ELECTRONIC COMMUNICATIONS SYSTEMS

KU-796-631

Online Services

Delmar Online

To access a wide variety of Delmar products and services on the World Wide Web, point your browser to:

> **http://www.delmar.com/delmar.html**
> or email: info@delmar.com

thomson.com

To access International Thomson Publishing's home site for information on more than 34 publishers and 20,000 products, point your browser to:

> **http://www.thomson.com**
> or email: findit@kiosk.thomson.com

A service of I(T)P®

ELECTRONIC COMMUNICATIONS SYSTEMS

Third Edition

Frank R. Dungan

President, FRADUN Enterprises

DELMAR

™

THOMSON LEARNING

Africa • Australia • Canada • Denmark • Japan • Mexico • New Zealand • Philippines
Puerto Rico • Singapore • Spain • United Kingdom • United States

NOTICE TO THE READER

Publisher does not warrant or guarantee any of the products described herein or perform any independent analysis in connection with any of the product information contained herein. Publisher does not assume, and expressly disclaims, any obligation to obtain and include information other than that provided to it by the manufacturer.

The reader is expressly warned to consider and adopt all safety precautions that might be indicated by the activities herein and to avoid all potential hazards. By following the instructions contained herein, the reader willingly assumes all risks in connection with such instructions.

The Publisher makes no representation or warranties of any kind, including but not limited to, the warranties of fitness for particular purpose or merchantability, nor are any such representations implied with respect to the material set forth herein, and the publisher takes no responsibility with respect to such material. The publisher shall not be liable for any special, consequential, or exemplary damages resulting, in whole or part, from the readers' use of, or reliance upon, this material.

Cover Illustration by Tom White

Delmar Staff

Publisher: Alar Elken
Acquisitions Editor: Paul Shepardson
Developmental Editor: Michelle Ruelos Cannistraci

Production Manager: Larry Main
Art and Design Coordinator: Mary Beth Voght
Editorial Assistant: Julie Waite

COPYRIGHT © 1998 Delmar, a division of Thomson Learning, Inc. The Thomson Learning™ is a trademark used herein under license.

Printed in the United States of America
  5 6 7 8 9 10 XXX 02

For more information, contact Delmar, 3 Columbia Circle, PO Box 15015, Albany, NY 12212-0515; or find us on the World Wide Web at http://www.delmar.com

International Division List

Asia
Thomson Learning
60 Albert Street, #15-01
Albert Complex
Singapore 189969
Tel: 65 336 6411
Fax: 65 336 7411

Japan:
Thomson Learning
Palaceside Building 5F
1-1-1 Hitotsubashi, Chiyoda-ku
Tokyo 100 0003 Japan
Tel: 813 5218 6544
Fax: 813 5218 6551

Australia/New Zealand:
Nelson/Thomson Learning
102 Dodds Street
South Melbourne, Victoria 3205
Australia
Tel: 61 39 685 4111
Fax: 61 39 685 4199

UK/Europe/Middle East
Thomson Learning
Berkshire House
168-173 High Holborn
London
WC1V 7AA United Kingdom
Tel: 44 171 497 1422
Fax: 44 171 497 1426

Latin America:
Thomson Learning
Seneca, 53
Colonia Polanco
11560 Mexico D.F. Mexico
Tel: 525-281-2906
Fax: 525-281-2656

Canada:
Nelson/Thomson Learning
1120 Birchmount Road
Scarborough, Ontario
Canada M1K 5G4
Tel: 416-752-9100
Fax: 416-752-8102

ALL RIGHTS RESERVED. No part of this work covered by the copyright hereon may be reproduced or used in any form or by any means—graphic, electronic, or mechanical, including photocopying, recording, taping, Web distribution or information storage and retrieval systems—without the written permission of the publisher.

For permission to use material from this text or product contact us by Tel (800) 730-2214; Fax (800) 730-2215; www.thomsonrights.com

Library of Congress Cataloging-in-Publication Data
Dungan, Frank R., 1924-
    Electronic communications systems / Frank R. Dungan.—3rd ed.
        p.   cm.
        Includes index.
        ISBN: 0-8273-8624-9
        1. Telecommunication. I. Title.
TK5101.D857   1997
621.382—DC21

97-31917
CIP

D
621.38
DUN

Contents

Dedication

This book is dedicated to my wife—Ruth Ann—
for thirty years of patience, encouragement,
and constructive criticism of my writing efforts.

Preface

Electronic Communications Systems, Third Edition, is designed for use in a one- or two-term course. The text is organized to allow complete flexibility in course construction.

Organization

Chapter 1 provides a brief overview of communications systems; Chapter 2 is a review of basic communications circuits; Chapter 3 details AM and SSB systems, with a section on stereophonics; Chapter 4 introduces angle modulation, including direct and indirect FM, and stereo FM receivers; Chapter 5 explains pulse modulation; Chapter 6 covers various transmission lines and techniques for their use; Chapter 7 contains a section on waveform characteristics and explains radio wave propagation; Chapter 8 offers extensive coverage of antennas; Chapter 9 introduces telephone systems and types of multiplexing; Chapter 10 covers data communications systems, codes, and equipment; Chapter 11 introduces microwave communications, waveguides, and equipment; Chapter 12 introduces TV transmission systems, standards, and applications and contains a section on CCD pickup devices; Chapter 13 covers TV reception circuits and methods; Chapter 14 provides satellite communications coverage; Chapter 15 offers extensive coverage of two-way communications services, stressing cellular telephone systems; Chapter 16 provides an overview of optical communications with fiber optics. New to this edition, Chapter 17 provides an introduction to the Information Superhighway; Chapter 18 covers Troubleshooting and Instrumentation.

New Features of the Third Edition

- New, updated, modern design and 2-color format makes book open and easy to read
- New Glossary of Key Terms for students' easy reference
- New Chapter on the Information Superhighway
- New Chapter on Troubleshooting and Instrumentation
- New and Expanded Coverage:

  Digital Data Communications   Advanced Satellite Technology and Systems
  Digital Satellite Systems   Concepts
  HDTV Technologies   Cellular Phones, Pagers
  Digital Television Systems   Fiber Optics and Testing

- An Extensive Supplements Package including:

Lab Manual   5 New Capstone Projects have been added. A great way to end the course and to test students on major concepts! Projects include:
Construct an AM Communications System
Construct an FM Communications System
Construct a Pulse Time Communications System
Construct an Optical Communications System
Construct a Time Division Multiplexing System

Instructor's Resource Kit   Available on CD-ROM, a New Instructor's Resource Kit includes everything an instructor needs to teach the course:

Lesson Plans.   Use them or adapt them to your course. Detailed chapter outlines, reference reading, and points of interest help guide lecture and discussion.

More Laboratory Projects Using Electronics Workbench.   Each project includes a theoretical discussion, a hands-on experiment, and a computer simulation integrating Electronics Workbench. Full pedagogical reinforcement: (1) Learn the theory, (2) Apply concepts using actual laboratory experiments, (3) Use Electronics Workbench to gather data, observe changes, and troubleshoot.

Computerized Testbank.   Over 800 Questions. Adapt tests by sorting by difficulty, adding your own questions or using as is. An Electronic Gradebook provided makes grading quick and easy.

Transparency Masters.   Use them to enhance your presentation!

Handouts.   Includes handy references, equations, and formulas.

Solutions to all Questions, Problems, and Laboratory Experiments.

- Software Simulation.   Use the new, stand-alone software simulation program, *Conquering Communications* by H. Paul Schuch. *Conquering Communications* is an interactive, highly visual, instructional software that acts as a "virtual laboratory." Great for students seeking additional instruction or to further their knowledge of electronic communications.

Recommendations

- For a one-term course: Chapters 1 through 11.
- For a two-term course: First term, Chapters 1 through 8; second term, Chapters 9 through 18.

Comprehensive end-of-chapter summaries provide an important study aid for students. Expanded end-of-chapter question and problem sets offer the instructor a wide choice for selecting and assigning problems to increase students' understanding of chapter contents.

For a student to gain the full benefits of the material in this text, recommended prerequisites are basic ac/dc electronics, semiconductor devices, and basic electronics mathematics.

Acknowledgments

The author and publisher wish to extend our thanks and appreciation to the following reviewers. We gratefully acknowledge their valuable contributions in reviewing the manuscript for this third edition.

Jim Allenbrand—ITT Portland, OR

Walter Banzhaf—University of Hartford, CT

Michael Benjamin—ITT Sylmar, CA

David Brett—ITT Youngstown, OH

Steve Brizendine—ITT Indianapolis, IN

Alan Green—ITT Austin, TX

Michael Guyote—Colorado Technical Univ., Colorado Springs, CO

David Jones—ITT Murray, UT

Frank Klonsky—ITT Framingham, MA

Clay Laster—San Antonio College, TX

Roger Lyons—ITT Jacksonville, FL

Kurt Osadchuk—National Institute of Technology, OH

Moises Rosado—ITT San Antonio, TX

Richard Segeberg—ITT Getzville, NY

Ken Somboon—ITT Van Nuys, CA

Frank R. Dungan

CHAPTER 1

Communications Systems—An Overview

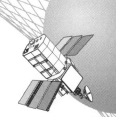

OBJECTIVES

When you complete this chapter, you should be able to:

- Define communication, carrier, intelligence, information, modulation, demodulation, heterodyne, noise, Signal-to-Noise Ratio, noise figure, transmission, reception, transducer, and bandwidth.
- Explain why an audio signal cannot be transmitted at its original frequency.
- Describe the basic types of modulation.
- Draw a block diagram of a basic communications system.
- List two levels of modulation.
- Discuss noise in terms of Signal-to-Noise Ratio and noise figure.
- List five general types of noise.
- List the frequencies, designations, and abbreviations for various ranges in the radio frequency spectrum.

INTRODUCTION

Since the beginning of civilization as we know it, the one thing that has distinguished *Homo sapiens* from other animals is the ability to communicate—that is, intelligently exchange ideas and information. Early communications took many forms: a glance, a particular body movement, a hand sign, a picture, even smoke signals. But as our civilization grew, so did our need for communications capacity. This need was filled by the discovery of electricity.

The field of electricity extended our communications range through the use of wires and telegraphy. Messages were sent over the wires by alternately turning electrical currents on and off, producing a series of "dots" and "dashes" in accordance with a telegraph code. Gradually, the telegraph system evolved into the telephone system, in which the electrical currents sent down the wires are varied at an audio rate, allowing the spoken word to be conveyed between two distant points. However, the requirement for wires remained in the telephone system, limiting its capabilities.

A significant advancement in communications was the development of radio, or wireless, communication. Radio extended the communications range to ships at sea and to remote areas of the world, providing for the first time instantaneous and worldwide communications. As a result, an entirely new branch of electronics was established.

Today, several of those early forms of communication are still used. However, modern electronic devices and techniques have contributed greatly to our communications capability. The transistor replaced the vacuum tube in many systems, resulting in smaller, less expensive, more efficient systems. More recently, the integrated circuit has provided highly reliable, low-power, high-performance functional electronic circuits that further reduce the size, cost, and weight of communications systems.

In this chapter, we will discuss the fundamentals of communications systems in general terms; examine the historical aspects and concepts of early communication; and define terms relating to communications systems, transmitters, modulation, noise, receivers, demodulation, and bandwidth. These subjects will be examined in more detail in later chapters.

1-1   COMMUNICATIONS AND SOCIETY

The ability to communicate has had a direct bearing on the progress of the human race. Communications among primitive humans was limited to gestures and facial expressions, and simple verbal utterings. The range of communication was very limited, and the rate of information transfer was very low. For these reasons, early social groups consisted of small groups of people banding together in packs, with little interaction between groups. The human social and economic situations improved as the ability to communicate improved.

Table 1-1 shows a historical time line for communications.

TABLE 1-1   Communications Historical Time Line

Year	Development
1440	Gutenberg printing press
1729	Discovery that electricity can be transmitted (Stephen Gray)
1746	Electrometer, first instrument for measuring electricity (Gralath)
1831	First electromechanical generator (Michael Faraday)
1844	Telegraph and Morse code (Samuel Morse)
1847	Boolean algebra, foundation for computer operation (George Boole)
1858	Transatlantic cable laid (Cyrus Field)
1876	Bell telephone system
1883	Edison effect (basis for vacuum tube discoveries)

TABLE 1-1 *(continued)*

Year	Development
1886	Verification of mathematical predictions for producing electromagnetic waves, opening field of practical radio
	Edison's carbon microphone
1889	Eastman photographic film
1895	Marconi's wireless telegraph (forerunner of radio)
1897	Ship-to-shore wireless demonstration (Marconi)
1899	Sound first recorded on magnetic wire
1900	First speech transmission by wireless
	Spark-gap transmitters developed
1901	First transatlantic wireless message
1906	First crystal detector
1915	First transatlantic radio telegraphy communication from the United States
1917	Campbell's electrical wave filter (made communication channels possible)
1920	First broadcast radio station (KDKA in Pittsburgh)
	Armstrong's superheterodyne circuit (forerunner of modern radio)
1923	Zworykin's iconoscope television camera tube
1925	Television demonstrated in England (John Baird)
1929	Zworykin's kinescope, first "modern" TV picture tube
1930	First basic radar system (Blair)
1931	Oscilloscope produced (Allen DuMont)
1934	Telecommunications Act of 1934
1937	Klystron tube (high-power microwave oscillator)
1939	Television broadcast demonstrated by NBC (using a new orthicon camera developed by RCA)
1941	Beginning of commercial FM broadcasting in the United States
	Commercial TV authorized by FCC
	First military application of radar (December 7, Pearl Harbor)
1942	Magnetic recording tape developed
	ENIAC, the first electronic computer
	First V-beam radar
1946	First mass-produced black and white TV (RCA, 10-inch tube)
	ENIAC computer, a digital-programmable machine, 5000 arithmetic calculations per second, weighed 30 tons, contained 18,000 vacuum tubes, required 130 kilowatts of power
	First all-electronic color television demonstrated (RCA)
1948	Transistors developed (begins modern electronics) (Bell Laboratories)
1951	UNIVAC 1 computer introduced
1952	Experiments on first all-transistorized TV set (RCA)
1953	Zeigler's beam master (forerunner of the laser)
1954	First transistorized consumer product (Regency Company 4-transistor radio)

TABLE 1-1 *(continued)*

Year	Development
1955	Varactor diode developed (used for electronic tuning with no moving parts in TVs and radios)
1956	All-electronic movie camera (Bell and Howell)
	Demonstration of feasibility of TV telephone (Bell Laboratories)
1957	Satellites (launch of Russian Sputnik)
	"Pill" transmitter (RCA's miniaturized FM transmitter, small enough to be swallowed, forerunner of modern versions used by astronauts to send physical data back to earth)
1958	First integrated circuits (Kilby of Texas Instruments and Noyce of Fairchild)
	Commercial field-effect transistors (General Electric and Crystalonics)
	Stereo radio broadcasting
1960	First IC products (Texas Instruments)
1961	Laser devices (Hughes, Bell Laboratories, Raytheon, and IBM)
1962	First light-emitting diodes (not commercially available until 1968)
	Telstar communications satellite (relays communications via telephone and television)
1963	First commercial mini-computer, 12-bit format, 1 K memory (Digital Equipment Corporation)
1965	Hearing aid, first consumer product using integrated circuits (Zenith)
1966	Magnet-bubble memory devices (Andrew Bobeck)
1969	Apollo 11 astronauts land on moon and return safely to earth
	First experimental network using Internet-like technology (consisted of four computers)
1970	Arpanet network established for Department of Defense (forerunner of today's Internet)
1971	First microprocessor (Intel)
	First hand-held calculators
1973	"Video Beam" projection TV introduced by Advent
1975	Home video tape recorder systems
	Microprocessor-based home computers
1977	40-channel citizens band radio approved by FCC
	First fiber optic communications system to provide regular telephone service placed in operation by General Telephone Company
1996	Telecommunications Act of 1996 passed
	Communications Decency Act passed

There was little progress in communications until the Gutenberg printing press was developed in 1440. The printing press made large amounts of information available over a wide range of territory; but because transportation was slow, delivery of the information was slow. However, there was a great impact on economic and social evolution because the printing press provided current commercial, governmental, and scientific knowledge. Only the rate and range of information transfer limited the effectiveness of communication.

The development of the Morse telegraph in 1844, the transatlantic cable in 1858, and the Bell telephone in 1876 dramatically changed communications. The range over which communication could occur within seconds reached thousands of miles; even continents were linked. International commerce, communications, and cooperation then began to expand rapidly. Subsequently, the Marconi wireless telegraph (radio) in 1895 and television in 1923 further accelerated the transition to global interaction between people and nations, with governments and business leading the way. With the development of satellite communications systems in 1957 and the more recent computer control of communications networks, the rate and range of information transfer today provides fast communications to any location on earth and beyond through global interconnection of telephone, television, and computer systems. For these reasons, it is now within our communicative capability to meet almost any requirement for the transfer of information.

The ultimate goal of any communications system is to extract intelligence from a transmitted signal. Therefore, it is vital that individuals keep up with the capabilities of communications systems and make the best possible use of alternatives as communications systems develop further.

1-2 METHODS OF COMMUNICATION

Because most communication in the past occurred between two people, information transfer was in a form compatible with the physical ability of a person to generate patterns and within the capabilities of a person's primary senses of hearing and sight and, in some cases, touch. The person sending the information had to generate patterns of light, sound, or texture. The person receiving the information had to decode the signals received.

Communications systems in the past used two broad techniques: aural and visual—the same techniques that are combined and used in today's television systems.

Aural Communication

In all aural communications, the transfer of information is in the form of sound waves generated by the sender. The reception of information is by the listener's ears, which convert sound waves to electrical impulses which are reconstituted into information by the brain. In this section, we will examine several forms of aural communication.

Speech   The most common and convenient method of aural communication is human speech. Speech communication began with basic utterings of grunts and growls to express anger or a threat. These utterings developed into specific sounds that represented specific things or needs. These sounds were the beginning of language in the form of simple words. As more simple words were used, a vocabulary grew into a complete language as we know it today.

Percussion   Spoken communication as it was originally developed is relatively slow and limited in range. The rate of information transfer is limited to about 100 words per minute; even with loud yells and under ideal conditions, the range is only a few hundred feet. The range limitation was overcome by the use of percussion devices, such as drums, to generate sound energy impulses that could carry messages thousands of feet. In this manner, a network of drummers could relay messages over great distances. However, the message rate was greatly reduced.

Telegraph   The telegraph, developed by Samuel F. Morse in 1844, was the earliest form of electrical communication. A digital signal, Morse code occupied less bandwidth than analog signals, such as voice. The code used by both the transmitter and the receiver was a series of dots and dashes representing letters of the alphabet.

With the laying of the transoceanic cable in 1858 and the use of the Morse code, continents were interconnected and the communications range was expanded to thousands of miles. Message delivery time was reduced to fractions of a second over this range, and the rate of information transfer was maintained at from 5 to 100 words per minute. However, only skilled telegraphers could use this system, so everyone who wanted to send a message had to go through a telegrapher.

Telephone   The Bell telephone system was developed in 1876. In this system, the speaker's voice was converted into electrical energy impulses that were sent over moderately long distances via wires to a receiver that converted those energy impulses back into the original sound waves for the listener. The telephone system provided many of the long-range communications capabilities of the telegraph, but it also offered the convenience of using speaking and hearing directly, so everyone could use the system, thus eliminating the need to go through a telegrapher. The rate of information transfer was limited to the speaker's rate of speech.

Wireless Telegraph   The next major change in sound communication was the approach to transmission. Wireless telegraphy, the forerunner of modern radio, was developed by Marconi in 1895. This system permitted telegraph communication over long distances without the need for running wires from the transmitter to the receiver. Messages could now be sent to mobile receivers on ships at sea, to military units, and to remote expeditions. The link between the transmitter and the receiver was electromagnetic waves.

Wireless telegraph transmission was not practical until the vacuum tube was perfected. The vacuum tube provided the high frequencies at the power needed to convert information to electromagnetic radiation.

Radio   After people learned how to encode and decode the human voice in a form that could be superimposed on electromagnetic waves and transmitted to a receiver, they used this communications method directly with human speech. Finally, the human voice could be transmitted thousands of miles, picked up by receivers, and converted back into speech by loudspeakers. Radio was the first complete electronic communications system. Speech could now be transferred from one part of the earth to another or from a point on earth to a point in space. Furthermore, the message delivery rate was increased to the speed of light, making possible international communications within fractions of a second, and space communications within seconds.

Visual Communication

Visual communication has undergone many changes since the first forms of body positions and facial expressions. Many early techniques are still used; others have evolved into more useful or entertaining forms. In this section, we will examine several forms of visual communication.

Sign Language   Initially, hand signals were used to indicate greeting, friendship, and peace. The hand signal evolved in many ways and is still evident in human communication. The military hand salute is used to indicate respect for an officer's rank. A wave of the hand can mean hello, good-bye, or simply that a person is recognized at a distance. Today, sign language is used as the primary form of communication with the deaf, with separate hand symbols representing a letter of the alphabet, a word, or a complex idea.

Flag Signals   An extension of the hand signal form of sign language is the semaphore flag code, used extensively by the navies of the world for communications between ships at sea. This form of communication is especially valuable during wartime when radio silence is necessary. A variation of the semaphore flag system is the paddle method used aboard aircraft carriers to land or to wave off aircraft.

Light Signals   Light signals in a variety of forms have been used throughout history for communication. A shiny piece of metal or a mirror reflecting the sun in a series of flashes has been used in many wars. The same approach has been used to save many persons stranded in remote land locations and at sea. Navies send coded messages between ships at sea by interrupting a light beam in the Morse code pattern. Traffic in its various forms is controlled by light patterns: street traffic is controlled by traffic lights, airport ground traffic is controlled by light color and pattern, aircraft without radios are controlled by different light colors and patterns from the airport control tower, and railroad flagmen use lanterns to signal the train engineer.

Today, information transfer can be achieved by the use of a light source, such as light-emitting diode or laser beam, and an optical fiber. The optical fiber is like a very thin cable, made of glass or plastic. The light signal is contained within this fiber, which reflects the signal from the source end to the receive end at very high speed.

Pictorial Language   Pictorial language is the most advanced form of visual communication. It evolved from early humans who left behind cave drawings depicting activities and events in their lives. Hieroglyphics is a picture language developed by early Egyptians. Native American Indian tribes used many forms of pictorial language, some of which may be seen in Indian ruins. Today, films and television provide entertainment, and television provides on-the-spot news from around the world.

The Alphabet and Written Language   The Egyptians began using sheets of papyrus around 2500 B.C., which led to written records of events. The Semites developed the first written alphabet around 1500 B.C. Over the next 2000 years, evolution led to a number system, a form of shorthand, parchment writing paper, books, and libraries. The use of ink on paper, block printing, and the present decimal number system had developed by around A.D. 1000. However, dissemination of available knowledge was still limited.

The Printing Press and Photography The development of the Gutenberg printing press in 1440 rapidly expanded the transfer of knowledge. Photographic film, developed by Eastman in 1889, made it possible to communicate through pictures. The typewriter, invented about the same time, allowed the efficient generation of written communication and information. All of these later developments enabled the important exchange of scientific information between people and greatly increased their understanding of the world around them. In turn, this resulted in rapid advances in developing more modern communications techniques.

Electronic Visual Communication The rapid advance of modern communications techniques led to the capability of transmitting printed pictures and written documents over long distances electronically. The first use of this approach was called *facsimile* (FAX), a process of sending still pictures or printed material by telephone from one location to another for reproduction. A later use for long distance electronic visual communication was television. Visual communication has been greatly enhanced by the development of satellite links, which provide a means for transmitting all types of information from one point to another anywhere in the world in a fraction of a second—a tremendous advantage over early methods of communication. Therefore, we can state that, primarily through the use of electronic communications techniques, the communicative capabilities of humanity have improved tremendously.

The *Communications Act of 1934* provided for the regulation of interstate and foreign communication by wire or radio. The Act created the *Federal Communications Commission* (FCC), thereby centralizing authority that had been previously granted to several agencies. The 1934 Act was updated and expanded by the *Telecommunications Act of 1996*. These Acts are discussed in detail in Chapter 17.

Another communications development has evolved from a collection of computer networks developed in the 1970s which started with *Arpanet,* a network sponsored by the United States Department of Defense. Arpanet has long since been expanded and replaced, and today its descendants form the global backbone of the *Internet.* This important communications medium is detailed in Chapter 17.

1-3 THE MEANING OF COMMUNICATION _____

Communication is defined as the transfer of meaningful information (intelligence) from one point (source) to another (destination), as shown in the block diagram of Figure 1-1. A *communications system* is a collection of individual communications

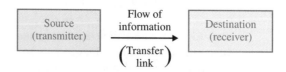

FIGURE 1-1 Information transfer in a communications system

networks consisting of intelligence origination, transmission systems, receiving systems, and intelligence reproduction.

Most communications systems use some form of coding and decoding in sending the information from the source to the destination. This is particularly true for electronic communications systems, where coding and decoding are referred to as modulation and demodulation, respectively. *Modulation* is the process of impressing information in the form of a low-frequency signal, called the *modulating signal,* onto a high-frequency signal, called the *carrier. Demodulation* is the process of stripping the carrier from the modulated signal, thus recovering the original information. A simplified block diagram demonstrating this concept is shown in Figure 1-2.

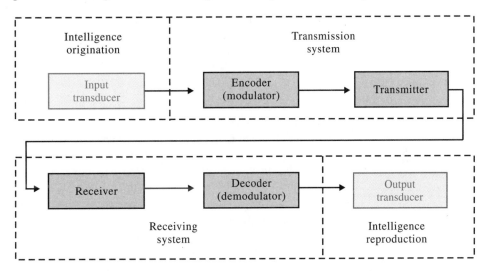

FIGURE 1-2 Electronic communications system

The purpose of a communications system is to communicate a *message.* The message consists of information (intelligence) from an originating source. Although written words and face-to-face spoken words are fundamental methods of conveying information, for communications systems in this text, *information* is defined as any knowledge or intelligence representing data that is conveyed or transmitted as an electrical signal. For demonstration purposes, we will use the concept of radio voice communications in the following discussion.

Processing the Message

The method chosen to process information depends on the method used for sending the information to its final destination. In radio voice communications, the message is voiced into a microphone, which serves as a transducer. A *transducer* is any device that converts one form of signal, or energy, into another form. In electronics, transducers convert ac or dc into sound, heat, light, radio waves, or other forms of energy; conversely, these other forms can be converted into ac or dc. In the case of radio, the mechanical audio sounds are converted into electrical voltages. These electrical voltages become the intelligence or modulating signal in the form of low *audio frequency* (AF) signals. The modulating signal is impressed onto the carrier, which is a fixed, high *radio frequency* (RF) signal. This process of impressing a low AF modulating signal onto a high RF carrier is modulation (coding).

Modulation is required because of the impracticality of transmitting low audio frequencies. The frequencies encountered in the intelligible AF range would necessitate an antenna of great length. For example, the length of a half-wave antenna for a 10 kHz AF signal would be almost 9 *miles*. But by impressing the AF signal onto a high RF carrier, say 10 MHz, the antenna length can be reduced to approximately 49 *feet*, making the transmission of the intelligence practical. The method used to determine these values is discussed later.

In a *transmitter*, a device that produces a signal for broadcasting or communications purposes, the modulated signal is amplified in a *power amplifier* (PA) to a level suitable to drive an *antenna*, a device for transmitting or receiving electromagnetic energy. The method used for modulation varies with different systems. Modulation may be *high level* or *low level*, depending on the point in the system at which the modulating signal is injected.

Basic Types of Communications Systems

There are three basic types of communications systems, each of which is discussed in separate chapters: (1) systems that use *amplitude modulated* (AM) carriers, (Chapter 3); (2) systems that use *angle modulation*, which includes *frequency modulated* (FM) carriers and *phase modulated* (PM) carriers (Chapter 4); and (3) systems that use digital techniques, normally referred to as *pulse modulation* (Chapter 5).

A carrier wave consists of a peak amplitude (X), a carrier frequency (f), and a phase angle (θ). If only X is changed by the message signal, the resulting modulation is AM. If only f is changed by the message signal, the resulting modulation is FM, and if only θ is changed by the message signal, the resulting modulation is phase modulation. Refer to Figure 1-3 for an illustration of these relationships. Examples for each of these modulation methods are presented in the appropriate chapters.

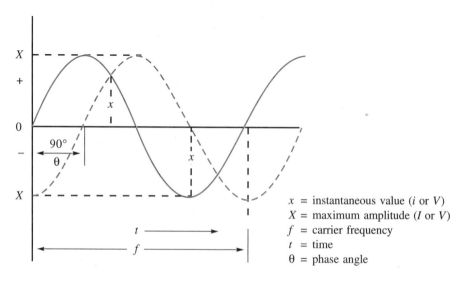

x = instantaneous value (i or V)
X = maximum amplitude (I or V)
f = carrier frequency
t = time
θ = phase angle

FIGURE 1-3 Electronic communications system

Sending the Message

The modulated signal is transferred from the power amplifier through transmission lines to the transmitting antenna. There it is radiated as a complex radio wave that *propagates* (travels) through the air to be picked up by a receiving antenna at some distant point. The paths followed by radio waves vary with environmental conditions, time of day, type of transmitter antenna used, frequencies and strength of the transmitted signal, and the terrain over which the signal is sent.

Receiving the Message

By the time the transmitted signal reaches its destination, much of its power has been dissipated by the various conditions just described, and several forms of noise (static, interference) have been added. A *receiver,* a device or circuit that intercepts a signal, processes it, and converts it to a form useful to a person, must filter out the unwanted noise signals, select the desired frequency, and amplify the weak signal enough to overcome any internal noise generated by the receiver. Then, through the process of demodulation (decoding), the carrier is stripped from the modulated signal, thus recovering the original message, which is then amplified and coupled to some type of output transducer, such as a radio loudspeaker.

There are many varieties of receivers in communications systems. The particular receiver used for an application depends on the modulation system used, the operating frequency, the transmitted signal power, and the type of output transducer required. Receivers will be discussed in detail later in the text. For present purposes, it is sufficient to state that most receivers are *superheterodyne. Heterodyning* is the frequency-converting (mixing) of the modulated signal with a signal provided by a *local oscillator* (LO). This produces an output signal which is at an entirely different frequency. This new frequency is called an *intermediate frequency* (IF). The term *super* was added to reflect the fact that this frequency conversion technique produces much greater tuning stability than the tuned circuits used previously. The superheterodyne receiver is shown in Figure 1-4. The IF signal is generally higher in frequency than the

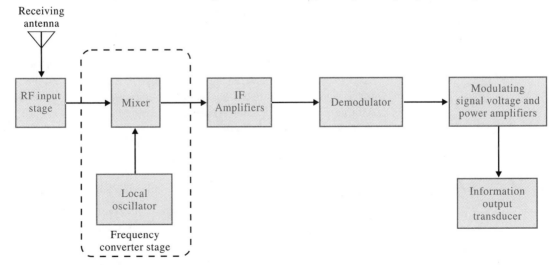

FIGURE 1-4 Standard superheterodyne receiver

intelligence but lower in frequency than the carrier. The IF signal is then demodulated, amplified, and sent to an output transducer.

A simplified block diagram of a complete communication system is shown in Figure 1-5. For such a system, the individual components must serve a common purpose, be technically compatible, employ common procedures, respond to some form of control and, in general, operate in unison. Such systems include all radio, wire, and other means used for voice, electrical or electronic, and visual transmission and reception of information or messages.

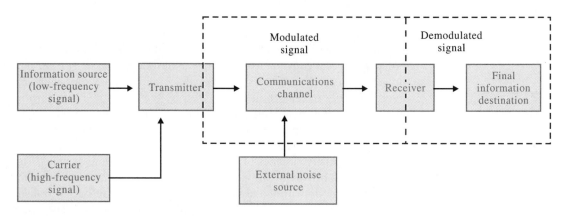

FIGURE 1-5 Radio communications system

1-4   NOISE—A LIMITING FACTOR

Several forms of noise affect the accurate reproduction and intelligibility of the transmitted signal. *Noise* is defined as any unwanted form of energy that tends to interfere with the reception and accurate reproduction of wanted signals. Noise is always present in electronic systems, and its effects can be devastating to the performance of the system. The term *bandwidth* (*BW*) defines the frequency occupied by a signal and required for effective transfer of the information carried by that signal. The receiver must have a *bandpass response* (the ability to pass a band of frequencies) at least as great as the signal bandwidth. If the receiver bandpass response is too narrow, the signal cannot be readily understood. Conversely, if the bandpass response is much wider than the bandwidth of the signal, a lot of noise enters the receiver along with the desired signal. Signals on nearby frequencies may also interfere with the desired signal. In general, received noise increases as the bandwidth of a receiver is increased.

External Noise
External noise is the noise generated outside a receiver. It may be caused by atmospheric conditions, including space, solar, and cosmic noise, or it may be man-made.

Atmospheric Noise *Atmospheric noise* is caused by electrical disturbances in the earth's atmosphere, and it manifests itself in what we commonly call *static*. Lightning is the most visible source of static. The random nature of lightning is such that it creates spurious radio signals, which are spread across the RF spectrum usually used for broadcasting. These spurious radio waves are propagated in the same manner as regular radio waves of the same frequency; therefore, static can occur at any point on earth, whether from a local or distant lightning strike. Atmospheric noise is most noticeable in the AM (550 kHz–1800 kHz) band, and it usually is more severe at night. It is much less a problem at frequencies above the AM band, although it can create severe disturbances in television picture signals.

Problems caused by *space noise* are about equally divided between the sun and all the other stars in the universe. Higher space noise levels are produced at intervals because of the cyclical nature of the sun's disturbances, including solar flares and sunspots. A *solar flare* is a violent storm on the sun's surface that is thousands of miles across and thousands of miles high and appears as a bright spot on the solar disk. A *sunspot* is a massive magnetic storm on the sun. Sunspots appear as dark regions on the sun's surface. A typical sunspot is several times the diameter of the earth. The periods of great magnetic disturbance caused by sunspots occur about every 11 years, with even higher maximum levels of disturbance about every 100 years. Solar flares can occur at any time, but they seem to take place most often near the peak of the 11-year sunspot cycle.

The amount of radio noise emitted by the sun is called *solar radio noise flux,* or simply the *solar flux,* an indicator of general solar activity. The solar flux varies with frequency. However, at any frequency, the level of solar flux increases abruptly when a solar flare occurs, which makes the solar flux useful for propagation forecasting. A sudden increase in the solar flux indicates that ionospheric propagation conditions will deteriorate within a few hours. Solar flux is often monitored at a frequency of 2800 MHz, which is correlated with the 11-year sunspot cycle. On the average, the solar flux is higher near the peak of the sunspot cycle and lower near a sunspot minimum.

Distant stars, which are also suns with high temperatures, radiate constant noise just as our own sun does. The noise generated by the stars is termed *cosmic noise.* Although these stars are more distant from earth than our sun, their vast numbers and additive effect contribute space noise about equal to that of the sun. Cosmic noise occurs at all wavelengths from the very low RF band to the x-ray band and above. Cosmic noise limits the sensitivity obtainable with receiving equipment because this noise cannot be eliminated.

Space noise is most troublesome at frequencies from about 8 MHz to 1.5 GHz. However, below 20 MHz, very little space noise penetrates the ionosphere, the outer layers of the earth's atmosphere, where electron and ion content tend to attenuate radio wave travel. At frequencies above 1.5 GHz, atmospheric absorption prevents most space noise from reaching earth.

Man-Made Noise   *Man-made noise* is any form of electromagnetic interference that can be traced to nonnatural causes. In particular, man-made noise refers to interference such as ignition and impulse noise, originating from internal combustion engines and electrical appliances. High-voltage line leakage and fluorescent lights are other sources.

Man-made noise is normally more intense than noise from any other source, whether external or internal to a receiver. It has become an increasing problem in radio communications at the Very Low, Low, Medium, and High Frequencies, because of the proliferation of household appliances that generate electromagnetic energy. (See Table 1-2.) At Very High Frequencies, man-made noise is not usually a problem except when extreme receiving sensitivity is required. For these reasons, most quality communications systems are established at reasonable distances from industrial areas.

TABLE 1-2   **Radio Frequency Spectrum**

Frequency Range	Designation	Abbreviation	Examples of Applications
30–300 Hz	Extremely Low Frequency	ELF*	Telephone
300 Hz–3 kHz	Voice Frequency	VF*	
3–30 kHz	Very Low Frequency	VLF*	
30–300 kHz	Low Frequency	LF	
300 kHz–3 MHz	Medium Frequency	MF	AM Broadcast
3–30 MHz	High Frequency	HF	Aircraft Radio
30–300 MHz	Very High Frequency	VHF	FM Broadcast
300 MHz–3 GHz	Ultra High Frequency	UHF	
3–30 GHz	Super High Frequency	SHF	Satellite/Ground
30–300 GHz	Extremely High Frequency	EHF	Microwave

*Within the ranges covered by ELF, VF, and VLF is an important subsegment:

30 Hz–15 kHz	Audio Frequency	AF	

Internal Noise

Random noises created by the passive or active devices inside a receiver are distributed about equally over the entire RF spectrum. The power of this random noise, termed *internal noise,* is proportional to the bandwidth over which it is measured. This type of noise can be divided into two broad categories: (1) thermal noise, generated in any resistance or the resistive component of any impedance; and (2) shot noise, created by the shot effect present in all active devices.

Thermal Noise   Noise termed variously as *thermal, thermal agitation, white,* or *Johnson noise* is generated by the random motion of the atoms and electrons in any electronic equipment. The power of this random noise is related directly to bandwidth. Therefore, it is advisable to limit the bandwidth of a receiver as much as feasible; the lower the bandwidth, the lower the noise power. The average noise power is also related directly to the absolute temperature of the resistance. *Noise power* (P_n) can be calculated by the following equation:

$$P_n = kT(BW) \tag{1-1}$$

where

k = Boltzmann's proportionality constant, or $1.38 \times 10^{-23} J/°K$

T = temperature in degrees kelvin ($°K = °C + 273°$)

BW = bandwidth of frequencies to be measured (in Hz)

Even when a resistor is not connected to a voltage source, the internal movement of electrons generates a random rms voltage across the resistor and a sensitive ac meter will register a reading. The value of the rms noise voltage e_n is

$$e_n = \sqrt{4\,kT\,(BW)\,R} \qquad (1\text{-}2)$$

where

k = 1.38 10^{-23}J/°K

T = absolute resistance temperature

BW = bandwidth of frequencies to be measured (in Hz)

R = resistor generating the noise

Shot Noise   *Shot noise* is present in all amplifying devices. It is a result of the *shot effect,* which is caused by random variations in the arrival of the majority current carriers (electrons or holes) at the output terminal of an amplifying device. The term *shot noise* is used because excessive randomly varying noise current, when superimposed on the dc of the output terminal and amplified, sounds like a shower of lead shot dropping onto a metal surface.

Both shot noise and thermal noise are present in amplifiers and are therefore additive. However, since shot noise is a current and thermal noise is a voltage, it is not possible to add them together in calculations, so equations for shot noise calculations are generally approximations. For simplicity, manufacturers' data sheets generally specify an *equivalent input noise resistance* (R_{eq}), which has a resistance value that will produce a noise level equivalent to that produced by the shot noise of an amplifying device. Purely a fictitious resistance, R_{eq} is provided solely to simplify shot noise calculations. In those calculations, R_{eq} is treated like any other noise-generating resistance, operating at the same temperature and electrically connected in series with the input terminal of the amplifying device. Table 1-3 summarizes noises and their causes.

Noise Evaluation

From the foregoing discussion on noise, it is obvious that an amplifier stage generates a certain amount of noise. Careful receiver design, particularly of the first amplifier stage, can minimize the internally generated noise output. However, it is possible that an RF amplifier can produce more noise than gain, especially at ultra-high frequencies.

Two approaches to noise evaluation in an amplifier or a receiver are the Signal-to-Noise Ratio and the noise figure.

Signal-to-Noise Ratio   The *signal-to-noise ratio (S/N)* is defined as the ratio of signal power to noise power at the same point. Mathematically, it is

$$S/N = \frac{\text{signal power}}{\text{noise power}} = \frac{P_s}{P_n} \qquad (1\text{-}3)$$

or, in decibel form,

$$S/N = 10\,\log\frac{P_s}{P_n} \qquad (1\text{-}4)$$

TABLE 1-3   Noise Sources and Causes

EXTERNAL	Any noise generated outside a receiver
Atmospheric	Electrical disturbances in the earth's atmosphere
static	Lightning strikes
space noise	Sun and star disturbances
solar flare	Violent storm on sun's surface
sunspot	Massive magnetic storm on sun
solar flux	Varies with frequency, increases when solar flare occurs
cosmic	From stars more distant than sun
Man-made	Any form of electromagnetic interference from nonnatural causes. Ignition and impulse noise from combustion engines and electrical appliances; high voltage lines; fluorescent lights
INTERNAL	Random noises from active or passive devices inside receiver
Thermal	Also termed thermal agitation, white, or Johnson; generated by rapid and random motion of atoms and electrons. Related directly to bandwidth
Shot	From shot effect, caused by random variations in arrival of majority carriers at the output of an amplifying device

Note: Log, as used here and throughout the text, is understood to represent the common log, or $\log_{10}$.

Usually, the sensitivity is specified as the signal strength in microvolts (μV) that is necessary to cause an S/N ratio of 10 decibels (dB). Modern communications receivers require 0.5 μV or less to produce an S/N ratio of 10 dB at the high frequencies in the continuous-wave (CW) or single-sideband (SSB) modes. For AM, the rating is 1 μV or less.

Noise Figure   More realistically, the *noise figure (NF)*, or the *noise factor (nf)*, is used in comparing the performance of receivers with respect to noise. It is defined as the ratio of the S/N at the input of the receiver to the S/N at the output of the receiver:

$$nf = \frac{\text{input } S/N}{\text{output } S/N} = \frac{S_{in}/N_{in}}{S_{out}/N_{out}} \tag{1-5}$$

or, in decibel form,

$$NF = 10 \log \frac{S_{in}/N_{in}}{S_{out}/N_{out}} \text{ or } 10 \log nf \tag{1-6}$$

Noise figures for microwave receivers range from 10 dB to 20 dB without the use of special input amplifiers. In the high and very high frequency ranges, noise figures from 3 dB to 12 dB are common. Typical noise figures for FM and television receivers vary from 3 dB to 10 dB.

EXAMPLE 1-1

Assume that a receiver has a first amplifier stage with a 5 kΩ input resistance, a gain of 300, an input audio signal of 20 µV, and an operating temperature of 17°C. When the amplifier is operating with a bandwidth first of 5 MHz and then of 1 MHz, find (a) the rms input noise levels, (b) the audio output levels, and (c) the rms output noise levels.

SOLUTION

Convert degrees Celsius to kelvins:

$$°K = 17°C + 273° = 290°K$$

For $BW = 5$ MHz,

(a) $e_n = \sqrt{4kT(BW)R}$

$$= \sqrt{(4)(1.38 \times 10^{-23} \text{ J/K})(290 \text{ K})(5 \text{ MHz})(5 \text{ k}\Omega)}$$

$$= 20 \text{ µV}_{rms}$$

(b) $e_{out} = (20 \text{ µV})(300) = 6.0 \text{ mV}_{rms}$

(c) $e_{n(out)} = (20 \text{ µV})(300) = 6.0 \text{ mV}_{rms}$

For $BW = 1$ MHz,

(a) $e_n = \sqrt{(4)(1.38 \times 10^{-23} \text{ J/K})(290 \text{ K})(1 \text{ MHz})(5 \text{ k}\Omega)}$

$$= 8.95 \text{ µV}_{rms}$$

(b) $e_{out} = (20 \text{ µV})(300) = 6.0 \text{ mV}_{rms}$

(c) $e_{n(out)} = (8.95 \text{ µV})(300) = 2.69 \text{ mV}_{rms}$

Example 1-1 graphically demonstrates the devastating effect that noise has on the output signal of an amplifier. It also demonstrates that a narrower operating bandwidth reduces the noise levels. We conclude that we should limit the bandwidth of the first amplifier in a receiver, and that we should use low-noise resistors wherever possible. The best low-noise resistors are the wirewound variety.

EXAMPLE 1-2

In the second part of Example 1-1, we had an amplifier with a 1 MHz bandwidth and a 20 µV input signal. We found that input noise = 8.95 µV, output signal = 6.0 mV, and output noise = 2.69 mV. For the same amplifier condition, find (a) S_{in}/N_{in}, (b) S_{out}/N_{out}, and (c) the noise figure for the amplifier.

SOLUTION

(a) $S_{in}/N_{in} = \dfrac{P_{s(in)}}{P_{n(in)}} = \left(\dfrac{V_{s(in)}}{V_{n(in)}}\right)^2 = \left(\dfrac{20 \text{ µV}}{8.95 \text{ µV}}\right)^2 = 4.99$

(b) $S_{\text{out}}/N_{\text{out}} = \dfrac{P_{\text{s (out)}}}{P_{\text{n (out)}}} = \left(\dfrac{V_{\text{s (out)}}}{V_{\text{n (out)}}}\right)^2 = \left(\dfrac{6.0 \text{ mV}}{2.69 \text{ mV}}\right)^2 = 4.98$

(c) $NF = \dfrac{S_{\text{in}}/N_{\text{in}}}{S_{\text{out}}/N_{\text{out}}} = \dfrac{4.99}{4.98} \cong 1.0$

Note: Using V^2 in the equations is valid only if impedances remain constant, which in this case they do. Impedance is a requirement for using the decibel form correctly, so *NF* is commonly used as a measure of noise since varied input/output impedances are effectively eliminated.

The result in Example 1-2 indicates that the amplifier is almost ideal because it did not generate noise internally. Theoretically, it is possible for such an amplifier to exist, but it is unrealistic to expect it.

EXAMPLE 1-3

An FM receiver has $S_{\text{in}}/N_{\text{in}} = 48$ and $S_{\text{out}}/N_{\text{out}} = 12$. Calculate the receiver's noise figure in decibels.

SOLUTION

$$NF = 10 \, \log \frac{S_{\text{in}}/N_{\text{in}}}{S_{\text{out}}/N_{\text{out}}}$$

$$= 10 \, \log \frac{48}{12} = 10 \, \log 4.0$$

$$= 6.0 \text{ dB}$$

The result in Example 1-3 is a typical noise figure for an FM broadcast receiver.

1-5 BANDWIDTH AND CHANNEL ALLOCATIONS

Many factors affect the performance of a communications system. However, the two predominant limitations are noise and bandwidth. We have already considered the effects of noise. This section examines bandwidth and its limiting effect on communications, and defines the frequency ranges in the RF spectrum and the bandwidths for the various communications systems as specified by the Federal Communications Commission (FCC).

Bandwidth

An ideal resonant circuit is resonant at only one frequency. Although this is true for the maximum resonant effect, frequencies slightly above and slightly below the maximum resonant frequency are also effective. We can therefore state that any resonant frequency has a band of frequencies associated with it that also provides resonance effects. This band lies between the points on a resonant frequency waveform at 70.7% of the maximum amplitude at resonance. These points are known as the *half-power points* or *−3 dB points,* as shown in Figure 1-6.

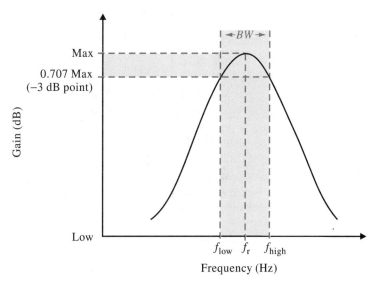

FIGURE 1-6 Resonant frequency curve

The width of the band of frequencies centered around the *resonant frequency (f_r)* depends on the *quality (Q)* of the tuned circuit and is called the *bandwidth (BW).* The bandwidth is directly related to the resonant frequency and inversely related to the quality of the tuned circuit:

$$BW = \frac{f_r}{Q} \tag{1-7}$$

EXAMPLE 1-4

A circuit with a *Q* of 250 is resonant at 5 MHz. What is its bandwidth?

SOLUTION

$$BW = \frac{f_r}{Q} = \frac{5 \text{ MHz}}{250} = 20 \text{ kHz}$$

Resonant circuits are discussed in detail in Chapter 2.

Channel Allocations

Communications systems are often designated by the frequency of the carrier or by the modulation method. The method of modulation determines the bandwidth required for a specific communications transmission. To conserve space in the available RF spectrum (see Table 1-2 and Figure 1-7), the FCC regulates the bandwidth of the various communications techniques. These allocated bandwidths, called *channels,* are listed in

Table 1-4. Note the wide variation in the bandwidth of the channels. In later chapters, specific channels are discussed in detail.

TABLE 1-4 FCC Channel Bandwidth Allocations

Designation	Frequency Range	Channels	Bandwidth
AM broadcast	535–1605 kHz	106	10 kHz
Citizens band	26.965–27.405 MHz	40	10 kHz
FM broadcast	88–108 MHz	100	200 kHz
TV broadcast*	54–806 MHz	67	6 MHz

*In the United States, television signals are transmitted into distinct frequency bands: (1) the VHF band (54–72 MHz for channels 2 through 4 and 76–216 MHz for channels 5 through 13) and (2) the UHF band (470–608 MHz and 614–806 MHz for channels 14 through 69.) In the VHF band, FM and aircraft navigation/ communications frequencies are allocated in the missing space, and in the UHF band, radio astronomy occupies the missing space.

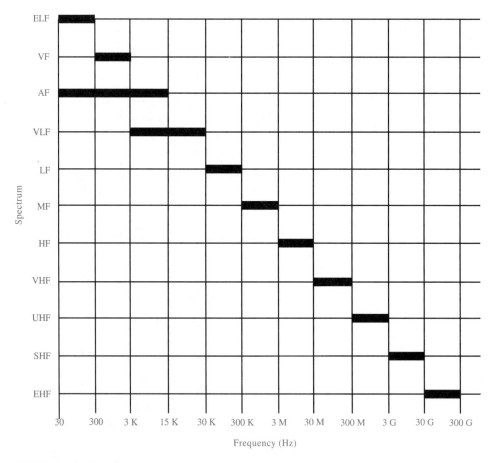

FIGURE 1-7 Graphic representation of radio frequency spectrum

1-6 IMPORTANT COMMUNICATIONS SYSTEMS CONSIDERATIONS

Several important interrelated requirements must be considered in any communications system. Although these requirements depend on one another, the order of importance changes for a given system. The requirements are as follows: (1) the rate at which information is transferred, (2) the reliability of the transfer of information, (3) the convenience of system use, and (4) the costs in money and in energy incurred in transferring the information.

Information Transfer Rate

In many cases, the rate at which information is transferred is the most important consideration. The rate of information transfer is determined by the amount of information that must be communicated from one point to another in a given amount of time. Furthermore, the rate of information transfer determines the physical form and techniques used to send and receive the information, thereby determining the system design and construction. If the information transfer rates are so low that any type of system meets the requirements, then this requirement takes a low priority.

Information Transfer Reliability

A communications system user must be able to rely on the system. That is, the system must work when needed, and it must be capable of transmitting and receiving error-free information.

Communications system reliability can be increased in several ways. First, to prevent or minimize the possibility of failure, the system builder should use modern electronic components, especially high-quality integrated circuits. Second, careful design of the modulating and demodulating devices can prevent or minimize errors in the communicated information. Finally, system design must incorporate every possible noise-limiting device available to minimize the effects of the various types of noise discussed previously.

System Convenience

For some people, the convenience of a communications system is its most important feature. For example, home telephones, pagers, and cellular telephones provide a convenient means whereby, in a matter of minutes, one person can hold a conversation with another person almost anywhere in the world. Radio and television receivers bring entertainment and the latest news into the home. Today, computers provide the convenience of data transfer between interested parties, and access to the plethora of information available on the Internet. We are told that, through the Telecommunications Act of 1996, we will be able to make purchases and pay bills via home computers.

System Costs

Although the cost of a system may be the most important consideration in some cases, it may sometimes be only one of the deciding factors. A system user always wants—and should receive—the most performance at the least cost, with good reliability and convenience.

In addition to monetary costs, energy and the raw materials used must also be considered. The key to determining the effectiveness and efficiency of the cost of a system is its cost per unit of information transferred. In this manner, effective comparisons of the cost of alternative systems can be made.

SUMMARY

- The ultimate goal of any communications system is to extract intelligence from a transmitted signal. Two broad techniques used for communications in the past were aural (speech, percussion, telegraph, telephone, wireless telegraph, and radio) and visual (sign language, flag signals [semaphore], light signals, pictorial language, the alphabet and written language, and televised signals).

- Communication is the transfer of meaningful information (intelligence) from a source to a destination. A communications system is a collection of individual communications networks consisting of intelligence origination, transmission systems, receiving systems, and intelligence reproduction. The purpose of a communications system is to communicate a message. All communications systems, with the exception of some local telephone lines, use some form of modulation (coding) and demodulation (decoding). Information is any knowledge or intelligence representing data that is conveyed or transmitted as an electrical signal from a source to a destination. Modulation is the impression of a low- (audio) frequency signal, called the modulating signal or intelligence, onto a high- (radio) frequency signal, called the carrier. This procedure eliminates the need for an extremely long antenna.

- The three basic types of modulation are amplitude modulation (AM), frequency modulation (FM), and phase modulation (PM). Both frequency modulation and phase modulation are categorized as angle modulation.

- At the transmitter, the modulated (coded) signal is amplified and fed to an antenna where it is radiated as radio waves and propagated through space. At the receiver, the radiated signal is picked up by a receiving antenna, fed to an amplifier, demodulated (decoded), and fed to an output transducer (speaker) as a reproduction of the original modulating signal.

- Noise is any unwanted form of energy that tends to interfere with the reception and accurate reproduction of wanted signals. In general, noise is external or internal. External noise is generated outside a receiver and consists of atmospheric or manmade noise. Internal noise consists of thermal noise and shot noise. Noise in an amplifier or a receiver is evaluated by the Signal-to-Noise Ratio (S/N) and the noise figure (NF).

- Bandwidth (BW) is the width of the band of frequencies centered around a resonant frequency (f_r) and depends on the quality (Q) of the tuned circuit. $BW = f_r/Q$.

- The FCC regulates the bandwidth of the various communications techniques and allocates specific bandwidths to specific channels.

- Important communications system considerations include information transfer rate, information transfer reliability, system convenience, and system costs.

QUESTIONS AND PROBLEMS

1. List five early forms of communication.
2. State the ultimate goal of any communications system.
3. Define aural communication.
4. Briefly describe the methods of aural communication.
5. Briefly describe the methods of visual communication.
6. Define communication.
7. Define communications system.
8. State the purpose of a communications system.
9. Define information.
10. Describe how a radio message is processed.
11. Define modulation and explain why modulation is required.
12. Explain the three basic types of communications systems.
13. Explain how a message is transmitted.
14. Define demodulation.
15. What type of receiver is most commonly used today?
16. Define heterodyning.
17. Briefly describe the types of external noise.
18. Briefly describe the types of internal noise.
19. The first amplifier stage of a receiver has a 1 kΩ input resistance, a gain of 350, an input signal of 25 μV, and an operating temperature of 25°C. For the two cases where the amplifier is operating at bandwidths of (1) 7.5 MHz and (2) 2.75 MHz, determine (a) the rms input noise levels, (b) the audio output levels, and (c) the rms output noise levels.
20. For the amplifier condition of part 2 of Problem 19, determine (a) S_{in}/N_{in}, (b) S_{out}/N_{out}, and (c) the noise figure for the amplifier.
21. For a receiver with $S_{in}/N_{in} = 38$ and $S_{out}/N_{out} = 12.4$, determine NF in decibels.
22. List the two predominant limitations on the performance of communications systems.
23. Define bandwidth.
24. Briefly describe the important considerations for communications systems.
25. What are the determining factors in the type of modulation a communications system is using?
26. Explain how the factors in Problem 25 determine (a) AM; (b) FM; (c) PM.
27. Assume $f = 101.9$ MHz, $t = 2$ μs, $X = 7.5$ V, $\theta = +20°$. (a) Find x and (b) state the type of modulation.
28. From Problem 27, find x if f is changed to 91.8 MHz.
29. What causes solar flare?
30. What type of noise is directly related to bandwidth?
31. Define shot effect.
32. What causes white noise?

CHAPTER 2

Communications Circuits

OBJECTIVES

When you complete this chapter, you should be able to:

- Identify the four basic types of filter circuits and discuss the uses of each.
- Define filter and discuss the two basic filter categories.
- List the filter classifications and explain how filters are classified.
- Recognize the basic characteristics of the filters used in communications circuits.
- Calculate the required component values for basic filter circuits.
- Define pass band, stop band, cutoff frequency, and roll-off.
- Define and discuss the transfer function of a filter.
- Define decade and octave.
- Explain how tuned circuits are used to select a desired frequency.
- Define series resonance and parallel resonance.
- Calculate the resonant frequency for any series RLC or parallel LC combination circuit.
- Determine the impedance, bandwidth, and Q for any series or parallel resonant circuit.
- Discuss the factors that influence selectivity.
- Explain the operation of three general classes of feedback oscillators.
- Recognize oscillators by observation.
- Determine the effects on frequency when reactive elements are varied.
- Determine if a crystal is operating in the series resonant or parallel resonant mode.
- Identify the frequency-determining components of various oscillators.
- Identify the three basic amplifier circuit configurations and describe the most important characteristics of each.

- Identify the basic amplifier coupling methods and discuss advantages and disadvantages of each.
- Identify basic biasing arrangements and explain how they operate.
- Determine the operating class of an amplifier.
- Identify the two basic types of amplifiers used in communications systems and explain the basic function of each.
- Explain how the three basic circuits operate together to form the basis of all communications systems.

INTRODUCTION

Although we assume you have studied the circuits discussed in this chapter, they are the building blocks and, as such, key elements in the understanding of any communications system. Therefore, this chapter reviews these key circuits: filters, tuned circuits and resonance, oscillators, and amplifiers.

2-1 FILTERS

Electrical filters are designed to attenuate unwanted frequencies or to amplify desired frequencies. Attenuation is accomplished by using passive filters that contain only resistors, capacitors, and inductors. Amplification is accomplished by using active filters that utilize resistors, capacitors, and an active device, usually an operational amplifier (op amp).

In this section, we will review the fundamentals, types, characteristics, and configurations of filters.

Filter Fundamentals

A *filter* is a device that screens out or passes electric current of only certain frequencies. The two basic categories of filters are passive and active.

Passive filters are constructed with resistors, capacitors, and inductors. They contain no active devices, such as vacuum tubes, transistors, or op amps. The inductors used in passive filters at low audio frequencies are bulky and relatively expensive, but at higher frequencies the inductors are smaller and less expensive.

Active filters are constructed with a network of resistors and capacitors around an active device, usually an op amp. Inductors are seldom used in active filters. Hence, active filters are more often used in low- to medium-frequency equipment. The limiting frequency response of op amps precludes their use at much higher frequencies.

Two basic techniques to attenuate or reject the passage of undesired frequencies are the following:

1. Place a high impedance in series with the signal path to oppose current flow at the undesired frequencies.
2. Place a low-impedance shunt across the signal path so that the undesired frequencies are by-passed to ground.

Both techniques are examined in the following discussion.

Filter Classifications

Filters are classified according to the range of frequencies they reject or allow to pass. They are classified as low-pass (LP), high-pass (HP), band-pass (BP), and band-reject (notch). The frequency pass band implied by the classifications is relative to the frequency band being encountered. For example, an LP filter for an audio system may pass frequencies from dc to 400 Hz, whereas an LP filter at RF frequencies may pass frequencies from dc to several megahertz.

Low-Pass Filters   *Low-pass* filters (Figure 2-1) pass frequencies below a given value and reject frequencies above that value. When an inductor is used for the series component, there is little or no opposition at low frequencies or dc but an increasing opposition to current at higher frequencies. The value of the capacitor placed across the applied signal voltage is chosen so that it will offer little opposition to the frequencies above cutoff but great opposition to the frequencies below cutoff. This capacitor bypasses the unwanted frequencies.

High-Pass Filters   *High-pass* filters (Figure 2-2) reject lower frequencies and pass only frequencies above a given value. These circuits operate in the same manner as LP filters, but the components are interchanged. The value for the series capacitor is chosen so that it offers little or no opposition to high frequencies but increases opposition to

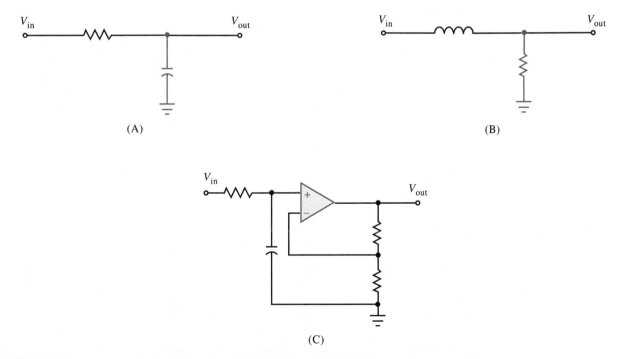

(A)

(B)

(C)

FIGURE 2-1   Low-pass filters (A) Passive *RC* (B) Passive *LC* (C) Active

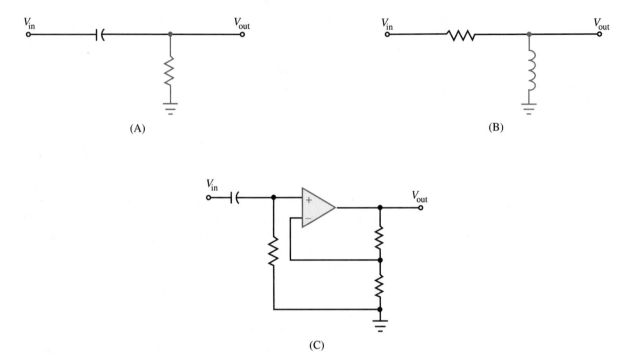

FIGURE 2-2   High-pass filters (A) Passive *RC* (B) Passive *LC* (C) Active

low frequencies. If an inductor is connected across the input signal voltage, inductive reactance decreases as frequency decreases, thus bypassing the unwanted frequencies.

Band-Pass Filters   *Band-pass* filters (Figure 2-3) pass a band of frequencies between a designated upper value and a designated lower value and reject frequencies above and below those designated values. They can be *narrow-band* or *wide-band*. The higher the Q of the circuit, the narrower the bandwidth. Figure 2-3(C) demonstrates the relationship between Q and bandwidth. *Bandwidth (BW)* can be calculated by:

$$BW = \frac{f_r}{Q} \qquad (2\text{-}1)$$

where

   f_r = resonant frequency of the tank circuit

   Q = quality of the circuit

Band-Reject Filters   *Band-reject* filters (Figure 2-4) prevent a band of frequencies between two designated values from passing and pass those frequencies above and below that band. The band-reject filter is also called a *band-stop* filter, or a *notch* filter. The band-reject filter is used extensively in noise elimination circuits.

Wave Traps   A form of band-reject filter called a *wave trap* is sometimes used in the antenna systems of receivers or HF equipment. The two commonly used wave traps,

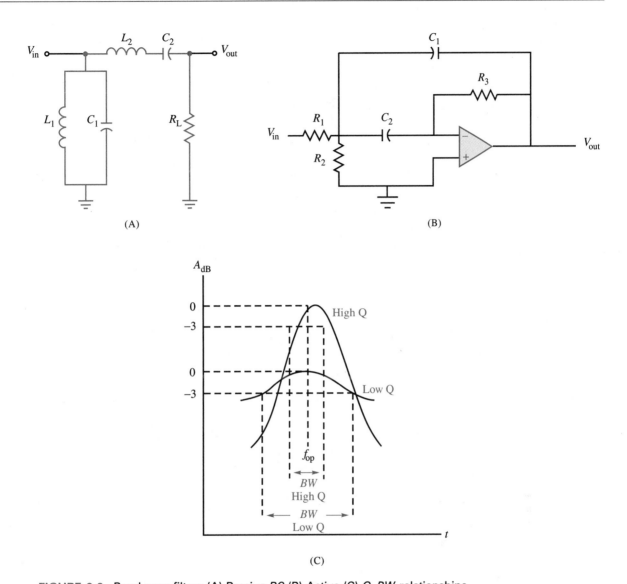

FIGURE 2-3 Band-pass filters (A) Passive *RC* (B) Active (C) *Q–BW* relationships

the *parallel-tuned* filter and the *series-tuned* filter, are shown in Figure 2-5. Parallel tuned circuits are often called *tank circuits*. These wave traps are used to prevent interference from unwanted signals or from harmonics of the desired frequency. The tank circuit in series with the antenna is tuned to resonance at the undesired frequency. The circuit presents a high impedance to currents at the undesired frequency but allows currents at other frequencies to pass with little or no opposition.

The series-tuned circuit connected between the antenna and the receiver's ground terminal is tuned to resonance at the undesired frequency, thus offering low impedance and bypassing the undesired signal to ground. The desired signal frequencies are unaffected and are passed to the receiver.

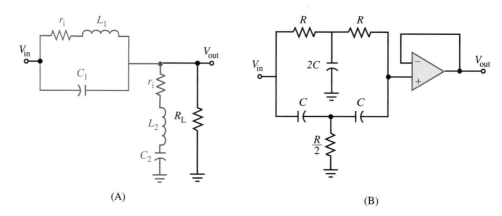

FIGURE 2-4 Band-reject filters (A) Simple passive circuit (B) Active twin-T circuit

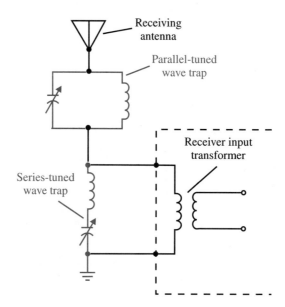

FIGURE 2-5 Wave traps

Decibels To provide a better understanding of the relationship between filter output and frequency, a discussion of the *decibel* is introduced here. Decibel figures are used extensively in electronics to indicate circuit gain, attenuator losses, and antenna power gain figures, each of which is discussed later in this text.

The *decibel (dB) is a means of measuring relative levels of current, voltage, or power.* A reference current I_o, or voltage E_o, or power P_o must first be established. Then the *ratio* (expressed as decibels or db) of an arbitrary current I to the reference current I_o is given by:

$$dB = 20 \ \log_{10}(I/I_o) \tag{2-2a}$$

For voltage, substitute E for I:

$$dB = 20 \ \log_{10}(E/E_o) \tag{2-2b}$$

A negative decibel figure indicates that I is smaller than I_o or that E is smaller than E_o. A positive decibel value indicates that I is larger than I_o, or that E is larger than E_o. A tenfold increase in current or voltage, for example, is a change of *+20 dB:*

$$dB = 20 \ \log_{10}(10 \ E/E_o) \tag{2-2c}$$

$$= 20 \ \log_{10}(10)$$

$$= 20 \times 1 = 20$$

For power, the ratio of an arbitrary wattage P to the reference wattage P_o is given by:

$$dB = 10 \ \log_{10}(P/P_o) \tag{2-2d}$$

As with current and voltage, a negative decibel value indicates that P is less than P_o; a positive value indicates that P is greater than P_o. A tenfold increase in power represents an increase of +10 decibels:

$$dB = 10 \ \log_{10}(10P/P_o) \tag{2-2e}$$

$$= 10 \ \log_{10}(10)$$

$$= 10 \times 1 = 10$$

The reference value for power is sometimes set at *1 milliwatt (0.001 watt)*. Decibels measured relative to 1 milliwatt, across a pure resistive load of 600 ohms, are abbreviated *dBm*. Decibels measuring relative levels of noise are abbreviated *dBa*. Decibels measuring the power gain of an antenna are abbreviated *dBd* primarily, but sometimes *dBi*. Each of these designations is discussed in the appropriate chapters.

Decibel Meter   A *decibel meter* is used to indicate the level of current, voltage, or power, in decibels, relative to some fixed reference value. The reference value may be arbitrary, or it may be some specific quantity, such as 1 milliwatt or 1 volt. In any case, the reference corresponds to *0 dB* on the meter scale. Levels greater than the reference level are assigned positive decibel values on the scale. Levels lower than the reference are assigned negative values. Figure 2-6 illustrates the scale of a typical decibel meter.

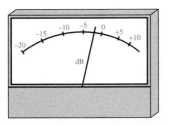

FIGURE 2-6   Decibel meter

Filter Output Relationships   Graphic representations of these various filters show the relationship between filter output in decibels (dB) and the frequency in hertz (Hz). Figure 2-7 illustrates generalized response curves of the filters discussed; we assume a constant amplitude for all input signal frequencies. Note that unwanted frequencies are not completely rejected but *roll off* in a sharp curve. The designation f_c is the *frequency cutoff* and is the limiting frequency for the filter.

The range of frequencies passing through a filter with maximum gain or minimum attenuation is called the *pass band*. The range of frequencies attenuated is called the *stop band*. The value of the cutoff frequency determines the bandwidth of the pass band.

The *transfer function* of a filter is considered to be its gain or amplitude response. The transfer function is represented by the relationship between the filter's output voltage and its input voltage at various frequencies and expressed as V_{out}/V_{in}, a function of

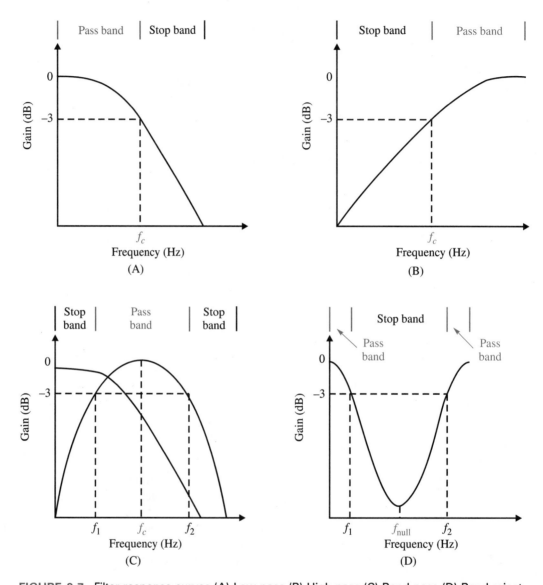

FIGURE 2-7 Filter response curves (A) Low-pass (B) High-pass (C) Band-pass (D) Band-reject

frequency. This ratio is more conveniently expressed in terms of decibels to eliminate the use of decimal voltage values:

$$A_{dB(max)} = 20 \log A_v \qquad \text{(2-3a)}$$

where A = a numerical value,

$$A_v = \frac{V_{out}}{V_{in}} \qquad \text{(2-3b)}$$

and V_{out}, V_{in} = volts

The gain A_v may be equal to, greater than, or less than (=, >, <) unity. A resistive network will attenuate a signal, which results in a loss, or a gain of less than unity. This result is shown as a −dB value. An amplifier in the system may provide a positive or greater than unity dB value. Unity gain is represented by 0 dB.

The frequency at which the gain of the filter circuit drops to 70.7% of its maximum value is the cutoff frequency f_c. The frequency or frequencies at which this occurs identify the filter's pass band limits. The cutoff frequency point is variously called the *0.707 point,* the *−3 db point,* the *break point,* and the *half-power point.* Expressed in equation form,

in dB

$$f_c \text{ point} = \text{AdB}_{max} - 3 \text{ dB} \tag{2-4a}$$

or in volts

$$f_c \text{ point} = 0.707 A_{v(max)} \tag{2-4b}$$

where

$$A_{v(max)} = \frac{V_{out}}{V_{in}} \tag{2-4c}$$

Filter performance is usually expressed by the number of decibels the signal is increased or attenuated per decade or per octave, as a function of frequency. This increase or decrease is called the *rolloff* or *fall-off* of the response curve. A *decade* represents a tenfold increase or decrease in frequency, and an *octave* represents a doubling or halving of a given frequency.

The steepness of the roll-off is a measure of the efficiency of the filter. At 20 dB/decade, we have a first-order filter with limited efficiency. A steeper roll-off, with a higher efficiency, is achieved with a higher-order filter. For example, a second-order filter provides 40 dB/decade, a third-order filter provides 60 dB/decade, and a fourth-order filter provides 80 dB/decade.

EXAMPLE 2-1

An active filter has $V_{in} = 1 \text{ V}_{p-p}$ and $V_{out} = 3 \text{ V}_{p-p}$. Determine (a) the gain of the circuit in decibels, and (b) the cutoff point in decibels. (c) Assuming the cutoff frequency is 1 kHz and the roll-off is 20 dB/decade, find the frequency at 20 dB down from $A_{dB(max)}$.

SOLUTION

(a) From Equation 2-3b,

$$A_v = \frac{V_{out}}{V_{in}} = \frac{3}{1} = 3$$

From Equation 2-3a,

$$A_{dB(max)} = 20 \log A_v = 20 \log 3$$

$$= 20(0.477) = 9.5 \text{ dB}$$

(b) From Equation 2-4a,

f_c point $= A_{dB(max)} - 3$ dB $= 9.5 - 3 = 6.5$ dB

(c) At 20 dB down from $A_{dB(max)}$, $9.5 - 20 = -10.5$ dB. Since one decade represents a tenfold increase and the roll-off is 20 dB/decade, the frequency at -10.5 dB will be $10(1 \text{ kHz}) = 10$ kHz.

EXAMPLE 2-2

A BP filter has a resonant frequency of 50 kHz and a Q of 100. Determine BW.

SOLUTION

From Equation 2-1,

$$BW = \frac{f_r}{Q} = \frac{50 \text{ kHz}}{100} = 500 \text{ Hz}$$

2-2   TUNED CIRCUITS AND RESONANCE

The inductive reactance (X_L) of a circuit varies directly as the frequency, and the capacitive reactance (X_C) varies inversely as the frequency. That is, inductive reactance will increase and capacitive reactance will decrease as the frequency is increased. Because of these opposite characteristics, for any LC combination there must be a frequency at which X_L equals X_C, since one increases while the other decreases. This case of equal and opposite reactances is called *resonance,* and the ac circuit is then a *resonant circuit.*

Any LC circuit can be resonant; it all depends on the frequency. The frequency at which the opposite reactances are equal is the *resonant frequency (f_r).* At the resonant frequency, an LC circuit provides the resonance effect. Above or below the resonant frequency, the LC combination is just another ac circuit and acts inductively or capacitively.

Tuning is the most common application of resonance in RF circuits. Tuned (resonant) circuits produce the phenomenon of resonance, the basis for frequency selectivity in communications systems. For example, resonant circuits allow a radio or television receiver to select a specific frequency transmitted by a particular station and to eliminate frequencies from all other stations. Also, resonant circuits are widely used in filter applications. Finally, resonant circuits are often combined with amplifier circuits in some manner. Both filters and amplifiers are covered elsewhere in this chapter.

Resonance occurs only under very specific conditions in circuits having combinations of inductance and capacitance. In general, we can say that large values of L and C provide a relatively low resonant frequency; smaller values of L and C provide higher values for the resonant frequency. The resonance effect is most useful for radio frequencies, where the required values of microhenries for L and picofarads for C are easily obtainable.

In this section, we will examine the conditions that produce resonance and review the characteristics of series and parallel resonant circuits.

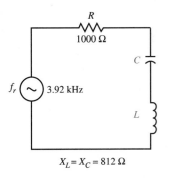

$X_L = X_C = 812\ \Omega$

FIGURE 2-8 Series resonant circuit

Series Resonance

Figure 2-8 illustrates the *series resonant* condition, the condition that occurs when the current in a series *RLC* circuit is in phase with the voltage across the circuit, the impedance is low at the resonant frequency and all other frequencies are highly attenuated, and inductive reactance is equal to capacitive reactance. In a series resonant circuit, the total impedance (Z_T) is purely resistive. The following equations fully define the series resonant condition:

$$f_r = \frac{1}{2\pi\sqrt{LC}} \tag{2-5}$$

$$X_L = X_C \tag{2-6a}$$

where

$$X_L = 2\pi f_r L \tag{2-6b}$$

and

$$X_C = \frac{1}{2\pi f_r C} \tag{2-6c}$$

$$Z_T = \sqrt{R^2 + (X_L - X_C)^2} \tag{2-7a}$$

Since $X_L = X_C$, the X terms cancel, and we obtain

$$Z_T = R \tag{2-7b}$$

Series Impedance   At the resonant frequency in a series *RLC* circuit, $X_L = X_C$, so the circuit impedance is purely resistive. At frequencies below f_r, $X_C > X_L$, and the circuit impedance is capacitive. At frequencies above f_r, $X_C < X_L$, and the circuit impedance is inductive.

From these statements, we conclude that the magnitude of *impedance is a minimum at resonance* $(Z_T = R)$ and increases above and below the resonant frequency. Figure 2-9 graphically illustrates how impedance changes with frequency.

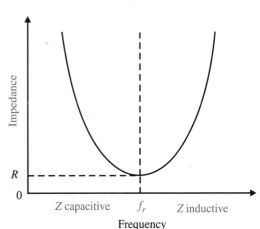

FIGURE 2-9 Impedance vs. frequency response curve for a series RLC circuit

EXAMPLE 2-3

For the series *RLC* circuit in Figure 2-8, determine Z_T, L, and C.

SOLUTION

$Z_T = R$ at the resonant frequency = 1000 Ω

$$X_L = 2\pi f_r L$$

and by algebraic manipulation, we have

$$L = \frac{X_L}{2\pi f_r} \tag{2-8}$$

$$= \frac{812\,\Omega}{(6.28)(3920\,\text{Hz})} = 33\,\text{mH}$$

$$X_C = \frac{1}{2\pi f_r C}$$

and by algebraic manipulation, we have

$$C = \frac{1}{2\pi f_r X_C} \tag{2-9}$$

$$= \frac{1}{(6.28)(3920\,\text{Hz})(812\,\Omega)} = 0.05\,\mu\text{F}$$

EXAMPLE 2-4

For a series *RLC* circuit, let $R = 10$ Ω, $L = 100$ mH, and $C = 0.01$ μF. Determine f_r, Z_T, X_L, and X_C. Then determine X_L, X_C, and Z_T at $f_r - 1$ kHz and at $f_r + 1$ kHz.

SOLUTION

$$f_r = \frac{1}{2\pi \sqrt{LC}} = \frac{1}{6.28\sqrt{(100\,\text{mH})(0.01\,\mu\text{F})}}$$

$$= 5.03\ \text{kHz}$$

Impedance at resonance is equal to *R*:

$$Z_T = R = 10\,\Omega$$

$$X_L = 2\pi f_r L = (6.28)(5.03\,\text{kHz})(100\,\text{mH})$$

$$= 3.16\ \text{k}\Omega$$

At resonance, $X_C = X_L$; therefore,

$$X_C = 3.16\ \text{k}\Omega$$

At $f_r - 1$ kHz, $f = 4.03$ kHz. Substituting f for f_r gives

$$X_L = (6.28)(4.03\ \text{kHz})(100\ \text{mH}) = 2.53\,\text{k}\Omega$$

$$X_C = \frac{1}{(6.28)(4.03 \text{ kHz})(0.01 \text{ }\mu\text{F})} = 3.95 \text{ k}\Omega$$

$$Z_T = \sqrt{R^2 + (X_L - X_C)^2}$$

$$= \sqrt{(10 \text{ }\Omega)^2 + (2.53 \text{ k}\Omega - 3.95 \text{ k}\Omega)^2}$$

$$= 1.42 \text{ k}\Omega$$

Notice that X_C is greater than X_L; therefore Z_T is more capacitive

At $f_r + 1$ kHz, $f = 6.03$ kHz. Substituting f for f_r gives

$$X_L = (6.28)(6.03 \text{ kHz})(100 \text{ mH}) = 3.79 \text{ k}\Omega$$

$$X_C = \frac{1}{(6.28)(6.03 \text{ kHz})(0.01 \text{ }\mu\text{F})} = 2.64 \text{ k}\Omega$$

$$Z_T = \sqrt{R^2 + (X_L - X_C)^2}$$

$$= \sqrt{(10 \text{ }\Omega)^2 + (3.79 \text{ k}\Omega - 2.64 \text{ k}\Omega)^2}$$

$$= 1.15 \text{ k}\Omega$$

Note that X_L is now greater than X_C; therefore, Z_T is now more inductive. The conclusion to be drawn from this example is verified by the graph in Figure 2-9: That is, at resonance, $Z_T = R$; at a frequency below the resonant frequency, the circuit impedance is capacitive; and at a frequency above the resonant frequency, the circuit impedance is inductive.

Series Current and Voltages   At *resonance* in the series *RLC* circuit, circuit impedance is a minimum, so *circuit current is a maximum ($I_{MAX} = V_S/R$)*. At frequencies above and below resonance, the current decreases because the impedance increases. Figure 2-10 illustrates the inverse relationship between current and impedance as a function of frequency.

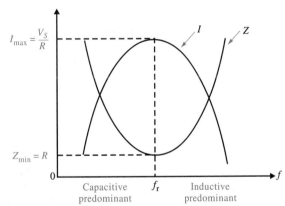

FIGURE 2-10 Inverse relationship between current and impedance for a series resonant circuit

The voltages across components in a series RLC circuit are maximum at resonance but drop off above and below f_r, as illustrated in Figure 2-11. Therefore, we can conclude that the circuit responds best to signals having f_r and tends to reject signals with frequencies above and below f_r. This type of response is basic to a band-pass filter, which was discussed earlier in this chapter.

Notice that the voltages across L and C at resonance, as shown in Figure 2-11, are exactly equal in amplitude. However, they are 180° out of phase, so they cancel. Therefore, the voltage across L and C together is zero, and the voltage across R is equal to V_S at resonance. Individually, V_L and V_C can be much greater than the source voltage, as we now show with an example.

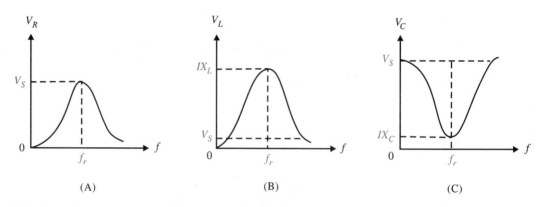

FIGURE 2-11   Response curves for components in a resonant circuit (A) Resistance voltage vs. frequency (B) Inductive current vs. frequency (C) Capacitive current vs. frequency

EXAMPLE 2-5

In a series resonance circuit, $V_S = 50$ V, $R = 25$ Ω, and $X_L = X_C = 100$ Ω. Calculate the circuit current and the voltage amplitudes across R, L, and C.

SOLUTION

Apply Ohm's law. At resonance, circuit current is a maximum and equal to V_S/R:

$$I = \frac{V_S}{R} = \frac{50 \text{ V}}{25 \text{ Ω}} = 2 \text{ A}$$

$$V_R = IR = (2 \text{ A})(25 \text{ Ω}) = 50 \text{ V}$$

$$V_L = IX_L = (2 \text{ A})(100 \text{ Ω}) = 200 \text{ V}$$

$$V_C = IX_C = (2 \text{ A})(100 \text{ Ω}) = 200 \text{ V}$$

You can see from Example 2-5 that all of the source voltage is dropped across the resistor. Notice also that V_L and V_C are equal in amplitude and at a much higher voltage than the source voltage. However, their 180° out-of-phase relationship causes their voltage to cancel, so the total voltage drop across the reactive components together is zero.

Parallel Resonance

Parallel resonance occurs at the frequency at which $X_L = X_C$ in a parallel *LC* circuit. The current in a parallel resonant *LC* circuit is in phase with the source voltage. The phenomenon of resonance, which we have described in terms of series *RLC* circuits, is applicable to parallel *LC* circuits. However, a different set of operating conditions exists because of the application of the circuit. The parallel resonant circuit is often called a *tank circuit* because energy is stored by the inductance and capacitance.

An Ideal Parallel Resonant Circuit   An *ideal* parallel resonant circuit, shown in Figure 2-12(A), has zero resistance. Ideally, at resonance the current in the inductor and the current in the capacitor are equal in magnitude, but they flow in opposite directions. The source voltage is across both reactive branches, and, since $X_L = X_C$, $I_L = I_C$. As illustrated in Figure 2-12(B), I_L lags V_S by 90°, and I_C leads V_S by 90°, which means the inductive and capacitive currents I_L and I_C are equal in magnitude but 180° out of phase with each other. Therefore, they cancel, and the total line current is ideally *zero,* as indicated in Figure 2-12(A). We can then conclude that in an ideal tank circuit, there is a current circulating back and forth, but no current is drawn from the source.

Practical Parallel Resonant Circuits   In practical parallel resonant circuits, there is always some small value of resistance (normally represented by the resistance of the inductor's winding) in series with the inductor. Figure 2-13 shows a practical tank circuit. Because of the small value of resistance, some energy is lost. For this reason, the currents in the circuits are not exactly equal, and the total line current has some small value, which means that there must be some small amount of current drawn from the source to replace the energy lost in the resistance.

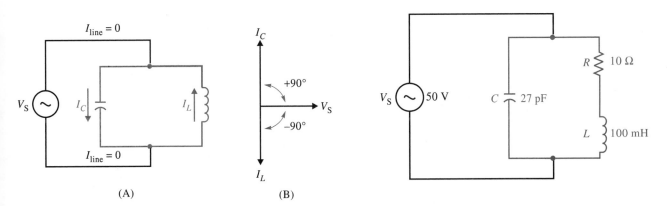

(A)   (B)

FIGURE 2-12 Parallel resonant circuit (A) Ideal circuit (B) Phasor relationship

FIGURE 2-13 Practical parallel resonant (tank) circuit

Parallel Circuit Resonant Frequency   We have stated that a condition for resonance in a parallel *LC* circuit is that $X_L = X_C$. Therefore, the equation for the resonant frequency is the same as for the series *RLC* circuit—that is, $f_r = 1/(2\pi\sqrt{LC})$—but only if $R \ll X_L$ at resonance so that there is negligible energy loss. For most cases where a high-*Q* inductor is used, the equation is valid.

Parallel Circuit Impedance and Current   At the parallel resonant frequency, the *total impedance is a maximum* and purely resistive, and the *total current is therefore a minimum*. The impedance-current relationship as a function of frequency is illustrated in Figure 2-14. Observe that impedance decreases and current increases above and below f_r. At frequencies below f_r, $X_L < X_C$, and because the smaller value in parallel circuits is predominant, the circuit impedance is inductive. At frequencies above f_r, $X_L > X_C$; therefore, the circuit impedance is capacitive. Notice that this characteristic is opposite to that for the series resonant circuit.

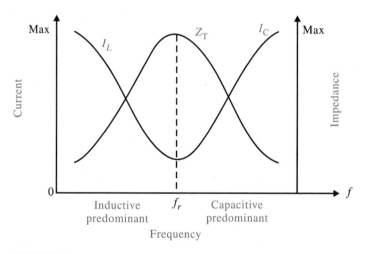

FIGURE 2-14   Inverse relationship between current and impedance for a parallel resonant circuit

The impedance of a parallel resonant circuit can be determined in several ways, as shown by the following equations. Impedance in terms of *R, L,* and *C* can be determined by

$$Z_T = \frac{L}{RC} \tag{2-10a}$$

When both the source voltage and the current are known, impedance can be determined by

$$Z_T = \frac{V_S}{I_S} \tag{2-10b}$$

At frequencies above and below resonance, with *R* ignored because it is so small compared to X_L, impedance can be determined by using the *j operator,* an imaginary number, mathematically defined as the square root of −1. While some mathematicians denote this quantity as the lower case *i,* the lower case *j* is used in electronics to distinguish the quantity from the *i* used to represent current.

The j operator is used to represent reactance. An inductive reactance of X ohms, where X is any positive real number, is written *jX,* or sometimes *Xj.* A capacitive reactance of *–X* ohms is written *−jX* or *−Xj.* The set of possible reactances is the set of imaginary numbers.

A complex impedance is represented by the sum of a pure resistance and an inductive (positive) or capacitive (negative) reactance. If the resistive component is *R,* then the impedance *Z* takes the form of $R + jX$ or $R − jX$.

Complex impedances can be added together, or even multiplied, in this form. The important thing to remember is that when *j* is multiplied by itself, the result is −1.

$$Z_T = \frac{(jX_L)(-jX_C)}{jX_L - jX_C} = \frac{X_L X_C}{jX_L - jX_C} \tag{2-10c}$$

At resonance, Z_T is purely resistive, so there will be no *j* terms. Therefore,

$$Z_T = \frac{R^2}{R} = R \tag{2-10d}$$

At frequencies above and below resonance, and assuming that the reactance value is more than 10 times greater than the resistance of the inductive branch, we have

$$Z_T = \frac{X_L^2}{R} \tag{2-10e}$$

EXAMPLE 2-6

For the practical tank circuit in Figure 2-13, calculate f_r; Z_T at resonance; and Z_T, I_C, I_L, and I_{line} at 2 kHz below resonance and then at 2 kHz above resonance.

SOLUTION

$$f_r = \frac{1}{2\pi \sqrt{LC}} = \frac{1}{6.28 \sqrt{(100 \text{ mH})(27 \text{ pF})}}$$

$$= 96.9 \text{ kHz}$$

At resonance,

$$Z_T = \frac{L}{RC} = \frac{100 \text{ mH}}{(10 \text{ }\Omega)(27 \text{ pF})} = 370 \text{ M}\Omega$$

At $f_r - 2$ kHz, $f = 94.9$ kHz. Substituting f for f_r gives

$$X_L = 2\pi f L = (6.28)(94.9 \text{ kHz})(100 \text{ mH})$$

$$= 59.6 \text{ k}\Omega$$

$$X_C = \frac{1}{2\pi f C} = \frac{1}{(6.28)(94.9 \text{ kHz})(27 \text{ pF})}$$

$$= 62.1 \text{ k}\Omega$$

Ignoring R, we obtain

$$Z_T = \frac{(jX_L)(-jX_C)}{jX_L - jX_C} = \frac{(j59.6 \text{ k}\Omega)(-j62.1 \text{ k}\Omega)}{j59.6 \text{ k}\Omega - j62.1 \text{ k}\Omega} = 1.48 \text{ M}\Omega$$

$$I_C = \frac{V_S}{X_C} = \frac{50 \text{ V}}{62.1 \text{ k}\Omega} = 805 \text{ μA}$$

$$I_L = \frac{V_S}{X_L} = \frac{50 \text{ V}}{59.6 \text{ k}\Omega} = 839 \text{ μA}$$

Since I_L and I_C are 180° out of phase with each other,

$$I_{\text{line}} = I_L - I_C = 839 \text{ μA} - 805 \text{ μA} = 34 \text{ μA}$$

At $f_r + 2$ kHz, $f = 98.9$ kHz. Substituting f for f_r gives

$$X_L = (6.28)(98.9 \text{ kHz})(100 \text{ mH}) = 62.1 \text{ k}\Omega$$

$$X_C = \frac{1}{(6.28)(98.9 \text{ kHz})(27 \text{ pF})} = 59.6 \text{ k}\Omega$$

By ignoring R and observing X_L and X_C, we can see that Z_T, I_C, I_L, and I_{line} are the same values as those calculated for $f_r - 2$ kHz except that, in this case, they are shifted by 180°. Figure 2-14 verifies this relationship.

Bandwidth and Selectivity of Resonant Circuits

The *bandwidth* for either series or parallel circuits is the range of frequencies between the cutoff frequencies for which the I and Z response curves are 70.7% of the maximum values. We can therefore state that the bandwidth is actually the *difference* between f_2 and f_1, as shown in Figure 2-15, and can be determined by

$$BW = f_2 - f_1 \tag{2-11}$$

Ideally, the resonant frequency is the *center frequency*, midway between f_2 and f_1. If BW and either f_2 or f_1 are known, f_r can be determined by either of the following equations:

$$f_r = f_2 - \frac{BW}{2} \tag{2-12}$$

$$f_r = f_1 + \frac{BW}{2} \tag{2-13}$$

The bandwidth of a resonant circuit determines the selectivity of the circuit. *Selectivity* defines how well a resonant circuit responds to a certain frequency while discriminating against all others. The smaller the bandwidth, the better the selectivity.

Another factor related to selectivity is the *sharpness* of the slopes of the response curve. The faster the curve drops off at the cutoff frequencies, the more selective the circuit is. Three general response curves with varying degrees of selectivity are compared in Figure 2-16(A).

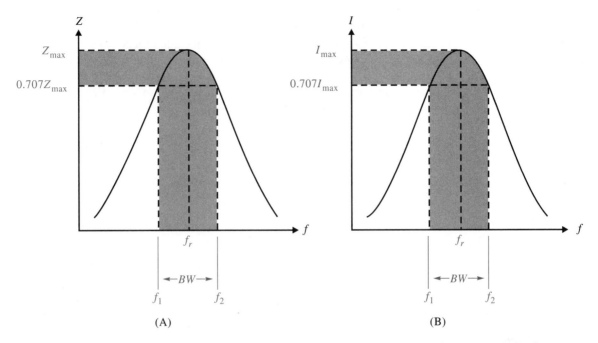

FIGURE 2-15 Bandwidth for resonant circuits (A) Impedance response curve for a parallel circuit
(B) Current response curve for a series circuit

We cannot assume that a resonant circuit accepts only those frequencies within its
bandwidth and completely rejects all other frequencies. This is not the case, because
frequencies outside the bandwidth are not completely eliminated. Their magnitudes
are greatly reduced, however, and the farther those frequencies are from the cutoff frequencies, the greater that reduction is. The shaded area in Figure 2-16(B) illustrates
this concept. Figure 2-16(C) represents an ideal selectivity curve.

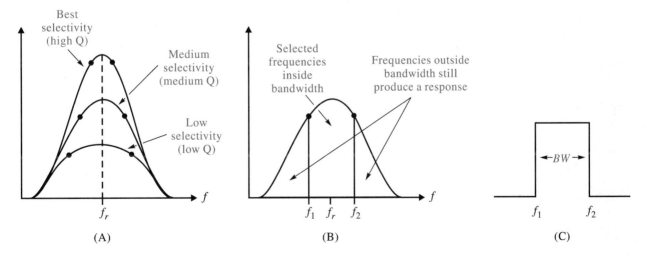

FIGURE 2-16 Representation of bandwidth and selectivity responses (A) Three selectivity response curves
compared (B) Model representing bandwidth (C) Ideal selectivity curve

Quality Factor (Q) of Resonant Circuits

The *quality factor* Q (a numerical value) is the ratio of reactive power (in either the inductor or the capacitor) to resistive power. Stated another way, Q is the ratio of energy stored to energy lost. The quality factor is important in both series and parallel resonant circuits and can be determined by either of the following equations:

$$Q = \frac{X_L}{R} \text{ or } \frac{X_C}{R} \tag{2-14}$$

$$Q = \frac{f_r}{BW} \tag{2-15}$$

where

X_L, X_C, R = ohms

f_r, BW = hertz

When the resistance is just the winding resistance of the inductor, as it is in most parallel resonant circuits, the Q of the circuit and the Q of the inductor are the same. Since Z_L varies with frequency, so does Q. Therefore, we are primarily interested in Q at resonance.

The voltages across L and C in a series *RLC* circuit depend on the circuit Q at resonance. At resonance, the combined voltages across X_L and X_C are zero, so the total source voltage is dropped across R. The following equations tell us that the circuit Q times the source voltage is the voltage across L or C at resonance:

$$V_L = V_C = QV_S \tag{2-16}$$

From these equations, we can conclude that if Q is high, V_L and V_C can be many times greater than V_S. However, since V_L and V_C are 180° out of phase and cancel, the net reactive voltage is zero.

Q, Bandwidth, and Selectivity

Comparing the three general response curves in Figure 2-16(A) shows that a higher value of Q results in a smaller bandwidth. Therefore, a higher Q also provides better selectivity. The bandwidth of a resonant circuit in terms of Q is

$$BW = \frac{f_r}{Q} \tag{2-17}$$

EXAMPLE 2-7

Refer to Figures 2-17(A) and (B). (a) Determine Q at resonance for the two circuits and (b) calculate the capacitor values for both circuits at resonance.

SOLUTION

For Figure 2-17(A),

(a) $X_L = 2\pi f_r L = (6.28)(2 \text{ kHz})(10 \text{ mH})$

$= 125.6 \ \Omega$

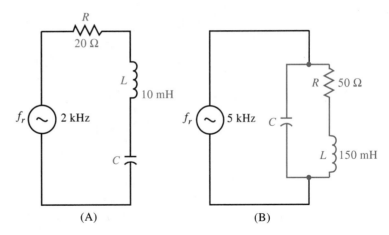

FIGURE 2-17 Figures for Examples 2-7 through 2-9 (A) Series resonant circuit (B) Parallel resonant circuit

$$Q = \frac{X_L}{R} = \frac{125.6\,\Omega}{20\,\Omega} = 6.28$$

(b) Since $X_C = X_L = 125.6\,\Omega$

$$C = \frac{1}{2\pi f_r X_C} = \frac{1}{(6.28)(2\ \text{kHz})(125.6\,\Omega)}$$

$$= 634\ \text{nF}$$

An alternate calculation for C:

$$C = \frac{1}{4\pi^2 f_r^2 L} = \frac{1}{(39.44)(4\ \text{MHz})(10\ \text{mH})}$$

$$= 634\ \text{nF}$$

For Figure 2-17(B),

(a) $X_L = 2\pi f_r L = (6.28)(5\ \text{kHz})(150\ \text{mH})$

$$= 4.71\ \text{k}\Omega$$

$$Q = \frac{X_L}{R} = \frac{4.71\ \text{k}\Omega}{50\,\Omega} = 94.2$$

(b) $C = \dfrac{1}{2\pi f_r X_C} = \dfrac{1}{(6.28)(5\ \text{kHz})(4.71\ \text{k}\Omega)}$

$$= 7\ \text{nF}$$

or,

$$C = \frac{1}{4\pi^2 f_r^2 L} = \frac{1}{(39.44)(25\ \text{MHz})(150\ \text{mH})}$$

$$= 7\ \text{nF}$$

EXAMPLE 2-8

For Figure 2-17(A), assume that $V_S = 10$ V, $X_L = X_C = 1$ kΩ, $R = 50$ Ω, and $f_r = 1.59$ MHz. Determine Q, V_L V_C and BW.

SOLUTION

$$Q = \frac{X_L}{R} = \frac{1 \text{ k}\Omega}{50 \text{ }\Omega} = 20$$

$$V_L = V_C = QV_S = (20)(10 \text{ V}) = 200 \text{ V}$$

$$BW = \frac{f_r}{Q} = \frac{1.59 \text{ MHz}}{20} = 79.5 \text{ kHz}$$

EXAMPLE 2-9

For Figure 2-17(B), assume that $V_S = 20$ V, $X_L = X_C = 1.41$ kΩ, $R = 100$ Ω, and $f_r = 22.5$ kHz. Determine L, C, Q, and BW.

SOLUTION

$$L = \frac{X_L}{2\pi f_r} = \frac{1.41 \text{ k}\Omega}{(6.28)(22.5 \text{ kHz})} = 10 \text{ mH}$$

$$C = \frac{1}{2\pi f_r X_C} = \frac{1}{(6.28)(22.5 \text{ kHz})(1.41 \text{ k}\Omega)}$$

$$= 0.005 \text{ }\mu F$$

$$Q = \frac{X_L}{R} = \frac{1.41 \text{ k}\Omega}{100 \text{ }\Omega} = 14.1$$

$$BW = \frac{f_r}{Q} = \frac{22.5 \text{ kHz}}{14.1} = 1.6 \text{ kHz}$$

2-3   OSCILLATORS

Oscillation is any repetitive motion, such as the swinging of a pendulum, the expansion and compression of a spring, or the continual movement of electrons in an electrical circuit. Oscillation may be short-lived, settling with time; or it may continue indefinitely if an outside source of energy is provided to sustain it.

In electronics, oscillation generally refers to the phenomenon caused by positive feedback. Circuits can be made to oscillate deliberately at a precise frequency. Sometimes oscillation occurs when it is not desired. This is likely to happen when an amplifier has too much gain or is improperly designed for the application intended.

Oscillation is always the result of periodic storage and release of energy. Pendulum energy is stored as the weight rises and released as the weight falls. Spring energy is stored with compression and released with expansion. In an inductance-capacitive circuit, energy is alternately stored and released as it is transferred between the inductive and capacitive reactances.

An *oscillator* is a circuit designed specifically to produce electrical oscillation. All oscillators use the feedback principle. Although there are many different types of oscillator circuits, they all consist basically of an amplifier by which some of the output is applied, in phase, to the input.

An oscillator requires an active amplifying device, such as a transistor, field-effect transistor, vacuum tube, or operational amplifier (op amp), and a power source. If the oscillator is to produce alternating currents at a specific frequency, some form of resonant circuit, such as an inductance-capacitance network or a piezoelectric crystal, is necessary. *Specialized devices* (discussed in detail later in this text), such as the Gunn diode and the klystron tube, produce oscillation because of negative-resistance effects.

Oscillators are important circuits found in almost every branch of electronics. Because of their unique properties, sine wave oscillators have a variety of applications in communications equipment. They are used as carrier frequency generators in the *exciter* (RF power driver) section of a transmitter, as *local oscillators* (LO) in superheterodyne receivers, and as *beat frequency oscillators* (BFO) to detect code signals.

The oscillator is an electronic generator that operates from a dc power source, has no moving parts, and can produce ac signal frequencies ranging into millions of hertz. A well-designed oscillator will have a uniform output, varying in neither frequency nor amplitude. This section reviews oscillator fundamentals and the three basic classifications of sine wave oscillators: *RC* oscillators, *LC* oscillators, and crystal oscillators.

Oscillator Fundamentals

In studying resonant circuits, we discovered that excitation of a tank circuit (such as Figure 2-17(B), where the circuit stores energy by exchanging it alternately between the two reactances) by a dc source tends to cause oscillations in the circuit. These oscillations, or pendulum-like back-and-forth oscillatory motions, are the result of circulating current flow inside the tank circuit. If there were no resistance within the circuit, the oscillations, once begun, would continue indefinitely. However, there is resistance, which dissipates the energy and damps the oscillations.

For continued oscillation, the energy lost due to resistance must be replaced. Simply stated, this is accomplished through *positive (regenerative) feedback,* whereby a small portion of the output signal is fed back to the input of the circuit, as shown in Figure 2-18(A). This feedback must be in phase with the original signal. To accomplish that, we may need a *phase-shift* network to shift the output signal the required 180°, assuming a 180° shift in the device itself. The amplifier gain will then replenish the lost energy.

For an oscillator to be useful, it must have a stable output frequency. Stability is achieved by carefully selecting components in the *frequency-determining network,* as shown in Figure 2-18(B). The parallel *LC* network in the positive feedback loop resonates at a frequency determined by component values. The desired 180° phase shift is produced by the reactance of the components, and the amplifier replenishes the energy lost in the tank circuit.

Figure 2-18(C) shows a frequency-determining *RC* network. The values of the components selected are such that the time constants determine the frequency of oscillation, and the desired phase shift is determined by the number of *RC* sections contained in the feedback loop.

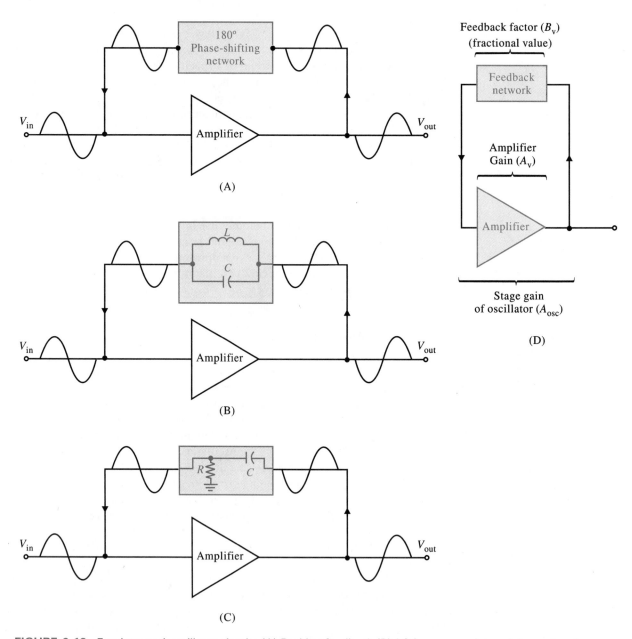

FIGURE 2-18 Fundamental oscillator circuits (A) Positive feedback (B) *LC* frequency-determining network (C) *RC* frequency-determining network (D) Oscillator circuit model for the Barkhausen criterion

Barkhausen Criterion

Examining the basic oscillator circuit in Figure 2-18(D) shows that certain conditions must be met for an oscillator to be self-starting and self-sustaining. For the oscillator to produce its own input signal continuously, the product of the amplifier gain (A_v) and the fractional feedback factor (B_v) (a small fraction of the output signal that is fed back to the input) must equal unity ($A_vB_v = 1$). This condition is called the *Barkhausen criterion.* Mathematically, the oscillator stage gain (A_{osc}) is

$$A_{osc} = \frac{A_v}{1 - A_vB_v} \qquad\qquad (2\text{-}18)$$

When the condition $A_vB_v = 1$ is met, the oscillator stage gain is theoretically infinite, which satisfies one of the requirements for an oscillator: An output signal must be present without an input signal.

Classifications of Oscillators

Oscillators are generally classified according to the components used in the frequency-determining networks. The three basic classifications for sine wave oscillators are resistor-capacitor (*RC*) oscillators, inductor-capacitor (*LC*) oscillators, and crystal oscillators.

Briefly summarized, the basic characteristics common to all oscillators are as follows:

1. *Amplification* is required to replace circuit losses.
2. A *frequency-determining network* is required to set the desired frequency of oscillation.
3. A *positive feedback signal* is required to sustain oscillation.
4. The oscillator is required to be *self-starting,* with no input signal.

RC Oscillators

The *RC oscillator* used resistance-capacitance networks to determine oscillator frequency. Resistors and capacitors are in plentiful supply and relatively inexpensive. Thus, the *RC* oscillator is an inexpensive, easily constructed, relatively stable circuit design for use in LF and AF ranges. The three basic types of sine wave–producing *RC* oscillators are the Wien-bridge oscillator, the phase-shift oscillator, and the twin-T oscillator.

Wien-Bridge Oscillator Figures 2-19(A) and (B) show two versions of a *Wien-bridge* oscillator, the industry-preferred oscillator circuit for low- to medium-range frequencies. It is widely used in commercial audio generators and many other LF applications.

In the Wien-bridge oscillator, the *RC* networks are part of a bridge circuit that provides both positive (regenerative) and negative (degenerative) feedback. The networks select the frequency at which the feedback occurs, but they do not shift the phase of the feedback signal.

In Figure 2-19(B), feedback is applied to both inputs of the op amp. The frequency-selective network, sometimes called the lead-lag network, consists of $C_1 - R_1$ and

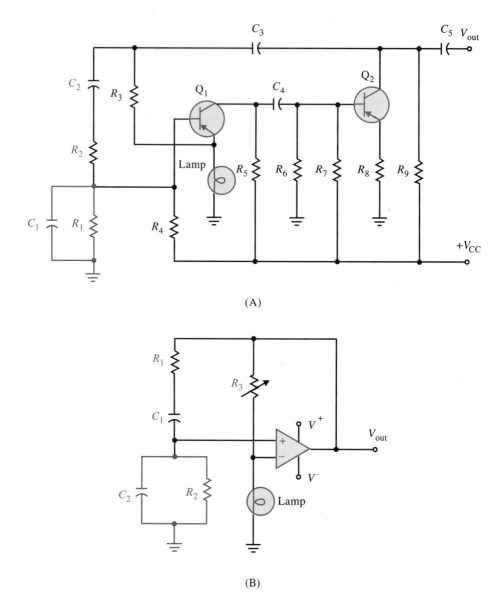

(A)

(B)

FIGURE 2-19 Wein-bridge oscillator (A) Discrete circuit (B) Op amp circuit

$C_2 - R_2$ and provides positive feedback to the noninverting input terminal. Negative feedback is developed across R_3 and the lamp resistance and is applied to the inverting input terminal. Resistor R_3 is made variable so that negative feedback can be reduced, because positive feedback must be greater than negative feedback for the circuit to sustain oscillation. The setting of R_3 is such that the circuit will start oscillating. The ratio of the lamp resistance to R_3 must be greater than 2:1 for proper operation. In most such circuits, the lamp can be replaced with a variable resistor, so that the required ratio can be retained.

Because resistance values in the negative feedback path do not change with frequency, negative feedback remains constant. However, positive feedback depends on the frequency response of the frequency-selective network, which is frequency sensitive. If oscillator frequency begins to increase, the reactance of C_2 will decrease and shunt some positive feedback to ground. Likewise, if the frequency begins to decrease, the reactance of C_1 becomes greater, causing less voltage to be developed across the $C_2 - R_2$ network, which reduces positive feedback. In this manner, the network forces the oscillator to stay on its operating frequency.

The circuit operating frequency is determined by the values of C_1, C_2, R_1 and R_2, and can be calculated as follows:

$$f_{op} = \frac{1}{2\pi \sqrt{R_1 R_2 C_1 C_2}} \qquad (2\text{-}19)$$

However, if $R_1 = R_2$ and $C_1 = C_2$, then

$$f_{op} = \frac{1}{2\pi R_1 C_1} \qquad (2\text{-}20)$$

EXAMPLE 2-10

Refer to Figure 2-19(B). Assume that $R_1 = R_2 = 20$ kΩ and $C_2 = 1$ nF. Calculate f_{op}.

SOLUTION

From Equation 2-20,

$$f_{op} = \frac{1}{2\pi R_1 C_1} = \frac{1}{(6.28)(20 \text{ k}\Omega)(1 \text{ nF})}$$

$$= 8 \text{ kHz}$$

EXAMPLE 2-11

Refer to Figure 2-19(B). Assume that $R_1 = 10$ kΩ, $R_2 = 20$ kΩ, $C_1 = 0.5$ nF, and $C_2 = 1$ nF. Calculate f_{op}.

SOLUTION

From Equation 2-19,

$$f_{op} = \frac{1}{2\pi \sqrt{R_1 R_2 C_1 C_2}}$$

$$= \frac{1}{6.28 \sqrt{(10 \text{ k}\Omega)(20 \text{ k}\Omega)(0.5 \text{ nF})(1 \text{ nF})}}$$

$$= 1.6 \text{ kHz}$$

Phase-Shift Oscillator As the name implies, the *phase-shift* oscillators shown in Figures 2-20(A) and (B) employ phase-shifting *RC* feedback networks that provide the 180° shift necessary to produce positive feedback. As shown, either the resistors or the capacitors can be used in the series feedback path. This type of oscillator is usually used only in fixed-frequency applications.

The phase difference in an *RC* circuit is a function of the capacitive reactance X_C and the resistance *R* of the network. Resistance does not change with frequency, so the capacitor is the frequency-sensitive component. By carefully selecting the components, we can control the amount of phase shift across an *RC* section. Stability can be improved by increasing the number of *RC* sections. Typically, three *RC* sections are used.

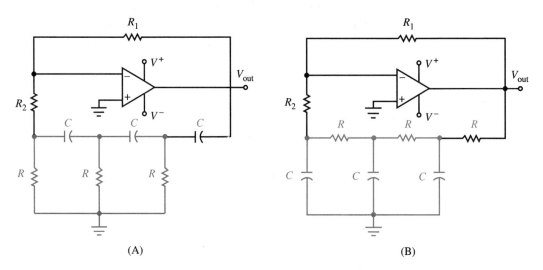

(A)

(B)

FIGURE 2-20 Phase-shift oscillator (A) Lead networks (B) Lag networks

The operating frequency for a phase-shift oscillator is

$$f_{op} = \frac{1}{2\pi RC\sqrt{6}}$$ (2-21)

where all resistors (*R*) are identical in each of the sections and all capacitors (*C*) are identical in each of the sections.

For oscillations to start, A_v in this circuit must be greater than 29 to satisfy the Barkhausen criterion. The gain A_v is set by the ratio of resistors R_1 and R_2. Resistor R_2 is often made variable to ensure that the exact required ratio is achieved.

EXAMPLE 2-12

Refer to Figure 2-20(A). Assume that each *RC* section has values of $R = 1$ kΩ and $C = 0.1$ µF. Calculate f_{op}.

SOLUTION

From Equation 2-21,

$$f_{op} = \frac{1}{2\pi RC\sqrt{6}} = \frac{1}{(6.28)(1\ k\Omega)(0.1\ \mu F)(2.45)}$$

$$= 650\ Hz$$

Twin-T Oscillator   The *twin-T* oscillator in Figure 2-21 provides a 180° phase shift from the active device output to its input, just as do the phase-shift oscillators in Figure 2-20. A positive feedback is through the voltage divider, composed of resistor R_1 and the variable resistance lamp, to the noninverting input. Negative feedback is through the twin-T filter network to the inverting input. When power is first applied to the circuit, lamp resistance is low and positive feedback is maximum. As oscillations increase, lamp resistance increases and positive feedback decreases. With decreasing feedback, oscillations level off and become constant. We can therefore state that the lamp stabilizes the output voltage level of the oscillator.

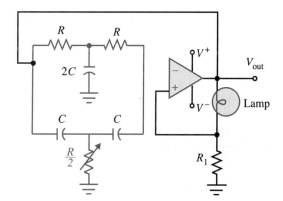

FIGURE 2-21
Twin-T oscillator

When compared with the phase-shift oscillator, the twin-T oscillator is much more stable and has less output distortion. The twin-T circuit is most effective when used with op amps. However, the frequency of oscillation of the circuit is limited by the op amp's frequency range. The frequency of oscillation for the twin-T oscillator is

$$f_{op} = \frac{1}{2\pi RC} \tag{2-22}$$

EXAMPLE 2-13

Refer to Figure 2-21. Assume that $R = 100\ k\Omega$ and $C = 10\ pF$. Calculate f_{op}.

SOLUTION

From Equation 2-22,

$$f_{op} = \frac{1}{2\pi RC} = \frac{1}{(6.28)(100\ k\Omega)(10\ pF)}$$

$$= 159.24\ kHz$$

LC Oscillators

The frequency-determining network in the *LC oscillator* is a tuned circuit consisting of inductors and capacitors connected either in series or in parallel. Component values determine the oscillating frequency. One or more of the components can be made variable if a range of frequencies is desired. This section examines Hartley, Colpitts, and Clapp *LC* oscillators.

Hartley Oscillator A widely used *variable-frequency oscillator* (VFO) is the *Hartley* oscillator. The identifying feature of the Hartley oscillator is the *tapped coil,* as shown in Figures 2-22(A) and (B). The capacitor placed across the coil makes the entire coil part of a tuned circuit. Current flowing through L_{1A} replaces energy lost in the tank, thus providing the positive feedback necessary for oscillation. The amount of feedback can be controlled by adjusting the position of the coil tap.

The frequency of oscillation for the Hartley oscillator is approximated by

$$f_{op} = \frac{1}{2\pi \sqrt{L_{eq}C}} \tag{2-23a}$$

with

$$L_{eq} = L_{1A} + L_{1B} + 2L_M \tag{2-23b}$$

and

$$L_M = k \sqrt{L_{1A}L_{1B}} \tag{2-23c}$$

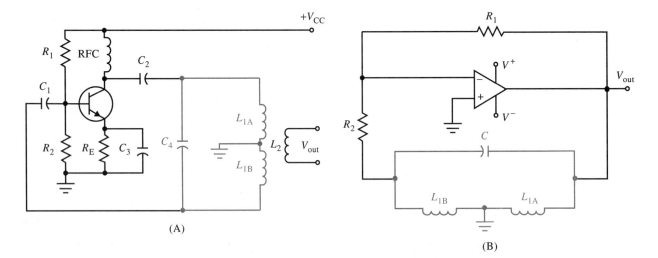

(A)

(B)

FIGURE 2-22 Hartley oscillator (A) Discrete circuit (B) Op amp circuit

where

L_M = mutual inductance

k = coefficient of coupling

For L_{1A} and L_{1B} on paper or plastic form, $k = 0.1$; for L_{1A} wound over L_{1B}, $k = 0.3$; and for L_{1A} and L_{1B} on the same iron core, $k = 1$.

EXAMPLE 2-14

Refer to Figure 2-22(B). Assume that $L_{1A} = L_{1B} = 10$ mH wrapped on paper form and $C = 10$ pF. Calculate f_{op}.

SOLUTION

From Equation 2-23c,

$$L_M = k\sqrt{L_{1A}L_{1B}} = 0.1\sqrt{(10 \text{ mH})(10 \text{ mH})}$$

$$= 1 \text{ mH}$$

From Equation 2-23b,

$$L_{eq} = L_{1A} + L_{1B} + 2L_M$$

$$= 10 \text{ mH} + 10 \text{ mH} + 2(1 \text{ mH})$$

$$= 22 \text{ mH}$$

From Equation 2-23a,

$$f_{op} = \frac{1}{2\pi\sqrt{L_{eq}C}} = \frac{1}{6.28\sqrt{(22 \text{ mH})(10 \text{ pF})}}$$

$$= 339.5 \text{ kHz}$$

There are two disadvantages to the Hartley configuration:

1. The coils tend to be mutually coupled, which makes the frequency of oscillation differ slightly from the calculated frequency.
2. The oscillating frequency cannot easily be varied over a wide range.

The second disadvantage stems from the limited variable range of an inductor. Some limited frequency variation can be made in the Hartley oscillator by making the capacitor variable.

Colpitts Oscillator   One of the most widely used VFO circuits is the *Colpitts* oscillator, shown in Figures 2-23(A) and (B). Note the similarity to the Hartley oscillator. The identifying feature of the Colpitts oscillator is the *tapped capacitor* arrangement. Assuming that the Q of the tank circuit is greater than 10, we can approximate the

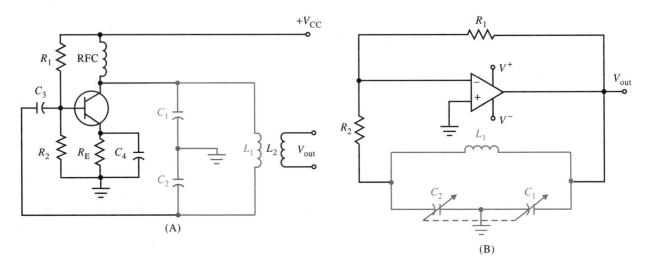

FIGURE 2-23 Colpitts oscillator (A) Discrete circuit (B) Op amp circuit

operating frequency by inductor L_1 and the series combination of capacitors C_1 and C_2, so that

$$f_{op} = \frac{1}{2\pi \sqrt{LC_{eq}}}$$ (2-24a)

with

$$C_{eq} = \frac{C_1 C_2}{C_1 + C_2}$$ (2-24b)

The disadvantages of the Hartley oscillator are overcome in the Colpitts oscillator. Variable capacitors are readily available. With a wide range of capacitance, a wide range of frequencies can be attained. Note that both capacitors in the Colpitts oscillator in Figure 2-23(B) are variable, allowing the user to vary the frequency over a wide range. However, the ratio of capacitances must remain constant to provide a constant feedback voltage ratio, thus ensuring the condition $A_v B_v = 1$.

The Colpitts oscillator is used extensively in AM and FM radio receivers. The capacitors are ganged (tied together) so that adjusting the tuning dial selects both the mixer/oscillator frequency and the resonant frequencies of the RF amplifier stages. For television reception, where mechanical switching is sometimes used, the two capacitors are fixed, and different values of inductance are switched into the circuit. This is accomplished by mounting a fixed-value inductor for each TV channel on a shaft that can be rotated. Each inductor is put into the oscillator tank circuit, one at a time, as the shaft is rotated, thereby changing the frequency of the circuit to the desired channel frequency.

EXAMPLE 2-15

Refer to Figure 2-23(B). Assume that $C_1 = 0.1$ nF, $C_1 = 1$ nF, and $L_1 = 10$ mH. Calculate f_{op}.

SOLUTION

From Equation 2-24b,

$$C_{eq} = \frac{C_1 C_2}{C_1 + C_2} = \frac{(0.1 \text{ nF})(1 \text{ nF})}{0.1 \text{ nF} + 1 \text{ nF}} = 90.9 \text{ pF}$$

From Equation 2-24a,

$$f_{op} = \frac{1}{2\pi \sqrt{L C_{eq}}} = \frac{1}{6.28 \sqrt{(10 \text{ mH})(90.9 \text{ pF})}}$$

$$= 167 \text{ kHz}$$

Clapp Oscillator   A refined version of the Colpitts oscillator is the *Clapp* oscillator, shown in Figures 2-24(A) and (B). In these circuits, the identifying features are the fixed-value capacitors, C_1, C_2, C_3, with C_3 placed in series with the inductive branch of the tank circuit. The feedback signal still depends on the relative values of C_1 and C_2. However, these capacitors are made much larger than C_3, resulting in capacitive reactances of $X_{C3} > 10X_{C1}$ or $10X_{C2}$, so that the frequency of the tank circuit depends essentially on the series resonance of C_3 and L_1. The resonant frequency f_r is approximated by

$$f_r = \frac{1}{2\pi \sqrt{L_1 C_3}} \tag{2-25}$$

Because of the low impedance of this series resonant circuit, and because of the large capacitance values of C_1 and C_2, the frequency of oscillation is relatively unaffected by changes in active device parameters, which provides added stability.

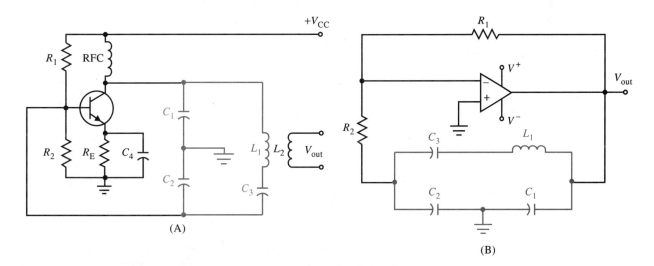

(A)

(B)

FIGURE 2-24   Clapp oscillator (A) Discrete circuit (B) Op amp circuit

EXAMPLE 2-16

Refer to Figure 2-24(B). Assume that $L_1 = 10$ mH and $C_3 = 10$ pF. Calculate the approximate resonant frequency.

SOLUTION

From Equation 2-25,

$$f_r = \frac{1}{2\pi \sqrt{L_1 C_3}} = \frac{1}{6.28 \sqrt{(10 \text{ mH})(10 \text{ pF})}}$$

$$= 504 \text{ kHz}$$

Crystal Oscillators

Oscillator instability is a problem common to all the oscillator circuits discussed thus far. This instability problem stems from several sources: temperature changes, aging of components, Q of the circuit, and circuit design. The use of crystals in the oscillator circuits provides the desired stability.

A crystal used in oscillator circuits must possess the property of *piezoelectricity*—that is, the qualities of (1) generating a voltage across its faces when subjected to mechanical pressure, and (2) compressing when a voltage is applied across its faces. A crystal has a *natural frequency of vibration* f_n, which provides an electrical signal from the crystal. This natural frequency of vibration is extremely constant, which makes the crystal ideal for oscillator circuits.

The natural frequency of a crystal is determined mostly by the crystal's thickness: The thinner a crystal, the higher its natural frequency; the thicker the crystal, the lower its natural frequency. There are, however, practical limits on how thin a crystal may be cut without becoming so fragile that it is easily fractured. In general, a crystal will have an upper limit on its natural frequency of around 50 MHz. For operation at higher frequencies, a tuned output circuit can be used to select a desired harmonic of the crystal's fundamental frequency.

Hartley Crystal-Controlled Oscillator In the Hartley oscillator in Figure 2-25(A), the crystal is connected in series with the feedback path. The *LC* network is tuned to the series resonant frequency of the crystal; therefore, the crystal operates at its series resonant frequency. When the oscillator is operating at the crystal frequency, the equivalent circuit offers minimum impedance. Therefore, there is minimum opposition to current flow, and the feedback is maximum. If the oscillator drifts away from the crystal frequency, the impedance increases, reducing feedback and thereby forcing the oscillator to return to the crystal frequency. We conclude that when the crystal is series connected, it controls the amount of feedback.

Colpitts Crystal-Controlled Oscillator In the Colpitts oscillator shown in Figure 2-25(B), the crystal is also connected in series with the feedback path. The *LC* network is tuned to the crystal frequency, and the crystal controls the amount of feedback.

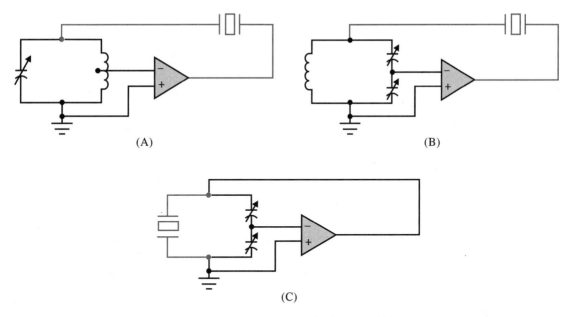

FIGURE 2-25 Crystal-controlled oscillators (A) Hartley (B) Colpitts (C) Pierce

Pierce Crystal-Controlled Oscillator A variation of the crystal-controlled Colpitts oscillator is the *Pierce* oscillator, shown in Figure 2-25(C). In this circuit, a crystal replaces the inductor in a standard Colpitts oscillator circuit. The crystal operates at its parallel resonant frequency, which is slightly higher than its series resonant frequency, and the crystal therefore appears as an inductance. The Pierce oscillator circuit is normally used in systems requiring a low voltage output. Hence, a buffer amplifier (used in many oscillator types) is usually required to increase the voltage to a suitable level. *Buffer* means that loading, such as would decrease the amplitude or affect the frequency of output voltage is not occurring.

The Pierce oscillator is used extensively in radio-frequency applications. The main advantage of the Pierce circuit is that the crystal acts as its own tuned circuit.

Other Oscillator Circuits

Many other oscillator circuits exist. Several are simply variations of those we have already considered. However, for some applications, such as higher-frequency operations, special oscillator circuits are required. This section discusses some of those circuits.

Electron-Coupled Oscillator An *electron-coupled oscillator* (ECO) is one in which one of the previously discussed VFOs is combined with an amplifier circuit. Figure 2-26 shows a junction field-effect transistor (JFET) circuit configured as a Hartley ECO. Capacitors C_3 and C_4 in the output LC circuit effectively complete the ac circuit from drain to ground, thus allowing the circuit to operate as a Hartley oscillator at the fundamental frequency.

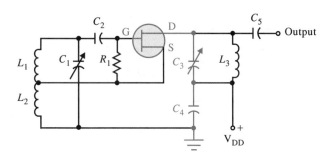

FIGURE 2-26
Electron-coupled
Hartley oscillator

Ultraudion Oscillator For operation at frequencies above about 100 MHz, an *ultraudion* oscillator, such as the series-fed Colpitts vacuum tube circuit shown in Figure 2-27, is used. Because of the interelectrode capacitances of the tube (C_{cp} and C_{cg}), the cathode is brought to the approximate center of the LC circuit capacitance. At such high frequencies, the coils are quite small, and the interelectrode capacitances are large enough to act as a voltage-dividing network across the LC circuit. The radio frequency choke (RFC) can be connected to the center of the coil (as shown) or to either the plate or the grid end of the coil. A radio frequency choke, because of its high X_L, is used to prevent the RF signal from being shorted to ground by the power supply.

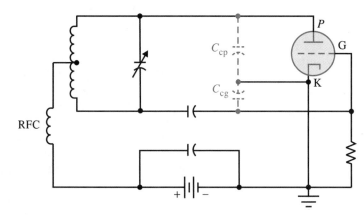

FIGURE 2-27
Ultraudion Colpitts
oscillator

2-4 AMPLIFIERS

One of the most basic, yet one of the most important, building blocks in electronics equipment is the *amplifier*. Various types of amplifiers are used, but they all perform the same basic function. They increase the amplitude, or level, of an electronic signal. Although some communications systems may use vacuum tubes, FETs, or op amps for amplification, we will consider only transistor circuits to demonstrate amplifier configurations, coupling methods, types of biasing, and classes. We will examine both voltage and power amplification in the two basic amplifier systems used for communications: the AF amplifier and the RF amplifier.

Amplifier Circuit Configurations

The three basic circuit configurations for amplifiers are *common-emitter* (CE), *common-collector* (CC), and *common-base* (CB). This section briefly reviews each circuit and its characteristics, advantages, disadvantages, and uses.

Common-Emitter Amplifier A CE amplifier is shown in Figure 2-28. Because the emitter is bypassed to ground, this circuit is sometimes called a *ground-emitter amplifier*. This term does not mean that the emitter is at dc ground, but it is at ac ground through the emitter bypass capacitor C_3. Because of its many desirable features, the CE circuit is the most commonly used of the three basic circuits.

In the CE circuit, the emitter is common to both the input and output signals, with the base and collector leads serving as input and output leads, respectively. A small ac signal applied to the base produces variations in the base current. Because of the transistor *ac current gain* A_i (more commonly known as the *ac beta*, β, or h_{fe}), the collector current is an amplified ac signal of the same frequency. The collector current flowing through the collector resistor R_C produces an *inverted*, amplified output voltage across R_L.

The ac current gain (A_i, β, or h_{fe}) of a transistor is defined as the ratio of the ac collector current i_c to ac base current i_b:

$$A_i = \beta = h_{fe} = \frac{i_c}{i_b} \tag{2-26}$$

The ac voltage gain A_v is defined as the ratio of ac collector voltage v_c to ac base voltage v_b, or as the ratio of ac output voltage v_{out} to ac input voltage v_{in}:

$$A_v = \frac{v_c}{v_b} = \frac{v_{out}}{v_{in}} \tag{2-27}$$

Other factors must be considered also when evaluating any amplifier—for example, the amplifier input impedance z_{in}, the input impedance to the base of the transistor, the

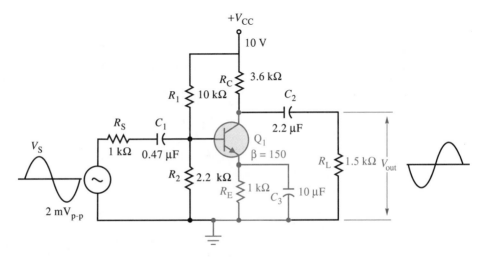

FIGURE 2-28 Common-emitter amplifier

ac resistance of the emitter r'_c, and the loaded and unloaded voltage gains of the stage. In the following example, we calculate these factors.

EXAMPLE 2-17

Refer to Figure 2-28. Calculate the loaded ac output voltage.

SOLUTION

First, calculate the ac emitter resistance. (Using calculus, we can prove that a very good approximation, at room temperature, for the ac emitter resistance for CE circuits equals 25 mV divided by the dc emitter current, but the derivation is beyond the scope of this text.)

$$V_B = \frac{R_2}{R_1 + R_2}(V_{CC}) \tag{2-28}$$

$$= \frac{2.2\,k\Omega}{10\,k\Omega + 2.2\,k\Omega}(10\,V) = 1.8\,V$$

$$I_E = \frac{V_B - V_{BE}}{R_E} \tag{2-29}$$

$$= \frac{1.8\,V - 0.7\,V}{1\,k\Omega} = 1.1\,mA$$

$$r'_e = \frac{25\,mV}{I_E} \tag{2-30}$$

$$= \frac{25\,mV}{1.1\,mA} = 22.7\,\Omega$$

Next, calculate the input impedance seen by the base of the transistor:

$$z_{in\,(base)} = \beta r'_e \tag{2-31}$$

$$= (150)(22.7\,\Omega) = 3.4\,k\Omega$$

Calculate the amplifier input impedance

$$z_{in} = R_1\,|\,R_2\,|\,\beta r'_e| \tag{2-32}$$

$$= 10\ k\Omega\,|\,2.2\ k\Omega\,|\,3.4\ k\Omega = 1.18\,k\Omega$$

Calculate the input voltage seen at the base:

$$v_{in\,(base)} = \frac{z_{in}}{R_s + z_{in}}(v_s) \tag{2-33}$$

$$= \frac{1.18\,k\Omega}{1\,k\Omega + 1.18\,k\Omega}(2\ mV)$$

$$= 1.08\ mV$$

Calculate the unloaded voltage gain. The minus sign is used to denote an inverted output:

$$A_{v(unl)} = \frac{-R_C}{r'} \tag{2-34}$$

$$= \frac{-3.6 \text{ k}\Omega}{22.7 \text{ }\Omega} = -159$$

Calculate the unloaded output voltage:

$$V_{out(unl)} = A_{v(unl)} \times V_{in(base)} \tag{2-35}$$

$$= (-159)(1.08 \text{ mV}) = -172 \text{ mV}$$

Finally, calculate the loaded ac output voltage:

$$V_{out} = \frac{R_L}{R_C + R_L}(V_{out(unl)}) \tag{2-36}$$

$$= \frac{1.5 \text{ k}\Omega}{3.6 \text{ k}\Omega + 1.5 \text{ k}\Omega}(-172 \text{ mV})$$

$$= -51 \text{ mV}$$

A CE circuit provides high current gain and, when suitable load resistances are used, high voltage gain and high power gain. It has low input resistance and high output resistance.

Common-Collector Amplifier A CC amplifier circuit is shown in Figure 2-29. The absence of a collector resistance places the collector at ac ground, so the CC amplifier is sometimes called a *grounded-collector amplifier.*

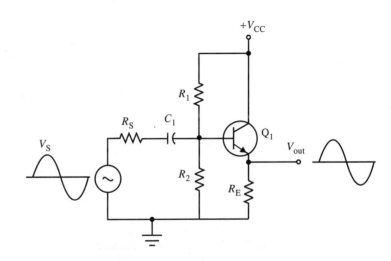

FIGURE 2-29
Common-collector amplifier

In the CC circuit, the collector is common to both the input and output signals, with the base and emitter leads serving as input and output leads, respectively. A small ac signal applied to the base develops a voltage across the emitter resistor that is in phase with, and approximately the same level as, the input voltage; that is, the output voltage follows the input voltage. For this reason, the CC amplifier is also called an *emitter follower.*

The unloaded voltage gain for the CC amplifier is

$$A_{v(unl)} = \frac{v_{out}}{v_{in}} = \frac{R_E}{R_E + r'_e} \tag{2-37a}$$

but R_E is so much larger than r'_e in most CC circuits and, because usually $R_E \ll R_L$, the unloaded voltage gain is approximately equal to the loaded voltage gain, so that the voltage gain approaches unity:

$$A_v \cong 1 \tag{2-37b}$$

The CC amplifier circuit provides high current gain, low (unity) voltage gain, and low power gain. Its major advantages are as follows:

1. It is inherently a low-distortion amplifier.
2. Because of its very high input resistance and very low output resistance, it is ideally suited for impedance matching between high output resistance and low input resistance stages so that efficient power transfer can occur.

Therefore, the CC circuit is used extensively for impedance matching, and thus it is sometimes called a *buffer amplifier.*

Common-Base Amplifier   A CB amplifier circuit is shown in Figure 2-30. In this circuit, the base is common to both the input and output signals, with the emitter and collector leads serving as input and output leads, respectively. Because the base is at ac ground, this circuit is also called a *grounded-base amplifier.*

The voltage gain for the CB amplifier is

$$A_v = \frac{v_{out}}{v_{in}} = \frac{i_c R_C}{i_e r'_e} \tag{2-38a}$$

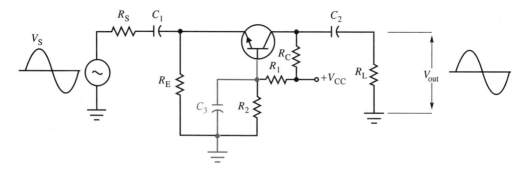

FIGURE 2-30 Common-base amplifier

but since i_c is approximately equal to i_e, the equation can be rewritten as

$$A_v \cong \frac{R_C}{r'_e} \tag{2-38b}$$

The input resistance of the circuit is so low that it overloads most signal sources. Thus, the CB circuit constructed with discrete components is seldom used for low frequencies. However, it is used for frequencies above about 10 MHz, and it is used extensively in integrated circuits (ICs) as part of the differential amplifier input stage.

The CB amplifier circuit produces low current gain (slightly less than unity), high voltage gain, and medium power gain. It has very low input resistance and very high output resistance.

Table 2-1 shows the characteristics and relationships of the three basic amplifier configurations.

TABLE 2-1 Amplifier Characteristic Approximations

Characteristic	CE	CC	CB
Input resistance r_{in} (Ω)	1 k	250 k	50
Output resistance r_{out} (Ω)	50 k	500	1 M
Power gain (with external R_L)	10 k	< 1	1 k
Phase inversion of signal	Yes	No	No
Advantage	High gain	High r_{in}	High r_{out}
Voltage gain	Yes	No	Yes
Current gain	Yes	Yes	No

Comments: The CC amplifier is not useful as a voltage amplifier, but it is good for impedance matching (high Z to low Z). The CB amplifier overloads most inputs, but it is useful at frequencies greater than 10 MHz.

Coupling Circuits

Coupling is a means of transferring energy from one stage of a circuit to another, or from the output of a circuit to a load. Four types of coupling circuits are used between amplifier stages: (1) resistance-capacitance (*RC*) coupling, (2) impedance (a form of capacitive) coupling, (3) transformer (inductive) coupling, and (4) direct coupling.

Resistance-Capacitance Coupling A common method of coupling ac signals from one voltage amplifier to the next is *resistance-capacitance* (*RC*) coupling, shown in Figure 2-31(A). A disadvantage of this type of coupling is that the gain of each stage is affected by its load resistance. For example, the low input impedance of each succeeding stage appears to the ac signal as being in parallel with the collector load resistance of the preceding stage. This effectively lowers the load resistance and reduces the voltage gain of each stage, resulting in an overall voltage gain for the system (the product of the individual stage gains) somewhat lower than expected. However, for audio amplifiers constructed with discrete devices, *RC* coupling is the least costly and most convenient approach.

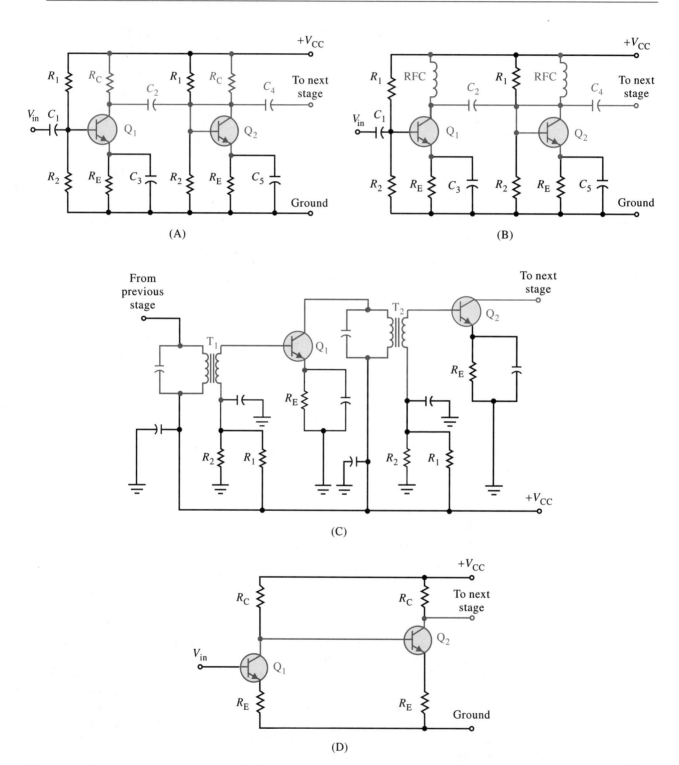

FIGURE 2-31 Coupling circuits (A) *RC* coupling (B) Impedance coupling (C) Transformer coupling (D) Direct coupling

Impedance Coupling In circuits where higher frequencies are encountered, *imped-ance* coupling may be used. As Figure 2-31(B) shows, collector resistors are replaced by inductors, called *RF chokes*. At dc and very low frequencies, the inductive reactance X_L is very low and the inductors appear to be short-circuit, passing dc voltages. However, as frequency increases, X_L approaches infinity and the inductors appear open, effectively blocking ac signals.

Replacing the power-consuming collector resistors with inductors provides the advantage of impedance coupling; inductors have very small power dissipation. Disadvantages are that RF chokes are more costly than resistors, and their impedance decreases at lower frequencies. In general, impedance coupling is used only in RF amplifiers.

Transformer Coupling A two-stage, *transformer*-coupled, based-biased audio amplifier is shown in Figure 2-31(C). The primary winding of coupling transformer T_1 serves as the collector load for the previous stage. The secondary of T_1 is connected in series with the base lead of transistor Q_1. The ac signal developed across the primary winding is inductively coupled to the secondary winding, thus affecting the base current of Q_1. Resistors R_1 and R_2 provide the proper no-signal input base current for Q_1 and Q_2. The output of Q_1 is inductively coupled from the primary winding to the secondary winding of coupling transformer T_2. Additional stages can be added if necessary for sufficient amplification.

One advantage of transformer coupling, which can be used in voltage or power amplifiers, is that the low resistance of the transformer primary winding results in a collector voltage almost equal to V_{CC}—a real advantage in power amplifiers. In addition, the turns ratio of the transformer can be made such that matching the input and output impedances of the two stages is possible, thus allowing maximum transfer of signal power between stages. Likewise, transformer coupling can result in higher gain per stage by using step-up transformers between stages.

Disadvantages of transformer coupling, compared with other types of coupling, are that it is costlier, bulkier, and heavier, and it requires good shielding to prevent picking up hum from nearby electrical power wiring. Less expensive transformers can respond only to a specific range of ac frequencies, thus introducing distortion into the signal. Distortion occurs because insufficient core iron prevents the transfer of low frequencies from primary to secondary, and the inherent characteristic capacitance in the transformer bypasses higher frequencies to ground. Finally, transformer coupling is not used in systems employing multistage audio amplifier ICs, because inductors cannot easily be constructed on the IC chips.

Direct Coupling In *direct* coupling, all bypass and coupling capacitors are eliminated, as shown in Figure 2-31(D). This approach eliminates a problem created at lower frequencies—the need for large capacitors, both electrically and physically. There is no lower frequency limit when using direct coupling; all frequencies down to dc, or zero hertz, are amplified.

A primary disadvantage of direct coupling is drift. *Drift* is an unwanted change in the final output, caused as transistor characteristics vary with temperature. These variations cause the collector currents and voltages to change, and, because of direct coupling, the voltage changes are coupled to succeeding stages and appear in the final output as an amplified voltage. This drift voltage cannot be distinguished from the desired amplified voltage of the input signal.

Biasing Circuits

Before an ac signal can be coupled into an amplifier stage, a *quiescent* (*Q*) point of operation must be established. The *Q* point is the *resting,* or *no-signal,* point on the dc load line, as shown in Figure 2-32. For linear operation, the *Q* point must be near the middle of the dc load line. In that way, the incoming ac signal can fluctuate above and below the *Q* point with minimum distortion. In the section on classes of amplifiers, we will discuss placement of the *Q* point at other than the middle of the dc load line.

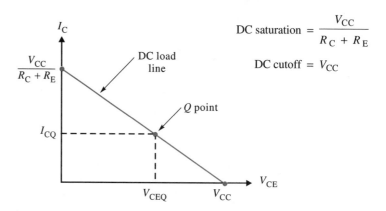

$$\text{DC saturation} = \frac{V_{CC}}{R_C + R_E}$$

$$\text{DC cutoff} = V_{CC}$$

FIGURE 2-32
dc load line
(*Q* point)

Base Bias   An example of *base bias,* also called *fixed bias,* is shown in Figure 2-33(A). Base bias is the least effective way to bias a circuit for linear operation (operation in the active region of load line) because the *Q* point is unstable. This instability occurs because the *dc beta,* β_{dc}, or h_{FE} of a transistor varies greatly with current and temperature. Thus base bias should never be used in linear circuits.

Emitter-Feedback Bias   An example of *emitter-feedback bias* is shown in Figure 2-33(B). Theoretically, in this biasing arrangement, an increased collector current produces more voltage across the emitter resistor, which reduces the base current and, consequently, the collector current, thus compensating for the variations in β_{dc}. However, for effective operation, the emitter resistor must be as large as possible, but to avoid collector saturation the emitter resistor must be relatively small. Because both situations obviously cannot occur simultaneously, in typical designs emitter bias is almost as sensitive to changes in β_{dc} as is base bias. Emitter-feedback bias is, therefore, not a preferred biasing method and should be avoided for linear operation.

Collector-Feedback Bias   An example of *collector-feedback bias,* also called *self-bias,* is shown in Figure 2-33(C). Note the distinguishing feature between collector-feedback bias and base bias: The base resistor is connected to the collector rather than to the power supply. In collector-feedback bias, when β_{dc} increases with temperature, collector current increases, which causes a greater voltage drop across R_C and decreases collector-emitter voltage. Hence, less voltage is across the base resistor, which

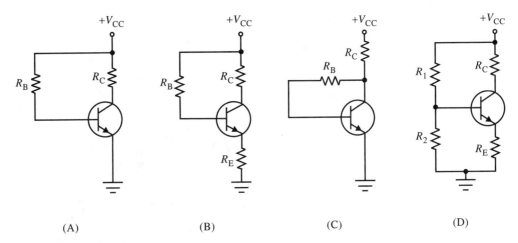

FIGURE 2-33   Biasing circuits (A) Base bias (B) Emitter-feedback bias (C) Collector-feedback bias (D) Voltage-divider bias

decreases the base current. This smaller base current offsets the increase in collector current caused by the change in β_{dc}.

Although collector-feedback biasing is somewhat sensitive to changes in β_{dc}, it is more effective than emitter-feedback bias and is used in practical linear circuits. Advantages offered by this form of bias are as follows:

1. The design, using only two resistors, is simple.

2. The transistor cannot be saturated.

Voltage-Divider Bias   An example of *voltage-divider bias,* also called *universal bias,* is shown in Figure 2-33(D). Resistors R_1 and R_2 form a voltage divider, hence the name. The circuit offers good thermal stability and is immune to changes in β_{dc}. Therefore, voltage-divider bias is the preferred and, by far, the most widely used form of bias in linear circuits.

By ratio, the voltage drop across R_2 should be substantially lower than that across R_1. Resistor R_2 increases thermal stability. As the resistance of the voltage-divider network decreases, the thermal stability increases, but the input impedance decreases and the signal loss increases, due to power dissipated in these resistors. A bypass capacitor across R_E can prevent signal degeneration while maintaining long-term or dc thermal stability.

Classes of Amplifiers

Many modes of operation are used in amplifier circuits. The more common are class A, class B, and class C. Each mode of operation has its own characteristics and, therefore, its own purpose for being.

Class A Operation   *Class A* is the common method of operation in audio voltage amplifier circuits. The circuit operates in the active region at all times, as shown in Figure 2-34. The current, therefore, flows for 360° of the ac signal cycle, thus producing minimum distortion. Class A leads to the simplest and most stable biasing circuits,

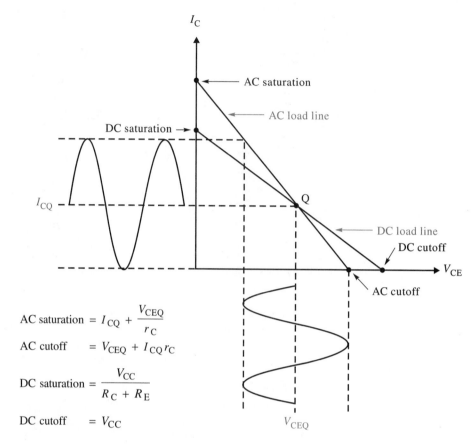

$$\text{AC saturation} = I_{CQ} + \frac{V_{CEQ}}{r_C}$$

$$\text{AC cutoff} \quad = V_{CEQ} + I_{CQ}r_C$$

$$\text{DC saturation} = \frac{V_{CC}}{R_C + R_E}$$

$$\text{DC cutoff} \quad = V_{CC}$$

FIGURE 2-34 Class A operation dc and ac load lines

but it is the least efficient mode in which to operate. Theoretically, in power amplifiers efficiency as high as 50% can be attained, but the normal operating efficiency is around 25%.

Class B Operation   In *class B* operation, shown in Figure 2-35(A), current flows for only 180° of the ac signal cycle. The *Q* point is located at approximately cutoff on both the dc and ac load lines. The advantages of class B operation are lower power dissipation, reduced current gain, and a higher efficiency of about 65%, with a possible efficiency of about 80%. A disadvantage is that, for audio purposes, because of the distortion resulting from the half-cycle current flow, we must use two transistors in *push-pull*. In a push-pull amplifier, illustrated in Figure 2-35(B), one transistor conducts during one half-cycle, and the other transistor conducts during the other half-cycle. The result of the push-pull arrangement is a class B amplifier circuit with large load power and low distortion.

Class C Operation   In *class C* operation, current flows for less than 180° of the ac signal cycle, implying that the circuit is biased below cutoff, as illustrated by

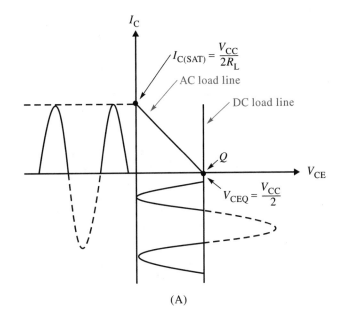

(A)

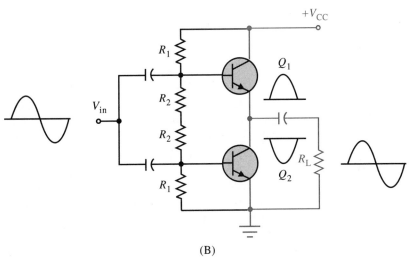

(B)

FIGURE 2-35 Class B operation (A) dc and ac load lines (B) Push-pull circuit

Figure 2-36. The result is a nonsinusoidal wave because current flows in pulses. When narrow current pulses like these drive a high-Q resonant circuit, the voltage across the tuned circuit is almost a perfect sine wave.

To avoid large inductors and capacitors in the resonant circuit, class C amplifiers normally operate only at radio frequencies—that is, above 15 kHz. Even though it is the most efficient of all classes, a class C amplifier is useful only when we want to amplify a narrow band of radio frequencies.

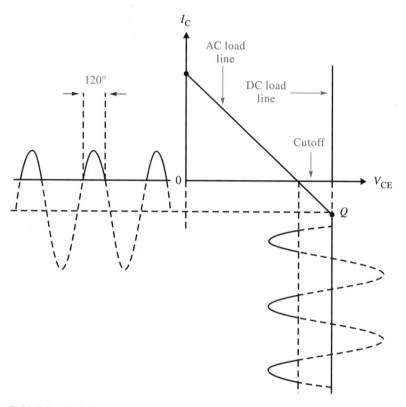

FIGURE 2-36 Class C ac and dc load lines

One important application for the high efficiency of a class C amplifier is an oscillator, which produces an output sine wave without an input signal. Tuned class C amplifiers are also used with *frequency multipliers.* In this circuit, the resonant tank circuit is tuned to a *harmonic,* or multiple, of the input frequency. However, the output power decreases when the circuit is tuned to higher harmonics; that is, the higher the harmonic, the lower the output power. For that reason, a tuned class C amplifier frequency multiplier is ordinarily used only for lower harmonics such as the second or third.

Class C amplifiers are the most efficient method of operation; thus, they produce the most load power. An efficiency of 95% or higher is possible with a well-designed class C circuit, but typical efficiency is about 85%.

Table 2-2 shows the characteristics and relationships for the three basic amplifier modes of operation.

Audio Frequency Amplifiers   The AF range covers all frequencies to which the human ear can respond. The range of frequencies is generally accepted to extend from approximately 20 Hz to 20,000 Hz. The frequencies developed by the human voice fall within this range, as do the various frequencies developed by musical instruments. Therefore, *audio amplifiers* are extremely important in any electronic equipment used

TABLE 2-2   **Typical Amplifier Characteristics**

Characteristic	Class A	Class B	Class C
Conduction (°)	360	180	Less than 180
Q point	Center	Cutoff	Below cutoff
Efficiency (%)	25–50	65–80	85–98
Distortion	Low	High	Severe
Normal use	Audio voltage amplifier	Push-pull audio output	RF power output

for the transmission, reception, and processing of audio signals. This equipment includes such communications systems as radio and television transmitters and receivers, stereo systems, two-way radio communications, telephone systems, public address systems, and all types of audio recorder and playback systems.

Transistor amplifier circuits are generally referred to as current amplifiers. However, they can provide current, voltage, and power amplification. Transistor audio amplifiers are generally designed to operate in the class A mode, which results in a signal with minimum distortion. We will now examine AF voltage and power amplifiers.

Audio Frequency Voltage Amplifiers   Audio amplifiers specifically designed to amplify low-voltage audio signals are broadly classified as *voltage amplifiers*. The output signals from microphones, turntables, and detector stages of receivers are normally in the millivolt or microvolt range and must be increased to a level high enough to drive a power amplifier stage. Therefore, one or more voltage amplifiers may be required in a system.

Voltage-amplifying devices are physically small and operate at very low current and power levels; hence, many such devices may be operated from the same power supply.

A single-stage, common-emitter audio voltage amplifier is shown in Figure 2-37(A). Resistor R_1 controls the input base current and provides the necessary forward bias to provide minimum distortion—that is, class A operation. Emitter feedback is provided by resistor R_3, greatly improving the thermal stability of the circuit. Bypass capacitor C_3 prevents ac input signal degeneration. Resistor R_2 is the collector load resistor. The ac input signal, applied through coupling capacitor C_1, appears between the base and circuit ground, effectively controlling the base-emitter current. The amplified ac output signal appears between the collector and circuit ground and is coupled to the next stage through coupling capacitor C_2. Such an amplifier provides substantial voltage gain over the relatively wide frequency range necessary for audio signals. It cannot amplify dc signals because of the coupling capacitors used in the signal path. Voltage gain is low at the lower end of the frequency spectrum owing to the increase in reactance of the capacitors, and the reduced beta of the transistor with increased frequency causes a similar low gain at the upper end of the spectrum. However, the overall frequency response gain can be increased substantially by careful design.

Two or more amplifier stages can be *cascaded* (connected together) to provide a higher overall gain. A two-stage CE amplifier is shown in Figure 2-37(B).

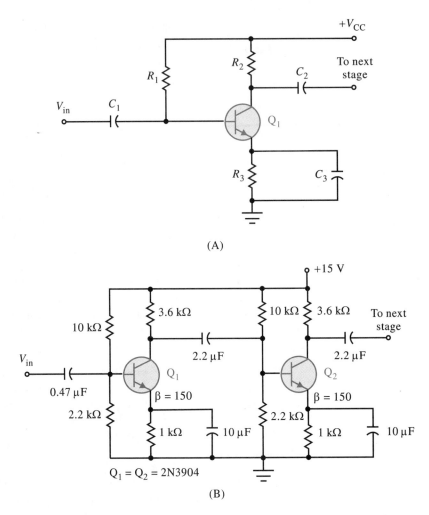

FIGURE 2-37 AF voltage amplifier circuits (A) Single-stage CE amplifier
(B) Two-stage CE amplifier

Audio Frequency Power Amplifiers   A *power amplifier* (PA) is normally the last
stage of amplification before an ac signal leaves the AF system, as shown in the block
diagram of Figure 2-38. The PA stage receives the amplified signal from the voltage
amplifiers, and its output is used to perform some type of work requiring high power,
such as operating loudspeakers in receivers or modulating transmitters.

Power amplifiers are rated in terms of *watts* (W), and the power can be calculated
with the basic power equation $P = V^2/R$, where P = power in watts, V = voltage in rms

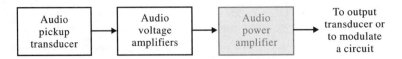

FIGURE 2-38 System block diagram

volts, and R = load resistance in ohms. Power amplifier circuits are designed to work into specific loads. For example, transistorized amplifiers or IC amplifiers in receivers can be safely operated into loads between 4 Ω and 16 Ω. Output load ratings, with safe operating ranges, are provided in manufacturers' specification sheets.

Radio Frequency Amplifiers

There are many similarities between RF amplifiers and AF amplifiers, but there are also many differences. *Radio frequency amplifiers* perform the same amplifying functions at radio frequencies as do AF amplifiers at audio frequencies. Voltage and power amplifiers are used in both types of systems. But since the signal frequencies are very different, and RF amplifiers require much higher selectivity than AF amplifiers, the RF amplifiers normally make use of tuned (resonant) circuits or selective filters to achieve the required selectivity.

Radio Frequency Voltage Amplifiers   *Radio frequency voltage amplifiers* are used in transmitters to amplify RF signals before modulation, in receivers to amplify received RF signals, and as frequency multipliers. The two types of RF voltage amplifiers are the standard broad-band, multiple-frequency RF amplifier (such as those found in transmitter outputs and receiver input circuits), and the *intermediate-frequency* (IF) amplifier, a fixed-frequency RF amplifier commonly used in superheterodyne receivers. The IF amplifier follows the mixer stage and precedes the detector stage. In the receiver, the IF amplifier amplifies a narrow band of frequencies between the AF and RF ranges.

Biasing arrangements used in RF and IF amplifiers are similar to those used in AF amplifiers. A typical RF amplifier circuit is shown in Figure 2-39. What distinguishes

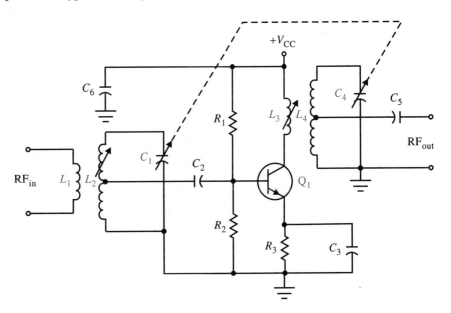

FIGURE 2-39   RF voltage amplifier circuit

RF amplifiers from AF amplifiers is that the input and output circuits are tunable in RF amplifiers. Tunable circuits provide improved selectivity and impedance matching for good transfer of power between stages. In Figure 2-39 transformers L_1L_2 and L_3L_4 are tuned to accept a broad band of RF frequencies. Adjustable capacitors C_1 and C_4 are ganged (tied together) to allow the selection of the desired incoming frequency. Adjusting the capacitors alters the frequency input and output simultaneously, thus providing maximum transfer of power. In receivers, RF voltage amplifiers are normally biased to operate class A, thus enabling 360° collector current flow with minimum distortion.

Intermediate-Frequency Amplifiers　*Intermediate-frequency amplifiers* are similar to RF amplifiers, except that IF amplifiers normally use special slug-tuned transformers in the tank circuits. A typical IF amplifier is shown in Figure 2-40. In this amplifier, the tank circuit transformers are pretuned to a specific intermediate frequency, so that both the primary and secondary windings are resonant at the same frequency. For alignment purposes, the tank resonant frequency can be changed slightly by adjusting slugs in the transformer.

The IF amplifier serves two purposes: to provide high gain and to provide excellent selectivity. For those reasons, transformers used in IF amplifiers have a narrower pass band than transformers used in other communications circuits.

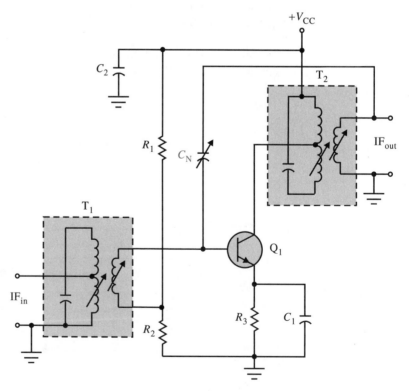

FIGURE 2-40　IF amplifier circuit with neutralization

Neutralizing Unwanted Oscillations   At the high operating frequencies of RF amplifiers, feedback paths are formed between the input and output tuned tank circuits. Because of this feedback, unwanted oscillations can occur at the resonant frequency of the tank circuits. These oscillations must be neutralized.

Neutralization, a method of reducing the possibility of those unwanted oscillations, is accomplished with a degenerative (negative) feedback circuit. A small amount of the output signal is fed back, 180° out-of-phase, to the input, normally through a small, variable capacitor, such as shown by C_N in Figure 2-40. The negative feedback signal effectively cancels the reactive feedback of the circuit.

Radio Frequency Power Amplifiers   *Radio frequency power amplifiers* are designed to deliver amplified current values to a load. In communications applications, RF power amplifiers are used to supply energy to another power amplifier or to a transmitting antenna.

The RF power amplifier can use either a *single-ended* arrangement, shown in Figure 2-41(A), or a *parallel* arrangement, illustrated in Figure 2-41(B). Single-ended arrangements are typically used as output stages in low-power RF devices. With small-value emitter resistors, these circuits can operate either class B or class C. However, the low output impedance of RF power transistors usually makes it necessary to insert an impedance-matching circuit between the output stage and the load. In the circuit of Figure 2-41(A), impedance matching is accomplished by the *pi network* formed by C_5, L_1, and C_6.

Maximum power output and efficiency are obtained when the amplifier is tuned to the input frequency. The correct frequency is selected in the pi network by filtering

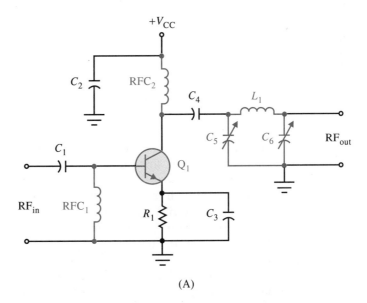

(A)

FIGURE 2-41   RF power amplifier circuits (A) Single-ended (B) Parallel

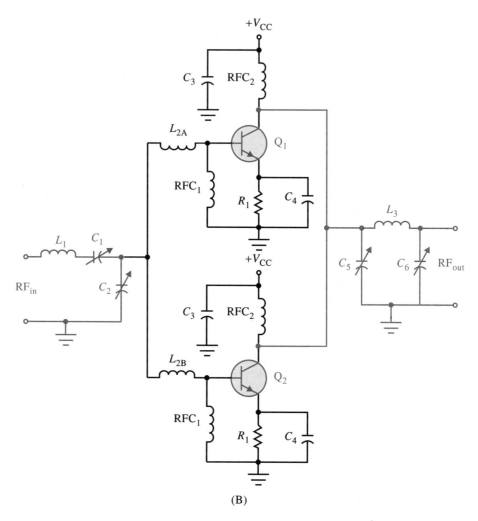

FIGURE 2-41 (*continued*)

out the fundamental frequency from the collector current pulses, thus permitting the circuit output to be sinusoidal. The RF signal is filtered out by RF chokes RFC_1 and RFC_2. Base current is kept constant by the filtering action of RFC_1, and the V_{CC} supply is isolated from the RF signal by the filtering action of RFC_2.

Operating the final output stage in parallel, as shown in Figure 2-41(B), provides greater RF power output. Each of the identical stages must deliver equal power to the load. Therefore, the transistors must be perfectly matched; that is, they must have identical characteristics, and the circuits must be exactly the same.

A final word of caution for working with power amplifiers: Transistors used in these circuits get extremely hot. To ensure proper heat dissipation, adequate heat sinks must be used with RF power transistors. In some very high-power operations, it may be necessary to install blower fans to provide cooling air for the transistors and heat sinks, or even liquid-cooled heat sinks. The heat, as well as the RF signal produced, can cause burns, which may appear minor on the surface but can severely damage deeper tissues.

SUMMARY

- Electrical filters are designed to attenuate unwanted frequencies or to amplify desired frequencies. Passive filters provide only attenuation, and active filters may provide amplification. Active filters are often used in LF and MF equipment. Filters are classified as low-pass, high-pass, band-pass, and band-reject (or notch). A wave trap is a form of notch filter sometimes used in antenna systems of receivers and HF equipment. The two commonly used wave traps are the parallel tuned and the series tuned. The pass band is the range of frequencies passed by a filter with minimum attenuation or maximum gain. The stop band is the range of frequencies attenuated. The cutoff frequency is the frequency at which the voltage gain of the filter drops to 70.7% of its maximum value. The value of the cutoff frequency determines the bandwidth of the pass band. The cutoff frequency point is variously called the 0.707 point, the −3 dB point, the break point, or the half-power point. The per decade or per octave increase or decrease above or below the cutoff frequency in a filter's response curve is called the rolloff or the fall-off and represents the performance of the filter. A decade represents a tenfold increase or decrease in a frequency, and an octave represents a doubling or halving of a frequency.

- Resonance is the basis for selectivity in communications systems. Resonant circuits are widely used in filter applications and are often combined with amplifier circuits. The selectivity of a resonant circuit is determined by the circuit bandwidth and the sharpness of the slopes of the response curve. The Q of the circuit determines the sharpness of the slopes. In series resonant circuits, impedance is minimum (equal to the circuit resistance), current is maximum, and $X_L = X_C$. Voltage across the reactive components is much higher than the source voltage but 180° out of phase with it; therefore, the voltages cancel. In parallel resonant (tank) circuits, impedance is maximum, line current is minimum, and $X_L = X_C$. In an ideal parallel tank circuit, current in the tank draws no current from the source.

- An oscillator is an electronic generator that operates from a dc power source, has no moving parts, and can produce ac signal frequencies. Fundamental requirements for all oscillators are amplification, a frequency-determining network, positive feedback, and the ability to be self-starting. The three basic classifications for sine wave oscillators are *RC, LC,* and crystal. The three basic *RC* oscillators are the Wien-bridge, the phase-shift, and the twin-T. The two basic *LC* oscillators are the Hartley, identified by a tapped coil arrangement, and the Colpitts, identified by a tapped capacitor arrangement. The Clapp oscillator is a refined version of the Colpitts, in which a third capacitor is added in series with the inductive branch of the tank circuit.

- Crystals are used in oscillator circuits to overcome an inherent oscillator instability. The crystals must possess the property of piezoelectricity. Crystals have a natural frequency of vibration, determined mostly by their thickness. Crystal-controlled oscillators may include the Hartley and the Colpitts. The Pierce oscillator is a variation of the crystal-controlled Colpitts, in which a crystal replaces the inductor in a standard Colpitts circuit.

- Communications systems use both AF and RF voltage and power amplifiers. The three basic configurations for amplifiers are common-emitter, common-collector, and common-base. The CE circuit is the most commonly used. The CC circuit, also called an emitter follower, is often used as an impedance-matching device; hence, it is sometimes called a buffer amplifier. The CB circuit, when constructed with discrete components, is normally used for frequencies above 10 MHz. It is also used in ICs as part of the differential amplifier input stage. The four types of coupling between amplifier stages are resistance-capacitance, impedance, transformer, and direct. The four amplifier biasing methods are base bias, emitter-feedback bias, collector-feedback bias, and voltage-divider, or universal, bias. The most commonly used mode of operation in audio voltage amplifiers is class A. Class A operation produces 360° of current flow, the least distortion, and the lowest efficiency. Class B operation produces 180° of current flow, lower power dissipation, and higher efficiency. When used in push-pull, it offers large load power, low distortion, and high efficiency. Class C operation produces less than 180° of current flow, resulting in an output of pulses. Class C amplifiers normally operate only at radio frequencies, offer the highest load power at the highest efficiency, and are used as oscillators and frequency multipliers.

- Audio frequency amplifiers are designed to operate in the AF range (20 Hz to 20 kHz). Voltage amplifiers are designed to amplify low-voltage audio signals. Power amplifiers are used as the final stage of amplification before an audio signal leaves the AF system. Radio frequency voltage and power amplifiers are designed to operate at frequencies above the audio range and are sometimes used as frequency multipliers. The distinguishing feature of RF amplifiers is that their input and output circuits are tunable, thus providing improved selectivity and impedance matching for good power transfer between stages. Intermediate-frequency amplifiers are used in receivers and are designed to operate at frequencies above the AF range but below the received RF range.

QUESTIONS AND PROBLEMS

1. List and define the four classifications of filters.

2. Explain where and why wave traps are used.

3. Define pass band, stop band, roll-off, and fall-off.

4. Define the cutoff frequency for a filter. List the other identifiers for the cutoff frequency.

5. Define decade and octave.

6. Why is resonance used in radio and television receivers?

7. For a series RLC circuit where $R = 1$ kΩ, $L = 2$ mH, and $C = 100$ nF, find f_r, Z_T, X_L, and X_C.

8. Determine X_L, X_C, and Z_T for the circuit in Problem 7 first at $f_r - 2$ kHz and then at $f_r + 2$ kHz.

9. What is the relationship between current and impedance in a series resonant circuit?

10. For a series circuit with $V_s = 100$ V, $R = 20$ Ω, and $X_L = X_C = 200$ Ω, find I, V_R, V_L, and V_C. Explain the results of your calculations for V_L and V_C.

11. For a parallel *LC* circuit operating at 5 MHz with $R = 50\ \Omega$, $L = 9\ \mu H$, and $C = 100$ pF, find X_L, X_C, and Z_T.

12. At what frequency will the circuit in Problem 11 resonate?

13. At resonance, find Q and BW for the circuit in Problem 11.

14. List the four basic characteristics common to all oscillators.

15. List and explain three uses of oscillators in communications equipment.

16. For a Wien-bridge oscillator with $R_1 = R_2 = 47$ kΩ and $C_1 = C_2 = 100$ pF, find f_{op}.

17. Assume that each section of a three-section phase-shift oscillator has values of $R = 10$ kΩ and $C = 0.01\ \mu F$. Find f_{op}.

18. For a twin-T oscillator with $R = 27$ kΩ and $C = 220$ pF, find f_{op}.

19. List the identifying features of the Hartley, Colpitts, Clapp, and Pierce oscillators.

20. For a Hartley oscillator with $L_{1A} = L_{1B} = 15$ mH, with L_{1A} wound over L_{1B}, and $C = 100$ pF, find f_{op}.

21. For a Colpitts oscillator with $C_1 = 100$ pF, $C_2 = 10$ pF, and $L = 100$ mH, find f_{op}.

22. For a Clapp oscillator with $L_1 = 100$ mH and $C_3 = 220$ pF, find f_r.

23. Define piezoelectricity.

24. What characteristic of a crystal makes it ideal for use in oscillator circuits?

25. List the three basic circuit configurations for amplifiers. Include their abbreviated designations.

26. Which amplifier circuit configuration is most commonly used? List its major features.

27. List the characteristics and major advantages of the CC amplifier.

28. List the characteristics of the CB amplifier. What is its major disadvantage? Where is it most often used?

29. List the four types of coupling circuits. State the advantages and disadvantages for each.

30. List the biasing circuits. Include any other names with which they may be identified.

31. What is the distinguishing feature of the self-bias circuit? What are its advantages?

32. List the three classifications of operation for amplifiers and identify the characteristics of each.

33. Where in an AF system is the PA stage situated? What does the PA do in an AF system?

34. Where and why are RF voltage amplifiers used?

35. What distinguishes RF amplifiers from AF amplifiers? What advantages does this offer?

36. In receivers, what class of operation is normal for RF amplifiers? Why?

37. What is the function of the IF amplifier in a radio receiver? What distinguishes the IF amplifier from the RF amplifier?

38. Explain how unwanted oscillations in RF amplifier circuits can be prevented.

39. What function do RF power amplifiers perform in communications applications?

40. What precautions should be observed when working with RF power amplifiers?

CHAPTER **3**

Amplitude Modulation

OBJECTIVES

When you complete this chapter, you should be able to:

- Define modulation.
- List three methods of generating AM signals and state the advantages and disadvantages of each method.
- Calculate the frequencies contained in an AM signal.
- Calculate the percent modulation of an AM signal.
- Calculate the power or current relations to each other of the frequencies contained in an AM signal.
- Define standard AM and list its disadvantages.
- List the standard AM detectors and the characteristics of each.
- Draw block diagrams of the tuned radio frequency (TRF) and superheterodyne AM receivers.
- Define receiver selectivity and sensitivity.
- List the advantages of superheterodyne receivers over TRF receivers.
- Identify and explain the functions of the stages of a typical superheterodyne receiver.
- Define image frequency and state methods for avoiding the problems it causes.
- Explain the purpose and operation of automatic gain control (AGC) circuits.
- State the differences between low-level and high-level modulation.
- State the differences between double-sideband, suppressed-carrier (DSSC) signals and SSB signals.
- List two methods of generating SSB signals.
- List the advantages of SSB over standard AM and suppressed-carrier AM.
- List two types of sideband filters.
- Draw block diagrams of standard AM and SSB transmitters.
- Draw a block diagram of an SSB receiver.

INTRODUCTION

The need for modulation was discussed in Chapter 1. We stated that modulation can be made to affect one of three factors: the amplitude, the frequency, or the phase. This chapter discusses the effect of varying the amplitude of the carrier, known as amplitude modulation (AM). We will also discuss single-sideband, suppressed-carrier (SSSC or, more simply, SSB) systems, used in many modern two-way voice communication systems.

3-1 AM SIGNAL GENERATION

Modulation is defined as the process of impressing a low-frequency intelligence (typically AF) signal upon a higher-frequency carrier (generally RF) signal. The HF signal is termed the *carrier,* and the LF signal is normally termed *intelligence,* but it may be called the *modulating signal, modulating wave, information signal,* or simply the *audio signal.*

If we vary the amplitude of the carrier at an audio rate, then we have *amplitude modulation* (AM). See Equation 3-1.

Amplitude modulation is generated by combining intelligence and carrier frequencies through a nonlinear device called a modulator. Many different circuits can amplitude modulate an RF carrier. For example, a diode has a nonlinear area and can be used as a modulator, but a diode modulator is impractical because it offers no gain. Most practical amplitude modulators use an amplifier stage. Although some high-power transmitters may still require vacuum tube modulator circuits, transistors offer nonlinear operation when properly biased, and they provide amplification; thus, they are ideal for this application. A sinusoidal function can be expressed as follows:

$$x(t) = X \sin(2\pi ft \pm \theta) \tag{3-1}$$

where

x = instantaneous value (i or v)

X = maximum amplitude (I or V)

f = carrier frequency

t = time

θ = phase angle

EXAMPLE 3-1

Assume f = 1 kHz, t = 100 µs, θ = +135°. Find x for (a), X = 5 V and (b) X = 10 V.

SOLUTION

(a) x = 5 V sin (6.28 × 1 kHz × 100 µs + 135°)
= 5 sin (135.63) = 5 (0.7) = 3.5 v

(b) x = 10 sin (135.63) = 10 (0.7) = 7 v

The 3.5 v and 7 v values are the instantaneous amplitudes of the modulated 1 kHz frequency at $t = 100$ μs.

Note that the maximum amplitude (X) is varied, with frequency (f) and phase angle (θ) held constant; therefore, this is an AM signal.

In transistor amplifier modulators, the intelligence signal can be injected into any of the three transistor pins, thus producing collector-modulation, base-modulation, or emitter-modulation circuits. The carrier is amplified while the intelligence signal varies the gain of the circuit. This technique is used in both high-level and low-level modulation, discussed in Section 3-3.

Although the following sections refer to transistor stages, the discussions apply equally well to vacuum tube and FET circuits. Figure 3-1 shows the schematic diagrams and elements of each of these devices.

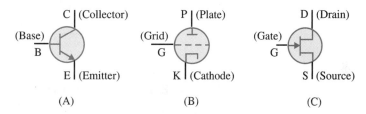

(A) (B) (C)

FIGURE 3-1 Amplitude modulator devices (A) Bipolar transistor (B) Vacuum tube (C) FET

Collector Modulation

The most commonly used AM technique is collector modulation, shown in Figure 3-2. Here, the modulating signal is applied to the collector in series with the dc collector supply voltage. This type of circuit, an RF class C amplifier, is often used for high-level modulation. The nonlinear amplification characteristics of class C amplifiers provide excellent mixing action. The modulating signal is shunt fed to the collector through transformer T_1. The modulating signal alternately aids and opposes the collector voltage, and the peak amplitude of the signal can approach V_{CC}.

The RF carrier is applied to the base of Q_1, which amplifies the signal. The gain of Q_1 is a function of the ratio of the collector impedance to the base impedance. The intelligence signal varies the collector impedance, and therefore varies the carrier gain.

Transistor Q_1 is a class C amplifier; therefore, only positive signal pulses will appear at its output. To produce a complete AM waveform, these pulses are coupled to the resonant circuit formed by L_4, C_5, and C_6.

In the circuit of Figure 3-2, 100% modulation is not possible because of saturation effects that prevent the transistor output voltage from rising in proportion to the increase in collector voltage. For high levels of modulation, it is necessary to modulate both a driver and the final amplifier, as shown in Figure 3-3. Modulating the driver provides extra drive to the final amplifier at the positive peak, allowing it to reach a full 100% modulation.

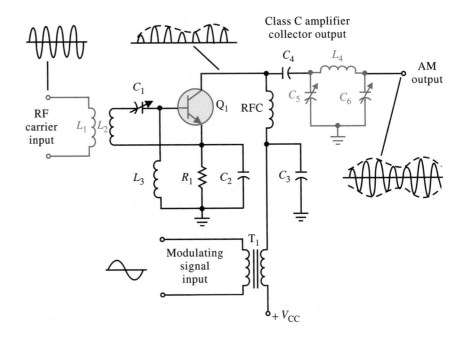

FIGURE 3-2 Basic collector-modulated circuit

High modulating signal power, equal to one half of the carrier power for 100% modulation, is required for collector modulation. However, collector modulation offers high collector efficiency, low distortion, and easy circuit adjustments.

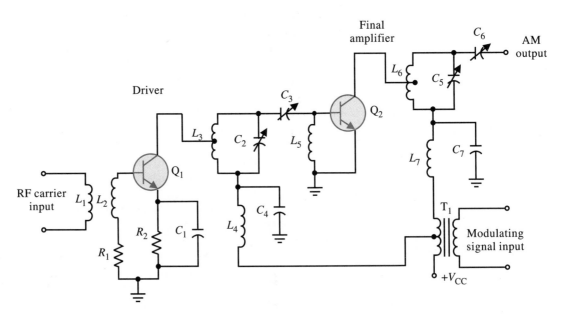

FIGURE 3-3 Collector-modulated circuit for high modulation levels

Base Modulation

Injecting the modulating signal into the base circuit of the transistor produces an appreciable reduction in the modulating power requirement over the collector-modulated circuit. The intelligence signal can be series fed (Figure 3-4(A)) or shunt fed (Figure 3-4(B)). In either case, RC or transformer coupling can be used, although a better impedance match can be achieved with transformer coupling.

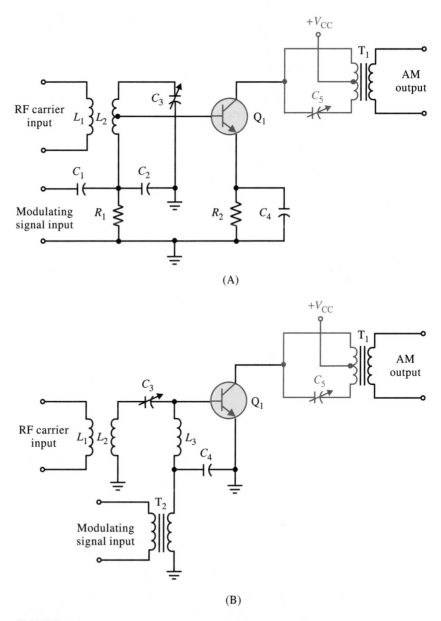

(A)

(B)

FIGURE 3-4   Base-modulated circuits (A) Series fed (B) Shunt fed

In Figure 3-4, the transistor conducts only when its base-emitter voltage swings above about 0.6 V, and the transistor is cut off for the negative half-cycle. A low-Q resonant tank circuit in the collector circuit of Q_1, formed by C_5 and the primary of T_1, restores the missing half-cycle. Each pulse of current causes the tank circuit to oscillate, and the low Q of the circuit causes the amplitude of the restored half-cycle to be equal to that of the original half-cycle. The result is a complete AM signal at the secondary of T_1.

Compared with collector modulation, base modulation (1) is more difficult to adjust, (2) has lower collector efficiency, (3) has lower power output (if the same transistor is used), and (4) has poorer linearity.

Emitter Modulation

Injecting the modulating signal into the emitter circuit of the transistor produces characteristics between those of base and collector modulation. As shown in Figure 3-5, the carrier is applied through C_1 to the base of the transistor. The modulating signal is applied to the emitter circuit through T_1, which also provides a good impedance match. Biasing is accomplished with $R_1 - C_2$.

To the carrier, the circuit appears as a common-emitter amplifier, the gain of which is adjusted by the intelligence signal applied at the emitter. In this type of modulator, the intelligence signal amplitude is usually much higher than the carrier amplitude. For this reason, emitter modulation is best suited for applications requiring low-level modulation.

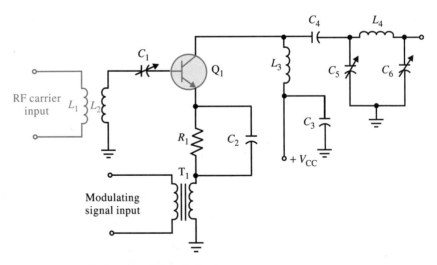

FIGURE 3-5 Emitter-modulated circuit

3-2 THE AM SIGNAL

Generally, AM is used for relaying messages by voice, television, facsimile, or other relatively sophisticated modes. The process is always the same: Audio or low frequencies are impressed upon a carrier wave of a much higher frequency.

The AM Waveform

The modulation of an AM signal may be considerable, or it may be very small. The intensity or degree of modulation is expressed as a percentage. Figure 3-6 illustrates the relationships between the unmodulated RF carrier (Figure 3-6(A)) and a sine wave or sinusoidal audio tone (Figure 3-6(B)). The modulated waveform (Figure 3-6(C)) shows that the amplitude of the carrier is greatest on positive peaks of the sinusoidal tone and smallest on negative peaks. Note that the modulated waveform is an AM signal at the carrier frequency whose amplitude varies at the same rate as does the modulating signal frequency.

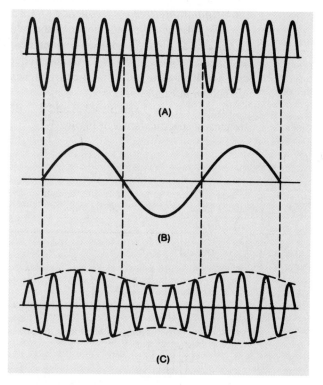

FIGURE 3-6 Relationships between unmodulated and modulated signals (A) RF carrier (B) RF modulating signal (C) Modulated RF Carrier

The dashed lines in Figure 3-6(C) represent the intelligence *envelope* of the waveform. The envelope is a replica of the intelligence signal and shows on both the top and bottom of the AM waveform. It is a depiction of the actual amplitude variations resulting from modulation. When the signal is viewed on an oscilloscope, the envelope may not be visible. However, it may be seen at lower frequencies, such as a 500 Hz sine wave modulating a 1 MHz carrier. The envelope is drawn here to demonstrate the concept of modulating a carrier with a sine wave.

The AM waveform appears to have only the carrier frequency whose envelope varies as a function of the intelligence signal. However, a mathematical analysis of the frequencies contained within the AM waveform will show that three frequencies exist: the original carrier frequency, and two additional frequencies, which are called *sidebands*. The frequency spacing of the sidebands around the carrier is a direct function of the intelligence frequency. These sidebands contain the information required to reconstruct the intelligence signal, and affect the bandwidth of the modulated signal.

The complex (modulated) waveform results from mixing the two signals through a nonlinear device called a *modulator.* Modulators and the methods used for mixing the signals were discussed in Section 3-1.

Combining the two signals through a nonlinear device produces the following frequency components: (1) a *dc* level, (2) the *original carrier frequency* (f_c) and the *original intelligence frequency* (f_i), (3) the *sum* of the two original frequencies ($f_c + f_i$), (4) the *difference* between the two original frequencies ($f_c - f_i$), and (5) the *harmonics* of the two original frequencies. However, filtering removes the dc level and the harmonics of the two original frequencies and leaves the two original frequencies and their sum and difference frequencies to form the AM signal. These concepts are shown in Figure 3-7.

Figure 3-8 and Examples 3-2 and 3-3 illustrate the concepts of harmonics and the frequency spectra shown in Figure 3-7.

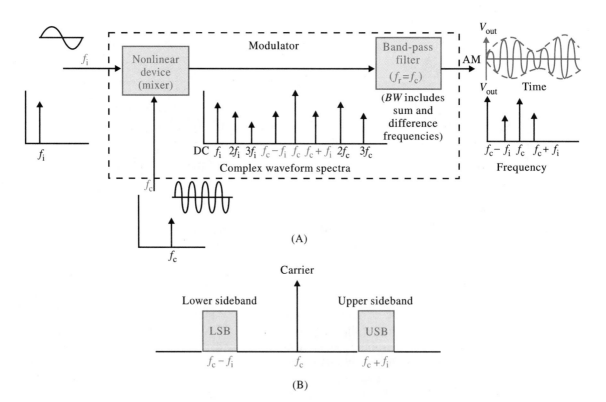

FIGURE 3-7 Modulator action (A) Combining signals to produce AM outputs and develop sidebands (B) Sideband representation

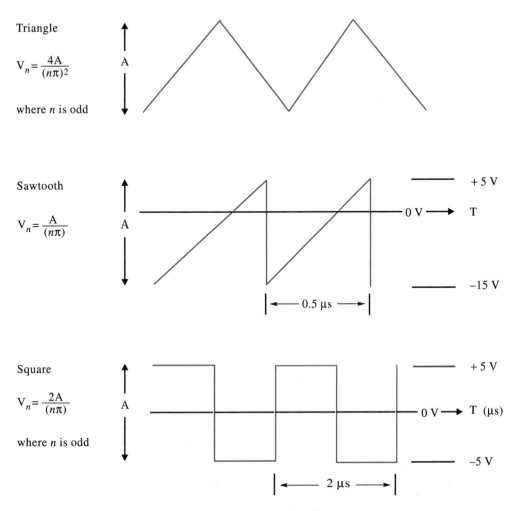

Triangle

$$V_n = \frac{4A}{(n\pi)^2}$$

where n is odd

Sawtooth

$$V_n = \frac{A}{(n\pi)}$$

Square

$$V_n = \frac{2A}{(n\pi)}$$

where n is odd

FIGURE 3-8   Periodic waveforms for Examples 3-2 and 3-3

EXAMPLE 3-2

For the square wave of Figure 3-8: (a) solve for the first three harmonics; (b) solve for the amplitudes of the three harmonics; (c) draw the frequency spectrum of the three harmonics.

SOLUTION

(a) $f_i(1) = \frac{1}{T} = \frac{1}{2}$ µs = 500 kHz or 0.5 MHz. A square wave consists of odd harmonics, therefore: $f_i(3) = 3(500 \text{ kHz}) = 1500$ kHz or 1.5 MHz; $f_i(5) = 5(500 \text{ kHz}) = 2500$ kHz or 2.5 MHz.

(b) Since $V_n = \frac{2A}{(n\pi)}$ where n is odd, $\pi = 3.14$, and $A = 10$ V;

$$V_1 = \frac{2(10\text{ V})}{(1\ \pi)} = \frac{20\text{ V}}{3.14} = 6.37\text{ V}$$

$$V_3 = \frac{2(10\text{ V})}{(3\ \pi)} = \frac{20\text{ V}}{9.42} = 2.12\text{ V}$$

$$V_5 = \frac{2(10\text{ V})}{(5\ \pi)} = \frac{20\text{ V}}{15.7} = 1.27\text{ V}$$

(c) See Figure 3-9.

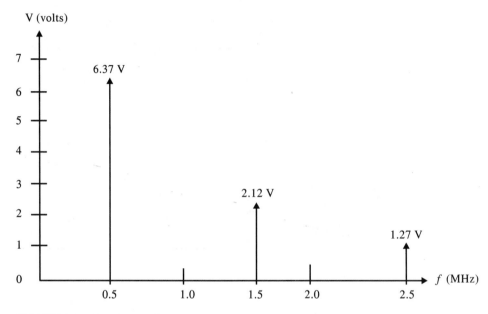

FIGURE 3-9 Solution to Example 3-2(c)

EXAMPLE 3-3

For the sawtooth wave of Figure 3-8: (a) solve for the first five harmonics; (b) solve for the amplitudes of the five harmonics; (c) draw the frequency spectrum of the five harmonics.

SOLUTION

(a) $f_i(1) = \dfrac{1}{T} = \dfrac{1}{0.5}\ \mu s = 2$ MHz. A sawtooth wave consists of all harmonics, therefore: $f_i(2) = 2(2 \text{ MHz}) = 4 \text{ MHz}$; $f_i(3) = 3(2 \text{ MHz}) = 6 \text{ MHz}$; $f_i(4) = 4(2 \text{ MHz}) = 8 \text{ MHz}$; $f_i(5) = 5(2 \text{ MHz}) = 10 \text{ MHz}$.

(b) Since $V_n = \dfrac{A}{(n\pi)}$ where n is odd, $\pi = 3.14$, and A = 20 V;

$$V_1 = \frac{(20\text{ V})}{(1\ \pi)} = \frac{20\text{ V}}{3.14} = 6.37\text{ V}$$

$$V_2 = \frac{(20 \text{ V})}{(2\pi)} = \frac{20 \text{ V}}{6.28} = 3.18 \text{ V}$$

$$V_3 = \frac{(20 \text{ V})}{(3\pi)} = \frac{20 \text{ V}}{9.42} = 2.12 \text{ V}$$

$$V_4 = \frac{(20 \text{ V})}{(4\pi)} = \frac{20 \text{ V}}{12.48} = 1.60 \text{ V}$$

$$V_5 = \frac{(20 \text{ V})}{(5\pi)} = \frac{20 \text{ V}}{15.7} = 1.27 \text{ V}$$

(c) See Figure 3-8.

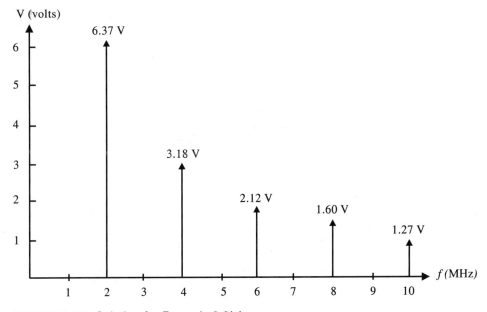

FIGURE 3-10   Solution for Example 3-3(c)

Modulation Factor

The variation in the AM signal compared with the unmodulated carrier is variously called the *modulation factor, index of modulation,* or *percent modulation.* The modulation factor m is a measure of the extent to which the carrier current (or voltage) is varied by the intelligence. The mathematical relationship between the relative amplitudes of the carrier and intelligence signals can be determined in several ways. A commonly used equation is

$$m = \frac{I_{\max} - I_{\min}}{I_{\max} + I_{\min}} \times 100\% \tag{3-2}$$

where I is the current of the modulated carrier. Another way to calculate m is as follows:

$$m = \frac{I_i}{I_c} \times 100\% \qquad (3\text{-}3)$$

where I_i is the intelligence current and I_c is the carrier current. The units used can be peak, peak-to-peak, or rms values as long as both are the same. If voltage is used instead of current, we simply substitute V for I in the equations.

Proof of the equations for m are demonstrated by using V and Figure 3-11.

$$V_{max} = V_C + V_I$$

$$V_{min} = V_C - V_I$$

$$m = \frac{V_{max} - V_{min}}{V_{max} + V_{min}} = \frac{V_C + V_I - V_C - V_I}{V_C + V_I + V_C - V_I} = \frac{2V_I}{2V_C} = \frac{V_I}{V_C}$$

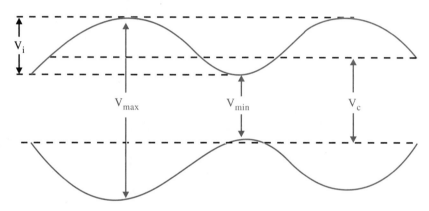

FIGURE 3-11 Reference for proof of m

Four examples of AM are shown in Figure 3-12. The varying amplitudes of the RF carrier wave result in a symmetrical envelope (for a sine wave modulating signal) that corresponds to the intelligence modulating signal.

EXAMPLE 3-4

Assume an unmodulated carrier current of 5 A. If the maximum current is 10 A and the minimum current is 0 A, what is the modulation factor?

SOLUTION

From Equation 3-2,

$$m = \frac{I_{max} - I_{min}}{I_{max} + I_{min}} \times 100\%$$

$$= \frac{10\,A - 0\,A}{10\,A + 0\,A} \times 100\%$$

$$= 1 \times 100\% = 100\%$$

Note that the RF signal for 100% modulation shown in Figure 3-12(B) varies from zero level to twice the value of the unmodulated carrier.

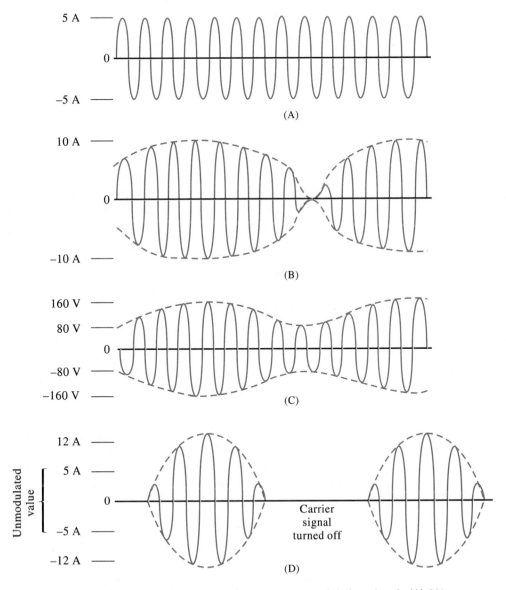

FIGURE 3-12   Relationships between various percent modulation signals (A) 0% modulation (unmodulated carrier) (B) 100% modulation (C) 33% modulation (D) Overmodulation (greater than 100%)

EXAMPLE 3-5

Assume that $V_c = 160$ V and $V_i = 80$ V. What is m?

SOLUTION

From Equation 3-3,

$$m = \frac{V_i}{V_c} \times 100\% = \frac{80\ \text{V}}{160\ \text{V}} \times 100\%$$

$$= 0.5 \times 100\% = 50\%$$

For the most intelligence to be recovered by a receiver circuit called the *demodulator,* the transmitter must have a high m. For that reason, modulation is generally maintained close to 100%, the maximum allowable value. However, the complexity of intelligence signals could possibly cause the modulation to exceed 100%, an undesirable condition called *overmodulation,* demonstrated by Figure 3-12(D). In this condition, the modulated carrier goes to more than twice its unmodulated value, and the signal is intermittently turned off, the effect of which is the generation of new frequencies that interfere with nearby channels and cause severe distortion of the transmitted intelligence, and lost information.

Sidebands

When a carrier is modulated in any manner, for conveying intelligence of any sort, *sidebands* are produced. Sideband signals occur above and below the carrier frequency.

Sideband signals are the result of mixing between the carrier and the modulating signal. The higher the frequency components of the modulating signal, the farther from the carrier the sidebands will appear (see Figure 3-7(B)).

When the percent modulation is known, the sideband power (P_{sb}) of an AM signal can be determined as a percentage of the carrier power (P_c). Total power (P_t) is the sum of the carrier power (a fixed value) and the sideband power (a variable value). The power in the upper sideband (P_{usb}) and the power in the lower sideband (P_{lsb}) are equal, with the total sideband power equaling one half of the carrier power (at 100% modulation). Thus, at 100% modulation (but only on the condition that the modulating frequency is a sine wave), sideband power will be equal to one third (33.3%) of the total power. The sideband power can be calculated as follows:

$$P_{sb} = \frac{m^2 P_c}{2} \tag{3-4a}$$

or at 100% modulation:

$$P_{sb} = \frac{P_c}{2} \tag{3-4b}$$

This equation demonstrates that sideband power equals one half of the carrier power at 100% modulation.

Since both sidebands have equal power, it can be stated that each sideband contains one quarter of the carrier power, calculated as follows:

$$P_{usb} = P_{lsb} = \frac{m^2 P_c}{4} \tag{3-5}$$

The total AM signal is the sum of the carrier power and sideband power:

$$P_t = P_c + P_{sb} = P_c \left(1 + \frac{m^2}{2}\right) \tag{3-6}$$

When the carrier voltage is known, the carrier power can be determined by the following methods:

$$P_c = \frac{V_c^2}{R} \quad \text{(using rms values)} \tag{3-7}$$

or

$$P_c = \frac{V_c^2}{2R} \quad \text{(using peak values)} \tag{3-8}$$

where R is the resistive load impedance.

Modulated and unmodulated currents are easily measured values, so it is helpful to have the following equation for calculating current:

$$I_t = I_c \sqrt{1 + \frac{m^2}{2}} \tag{3-9a}$$

or

$$\left(\frac{I_t}{I_c}\right)^2 = 1 + \frac{m^2}{2}$$

$$\frac{m^2}{2} = \left(\frac{I_t}{I_c}\right)^2 - 1$$

$$m = \sqrt{2\left[\left(\frac{I_t}{I_c}\right)^2 - 1\right]} \tag{3-9b}$$

EXAMPLE 3-6

A 1 MHz RF carrier signal is modulated with a 3 kHz AF sinusoidal tone. The modulated carrier voltage is 30 V_{max} and 15 V_{min} across a 100 ohm resistive load impedance. Assume rms values. Determine the (a) RF carrier voltage when the percent of modulation is zero, (b) modulation factor, (c) carrier power, (d) sideband power, (e) total power, (f) various frequencies, and (g) bandwidth.

SOLUTION

(a) The zero-modulation RF carrier voltage is the average value of the maximum and minimum voltages of the modulated signal:

$$V_C = \frac{V_{max} + V_{min}}{2} = \frac{30 \text{ V} + 15 \text{ V}}{2} = 22.5 \text{ V}$$

(b) The modulation factor can be determined by using Equation 3-2:

$$m = \frac{V_{max} - V_{min}}{V_{max} + V_{min}} \times 100\%$$

$$= \frac{30 \text{ V} - 15 \text{ V}}{30 \text{ V} + 15 \text{ V}} \times 100\%$$

$$= \frac{15 \text{ V}}{45 \text{ V}} \times 100\% = 0.33 \times 100\% = 33\%$$

(c) Carrier power can be calculated by Equation 3-7:

$$P_c = \frac{V_c^2}{R} = \frac{22.5^2}{100} = \frac{506.25}{100}$$

$$= 5.0625 \text{ W}$$

(d) Sideband power can be calculated by Equation 3-4a:

$$P_{sb} = \frac{m^2 P_c}{2} = \frac{(0.33^2)(5.0625)}{2}$$

$$= 275.7 \text{ mW}$$

(e) Total power can be calculated by Equation 3-6:

$$P_t = P_c + P_{sb} = 5.0625 + 0.2757 = 5.338 \text{ W}$$

or

$$P_t = P_c\left(1 + \frac{m^2}{2}\right) = 5.0625\left(1 + \frac{0.33^2}{2}\right)$$

$$= 5.338 \text{ W}$$

(f) The various frequencies are determined in the following manner:

upper side frequency $= f_c + f_i$
$= 1 \text{ MHz} + 3 \text{ kHz}$
$= 1.003 \text{ MHz}$

carrier frequency $= f_c$
(given) $= 1 \text{ MHz}$

lower side frequency $= f_c - f_i$
$= 1 \text{ MHz} - 3 \text{ kHz}$
$= 997 \text{ kHz}$

(g) Bandwidth is the difference between the upper side frequency and the lower side frequency: $1.003 \text{ MHz} - 0.997 \text{ MHz} = 6 \text{ kHz}$

EXAMPLE 3-7

A 100 W carrier is to be modulated to a level of 80%. Determine the (a) carrier power after modulation and (b) sideband power.

SOLUTION

(a) The carrier power is not affected; therefore,

$$P_c = 100 \text{ W}$$

(b) By Equation 3-4a,

$$P_{sb} = \frac{m^2 P_c}{2} = \frac{(0.8^2)\,(100\text{ W})}{2} = 32 \text{ W}$$

EXAMPLE 3-8

An unmodulated carrier current is 8 A. Current becomes 8.95 A when the carrier is modulated with a sine wave. Determine the (a) index of modulation and (b) total current when the signal is modulated to a level of 70%.

SOLUTION

(a) By Equation 3-9b,

$$m = \sqrt{2\left[\left(\frac{I_t}{I_c}\right)^2 - 1\right]} = \sqrt{2\left[\left(\frac{8.95}{8}\right)^2 - 1\right]}$$

$$= \sqrt{2[1.252 - 1]} = \sqrt{0.504} = 0.71 = 71\%$$

(b) By Equation 3-9a,

$$I_t = I_c \sqrt{1 + \frac{m^2}{2}} = 8\sqrt{1 + \frac{0.7^2}{2}} = 8\sqrt{1.245}$$

$$= 8(1.116) = 8.93 \text{ A}$$

3-3 STANDARD AM TRANSMITTERS

The individual circuits discussed thus far are used in the various stages of a basic AM transmitter. We will now put those necessary stages together in block diagram form to see how the standard AM signals are processed and transmitted. In a typical transmitter, a carrier oscillator is followed by a number of amplifier stages and a final power amplifier (FPA). The output of the FPA is fed to the antenna.

The modulating signal can be applied to any of the RF amplifier stages. In general, AM transmitters are classified as employing *high-level* or *low-level* modulation according to the stage of application of the intelligence signal. In *high-level* modulation, the modulated RF stage is collector modulated and its output feeds the antenna. In *low-level* modulation, the modulated RF stage is collector, emitter, or base modulated and is followed by a linear amplifier that feeds the amplified signal to the antenna.

High-Level Modulation

A basic high-level AM transmitter is shown in the block diagram of Figure 3-13. Most commercial transmitters use high-level modulation. Note that the modulating signal is applied to FPA where maximum power amplification can occur. The FPA in high-level-modulated transmitters is generally a highly efficient class C stage.

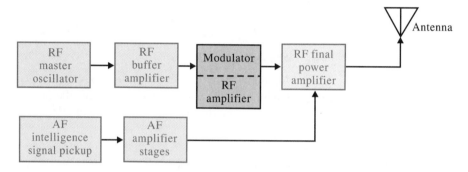

FIGURE 3-13 Block diagram of a high-level AM transmitter

The carrier frequency is generated by the RF oscillator. In general, a crystal oscillator is used for stability if the transmitter is to be used at a standard fixed frequency. However, if the operating frequency is high, better stability is obtained by using a lower-frequency oscillator, which then feeds frequency multipliers to obtain the desired frequency.

For highest stability, the load on the oscillator circuit must be constant, with no feedback allowed from following stages. Stability is accomplished by using a *buffer* amplifier between the oscillator stage and the following stages, which are normally operated class C for maximum efficiency.

Class C operation of the RF amplifiers is possible because they are unmodulated. This is the major advantage of high-level modulation. The number of stages used depends on the power required to drive the FPA. If multiplication of the oscillator frequency is necessary, any of these stages can be operated as frequency doublers or triplers.

The initial audio or intelligence input is through a pickup device called a *transducer,* which transforms one form of energy or disturbance into another. Since a wide range of frequencies must be amplified by the same amount without any distortion, the AF amplifiers must be linear. For standard voice or music transmission, the transducer would be a microphone, followed by audio amplifiers. For television transmission, the transducer would be a camera, followed by video amplifiers. For telemetry operation, the transducer would be a strain gauge, a pressure gauge, or a thermocouple, followed by amplifiers having an appropriate bandwidth.

The carrier and intelligence signals are input to the FPA where modulation occurs. The FPA is operated class C for maximum efficiency and good mixing action, using collector (plate for vacuum tubes or drain for FETs) modulation, where the intelligence signal is applied to the collector (plate, drain) in a series with the dc supply voltage. The output of the FPA is fed through a transmission line to the antenna.

Low-Level Modulation

A basic low-level AM transmitter is shown in the block diagram of Figure 3-14. In this system, the intelligence signal is applied to any of the amplifier stages preceding the FPA. The earlier the stage, the lower the carrier power level required. Consequently, the power requirement for the modulating signal is much lower. However, with the RF amplifiers modulated, all of the following stages and the FPA must be operated in their linear regions, generally either class A or class B push-pull, to prevent distortion of the signal. Because of this, efficiency and power handling capability are reduced. Low-level modulation usually is the most economical method for low-power transmitters.

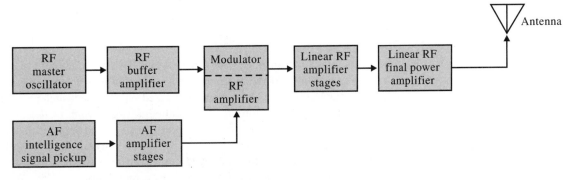

FIGURE 3-14 Block diagram of a low-level AM transmitter

3-4 AM RECEIVERS

We have examined the process of impressing an intelligence signal in the form of amplitude variations onto a carrier wave, where it is then amplified and applied to a transmission antenna. The modulated signal is then radiated into the air, where an antenna at a receiver picks up the signal. The ultimate goal of any communications system is to extract the intelligence from the transmitted signal, process it, amplify it, and reproduce it into a load as intelligible information. Typical loads for received AM signals are sound system loudspeakers and television video screens. In this section, we will study how the received signal is processed in the receiver and will examine the original tuned radio frequency (TRF) receiver and modern superheterodyne receivers. Before examining the different types of receivers, we discuss characteristics common to all receivers.

AM Frequency Spectrum

As we saw in Table 1-4, the AM broadcast system covers the frequency range from 535 kHz to 1605 kHz. It consists of 106 channels, each with a bandwidth of 10 kHz. Therefore, all AM radio receivers must be capable of receiving transmitted signals at frequencies within that frequency range. Also, receivers must be sensitive enough, and selective enough, to select only one desired frequency from among the signals received while rejecting all others.

Sensitivity and Selectivity

The *sensitivity* of a receiver is a measure of its ability to receive and amplify weak signals. The amount of gain a receiver has determines its sensitivity. Gain is often defined in terms of the signal voltage amplitude that must be applied to the receiver's input to provide a standard output power, measured at the receiver's output terminals. Therefore, the more gain produced in the input (RF) stages, the greater the sensitivity of the receiver.

The *selectivity* of a receiver is a measure of its ability to select one signal while rejecting all others at nearby frequencies. Selectivity determines the adjacent-channel rejection of a receiver. Selectivity is increased as bandwidth is reduced. However, selectivity varies with the received frequency, decreasing when the received frequency increases.

Demodulation

The modulated signal is transmitted into space where a small fraction of it is captured by the receiving antenna. At the receiver, the information contained in the modulated signal must be separated from the carrier and passed on to a load, such as a speaker or television screen. This separation process is called *demodulation,* or *detection,* and it involves stripping the carrier wave from the intelligence (audio signal) and amplifying the original intelligence signal to some usable level. The circuits used for this process depend on the form of modulation employed in the transmission of the signal. For AM signals, the circuit is called an *envelope detector* or, more simply, a *detector.*

The Diode Detector   The simplest and most widely used AM demodulator is the *diode detector* shown in Figure 3-15(A). The input to the circuit is the 455 kHz IF signal that has been selected and amplified by earlier stages in the receiver. The signal is then applied to diode D, which acts as a half-wave rectifier. An LP filter, composed of the RC parallel network, filters out the 455 kHz IF signal but passes the desired audio frequencies to the audio amplifier.

The Transistor Detector   Transistors are sometimes used to provide detection and amplification in a single stage. *Transistor detectors* are generally used only in low-cost, all-transistor receivers, whereas diode detectors are chosen for many kinds of equipment because of their simplicity and excellent performance.

A typical transistor detector is shown in Figure 3-15(B). Resistors R_1 and R_2 bias the transistor for precise class B operation, a requirement for simultaneous rectification and amplification. Resistor R_3 serves as the collector load, and RF components are filtered out by capacitor C_2.

Types of Radio Receivers

Various types of radio receivers have been proposed, but only two types have survived the test of time: the *tuned radio frequency* (TRF) receiver and the *superheterodyne*

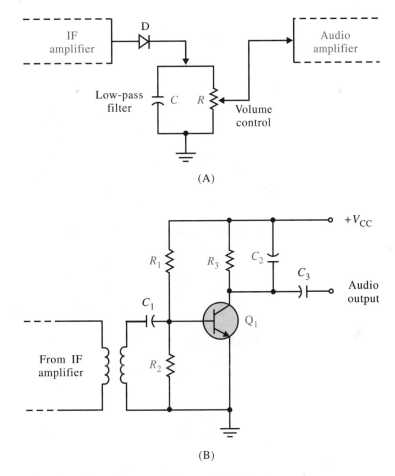

FIGURE 3-15 AM demodulators (A) Diode detector
(B) Transistor detector

receiver. Today, only the superheterodyne is in general use, although the TRF may be found in some fixed-frequency applications.

The TRF Receiver   Used today primarily as a fixed-frequency receiver in special applications, the TRF receiver offers simplicity and high sensitivity. These virtues represented a great improvement over other types of receivers previously used.

Figure 3-16 shows a block diagram of a TRF receiver. If a single RF amplifier is used, the bandwidth is too wide to reject all frequencies except the desired one. More typically, two or more RF amplifiers, all tuned together, are used. This configuration decreases the bandwidth to a point where only the one desired frequency will pass to the detector. In a TRF receiver, selectivity is determined by the bandwidth of the RF amplifiers. The RF amplifiers also provide amplification for the desired signal. The two characteristics important in all receivers—sensitivity and selectivity—are increased by additional RF amplifiers.

Assume that the receiver represented by Figure 3-16 has sufficient RF amplifier stages to make its sensitivity and selectivity adequate for the AM broadcast band,

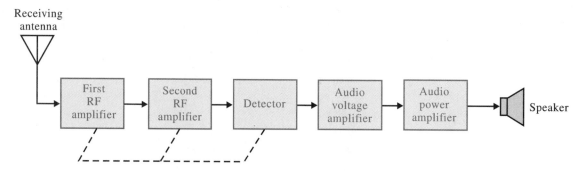

FIGURE 3-16 Block diagram of a TRF receiver

ranging from 535 kHz to 1605 kHz. It would be extremely difficult to get all the RF amplifiers to tune to exactly the same frequency across the entire broadcast band. Also, because the bandwidth of the RF amplifiers increases as the operating frequency increases ($BW = f_r/Q$), selectivity decreases at the high end of the band, causing insufficient adjacent-channel rejection.

Another difficulty presented at the higher frequencies is the risk of instability associated with high gain being achieved at one frequency by the several stages of RF amplification. This instability can cause feedback from the output to the input through some stray path and produce unwanted oscillations. The inherent problems of the TRF receiver resulting from bandwidth variation, insufficient adjacent-channel rejection, and instability are all solved by the superheterodyne receiver.

The Superheterodyne Receiver We learned earlier that a TRF receiver works well when it is fixed-tuned to a single frequency. The principle of fixed tuning is used in the superheterodyne receiver, which mixes the incoming RF signal with a signal generated within the receiver. This mixing produces a difference frequency, which is usually a lower frequency than the received RF signal. This lower difference frequency signal contains the same modulation as the original carrier, and it is amplified, demodulated, and reproduced at the output.

A block diagram of a superheterodyne receiver for use in the AM broadcast band is shown in Figure 3-17. The AM signal that operates in the 535–1605 kHz range is received by the antenna and coupled into a *tunable-circuit RF section,* which must be capable of tuning over the entire broadcast band. Except in some entertainment receivers used in high-signal-strength areas, the output of the tuned RF circuit is fed into an RF amplifier to provide gain. The number of RF amplifiers used in a particular receiver is determined by system design and system requirements.

The section shown within the dashed-line area is the key to the superheterodyne receiver. It is the *frequency-conversion* section, more commonly called the *mixer stage,* where mixing (heterodyning) of the received RF signal and the LO signal occurs. Note that the RF, mixer, and LO stages are *ganged* (interconnected) for simultaneous tuning. The mixer circuit is tunable over the entire broadcast band, and it is tuned to the same frequency as the RF stage for any setting on the selector dial. The LO is also a variable-frequency stage, the frequency of which is always a fixed amount higher than the RF frequency of the other two ganged stages. The output of the mixer stage is the

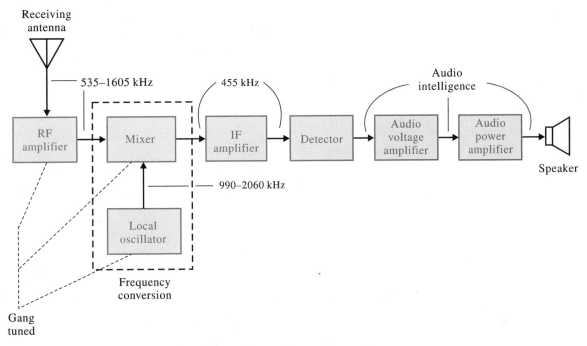

FIGURE 3-17 Block diagram of a AM superheterodyne radio receiver

difference frequency between the RF signal and the LO frequency. This difference frequency is a constant value because of the relation between RF and LO tuning. For AM, the standard difference frequency is 455 kHz. It is still a radio frequency, but to distinguish it from the received RF signal, and because it lies between the original RF carrier and AF modulating frequencies, it is termed the *intermediate frequency* (IF). In the process, the modulating signal contained in the original carrier signal is converted from a higher region in the RF spectrum to a lower IF region.

The IF section is designed for optimum results at the single, fixed frequency of 455 kHz. For this reason, there is no tracking problem. It can contain any number of amplifier circuits. The IF section primarily determines the sensitivity and selectivity characteristics of the superheterodyne receiver.

The amplified IF signal is coupled to the detector where the original modulating information is recovered. The detected audio signal is coupled to suitable voltage and power amplifiers, and finally to the loudspeaker load.

Image Frequency   In a superheterodyne receiver, frequency conversion produces a constant IF, which simplifies the problem of obtaining good selectivity and gain over a wide range of input frequencies. However, it is possible for input signals on two different frequencies to produce an output at the IF. One signal is desired; the other is not. The undesired response frequency is called the *image frequency*. For example, if the desired signal is at 1045 kHz, the LO will oscillate at 1500 kHz—that is, 455 kHz above the desired frequency. A signal on either 1045 kHz (the desired frequency) of 1955 kHz (an undesired frequency 455 kHz above the LO) will mix with the LO and produce the 455 kHz IF. The undesired signal on 1955 kHz is the image frequency.

Assuming the standard method where the LO is a fixed frequency above the incoming signal frequency, we use the following equations to demonstrate the relations between the image frequency F_I, the LO, the IF, and the signal frequencies F_S:

$$F_I = LO + IF \tag{3-10a}$$

or

$$F_I = F_S + 2(IF) \tag{3-10b}$$

$$LO = F_S + IF \tag{3-11}$$

$$IF = LO - F_S \tag{3-12a}$$

$$F_S = LO - IF \tag{3-12b}$$

EXAMPLE 3-9

Determine the image frequency for a standard broadcast AM receiver tuned to a station at 1320 kHz.

SOLUTION

First, determine the frequency of the LO:

$$LO = F_S + IF$$
$$= 1320 \text{ kHz} + 455 \text{ kHz}$$
$$= 1775 \text{ kHz}$$

Next, determine the image frequency:

$$F_I = LO + IF$$
$$= 1775 \text{ kHz} + 455 \text{ kHz}$$
$$= 2230 \text{ kHz}$$

Alternatively,

$$F_I = F_S + 2(IF)$$
$$= 1320 \text{ kHz} + 2(455 \text{ kHz})$$
$$= 2230 \text{ kHz}$$

In standard AM superheterodyne receivers, greater image rejection can be obtained by adding a tuned RF amplifier before the mixer. However, at the higher frequencies encountered in many communications receivers, tuned RF amplifiers may not adequately reject image frequencies. At frequencies around 30 MHz, dual conversion is used to solve the image rejection problem. Dual conversion is discussed in the section on SSB receivers.

3-5 DISADVANTAGES OF STANDARD AM

The major advantage of the standard AM system we have been discussing is that it uses straightforward and inexpensive transmitting and receiving equipment. However, it has several disadvantages. The three most important are as follows.

1. *Power is wasted in the transmitted signal.* Most of the transmitted power is in the carrier, which does not contribute to the transmitted intelligence. The carrier contains no intelligence. It merely acts as a reference frequency, and all intelligence being transmitted is in the sidebands. For example, if an unmodulated output of a standard AM transmitter is 100 W, the entire 100 W is contained in the unmodulated carrier. At 100% modulation, the transmitted power increases to 150 W. The 100 W carrier is still being produced by the transmitter, with the additional 50 W representing the power in the sidebands. Each sideband contributes only 25 W to the total transmitted power, because the upper sideband power and the lower sideband power are equal. At lower levels of modulation, the power contribution of the sidebands is even lower. We can correctly deduce, therefore, that in a standard AM system two thirds of the total transmitted power is in a signal that contains no intelligence.

2. *The transmitted signal requires twice the bandwidth of the transmitted intelligence.* In Section 3-2, we deduced that the bandwidth is determined by the intelligence. In Example 3-6, we showed that when a 1 MHz carrier is modulated with a 3 kHz AF sine wave, a 1.003 MHz upper sideband and a 0.997 MHz lower sideband are produced. The resulting bandwidth is 1.003 MHz − 0.997 MHz, or 6 kHz. Since both sidebands are transmitted along with the carrier in standard AM, the transmitted bandwidth is 6 kHz, twice the modulating frequency signal.

3. *Very precise amplitude and phase relationships between the sidebands and carrier are required.* In a standard AM system, an undistorted received signal must be exactly as transmitted. Under certain propagation conditions, these relationships are difficult, if not impossible, to maintain. Each sideband component of the AM signal has different frequencies and can therefore be affected differently by poor propagation conditions. Fading and interference are the major problems with standard AM transmissions.

3-6 SINGLE-SIDEBAND AM TRANSMISSION

The three disadvantages of standard AM discussed in Section 3-7 can be overcome by a system called *single-sideband, suppressed-carrier* (SSSC) or, more simply, *single-sideband* (SSB) transmission. Single-sideband emission is a form of AM in which only one sideband is transmitted. The remaining sideband and the carrier are suppressed at the transmitter. Single-sideband, suppressed-carrier emission provides greater communications efficiency than standard AM, because two-thirds of the power in a standard AM signal is taken up by the carrier wave, which conveys no intelligence. In an SSB signal, all of the power is concentrated into one sideband. An SSB signal has a bandwidth about half that required for standard AM. Therefore, it is possible to get twice as many SSB signals as standard AM signals into a given amount of spectrum space.

Most voice communication at the low, medium, and high frequencies is carried out via SSB. Generally, the lower sideband is used at frequencies below about 10 MHz. The upper sideband is preferred at frequencies above 10 MHz. These preferences are simply a matter of convention; either sideband will provide equally good communication at a given frequency.

Compared with standard AM transmitters, SSB transmitters are more complicated and consequently more expensive. Different modulation techniques are used, and sharply tuned filters are necessary to suppress the carrier and unwanted sideband. Also, SSB receivers are more complicated, because they must add a substitute carrier back into the received signal before detection of the intelligence can be achieved. However, in most cases the advantages of SSB outweigh the additional design time and cost.

Single-sideband transmitters are essentially amplitude modulated transmitters, in that an RF carrier and two sidebands are produced. Recall that in standard AM, all of the intelligence is contained in the sidebands, but the sidebands contain only one-third of the total power. The other two thirds of the power is in the carrier. Hence, a great amount of power is wasted during transmission. The basic principle of SSB is to eliminate or greatly suppress the high-energy RF carrier without jeopardizing the integrity of the intelligence. This is possible because both upper and lower sidebands contain the same intelligence. Therefore, in the SSB transmitter, one sideband is suppressed and all or part of the carrier is suppressed, thus converting the power lost in the carrier into useful power to transmit the intelligence contained in the remaining sideband. Also, since only one sideband is transmitted, the bandwidth required for transmission is cut in half.

The SSB transmitter is rated in terms of *peak envelope power* (PEP), which is the product of rms voltage V_{rms} and rms current I_{rms}:

$$PEP = V_{rms}I_{rms} = \frac{V_{rms}^2}{R} \tag{3-13a}$$

where V_{rms} is the rms value of the carrier amplitude during its highest modulation (for example, on a voice peak), rms equals $0.707V_p$, and R is the resistive load impedance. The PEP is also equal to the average power (P_{avg}) of the SSB signal.

$$PEP = P_{avg} \tag{3-13b}$$

The dc power equations for the PEP are as follows:

$$PEP = \frac{V^2}{R} = I^2R \tag{3-13c}$$

EXAMPLE 3-10

An oscilloscope display shows a peak SSB envelope voltage (V_p) of 300 V when measured across a 75 Ω line. What is the PEP?

SOLUTION

Use Equation 3-13a:

$$V_{rms} = 0.707V_p = 0.707(300) = 212 \text{ V}$$

$$PEP = \frac{V_{rms}^2}{R} = \frac{212^2}{75} = 599 \text{ W}$$

EXAMPLE 3-11

The rms values for an SSB signal are measured as 30 V and 15 A. What is the PEP?

SOLUTION

Use Equation 3-13a:

$$PEP = V_{rms} I_{rms} = (30)(15) = 450 \text{ W}$$

EXAMPLE 3-12

The average power of an SSB signal is 375 W. What is the PEP?

SOLUTION

Use Equation 3-13b:

$$PEP = P_{avg} = 375 \text{ W}$$

3-7   SSB GENERATION

Single-sideband transmission is used extensively in HF communications systems. Therefore, SSB receivers must be capable of receiving signals in the HF band (3–30 MHz).

To obtain SSB emission, we must use a balanced modulator followed by a filter or phasing circuit. Thus, we have two SSB generation methods: the *filter* method and the *phase-cancellation* method. Both methods use a balanced modulator to produce a double-sideband signal with a suppressed carrier. The filter or phasing network then removes either the upper sideband or the lower sideband. If the upper sideband is removed, the resulting SSB signal is called a lower sideband (LSB) signal. If the lower sideband is removed, the resulting SSB signal is called an upper sideband (USB) signal.

Balanced Modulators

The balanced modulator (a symmetrical circuit that delivers an output signal containing the frequency sum and frequency difference of two input signals) is the key circuit in SSB generation. A commonly used SSB generator consists of a balanced modulator followed by an extremely selective filter, as shown in Figure 3-18. The filter can be crystal, ceramic, or mechanical. In general, two output frequencies are produced by the balanced modulator: (1) a USB frequency equal to the injected carrier plus the input audio frequency and (2) an LSB frequency equal to the injected carrier minus the input audio frequency. In theory, the injected carrier is balanced out so that it does not appear in the output. Practical design characteristics determine the extent to which the carrier can be balanced out. Existing balanced modulators suppress the carrier from 30 dB to 40 dB below the PEP of the sidebands. The SSB filter further suppresses the carrier by an additional 20 dB. Thus, total carrier suppression of from 50 dB to 60 dB can reasonably be expected from SSB transmitter system.

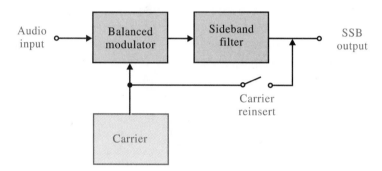

FIGURE 3-18 Simplified block diagram of a filter-type SSB generator

Dual-FET Balanced Modulator   A relatively simple balanced modulator using FETs is shown in Figure 3-19. Note the similarity to a conventional push-pull amplifier. In this simple circuit, the carrier input voltage is applied through L_1–L_2 and is in phase at the gates of the FETs, causing equal but opposite drain currents to flow in the primary of T_2. The magnetic fields caused by the opposing drain currents cancel, resulting in zero output of the carrier at T_2. Of course, good suppression of the carrier can occur only if the FETs and all other circuit components are perfectly matched.

When a modulating input voltage is applied across center-tapped transformer T_1, it is 180° *out of phase at the gates*—a push-pull arrangement. Drain currents flow on both the positive and negative signal swings, first through one FET then through the other, causing nonlinear mixing across the FETs. This mixing results in sum and difference frequency sidebands in push-pull across the primary of T_2. The output from the tuned circuit consists of the upper and lower sidebands only. Stated another way, the output of a properly balanced modulator is a *double-sideband, suppressed-carrier* (DSSC) signal.

Diode-Ring Modulator   A widely used balanced modulator, sometimes called a *balanced ring modulator,* uses diodes as the nonlinear elements, as shown in Figure 3-20. This circuit is capable of suppressing the carrier by 60 dB, but the diodes must be carefully matched to achieve such good carrier suppression.

The carrier balancing action of the diode ring modulator is analyzed by following the current flows of the carrier signal alone through the circuit. Assume that the carrier generator voltage is such that D_3 and D_4 conduct. As shown by the solid arrows in Figure 3-20, the current will flow through T_1, D_3, and D_4, T_2, and back to the RF input. The currents through the two windings of output transformer T_2 are out of phase and

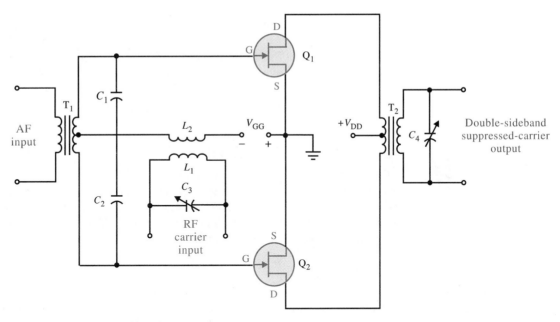

FIGURE 3-19   Dual-FET balanced modulator

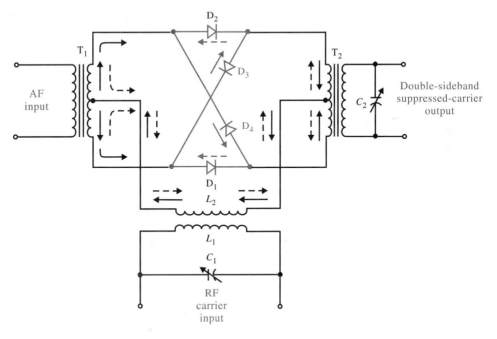

FIGURE 3-20   Diode ring modulator

will cancel. On the next carrier half-cycle, as shown by the dashed arrows, D_1 and D_2 conduct and the phases of all currents are changed by 180°. The output currents are again out of phase. Therefore, no carrier voltage appears across the secondary of T_2.

For balanced modulator action to occur, both carrier and modulating signals must be applied. The carrier must be several times stronger than the modulating signal. Conduction is determined by the polarity of the carrier, as discussed in the preceding paragraph. The modulating voltage will either aid or oppose the conduction of the diodes, thus upsetting the current balance in the primary of T_2. An output at the sideband frequencies is, therefore, developed in the secondary of T_2.

3-8 SSB SUPPRESSION

Both basic methods of SSB generation use some form of balanced modulator to suppress the carrier, but they differ in the suppression of the unwanted sideband. Depending on system requirements and circuit configuration, either method will suppress either the upper or the lower sideband with equal ease. This section discusses both the filter and the phase-cancellation methods of SSB suppression.

Filter Method

The most widely used technique for suppressing an SSB signal is the *filter* method. The filter method uses a BP filter having sufficient selectivity to pass one sideband and reject the other. Although transmitter stability and accuracy are determined primarily by the carrier oscillator, the balanced modulator and the sideband filter circuits are more important for obtaining the desired SSB output.

A block diagram of a basic filter system SSB transmitter is shown in Figure 3-21. Here, the carrier and the amplified intelligence signal are applied to the balanced modulator. The DSSC output of the balanced modulator is then applied to the sideband filter. The filter is designed to pass the desired sideband and to block the unwanted sideband. Therefore, the SSB output is either the upper sideband or the lower sideband, depending on the filter pass band. However, the two sidebands may be only a

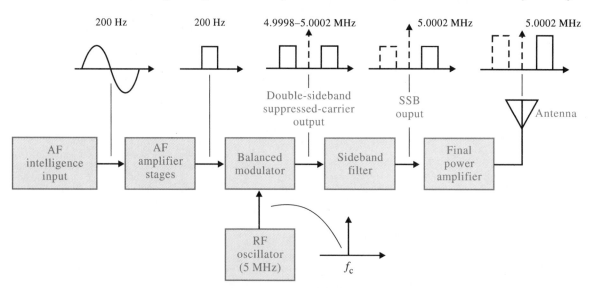

FIGURE 3-21 Simplified block diagram of a filter-type SSB transmitter

few hertz apart, which places a design burden on filters. For example, at an audio frequency of 50 Hz, the side frequencies are the carrier frequency $f_c \pm 50$ Hz—only 100 Hz apart. To ease filter design burdens, voice frequencies below 100 Hz are suppressed. Suppressing these lower frequencies does not affect the intelligibility of the intelligence.

The carrier frequency also affects the degree of sideband separation. For example, assume a carrier frequency of 5 MHz and a modulating frequency of 200 Hz. The two side frequencies would then be 5.0002 MHz and 4.9998 MHz. If the upper sideband is selected, the filter then must pass the upper side frequency and reject the lower side frequency, which is only 400 Hz away. To gain a better perspective, let us look at a percentage. The percent difference in frequency is $(0.0002/5.0002) \times 100\% = 0.004\%$. If we change the carrier frequency to 100 kHz, the side frequencies are still only 400 Hz apart, but the percent difference in frequency is $(0.2/100.2) \times 100\% = 0.2\%$. This example demonstrates that circuit design can be eased considerably by the use of LF carrier oscillators when using the filter method of SSB generation. Because excellent crystal oscillator stability can be obtained at 100 kHz, this frequency is commonly used in SSB transmitter carrier oscillators.

A problem arising from the use of LF carrier oscillators is that of raising the signal of a desired RF level. Frequency multipliers cannot be used because they alter the intelligence signal and increase the bandwidth. The block diagram in Figure 3-22 shows a *dual-conversion,* filter-type SSB transmitter, a typical way to preserve the intelligence signal and bandwidth by using frequency translation (conversion) to raise the

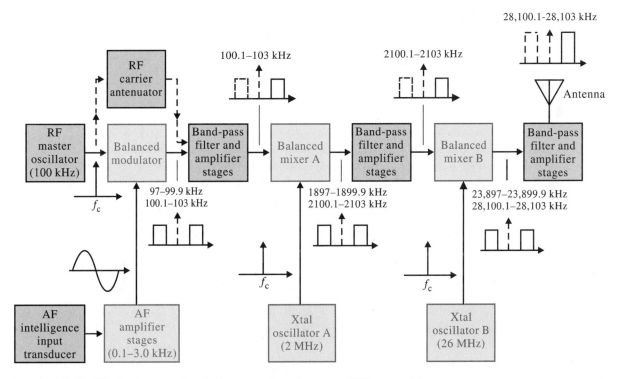

FIGURE 3-22 Block diagram of a dual-conversion, filter-type SSB transmitter

carrier frequency. This preservation is achieved by using two frequency-conversion stages. Modulating the 100 kHz carrier with voice frequencies of 0.1–3 kHz, we see that the two sidebands at the balanced modulator output are only 200 Hz apart (99.9–100.1 kHz). At the output of mixer A, the separation is 200 kHz (1899.9–2100.1 kHz), and at the output of mixer B, the sidebands are separated by 4.2 MHz (23.8999–28.1001 MHz). The spacing provided by the double conversion makes it easy to reject one sideband without affecting the other, because filtering becomes more efficient.

Sideband Filters

Voice transmission requires audio frequencies from about 100 Hz to 3000 Hz—all that is required for adequate speech intelligibility. In SSB transmitters, once the carrier has been suppressed, one sideband must be rejected without affecting the other. At audio frequencies, the separation between the upper and lower sidebands is only 200 Hz, as shown in Figure 3-23(A). Because of the closeness of the two sidebands, high-Q filters are necessary. For example, a 5 MHz carrier and a lower audio frequency of 100 Hz would require a filter Q of 5 MHz/200 Hz = 25,000. On the other hand, if we use a

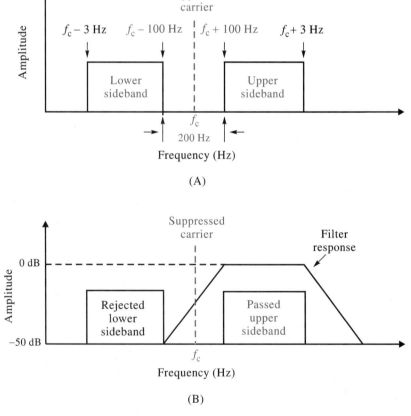

FIGURE 3-23   SSB transmitter action (A) Carrier suppression (B) Filter pass band

typical 100 kHz carrier, the Q of the filter is 100 kHz/200 Hz = 500, a significant reduction in the high-Q requirement.

The filters used in filter-type SSB transmitters must have sharply defined skirts (steep frequency rise and frequency fall-off slopes) and flat BP characteristics to pass only those frequencies in the desired sideband and to reject all others. For example, in Figure 3-23(B) we see that to pass the upper sideband and reject the lower sideband, the filter must rise from about −50 dB attenuation to 0 dB attenuation within the 200 Hz frequency range that separates the two sidebands. Crystal, ceramic, or mechanical filters can meet these specifications and therefore are used in SSB transmitters.

Crystal Filters   *Crystal* filters are commonly used in SSB systems because of their very high Q values. Piezoelectric crystals have Q values in the thousands in the HF range. When used as elements in a filter, they produce the narrow pass bands and sharply defined skirts required to properly filter out the unwanted sideband. Figure 3-24 shows a *crystal lattice* filter that uses two matched pairs of crystals in a bridge circuit. Crystals X_1 and X_2 are identical to each other and are in series with the signal path. They are selected so that their series resonant frequency falls within the desired pass band. Crystals X_3 and X_4, also identical to each other but slightly different from crystals X_1 and X_2, are in parallel with the signal path. Recall that crystals have both series resonant and parallel resonant frequencies. The high impedance of the parallel resonant frequencies corresponds to the low impedance of the series resonant frequencies. Therefore, at their parallel resonant frequency, crystals X_3 and X_4 will not bypass the desired sideband. However, their series resonant frequency will bypass the unwanted sideband to ground, while the parallel resonant frequency of crystals X_1 and X_2 will block passage of the unwanted sideband. If crystals with the proper characteristics of series and parallel resonant frequencies and Q are chosen and carefully designed, crystal lattice filters can produce the almost ideal results of very sharp skirts, flat top, and small band-pass.

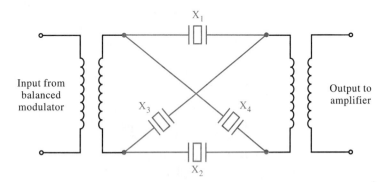

FIGURE 3-24   Crystal lattice filter

Ceramic Filters   The piezoelectric effect of crystals is also utilized in modern *ceramic* filters constructed from lead zirconate titanate. Figure 3-25 shows the equivalent circuit of a typical *ceramic ladder* filter. The ceramic material behaves like quartz crystals; therefore, each ceramic disc has a series and a parallel resonant frequency.

Likewise, the piezoelectric characteristic of the ceramic material makes input and output transducers unnecessary.

Ceramic filters are available at 455 kHz or 500 kHz center frequencies, with bandwidths ranging from 2 kHz to 50 kHz. They have Q values significantly lower than those of crystal filters. However, the lower cost, smaller size, and ruggedness of ceramic filters make them more practical than crystal filters for portable SSB equipment applications.

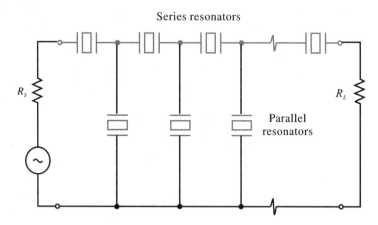

FIGURE 3-25   Ceramic ladder filter

Mechanical Filters   *Mechanical* filters are mechanically resonant devices that use a double-conversion technique for better unwanted frequency suppression. An input transducer receives electrical energy and converts it into mechanical vibrations, which are passed through a series of mechanical resonators in the form of thin, machined nickel discs with wire couplers. An output transducer then converts the mechanical vibrations back into electrical energy.

The most commonly used transducers are *magnetostrictive* devices. These devices are constructed with nickel and ferrites, which will lengthen or shorten in the presence of a magnetic field. The result is the mechanical vibrations that drive the mechanically resonant elements of the filter.

The construction and electrical equivalent circuit of a typical mechanical filter are shown in Figure 3-26. In the equivalent circuit, the discs of the mechanical filter are represented by the parallel-tuned *LC* circuits, the coupling wires by the series inductances, and coil losses by the series resistances.

The center frequency of the mechanical filter is determined by the resonators, represented by the parallel resonant *LC* circuits. Practical SSB mechanical filters are available with center frequencies from about 60 kHz to 500 kHz, and bandwidths between 3 kHz and 4 kHz. Increasing the number of discs in the mechanical filter improves its selectivity, defined as its *shape factor,* which is determined by dividing the bandwidth at 60 dB attenuation by the bandwidth at 6 dB attenuation:

$$\text{shape factor} = \frac{BW_{60\,dB}}{BW_{6\,dB}} \tag{3-14}$$

where BW = Hz

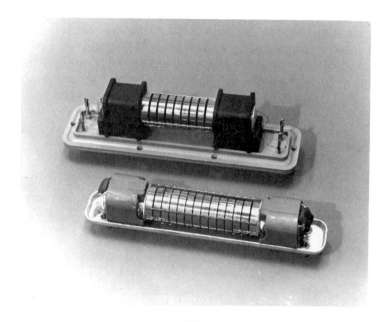

(A)

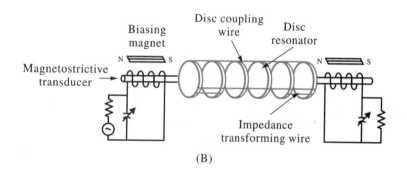

(B)

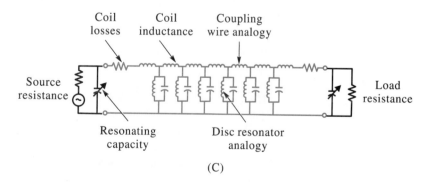

(C)

FIGURE 3-26 Mechanical filter (A) Top, seven-disc resonator; bottom, twelve-disc resonator (B) Physical structure (C) Electrical analogy

EXAMPLE 3-13

A mechanical filter has a bandwidth of 3.5 kHz at 60 dB attenuation and 1.75 kHz at 6 dB attenuation. What is the shape factor?

SOLUTION

Using Equation 3-14 gives

$$\text{shape factor} = \frac{BW_{60\,dB}}{BW_{6\,dB}}$$

$$= \frac{3.5 \text{ kHz}}{1.75 \text{ kHz}} = 2:1$$

The response characteristics illustrated here also apply to the crystal and ceramic filters discussed earlier.

EXAMPLE 3-14

A crystal filter has a bandwidth of 6 kHz at 60 dB and a bandwidth of 4.2 kHz at 6 dB. What is the shape factor?

SOLUTION

Using Equation 3-14 gives

$$\text{shape factor} = \frac{6 \text{ kHz}}{4.2 \text{ kHz}} = 1.43:1$$

The frequency response showing the attenuation provided by the filter in Example 3-14 is shown in Figure 3-27. The variations seen at the top of the response curve are termed *ripple amplitude* or *peak-to-valley ratio,* the ratio of maximum to minimum attenuation within the useful pass band of the filter. Ripple amplitudes from 1 dB to about 3 dB are common.

The insertion loss shown in Figure 3-27 is a reduction in the output of the filter. For a selective filter, the insertion loss is normally specified for the frequency or band of frequencies at which attenuation is the least. Most selective filters have low insertion loss at the frequencies to be passed, and high loss at other frequencies. Insertion loss is usually expressed in decibels. In Figure 3-27, the insertion loss is about 2.5 dB.

Phase-Cancellation Method

The principle involved in the *phase-cancellation* method of SSB generation is shown in Figure 3-28. This system uses two balanced modulators. The audio signal is split into two components that are identical except for a phase difference of 90°. The output of the carrier oscillator is also split into two separate components having a 90° phase difference. The direct carrier component and the direct audio component are combined in balanced modulator A. The 90° phase-shifted carrier and audio components are combined in modulator B. Both balanced modulators suppress the carrier and have a double-sideband output. When the two signals are exactly 180° out of phase, they are

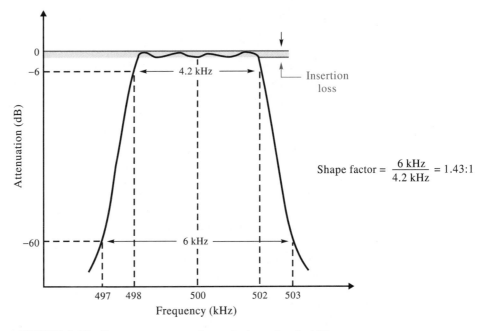

FIGURE 3-27 Response curve of a typical mechanical filter

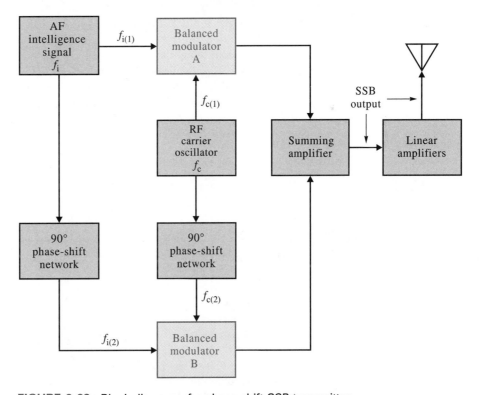

FIGURE 3-28 Block diagram of a phase-shift SSB transmitter

said to be in phase opposition. This condition results in a net amplitude that is the absolute value of the difference in the amplitudes of two signals having identical frequency. The phase of the resultant signal is the same as the phase of the stronger of the two signals. If the two signals have the same amplitude, they combine in phase opposition to produce no signal; that is, phase-cancellation.

This phenomenon results when the two sideband signals are combined in a summing circuit, where the relative phases of the sidebands are such that one sideband is canceled out while the other sideband is doubled, leaving only the desired sideband at the output of the summing circuit. If the output of the summing circuit is of sufficient amplitude, such an SSB generator can operate directly into the antenna. If not, the summing circuit output is applied to a linear amplifier for power amplification.

An advantage of the phase-cancellation method is that it is possible to generate the SSB signal at the operating frequency without frequency conversion, because no selective filter is required. A disadvantage is that it is necessary to maintain accurately the phase shifts and amplitudes of the signals applied to the summing circuit. Also, changing the phase angles of all possible intelligence frequencies by 90° is difficult. For these reasons, filter-type SSB transmitters are still very popular. However, the availability of special IC devices, which results in lower cost and higher reliability, has made the phase-cancellation method of SSB generation much more practical.

3-9 SSB RECEIVERS

Figure 3-29 shows a block diagram of a typical SSB receiver. It has three main sections: an RF section, an IF section, and an AF section. The principal requirement of the RF section is to select the desired signal in the antenna and convert this RF signal to a lower IF signal with minimum distortion and minimum generation of spurious signals. The IF section provides selectivity and amplification. The AF section recovers the AF intelligence and provides necessary AF amplification.

The RF section consists of an RF amplifier and one or more mixers that convert the RF signal to the IF signal. Using an RF amplifier as the first stage of a receiver has

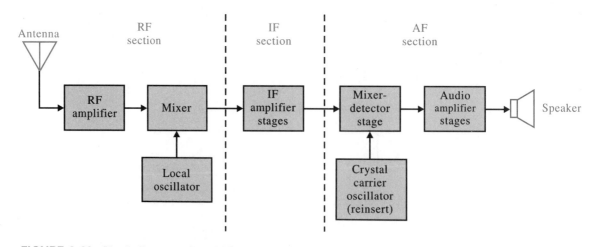

FIGURE 3-29 Block diagram of an SSB receiver

two advantages: (1) increased sensitivity resulting from the lower inherent noise of amplifiers compared with mixers and (2) a reduction of spurious responses because increased RF filtering can be used without degrading the *S/N* ratio. The amplification provided by the RF amplifier offsets the losses inherent in the passive filter circuits.

The RF signal is converted from the operating frequency to the IF by modulation in mixer circuits. Careful selection of frequencies used in the IF amplifier is necessary to avoid spurious responses. These responses occur whenever the spurious response frequency coincides with the desired frequency. Spurious responses are minimized if the IF is kept as low as possible consistent with good image rejection.

The Synchronous Detector   The envelope of a DSSC amplitude modulated signal does not duplicate the modulating signal. Because of the absence of the carrier, the two sidebands combine and the envelope has a frequency twice that of the modulating frequency. When DSSC signals are transmitted, the received signal cannot be demodulated by the diode detector or the transistor detector. A special circuit called a *synchronous detector* must be used.

A typical synchronous detector is shown in Figure 3-30(A). It uses two diodes and has a separate carrier input. Recall that the carrier was suppressed at the transmitter. The carrier input in this circuit is a signal that is *regenerated* by earlier stages of the receiver. The receiver-generated carrier must have exactly the same frequency as, and must be exactly in phase with, the original suppressed carrier.

The carrier input to the detector causes a switching action in diodes D_1 and D_2. On the half-cycle of the input that results in a negative voltage at the top of the secondary of T_1 and a positive voltage at the bottom, both diodes conduct. All of the current developed stays within the transformer-diode loop, which means that the carrier is effectively balanced out and does not reach the output. On the other half-cycle of the input, both diodes are reverse biased and do not conduct; hence, no current is developed. To ensure that the diodes are switched on for only a brief period of each input cycle, diode loads composed of the combinations of R_1–C_1 and R_2–C_2 are inserted. Any RF components that reach the output are filtered out by C_4.

Because the modulated signal input level is of such low value, the only time the modulated signal can reach the output is when the carrier input forward biases the diodes. The carrier input samples small portions of the positive peaks of the modulated signal during one half-cycle. Every time the carrier input goes positive, the diodes conduct and the suppressed-carrier signal passes to the output, resulting in an accurate duplication of the modulating signal at the output.

The greatest disadvantage of DSSC amplitude modulation is the need for extensive circuitry in the receiver to reinsert the carrier. Accuracy in frequency and phase in the receiver-generated carrier is important because any slight variation in either one will shift the sampling points and cause severe distortion.

The Product Detector   Single-sideband receivers usually use a *product detector.* Any nonlinear mixer circuit can be operated as a product detector. A typical product detector circuit is shown in Figure 3-30(B). It is a mixer stage, except that the IF input plus the *beat frequency oscillator* (BFO) input produces an AF output instead of an IF output. The BFO is used to produce a tone to beat (mix) with the IF signal, thus producing a heterodyne action. Because either the upper or lower sideband may

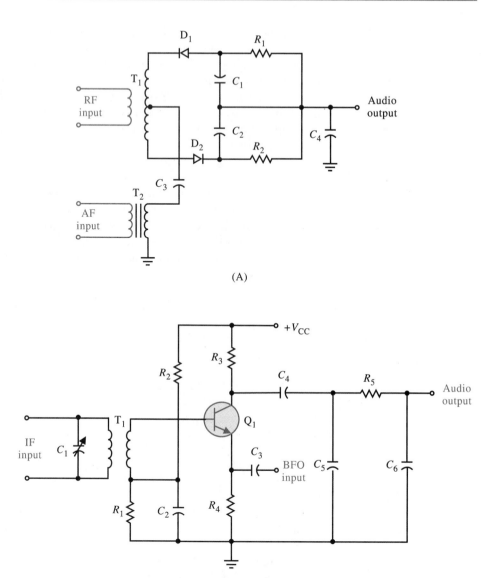

FIGURE 3-30 SSB detectors (A) Synchronous detector (B) Product detector

be transmitted, the BFO must be capable of tuning at least 1.5 kHz above and below the center of the IF pass band. In SSB receivers, the BFO is sometimes called a *carrier oscillator.* Although this circuit is most often used for SSB, it can be used for demodulating all other forms of AM.

In the circuit shown, the SSB signal input is applied to the base of Q_1 through fixed-frequency IF transformer T_1. The BFO signal from a crystal oscillator is applied to the unbypassed emitter. If the BFO frequency is within approximately 50 Hz of the original suppressed-carrier frequency, the difference frequencies will be similar to the original modulating signal. These signals range from about 100 Hz to 3000 Hz and are the desired frequencies. An LP filter composed of C_5–R_5–C_6 removes any RF components and allows only the desired audio signal to pass to the output.

Receiver Similarities

Many similarities exist between the superheterodyne broadcast and SSB receivers. In the SSB receiver, the received SSB radio frequency signal is amplified, converted to a lower IF value, demodulated, and amplified to a usable AF signal level. For high sensitivity and selectivity, SSB receivers use superheterodyne techniques. The absence of a carrier in the received SSB signal is the principal difference between SSB and AM. To recover the intelligence from the SSB signal, we must first restore the carrier. This locally restored carrier must have the same relationship with the sideband components as the original carrier used in the SSB transmitter modulator. Therefore, it is a stringent requirement of SSB receivers that the oscillator producing the reinserted carrier have extremely high frequency accuracy and stability. The total frequency error of the system must be less than 100 Hz to prevent the intelligibility of the received signal from being degraded.

Dual Conversion

As the range over which the receiver must operate is increased, it becomes more difficult to avoid spurious responses. A high degree of spurious response attenuation is maintained when covering the HF range (3–30 MHz) by using *dual conversion* (double conversion), or the use of two IF sections, as shown in Figure 3-31. Single conversion is used at the lower frequencies, dual conversion at the higher frequencies.

Dual conversion improves frequency stability through a crystal-controlled HF oscillator. Ganging a variable first IF stage to the tunable LF oscillator provides tuning. The

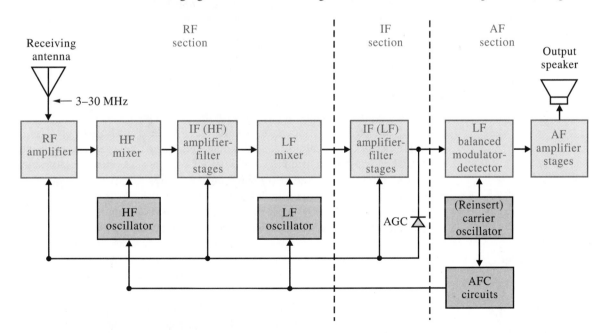

FIGURE 3-31 Block diagram of a dual-conversion SSB receiver

crystal-controlled oscillator ensures that the tuning rate remains the same on the HF bands as on the LF bands.

The IF section contains the frequency-selective filter elements and the principal amplifier stages. It is desirable to place the selective filter in the circuit ahead of the amplifier stages so that strong adjacent-channel signals are attenuated before they can drive the amplifiers. These filters are similar to, and of the same general construction as, the filters used in the transmitters for selecting the desired sideband while rejecting the unwanted sideband. However, for the receiver to provide good rejection of strong adjacent-channel signals, it must have an attenuation of 60 dB or more.

The amplifier portion of the IF section contains the amplifiers necessary to increase the signal to a level suitable for the demodulator. This portion usually consists of cascaded class A linear amplifier stages. Tuned circuits can provide the load resistance for these stages.

Automatic Gain Control

A factor to be carefully considered in SSB receiver design is *automatic gain control* (AGC). The basic function of AGC is to keep the signal output of the amplifier constant and thus maintain a constant audio output for changing signal levels. As shown in Figure 3-31, this AGC is also applied to amplifiers in the RF section. However, it is necessary to delay the application of AGC to the RF amplifier until a suitable *S/N* ratio is reached. Conventional AM automatic gain control systems are generally not usable in SSB receivers because they operate on the level of the carrier, and the carrier is suppressed in SSB. Therefore, we must use AGC systems for SSB that obtain their information directly from the modulation envelope. This AGC system can be achieved with conventional diode rectifiers and additional amplification. Special care must be taken to isolate the AGC system from the reinserted carrier because the carrier is a large signal of the same frequency as the IF signals. This problem can be avoided if the AGC voltage is developed from the audio signal.

The intelligence carried by the SSB signal is recovered and amplified to a level suitable for the audio output circuits. The circuits for recovering the audio intelligence perform the same function in the receivers as the modulator does in the transmitter; therefore, the same circuits can be used. In the demodulator, the SSB signal modulates the local carrier. Product detector circuits (see Figure 3-30(B)) are preferred in SSB reception because they minimize intermodulation distortion products in the audio output signal and do not require large local carrier voltages. The demodulator output consists of an audio signal and several RF outputs. These signals are easily filtered by passing the output of the demodulator through an audio LP filter.

The proper frequency relationship between the sideband signal and the carrier must be maintained. If the received signal is an upper sideband, the carrier frequency is below the sideband signal; if the received signal is the lower sideband, the carrier frequency is above the sideband signal. If a receiver is to be used to receive either the upper sideband or the lower sideband, we must provide a means of changing the relative position of the carrier with respect to the sideband. One way of doing this is to use two sideband filters in the IF section and a single carrier at the demodulator. The desired sideband is then selected by switching in the proper filter. Another approach for dual-sideband reception is to use a single filter and provide a means of shifting the local carrier from one side of the IF filter pass band to the other and then retuning the oscillators in the RF section.

The ISB Receiver

Another type of SSB receiver is the *independent-sideband* (ISB) *receiver* shown in the block diagram of Figure 3-32. The ISB receiver must have two audio outputs to reproduce the two separate audio signals originating at the transmitter. To separate the two transmitted sidebands, we place a selective crystal filter, capable of shifting the pass band slightly above or below the IF frequency, in the IF section.

The two conversion sections of the ISB receiver are similar to those of other SSB receivers. However, in the ISB receiver a hybrid circuit separates the upper and lower sidebands before application to their respective channels.

An LF oscillator output is mixed with each of the separated sidebands in their own demodulators. The outputs of these two demodulators are then applied to their respective audio amplifier sections, where the signals are increased to a level suitable to drive the output loudspeaker loads.

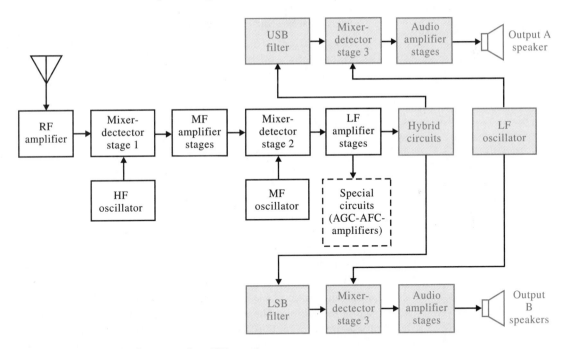

FIGURE 3-32 Block diagram of an ISB receiver

3-10 SSB ADVANTAGES, DISADVANTAGES, AND VARIATIONS

As stated earlier, SSB transmission overcomes the disadvantages of standard AM. However, SSB emissions have their own advantages and disadvantages, and several variations are found in SSB systems. This section discusses the advantages, disadvantages, and variations of SSB transmission.

Advantages of SSB

The need for SSB communications systems has arisen because modern radio communications require faster, more reliable, spectrum-conservative systems. Today, SSB communication plays an increasingly important role in radio communications because of its many advantages over standard AM, several of which follow.

1. *All the power transmitted represents intelligence.* In our discussion of standard AM, we found that 100 W, or two thirds of the 150 W transmitted, was power wasted in the carrier. An SSB transmitter radiating 50 W produces the same level of intelligence at the receiver as a standard AM transmitter radiating 150 W.

2. *Bandwidth requirements are reduced.* Bandwidth reduction satisfies the need for spectrum conservation. Because only one sideband is transmitted, the bandwidth is only one half that of the standard AM system. Thus, for a given frequency spectrum, SSB offers twice as many channels as standard AM.

3. *Transmission reliability is improved.* Improved reliability means less noise at the receiver and better *S/N* ratios. In long-range communications, SSB transmission reduces distortion and fading caused by multipath reception.

4. For equal performance, the *size, weight, power output,* and *peak antenna voltage* of the SSB transmitter *are significantly less* than for the standard AM transmitter.

Disadvantages of SSB

As with all systems, there are certain problems that arise in SSB operation. The major disadvantages are the added cost and system complexity involved in providing better frequency stability. Because standard AM transmitters offer a relatively high degree of stability, standard AM receivers do not require any special circuitry to maintain LO stability. However, in SSB operation, frequency stability is very important. The same high stability must be maintained in both transmitters and receivers. Precision frequency control circuitry or complex *automatic frequency control* (AFC) circuitry is used in high-quality receivers. Because the carrier is not transmitted, it must be generated locally at the receiver. For maximum fidelity, this locally generated carrier must be identical, in both frequency and phase, to the original transmitter carrier. In voice communication, small phase shifts are tolerable, but small frequency shifts can make the signal unintelligible. The amount of frequency and phase shift that can be tolerated will vary, depending on the *S/N* ratio. The requirements for transmitting digital information are much more critical, however, because phase shifts under these conditions cannot be ignored.

SSB Variations

Several types of systems in the general classification of SSB transmission have been developed. Each variation has its own particular purpose. The definition, description, and application of the various forms of SSB follow.

1. *Single-sideband, suppressed-carrier* (SSSC) is another designation for the SSB systems we have been discussing. The carrier is suppressed by about 50 dB at the transmitter. Receivers using *frequency synthesizers* (devices that generate highly precise frequency signals by means of a single crystal oscillator in conjunction with frequency dividers and multipliers) for reinsertion of the carrier have made SSSC the standard form of SSB for HF mobile communications.

2. *Single-sideband, reduced-carrier* (SSRC) is a *pilot carrier* system in which we reinsert an attenuated carrier into the SSB signal, after removal of the unwanted sideband, to facilitate receiver tuning and demodulation. The pilot carrier consists of a small amount of ac signal drawn from the carrier oscillator, equivalent to about 10% modulation. The receiver separates the pilot carrier from the SSB signal. The pilot carrier is then amplified, and the signal is used to reinsert the original carrier. However, the pilot carrier can instead be used as a reference in the receiver's AFC circuit, where it is compared with the insertion oscillator frequency. It can also provide *automatic volume control* (AVC) voltages. This system is widely used in transmarine point-to-point radiotelephony, and in maritime mobile communications. Primarily, it is used on the maritime mobile distress frequencies.

3. *Single-sideband, controlled-carrier* (SSCC) is another type of pilot carrier. The controlled carrier rises to almost full amplitude during brief inactive periods of transmission but is reduced to very low levels during actual modulation, thus keeping the average power output of the transmitter constant. (This constant transmitter power output is similar to squelch circuit operation in two-way communications receivers, which are discussed in detail in Chapter 15.) For this reason, power is independent of the presence or absence of modulation. Slow-acting AFC and AGC (automatic gain control) circuits in the receivers maintain proper frequency during demodulation.

4. *Single-sideband, full-carrier* (SSFC) can be used as a compatible AM broadcasting system, because distortion is very low. The transmitted signal can be received and demodulated by standard AM and SSB receivers. However, since the full carrier is transmitted (actually, at a level about 5 dB below the peak power level of the transmitter), the reduced power advantages of SSB are lost.

5. *Independent-sideband* (ISB), called *twin-sideband, suppressed-carrier* (TSSC), transmits two independent sidebands, each containing different intelligence, with the carrier suppressed by about 50 dB. It is frequently used for HF point-to-point radiotelephony, where more than one channel is required. Its principal application is in military communications.

6. *Vestigial sideband* is used throughout the world for television video transmissions. In this system, a trace, or *vestige*, of the unwanted sideband is transmitted, usually with the carrier and the other sideband left intact. A detailed discussion of television is presented in Chapters 12 and 13.

3-11 STEREOPHONICS

Stereophonics (stereo) is the technology of two-channel sound reproduction. When high-fidelity music is reproduced in stereo, the music sounds more realistic because you get a sense of sounds originating from different sources.

A stereo recording has two independent channels (left channel and right channel), recorded with separate microphones and played back through separate speakers or earphones. This requires two sets of audio recording and reproduction systems.

Stereo is almost universally used in the recording of commercial music today. A wide variety of stereo systems and accessories are manufactured commercially.

Quadriphonics

Quadriphonics, or quad system, is now an outdated four-channel, high-fidelity recording and reproduction system. Where an ordinary stereo system has two channels, a quad system has four channels, usually designated left front, right front, left rear, and right rear. In a true quad system, the four channels are entirely independent, providing true 360-degree reproduction of sound.

Stereo Disk Recording

Stereo sound is obtained by two independent sound channels being impressed onto a single groove in a disk. This is done by a stereo disk recorder. The recorder cuts a groove by vibrating in a plane perpendicular to the groove. Left-channel information is impressed in the groove by vibrations that occur at a 45-degree angle with respect to the surface of the disk; the right-channel by vibrations at a right angle (90 degrees) to those of the left channel.

When a stereo disk is played back, the stylus (needle) follows the groove, vibrating in the same way as did the original cutting stylus. Because the left-channel and right-channel vibrations occur at right angles to each other, their instantaneous amplitudes are entirely independent.

Stereo AM Broadcast

Figure 3-33 shows a typical stereo AM broadcast transmitter. FM stereo is discussed in Chapter 4. The RF oscillator provides the carrier signal that is fed into the left-channel AM modulator and a phase shifter. The output of the phase shifter provides a second carrier signal to the right-channel AM modulator. The outputs of the two modulators are mixed and fed into the power amplifier, thence to the antenna. A 25-Hz signal is mixed with one of the modulator inputs to provide a signal for an indicator light on a receiver, that lights up to indicate the reception of stereo.

Stereo Multiplex

Broadcasting the two independent channels of sound over a single RF carrier to obtain stereo sound in a specially designed receiver is accomplished by stereo multiplex.

Stereo multiplex is achieved by modulating two low-frequency subcarriers with audio signals from the left and right channels. The two subcarriers have different frequencies. The main RF carrier is then modulated by both subcarrier signals simultaneously. To maintain separation between channels, linearity in the circuits is required. Figure 3-34 shows a block diagram of a stereo-multiplex transmitter.

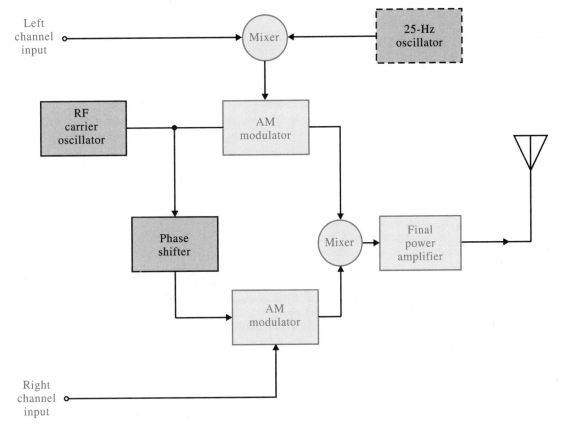

FIGURE 3-33   Typical stereo AM transmitter

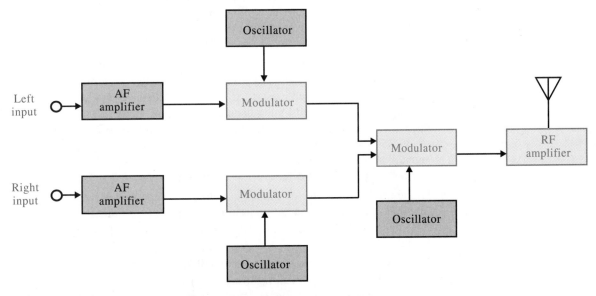

FIGURE 3-34   Block diagram of a stereo-multiplex transmitter

Stereo Receiver

Figure 3-35 shows a typical stereo-multiplex receiver. The demodulator (detector A) strips the two subcarrier signals from the main RF carrier. A pair of selective circuits (filters left and right) separates the subcarrier channels. A detector is then used in each channel to obtain the original left-channel and right-channel audio signals.

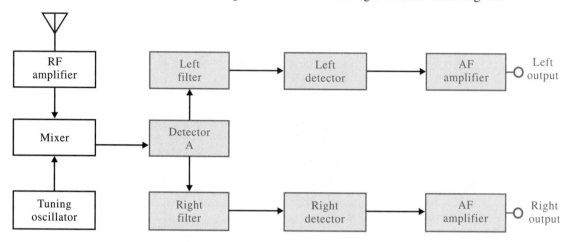

FIGURE 3-35 Typical stereo-multiplex receiver

SUMMARY

- Amplitude modulation (AM) is accomplished by the effect of varying the amplitude of an HF carrier with an LF intelligence signal. The AM signal is a complex waveform resulting from mixing the carrier and intelligence signals through a modulator. Mixing the carrier and intelligence signals produces frequency components, including a dc level and harmonics, which are removed by filtering. An intelligence envelope is a replica of the intelligence signal.

- The modulation factor m is a measure of the extent to which the carrier current or voltage is varied by the intelligence. Maximum allowable modulation is 100%. Modulation is maintained close to 100% so that the maximum amount of the intelligence can be recovered by the demodulator circuit in a receiver. Overmodulation occurs when the modulated carrier goes to more than twice its unmodulated value (over 100% modulation). At 100% modulation, and assuming that the modulating signal is a sine wave, sideband power is always equal to one half (50%) of the carrier power, and one third (33.3%) of the total power.

- Diode modulators offer no amplification. Some high-power transmitters may still require vacuum tube modulator circuits. Properly biased transistors can operate in a nonlinear mode and offer amplification, making the transistor ideal for use as modulators.

- In transistor amplifier modulators, the intelligence signal can be injected into any of the three transistor elements, resulting in collector modulation, emitter modulation, or base modulation. In collector modulation, the most commonly used technique, the disadvantage of a high modulating signal power requirement is outweighed by the advantages of high collector efficiency, low distortion, and easy circuit adjustments. In base modulation, the advantage of a lower modulating power requirement is overshadowed by the disadvantages of lower collector efficiency, lower power output, poorer linearity, and difficulty of circuit adjustments. In emitter modulation, the intelligence signal amplitude is usually higher than the carrier amplitude, which makes emitter modulation best suited for low-level modulation applications.

- The level of modulation is determined by where in the amplification system the modulating signal is applied. Applying the modulating signal to the last possible amplifier stage, usually to the collector of the final power amplifier, produces high-level modulation. Low-level modulation is the result of applying the modulating signal at some earlier amplifier stage. The major advantage of high-level modulation is that the RF amplifiers in a transmitter can be operated class C. In low-level modulation, all amplifier stages following application of the modulating signal must be operated either class A or class B push-pull to prevent distortion of the signal. A disadvantage of low level modulation is that linear operation reduces efficiency and power-handling capabilities. Low-level modulation is the most economical method for low-power transmitters.

- To provide the highest stability in a standard AM transmitter, we use a buffer amplifier between the oscillator stage and the following stages to provide a constant load on the oscillator and to prevent feedback to the oscillator from the following stages.

- Demodulation, or detection, involves stripping the carrier wave from the intelligence signal in a circuit of the receiver. In AM receivers, the circuit used for demodulation is called the envelope detector or, more simply, the detector. The diode detector is the simplest and most widely used AM demodulator, but it provides no amplification. Transistors can provide detection and amplification in a single stage, and transistor detectors are generally used in low-cost, all-transistor receivers. The synchronous detector is a special circuit required to demodulate the DSSC signal. The circuit input is a carrier signal regenerated by earlier stages of the receiver. Single-sideband receivers usually use some form of product detector, which can be any nonlinear mixer circuit. The product detector can demodulate all forms of AM. The beat frequency oscillator (BFO) used in product detectors is an oscillator whose output frequency is beat (mixed) with the IF input to produce an AF output. It is sometimes called a carrier oscillator.

- The tuned radio frequency (TRF) receiver offers simplicity and high sensitivity. Its greatest use today is in fixed-frequency applications.

- The superheterodyne receiver uses the principle of fixed tuning. Fixed tuning is accomplished by mixing the incoming RF signal with a locally generated signal, which produces an IF of constant value (455 kHz for standard AM). Heterodyning (mixing) of the RF signal and the LO occurs in the mixer stage of the receiver. The mixer output is the IF. The LO is a variable-frequency stage whose frequency is always a fixed amount higher than the RF signal. Image frequency is a frequency other than the desired frequency that produces the IF in a receiver. Image frequency rejection is accomplished by adding tuned RF amplifiers before the mixer or by dual conversion.

- The major advantage of the standard AM system is that it uses straightforward and inexpensive transmitting and receiving equipment. The three most important disadvantages in the standard AM system are as follows: (1) Power is wasted in the transmitter. (2) The transmitted signal requires twice the bandwidth of the transmitted intelligence. (3) Very precise amplitude and phase relationships between the sidebands and carrier are required.

- Single-sideband (SSB) transmission is a form of AM in which only one sideband is transmitted; the other sideband and the carrier are suppressed at the transmitter. Single-sideband transmitters are more complicated, hence more expensive, than standard AM transmitters. Some advantages of SSB over standard AM are as follows: (1) All of the power transmitted represents intelligence. (2) Bandwidth requirements are reduced. (3) Transmission reliability is improved. (4) Size, weight, power output, and peak antenna voltage are reduced. The major disadvantage of SSB operation is the added cost and system complexity involved in providing better frequency stability in the transmitter and the receiver. Several variations of SSB operations are: (1) single-sideband, suppressed-carrier (SSSC), used for HF mobile communications; (2) single-sideband, reduced-carrier (SSRC), a pilot carrier system widely used in transmarine point-to-point radiotelephony and in maritime mobile communications using primarily the maritime distress frequencies; (3) single-sideband, controlled-carrier (SSCC), another pilot carrier system, in which power is independent of modulation and the transmitter average power output is constant; (4) single-sideband, full-carrier (SSFC), a low-distortion, compatible AM broadcasting system whose transmitted signal can be received by standard AM and SSB receivers; (5) independent-sideband (ISB), also called twin-sideband, suppressed-carrier (TSSC), which transmits two independent sidebands, each containing different intelligence, frequently used for HF point-to-point radiotelephony, with primary application in military communications; and (6) vestigial sideband, used worldwide for television signal transmissions.

- Two basic methods for SSB generation are the filter method and the phase-cancellation method. Both methods use some form of balanced modulator to suppress the carrier. Commonly used SSB generators consist of a balanced modulator followed by an extremely selective crystal, ceramic, or mechanical filter.

- The output of a dual-FET balanced modulator is a DSSC amplitude modulated signal. The greatest disadvantage of DSSC amplitude modulation is the need for extensive circuitry in the receiver to reinsert the carrier. The diode ring modulator, also called a balanced ring modulator, uses carefully matched diodes as the nonlinear elements of the circuit.

- The most widely used technique for SSB generation is the filter method, which uses a BP filter to pass one sideband and reject the other. In some SSB transmitters using LF carrier oscillators, dual conversion is used as a way to preserve the intelligence signal and bandwidth while raising the carrier frequency to a desired RF level. The filters used in filter-type SSB transmitters must have very fast rise and fall slopes and flat BP characteristics. Crystal filters are commonly used in SSB systems because of their very high Q values. The crystal lattice filter uses two matched pairs of crystals in a bridge circuit. Ceramic filters have lower Q values than crystal filters, but the lower cost, smaller size, and ruggedness of ceramic filters make them more practical than crystal filters for portable SSB equipment applications. Mechanical filters are mechanically resonant devices that use a double-conversion technique for better suppression of unwanted frequencies.

- The phase-cancellation method of SSB generation uses two balanced modulators, both of which suppress the carrier and have a double-sideband output. The two sideband signals are combined in a summing circuit, the output of which is only the desired sideband. An advantage of the phase-cancellation method is that the SSB signal at the operating frequency can be generated without frequency conversion. A disadvantage is that the phase shifts and amplitudes of the signals applied to the summing circuit must be accurately maintained.

- Single-sideband receivers are superheterodyne but differ from standard AM receivers in that additional circuitry must be used to restore the suppressed carrier. A high degree of spurious response attenuation in HF receivers is maintained by dual conversion. Two IF sections and two local oscillators are required.

- The ISB receiver requires two conversion sections. It must have two audio outputs to reproduce the two separate audio signals originating at the transmitter. The two transmitted sidebands are separated in an ISB receiver by a selective crystal filter in the IF section.

QUESTIONS AND PROBLEMS

1. Define (a) modulation, (b) carrier, and (c) intelligence.

2. List the frequency components resulting from mixing the carrier and intelligence signals through a modulator.

3. Define modulation factor.

4. An AM signal has a carrier current of 6 A, a maximum current of 9 A, and a minimum current of 3 A. What is m?

5. An AM carrier voltage of 60 V is modulated with an intelligence voltage of 15 V. What is the index of modulation?

6. A carrier voltage of 100 V is modulated with an intelligence voltage of 120 V. What is the modulation factor?

7. At 100% modulation, what percentage of the total power is contained in the sidebands?

8. A 500 kHz RF signal is modulated with a 5 kHz AF signal. The modulated carrier voltage is 25 V maximum and 15 V minimum across a 300 Ω resistive load impedance. Determine the (a) zero-modulation RF carrier voltage, (b) modulation factor, (c) carrier power, (d) sideband power, (e) total power, (f) upper

side frequency, (g) carrier frequency, and (h) lower side frequency.

9. A 50 kW carrier is modulated to a level of 90%. Determine the (a) carrier power, (b) sideband power, and (c) total power.

10. An unmodulated carrier current is 5 A. Current increases to 5.75 A when modulated. Determine the (a) modulation factor and (b) total current when the signal is modulated to a level of 60%.

11. When transistors are used as modulators, what advantage do they have over diodes?

12. List the ways in which a transistor amplifier modulator may be modulated.

13. What is the most commonly used AM technique? Explain how this technique works.

14. What type of amplifier is the circuit in Problem 13? In what level of modulation is it often used?

15. What must be done to the circuit in Problem 13 to obtain 100% modulation?

16. How much power is required to obtain 100% modulation in a collector-modulated circuit?

17. What are the advantages of collector modulation?

18. Compared to collector modulation, what are the disadvantages of base modulation?

19. For what level of modulation is emitter modulation best suited? Why?

20. Draw a block diagram of a standard AM transmitter. Explain the purpose of each block.

21. What is the purpose of a buffer amplifier in an AM transmitter?

22. What is the major advantage of high-level modulation?

23. List the major advantage and the most important disadvantages of standard AM.

24. Define SSB and explain why it is needed.

25. List four advantages SSB has over standard AM.

26. List the major advantages of SSB.

27. Briefly describe five variations of SSB transmission.

28. The peak SSB envelope value is 125 V when measured across a 50 Ω line. Calculate the PEP.

29. The rms values for an SSB signal are 45 V and 12 A. Calculate the PEP.

30. The average power of an SSB signal is 250 W. What is the PEP?

31. Briefly describe the operation of a balanced ring modulator.

32. What are the two basic methods used for SSB suppression? Which is the most widely used?

33. Draw a block diagram of a basic filter system SSB transmitter. Briefly describe its operation.

34. Draw a block diagram of a dual-conversion, filter-type SSB transmitter. Briefly explain why dual conversion is used and why it is necessary.

35. Briefly describe the operation of a crystal lattice filter.

36. What technique is used in mechanical filters to obtain better sideband suppression?

37. In mechanical filters, how can selectivity be improved? How can bandwidth be increased?

38. Draw a block diagram of a phase-shift SSB transmitter. Briefly describe its operation.

39. List an advantage and a disadvantage of the phase-shift SSB method.

40. What is the simplest and most widely used AM demodulator? What type of demodulator is generally used in low-cost, all-transistor receivers?

41. What type of device is used to demodulate a DSSC amplitude modulated signal? Briefly describe its operation.

42. What is the greatest disadvantage of DSSC amplitude modulation?

43. What type of device is used to demodulate SSB signals?

44. What is the function of a BFO? What is another name for the BFO?

45. Draw a block diagram of a TRF receiver. Briefly describe its operation.

46. Define sensitivity of a receiver. How is it determined?

47. Define selectivity of a receiver. Explain what it does.

48. Define heterodyne.

49. Draw a block diagram of a superheterodyne radio receiver. Describe the function of each block.

50. Calculate the image frequency for a standard AM receiver tuned to a station at 1105 kHz.

51. A standard AM receiver receives an image frequency of 1205 kHz. What is the desired frequency?

52. A receiver has an IF of 670 kHz and is tuned to a station at 830 kHz. What is the LO frequency?

53. Draw a block diagram of an SSB receiver. Briefly describe the function of each section.

54. Draw a block diagram of an ISB receiver. Briefly describe its operation.

55. Define (a) stereophonics and (b) quadriphonics.

56. Describe how a stereo recording is made.

57. Describe how stereo multiplex works for stereo broadcasting.

58. Describe the operation of a stereo-multiplex receiver.

CHAPTER 4

Angle Modulation

OBJECTIVES

When you complete this chapter, you should be able to:

- Describe the two types of angle modulation.
- Discuss the determining factors for the amount and rate of carrier deviation.
- Define deviation, frequency swing, preemphasis, deemphasis, significant sidebands, guard bands, reinsertion, wide-band, and narrow-band.
- Define the direct and indirect classes of FM and describe the differences between them.
- Identify the schematic diagrams of different types of modulator circuits and describe their operation.
- Determine the multipliers required to achieve the transmitter output frequency when the frequency of the master oscillator and the station operating frequency are known.
- Determine the frequency of the master oscillator when the output frequency and transmitter multiplication are known.
- Determine the required bandwidth for an FM signal when the deviation and the intelligence frequency are known.
- Identify the schematic diagrams of the different types of demodulator circuits and describe their operation.
- Draw block diagrams of angle modulation transmitters and receivers and describe the function of each stage.

INTRODUCTION

There are two distinct types of angle modulation: (1) *frequency modulation* (FM), where the amplitude of the modulated carrier remains constant and its frequency is varied around the center frequency in accordance with the variations of the modulating signal; and (2) *phase modulation* (PM), where the amplitude of the modulated carrier also remains constant but the phase of the carrier is varied instead of its frequency. In

135

effect, PM is a modified version of FM, because any change in phase is related to a change in frequency. This chapter examines both types of modulation and the circuits and methods used for modulation and demodulation.

4-1 FREQUENCY MODULATION

The FM system was developed as an alternative to AM in an effort to make radio transmissions less susceptible to noise interference. In FM, the frequency of the carrier is made to fluctuate in accordance with the modulating signal.

Frequency modulation results in the generation of sidebands similar to those in an amplitude modulated system. However, as the deviation is increased, sidebands appear at greater and greater distances from the main carrier. The amplitude of the main carrier also depends on the amount of deviation.

The amount by which the signal frequency varies above and below the center of the main carrier (*rest* frequency) is called the *deviation*. The *amount* of deviation is determined only by the amplitude of the modulating signal; that is, all modulating signals having the same amplitudes will deviate the carrier frequency by the same amount. The amplitudes of the sidebands, which appear at integral multiples of the modulating signal frequency above and below the carrier, as well as the amplitude of the carrier itself, are a function of the ratio of the deviation to the modulating frequency. The function is rather complicated, but, in general, the greater the deviation, the greater the bandwidth of the signal. The ratio of the maximum frequency deviation to the highest modulating frequency is called the *modulation index*.

The circuits used for processing the signals, both at the transmitter and at the receiver, play an important role in noise immunization in FM systems. This section discusses these circuits, along with methods of modulation and demodulation.

The FM Signal

Assuming that the phase relationships of a complex sinusoidal modulating signal are preserved, we have only the two remaining important parameters of amplitude and frequency to consider. In FM, the amplitude remains constant, so we have only the frequency to consider (see Equation 3-1).

In FM, the modulating frequencies have no effect on the amount of deviation. Conversely, the *rate* of carrier deviation is determined only by the *frequencies* of the modulating signals. The individual amplitudes of the modulating signals have no effect on the rate of deviation.

As stated earlier, the amplitude of the FM wave remains constant. This characteristic provides the greatest advantage of FM over AM—that of superior noise immunity. Therefore, we can fully define FM as *the amount by which the instantaneous deviation is made proportional to the instantaneous value of the modulating signal*. Figure 4-1 demonstrates these concepts. For comparison purposes, an AM signal is shown. For ease of understanding, the FM signal is shown in greatly exaggerated form. The total frequency deviation above and below the rest frequency (100 kHz) is called the *frequency swing*. In the FM signal shown, the frequency swing is 130 kHz–70 kHz, or 60 kHz.

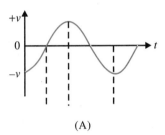

(A)

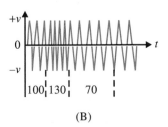

|100|130| 70 |

(B)

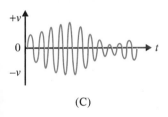

(C)

**FIGURE 4-1   FM signal
(A) Modulating signal
(B) FM wave (C) AM wave**

The FM signal may be mathematically analyzed as follows:

$$x = X \cos(\omega_c t + m_f \sin \omega_i t) \tag{4-1}$$

where

x = instantaneous value (voltage or current)

X = maximum value of carrier wave (voltage or current)

ω = $2\pi f$

ω_c = carrier angular velocity in radians per second (rad/s) ($2\pi f_c$)

m_f = FM modulation index

ω_i = intelligence signal angular velocity in radians per second ($2\pi f_i$)

The FM modulation index m_f is defined as

$$m_f = \frac{\text{maximum carrier deviation}}{\text{intelligence frequency}} = \frac{\delta}{f_i} \tag{4-2}$$

EXAMPLE 4-1

An AM signal has a rest frequency of 88 MHz and is swinging between 87.998 MHz and 88.002 MHz, a frequency swing of 4 kHz, at a rate of 2000 times per second. Determine the (a) intelligence frequency f_i and (b) effect on the signal if the intelligence signal amplitude is doubled.

SOLUTION

(a) The rate of deviation is 2000 times per second; therefore, $f_i = 2$ kHz.

(b) Doubling the amplitude of the modulating signal doubles the amount of deviation; therefore, the FM signal will swing between 87.996 MHz and 88.004 MHz—a frequency swing of 8 kHz. However, the rate of deviation will not be affected by the change in amplitude, so f_i remains at 2 kHz.

EXAMPLE 4-2

An FM signal has a deviation of 3.6 kHz when the modulating signal voltage (amplitude) is 2.3 V and f_i is 750 Hz. (a) The intelligence voltage is increased to 4.6V. What is the new deviation? (b) The frequency is decreased to 250 Hz. What is the new deviation? (c) Determine m_f for each condition.

SOLUTION

(a) When intelligence amplitude doubles, deviation doubles:

$\delta = 2(3.6 \text{ kHz}) = 7.2 \text{ kHz}$

(b) The frequency is reduced by two-thirds, but the deviation is unaffected by frequency changes. So

$\delta = 3.6 \text{ kHz}$

(c) For the original condition,

$$m_f = \frac{\delta}{f_i} = \frac{3.6 \text{ kHz}}{0.75 \text{ kHz}} = 4.8$$

For condition (a),

$$m_f = \frac{\delta}{f_i} = \frac{7.2 \text{ kHz}}{0.75 \text{ kHz}} = 9.6$$

For condition (b),

$$m_f = \frac{\delta}{f_i} = \frac{3.6 \text{ kHz}}{0.25 \text{ kHz}} = 14.4$$

Refer to Equation 3-1 to solve Examples 4-3 and 4-4.

EXAMPLE 4-3

Assume $X = 8$ V, $t = 10$ μs, $\theta = -90°$. Find x for (a) $f = 90$ MHz and (b) $f = 102$ MHz. What type of modulation is this?

SOLUTION

(a) $x = 8$ V sin (6.28 × 90 MHz × 10 μs − 90°)
    $= 8$ sin (5562) = 8 (0.31) = 2.48 v

(b) $x = 8$ V sin (6.28 × 102 MHz × 10 μs − 90°)
    $= 8$ sin (6316) = 8 (−0.27) = −2.16 v

The 2.48 v and −2.16 v values are the instantaneous amplitudes of the modulated 90 MHz and 102 MHz frequencies, respectively, at $t = 10$ μs. The frequency (f) is varied, with maximum amplitude (X) and phase angle (θ) held constant; therefore, this is an FM signal.

EXAMPLE 4-4

Assume $X = 10$ V, $t = 5$ μs, $f = 98$ MHz. Find x for (a) $\theta = +90°$ and (b) $\theta = -45°$. What type modulation is this?

SOLUTION

(a) $x = 10$ V sin (62.8 × 98 MHz × 5 μs + 90°)
    $= 10$ sin (3167) = 10 (−0.9553) = −9.533 v

(b) $x = 10$ V sin (6.28 × 98 MHz × 5 μs − 45°)
    $= 10$ sin (3032) = 10 (0.4664) = 4.664 v

The −9.553 v and 4.664 v values are the instantaneous amplitudes of the modulated 98 MHz frequency at $t = 5$ μs. The phase angle (θ) is varied, with maximum amplitude (X) and frequency (f) held constant; therefore, this is a PM signal.

Because f and θ are related, it actually is impossible to change one without changing the other. However, the one that is directly varied distinguishes between frequency modulation and phase modulation. These relationships are discussed in detail in Section 2.

Forms of Interference

A major advantage of angle modulation over amplitude modulation is its relative immunity to noise and other interference signals. Among the three modulating modes, FM has the greatest immunity to noise. However, noise or interference still can be introduced into any system at any point where amplitude-controlled phase- or frequency-sensitive circuits are used. The reason is that any change in amplitude in a reactive (inductive or capacitive) circuit produces a change in phase. Although amplitude-controlled signals are much smaller in FM systems than in AM systems, they can become large enough to produce a noticeable "hiss" in a received FM signal.

Noise Effects   Noise is generated by random electron motion in all electronic devices and components, whether passive or active. The higher the operating temperatures, the greater the amount of noise produced. This internal thermal noise increases with frequency, degrading the intelligence signals coming through any amplifier.

Frequency modulation has noise immunity properties that reduce the effects of noise. First, consider how noise affects the phase of the RF carrier. The phase deviation and the relative FM modulation index increase as the noise signal frequency increases. Increased phase deviation occurs because the phase shift is directly proportional to the amplitude of the incoming signal, and the amplitude of a noise pulse is usually much higher than that of the modulating signal. The relative FM modulation index increases because any change in the instantaneous phase of a carrier produces an instantaneous fluctuation in the frequency. The result is an increased noise level with increased frequency, but no change in the modulation signal level as its frequency increases. A consequence of this phenomenon is a smaller S/N ratio (greater noise level) at the higher frequencies. However, the increased modulation index resulting from the higher frequencies also results in a lower noise sideband distribution.

Not only do noise levels increase in the FM detection process at higher intelligence frequencies, but the amplitudes of higher-frequency intelligence are normally smaller than lower-frequency intelligence signals, so the signals sound normal and pleasing to the listener. Thus, at higher intelligence frequencies, the resultant S/N ratio will be lower, making the noise levels appear to be higher. Naturally, the best S/N ratios occur when noise signals are at their minimum levels and intelligence signals are at their maximum levels. In FM systems, maximum intelligence signal levels occur with maximum deviation of the RF carrier.

Comparative Noise Suppression   The degree to which noise is suppressed increases directly with the modulation index produced by the signal. Thus, the S/N ratio in a wide-band (±75 kHz deviation) FM system used in broadcasting is usually superior to that in a narrow-band (± 10 kHz deviation) system. In the wide-band system, thermal, circuit, and pulse noises weaker than the desired signal will be more completely suppressed than in the narrow-band system. However, the amount of thermal and circuit noise energy accepted by the receiver is proportional to the bandwidth of the receiver, which means that a wide-band system must deliver a greater signal strength to a receiver for the signal to be greater than the noise. Thus, wide-band systems are used where high-quality signals are to be transmitted. Conversely, where voice intelligibility is the primary purpose of the transmission, the narrow-band system is superior because the narrow bandwidth of the receiver limits to a greater extent the amount of thermal and circuit noise that will get into the signal circuits. A natural property of the FM system is to suppress the weaker of two signals. Therefore, it is essential in all cases to keep the noise level well below that of the desired intelligence level.

FM Waveform Analysis

In theory, an FM wave has an infinite number of sideband frequencies, referred to simply as *sidebands*, above and below the carrier and spaced at multiples of the intelligence frequency, indicating an infinite bandwidth requirement. However, the larger the multiple of these sidebands, the lower their amplitudes, so beyond a certain spread the amplitudes become negligible. What this means in practical terms is that the number of significant sidebands is limited and dependent on the modulation index. Significant sidebands are those that have a minimum amplitude of 1% of the amplitude of the unmodulated carrier.

Mathematically, analysis of an FM signal modulated with a sine wave requires a set of advanced calculus equations called *Bessel functions*. Such high-level mathematics is beyond the scope of this text; therefore, calculations for each pair of sidebands will not be shown. Separate calculations are not necessary because information is readily available in table form, shown by the Bessel functions in Table 4-1. Before showing how to use the table, a few comments are in order. When an intelligence frequency f_I modulates a carrier f_c, it creates sidebands, much as does AM. However, these sidebands are at the carrier frequency plus/minus integer multiples of f_I. The strength of each of these sidebands is a function of the modulation index, which in turn depends upon the *amplitude* of the intelligence frequency (as was shown in Equation 4-2). Refer to Table 4-1 (Bessel Functions) and note that the vertical axis contains values of the modulation index (m_f). The horizontal axis contains values of each Bessel function (J_n) where the n refers to the integer multiple of the intelligence frequency. Each table value is the amplitude of a particular sideband. Negative values indicate $180°$ phase shifts in the sideband frequency. Figure 4-2 shows the spectrum produced by modulation indices of 0.25, 1.0, 2.0, and 5.0. Note that as the modulation index increases, the number of significant sidebands also increases. This is due to the fact that a high modulation index produces greater frequency deviation, which in turn increases the number of sidebands.

The concepts of bandwidth, sideband, and carrier amplitudes are demonstrated by Figure 4-2. The sidebands at equal distances from f_c have equal amplitudes, and the distance between each successive J term is equal to f_i. We see that as sideband amplitudes increase, carrier amplitude decreases. From Table 4-1, we can determine that for values of m_f of 2.4, 5.5, and 8.65, the carrier component is at zero amplitude and all

TABLE 4-1 **Bessel Functions**

x (m_f)	J_0	J_1	J_2	J_3	J_4	J_5	J_6	J_7	J_8	J_9	J_{10}	J_{11}	J_{12}	J_{13}	J_{14}
0.00	1.00	—	—	—	—	—	—	—	—	—	—	—	—	—	—
0.25	0.98	0.12	—	—	—	—	—	—	—	—	—	—	—	—	—
0.5	0.94	0.24	0.03	—	—	—	—	—	—	—	—	—	—	—	—
1.0	0.77	0.44	0.11	0.02	—	—	—	—	—	—	—	—	—	—	—
1.5	0.51	0.56	0.23	0.06	0.01	—	—	—	—	—	—	—	—	—	—
2.0	0.22	0.58	0.35	0.13	0.03	—	—	—	—	—	—	—	—	—	—
2.4	0	0.52	0.43	0.20	0.06	—	—	—	—	—	—	—	—	—	—
2.5	−0.05	0.50	0.45	0.22	0.07	0.02	—	—	—	—	—	—	—	—	—
3.0	−0.26	0.34	0.49	0.31	0.13	0.04	0.01	—	—	—	—	—	—	—	—
4.0	−0.40	−0.07	0.36	0.43	0.28	0.13	0.05	0.02	—	—	—	—	—	—	—
5.0	−0.18	−0.33	0.05	0.36	0.39	0.26	0.13	0.05	0.02	—	—	—	—	—	—
5.5	0	−0.34	−0.12	0.26	0.40	0.32	0.19	0.09	0.03	0.01	--	—	—	—	—
6.0	0.15	−0.28	−0.24	0.11	0.36	0.36	0.25	0.13	0.06	0.02	—	—	—	—	—
7.0	0.30	0.00	−0.30	−0.17	0.16	0.35	0.34	0.23	0.13	0.06	0.02	—	—	—	—
8.0	0.17	0.23	−0.11	−0.29	−0.10	0.19	0.34	0.32	0.22	0.13	0.06	0.03	—	—	—
8.65	0	0.27	0.06	−0.24	−0.23	0.03	0.26	0.34	0.28	0.18	0.10	0.05	0.02	—	—
9.0	−0.09	0.24	0.14	−0.18	−0.27	−0.06	0.20	0.33	0.30	0.21	0.12	0.06	0.03	0.01	—
10.0	−0.25	0.04	0.25	0.06	−0.22	−0.23	−0.01	0.22	0.31	0.29	0.20	0.12	0.06	0.03	0.01

the transmitted power is then in the sidebands. The required bandwidth for an FM signal can be calculated by using Table 4-1 to see which J coefficient is the last one shown for a specific modulation index.

EXAMPLE 4-5

An FM signal has a maximum deviation of 10 kHz and is modulated with an intelligence frequency of 2.5 kHz. What is the required bandwidth?

SOLUTION

First, determine the modulation index:

$$m_f = \frac{\delta}{f_i} = \frac{10 \text{ kHz}}{2.5 \text{ kHz}} = 4$$

Then, from Table 4-1, observe that the highest J coefficient included for $m_f = 4$ is J_7. All higher values of Bessel functions for $m_f = 4$ have values less than 0.01 (1%) and can be ignored. Therefore, the pair of sidebands farthest from the carrier is the seventh pair. The bandwidth is then

$BW = f_i \times$ highest needed sideband $\times 2$
$= (2.5 \text{ kHz})(7)(2) = 35 \text{ kHz}$

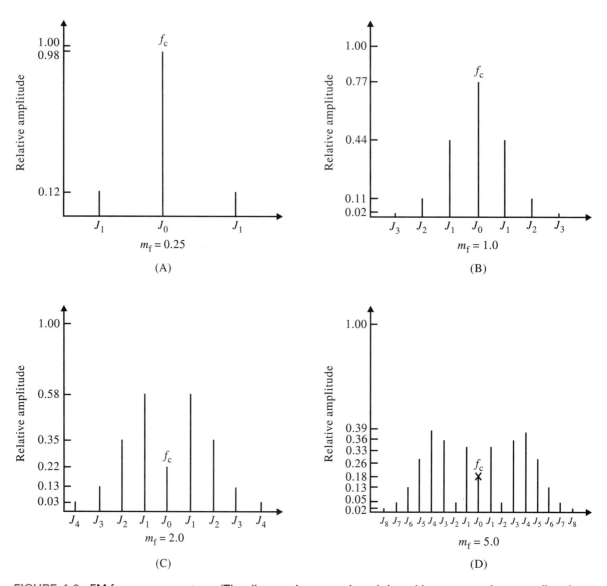

FIGURE 4-2 FM frequency spectrum (The distance between J_0 and J_1 and between each succeeding J term is equal to f_i in all cases.)

Standard Broadcast FM

The FCC has allocated each standard broadcast FM station a bandwidth of 200 kHz. This FCC allocation, shown in Figure 4-3, allows a maximum carrier deviation of ±75 kHz and provides 25 kHz *guard bands* at the upper and lower ends. These guard bands are provided to prevent interference from adjacent stations.

Broadcast FM uses modulating signals up to 15 kHz. These frequencies are within the upper limits of the audible range and provide true high-fidelity reproduction of the original intelligence, a major advantage of FM over AM. For broadcast FM, the

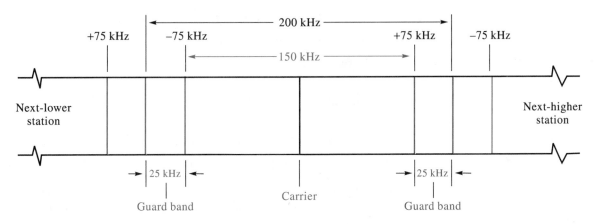

FIGURE 4-3 FCC allocation for a standard broadcast FM station

modulation index for the *standard deviation ratio* occurs when deviation δ and intelligence frequency f_i are at their maximum permissible values—that is, 75 kHz and 15 kHz, respectively. This produces the specific value of m_f that results in a maximum bandwidth.

EXAMPLE 4-6

A broadcast FM station has modulating frequencies ranging from 30 Hz to 15 kHz. Determine the FCC allowable range in maximum modulation index.

SOLUTION

Broadcast FM has a maximum deviation of 75 kHz; hence,

$$m_{f(max)} = \frac{\delta}{f_{i(min)}} = \frac{75 \text{ kHz}}{30 \text{ Hz}} = 2500$$

and

$$m_{f(min)} = \frac{\delta}{f_{i(max)}} = \frac{75 \text{ kHz}}{15 \text{ kHz}} = 5$$

Note that the maximum permissible values in Example 4-6 provide a result of 5, which is the modulation index for the standard deviation ratio for broadcast FM.

However, from the Bessel function table, we see that the result of this calculation gives J_8 as the last significant sideband. This would give a BW of:

$$BW = f_{i(max)} \times J_n \times 2 = 15 \text{ kHz} \times 8 \times 2 = 240 \text{ kHz}$$

This result exceeds the maximum allowable 200 kHz bandwidth (with guard bands). To correct this problem, either the maximum deviation of 75 kHz must be reduced, or the intelligence frequency of 15 kHz must be increased. One possibility is to reduce the maximum deviation frequency to 50 kHz, resulting in:

$$m_f = \frac{50 \text{ kHz}}{15 \text{ kHz}} = 3.33$$

and a significant sideband between J_6 and J_7, with a modulation index of between 3.0 and 4.0:

with $m_f = 3.0$: BW = 15 kHz × 6 × 2 = 180 kHz

with $m_f = 4.0$: BW = 15 kHz × 7 × 2 = 210 kHz

Splitting the difference here results in 210 kHz − 180 kHz = 30 kHz and 30 kHz/3 = 10 kHz, so the resultant BW = 180 kHz + 10 kHz = 190 kHz, is well within the allowable BW.

FM for Communication Use

Standard broadcast is not the only use for FM. Frequency modulation techniques are widely applied to communication networks used by police, military, aircraft, weather services, taxis, home service companies, and private industry. The FCC allocates bandwidths of 10–30 kHz to such systems. For this reason, these systems are designated *narrow-band*. While voice transmissions in these systems are limited to maximum modulating frequencies of 3 kHz, typical industrial uses include instrumentation and telemetry.

EXAMPLE 4-7

A narrow-band FM system has modulating frequencies ranging from 30 Hz to 3 kHz. Determine the FCC allowable range in maximum modulation index for (a) a maximum deviation of 10 kHz and (b) a maximum deviation of 30 kHz.

SOLUTION

$$m_f = \frac{\delta}{f_i}$$

(a) $\dfrac{10 \text{ kHz}}{30 \text{ Hz}} = 333.3$

and

$\dfrac{10 \text{ kHz}}{3 \text{ kHz}} = 3.3$

(b) $\dfrac{30 \text{ kHz}}{30 \text{ kHz}} = 1000$

and

$\dfrac{30 \text{ kHz}}{3 \text{ kHz}} = 10$

4-2   FM SIGNAL GENERATION

Frequency modulation systems are grouped into two classes: (1) *direct* FM, in which the carrier is modulated at the point where it is generated, in the master oscillator (MO); and (2) *indirect* FM, in which the MO is not modulated but instead modulation is applied in some following stage.

The frequency of a wave can be directly varied only at the point where that wave is being generated. This would seem to indicate that a wave could be frequency modulated only in the MO. In a sense, this is true. However, the modulating signal can shift the phase of a carrier wave current or voltage after the carrier has been generated; that is, the carrier can be *phase modulated* at any point in the transmitter. This phase modulation is then easily converted into frequency modulation, so we can say that the frequency modulation is produced indirectly, and we call the result indirect FM. Indirect FM can also be produced by first amplitude modulating the carrier and then converting the resultant AM into FM.

The Crosby method was the first popular method used to produce direct FM, and the Armstrong method was the first popular method used to produce indirect FM. Today, direct FM and indirect FM systems are both in common use.

Direct FM Generation

A simplified circuit for direct FM generation is shown in Figure 4-4. Although the circuit is not practical for commercial FM systems, we use it for demonstration purposes because it is easily understood. In this circuit, a capacitor microphone (C_M) is connected in parallel with the *LC* tank circuit of an RF oscillator. Sound waves applied to the microphone cause the microphone's capacitance to vary in step with the sound wave vibrations. Since the microphone is connected in parallel with the oscillator's tank circuit, the frequency of the RF wave generated by the oscillator also varies (is frequency modulated) in step with the sound wave vibrations. In this manner, the sound wave vibrations directly control the frequency variations in the RF wave—hence the term *direct FM generation*.

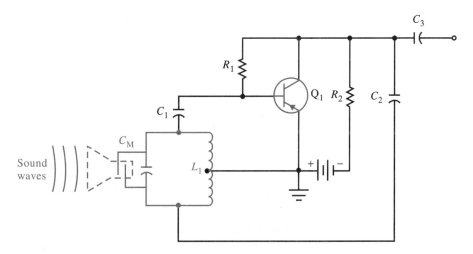

FIGURE 4-4   Simplified circuit for direct FM generation

The shunt-fed Hartley oscillator circuit used for this discussion is a practical operative circuit, but it is not used in practice for several reasons:

1. The carrier frequency stability is poor.
2. The type of microphone to be used is limited to the capacitance type.
3. The magnitudes of the frequency deviations produced are too low for normal modulation requirements, which produces a low modulation index and consequently a poor *S/N* ratio.

Reactance Modulator A practical method for producing direct FM is the *reactance modulator*, which provides a varying capacitance or inductance in a resonant oscillator circuit to achieve frequency deviation. Depending on the specific circuitry used, the reactance modulator circuit acts as a variable inductance or as a variable capacitance in parallel with the resonant *LC* circuit of an RF oscillator. A critical requirement of the reactance modulator circuit is that it be totally reactive. For proper operation, there must be either a +90° or a −90° phase angle, which means there must be no resistive component in the modulator circuit. Resistive components will produce undesirable AM signals at the output. Removing these unwanted AM signals at the receiver requires additional limiter processing circuitry.

Four fundamental arrangements of reactance modulators will produce satisfactory results. These arrangements are shown in Figure 4-5. Although the circuits shown use a transistor as the active device, other devices, such as vacuum tubes, JFETs, MOSFETs, and ICs may also be used.

In the commonly used circuit in Figure 4-6, an *RC* capacitive transistor reactance modulator operates on the tank circuit of the MO. Any reactance modulator may be connected across the tank circuit of a non-crystal-controlled *LC* oscillator, provided that the oscillator used does not require two tuned circuits for its operation. Most often used are isolated (buffered) Hartley, Colpitts, or Clapp oscillators. The Clapp oscillator is used in the circuit shown, and the RF chokes provide circuit isolation for ac while providing dc paths.

Varactor Diode Modulator A voltage-variable capacitance diode, called a *varactor* (or *varicap*) diode, is one in which the junction capacitance is easily varied electronically. This variation is accomplished simply by changing the reverse bias on the diode. A varactor diode can be connected in various ways to affect a resonant circuit. For this reason, a varactor diode modulator is one of the simplest means of generating an FM signal.

Typical varactor diode modulators are shown in Figure 4-7. In Figure 4-7(A), the total capacitance in the tank circuit of a Hartley oscillator is provided by a varactor diode. In the Colpitts oscillator of Figure 4-7(B), a varactor diode is one of the capacitive components in the tank circuit. In these two circuits, the RF oscillator frequency is determined by the reverse bias applied to the varactor diodes, and the frequency modulation results from the AF modulating signal applied through RFC_1.

The circuit in Figure 4-7(C) offers simplicity, reliability, and the stability of a crystal oscillator. In this circuit, a varactor diode is connected in a Pierce oscillator circuit. The oscillator's center frequency is established by the nominal capacitance value of the varactor diode, which is set by the reverse bias provided by resistors R_1 and R_2. An AF modulating signal applied through RFC_1 varies the reverse bias of the varactor,

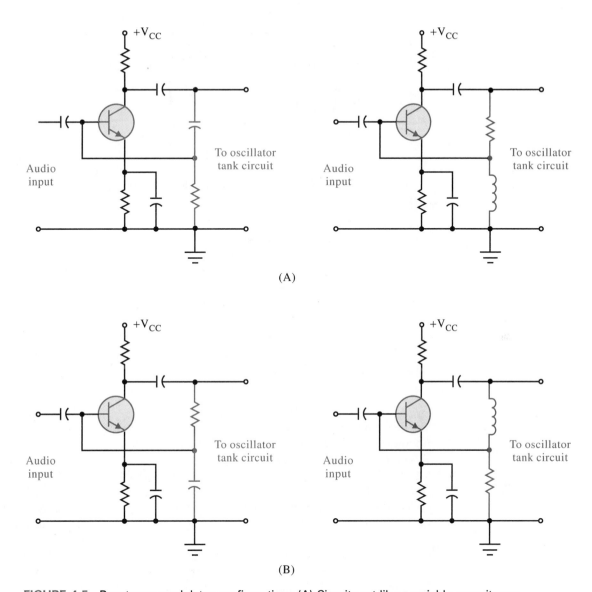

FIGURE 4-5 Reactance modulator configurations (A) Circuits act like a variable capacitance (B) Circuits act like a variable inductance

changing the capacitance of the varactor and, as a consequence, the frequency of the oscillator.

Automatic Frequency Control In many narrow-band applications, a crystal-controlled MO may be directly modulated. However, the high Q of crystal oscillators results in a limited deviation, thus preventing them from being directly frequency modulated for wide-band applications. Therefore, a direct FM transmitter is subject to poor carrier frequency stability. In general, in direct FM generation the MO is not

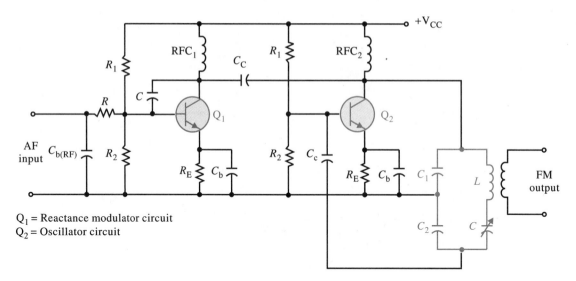

FIGURE 4-6 Transistor reactance modulator

Q_1 = Reactance modulator circuit
Q_2 = Oscillator circuit

crystal controlled. Carrier frequency stability in a communications system is tightly controlled by the FCC. Therefore, meeting FCC standards requires a means of *automatic frequency control* (AFC) to correct for any carrier drift.

The Crosby System

The most common method of frequency stabilization involves a *phase discriminator*. A typical frequency-controlled system is the *Crosby* direct FM system in Figure 4-8. To aid in understanding how the system works, we will trace the signals through the system. The audio signal feeds into the reactance modulator where it frequency modulates the MO. The center frequency of the MO is the frequency produced when no audio signal is present at the input to the reactance modulator. The output of the frequency multiplier (which has the center frequency as its average frequency) is compared with the ideal center frequency provided by the CCO and the frequency multiplier. If frequency drift has occurred, the LP filter provides a slowly varying signal which helps to drive the reactance modulator back to the desired center frequency. The LP filter/discriminator combination ignores the quickly varying intelligence signal and only looks at the long-term frequency average. Note that the mixer/LP filter combination functions in the same manner as in a superheterodyne receiver.

The CCO is used as a frequency standard against which the multiplied frequency of the MO is compared. The heterodyne frequency produced in the mixer reflects this comparison. If no difference exists between the two frequencies being compared, no heterodyne frequency will be produced, and we have what is called the *zero-beat* frequency. However, if the carrier center frequency drifts above or below its correct value, the mixer output will contain the difference frequency generated by the mixer's heterodyning action.

In most systems, the CCO used as the frequency standard is operated at a specified number of hertz from the transmitter center frequency. The discriminator is then adjusted so that its output voltage will be zero when its input frequency is the exact difference frequency specified for the CCO. For example, assume that the correct transmitter center frequency at the point where the sample is taken off and fed to the

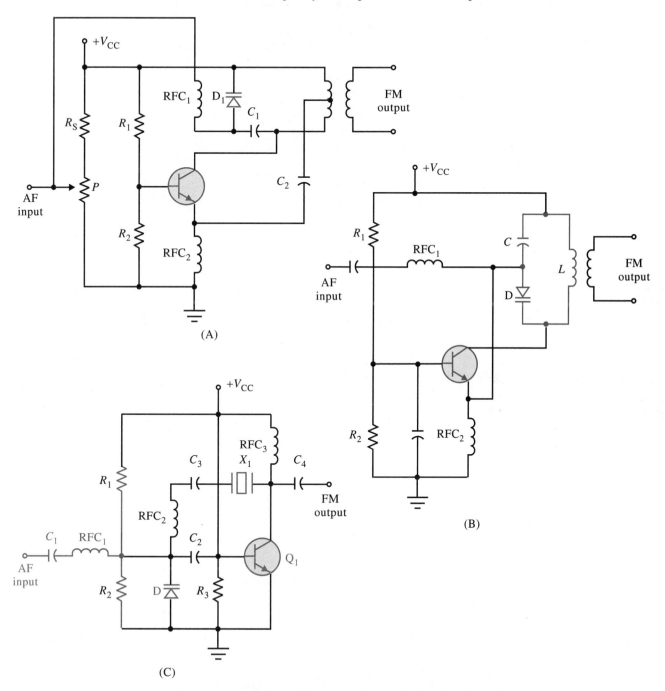

FIGURE 4-7   Varactor diode modulators (A) Hartley oscillator (B) Colpitts oscillator (C) Pierce oscillator

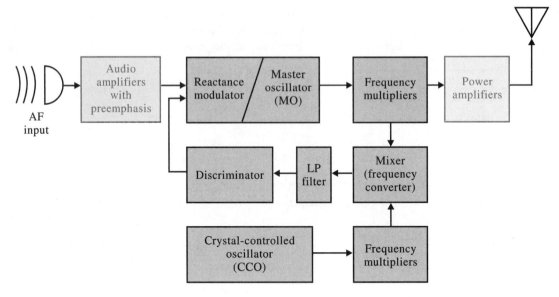

FIGURE 4-8   Crosby direct FM system

mixer is 100 MHz. Further, assume that the input to the mixer from the CCO is at 98 MHz. When the carrier is exactly *on frequency*, the difference frequency generated in the mixer by heterodyning action is 2 MHz. The discriminator for this system is adjusted so that an input frequency of 2 MHz causes the output voltage to be zero. Hence, when the transmitter is operating exactly on frequency, no correction voltage is delivered to the reactance modulator. On the other hand, if, for example, the transmitter frequency at the sampling point drifts down to 99.95 MHz, the difference frequency produced in the mixer and fed to the discriminator becomes 1.95 MHz. In this case, the discriminator would deliver an output voltage to the reactance modulator of the polarity and amplitude necessary to increase the MO frequency to such value that the center frequency at the sampling point returns to 100 MHz.

The polarity of the discriminator's output voltage is determined by whether the transmitter center frequency drifts up or down, and the amplitude of the voltage is determined by how far the transmitter center frequency drifts away from its correct value.

4-3   PHASE MODULATION

Both phase modulation and frequency modulation fall into the category of angle modulation. In simple terms, the difference between PM and FM is that the modulation index (m_p for PM, m_f for FM) is defined differently in each system. In the equation $x = X \sin(2\pi f_c t + \theta)$, a PM wave is developed by varying the phase angle θ so that its magnitude is proportional to the instantaneous amplitude of the modulating signal. In an FM wave, the frequency f_c is varied so that its magnitude is proportional to the instantaneous amplitude of the modulating signal.

Signals developed with PM are not used in practical analog transmission systems. Instead, PM techniques are used to generate FM signals for transmission. The methods used for this FM signal generation are called indirect FM generation.

Indirect FM Generation

Indirect FM generation is the term often used for phase modulation, because the technique develops an FM signal from a fixed-frequency signal source, such as a crystal oscillator. In PM, the crystal oscillator stage is not modulated. The modulation is applied to the following stages. In other words, PM is accomplished by modulating the output of a crystal oscillator. In this way, the phase of a signal is changed, resulting indirectly in a frequency change. With PM, deviation is not only directly proportional to modulating audio voltage amplitude, as in FM, but also directly proportional to the audio voltage frequency. To produce equal deviation for all modulating frequencies, a low-pass *RC* network (integrator) must be used in one of the AF amplifier stages. This integration makes the audio output amplitude inversely proportional to the input frequency.

Phase Modulator   A simple FET *phase modulator* is shown in Figure 4-9. In this circuit, the RF input is from a very stable crystal oscillator, usually through an RF buffer amplifier. The RF signal is applied simultaneously to the gate of the FET and through C_2 to the output tank circuit. When a modulating signal is applied to the circuit, both gate bias and stage gain vary, resulting in amplitude variations and a phase shift at the output of the circuit. The output of the phase modulator is, therefore, both amplitude modulated and phase modulated. By operating succeeding stages in the class C mode, the clipping action removes the amplitude variations but does not affect the phase modulation.

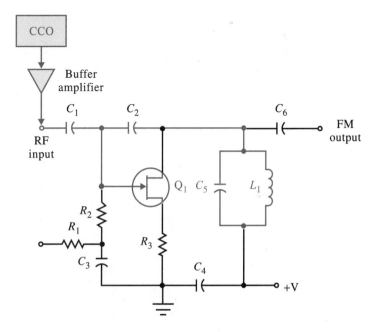

FIGURE 4-9   FET phase modulator

The phase modulator offers a major advantage over direct modulation techniques in that the MO is unaffected by modulation; thus, it offers increased frequency stability. However, a disadvantage is that it is difficult to obtain wide frequency deviations with a phase modulator, and therefore its primary application is in narrow-band, two-way FM communication systems, such as the voice modulated PM system in Figure 4-10.

In this circuit, the oscillator output is applied simultaneously to the base of Q_2 across C_4–R_6 and to the tank circuit C_6–L_1. The frequency across R_6 and across L_1 is the same, but because one circuit is an RC circuit and the other an LC circuit, the voltage across L_1 will be leading the voltage across R_6. This out-of-phase voltage on the base of the modulator produces an out-of-phase current in the collector circuit through L_1. The current flowing through L_1 has two out-of-phase components—one due to the voltage applied to the base of Q_2 and one due to the voltage from the oscillator. Therefore, the voltage drop across L_1 is not in phase with either voltage but is at some resultant phase. When a voice signal is applied to the base circuit of the modulator through C_3–RFC_1, the modulator's collector current will retain a steady oscillator component, but now the voltage at the base of Q_2 varies in amplitude. The result is a phase shift that varies with the applied voice frequency. In this way, PM is developed across L_1 and coupled through L_2 to the multiplier stages.

In the first multiplier stage, the audio signal is fed through an LP filter network that does two things. It tends to round off any square-wave-shaped signals that may have been developed by limiter action, thereby reducing higher-frequency audio harmonics. It also provides the integration of the audio signal necessary for PM systems to produce FM.

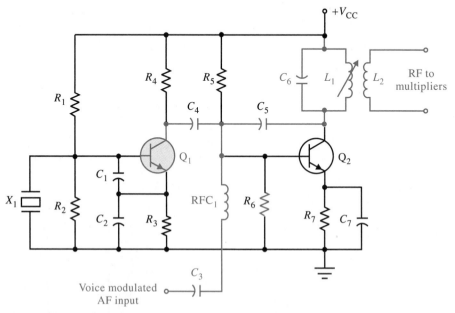

FIGURE 4-10
Voice-modulated PM system used in mobile FM transmitters

Voice modulated AF input

Q_1 = Crystal oscillator circuit
Q_2 = Phase modulator circuit

The Armstrong System An early system used for FM broadcast transmitters was the *Armstrong system,* shown in the block diagram of Figure 4-11(A). The Armstrong system requires a phase shifter and a balanced modulator. The RF signal from the crystal oscillator is applied to both the phase shifter and the balanced modulator. The only thing that occurs in the phase shifter is a 90° phase shift of the unmodulated signal. Both the modulating AF signal and the unmodulated RF signal are applied to the input of the balanced modulator. In the balanced modulator, the RF signal is amplitude modulated by the AF signal and upper and lower sideband frequencies are produced. However, recall that the action of a balanced modulator is such that the RF carrier is suppressed or canceled, leaving only the sideband energy at the output. The sideband energy from the balanced modulator is then combined with the carrier frequency output of the phase shifter. The result is a PM signal, which will result in FM if the audio signal was first integrated. This narrow-band FM signal is then further processed by circuits that will produce wide-band FM.

In the Armstrong system, AM is used to generate sidebands. The carrier is then removed from the AM signal, and a new carrier, shifted 90° from the original, is substituted for the original carrier. This process of carrier substitution, or the reinserting of a new carrier to replace the old carrier, is called *reinsertion*. See Figure 4-11(B).

As stated earlier, it is difficult to obtain very wide frequency deviations from PM systems. For example, a frequency deviation of 50 Hz per 1 MHz signal is typical. A 100 MHz signal with a multiplication factor of 100 gives a deviation of 100×50 Hz, or 5 kHz. This is adequate for FM voice communication with its 10 kHz bandwidth, but it could not be used for FM broadcasting with its 200 kHz bandwidth requirement.

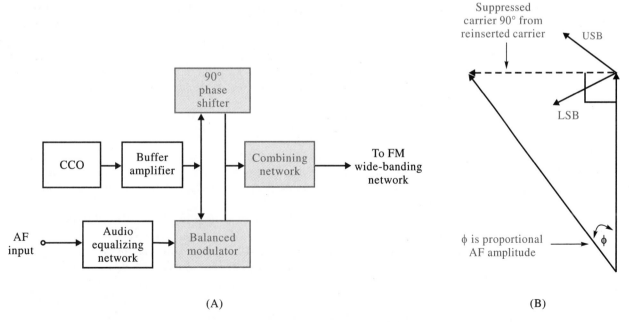

(A)   (B)

FIGURE 4-11   (A) Block diagram of an Armstrong PM system (B) Vector diagram representation of the Armstrong system

Therefore, multiplication alone may be sufficient for FM mobile communication, but an FM broadcast station requires a higher maximum deviation with a lower center frequency; so both mixing and multiplication are required. However, methods have been developed to obtain relatively wide-band deviation from Armstrong PM systems. One such method is shown in the partial block diagram of Figure 4-12.

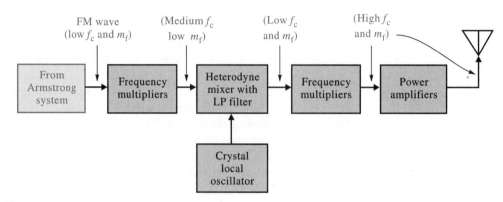

FIGURE 4-12 Wide-band FM from the Armstrong system

Frequency Synthesizers
Instead of individual oscillators, modern communications equipment uses *frequency synthesizers:* circuits that generate precise frequency signals by means of a single crystal oscillator in conjunction with frequency dividers and multipliers.

Some frequency synthesizers are wholly digital devices, characterized by discrete steps in frequency instead of being adjustable over a continuous range. The steps can be large or small, depending upon the application in which the device is to be used.

Figure 4-13 shows the block diagram of the HEF4750V frequency synthesizer, a device using a combination of analog and digital techniques. Used in combination with the HEF4751V universal divider, Figure 4-14, the HEF4750V offers a variety of applications choices.

Frequency Multiplication
A *frequency multiplier* is a circuit that produces a whole numerical multiple, or many whole numerical multiples, of a given input signal. A frequency multiplier is sometimes called a *harmonic generator* because the output of the circuit is a harmonic of the fundamental input frequency.

The frequency stability of MF oscillators is superior to that of VHF and UHF oscillators. Therefore, it is desirable to use an MF oscillator as the carrier wave generator at the transmitter. However, because FM broadcast stations operate at frequencies above the 30 MHz MF upper limit, frequency multipliers must be used in transmitter stages following the MO to obtain the assigned carrier frequency to be transmitted.

Frequency multipliers used in FM transmitters increase the deviation but not the modulation index. This occurs because when an FM signal is passed through a frequency multiplier, the carrier frequency and the frequency deviation are multiplied. The multiplication ratio is equal to the harmonic number to which the output is tuned.

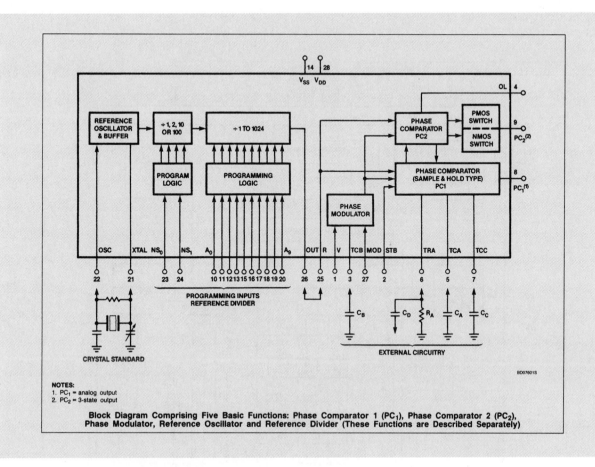

FIGURE 4-13 Block diagram of the HEF4750V frequency synthesizer

In Figure 4-12, the required deviation is obtained through multipliers and mixing. Note the effect on both the carrier frequency f_c and the modulation index m_f as the signals progress through the multipliers and the mixer. The mixer converts the carrier back to its original frequency, with the deviation multiplied by the harmonic number of the multipliers. By carefully selecting the multipliers and the oscillator frequency for the mixer, we will obtain the desired output carrier and intelligence frequencies. The following examples clarify how frequency multiplication is achieved.

EXAMPLE 4-8

An FM station is assigned a transmitter carrier frequency of 101.7 MHz. The MO at the transmitter oscillates at 2.825 MHz. (a) Determine the number of times the MO frequency must be multiplied to obtain the assigned 101.7 MHz carrier. (b) Determine what combination of frequency multiplier stages will give the desired frequency. (Caution: Multipliers of $\times 5$ or less work best.) (c) Draw a block diagram of the FM transmitter showing the multipliers between the buffer and the final power amplifier (FPA) and the output frequency for each stage.

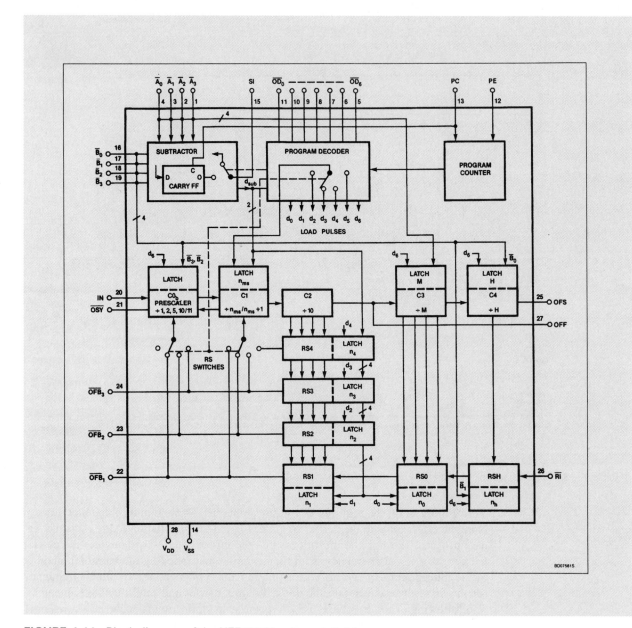

FIGURE 4-14   Block diagram of the HEF4751V universal divider

SOLUTION

(a) To determine the multiplication total M_T, divide the assigned carrier frequency f_c by the MO frequency f_{mo}:

$$M_T = \frac{f_c}{f_{mo}} = \frac{101.7 \text{ MHz}}{2.825 \text{ MHz}} = 36 \tag{4-3}$$

(b) Find several small numbers that can be combined to produce a product of 36. Some possible choices are (6)(6), (4)(9), (3)(12), (2)(18). However, keeping in mind the cautionary note, we should select numbers below 5. Thus (6)(6) could be changed to (3)(2)(3)(2), (4)(9) could be changed to (4)(3)(3), (3)(12) could be changed to (3)(3)(4), and (2)(18) could be changed to (2)(3)(3)(2). From the possibilities, we have the least number of multipliers with the (4)(3)(3) or the (3)(3)(4) combination. The two possibilities are the same, so it is immaterial which is used. However, another rule of thumb is to use the small multipliers first. Therefore, we will use the (3)(3)(4) combination.

(c) Figure 4-15 is a block diagram showing the result of the multipliers.

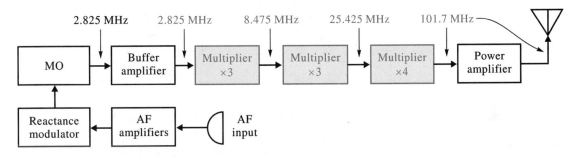

FIGURE 4-15 Block diagram of an FM transmitter with multipliers

EXAMPLE 4-9

An FM transmitter operating on 100.1 MHz has multipliers (2)(3)(4). What is the MO oscillating frequency?

SOLUTION

To determine the MO oscillating frequency, divide the carrier frequency by the multiplication total:

$$f_{mo} = \frac{f_c}{M_T}$$

$$= \frac{100.1 \text{ MHz}}{24} = 4.1708 \text{ MHz} \tag{4-4}$$

4-4 FM RECEIVERS

The FM receiver in Figure 4-16 is very similar to the AM receiver. They both use the super heterodyne principle. However, the FM receiver has specific differences.

1. It has much higher operating frequencies.

2. It requires some method of limiting and deemphasis.

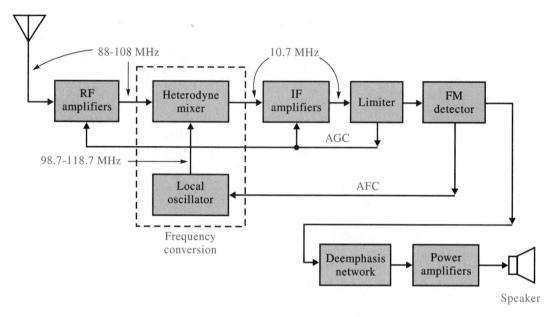

FIGURE 4-16   Block diagram of a broadcast FM receiver

3. It uses very different methods of demodulation.
4. It uses different techniques for obtaining AGC.

This section examines these differences.

Operating Frequencies

An FM receiver operates at frequencies of 88–108 MHz. The inherent noise reduction capability of FM means that FM receivers can operate satisfactorily with much lower incoming signal amplitudes than can AM receivers. The sensitivity is such that the receiver can accept input signals as low as 1 μV. But to prevent mixer noise from overpowering the intelligence, this small signal must be amplified before being fed to the mixer. The required amplification is performed by the receiver's front end RF amplifier, which also reduces image frequency problems and prevents LO feedback to the antenna.

The LO/mixer stage in FM receivers produces an IF of 10.7 MHz. This standard IF was chosen for FM for two reasons: (1) to provide good image frequency rejection, and (2) to provide the wider bandwidth required for FM broadcast signals. Recall that maximum transmitter deviation in broadcast FM is ±75 kHz. This means that minimum FM receiver bandwidth must be 150 kHz. However, the true bandwidth of an FM signal depends on the Bessel function, not just twice the deviation. In fact, most FM receivers have bandwidths of 180–200 kHz.

The LO in the broadcast FM receiver must be extremely stable. For example, a frequency drift of just 0.1% at 108 MHz will cause the IF to drift 108 kHz. The result of this drift is serious distortion of the output signal. Although crystal oscillators are used in many two-way FM communications systems where only a very few frequencies must be selected, they are not practical for broadcast FM receivers because many different frequencies must be selected. Thus, an AFC circuit is used in some FM receivers to lock the LO frequency onto the incoming signal. The basic principle of AFC was discussed in Section 4-2. Most of the amplification for the mixed signal is provided by the IF amplifier, which also reduces the received bandwidth to select the correct desired signal while rejecting all others.

Limiter Action

The limiter removes any amplitude variations, such as noise, from the received signal. To perform its function, the limiter must be overdriven; that is, the input level must be high enough to drive the limiter into saturation and cutoff. Limiter action is accomplished by going into saturation when the input signal is positive and going into cutoff when the input signal is negative. In this way, both the positive and negative peaks of the waveform are clipped, thus removing any amplitude variations caused by noise pulses. The frequency variations of the original signal are not affected by this clipping action.

Demodulation

There are many different approaches to demodulating angle modulated signals. *Demodulators* are also called *discriminators*, or *detectors*. The detector converts into audio voltages the frequency (or phase) variations brought into the receiver on the FM or PM wave. The amplitude of the detected audio signal depends on instantaneous carrier frequency deviation, and its frequency depends on the carrier's rate of frequency deviation. Detection occurs following the IF amplifiers, where the carrier has been translated (converted) to 10.7 MHz, but where the original modulating frequency deviation remains.

The detector used in FM receivers is more properly called a *discriminator*. The discriminator circuit produces an output voltage that depends on the frequency of the incoming signal. In this way, the circuit detects the frequency-modulated waveform.

When a signal is at the center of the pass band of the discriminator, the voltage output of the circuit is zero. If the signal frequency drops below the channel center, the output voltage becomes positive. The output voltage is proportional to the deviation of the signal frequency; that is, the greater the deviation of the signal frequency below the channel center, the greater the positive voltage at the output of the discriminator. Conversely, if the signal frequency rises above the channel center, the discriminator output voltage becomes negative, and the output voltage is again proportional to the deviation of the signal frequency. The amplitude of the voltage at the output of the circuit is linear and directly proportional to the frequency of the signal. Therefore, an undistorted output is ensured.

Figure 4-17 shows a simple discriminator circuit suitable for use in an FM receiver. A shift in the input signal frequency causes a phase shift in the voltages on either side of the transformer. When the signal is at the center of the channel, the voltages are equal and opposite, and the output is zero.

Foster-Seeley Discriminator   The *phase discriminator*, more popularly known as the *Foster-Seeley discriminator* (Figure 4-18(A)), converts into audio voltages the frequency or phase variations brought into the receiver on the FM or PM waves. Because the circuit is also sensitive to amplitude variations in the FM wave, a limiter stage

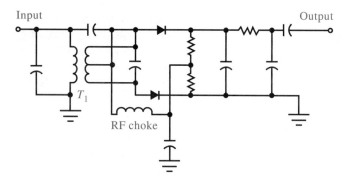

FIGURE 4-17 Simple discriminator circuit for FM receiver

immediately preceding the discriminator is required. Both primary and secondary windings of T_1 are tuned to the IF center frequency. This method of tuning greatly simplifies circuit alignment and provides increased linearity. The output of the circuit has an S-shaped response curve as shown in Figure 4-18(B).

The Foster-Seeley operates on the principle that two ac voltages in series add vectorially. This means that the phase relationship between the two voltages is an important consideration in determining the combined voltage. For example, if two ac voltages of 5 V each are placed in series, the combined voltage of the two will be 10 V *only* if the two voltages are exactly *in phase* with each other. On the other hand, if the two voltages are 90° out of phase, the resultant voltage will be about 7 V, and if they are 180° out of phase, the result will be 0 V. The important point is that the resulting total voltage of two in-series ac voltages is determined by the phase relationship between the two voltages.

The input to the Foster-Seeley circuit is an IF signal, through a limiter stage, which is varying ±75 kHz at an audio rate. The circuit output is the detected audio signal. When both diodes conduct, equal but opposite-polarity voltages are developed across R_1 and R_2, the voltages tend to cancel, and the output is 0 V. However, if D_1 conducts harder, the output is a positive voltage, and if D_2 conducts harder, the output is a negative voltage. Therefore, the audio output signal can be recovered when the following conditions exist:

1. Both diodes conduct exactly equally at f_c.
2. Diode D_1 conducts harder at frequencies above f_c.
3. Diode D_2 conducts harder at frequencies below f_c.

What determines how each diode conducts is explained in the following discussion.

Through transformer action, the incoming IF signal is coupled from primary coil L_1 to center-tapped secondary coil L_2–L_3. Because of the center-tapped configuration, the voltage V_2 developed across L_2 is 180° out of phase with the voltage V_3 developed across L_3. The conduction of D_1 is controlled by V_2 and V_3 controls the conduction of D_2. Remember that these two voltages are equal in amplitude but 180° out of phase at f_c.

The incoming IF signal is also capacitively coupled through C_2 to L_4. Voltage V_4 is developed across L_4, and this voltage also controls the conduction of both diodes. The circuit configuration is such that V_4 leads V_3 by 90° and lags V_2 by 90°. However, this is true only when the IF signal is at its center frequency. For this reason, the amount that D_1 conducts is determined by V_2 and V_4, and the amount that D_2 conducts is determined by V_3 and V_4. The result is zero output voltage for an input frequency of f_c.

Since the parallel resonant circuit resonates at f_c, X_L equals (and cancels) X_C, and the resonant circuit appears resistive. However, above f_c, X_L is larger than X_C. Thus, the net reactance shifts the phase of V_2 more in phase with V_4, whereas V_3 is shifted

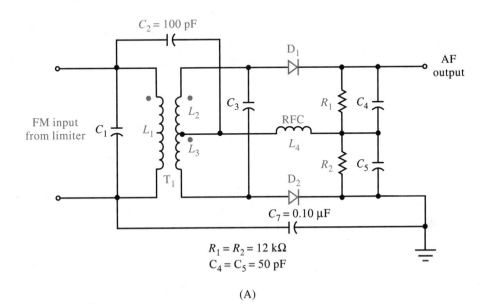

(A)

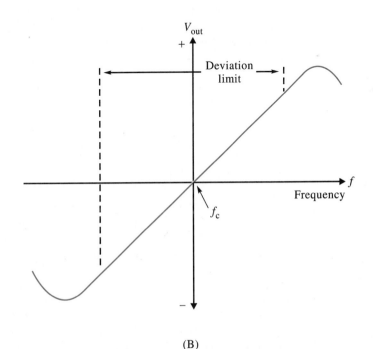

(B)

FIGURE 4-18
(A) Foster-Seeley
discriminator
(B) Response curve

more out of phase with V_4. This phase shift means that V_4 tends to add to V_2 and subtract from V_3, which makes D_1 conduct harder than D_2. This action produces a positive output voltage swing each time the IF signal swings above f_c.

At frequencies below f_c, X_C is larger than X_L, and the net reactance shifts the phase of V_2 and V_3 in the opposite direction. Now, V_4 tends to add to V_3 and subtract from V_2. Diode D_2 conducts harder than diode D_1, producing a negative output voltage. Thus, each time the IF signal swings below f_c, the output voltage swings negative.

It is through this positive-negative output voltage swing that the discriminator produces an output sine wave. Thus, the recovered signal is a reproduction of the original modulating signal.

Ratio Detector An improvement over the Foster-Seeley discriminator is the *ratio detector* in Figure 4-19. The primary advantage of the ratio detector is its *self-limiting action*; that is, it requires no preceding limiter stage.

A quick glance may leave the impression that the ratio detector and the Foster-Seeley are identical, but note that diode D_1 is reversed and that the output configuration is different. The reversal of D_1 puts both diodes in series across the entire transformer secondary. Conduction of the two diodes is controlled by the same factors as in the Foster-Seeley; that is, at f_c, the two diodes conduct equally. However, in this circuit, instead of opposing voltages, the voltages build up across C_4 and C_5 in series.

A large-value capacitor, C_6, is the key to the unique action of the ratio detector. A few cycles of input signal voltage charge C_6 to a voltage that is proportional to the average input signal voltage. The capacitor is large enough that the voltage across R_2, and across the series combination of C_4 and C_5, is held constant, even if there are momentary amplitude variations. The constancy of C_6 is the reason the ratio detector is relatively immune to noise and thus requires no prelimiting.

At f_c, the two diodes conduct equally, and the voltage across C_4 and C_5 is equal. The voltage across C_4 plus the voltage across C_5 must always equal the voltage across C_6.

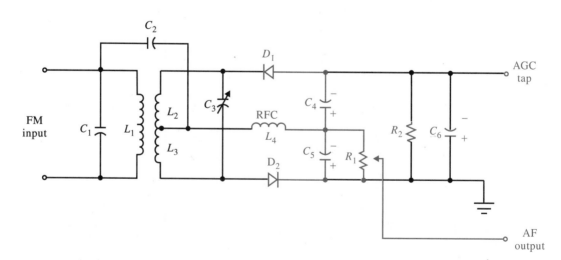

FIGURE 4-19 Ratio detector

The output audio signal is a negative dc voltage provided by a sample taken across C_5 and tapped from R_1.

Diodes D_1 and D_2 alternately conduct harder as the input signal swings above and below f_c. When the diodes conduct unequally, the difference current flows through L_4. When D_1 conducts harder, the voltage across C_4 is larger than that across C_5. Because the sum of these two voltages must remain constant, the voltage across C_5 must decrease, thus decreasing the output voltage. When D_2 conducts harder, the voltage across C_4 decreases, the voltage across C_5 increases, and the output voltage increases. Thus, we see that the output voltage swings respond to the frequency swings in the input signal; that is, when the frequency swings above f_c, the output voltage increases, and when the frequency swings below f_c, the output voltage decreases.

Automatic Gain Control Because the voltage across C_6 is directly proportional to the average input signal voltage, C_6 offers another valuable service: It provides a convenient source for tapping an *automatic gain control* (AGC) voltage.

Simple AGC is used in most domestic receivers and many cheaper communications receivers. In simple AGC receivers, the AGC bias starts to increase as soon as the received signal level exceeds the background noise level, and the receiver immediately becomes less sensitive.

Delayed AGC is used in most of the better broadcast and communications receivers. Delayed AGC is obtained when the generation of the bias is prevented until the signal level exceeds a preset threshold. Then, as the signal strength continues to increase, the delayed AGC provides greater and greater attenuation. The threshold may be fixed by the circuit design, or it may be adjustable. It is usually set to start taking effect when the signal has risen nearly to the level that produces the receiver maximum output under full gain (maximum sensitivity) conditions. Signals below the threshold level will pass through the receiver with maximum gain.

Although the Foster-Seeley and ratio detector discriminators just discussed can give good results, a major drawback to these two designs is that they require transformers for coupling and balance. Modern designs prefer to use ICs for lower cost, higher performance, and greater reliability, yet a transformer cannot be fabricated into an integrated circuit chip design. This has led to increased use of the quadrature detector and phase-locked loop, discussed next.

Quadrature Detector The term *quadrature* refers to the use of signals whose phases are at right angles (90°) to each other. The *quadrature detector* requires no transformer and is often built as part of the overall FM receiver IC. The idea of a quadrature detector is to compare the input signal to be demodulated with its phase-shifted version. The phase shifting is done by an LC combination which is resonant at the nominal IF frequency and so provides a shift of precisely 90° via the phase-shift capacitor at that frequency only. The amount of phase shift will vary from exactly 90° as the received signal frequency differs from the LC resonant frequency.

In operation, the quadrature detector uses the modulated IF signal and its phase-shifted version to control the flow of current through the two transistor switches (Figure 4-20(A)). When the two signals are in perfect quadrature (exact 90° phase difference) because the signal to be demodulated is at the carrier frequency, the on/off ratio of the switching action is 1:1, and a series of squarewave pulses is produced

(Figure 4-20(B)). The average value of these square waves is 50% of full scale. As the received signal frequency differs from the nominal value, the amount of phase shifting varies above and below 90°. The switching action now produces pulses that are not square waves (Figure 4-20(C)) but have on/off ratios of less than 1:1 or greater than 1:1. The average value of the pulses then goes above or below half scale, in direct proportion to the phase shift that resulted from the frequency variations.

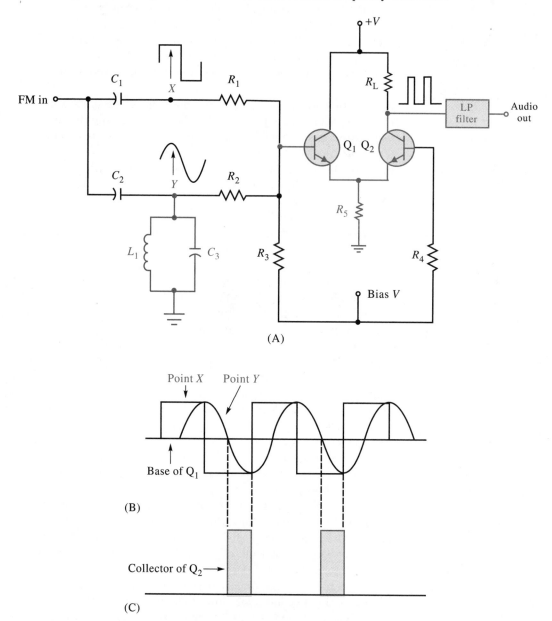

FIGURE 4-20 Quadrature detector (A) Simplified circuit (B) Combined waveforms at base of Q_1 (C) Detector output at collector of Q_2

As shown in Figure 4-20(A), the quadrature detector circuit is basically a difference amplifier formed by transistors Q_1 and Q_2, with the output taken at the collector of Q_2. The current in resistor R_5 is constant, and any change in the current of Q_1 produces an equal but opposite change in Q_2. The quadrature circuit provides the audio output when an FM carrier signal is applied to the input.

Capacitor C_1 is large enough in value that no phase shift of the input signal is produced at point X. The value of C_2 is small enough that its reactance to the carrier frequency is large when compared with the tuned circuit (L_1–C_3), resulting in a phase shift at point Y. The waveform at point Y is the FM carrier sine wave, and the waveform at point X is the squared-off constant amplitude from the limiter. The two waveforms are combined, as shown in Figure 4-20(B), at the base of Q_1. When both waveforms are negative, the detector output is high, as shown in Figure 4-20(C).

With modulation, the carrier frequency shifts and causes a phase shift of the signal at point Y. The shift occurs because the tuned circuit goes off resonance, which causes a change in the phase relationship between the tuned circuit and capacitor C_2. The phase change at point Y causes a variance in interval of the output signal when both waveforms are negative. The variance in interval has the effect of varying the output pulse width and, therefore, the average value of the output voltage. The change in the pulse average voltage is proportional to the amplitude of the original modulation, since it is the amplitude of the modulating signal voltage that causes the frequency change. The audio can now be separated from the pulse output through an LP filter.

ICs with Quadrature Detectors

A complete quadrature detector circuit requires some additional components to give performance that is linear and consistent over the desired range, with output voltage proportional to frequency deviation, but this is not a problem with ICs. In IC design, it is relatively simple and usually desirable to add extra transistors and related components to compensate for nonlinearities, temperature effects, and other imperfections, instead of requiring costly external components such as transformers plus manual trimming and adjustment. For these reasons, the quadrature detector is often used in modern IC FM demodulators. Figures 4-21 and 4-22 show two such ICs. The CA2111A (Figure 4-21) is an extremely simple device, containing an amplifier that combines limiting and amplification, a quadrature detector, and an emitter-follower output stage that provides a low-impedance output for driving an external audio amplifier. External coil L_1 allows easy tuning of the internal detector circuit.

A more advanced device is the LM3189N (Figure 4-22). This IC performs all the functions of a comprehensive FM–IF system. It includes a three-stage FM–IF amplifier–limiter configuration with level detectors for each stage, a double-balanced quadrature detector, and an audio amplifier that features the optional use of a *muting* (squelch) circuit.

The advanced design of the LM3189N provides such desirable features as programmable delayed AGC for the RF tuner, an AFC drive circuit, and an output signal to drive a tuning meter or to provide stereo switching logic. In addition, internal power supply regulators maintain a nearly constant current drain over the voltage supply range of 8.5 to 16 V. This IC device is ideal for high-fidelity operation; single-coil tuning capability is provided by the external detector coil.

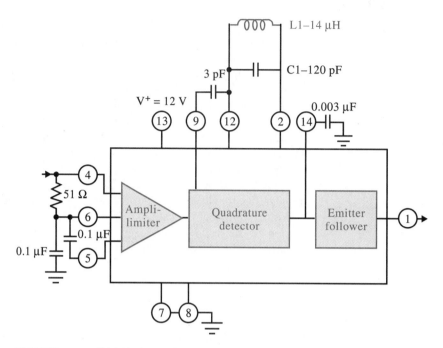

FIGURE 4-21 CA2111A quadrature detector IC

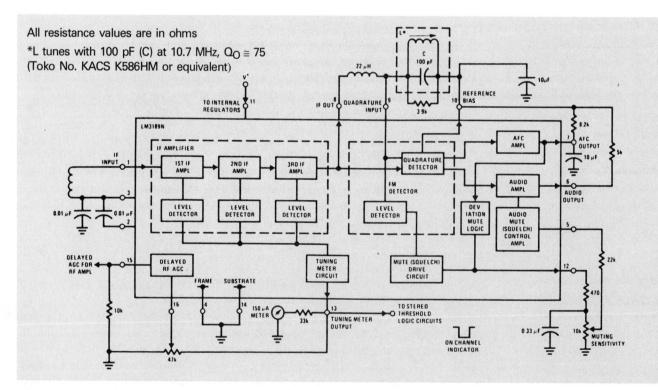

FIGURE 4-22 LM3189N FM–IF system

Phase-Locked Loop Demodulator   The development of ICs has made the *phase-locked loop* (PLL) increasingly popular as an FM demodulator. The PLL offers many advantages over the circuits previously discussed:

1. It requires no costly inductors or transformers, eliminating the need for intricate and time-consuming coil adjustments.

2. It produces excellent performance at low cost with a minimum of external components.

3. It can be used in many other electronics system applications.

A basic PLL (Figure 4-23) consists of a phase detector, a dc amplifier, an LP filter, and a voltage controlled oscillator (VCO). The VCO operates at the input frequency. The phase detector compares the input and VCO frequencies. The phase detector then develops an error voltage proportional to the amount and direction of the frequency difference. The dc amplifier increases the error voltage to a level needed to drive the VCO. The error signal is then coupled to the LP filter. The filter sets many of the dynamic characteristics of the PLL. It determines the frequency range over which the loop will capture and hold its phase lock, and it determines the speed with which the loop will respond to variations of the input frequency.

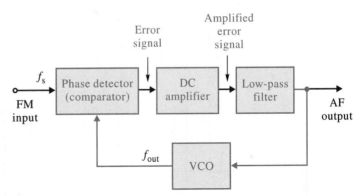

FIGURE 4-23
PLL demodulator

The error voltage from the filter is used to control the VCO. For example, if the input frequency swings above f_s (source frequency), an error voltage generated by the phase detector is amplified, fed to the filter, and applied to the VCO. The error voltage will cause the VCO frequency to increase in an exact lock with the input frequency. When the input signal is frequency modulated, the VCO tracks the FM deviation exactly, and the resulting error voltage is an exact reproduction of the intelligence signal.

Figure 4-24 shows the block diagram for the TDA7000 FM radio IC, a device that can be used for narrow-band FM reception.

 ## 4-5   UNDERSTANDING PHASE-LOCKED LOOPS ___

The PLL (Figure 4-23) is an electronic feedback loop. The purpose of the PLL circuit is to make a variable-frequency oscillator (the VCO) lock in at the frequency and phase angle of a standard frequency (f_s) used as a reference. The VCO will then have the same frequency accuracy as the referenced standard.

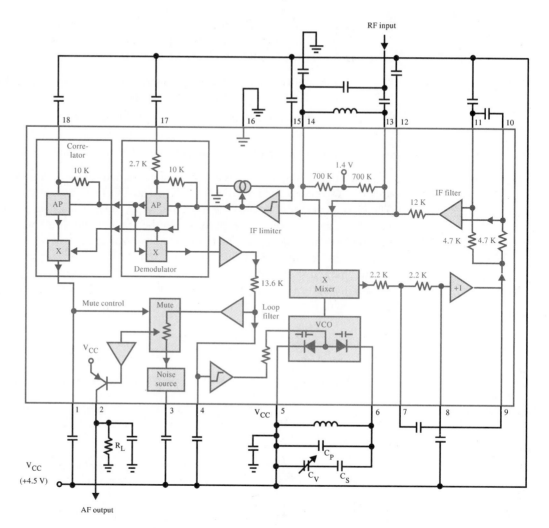

FIGURE 4-24   Block diagram of a TDA7000 FM radio IC

PLL Functional Blocks

To better understand the operation of the PLL, let us examine each of the four functional blocks: (1) the phase detector, (2) the dc amplifier, (3) the LP filter network, and (4) the VCO circuit.

The *phase detector*, or *comparator*, is used by all PLL systems, whether analog or digital, to generate the dc control voltage. The basic difference between the analog and digital type of PLL is the type of phase detector used. In general, digital systems use either an exclusive-OR gate or some type of edge-triggered phase detector, whereas the analog system uses a double-balanced mixer. Basically, the analog phase detector uses two diodes in a balanced rectifier circuit. The phase detector output voltage V_{out} is proportional to the phase difference between the two inputs:

$$V_{out} = K_c \Delta\phi \tag{4-5}$$

where

K_c = phase detector conversion gain in volts per radian (V/rad)

$\Delta\phi$ = input phase difference in radians (1 rad = 57.3°)

EXAMPLE 4-10

A phase detector has a conversion gain of 20 mV and an input phase difference of 0.15 rad. Calculate the phase detector output voltage.

SOLUTION

From Equation 4-5,

$$V_{out} = K_c\Delta\phi = (20 \text{ mV/rad})\ (0.15 \text{ rad}) = 3 \text{ mV}$$

EXAMPLE 4-11

Calculate the phase detector output for a conversion gain of 15 mV and an input phase difference of 20°.

SOLUTION

First, convert the phase difference from degrees to radians:

$$\frac{20°}{57.3°/\mathrm{rad}} = 0.349 \text{ rad}$$

Next, from Equation 4-5,

$$V_{out} = (15 \text{ mV/rad})\ (0.349 \text{ rad}) = 5.2 \text{ mV}$$

The *dc amplifier* circuit amplifies the filtered dc control voltage to the desired level for better control and in the polarity needed for the varactor in the VCO.

The *LP filter* network can be passive or active and serves two major functions:

1. It removes the ac signal variations of the two compared frequencies, or traces of higher-frequency noise, from the rectified dc output voltage of the phase detector.
2. It controls the lock, capture, bandwidth, and transient response of the loop; that is, the filter determines the dynamic performance of the loop.

The *VCO circuit*, also termed a *voltage-to-frequency converter*, is used for electronic tuning of the oscillator frequency. It uses a varactor, a semi-conductor capacitive diode that operates on the principle of a varying capacitance inversely proportional to the amount of reverse dc voltage applied. The reverse voltage applied can be either negative at the anode or positive at the cathode of the diode. The more reverse voltage applied, the wider the depletion area of the PN junction, and the effect is equivalent to an increased distance between capacitor plates, which produces less capacitance. In other words, the output frequency of the VCO is directly proportional to the input control voltage, which keeps the VCO locked into the frequency and phase of the referenced standard frequency.

In an actual implementation of a PLL, the VCO and the input are considered matched when they are at quadrature with a 90° phase difference between them. At quadrature, the control voltage is at the nominal value. As the phase difference shifts from 90° toward 0° the control voltage increases from nominal; as the phase difference shifts toward 180° the control voltage decreases from nominal toward its minimum value.

The error voltage is the key to the usefulness of the PLL. It represents the difference between the actual input signal frequency (and phase), compared to the VCO's last estimate of the input. By using the error signal in different ways, the input signal can be demodulated and manipulated. Note that the amplitude of the input versus the VCO amplitude is not involved. As long as the input voltage is within the range of the phase detector capability, the PLL will function properly. This is another advantage of the PLL: It is relatively insensitive to variations in signal level. Only the phase and frequency really affect the overall operation.

To examine a PLL used as an FM demodulator, consider a carrier at 1 MHz that has been frequency modulated by a 3-kHz voice signal with a maximum deviation of 30 kHz. The VCO is set to free-run at the carrier frequency, while the error voltage from the phase detector and low-pass filter can span +0.5 V and cause the VCO to swing −30 kHz around the 1-MHz point. As the modulated signal is received, the closed-loop action of the PLL will cause the error voltage to constantly correct the VCO to track the input frequency at any instant. The amplitude of the error voltage, therefore, is exactly proportional to the frequency modulation of the carrier. As the received frequency changes due to the FM, the phase detector determines the amount of difference, an error voltage is formed, and the error forces the VCO to correct itself (Figure 4-23).

There are some critical elements in a good PLL circuit. First, the phase detector (multiplier) must produce a linear output (multiplication result compared to actual phase difference) over the range of both inputs. The low-pass filter is critical. It determines the bandwidth and stability of the PLL and its ability to lock quickly onto the input and follow it. A PLL with the wrong filter parameters will be sluggish in following the input changes, possibly overshoot the correct error value, or cause the entire PLL to hunt constantly but never settle to the proper value. A narrow bandwidth is preferred to minimize noise, while a wider bandwidth is needed for faster response and higher data rates. Virtually all IC PLLs use external components which are not part of the IC for the filter so that the designer can tailor the filter to the intended application.

Finally, the VCO must be linear and produce a constant change in frequency per volt of control signal. Different PLLs offer a variety of VCO ranges. One model, for lower-frequency operation, has a VCO that can operate from 0 to 300 kHz (LM565), while the LM568 VCO operates up to 150 MHz and provides highly linear operation for deviations up to 10% around a 70 MHz center frequency.

A General-Purpose VCO

A popular monolithic IC voltage-controlled oscillator is the LM566C in Figure 4-25. It is a general-purpose VCO that may be used to generate square and triangular waves, the frequencies of which are very linear functions of a control voltage and of an

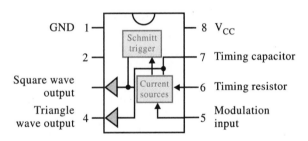

FIGURE 4-25
Block diagram of
the LM566C VCO

external resistor and capacitor. The LM566C operates over a supply voltage range of 10–24 V and provides very linear modulation characteristics, high temperature stability, and an excellent power supply rejection ratio (PSRR). It has a 10:1 frequency range with a fixed capacitor. Frequency is programmable by means of current, voltage, resistor, or capacitor. It can be operated from either a single supply or a split (±) power supply. Its applications include FM modulation, signal generation, function generation, frequency shift keying (FSK), and tone generation.

PLL Operating Modes

The operating modes for a PLL are *free-running, capture*, and *phase-lock* (sometimes called *lock-in* or *tracking*). If the VCO output frequency (f_{out}) is too far from the standard frequency f_s, the PLL cannot lock in the oscillator. Without such lock-in, the VCO is in the free-running mode. Once the control voltage from the dc amplifier starts to change the VCO frequency, the oscillator is in the capture mode. When f_{out} is exactly the same as f_s, the VCO is in lock-in and the PLL is in the phase-lock mode. The PLL will remain in the phase-lock mode as long as the dc control voltage is applied.

Lock Range and Capture Range

The frequency range over which the PLL can follow the incoming signal is called the *lock range*. The PLL is locked-in when the VCO output frequency is exactly the same as the standard frequency. The bandwidth over which capture is possible is called the *capture range*. Capture occurs when control voltage from the dc amplifier starts to change the VCO frequency. The capture range can never be wider than the lock range. To accomplish the desired result, we use the phase detector to compare the two frequencies. Any difference in phase causes an error signal at the output of the phase detector. This error signal is a dc voltage proportional to the difference in frequency and phase of the standard oscillator and the VCO. The dc error voltage is used to correct the VCO frequency by forcing it to change in a direction that reduces the frequency difference between the input signal and the VCO.

The LP filter circuit removes the ac variations of the two oscillators or traces of higher-frequency noise from the rectified dc output of the phase detector. The output of the comparator is a dc control voltage that is fed to the dc amplifier, where it is increased in amount to provide better control. The amplifier output provides the desired dc level and polarity for the control voltage needed for the varactor in the VCO.

The VCO uses the varactor to set the oscillator frequency. Input from the dc amplifier keeps the VCO locked into the frequency of the reference oscillator. Although locked in, there is always a finite phase difference between the input and output.

FM Stereo

Most FM broadcast stations use stereo transmission. Some FM broadcast receivers are capable of reproducing the stereo sound; others are not. For this reason, FM stereo broadcasting must be accomplished in a special way, so that it will be compatible with both types of receivers. To obtain FM transmission of the two independent stereo channels, modulated subcarriers are used (see Section 3-11).

An FM stereo signal is comprised of a main audio channel and two independent stereophonic channels. The main-channel audio consists of the combined audio signals from the left and right stereo channels. This signal frequency modulates the main carrier between 15 Hz and 15 kHz, thus allowing reception of the broadcast by receivers without stereo capability. Subcarriers impress the left and right stereo channels onto the main carrier signal. A pilot subcarrier frequency modulates the main carrier at 19 kHz. At twice this frequency (38 kHz) the stereo subcarrier is amplitude-modulated by the difference signal between the left and right channels. The pilot subcarrier maintains the frequency of the stereophonic subcarrier in the receiver.

The channel-difference sidebands occupy a modulating frequency range from 23 to 53 kHz, centered at the frequency of the stereophonic subcarrier. The different sidebands contain all the information necessary to reproduce a complete stereo signal in the FM receiver.

Subsidiary Communications   A subsidiary communications signal can be impressed on the main carrier in the modulating-frequency band from 53 to 75 kHz (the maximum modulating frequency allowed by law is 75 kHz). This subsidiary subcarrier is used for the transmission of background music, telemetry, and station-to-station information.

In order to use the subsidiary communication subcarriers in the United States, a broadcast station must obtain a Subsidiary Communications Authorization (SCA) from the Federal Communications Commission (FCC).

The Phase-Locked Loop and Stereo Demodulation

The phase-locked loop is one of the most effective circuits for demodulating an FM signal. The PLL is capable of outstanding performance under many difficult conditions. It is used not only in demodulation, but also in local oscillators that track received carrier drift (sometimes unavoidable), in demodulation of signals with very low (or even negative) signal-to-noise ratios, in frequency synthesizers, in tracking filters, and in many aspects of communications system design. The PLL principle and operation work in FM demodulation for both mono and stereo.

Preemphasis

Because lower frequencies carry better than higher frequencies, we must boost the amplitudes of the HF intelligence signals as compared to the amplitudes of the LF intelligence signals. To this end, we make the modulating signal equipment in the transmitter nonlinear to deliberately produce frequency distortion, with the higher frequencies being boosted to unnaturally high amplitudes as compared to the lower frequencies. The process of deliberate frequency response distortion of the intelligence signal is called *preemphasis*. Preemphasis improves noise suppression by increasing the modulation index at higher modulating frequencies.

Deemphasis

The signal produced by preemphasis at the transmitter is not desirable at the receiver in its distorted form. The receiver circuits following the detector must be designed to that they will restore the HF intelligence signals to their proper relative amplitude so that a linear signal is presented to the speaker. The process of reducing (attenuating) the HF amplitudes in the receiver by the same degree that they were boosted in the transmitter is called *deemphasis*. The end result of using preemphasis in the transmitter and deemphasis in the receiver is an improved *S/N* ratio at the higher frequencies of the audio spectrum. Figure 4-26 shows the response curves associated with preemphasis and deemphasis using the U.S. standard.

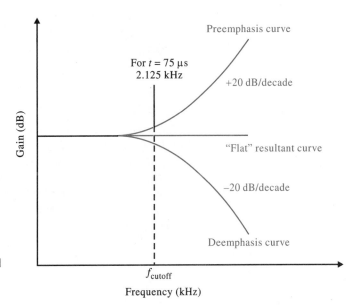

FIGURE 4-26
Preemphasis and deemphasis response curves

Standards for Preemphasis and Deemphasis

Preemphasis and deemphasis networks are simple and easily understood. In the United States, the FCC rules for broadcast FM require that such circuits have a time constant of 75 µs ($t = RC$ or L/R = 75 µs). The European standard is 50 µs, and the Dolby noise reduction system requires a 25 µs time constant. Among these three standards, the Dolby system offers the best noise reduction, which leads to the conclusion that the lower the time constant, the better. Figure 4-27(A) shows a typical preemphasis circuit, consisting of a series capacitor and a shunt resistor. If an inductor is used, then the resistor is the series component and the inductor is the shunt component, as shown in Figure 4-27(B). As you would expect, typical deemphasis circuits, shown in Figures 4-27(C) and (D), are simply the inverse of the preemphasis circuits; that is, the placement of the components is reversed. However, for the reproduced signal to be the same as before preemphasis and deemphasis, the time constants of the two circuits must be equal.

In general practice, *RC* networks are more prevalent than *RL* networks in preemphasis and deemphasis circuits. There are several reasons for this:

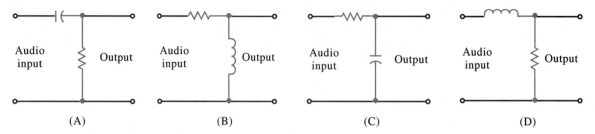

FIGURE 4-27 Simplified preemphasis and deemphasis circuits (A) *RC* preemphasis (B) *RL* preemphasis (C) *RC* deemphasis (D) *RL* deemphasis

1. In circuits constructed with discrete components, inductors are more expensive than capacitors.
2. The trend is toward miniaturization.
3. Inductors are not built into ICs used in modern miniaturized equipment.

Dolby Sound System

The *Dolby* system uses a 25 μs time constant, and the FCC has ruled that any FM station desiring to do so can use the 25 μs time constant rather than the prescribed standard 75 μs time constant. Dolby techniques are sometimes categorized as Dolby A, the original Dolby method, and Dolby B, a simpler system that works like preemphasis and deemphasis. The Dolby B system provides HF boost, like preemphasis—but of a dynamic nature. The amount of boost varies with signal loudness levels, as demonstrated by the response curves in Figure 4-28. But the real advantage of the

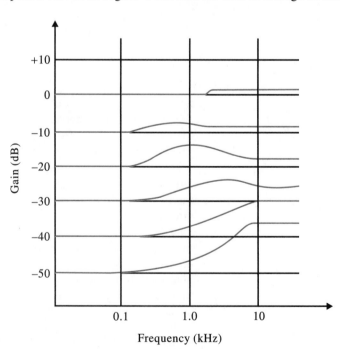

FIGURE 4-28
Dolby response

Dolby system over the preemphasis and deemphasis system is provided by the Dolby A system. This system has four independent noise reduction circuits, operating at the bass, midrange, treble, and HF portions of the audible sound spectrum.

The greatest beneficiaries of the Dolby system are listeners living in fringe reception areas, where noise effects are most disturbing. For maximum benefit, of course, the listener must have a Dolby receiver—that is, one that reverses the characteristics of the transmitted signal. Dolby receivers are more expensive than standard receivers because they require more complex circuitry to determine what equal attenuation is required. However, reception of Dolby signals by a standard FM radio receiver, although not perfect, is satisfactory if the treble control is properly adjusted.

SUMMARY

- The two distinct types of angle modulation are FM and PM. In FM, carrier amplitude remains constant, while carrier frequency varies around the center frequency. In PM, both carrier amplitude and frequency remain constant while carrier phase is varied.

- In FM, the amplitudes of the modulating frequencies determine the amount of carrier deviation. The modulating frequencies determine the rate of carrier deviation. Deviation is the amount by which the carrier frequency is varied above or below its center frequency. Frequency swing is the total deviation above and below the center carrier frequency.

- Frequency modulation is fully defined as the amount by which the instantaneous deviation is made proportional to the instantaneous amplitude of the modulating signal. Maximum intelligence signal levels occur with maximum deviation of the carrier signal.

- The Dolby noise reduction system uses a 25 μs time constant. Any broadcast FM station in the United States is allowed to use the 25 μs Dolby time constant instead of the standard 75 μs.

- Both the overall amplitude of an FM wave and the total transmitted power in FM always remain constant.

- Standard broadcast FM stations, operating in the 88–108 MHz frequency band, are allocated a bandwidth of 200 kHz. The bandwidth ensures that no significant sidebands are lost when the FM signal receives maximum deviation at the highest modulating frequency. Significant sidebands have minimum amplitudes of 1% of the unmodulated carrier. The prescribed modulation index (standard deviation ratio) of 5 for broadcast FM occurs when the deviation and the intelligence frequencies are at their maximum permissible values.

- Narrow-band FM communications system networks typically transmit voice only, where modulating frequencies ranging from 30 Hz to 3 kHz are normal. Allocated bandwidths for such systems are typically 10–30 kHz.

- The two classes of FM generation are direct and indirect. In a direct FM system, the carrier is modulated in the MO. In an indirect FM system, modulation is applied at some stage following the MO. Frequency modulation can be produced indirectly by shifting the phase of the carrier current or voltage after the carrier has been generated, or by first amplitude modulating the carrier and then converting the resultant AM wave to FM. Direct FM can be produced by reactance modulators or by varactor diode modulators. A popular and practical method is the Crosby system.

- Indirect FM is the term often used for phase modulation. By modulating the output of the MO, the phase of a signal is changed, resulting indirectly in a frequency change. Phase modulators used to produce indirect FM offer a major advantage over direct FM techniques in that the MO is unaffected, thus providing increased frequency stability. A disadvantage is that it is difficult to obtain wide frequency deviations. The primary application for the phase modulator is in narrow-band, two-way FM communications systems.

- For increased frequency stability in FM systems, MF oscillators are used as carrier generators. The carrier frequency and the deviation are then increased through frequency multipliers in the transmitter stages following the MO.

- Frequency modulation receivers use the same superheterodyne principle as do AM receivers, but they have higher operating frequencies, require some method for limiting and deemphasis, use different methods for demodulation, and employ different techniques for obtaining AGC. They operate on the 88–108 MHz frequency range, with an IF of 10.7 MHz, and require a minimum bandwidth of 150 kHz. Most FM receivers have a bandwidth of 180–200 kHz. The limiter stage removes amplitude variations from the received signal. Limiter action is accomplished by the stage operating in the saturation and cutoff modes. Demodulation can be obtained by using any of the following circuits; slope detector, Foster-Seeley discriminator, ratio detector, PLL demodulator, or quadrature detector. The ratio detector requires no prelimiting. The quadrature detector is used in most modern FM receivers having ICs as a major part of their design.

- Higher modulating frequencies in FM result in increased noise levels and decreased intelligence signal levels. However, FM systems provide better immunity from noise interference than AM systems do. Preemphasis at the transmitter and deemphasis at the receiver improve the *S/N* ratio at the higher intelligence frequencies. Preemphasis is the process of deliberate frequency response distortion of the intelligence signal at the transmitter, where the higher frequencies are boosted to unnaturally high amplitudes as compared to the lower frequencies. Deemphasis is the process of restoring the distorted HF intelligence signals to their proper relative amplitudes in the receiver. Standard preemphasis and deemphasis networks for broadcast FM systems are required to have time constants of 75 μs in the United States and 50 μs in Europe.

QUESTIONS AND PROBLEMS

1. List the two types of modulation included in angle modulation. Briefly describe each type.
2. What effect do modulating frequencies have on the amount of carrier deviation in FM?
3. What determines the rate of carrier deviation in FM?
4. What effect do the individual amplitudes have on the rate of carrier deviation in FM?
5. Define deviation.

6. Fully define FM.

7. Define frequency swing.

8. Define modulation index for FM.

9. An FM signal at rest frequency equals 100 MHz. It swings between 99.997 MHz and 100.003 MHz at a rate of 5000 times a second. Determine the (a) frequency swing, (b) intelligence frequency f_i and (c) effect on the signal if the amplitude of f_i is tripled.

10. An FM signal deviation is equal to 4.8 kHz when f_i is 1 kHz at an amplitude of 3.2 V. (a) If the amplitude of f_i is increased to 6.4 V, what is the new deviation? (b) If f_i is decreased to 500 Hz, what is the new deviation? (c) Determine m_f for each condition.

11. Why is a narrow-band system superior to wide-band for voice transmissions?

12. In theory, what is the number of sideband frequencies contained in an FM wave?

13. What is meant by significant sidebands?

14. Why is a Bessel function table used?

15. From Table 4-1, at which values for m_f is all the transmitted power in the sidebands?

16. Determine the required bandwidth for an FM signal with a maximum deviation of 7.5 kHz when modulated with an intelligence frequency of 3.75 kHz.

17. Determine the required bandwidth for an FM signal with a maximum deviation of 75 kHz when modulated with an intelligence frequency of 15 kHz.

18. Determine the required bandwidth for an FM signal with a maximum frequency swing of 50 kHz when modulated with an intelligence frequency of 5 kHz.

19. For a standard broadcast FM station, what is the (a) FCC-allocated bandwidth, (b) allowed maximum deviation, and (c) frequency swing?

20. What are guard bands? What are they used for? How wide are they?

21. Determine the FCC allowable range in maximum m_f for a broadcast FM station with modulating frequencies ranging from (a) 100 Hz to 12 kHz and (b) 30 Hz to 15 kHz.

22. A narrow-band FM system has modulating frequencies ranging from (a) 20 Hz to 2500 Hz and (b) 50 Hz to 3 kHz. Determine the FCC-allowable range in maximum modulation index for (a) and (b) for maximum deviations of 10 kHz and 30 kHz, respectively.

23. What is a practical device for producing direct FM? Briefly describe its operation.

24. What is one of the simplest means of generating an FM signal?

25. Draw a block diagram of the Crosby direct FM system. Briefly describe its operation.

26. In simple terms, what is the difference between PM and FM?

27. Why are PM techniques used? What are these methods called?

28. List the major advantage and disadvantage of the phase modulator over the direct modulation technique.

29. Draw a block diagram of the Armstrong PM system. Briefly describe its operation.

30. Define reinsertion.

31. A 4 MHz MO is used in a transmitter assigned a carrier frequency of 96 MHz. (a) Determine the number of times the MO output frequency must be multiplied to obtain the assigned frequency. (b) Determine the combination of frequency multipliers required. (c) Draw a block diagram of the resulting FM transmitter, showing its multipliers and the frequencies at each stage.

32. An FM transmitter assigned on a carrier frequency of 90 MHz has (2)(2)(3)(3) multipliers. What is the oscillating frequency of the MO?

33. Over what range of frequencies does an FM receiver operate? What is the FM receiver IF? Why was this IF chosen?

34. A receiver is tuned to a station at 88.1 MHz, and the LO is oscillating at 98.8 MHz. What is the IF? the image frequency?

35. Draw a block diagram of an FM receiver. Briefly describe the function of each block.

36. Why is a limiter stage used? How is limiter action accomplished?

37. List three FM demodulators. Which is the simplest? Which is most often used in modern FM receivers?

38. Briefly describe frequency shift keying. State which FM demodulator is used with FSK.

39. List the advantages offered by the PLL demodulator.

40. Draw the schematic diagram for a quadrature detector. Briefly explain its operation.

41. What is the prime characteristic of the (a) Foster-Seeley and (b) ratio detectors?

42. What is the main difference between the Foster-Seeley and ratio detectors?

43. Describe the general operation of the (a) Foster-Seeley and (b) ratio detectors.

44. Explain why the quadrature detector is favored over Foster-Seeley and ratio types?

45. How does the quadrature detector convert frequency differences to a series of pulses and then to a corresponding amplitude variation?

46. Define preemphasis. Explain its process.

47. Define deemphasis. Explain its process.

48. What is the result of using preemphasis and deemphasis?

49. For broadcast FM, preemphasis and deemphasis circuits have what time constants (a) in the United States, (b) in Europe, and (c) for the Dolby noise reduction system?

CHAPTER 5

Pulse Modulation

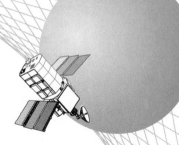

OBJECTIVES

When you complete this chapter, you should be able to:

- List the two broad categories of pulse modulation and explain the differences between the two.
- List the major forms of modulation within the two broad categories and describe the operation of each.
- Explain sampling.
- Define coding, Nyquist rate, quantization, and quantization noise.
- Define compander and explain its purpose.

INTRODUCTION

In previous chapters we discussed the use of analog signals for modulating a carrier wave. Systems using this technique are widespread and will continue to be so. But the use of modulation systems using digital or pulse signals has led to the alternative form of modulation called *pulse modulation.*

Pulse modulation falls into two broad categories: (1) *analog pulse modulation,* which includes pulse amplitude modulation and pulse time modulation; and (2) *digital pulse modulation,* which includes pulse code modulation and delta modulation. We will note other names and subcategories as we examine these two broad groupings.

A process used in the transmission of pulse modulation signals is called *companding.* We will examine the companding process relating to both the analog and digital pulse modulation categories.

5-1 ANALOG PULSE MODULATION

Analog pulse modulation includes *pulse amplitude modulation* (PAM) and *pulse time modulation* (PTM), with some subforms. To better understand how analog signals can be transmitted in the form of pulses, we will first examine sampling techniques.

In the analog communications systems studied thus far, the entire modulating waveform has been transmitted. However, in analog pulse modulation, periodic samples of the modulating waveform are taken, and only those samples, in the form of a pulse train (positive and negative spikes or pulses), are transmitted. If the sampling frequency is high enough—that is, if enough samples are transmitted—the modulating wave can be recovered at the receiver. We can state, therefore, that the higher the sampling frequency, the more accurate the reproduced signal at the receiver.

To demodulate the signal, an LP filter is used. If the number of samples is low, the pulse frequency will be low, and the filter may remove some of the intelligence. It is important, then, to have an adequate number of samples.

The minimum sampling frequency required for reliable reproduction of the modulating signal is defined by the following theorem: *To ensure that all of the information contained in the original signal is transferred, the sampling frequency in any pulse modulation system must not be less than twice the highest signal frequency.*

The minimum sampling frequency of twice the highest modulating signal frequency is known as the *Nyquist rate*. If the sampling rate is less than the Nyquist rate, there will be an overlap between the modulating signal and the lower sideband of the sampling frequency. With such an overlap, correct filtering cannot occur, and the modulating signal cannot be accurately reproduced.

Increasing the sampling rate to a frequency higher than the Nyquist rate provides a guard band that permits easier filtering and therefore improves the accuracy of the reproduced signal. A disadvantage is that the wider the guard band, the greater the pulse-modulated signal bandwidth must be. Therefore, a careful balance must be maintained to allow for an adequate guard band for easy filtering at a minimum signal bandwidth. For this reason, the sampling frequency is always slightly greater than twice the highest modulating frequency.

A simple example of these concepts is the telephone network. The voice modulating frequency in a telephone system is limited by filters to about 3.3 kHz. The standard sampling frequency is about 8 kHz, which is well above the Nyquist rate. The guard band (*gb*) is determined by the following:

$$gb = f_s - (2 \times f_m) \tag{5-1}$$

where

gb = guard band

f_s = sampling frequency

f_m = highest frequency in the modulating signal

Therefore, for the telephone network, we have a guard band of 8 kHz − 2(3.3 kHz) = 1.4 kHz, which does not significantly increase the bandwidth yet allows the use of simple LP filters for demodulation.

Pulse Amplitude Modulation

Pulse amplitude modulation (PAM), illustrated in Figure 5-1, is the simplest form of analog pulse modulation. It is the result of a train of pulses (Figure 5-1(A)) occurring at a fast pulse repetition rate (PRR), the number of pulses per unit time, usually pulses per second (pps), that are made to vary in amplitude in proportion to the amplitude of a lower-frequency modulating signal (Figure 5-1(B)). The resultant dual-polarity waveform is a series of pulses whose amplitudes correspond to the original modulating signal (Figure 5-1(C)). If the modulating signal is offset by a dc level sufficient to ensure that the pulses are always positive (Figure 5-1(D)), the result is a single-polarity waveform (Figure 5-1(E)). In either case, to recover the original modulating signal, we need only pass the PAM waveform through an LP filter.

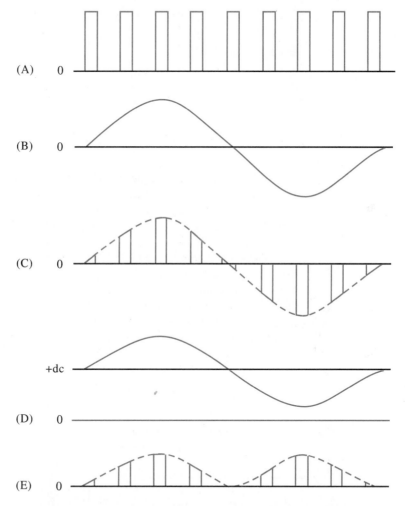

FIGURE 5-1 Pulse amplitude modulation (A) Uniform pulse train (B) Modulating signal (C) Dual-polarity PAM waveform (D) Modulating signal with dc offset (E) Single-polarity PAM waveform

Dual-Polarity PAM   Dual-polarity PAM can easily be generated by using a 4016 integrated circuit CMOS sampling switch, as shown in Figure 5-2. In this simple PAM generator, a positive sampling pulse closes the switch and the modulated signal input appears at the output. When the sampling pulse drops to zero, the switch opens and the output goes to zero.

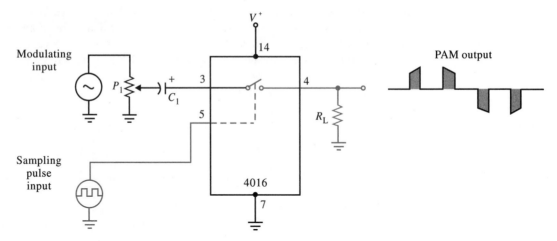

FIGURE 5-2 Dual-polarity PAM generator

Single-Polarity PAM Single-polarity PAM can be generated by simply adding two resistors to the circuit of Figure 5-2 to form a voltage divider, as shown in Figure 5-3. The voltage divider adds a dc level to the modulating input signal, causing a positive dc offset of the modulating single waveform. When the modulating input signal is sampled by the pulse train, the result is the single-polarity PAM output waveform shown.

Because PAM signals are more easily distorted by noise, crosstalk (undesired transfer of signals between systems or parts of systems), and other distortions than other forms of pulse modulation are, they are rarely used to transmit information directly. However, PAM is used frequently as an intermediate step in other pulse-modulating methods. When PAM is used, it normally is sent over cable or wire. It can be used to modulate an RF carrier, in which case it normally acts as a frequency modulator rather than as an amplitude modulator.

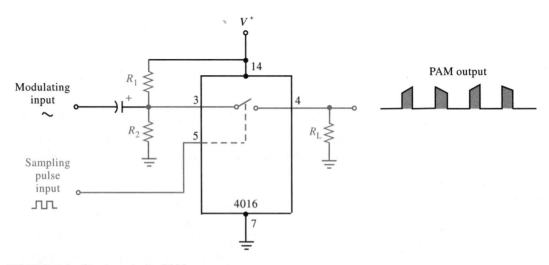

FIGURE 5-3 Single-polarity PAM generator

Pulse Time Modulation

Pulse time modulation (PTM) uses constant-amplitude pulses for sampling. The pulses vary in timing rather than in amplitude. The timing variable may be the duration or the position of the pulses, producing four distinct forms of PTM: pulse duration modulation, comprising three separate forms; and pulse position modulation.

Because the amplitudes of the sampling pulses are constant in PTM, limiters in a receiver can be used to remove noise spikes. This feature results in PTM having the same advantage over PAM as FM has over AM—that is, greater noise immunity.

Pulse Duration Modulation   When the duration of the pulses used for modulation is varied, the result is *pulse duration modulation* (PDM), also called *pulse width modulation* (PWM), or *pulse length modulation* (PLM). There are three different ways to achieve PDM: symmetrical PDM, leading-edge PDM, and trailing-edge PDM (Figure 5-4).

The modulating signal (Figure 5-4(A)) is sampled and the pulse width is varied in different ways, depending on the method chosen and the system being used. In *symmetrical* PDM (Figure 5-4(B)), both leading and trailing edges of the sampling pulses are varied in accordance with the amplitude of the sampling points of the modulating signal. The average (at-rest) pulse width occurs at sampling points 1, 4, and 7,

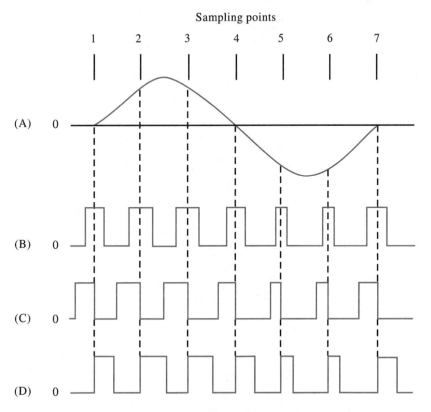

FIGURE 5-4 Pulse duration modulation (A) Modulating signal (B) Symmetrical PDM (C) Leading-edge PDM (D) Trailing-edge PDM

when the modulating signal is at zero amplitude. When the modulating signal is at a high positive amplitude (points 2 and 3), sampling pulse widths increase. When the modulating signal amplitude is at a high negative amplitude (points 5 and 6), sampling pulse widths decrease. Note that the spacing between pulse centers remains constant although the spacing between pulse edges varies.

In *leading-edge* PDM (Figure 5-4(C)), the trailing edge of the sampling pulses remains fixed and the leading edge varies in accordance with the modulating signal amplitude. The sampling pulses are at-rest width when the modulating signal amplitude is zero. At a high positive amplitude, sampling pulse widths increase, and at a high negative amplitude, sampling pulse widths decrease.

In *trailing-edge* PDM (Figure 5-4(D)), the leading edge of the sampling pulses remains fixed and the trailing edge varies in accordance with the modulating signal amplitude. As shown, at-rest pulse widths occur when the modulating signal amplitude is zero. At a high positive amplitude, sampling pulse widths increase, and at a high negative amplitude, sampling pulse widths decrease.

Caution: The important thing to remember when referring to leading-edge PDM and trailing-edge PDM is that *the name identifies the edge that varies, not the fixed edge.*

A typical approach to PDM generation is shown in the simplified block diagram of Figure 5-5(A). The sampling signal input and the modulating signal input are mixed in a summing network. The mixed-signal summing network output is fed to a clipper circuit, where all of the waveform below the clipping level is removed. The output of the clipper is a sloping-edge waveform of varying amplitude and width. This wave is then

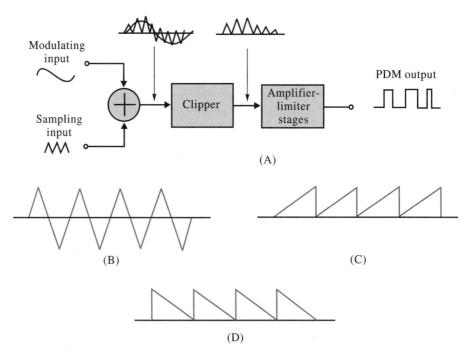

FIGURE 5-5 Generation of PDM (A) Simplified PDM generator (B) Sampling signal for symmetrical PDM (C) Sampling signal for leading-edge PDM (D) Sampling signal for trailing-edge PDM

fed into the number of amplifier–limiter stages necessary to result in all pulses having vertical edges and equal amplitudes. Depending on the type of sampling signal input (shown in Figures 5-5(B) through (D)), the output of the generator will be a symmetrical PDM wave, a leading-edge PDM wave, or a trailing-edge PDM wave.

A PDM generator using a 565 phase-locked loop (PLL) is shown in Figure 5-6. This circuit produces both PDM and pulse position modulation (PPM). (PPM is discussed later in this chapter.) A PPM wave is generated at the junction of pins 4 and 5 of the 565 PLL and amplified by transistor Q_1. Comparing the sampling pulse input signal and the amplified PPM signal in an exclusive-OR (XOR) gate results in a PDM output wave. The output of an XOR gate is *high* only when one or the other, but not both, of its inputs is high. Therefore, the PDM output wave is twice the frequency of the original sampling pulse input signal.

Potentiometer P_1 varies the center frequency of the VCO internal to the 565 PLL. Potentiometer P_2 is a phase adjustment used to establish the quiescent (at-rest) PDM duty cycle.

If desired, the PPM wave can be tapped off, as shown. Either the PDM or the PPM output signals can be used to modulate a carrier.

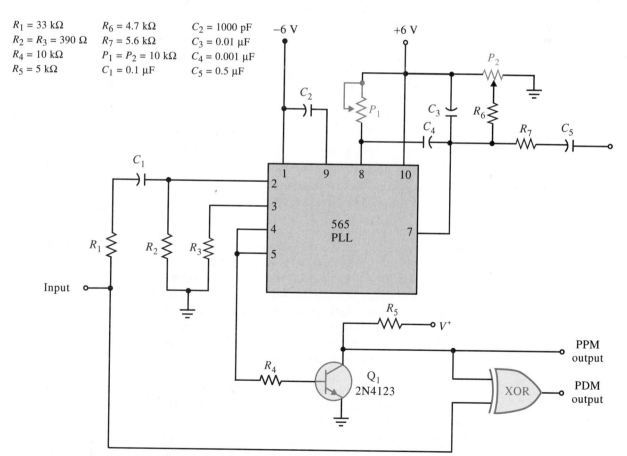

$R_1 = 33 \text{ k}\Omega$

$R_2 = R_3 = 390 \text{ }\Omega$

$R_4 = 10 \text{ k}\Omega$

$R_5 = 5 \text{ k}\Omega$

$R_6 = 4.7 \text{ k}\Omega$

$R_7 = 5.6 \text{ k}\Omega$

$P_1 = P_2 = 10 \text{ k}\Omega$

$C_1 = 0.1 \text{ }\mu\text{F}$

$C_2 = 1000 \text{ pF}$

$C_3 = 0.01 \text{ }\mu\text{F}$

$C_4 = 0.001 \text{ }\mu\text{F}$

$C_5 = 0.5 \text{ }\mu\text{F}$

FIGURE 5-6 PDM generator using a 565 PLL

PDM Demodulation An LP filter is the simplest method for PDM demodulation. A disadvantage of this method is severe distortion. A more practical approach is the demodulator shown in block diagram form in Figure 5-7. Although this approach is much more complex than the simple LP filter method, it has the advantage of relatively low distortion.

In this demodulator, a constant current source (CCS) receives a PDM input signal from a PDM generator. When the input signal goes positive, the CCS linearly charges capacitor C_1 until the PDM input signal returns to zero. When the input drops to zero, the CCS is turned off. Holding diode D_1 prevents any capacitor discharge, thus holding the voltage across C_1. The sampling circuit is turned on by a sampling pulse from the PDM input signal. Concurrently with the taking of the sample, a *dump* circuit (discharge path) discharges the capacitor. The sampling circuit output is a PAM wave, which is applied to a simple LP filter for demodulation and recovery of the original modulating signal.

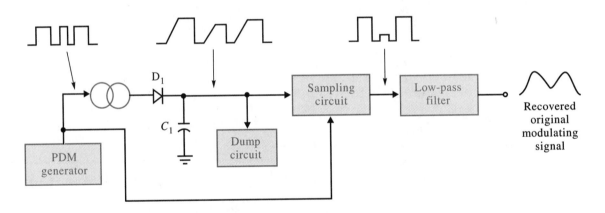

FIGURE 5-7 PDM demodulator

Pulse Position Modulation In *pulse position modulation* (PPM), the second form of analog PTM, both amplitude and width of the pulse remain constant. Figure 5-8(C) illustrates how PPM output pulses, relative to the reference pulses (Figure 5-8(B)), are varied in accordance with the modulating signal (Figure 5-8(A)). As shown at sampling points 1, 4, and 7, the PPM pulse is at rest. When the modulating signal goes positive (points 2 and 3), the PPM pulse lags the reference pulse by an amount of time proportional to the amplitude of the sample point. Conversely, when the modulating signal goes negative (points 5 and 6), the PPM pulse leads the reference pulse by a proportional amount of time.

Although there are similarities between PDM and PPM, PPM has the advantage of superior noise immunity. A typical PPM generator, with waveforms for the various stages, is shown in Figure 5-9. In this circuit, a PDM signal is used to develop the PPM output signal.

If we use a sawtooth sampling wave with a vertical leading edge and a sloping trailing edge along with the PDM generator of Figure 5-5(A), we produce trailing-edge

PDM, which we will use in the PPM generator. Recall that the leading edge in this form of PDM is fixed. The output of the PDM generator is differentiated. The positive spikes of the differentiated wave, of fixed time and duration, are locked to the leading edge of the PDM wave and are therefore of no further use. The positive spikes are

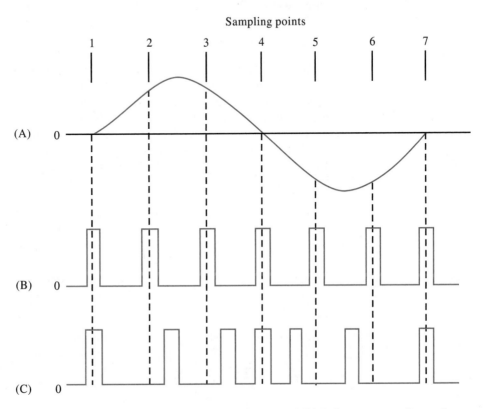

FIGURE 5-8 PPM waveforms (A) Modulating signal (B) Reference sampling pulses (C) PPM output pulses

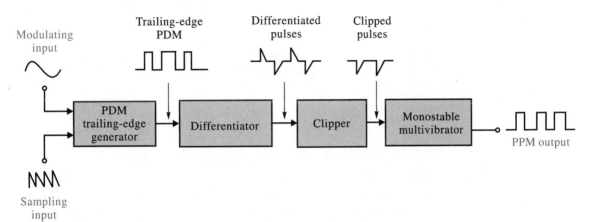

FIGURE 5-9 Generation of PPM

clipped off, and the remaining negative spikes are synchronized to the trailing edges of the PDM wave. The result is position-modulated negative spikes, which are used to trigger the monostable multivibrator. The multivibrator generates a fixed-width pulse for each input trigger, and since the negative spikes are position sensitive, the resultant output wave is PPM.

One approach to demodulation of a PPM signal is shown in Figure 5-10. A limiter removes any noise spikes and regenerates the incoming PPM signal. A synchronizing circuit removes synchronizing pulses transmitted with the PPM signal. These pulses are then used to lock a local reference pulse generator to the phase and frequency necessary to demodulate the PPM signal. The reference pulses and the PPM signal are applied to a flip-flop, with the PPM pulses lagging the reference pulse by an amount proportional to the modulation position. The reference pulse turns the flip-flop on, and the PPM pulse turns it off. The output of the flip-flop is a PDM wave with a pulse width that is directly proportional to the position of the PPM input pulse. This PDM signal is converted to a PAM signal, which is then applied to an LP filter, the output of which is a reproduction of the original modulating signal.

A disadvantage of the PPM demodulator, compared with the PDM and PAM demodulators, is the requirement in the PPM demodulator for a synchronized reference pulse generator. This requirement makes the PPM demodulator more complex and more costly, but the improved noise immunity of PPM offsets these problems.

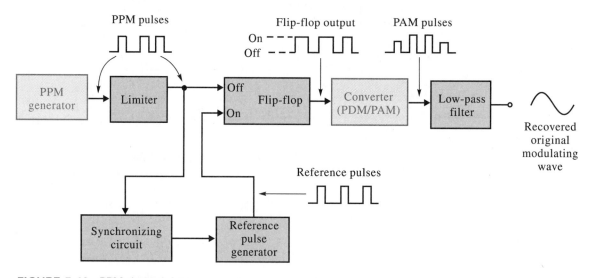

FIGURE 5-10   PPM demodulator

5-2   DIGITAL PULSE MODULATION

Although PTM has a noise immunity advantage over PAM, both forms of analog modulation are susceptible to noise distortion. Once noise is present on a PAM signal, removing the noise and recovering the original signal becomes extremely difficult. A limiter can be used with PTM signals to remove most of the noise, but the edges of PTM pulses must be precisely recorded, otherwise they can be distorted by noise. To help overcome these noise problems, *digital pulse modulation* (DPM) was developed. In DPM, pulse trains are transmitted in groups that represent binary numbers corresponding to the modulating signal voltage amplitudes. Demodulation of DPM signals does not depend on pulse amplitude, width, or position. Instead, it depends only on the presence or absence of pulses; thus, it is relatively easy to recover the pulses, even if high noise levels are present. Also, it is relatively easy to regenerate DPM signals. For these reasons, DPM is widely used in long distance communications. Digital pulse modulation includes pulse code modulation and delta modulation.

Pulse Code Modulation

The major form of DPM is *pulse code modulation* (PCM). Its primary advantage is much better noise and interference immunity. In PCM, the modulating signal is sampled, the sample amplitude is converted into a binary code, and the binary code is transmitted in groups as a train of pulses. A major difference is that, in PCM, the sampled amplitude must be transmitted as a specific binary number out of a limited range of binary numbers. To accomplish this, each sample must first be converted to the nearest standard amplitude, called the *quantum*. This process of sample conversion is called *quantizing the signal*.

A model for quantizing a modulating waveform is shown in Figure 5-11. Here, we use eight quantization levels, ranging from 0 to 7, where each level represents one volt. Table 5-1 shows the binary number and the 3-bit pulse code represented by each of the quantization levels.

In Figure 5-11(A), we see that many of the sampling points are not at a quantum level. In those instances, the sample amplitudes are represented by the nearest quantum level. For example, the amplitude of sampling pulse 1 is about 3.7 V. There is no quantum level at that voltage, so sampling pulse 1 is represented by quantum level 4. Another example is sampling pulse 6, where the amplitude is about 0.4 V, which is represented by quantum level 0. The error—that is, the difference between sampling

TABLE 5-1   Binary and Pulse Code for 8-Bit Quantization

Quantization Level (v)	Binary Number	Pulse Code Waveform
0	000	
1	001	
2	010	
3	011	
4	100	
5	101	
6	110	
7	111	

point amplitudes and quantum levels—is a distortion called *quantization noise,* so called because the errors are random. Stated another way, the difference between any quantum level and the amplitude of the signal at any instant is unpredictable.

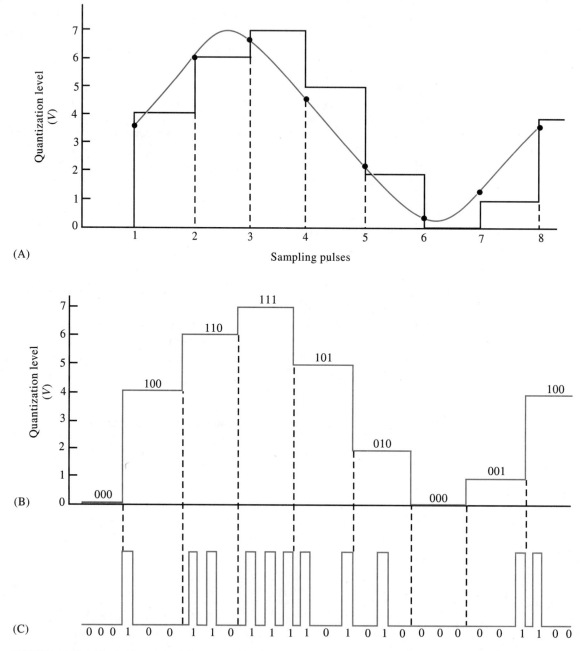

FIGURE 5-11   Model for quantizing a modulating signal (A) Quantized waveform (B) Coded waveform (C) Three-bit PCM pulse train

Quantization noise can be reduced by increasing the number of quantization levels. However, a disadvantage of increasing the number of levels is an increased transmission bandwidth requirement. Therefore, a compromise must be made between an acceptable transmission bandwidth and acceptable quantization noise. For example, the noise generated by a 3-bit code may be acceptable for the narrow bandwidth used for voice transmission but not for the much greater bandwidth required for television transmission.

After quantization and before transmission as a PCM signal, each sample is coded as a binary number. Coding the quantized waveform of Figure 5-11(A) is shown in Figure 5-11(B). The eight quantum levels used are represented by 3-bit binary words, as shown in Table 5-1. After the quantized waveform is coded, each sequential sample is transmitted as a pulse code, shown as the 3-bit PCM pulse train in Figure 5-11(C).

In practical PCM systems, the 3-bit word for this quantizing model is seldom used because of the quantization noise introduced. Instead, 8-bit words are more common because they provide 256 quantum levels, and therefore allow much better reproduction of the modulating signal with very little quantization noise. Also, in practical PCM systems, synchronizing pulses are transmitted with the pulse train to ensure that the receiver decodes the information pulses correctly.

5-3 DELTA MODULATION

Delta modulation (DM), sometimes called *slope modulation,* is a process of modulation in which a train of fixed-width pulses is transmitted. Figure 5-12 shows a model for the simplest form of DM, in which just one bit (binary digit) per sample is transmitted. Pulse polarity is determined by whether the modulating signal sample is rising or falling. The input is made to rise or fall by a fixed step height at each pulse. The polarity of the pulse indicates whether the demodulator output should rise or fall.

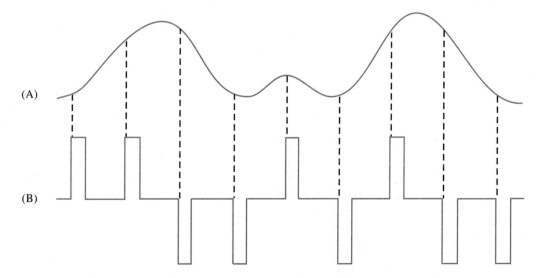

(A)

(B)

FIGURE 5-12 Model for delta modulation (A) Quantized waveform (B) DM waveform

In the block diagram of Figure 5-13, a modulating signal is applied to the noninverting input of a high-gain differential comparator, and a digitally reconstructed version of the modulating signal is applied to the inverting input. The saturated output of the comparator, either positive or negative, will depend on the polarity of the difference voltage between the input signals. The output will thus represent ± binary 1.

A train of pulses from the pulse generator is applied to the modulator at the desired sampling rate. The modulator either transmits these pulses directly for a +1 or inverts their polarity for a −1. The resultant signal is simultaneously applied to an integrator and transmitted as the output signal. The output signal of the integrator causes the comparator output to rise or fall by a fixed step height for each positive or negative pulse applied to the inverting input, thus causing a comparable change in the modulator.

The integrator cannot follow a signal that has a very fast rise time. When voice signals are being modulated, this limitation does not present a problem because voice signals do not change abruptly and there is generally little level change from one sample to the next. However, if high-frequency ramp signals are to be modulated by DM, a variable-gain amplifier is placed in the feedback path at the input to the integrator. The amplifier is essential for proper operation with high-frequency input signals.

At the receiver, the DM signal is reshaped in a regenerator circuit. Most of the noise that may have been picked up in the modulation/transmission process is removed before the regenerated DM signal is applied to a detector. The detector reconstructs the step waveform and feeds it through an LP filter to remove quantizing noise. The output of the LP filter is a reproduction of the original modulation signal.

Compared to PCM, the DM system has the advantages of greatly simplified encoding, decoding, and quantization, all of which result in simpler and less costly hardware.

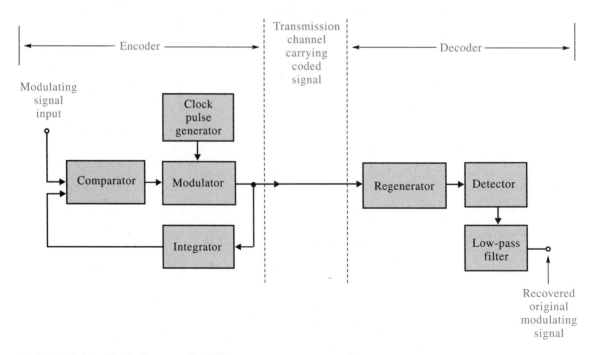

FIGURE 5-13 Block diagram of a DM system

However, because the DM system cannot follow fast-rising signals, it has the disadvantage of requiring a higher sampling rate and a greater bandwidth than PCM.

5-4 COMPANDING

One of the fundamental measures of quality for a quantizer is the *signal-to-quantizing-noise ratio* (SQR). For a linear system, the SQR is the ratio of the size of the input signal to $\frac{1}{4}$ (0.25) the size of the quantization level. This means that the SQR increases with signal level in a linear coder, so that large signals will have a higher SQR (better quality) than small signals.

Comparing two signals in a coder demonstrates the SQR concept. A small signal at quantization level 1 has an SQR of 4 (1/0.25 = 4), whereas a large signal at quantization level 6 has an SQR of 24 (6/0.25 = 24). This is an undesirable condition because small signals are more likely to occur than large signals, and the large signals tend to mask any noise present. The remedy for this condition is to adjust the size of the quantization levels in relation to the input signal amplitude so that the levels are smaller for large signals and larger for small signals. This adjustment produces a nonlinear output vs. input relationship and results in the output being compressed with respect to the input.

At the receiver end of the transmission, the decoder has a complementary expansion characteristic to restore linearity to the signal. The combination of characteristics in the *codec* (coder/decoder) is called a *compander,* and acronym for *com*pressor-ex*pander,* the circuit that compresses the dynamic range of an input signal in the coder and then expands it to almost original form in the decoder. When a compander is used, the SQR is about the same across the range of input signal levels. The devices at each end of the digital transmission channels in the public switched telephone network that perform the sampling, quantization, and coding to transform the speech signals to bits are called *channel banks.*

The companding process is illustrated in the block diagram of Figure 5-14(A). Signal levels in a complete compander system are shown in Figure 5-14(B). The compander works on a 2:1 compression ratio and a 1:2 expansion ratio. Noise inherent in the transmission medium is kept below the level of the weakest signal.

Companding itself does not produce any distortion in the recovered modulating signal, but it does reduce the level of quantizing noise during periods of very weak signals. As shown, the weak portion of the signal is made nearly equal to the strong portion of the signal. This is accomplished in the compressor amplifier, which amplifies the low-level signals more than it does the high-level signals, thus compressing the input voltage into a smaller span. The steps then transmitted have equal amplitudes. After detection, the signal levels are restored to their proper amplitudes in the expander. The result is lower quantizing noise levels during periods of weak signals. Companding is essential to quality transmission of digital modulation signals.

U-Law Compander

In large telephone systems, large numbers of digital channels are interconnected with one another. Therefore, all channel banks must use a common scheme for the analog-to-digital conversions (ADCs) and companding required. This scheme, standard for the

North American and Japanese telephone networks, is the "$u = 255$ law companded PCM digital coding standard," commonly called the *u-law*. The companding circuits operate on a logarithmic curve using the relationship

$$Fu(x) = sgn(x)\left[\frac{\ln(1 + u|x|)}{\ln(1 + u)}\right] \qquad (5\text{-}2)$$

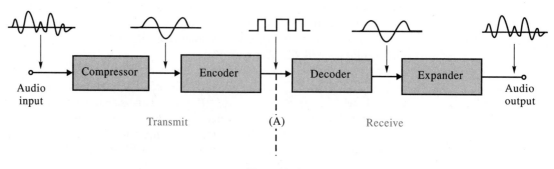

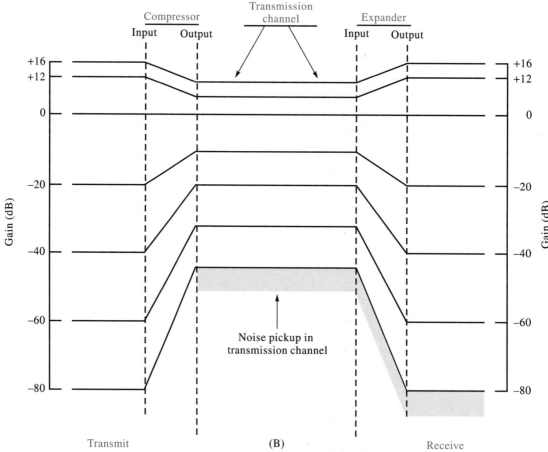

FIGURE 5-14   Companding process (A) Block diagram and waveforms (B) Compander signal levels

where

$Fu(x)$ = compressed output value

x = normalized input signal (between -1 and 1)

$sgn(x)$ = sign ($\pm$) of x

u = compression ratio, set at 255 for the North American and Japanese networks

$\ln$ = natural log

The encoder operates on a segmented linear approximation to the true logarithmic curve. It produces an 8-bit output: seven bits for level plus one bit for sign, with the most significant bit (MSB) (leftmost bit) representing the sign bit. The sign bit is 1 for positive input values and 0 for negative input values. The remaining seven bits of the code indicate the absolute value of the input signal. Since the sampling rate is 8000 samples per second, as determined by the Nyquist rate, the data rate for an individual voice channel when encoded using the u-law technique is 8000 samples per second $\times$ 8 bits per sample = 64,000 bits per second (bps). The transmission rate on most digital facilities is much higher than this because many channels are multiplexed together.

A-Law Compander

The companding standard for the European telephone network is the *A-law* characteristic, whose compression characteristics are defined as follows:

$$F(x) = sgn(x)\left[\frac{A|x|}{1 + \ln(A)}\right] \tag{5-3a}$$

when

$$0 \le |x| < \frac{1}{A}$$

and

$$F(x) = sgn(x)\left[\frac{(1 + \ln A|x|)}{(1 + \ln(A))}\right] \tag{5-3b}$$

when

$$\frac{1}{A} \le |x| \le 1$$

where

$F(x)$ = compressed output value

$sgn(x)$ = sign ($\pm$) of x

A = compression parameter, set at 87.6 for the European network

The *A*-law compander also produces eight bits per input sample in the same format as the *u*-law compander and a rate of 64,000 bps for each channel. The *A*-law scheme produces a slightly better SQR for small signals, but the *u*-law scheme exhibits lower noise on an idle channel.

The compander digitized schemes just described take enough samples and send enough bits to encode the complete waveform, so there is exact reproduction at the destination. Therefore, they are suitable for encoding and transmitting any waveform so long as its bandwidth is limited to what the chosen sampling rate can encode without error. The capability to accurately transmit any waveform is achieved at the cost of sending enough bits to encode the entire sample at every sample interval.

SUMMARY

- Nyquist rate, the minimum sampling frequency required for reliable reproduction of the modulating signal, is twice the highest modulating signal frequency. Increasing the sampling rate to a frequency higher than the Nyquist rate provides a guard band that allows easier filtering and improved accuracy of the reproduced signal.

- Analog pulse modulation includes pulse amplitude modulation (PAM) and pulse time modulation (PTM). PAM is the simplest form of analog pulse modulation. It is the result of a train of pulses occurring at a fast repetition rate that are made to vary in amplitude in proportion to the amplitude of a lower-frequency modulating signal. PAM waveforms can be single-polarity or dual-polarity. PAM signals are rarely used to transmit information directly because they are easily distorted by noise, crosstalk, and other forms of interference. PAM is frequently used as an intermediate step in other pulse-modulating methods. PTM uses constant-amplitude pulses varying in timing for sampling. The timing variable may be duration or position, resulting in four distinct forms of PTM: pulse duration modulation (PDM), comprising three separate forms; and pulse position modulation (PPM). PDM, also called pulse width modulation (PWM), or pulse length modulation (PLM), is the result of varying the duration of the pulses used for modulation. The three types of PDM are symmetrical, leading edge, and trailing edge. In PPM, both the amplitude and width of the pulse remain constant, but the position of the pulse varies in relation to a reference pulse, either leading or lagging, in accordance with the polarity of the modulating signal. The PPM demodulator requires a synchronized reference pulse generator.

- Digital pulse modulation (DPM) includes pulse code modulation (PCM) and delta modulation (DM). Digital pulse trains are transmitted in groups representing binary numbers that correspond to the modulating signal voltage amplitudes. The principal form of DPM is PCM, which provides greater immunity to noise and interference. Samples for PCM transmission must be quantized—that is, converted to the nearest standard amplitude, called the quantum. The error, or difference between sampling point amplitudes and quantum levels, is a distortion called quantization noise. After quantization and before transmission as a PCM signal, each sample is coded as a binary number. One of the fundamental measures of quality for a quantizer is the signal-to-quantizing-noise (SQR). In DM, also called slope modulation, a train of fixed-width pulses is transmitted. In the simplest form, just one bit per sample is transmitted. Pulse polarity is determined by whether the signal sample is larger or smaller than the previous sample. Compared to PCM, DM has the advantages of greatly simplified encoding, decoding, and quantization, but the disadvantages of a higher required sampling rate and a greater required bandwidth.

- A compander is a circuit that compresses the dynamic range of an input signal in a coder and expands it to almost original form in a decoder. The coder/decoder combination is called a codec. When the compander is used, the SQR is about the same across the range of input signal levels. North American and Japanese telephone systems use the u-law compander; the European telephone system uses the A-law compander.

QUESTIONS AND PROBLEMS

1. List the two broad categories of pulse modulation, what types each includes, and other names and subcategories.

2. What advantage does PTM have over PAM? How is this advantage achieved?

3. List three ways to achieve PDM and explain their differences.

4. What advantages does PPM have over PDM?

5. What disadvantage does the PPM demodulator have?

6. Compare the advantages and disadvantages of DM and PCM.

7. Define Nyquist sampling rate.

8. What is the effect of a sampling rate (a) less than the Nyquist rate and (b) higher than the Nyquist rate?

9. Find the guard band for a 10 kHz sampling frequency and a 3 kHz highest modulating frequency.

10. Find the highest modulation frequency for a 2 kHz guard band and a 7 kHz sampling frequency.

11. Find the sampling frequency for a 2.5 kHz highest modulating frequency and a 1.2 kHz guard band.

12. Defined quantization noise, explain how it can be reduced, and indicate the disadvantage of the reduction method.

13. Define compander and explain its operation.

14. Define channel banks.

CHAPTER 6

Transmission Lines and Techniques

OBJECTIVES

When you complete this chapter, you should be able to:

- List and describe several types of transmission lines.
- Distinguish between electrical and physical line length.
- Calculate line signal propagation for different types of lines.
- Calculate wavelength in meters and in feet for open-air propagation.
- Calculate propagation velocity and time delay for transmission lines.
- Calculate characteristic impedance for different transmission lines.
- Determine sources and loads in a system.
- Define nonresonant and resonant transmission lines.
- Understand and explain incident and reflected signals, standing waves, traveling waves, and wavelength.
- Recognize open and shorted lines and explain their voltage and current waves.
- Define and calculate VSWR, SWR, and ρ.
- List and explain three major transmission line losses.
- List and explain several transmission line applications.

INTRODUCTION

In communications systems, energy in the form of varying voltage and current must be transferred from one point to another. This transfer can be accomplished in several ways, but it is through transmission lines that the signals are initially transferred. The primary purpose of a transmission line is to transfer energy from a generator to a load. For example, the audio signals from a studio are fed through transmission lines to a transmitter at one location, and the RF signals from the final power amplifier of the transmitter are fed through transmission lines to an antenna at some remote location.

Likewise, at the receiver location, the received signals are fed through transmission lines from the receiving antenna to the receiver. At dc levels or very low frequencies, such as power lines and voice frequencies, few problems are encountered with the lines. However, at high frequencies, even short lines acquire very different characteristics. Because of these high-frequency characteristics, transmission lines are widely used for impedance matching, as measuring devices, and as resonant circuits. A good understanding of transmission line theory is essential for understanding the chapters that follow.

6-1   TYPES OF TRANSMISSION LINES

The two basic types of transmission lines are the two-wire line and the coaxial line. Transmission lines may be balanced or unbalanced with respect to ground. A *balanced* line is one in which both wires of the line carry RF current and the current in each wire is 180° out of phase with that in the other wire. An *unbalanced* line is one in which one wire is at ground potential and the other wire carries the RF current. In this section, we will consider each type of line.

Two-Wire Lines

Two-wire lines, usually operated as balanced lines, include two-wire open-air, two-wire ribbon (twin-lead), twisted-pair, and shielded-pair lines. All of these lines have characteristics that make them useful for specific applications.

Two-Wire Open-Air Line   In earlier times, *two-wire open-air* lines (Figure 6-1(A)) commonly consisted of a pair of No. 6 copper wires spaced 12 inches apart, giving a characteristic impedance of 600 Ω. Today, these lines more commonly consist of a pair of wires spaced from $\frac{1}{4}$ inch to 6 inches apart. The parallel lines are maintained at a fixed separation by insulated spacers placed at intervals along the length of the line. The principal uses for this type of line are power distribution, telephone, and telegraph lines. Less often, they may be used between a transmitter and an antenna or between an antenna and a receiver. Simple and inexpensive construction is the greatest advantage of this type of line. However, the lack of shielding causes high signal radiation losses and allows easy pickup of noise.

Two-Wire Ribbon Line   The *two-wire ribbon* line, commonly called *twin-lead* (Figure 6-1(B)), consists of two parallel wires embedded in a thin ribbon of low-loss dielectric, usually polyethylene. The ribbon ensures uniform spacing along the entire length of the line. The primary use of this type of line is to provide a 300 Ω feeder line between a television receiving antenna and a television receiver. The advantages and disadvantages of twin-lead are essentially the same as for the two-wire open-air line.

Twisted-Pair Line   As implied by the name, the *twisted-pair* line (Figure 6-1(C)) consists of two rubber-insulated wires twisted together. No spacers are used, and the result is line flexibility—its primary advantage. The major disadvantage of this type of line is high-frequency loss in the rubber insulation. This loss increases significantly if the line is wet. Thus, the twisted-pair line is commonly confined to protected short-run, low-frequency use, such as ac power cables.

Shielded-Pair Line   The *shielded-pair* line (Figure 6-1(D)) consists of two separated parallel wires surrounded by solid, low-loss dielectric. The insulating dielectric containing the parallel wires is enclosed in a metal braid for shielding, then covered with a flexible coating, usually rubber, to protect the line from moisture.

   The advantages offered by the shielded-pair line are reduced signal radiation losses and reduced noise pickup. The grounded shield provides a balance between the conductors and ground, so this type of line is ideally suited to balanced line feeds. The disadvantage is that the dielectric and leakage losses are higher than for the other line types.

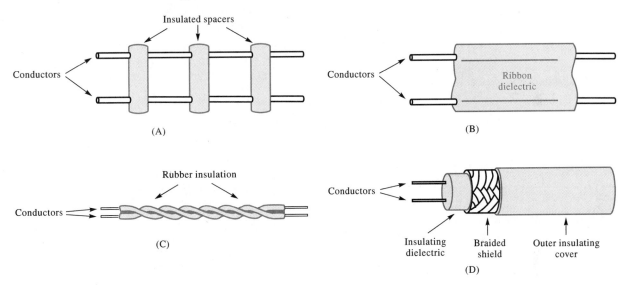

FIGURE 6-1   Two-wire transmission lines (A) Two-wire open-air (B) Two-wire ribbon (twin-lead) (C) Twisted-pair (D) Shielded-pair

Coaxial Lines

Coaxial (or simply *coax*) lines, usually operated as unbalanced lines, include the rigid (air, or concentric) line and the flexible (solid) line. The two types of coaxial transmission lines are represented in Figure 6-2. These lines are unbalanced because they consist of an inner conductor and an outer shield that is at ground potential. The center conductor carries the RF energy, and the shield prevents this energy from being radiated into space. The ability to minimize signal radiation loss is the chief advantage of coaxial lines. The electric and magnetic fields are confined to the space between the inner and outer conductors, and the grounded shield prevents noise pickup from external sources.

Rigid Coaxial Line   The *rigid,* or *concentric,* line shown in Figure 6-2(A) consists of an inner conductor maintained at a fixed distance from an outer conductor by spacers placed at regular intervals throughout the length of the line. The spacers are made of materials that provide good insulating qualities and low loss at high frequencies. However, the line is expensive to construct, and difficult to work with because of its rigidity. To prevent excessive leakage between the two conductors, the lines must be kept dry. Dryness is accomplished upon installation by insertion into the line, under pressure, an inert gas such as helium, argon, or nitrogen. The pressure is maintained to ensure that no moisture enters the line. In addition, the practical length of the line is limited by high-frequency losses over distance.

Flexible Coaxial Line   The *flexible* line in Figure 6-2(B) consists of an inner conductor of flexible wire insulated from the outer conductor by a solid, flexible insulating polyethylene plastic extending continuously for the length of the line. Flexibility is maintained by making the outer conductor with a braided copper wire. An outer polyethylene cover completes the construction of the line.

Although high-frequency losses are greater with the polyethylene insulator than with air, this is a slight disadvantage that is outweighed by other considerations. Polyethylene is flexible over wide temperature ranges and is highly resistant to the effects of such caustic substances as gasoline, oil, and salt water. For these reasons, the solid flexible coaxial line is the most commonly used type of transmission line.

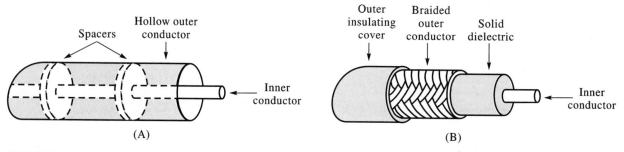

FIGURE 6-2   Coaxial transmission lines (A) Rigid (air or concentric) (B) Flexible (solid)

Long and Short Lines

Transmission lines are divided into two general categories: long lines and short lines. These designations refer to the *electrical length* of the line, not to the physical length. In general terms, *long lines* are one quarter wavelength (λ/4) long or longer; *short lines* are appreciably shorter than λ/4. The wavelength λ in open air is calculated by dividing velocity v by frequency f:

$$\lambda = \frac{v}{f} \tag{6-1}$$

where

v = velocity = 3×10^8 meters/second (m/s) or 9.84×10^8 feet/second (ft/s)

f = frequency in hertz (Hz)

For example, the wavelength of a 100 Hz signal is approximately 1860 miles long. Therefore, at the 100 Hz frequency, any line with a *physical* length of less than around

400 miles is classified as a short line, because 400 miles is appreciably less than $\lambda/4$. On the other hand, the wavelength of a 30 MHz signal is approximately 33 feet, so a line used to transfer the energy at this frequency that has a *physical* length of over 9 feet is a long line. Actually, the physical length of the lines discussed here would be somewhat shorter because of the resistance found in any conductor.

Most dc and low-frequency transmission lines are classified as short lines. Because short lines have no standing waves (stationary distribution of current or voltage along the line) to be considered, voltage, current, and impedance calculations are made by following Ohm's law. On the other hand, because standing waves appear on RF transmission lines, RF lines are classified as long lines and must be treated as resonant circuits. However, under specific conditions, RF lines can be made nonresonant. Nonresonant lines are discussed in Section 6-4.

Propagation Velocity

The *propagation velocity* v_P (the speed at which energy travels from a source) for transmission lines is a function of distance D and time T per unit length. Propagation velocity is important for those applications where a specific signal time delay is desired. An example is the construction of a delay line from a length of coaxial cable. In equation form, we write

$$v_P = \frac{D}{T} \qquad\qquad (6\text{-}2)$$

where

v_P = propagation velocity

D = travel distance

T = time = $\sqrt{LC}$

EXAMPLE 6-1

Assuming open-air transmission, calculate the wavelength for a signal frequency of 25 MHz in (a) meters and (b) feet.

SOLUTION

(a) $\dfrac{3 \times 10^8 \text{ m/s}}{25 \text{ MHz}} = 12$ m

(b) $\dfrac{9.84 \times 10^8 \text{ ft/s}}{25 \text{ MHz}} = 39.36$ feet

Or, to convert from meters to feet, multiply the number of meters by 3.28:

(12 m) (3.28) = 39.36 feet

EXAMPLE 6-2

Assume that the 25 MHz signal frequency in Example 6-1 is transmitted through an RG-63B/U coaxial cable, where capacitance is 10 pF/ft and inductance is

156.25 nH/ft. Calculate (a) the time delay for 1 foot of cable and (b) the propagation velocity.

SOLUTION

(a) $T = \sqrt{LC} = \sqrt{(10 \times 10^{-12})(156.25 \times 10^{-9})}$

   $= 1.25$ ns

(b) $v_P = \dfrac{1 \text{ foot}}{1.25 \text{ ns}}$

   $= 8 \times 10^{8}$ ft/s

Electrical Characteristics

In basic electronics studies, we were taught to think of the movement of electrical energy in ordinary low-frequency circuits as electrons flowing in one direction at any given instant throughout the entire circuit and then reversing direction in the entire circuit at a later instant, so that all the electrons are moving in only one direction at any given instant. However, this concept does not hold for electrically long circuits. For example, if the physical length of a two-wire transmission line is greater than λ/4, *electrons may flow in opposite directions at the same instant in the same wire.* This phenomenon occurs when the transmission line is not terminated in an impedance of the same value as the characteristic impedance of the line. When this happens, Ohm's law for voltage, current, and impedance of short lines cannot be applied. This means that new concepts must be dealt with.

Distributed Elements

A constant variable such as resistance, capacitance, or inductance is usually thought to exist in a discrete form in a component especially designed to have certain electrical properties. However, there is always some resistance, capacitance, and inductance in even the simplest circuits. The equivalent resistance, capacitance, and inductance in the wiring of a circuit are called *distributed elements* or *distributed constants.*

The effects of distributed elements in circuit design and operation are not usually significant at low frequencies. However, in the VHF and higher frequency ranges, distributed elements are an important consideration. At a higher frequency the lead length of a discrete component becomes an appreciable fraction of the wavelength causing lead inductance and capacitance to affect the operation of the circuit. A resistor has distributed inductance and capacitance. An inductor has distributed resistance because of the ohmic loss in the conductor, and distributed capacitance because of interaction between the windings.

In a transmission line, these three constants are distributed uniformly, as shown in Figure 6-3. The resistance and inductance are in series with the conductors. The capacitance appears across the conductors.

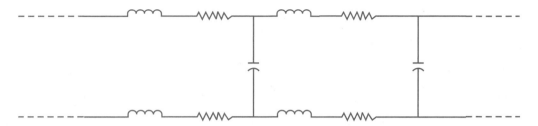

FIGURE 6-3   Equivalent circuit for distributed reactances along a transmission line

6-2   CHARACTERISTIC IMPEDANCE

A transmission line's *characteristic impedance* (often called *surge impedance,* or *surge resistance*) is a special property that is not influenced by the line's length, by the frequency at which the line is operated, or by the ohmic value of the load resistance. The characteristic impedance of any transmission line is determined by the distributed inductance (of the conductors) and the distributed capacitance (between the conductors) of the line. Expressed in ohms, the characteristic impedance is approximately equal to the square root of the ratio of inductance to capacitance, with the ratio taken for any given length of line. As an equation, in ohms per unit length, we have

$$Z_0 = \sqrt{\frac{L}{C}} \qquad\qquad (6\text{-}3)$$

Remember, the length of the transmission line does not affect the characteristic impedance. For example, if the length of a line is doubled, the inductance and capacitance in that line also double, but the value of the L/C ratio as a whole does not change. Therefore, the square root of the ratio is not changed by changing the length of the line. The following example demonstrates this fact.

EXAMPLE 6-3

A commonly used television receiver input transmission line is a coaxial cable, RG-59B/U, with a capacitance of 21 pF/ft and an inductance of 111.91 nH/ft. What is the characteristic impedance for (a) a 10 foot section and (b) a 500 foot section?

SOLUTION

From Equation 6-3,

$$\text{(a)} \ \ Z_0 = \sqrt{\frac{L \times 10}{C \times 10}} = \sqrt{\frac{111.91 \times 10^{-9} \times 10}{21 \times 10^{-12} \times 10}}$$

$$= \sqrt{5329} = 73 \ \Omega$$

(b) $Z_0 = \sqrt{\dfrac{L \times 500}{C \times 500}} = \sqrt{\dfrac{111.91 \times 10^{-9} \times 500}{21 \times 10^{-12} \times 500}}$

$= \sqrt{5329} = 73 \ \Omega$

The ohmic value of the pure resistance required to terminate a given transmission line in such a manner that no energy will be reflected back into the line from the terminating load is a specific characteristic of the line itself. This impedance value is characteristic of the line acting as a source, and it must be matched by the load's resistance if the load is to accept all and reflect none of the power fed to it by the line. We can therefore define the characteristic impedance of any transmission line as the exact number of pure resistance ohms required to properly terminate that particular transmission line.

The characteristic impedance, in ohms, of a two-wire open-air transmission line is determined by

$$Z_0 = 276 \log \frac{s}{r} \tag{6-4}$$

where

s = spacing between wire centers

r = radius of conductors

Quantities s and r must be in the same units.

The characteristic impedance, in ohms, of a coaxial line is determined by

$$Z_0 = 138 \log \frac{d_1}{d_2} \tag{6-5}$$

where

d_1 = inside diameter of outer conductor

d_2 = outside diameter of inner conductor

Note that Equations 6-4 and 6-5 do not include length, further substantiating our earlier statement.

EXAMPLE 6-4

A transmission line is made with two parallel lengths of No. 20 hookup wire, where $L = 500$ nH and $C = 2$ pF. Calculate the characteristic impedance of the line.

SOLUTION

From Equation 6-3,

$$Z_0 = \sqrt{\frac{L}{C}} = \sqrt{\frac{500 \times 10^{-9}}{2 \times 10^{-12}}}$$

$$= \sqrt{25 \times 10^4}$$

$$= 500 \, \Omega$$

EXAMPLE 6-5

A two-wire open-air line consists of No. 10 wire spaced 300 mm between centers. The diameter of No. 10 wire is 2.588 mm. What is the characteristic impedance of the line?

SOLUTION

First, solve for r:

$$r = \frac{d}{2} = \frac{2.588 \, \text{mm}}{2} = 1.294 \, \text{mm}$$

From Equation 6-4,

$$Z_0 = 276 \log \frac{s}{r}$$

$$= 276 \log \frac{300}{1.294}$$

$$= 276 \log 231.8$$

$$= (276) \, (2.365)$$

$$= 653 \, \Omega$$

EXAMPLE 6-6

A rigid coaxial line is made of an outside copper tubing 1.6 mm thick with an outside diameter of 25 mm. The copper tubing of the inner conductor is 0.8 mm thick with an outside diameter of 6 mm. Calculate the characteristic impedance of the line.

SOLUTION

First, solve for d_1:

$$d_1 = 25 - (2 \times 1.6) = 21.8 \, \text{mm}$$

From Equation 6-5,

$$Z_0 = 138 \log \frac{d_1}{d_2}$$

$$= 138 \log \frac{21.8}{6}$$

$$= 138 \log 3.633$$

$$= (138) \, (0.5603)$$

$$= 77.3 \, \Omega$$

Source or Load?

Since we know that the purpose of a transmission line is to transfer energy from one point to another, we can conclude that the line must work out of a source into a load. A *source* is an energy generator in any form. A *load* is a device that absorbs or consumes the energy fed into it. For example, if a transmission line is used to feed energy from a transmitter to a transmitting antenna, the transmitter is the source and the transmitting antenna is the load. The antenna absorbs the energy from the transmission line, but then the antenna itself becomes a source for radio waves by radiating the absorbed energy into space. Therefore, at the transmitter, the antenna acts both as a load (to the transmission line) and as a source (for radio waves). Conversely, at the receiver, the antenna acts as a load to the radio waves and as a source for the transmission line. The receiver absorbs the energy fed into it by the transmission line and thus acts as the load for the transmission line.

Let us take a look at what happens in these lines. Assume that a 3 kHz signal is applied to the two-wire line shown in Figure 6-1(A). The instantaneous electron flow is in opposite directions in successive 31 mile long sections of each wire. If the frequency is increased to 3 MHz, each successive 164 foot length will have opposite electron flow. Increasing the frequency to 300 MHz will cause opposite-direction current flow in about each 1.64 foot section. This discussion shows that as frequency is increased, the flow of electrons in the wires is in opposite directions within a more limited range, and the length of that range is one half the wavelength of the applied frequency.

Traveling Waves

It may be difficult to visualize the transfer of energy from one end of the line to the other under high-frequency conditions in which the electron flow in different sections of the same conductor is in opposite directions at the same instant. It is easier to visualize if we think of the energy as a *wave* moving down the line, in which an oscillatory electron motion is set up in sections as it moves down the line. This wave is known as a *traveling wave,* traveling at about the speed of light (186,000 miles/s or 3×18^8 m/s) along and between the two wires of the transmission line and setting up associated electron movement in sections of the wires as it moves along.

Moving at about the speed of light, the wave travels out from the source along the line for a distance of $\lambda/2$ during the time it takes the source to produce one alternation of the applied signal, or for a distance of one full wavelength during the time it takes the source to produce one full cycle of the applied signal. Therefore, the wave and its associated line current at any point along the line are the results of an input signal voltage applied to the line at some earlier instant. The input voltage applied to a particular wire of the line is alternately positive and negative, so the polarity of the electron flow in the wire will be opposite for each successive $\lambda/2$ of the wire, measured along the length of the wire. For this reason, in either of the two conductors, the electron flow in each half-wave section is opposite in direction to the electron flow in both the preceding and following half-wave sections of that same conductor.

Instantaneous current waveforms in a two-wire transmission line are illustrated in Figure 6-4. Note that the polarity reverses with each half-wave section in each conductor, and that the polarity of the wave in one conductor is reversed from that in the other conductor. These waveforms show instantaneous conditions only.

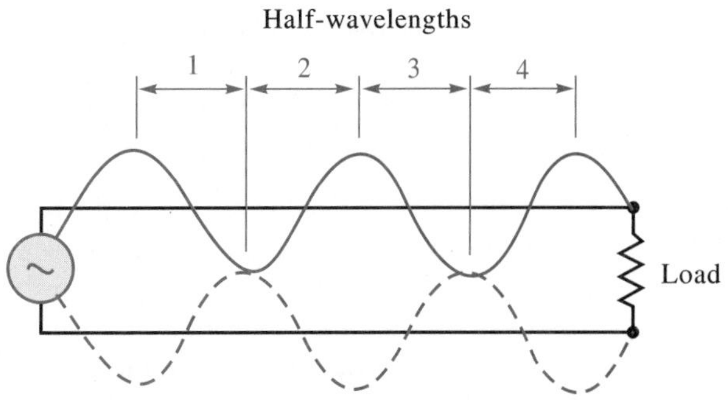

Half-wavelengths

FIGURE 6-4
Instantaneous
current waveforms
in a two-wire line

Load

6-3   RESONANT TRANSMISSION LINES

A transmission line that has a finite length, or one that is not properly terminated (that is, a line terminated in an impedance not equal to its characteristic impedance) is called a *resonant (unmatched) line.* A resonant circuit is one in which capacitive and inductive reactances cancel each other. An example of such a circuit, at a specific frequency, is a parallel or series *LC* circuit. Antennas and transmission lines are resonant circuits at many different frequencies.

All resonant circuits exhibit variable attenuation, depending on frequency. To obtain selectivity, notching, and impedance matching, resonant circuits are widely used in audio frequency and radio frequency design. A resonant line may be terminated in an open, in a short, or in some resistive value other than the characteristic impedance of the line. In any such case, the line cannot deliver full energy to a load. Some of the energy is reflected back to the source and forms *standing waves* on the line.

Standing Waves

On a resonant line, some of the energy sent down the line will be reflected back to the source, resulting in standing waves. Every $\lambda/2$ along a resonant line, high voltage and low current points appear. Halfway between these points, the opposite is true—low voltage and high current points. The ratio of the high voltage points to the low voltage points is called the *voltage standing wave ratio* (VSWR). The ratio of the high current points to low current points is known more simply as the *standing wave ratio* (SWR). As equations, we write

$$\text{VSWR} = \frac{V_{\max}}{V_{\min}} \tag{6-6}$$

and

$$\text{SWR} = \frac{I_{\max}}{I_{\min}} \tag{6-7}$$

The ratio of voltage (or current) reflected back down the line to that delivered to a load is the *reflection coefficient* ρ. In equation form, we write

$$\rho = \frac{V_r}{V_i} = \frac{I_r}{I_i} \tag{6-8a}$$

Related to SWR, we have

$$\rho = \frac{\text{SWR} - 1}{\text{SWR} + 1} \qquad\qquad (6\text{-}8b)$$

The points at which the two waves aid or oppose each other are determined by the speed and the length of the waves. Therefore, if the frequency (and, consequently, the wave-length) of the energy from the source remains constant, the points at which maximum and minimum currents (or voltages) appear along the line remain at fixed distances back from the open end of the line. In other words, these points "stand" in a fixed position—hence the term *standing wave*. The points of maximum strength are called *loops,* and the points of minimum strength are called *nodes.*

Open Line   When a transmission line of finite length is terminated in an *open,* current at the open end of both conductors must be zero, just as with Ohm's Law. This zero current occurs because the wave traveling out from the source comes to an infinite impedance at the open. The wave cannot continue to move in the outward direction, and there is no current flow at this point to dissipate the energy carried by the wave. Therefore, the only possible direction for the wave to take is back down the line—that is, to be reflected back to the source. The reflected wave travels back toward the source at the same speed as the original wave, which is still moving toward the load end of the line. The original wave moving outward from the source is called the *initial wave* (or *incident wave*), and the wave moving back toward the source is called the *reflected wave.*

The initial wave and the reflected wave are traveling at the same speed. The currents induced by each of these waves vary in value throughout $\lambda/2$, and the polarity changes with each successive $\lambda/2$. Since both of these waves are traveling on the same line but in opposite directions, their relative phases are such that at some points along the line the currents induced by them tend to cancel each other, and at other points along the line the currents aid each other. If the signal frequency (and therefore its wavelength) and the speed of the wave remain constant, the phase relationship between the two waves also remains constant. For these reasons, the points along the line at which the current of the initial wave is aided by the current of the reflected wave remain fixed. Stated another way, the points of maximum current do not move along the line as the two separate waves do but, instead, stand at fixed positions along the line. Also standing at fixed positions along the line are the points of minimum current.

In an open-ended, no-loss line, all of the energy is reflected. The strength of the reflected wave is equal to that of the initial wave. When the two waves are of equal strength, the points at which they are exactly in phase will have current values equal to twice the value of the initial wave alone. On the other hand, the points at which the two waves are exactly 180° out of phase will have a current value of zero. At points in between, the current will have a value between twice the initial current wave value and zero.

The distribution relationships for current, voltage, and impedance along an open-ended line are illustrated in the ladder diagram of Figure 6-5. As shown, the open end of the line represents maximum impedance, and hence, maximum voltage and minimum current. In other words, the open end of the line represents a current node, since there is no place for the current to go except back down the line (reflection). Note that each current node is located exactly λ/2 back down the line. Also observe that voltage standing waves are 90° out of phase with the current standing waves; that is, voltage loops occur when current nodes occur, and vice versa. We know that impedance represents a certain relationship between current and voltage; that is, current is higher when impedance is lower, and voltage is higher when impedance is higher. Because of those relationships, with reference to standing waves on a transmission line, the points of minimum and maximum impedance can be determined. Figure 6-5 shows those impedance relationships.

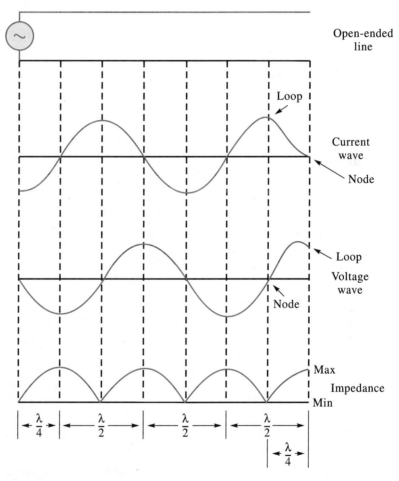

FIGURE 6-5 Relationships for current, voltage, and impedance on a two-wire open-ended line

Shorted Line In a *shorted* transmission line, the two wires are connected by a non-resistive conductor so that energy can flow freely from one wire of the line to the other. In the conductor connecting the two wires together, the currents from the two wires are in phase, and therefore the total current will be maximum, just as in Ohm's

Law. As shown in Figure 6-6, standing waves on a shorted transmission line act like those in an open line. The major difference is the location of the loops and nodes. Note that the output end of the shorted line has a current loop, and λ/4 back is a current node. The pattern repeats itself every λ/2.

Comparing the waveforms for the open-ended and shorted transmission lines shows that the standing waves on the open-ended line are displaced from those on the shorted line by exactly λ/4. Also, note that the condition of the standing wave λ/2 back down the line from the end is, in both cases, the same as that at the end of the line.

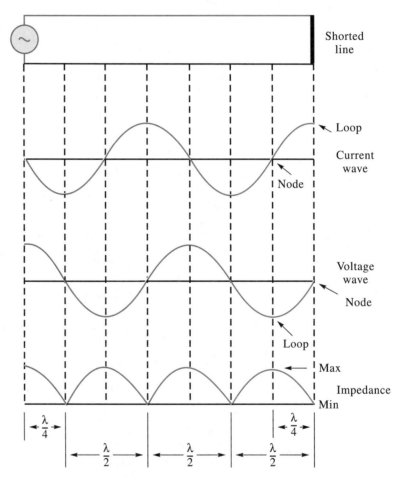

FIGURE 6-6 Relationships for current, voltage, and impedance on a two-wire shorted line

Improper Resistive Load Standing waves may be produced on transmission lines other than those terminated in opens or shorts. There is only one value of resistance that will properly terminate a line—that is, a line that will not produce standing waves. That resistance value, as stated earlier, is the resistance that is exactly correct to dissipate all of the energy delivered to it by the line. That is, the terminating resistance must be exactly equal to the characteristic impedance of the line. Any other resistance value will dissipate only part of the energy delivered to it and, therefore, will reflect the remaining energy back down the line to the source. The amplitude of the standing waves is determined by the percentage of the total energy input that is reflected. In practice, it is almost impossible to eliminate standing waves, but they can be reduced to a negligible level.

Measuring SWR   The standing wave ratio in a coaxial cable can be measured with an *SWR meter,* or *reflectometer.*

Also called a *directional coupler, directional power meter,* or *monimatch,* a reflectometer is an instrument installed in a transmission line for measuring the standing-wave ratio. Some reflectometers are calibrated in both forward watts and reflected watts.

Many different reflectometers for use in coaxial transmission lines are available from commercial manufacturers. A simple reflectometer might consist of a short length of coaxial line with a short piece of wire placed inside it through a directional coupler parallel to the center conductor. This wire feeds an SWR-calibrated dc meter through a diode, shown in Figure 6-7. When SWR is 1:1, there is no deflection on the meter, but as SWR increases, the meter reads higher. Because of slight frequency variations and environmental changes, in practice an SWR of 1.2:1 is about the best that can be achieved. Impedance matching then becomes necessary.

A directional coupler placed in the opposite direction will allow a reading of the reflected signal. If two directional couplers are connected back to back, both the signal moving forward and the reflected signal can be sampled. The dashed lines in Figure 6-7 indicate how the reflected signal is sampled. If the same type of device is calibrated in watts and some means of indicating a reference level is provided, it can be used as an *RF wattmeter.*

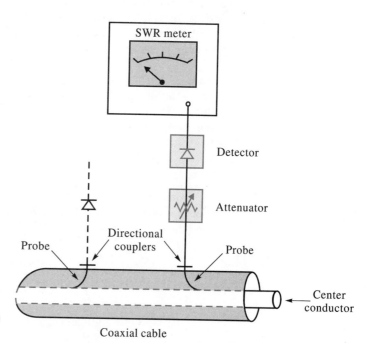

FIGURE 6-7
Reflectometer for measuring SWR on a coaxial cable

Slotted Lines   Another method for measuring SWR is the *slotted line*. The slotted line is simply a section of coaxial transmission line with a lengthwise slot cut into the outer conductor. A pickup probe is then inserted into this slot. The pickup probe, connected to a calibrated scale, shows that the magnitude of a signal picked up is proportional to the voltage between the conductors at the insertion point. The resulting measurement can be used to plot the SWR pattern as related to distance and to determine generator frequency, load impedance, and VSWR.

Effect on Input Impedance

The physical length of a resonant line is critical. As stated earlier, unless a load is resistive and equal to the characteristic impedance of the line to which it is connected, standing waves of voltage and current are set up along the line. When a line is not properly terminated, the input impedance of the line will be affected by the physical length of the line and by the signal frequency. That is, the impedance seen by the source will be affected by any fraction of a wavelength left over at the end of the line in addition to any whole number of wavelengths in the entire physical length of the line. For example, if the line were either slightly shortened or slightly lengthened, the input impedance of the open line in Figure 6-5 would be increased. Either of those changes would reduce the input impedance of the *shorted* line in Figure 6-6.

Quarter-Wave Transformer   For maximum energy transfer, the load must be matched to the line itself. As a general rule, this involves the tuning out of the unwanted load reactance (if any) and the transformation of the resulting impedance to the ohmic value required. Although ordinary RF transformers may be used up to the middle of the VHF range, their performance is not good enough at much higher frequencies because of excessive leakage inductance and stray capacitances. The *quarter-wave transformer* provides the solution. The quarter-wave transformer is compatible with transmission lines and provides unique opportunities for impedance transformation up to the highest frequencies.

Impedance Matching   Figure 6-8 shows a load (Z_L) connected to a length (s) of transmission line. When the length s is exactly $\lambda/4$ (or an odd number of quarter-wave lengths) and the line is lossless, the source impedance (Z_S), seen when looking toward the load, is given by

$$Z_S = \frac{Z_0^2}{Z_L} \tag{6-9}$$

Equation 6-9 shows that the impedance at the input of a quarter-wave transformer depends on the load impedance Z_L, which is fixed for any load at a constant frequency, and the characteristic impedance Z_0 of the interconnecting transmission line. If Z_0 can

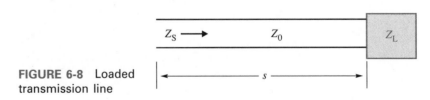

FIGURE 6-8   Loaded transmission line

be varied, the impedance seen at the input to the quarter-wave transformer will be varied accordingly, and the load may thus be matched to the characteristic impedance of the main line. Varying Z_0 is similar to varying the turns ratio of a transformer to obtain a required value of input impedance for a given value of load impedance. The following practical example will demonstrate the procedure.

EXAMPLE 6-7

The SWR along a television lead-in line needs to be reduced to 1:1. The television set provides a 300 Ω load to a 75 Ω transmission line. If the quarter-wave transformer is connected directly to the load, what must its characteristic impedance be?

SOLUTION

Since the condition SWR = 1:1 is desired along the main line, the source impedance Z_S at the input to the quarter-wave transformer must equal the characteristic impedance Z_0 of the main line. Let Z_t be the transformer characteristic impedance. Then manipulate Equation 6-9 so that

$$Z_S = \frac{Z_t^2}{Z_L} = Z_0 \quad \text{(of the main line)}$$

$$Z_t = \sqrt{Z_0 Z_L}$$

$$= \sqrt{(75)(300)}$$

$$= 150 \ \Omega$$

It must be understood that the transformer used in this example is λ/4 at only one frequency. It is thus highly frequency dependent. This property of the quarter-wave transformer makes it very useful as a filter to prevent unwanted frequencies from reaching the load. On the other hand, if broadband impedance matching is required, the transformer must be constructed of high-resistance wire to increase losses and lower the line Q, thereby increasing bandwidth.

The procedure becomes somewhat more involved if the load is complex rather than purely resistive. The quarter-wave transformer can still be used, but it must now be connected at some precalculated distance from the load. It is generally connected at the nearest resistive point to the load, whose position is found with the aid of a transmission line calculator.

6-4   NONRESONANT TRANSMISSION LINES

A transmission line is terminated in the load to which it feeds its energy. A *nonresonant* (matched) line is one of infinite length or one that is terminated with a resistive load equal to the characteristic impedance of the line. A transmission line is said to be *properly terminated* when the resistance of the load is the exact value required to dissipate energy at the same rate at which the line feeds energy to the load.

A circuit is nonresonant if: (a) the circuit is purely resistive over a continuous range of frequencies, or (b) the circuit contains capacitive and inductive reactances that do not cancel each other.

Resistive networks exhibit no significant reactance over a wide range of frequencies. Also, these circuits do not exhibit resonance. Therefore, resistive networks are nonresonant circuits.

Several examples of nonresonant circuits are band-pass filters, band-reject filters, high- and low-pass filters, piezoelectric crystals, antennas, and transmission lines. In these examples, either the capacitive reactance or the inductive reactance predominates. These reactances cancel only at specific frequencies or bands of frequencies.

The voltage and current waves in a nonresonant line move in phase with one another from the source to the load. These are the traveling waves discussed in Section 6-2. Since they move down the line as a wavefront, all of the instantaneous waves produced by the source travel down the line in the order in which they are produced. At any point along the line, the waveform will be an exact duplicate of the source waveform. The physical length of the nonresonant line is not critical. It can be infinite in length, or it can be terminated in its characteristic impedance. In either case, all of the energy produced by the source will be absorbed by the load impedance.

If we assume a transmission line of infinite length, the electromagnetic wave would travel along the line until all of the energy delivered to the line was dissipated. However, if we also assume that the line has absolutely no losses of any kind, the energy wave moving out from the source along the line would never weaken, and the current at some distant point would be the same as that at the source end. Also, the energy wave moving along such a line would never be reflected, and, therefore, no standing waves would be produced. Combining the facts that the current at some distant point and the current at the source are equal and that no standing waves are produced, we can deduce that the average current at any point along the line is the same as that at any other point. Stated another way, in an infinitely long, no-loss line, an ammeter inserted in either wire anywhere along the length of the line would read the same current as another ammeter inserted at any other point in either line.

Since there are no infinitely long lines, we must adopt a technique that produces the effect of such a line. That technique is quite simple: Ensure that the transmission line feeds into a load that properly matches the line's characteristic impedance. In such a properly terminated line, the line acts like an infinitely long line, and the current will be the same in all parts of the line because there will be no standing waves.

6-5   TRANSMISSION LINE LOSSES _____

There are three major types of losses that commonly occur in practical transmission lines: copper loss, dielectric loss, and radiation loss. Because of the standing waves of current and voltage, the losses in a resonant line are much greater than in a nonresonant line.

Copper Loss

Copper loss is the I^2R power dissipation due to heating that occurs in the pure resistance of the conductors. Because copper loss is directly proportional to the square of current, power dissipation at the current loops is much greater than if there were no current loops on the line—that is, if the line were properly terminated so that no standing waves were present.

In general, copper loss is greater in a line having a low characteristic impedance than in a line having the same resistance but a high characteristic impedance. Why is this so? For the same termination characteristics, lower impedance allows higher current, and, by definition, power dissipation in the line's resistance increases directly with the square of the line current. Less current is required in a high-impedance line to deliver a given amount of power to the antenna. We can state, therefore, that the reduced current in a high-impedance line results in reduced copper loss without causing a reduction in transmitted power.

Skin Effect   Another type of copper loss is called *skin effect*. The movement of electrons through the cross section of a conductor is uniform when a dc current is applied to the conductor. However, when an ac current is applied, the phenomenon known as *self-induction* takes over, retarding the free movement of electrons in the center of the conductor. As the frequency of the applied current is increased, more of the electron flow is on the surface (skin) of the conductor. At frequencies above about 100 MHz, almost all of the electron flow is on the surface of the wire. Skin effect, then, reduces the effective cross-sectional area of a conductor as frequency increases, and, because resistance is inversely proportional to the cross-sectional area, the resistance increases with increased frequency. This increased resistance results in increased power losses as frequency increases.

Crystallization   The copper loss in a line may increase as the line ages. This increased loss occurs in transmission lines subjected to high winds, high temperatures, and moisture. The bending of the line back and forth as it swings in the breeze causes the metal of the conductors to become brittle, and small cracks or breaks appear. The result is known as *crystallization* of the conductors. Crystallization increases resistance in the conductors, which in turn increases copper loss.

Dielectric Loss

Dielectric loss is the I^2R power dissipation due to heating that occurs in the dielectric (insulating material) between the conductors in a transmission line. Dielectric loss is proportional to the voltage across the dielectric, and, for this reason, standing waves of voltage on a line increase it.

Table 6-1 shows dielectric constants for various materials, many of which may be used in making up transmission lines. Note particularly the low dielectric constant values for gases.

The properties of a solid dielectric medium gradually worsen with increasing frequencies. However, for air, dielectric loss is negligible. Therefore, solid dielectric transmission lines should be limited to applications in which dielectric losses can be tolerated, with air dielectric lines being used for high-frequency applications.

Radiation Loss

Radiation loss is the unwanted radiation of energy as radio waves from a transmission line. This loss occurs because a line can act as an antenna if the spacing between the

TABLE 6-1 **Dielectric Constants**

Material	Dielectric Constant (K_e)	Gases	Dielectric Constant (K_e)	Liquids @ 25° C	Dielectric Constant (K_e)
Mica	2.5–7.0	Air	1.000590	Ammonia (liquid)	22
Nylon	3.5	Ammonia	1.0072	Methanol	32.63
Neoprene	4.1	Argon	1.000545	Benlene	2.283
Polyethylene	2.3	Carbon dioxide	1.000985	Carbon tetrachloride	2.24
Polyester	4.1–5.2	Helium	1.0000684	Oil, petroleum	2.13
Polystyrene moulding	2.4–2.5	Hydrogen	1.000764	Oil, transformer	2.24
Rubber, hard	2.8	Neon	1.000127	Oil, turpentine	2.23
Shellac	—	Oxygen	1.000523	Water	78.54
Vinyl chloride	3.5–4.5				
Alumina	4.5–8.4				
Porcelain	6.0–8.0				
Porcelain, zircon	7.1–10.5				
Titonote, (Be, Sr, Ca, Mg, Pb)	15–12,000				
Paraffin	2.0–2.5				

conductors is a relatively large fraction of a wavelength. Radiation losses are, therefore, more common in parallel-wire lines than in coaxial lines. Increased frequencies produce much shorter wavelengths, resulting in increased radiation loss. In a practical sense, at some high frequencies a particular transmission line becomes useless.

SUMMARY

- The primary purpose of transmission lines is to transfer energy from a generator to a load. Because of their high-frequency characteristics, transmission lines are widely used for impedance matching, as measuring devices, and as resonant circuits.

- The two basic types of transmission lines are two-wire and coaxial. The lines may be balanced or unbalanced. In balanced lines, both wires carry RF current, with the current in each wire 180° out of phase with that in the other wire. In unbalanced lines, one wire is at ground potential and the other wire carries RF current.

- Two-wire lines, usually operated as balanced lines, include two-wire open-air, two-wire ribbon (twin-lead), twisted-pair, and shielded-pair lines. The principal uses for two-wire open-air lines are power distribution, telephone, and telegraph lines. These lines may occasionally be used between a transmitter and an antenna or between an antenna and a receiver. Their greatest advantages are simplicity and inexpensive construction. Their disadvantage is lack of shielding, which causes high signal radiation losses and allows easy pickup of noise. The primary use of the twin-lead line is to provide a 300 Ω feeder line between a television receiving antenna and a television receiver. Advantages and disadvantages of the twin-lead line are essentially the same as for the two-wire open-air line. The twisted-pair line is commonly confined to protected short-run, low-frequency use, such as ac power cables. Its primary advantage is flexibility, and its major disadvantage is high-frequency loss in the rubber insulation. The shielded-pair line is ideally suited to balanced line feeds. Its advantages are reduced signal radiation losses and reduced noise pickup. Disadvantages are high dielectric and leakage losses.

- Coaxial lines, usually operated as unbalanced lines, include rigid (air, or concentric) and flexible (solid) lines. The lines consist of an RF energy-carrying inner conductor and a grounded outer shield. The major advantages are that the grounded shield minimizes signal radiation loss and prevents noise pickup from external sources. Disadvantages of the rigid line are that it is expensive to construct; it is difficult to work with because of its rigidity; it must be kept dry to prevent excessive leakage between the two conductors; and its practical length is limited by high-frequency losses over distance. Solid (flexible) coaxial lines are the most commonly used type of transmission lines.

- The terms *long line* and *short line* refer to the electrical length of transmission lines, not the physical length. In general, long lines are λ/4 or longer; short lines are appreciably shorter than λ/4 long. Most dc and low frequency transmission lines are classified as short lines, and most RF lines are classified as long lines. In electrically long lines, electrons may flow in opposite directions at the same instant in the same wire.

- A line's characteristic impedance, also called surge impedance or surge resistance, is a special property of the line itself. It is not affected by line length, signal frequency, or terminating impedance. It is determined by the distributed inductance and the distributed capacitance of the line and is expressed in ohms.

- Whether a device is a source or a load depends on whether it is generating or receiving a signal. A device may be acting as both a source and a load.

- A traveling wave is a wave of energy moving down a line at the speed of light, setting up an oscillatory electron motion in sections of the wires as it moves along.

- A resonant line is one that has a finite length or one that is not properly terminated. It may be terminated in an open, a short, or some impedance other than the line's characteristic impedance. Under each of these terminating conditions, some of the energy is reflected back toward the source, resulting in standing waves on the line. Voltage standing wave ratios are designated VSWR, and current standing wave ratios are designated simply SWR. The reflection coefficient is designated ρ. The physical length of a resonant line is critical. The line's input impedance will be affected by line length, signal frequency variations, and terminating impedance.

- The amplitude of standing waves is determined by the percentage of the total energy input that is reflected. The points at which the reflected waves aid or oppose one another is determined by the speed of the radio waves and the length of the waves. Points of maximum strength (aiding) are loops, and points of minimum strength (opposing) are nodes. In an open-ended, no-loss line, all of the energy is reflected. This reflection creates minimum current, maximum voltage, and maximum impedance at the output end of the line. In a shorted line, the output end has maximum current, minimum voltage, and minimum impedance.

- A nonresonant line is one of infinite length or one that is properly terminated in a resistive load equal to the characteristic impedance of the line. The physical length of a nonresonant line is not critical. A properly terminated line is one in which all of the energy produced by the source will be absorbed by the load impedance. In this case, the line acts like an infinitely long line, and the current will be the same in all parts of the line because there will be no reflected waves.

- A reflectometer can be used to measure SWR in a coaxial cable. The same device can be used as an RF wattmeter if calibrated in watts and if some means of indicating a reference level is provided. Another method for measuring SWR is the slotted line.

- Three major types of transmission loss are copper, dielectric, and radiation. Losses are greater in resonant lines than in nonresonant lines. Skin effect is a type of copper loss in which self-inductance retards free electron movement in the center of the conductor. Crystallization is caused in a line by aging, high winds, high temperatures, and moisture, resulting in higher conductor resistance and increased copper loss.

 QUESTIONS AND PROBLEMS

1. State the purpose of a transmission line.
2. Transmission lines are widely used for what purposes other than transfer of signals? Why?
3. What are the two basic types of transmission lines? List the subtypes for each.
4. With respect to ground, how may transmission lines be operated?
5. Define balanced and unbalanced lines.
6. Describe the construction, list the advantages and disadvantages, and state the uses for each of the following transmission lines: (a) two-wire open-air, (b) two-wire ribbon, (c) twisted-pair, (d) shielded-pair, (e) rigid coaxial, (f) solid coaxial.
7. Which of the lines in Problem 6 is the most commonly used? Why?
8. Define long lines and short lines in terms of electrical length.
9. Calculate the wavelength for lines 1 km long operated at (a) 1 kHz, (b) 50 kHz, and (c) 7.5 MHz.
10. State whether each line in Problem 9 is long or short. Explain your answers.
11. Explain why RF lines are normally considered long lines. How must they be treated?
12. Explain the concepts of source and load.
13. Explain the concept of traveling waves.

14. Define nonresonant line.
15. What is meant by a properly terminated line?
16. An ammeter inserted 1000 feet from the source end of an infinitely long, no-loss line measures 5 A. Another ammeter inserted 10 miles from the source end would measure how much current? Explain your answer.
17. Define resonant line.
18. Explain how standing waves are formed.
19. Calculate SWR and ρ for a transmission line where $I_{max} = 10$ A and $I_{min} = 2$ A.
20. Calculate VSWR for a transmission line where $V_{max} = 80$ V and $V_{min} = 10$ V.
21. Define (a) loop and (b) node.
22. What device is used to measure SWR on a coaxial cable? Describe how it is constructed.
23. Draw a ladder diagram showing the distribution relationships for current, voltage, and impedance along (a) an open-ended transmission line of finite length and (b) a shorted line.
24. Define (a) initial wave and (b) reflected wave.
25. Define characteristic impedance.
26. Calculate the characteristic impedance for a 75 foot section of RG-59B/U coaxial cable.

27. Calculate the characteristic impedance for a 1000 foot section of RG-11A/U coaxial cable, where $C = 20.5$ pF/ft and $L = 115.31$ nH/ft.

28. Could the coaxial cables from Problems 26 and 27 be interchanged without appreciable effect on the quality of transmission? Why?

29. Calculate the characteristic impedance for a two-wire open-air line made of two parallel lengths of No. 12 wire spaced 100 mm between centers. The diameter of No. 12 wire is 2.05 mm.

30. Calculate the characteristic impedance for a concentric coaxial line made of an outside copper tubing 1.4 mm thick with an outside diameter of 20 mm. The inner conductor tubing is 0.6 mm thick with an outside diameter of 4 mm.

31. List and describe the three major types of losses for transmission lines. Include any subtypes for each.

CHAPTER 7

Radio Wave Propagation

OBJECTIVES

When you complete this chapter, you should be able to:

- Describe the characteristics of radio waves.
- Name the three broad classifications of radio wave propagation and describe each.
- Find the space wave radio horizon when given antenna height.
- When given the height of a transmitting antenna, determine the required receiving antenna height to achieve a given communications travel distance.
- Name the various layers of the ionosphere.
- List the variations of the ionosphere and their causes.
- Define and explain each of the following: critical frequency, critical angle, skip distance, skip zone, and multiple-hop transmission.
- Define maximum usable frequency, lowest usable frequency, and optimum usable frequency.
- State the primary cause of fading.
- List and explain the two methods for extending the range of high-frequency communications.

INTRODUCTION

Of the many technical subjects communications technicians are expected to know, probably the one that is least susceptible to change is the theory of radio wave propagation. The basic principles that enable radio waves to be propagated through space are the same today as they were 100 years ago. A thorough understanding of these principles is a relatively simple task, although some people view radio wave propagation as something complex and confusing, undoubtedly because it is an invisible force that the senses of sight and touch cannot detect. It is left up to individual imagination to visualize the sequence of concepts and their relationship to practical application.

The advent of the Marconi wireless in 1895 made possible transmission of messages in the form of radio waves. You have already gained a basic knowledge of the generation, transmission, and reception of radio waves. You learned that the signal is fed to an antenna, which converts that signal to electromagnetic energy. The electromagnetic energy radiated from a transmitting antenna travels (radiates) outward in many directions. As the distance from the antenna increases, the energy field spreads out and the field strength decreases. The path, known as wave propagation, by which the signal reaches the receiving location also affects the field strength. There are three broad classifications for the signal path, or wave propagation: the ground wave, the space wave, and the sky wave. In this chapter, we examine each of these wave propagation methods and the effects various atmospheric conditions have on them.

7-1   RADIO WAVE CHARACTERISTICS

The radiation concept of radio waves can be visualized by dropping a pebble into a pool of water. As the pebble enters the water, a surface disturbance is created, causing the water to move up and down. From this point, the disturbance is transmitted on the surface of the pool in the form of expanding circles of waves. It should be noted that the water is not moving away from that point, for if a leaf or small stick were placed on the surface of the pool, it would have no sideways movement, but merely an up and down motion as each wave passes under it. The type of wave produced by the water is called a *transverse* wave—that is, the wave occurring in a direction or directions perpendicular to the direction of wave propagation. The wave is more simply called a traveling wave. The electromagnetic waves radiated by a transmitting antenna are examples of transverse waves.

The basic shape of the carrier wave generated by a transmitter is that of a sine wave. The transverse wave that is radiated out into space, however, may or may not retain the characteristics of a sine wave, depending on the type of modulation of the carrier. For our purposes, a knowledge of the characteristics of the basic sine wave is sufficient to understand the subjects of this chapter without getting into the complex variations involved in transverse waves. Those characteristics of a sine wave that are considered important to our discussion are covered in this section.

Cycle

The sine wave shown in Figure 7-1 is just like that used to depict the alternating current (ac) found in the home. The dotted line represents zero current and is used as a reference line. From point A, the cycle of current alternately rises above the reference line to a maximum value then falls below the reference line to another maximum value, then returns to zero at point C. That portion above the reference line (between points A and B) is called the *positive alternation,* and the portion below the reference line (between points B and C) is called the *negative alternation.* Thus, one cycle of a sine wave is the combination of one positive and one negative alternation (points A to C). At point C, the wave begins to repeat itself, with a second cycle completed at point D.

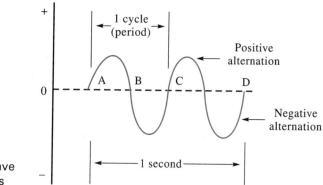

FIGURE 7-1
Basic sine wave
characteristics

Frequency

The *frequency (f)* of the wave is the number of cycles of a sine wave completed in one second. In the case of moving waves, such as radio waves, the frequency can be thought of as the number of cycles of the wave that pass a given point in one second. For example, Figure 7-1 shows two cycles occurring in one second; hence, that sine wave is said to have a frequency of two cycles per second (2cps).

In 1967, in honor of the German physicist Heinrich Hertz, the term *Hertz* was designated for use in lieu of the term 'cycles per second' when referring to the frequency of radio waves. It may seem confusing that in one place the term 'cycle' is used to designate the positive and negative alternations of a wave, but in another instance the term 'Hertz' is used to designate what appears to be the same thing. The key is the time factor; cycle refers to any sequence of events, such as the positive and negative alternations comprising one cycle of electric current; Hertz refers to the number of occurrences that take place in one second.

Hertz is abbreviated Hz, thousands of Hertz (kiloHertz) as kHz, millions of Hertz (megaHertz) as MHz, and billions of Hertz (gigaHertz) as GHz. Currently, the usable frequency range extends from approximately 15 Hz to about 300 GHz.

Audio Frequencies Audio frequencies (AF) are in the frequency range between about 15 Hz and 20 kHz. These frequencies are audible to the human ear and include all those sounds heard during everyday routine. As an example, the average speaking voice has an audible frequency of about 128 Hz; however, the singing voice of a high soprano may go as high as 1300 Hz. Extremely high-pitched sounds of some musical instruments and whistles approach, and may even exceed, the upper limits of the AF range.

Radio Frequencies The frequencies falling between 3 kHz and 300 GHz are called radio frequencies (RF) since they are commonly used in radio communication. Table 7-1 (Section 7-3) shows how the RF spectrum is divided into eight frequency bands, each of which is ten times higher in frequency as the one immediately below it. This arrangement serves as a convenient way to remember the range of each band.

Harmonics

Any frequency that is a whole number multiple of a basic frequency is known as a *harmonic* of that basic frequency. The basic frequency itself is called the first harmonic or, more commonly, the *fundamental frequency*. A frequency that is twice as great as the fundamental frequency is called the second harmonic; a frequency three times as great is the third harmonic, and so on. For example, the third harmonic of a 3 MHz fundamental frequency is 9 MHz; the fifth harmonic is 15 MHz.

Radio operators sometimes encounter signals being radiated simultaneously on both the fundamental frequency and one of its harmonics. Usually the harmonic is an unintentional radiation resulting from a transmitter malfunction. It is also the quality of harmonics that the radiated energy level of successive harmonics is progressively decreased—that is, the fundamental frequency's energy level is higher in value than the second harmonic, and so on.

Period

The *period (p)* of a radio wave is simply the amount of time required for the completion of one full cycle of its frequency. For example, the sine wave in Figure 7-1 has a frequency of 2 Hz; therefore, each cycle has a duration, or period, of one-half second. If the frequency was 10 Hz, the period of each cycle would be one-tenth second. Since the frequency of a radio wave is the number of cycles that are completed in one second, you can see that the higher the frequency is, the shorter the period will be.

Wavelength

A *wavelength* (symbolized by the Greek lambda, λ) is the space occupied by one full cycle of a radio wave at any given instant. If, for example, a radio wave could be frozen in place and measured, its wavelength would be the distance from any one point on a cycle to the corresponding point on the next cycle. This concept is illustrated in Figure 7-2.

Wavelengths vary from a few hundredths of an inch at the extremely high frequencies to many miles at the very low frequencies. In general practice, wavelengths are expressed in meters (one meter is equal to 3.28 feet). A thorough understanding of frequency and wavelength is necessary when considering the correct types of antennas to be used for successful communications.

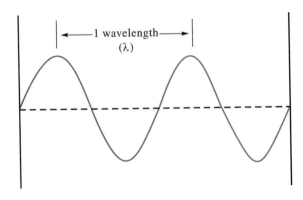

FIGURE 7-2
Concept of a
wavelength

Velocity

The *velocity* of a radio wave that is radiated into space by a transmitting antenna is simply that speed at which the wave travels. Radio waves travel in free space at about the speed of light, 186,000 miles per second (300,000,000 meters per second). Radio waves traveling inside the earth's atmosphere travel at a slightly lesser speed due to various factors such as barometric pressure, humidity, molecular content, and so on. Normally, when discussing the velocity of radio waves, we are referring to the free space velocity.

The frequency of a radio wave has nothing to do with the wave's velocity. A 5 MHz wave travels through space at the same speed as a 10 MHz wave. The velocity of radio waves is an important factor in making wavelength-to-frequency conversions.

Wavelength-to-Frequency Conversions   Radio waves are often referred to by their wavelength in meters instead of by frequency. At one time or another while listening to a commercial radio station you may have heard an expression similar to the following: "Station WXYZ operating on 240 meters ..." If you wish to tune receiving equipment that is calibrated in frequency to such a station, you must first convert the designated wavelength to its equivalent frequency.

As previously stated, a radio wave travels (in free space) 300,000,000 meters in one second; therefore, a radio wave of 1 Hz would have traveled a distance (or wavelength) of 300,000,000 meters. By doubling the frequency of the wave to 2 Hz per second the wavelength would be cut in half, or to 150,000,000 meters. This illustrates the principle that *the higher the frequency, the shorter the wavelength*. Thus, both wavelength and frequency are reciprocals and either one divided into the velocity of a radio wave will yield the other, as illustrated by the following equations:

For conversion to wavelength (in meters):

$$300,000,000/f \text{ (in Hz)} \qquad\qquad (7\text{-}1a)$$

or

$$300,000/f \text{ (in kHz)} \qquad\qquad (7\text{-}1b)$$

or

$$300/f \text{ (in MHz)}$$

For conversion to frequency:

$$300,000,000/\lambda \text{ (in meters)} = f \text{ (in Hz)} \qquad\qquad (7\text{-}2a)$$

$$300,000/\lambda \text{ (in meters)} = f \text{ (in kHz)} \qquad\qquad (7\text{-}2b)$$

$$300/\lambda \text{ (in meters)} = f \text{ (in MHz)} \qquad\qquad (7\text{-}2c)$$

To convert wavelength into feet, simply divide the meters by 3.28.

Electromagnetic Radiation

The radio wave radiated into space by the transmitting antenna is a very complex form of energy containing both electric and magnetic fields, hence the term *electromagnetic radiation*. A moving electric field always creates a magnetic field, and a moving magnetic field always creates an electric field. They are inseparable. The lines of force of these fields are perpendicular to each other, as shown in Figure 7-3.

In any technical discussion of radio waves, you will frequently encounter the interchangeable usage of the terms radio wave and RF energy. Just remember that both are one and the same.

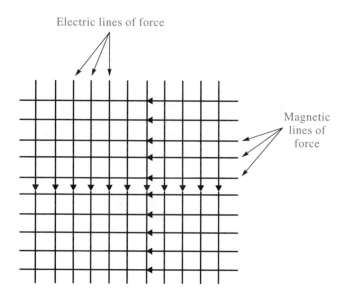

FIGURE 7-3
Electromagnetic
lines of force

Wave Polarization

Polarization of the radio wavefront is an important consideration in the efficient transmission and reception of radio signals. The polarization of a radio wave is determined by the direction of the electric field of the wave with respect to earth. If the electric field of the wave is vertical to the earth, as shown in Figure 7-4(A), the wave is *vertically polarized*. If the electric field is horizontal to the earth as in Figure 7-4(B), the wave is *horizontally polarized*. As shown in Figure 7-4, the position of the transmitting antenna determines whether the wave will be vertically or horizontally polarized. The polarization of the wave may be altered somewhat (by atmospheric conditions, hills, buildings, and so on) during its travel to the receiving site. However, regardless of the wave's position with respect to earth, the electric and magnetic fields will always be perpendicular to each other.

7-2 PROPAGATION CHARACTERISTICS

When an electromagnetic wave is sent out from an antenna, part of the radiated energy travels along or near the surface of the earth, and another part travels from the antenna upward into space. The energy that stays close to the surface of the earth is called the ground wave. The energy that travels upward into space is called the sky wave. Energy that travels directly from the transmitting antenna to the receiving antenna, without following the surface of the earth or radiating toward the sky, is called the space wave.

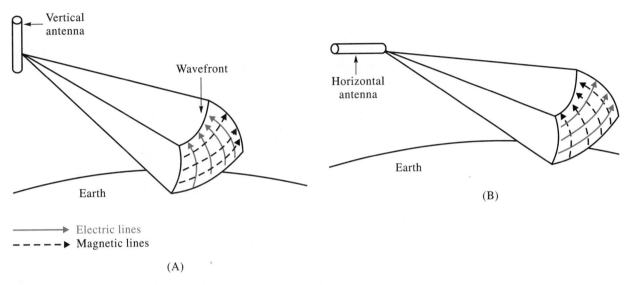

Electric lines
Magnetic lines

(A)

FIGURE 7-4 Wave polarization (A) Vertically polarized wave (B) Horizontally polarized wave

Ground Waves

As just stated, the *ground wave* is a radio wave that travels along the surface of the earth. Ground waves are the primary mode of propagation in the LF and MF bands. The longer wavelengths in these bands tend to follow the curvature of the earth and actually travel beyond the horizon, as shown in Figure 7-5. However, as the frequency increases, the ground wave is more effectively absorbed by the irregularities on the earth's surface, because hills, mountains, trees, and buildings become significant relative to the transmitted wavelength. For example, at 30 kHz, the wavelength is 10 km (about 6.2 miles), and even mountains are relatively insignificant compared to this value. But, at 3 MHz, the wavelength is only about 100 m, short enough that hills, trees, and large buildings break up and absorb the ground wave.

One way of greatly improving ground wave coverage is to use vertical polarization. When horizontal polarization is used, the electric field is parallel to the earth's surface, and any ground wave is effectively short-circuited by the conductivity of the earth. However, since ground wave propagation is limited to the LF and MF bands, constructing full-size quarter-wave antennas does present problems. For this reason and others, there are few communications services in the LF band. Most ground wave communications are in the MF band where antenna sizes are more practical.

FIGURE 7-5
Ground wave
propagation

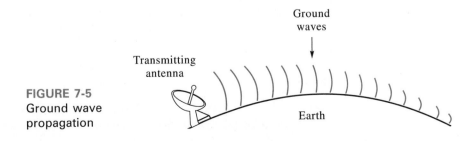

Space Waves

When the transmitted signal is increased above 4 or 5 MHz, the usable ground wave signal is limited to a few miles. Therefore, at these frequencies and above, signals can be transmitted farther using the *space* or *direct wave*. Space wave propagation is used primarily in the VHF, UHF, and higher-frequency bands. Figure 7-6 illustrates the concept of space waves.

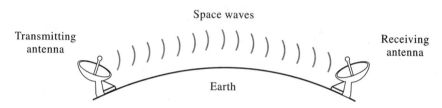

FIGURE 7-6   Space wave propagation

In general, space wave propagation is limited to line-of-sight distances. Most of the energy in radio waves at frequencies above 30 MHz moves through space in straight lines like light waves, although some of the energy at these higher frequencies follows the curvature of the earth, and good VHF reception is often obtained considerably beyond line-of-sight distances. At still higher frequencies, such as those in the UHF and SHF ranges, normal transmissions from conventional antennas are limited to line-of-sight.

When the transmitting and receiving antennas are within sight of each other, the signal is considered to be a space wave. It follows, then, that to increase the range of space wave propagation, you need only to increase the height of either or both the transmitting and receiving antennas. Examples of increased antenna height are satellites and aircraft antennas. In these circumstances, propagation ranges are greatly extended.

Radio Horizon

In space wave transmission, there is a phenomenon called the *radio horizon*, the distance of which is about one third greater than that of the optical horizon. This phenomenon is caused by refraction in the earth's lower atmosphere. The refraction occurs because the density of the earth's atmosphere decreases linearly as height increases. As a result, the top of the wave travels slightly faster than the bottom of the wave, effectively bending the wave slightly downward, where it follows the curvature of the earth beyond the optical horizon.

The radio horizon for both transmitting and receiving antennas can be calculated by the following equation:

$$D_t = 4\sqrt{H_t} \text{ or } D_r = 4\sqrt{H_r} \tag{7-3}$$

where

D_t and D_r = radio horizon distance in kilometers

H_t (H_r) = height of transmitting (receiving) antenna in meters

When measured in miles, substitute the following:

$$D = \sqrt{2H} \tag{7-4}$$

where

D = horizon distance in miles

H = height of antennas in feet

The maximum space wave communications distance is the sum of the numbers obtained by using Equation 7-3 for both antennas.

$$D_{max} = 4\sqrt{H_t} + 4\sqrt{H_r} \tag{7-5a}$$

or

$$D_{max} = D_t + D_r \tag{7-5b}$$

EXAMPLE 7-1

Find the radio horizon distance of a transmitting antenna with a height of 80 m.

SOLUTION

From Equation 7-3,

$$D_t = 4\sqrt{H_t} = 4\sqrt{80}$$

$$= 35.8 \text{ km}$$

EXAMPLE 7-2

Find the radio horizon distance of a receiving antenna with a height of 40 m.

SOLUTION

From Equation 7-3,

$$D_r = 4\sqrt{H_r} = 4\sqrt{40}$$

$$= 25.3 \text{ km}$$

EXAMPLE 7-3

Using the antennas from Examples 7-1 and 7-2, find the maximum space wave communications distance for signals.

SOLUTION

From Equation 7-5b,

$$D_{max} = D_t + D_r = 35.8 + 25.3$$

$$= 61.1 \text{ km}$$

EXAMPLE 7-4

A receiver is located 80 km from the transmitter. The transmitting antenna is 100 m high. Find the required height of the receiving antenna.

SOLUTION

Algebraic manipulation of Equation 7-5a gives

$$H_r = \left(\frac{D_{\max} - 4\sqrt{H_t}}{4} \right)^2$$

$$= \left(\frac{80 - 4\sqrt{100}}{4} \right)^2$$

$$= 100 \text{ m}$$

EXAMPLE 7-5

Find (a) the radio horizon distance for a transmitting antenna with a height of 300 feet; (b) the radio horizon distance of a receiving antenna with a height of 140 feet; and (c) the maximum space wave communications distance for the signals on these antennas.

SOLUTION

From Equation 7-4,

(a) $D_t = \sqrt{2(300)} = \sqrt{600} = 24.5$ miles

(b) $D_r = \sqrt{2(140)} = \sqrt{280} = 16.73$ miles

(c) $D_{\max} = D_t + D_r = 24.5 + 16.73 = 41.23$ miles

Sky Waves

Ionization is the loss or gain of electrons by an atom. There are different degrees of ionization, forming several recognizable *ionized* layers, known collectively as the *ionosphere*. These ionized layers of the atmosphere between 30 and 250 miles above the surface of the earth make *sky wave* reception possible. All of the energy radiated upward in the sky wave would be wasted, so far as earthbound radio communication is concerned, if it continued to travel in a straight line off into space. However, at certain frequencies and radiation angles, the ionosphere reflects radio waves much the same way a mirror reflects light. Also, radio waves at other frequencies and angles are re- fracted (bent) in such a manner that they return to earth. This concept is illustrated in Figure 7-7.

The amount of refraction depends on three factors: (1) the frequency of the wave, (2) the density of the ionized layer, and (3) the angle at which the wave enters the ionosphere. When conditions are just right, the wave will be refracted enough to return to earth. The refracted radio waves may return to earth very far from the point where they originated. Long distance communications often makes use of this characteristic

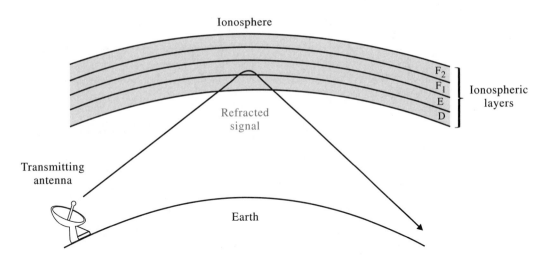

FIGURE 7-7 Sky wave propagation

by using carrier frequencies in the MF and HF bands, because waves radiated at these frequencies can be refracted back to earth with comparative ease. Almost all HF propagation, and nighttime long distance MF propagation, is by sky wave. However, waves at frequencies above 30 MHz are much more likely to penetrate the ionosphere and continue moving out into space. In the UHF and SHF bands, a very small percentage of the wave's energy is refracted back to earth.

7-3 ATMOSPHERIC CONDITIONS AND COMMUNICATIONS

Atmospheric conditions are in a continuous state of change. Variations may occur hourly, daily, monthly, seasonally, yearly, and even from decade to decade. Among the various conditions affecting communications are layers in the ionosphere, the aurora borealis, and sunspots. Some undesirable results of these phenomena are signal absorption, signal dispersion, and signal fading. However, under certain conditions, a greatly increased communications range and reliability may result. Atmospheric conditions have their greatest effect on the ionosphere, which primarily affects the sky wave.

The Ionosphere

The behavior of the sky wave is a result of the ionosphere. A graphic illustration of the designations of the ionospheric layers and their approximate altitudes is shown in Figure 7-8.

The D layer, 30–60 miles above the earth, is the lowest layer and exists only in the daytime. Since this layer is the farthest from the sun, its ionization is relatively weak; consequently, it does not affect the travel direction of radio waves. However, the D layer does absorb energy from the electromagnetic wave and, when present, attenuates the sky wave in both directions—on the way up to the other layers and on the way back down. Signals in the MF band are completely absorbed by the D layer, thereby limiting

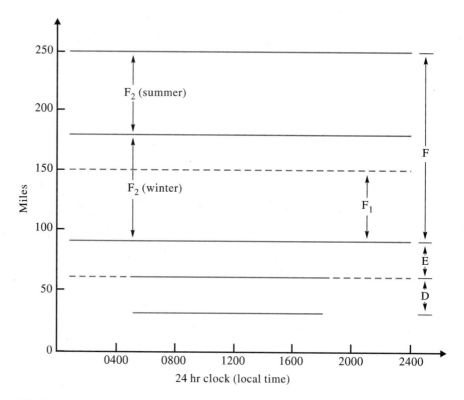

FIGURE 7-8 Ionization layers

these signals to ground wave propagation during daylight hours. At night, however, the D layer disappears, and long distance MF transmissions via sky wave propagation are possible.

The E layer, 60–90 miles above the earth's surface, reaches its maximum density at noon. Its ionization is so weak at night that the layer may disappear.

The F layer, 90–250 miles above the earth's surface, splits into F_1 and F_2 in the daytime, with F_2 varying from summer to winter. The F_1 layer ranges from about 90 to 150 miles above the surface of the earth. The F_2 layer is the closest to the sun and ranges in height from about 90 to 180 miles on a winter day, and from about 180 to 250 miles on a summer day.

The different layers of ionization are the result of the different gases that make up the earth's atmosphere. These gases ionize at different pressures, and they are affected differently by ultraviolet radiation and cosmic ray bombardment from the sun. Since the major influence on the ionosphere is radiation from the sun, the number of layers present, their heights above the earth, and the amount they refract the sky wave all vary over time. The hourly and seasonal variations shown in Figure 7-8 are the result of solar radiation which increases the density of the ionospheric layers and lowers their effective heights.

The *aurora borealis*, more popularly called the *northern* (or *southern*) *lights*, is a luminous atmospheric phenomenon ascribed to electricity, that occurs near, or radiates from the earth's northern or southern magnetic pole and is visible at different times and over varying portions of the earth's surface. Because of the ionic turbulence produced, signals reflected from the aurora borealis have a rapid fluttering sound. For that reason, during strong aurora borealis activity, voice communications are poor, and continuous wave (CW) is the best means of communicating.

Changes in the ionospheric layers can also result from *sudden ionospheric disturbances* and *ionospheric storms*, which are believed to be caused by solar flares. These ionospheric disturbances weaken signal levels in the HF band, either suddenly or gradually, and may cause signals to disappear completely. Conversely, some of these ionospheric disturbances can produce long distance communications contacts in the VHF band. One such disturbance is known as *sporadic E-layer ionizations*, caused when high-density ionization occurs in the E layer. The usual result is ionospheric refraction of signals high into the VHF band, thus allowing extremely long distance communications in this normally line-of-sight frequency band.

The sudden ionospheric disturbance usually starts suddenly and lasts from about 15 minutes to two hours. The ionospheric storm may develop suddenly or gradually and may last from one day to one week. When these disturbances are at their peak, ionospheric absorption is high enough to cause a radio blackout, especially at higher frequencies.

Ionospheric Variations

The existence of the ionosphere depends on solar radiation; therefore, any variations in that radiation must influence the ionosphere. The amount of solar radiation reaching the earth and thus influencing the ionosphere is determined, in part, by the earth's rotation and its revolution around the sun. The variations of the earth, sun, and ionosphere occur regularly and are predictable. They are divided into four categories:

1. *Diurnal* variations—the hour-to-hour changes in the various ionospheric layers caused by the rotation of the earth around its axis.
2. *Seasonal* variations—caused by the constantly changing position of any point on earth relative to the sun as the earth orbits the sun.
3. *Geographical* variations—caused by the varying intensity of the solar radiation striking the ionosphere at different latitudes.
4. *Cyclical* variations—caused by sunspot activity over an 11-year cycle.

The 11-year sunspot cycle has the most influence on the ionosphere. The greater the number of sunspots, the greater the intensity of solar radiation. The critical frequency during a sunspot maximum, which occurs every 11 years, is about twice that of a sunspot minimum and results in greatly increased communications range and reliability during years of high sunspot activity.

Time of day may also affect transmissions. Ground waves remain the same during the day and the night, as do space waves. However, there can be radical changes in the sky wave during these periods. At nightfall, the sun can no longer ionize the atmosphere above the darkened part of the earth, producing thinner ionized layers in these parts of the world. The thinly ionized layers refract sky waves back to earth over a wider area; thus, sky waves generally return to earth farther away at night than during the day. However, different frequency bands can be affected differently.

In the VLF and LF bands, there is little difference between day and night transmissions because signals are received primarily by ground wave. In general, however, at distances of several thousand miles the signals at night will be stronger.

Beyond the ground wave range, signals in the MF band improve significantly at night. The distance these signals travel may increase from hundreds of miles in the daytime to thousands of miles at night.

During the daytime, frequencies at the lower end of the HF band may return to earth 20–500 miles away, whereas at night, these same frequencies may be returned to earth 200 to many thousands of miles away. Frequencies at the high end of the band may return to earth 200–5000 miles away during the day, but at night they may penetrate the ionosphere and not return to earth at all. For these reasons, to maintain signals at usable strengths, HF communications systems may have to shift from one frequency during the day to another during the night.

Frequencies at the lower end of the VHF band may sometimes be refracted during the day, but rarely at night. Although unreliable for long distance communications, these lower-end frequencies are useful for communications up to about 50 miles. Frequencies above 100 MHz are rarely refracted by the ionosphere. The UFH, SHF, and EHF bands are used primarily for line-of-sight communications both day and night, so there is rarely any change in travel distance between day and night.

Table 7-1 shows the frequency bands and the major systems they service. Note the overlapping services in some of the bands.

TABLE 7-1 Frequency Bands and Major Services

Band	Range	Major Services
VLF LF	10–30 kHz 30–300 kHz	Radio navigation; time and frequency broadcasts; maritime mobile communications; aeronautical communications
MF	300 kHz–3 MHz	AM broadcasting; amateur communications; time and frequency broadcasts; fixed and mobile communications; maritime and aeronautical aids and communications
HF	3–30 MHz	Shortwave broadcasting; time and frequency broadcasts; point-to-point communication; amateur communications; land, maritime, and aeronautical communications
VHF	30–300 MHz	Land and aeronautical mobile communications; industrial and amateur communications; FM and TV broadcasting; space and meteorological communications; radio navigation
UHF	300 MHz–3 GHz*	TV broadcasting; aeronautical and land mobile communications; radioastronomy; telemetry; satellite communications; amateur communications
SHF EHF	3–30 GHz* 30–300 GHz*	Microwave relay; satellite and exploratory communications; amateur communications

*Frequencies above about 900 MHz are considered to be microwaves.

Critical Frequency and Critical Angle

The highest frequency returned to earth when radiated upward in the vertical direction is called the *critical frequency*. The value of the critical frequency depends on the condition of the ionosphere. Since ionospheric conditions change from hour to hour, day to day, month to month, season to season, and year to year, the critical frequency also changes constantly.

Lowering the radiation angle from the exact vertical direction allows the wave to travel longer through the ionized layer, causing a greater degree of refraction. In practical terms, this means that, with one limitation, signal frequencies higher than the critical frequency can be refracted back to earth. The limitation is that, for any given frequency, there is a *critical angle* above which the signal will not be refracted enough to return to earth. Figure 7-9 shows how signals above the critical angle penetrate the ionospheric layer while signals below the critical angle return to earth.

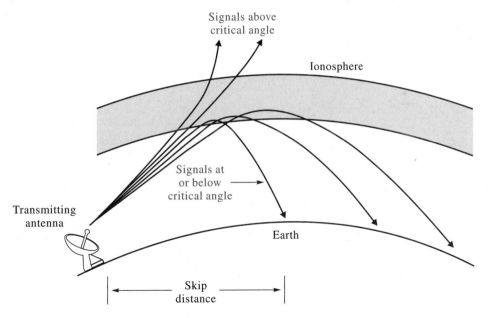

FIGURE 7-9 Effects of lowering the radiation angle

Multiple-Hop Transmission

In Figure 7-9, note that as the radiation angle decreases, the distance the signal wave travels over the earth increases. This distance is known as the *skip distance* and is an important consideration in long distance communications. At high frequencies, there may be long distances between the end of the usable ground wave signal and the reappearance of the reflected sky wave, as illustrated in Figure 7-10. This is a no-signal zone known as the *skip zone*.

The skip distance can be maximized by using the lowest radiation angle possible and the highest frequency that will be refracted at that angle. If the sky wave returns to earth and strikes a good conducting surface such as salt water, it can be reflected back into the ionosphere and take a double hop. If the signal strength is sufficient, the

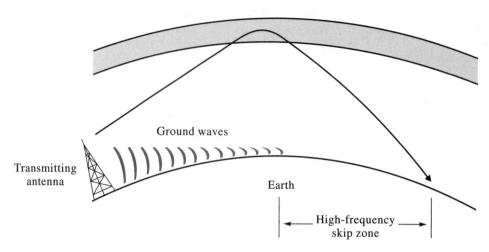

FIGURE 7-10 Skip zone

wave will be reflected again and take another hop, as shown in Figure 7-11. This concept is known as *multiple-hop transmission* and is used extensively for long-distance communications.

Under the best conditions, the maximum distance of a single hop is about 2000 miles. Limiting factors are the frequency used and the radiation angle, and since the radiation angle cannot be reduced below the horizon, multiple-hop transmission is used. Attenuation of the signal from the ionosphere and the ground reflection points are the primary determining factors for the distance that multiple-hop sky wave signals can travel.

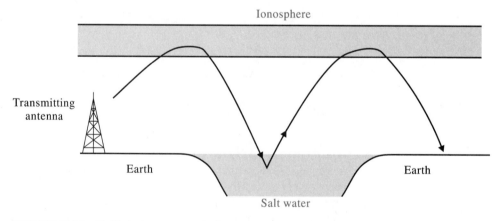

FIGURE 7-11 Multiple-hop transmission

Usable Frequencies

The *maximum usable frequency* (MUF) is the highest frequency that can be used for transmission between two points via sky wave—that is, the highest frequency at which a signal radiated into the ionosphere will be refracted back to earth with usable strength. The MUF varies between 8 MHz and 30 MHz with time of day, distance, direction, season, and solar activity.

A frequency lower than the MUF can actually be used, because it will also be refracted. However, as the signal frequency decreases, the signal energy absorbed by the ionosphere increases, drastically reducing the usable signal strength. Below a certain frequency, sometimes called the *lowest usable frequency* (LUF), the RF signal will be totally absorbed by the ionosphere. Therefore, operating at the MUF will produce the maximum received signal strength.

The MUF changes constantly with atmospheric conditions, and the LUF is borderline for refraction. Therefore, to obtain the most reliable sky wave propagation, the *optimum usable frequency* (OUF) is used. The OUF is well above the LUF but far enough below the MUF that the signal is not appreciably affected by the constant atmospheric changes, thus providing relatively reliable sky wave communications.

Because the usable frequencies are constantly changing, charts are available that predict them for every hour of the day, over any path on the earth, and for any month of the year. The predictions on these charts are based on extensive solar observations and can be used to optimize sky wave communications.

Fading

When sky wave propagation is used, variations in the ionosphere may refract more or less energy to the receiving point at different instants, causing periodic increases and decreases in the received signal strength. If the changes in signal strength are small, the receiver's automatic gain control (AGC) may compensate. On the other hand, if the changes are large, the signal may be lost completely. This variation in signal strength is called *fading* and may be the result of multipath reception.

When the receiving antenna is within range of both the ground wave and the sky wave, as shown in Figure 7-12(A), the two signals may arrive at the receiver in phase or out of phase. If the signal paths vary, alternate cancellation and reinforcement will occur because the signal received at any instant is the phasor sum of all waves received. When arriving in phase, the waves add to each other, producing a strong signal. When arriving 180° out of phase, the waves tend to cancel each other. Changing ionospheric conditions can change the sky wave travel distance and therefore the phase relationship of the ground and sky waves. This condition occurs primarily in the MF band.

If the received signal arrives via both single- and double-hop sky waves paths, as shown in Figure 7-12(B), the lengths of the paths, the signal strength, and the phasing of the signals will vary with ionospheric conditions. The result is the same as for combined ground and sky wave reception; that is, fading occurs.

In the VHF and UHF bands, the MUF seldom rises above 30 MHz except during maximum sunspot activity, when it may reach as high as 50–60 MHz. Therefore, the space or direct wave is used, limiting communication in these bands to line-of-sight paths. Limiting factors for line-of-sight transmissions are local terrain, curvature of the earth, and buildings or overflying airplanes. For example, an airplane flying overhead can act as a reflector of the direct wave. Any receiving antenna below the airplane will receive the direct wave from the transmitter plus the reflected wave from the airplane. The transmitter–reflector–receiver distance changes constantly because the airplane is moving. Therefore, the relationship between the direct and reflected waves at the

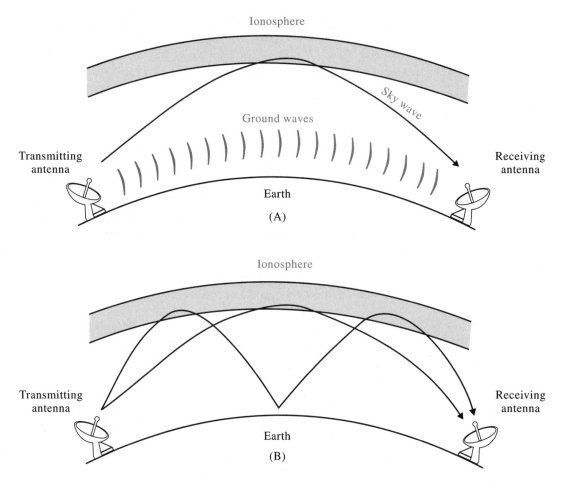

FIGURE 7-12   Multipath reception (A) Ground plus sky wave reception (B) Multipath sky wave reception

receiver is alternately in phase and out of phase, producing a varying fading signal. This phenomenon is particularly troublesome for television viewers.

Fading signals have always been a problem in long distance communications. In amplitude modulation, the AM signals fade up and down in strength, and the carrier may fade at different rates and times in either sideband, the combination of which produces varying signal strengths and a characteristic rolling distortion. The result of the carrier and sidebands fading at different times may be voice transmissions that are sometimes completely unintelligible. This fading can be eliminated by using single-sideband transmission—that is, by balancing out the carrier at the transmitter, filtering out one sideband, and transmitting only the remaining sideband.

Amplitude variations can also cause fading in mobile communications because of the shielding effect of buildings, hills, or heavily wooded areas, or because of reflected or multipath signals. Multipath signals can also create problems in frequency modulation. When an FM signal is reflected, it may undergo a 180° phase reversal. If the reflected path is a number of full wavelengths longer than the direct path, the signals at the receiving antenna are out of phase. The strength of the signal at the antenna then depends on the relative strengths of the two signals, and total signal cancellation can occur if the signals are exactly 180° out of phase and of equal strength.

If the difference in reflected and direct path lengths is an odd number of half-wavelengths, the signals will be in phase and will produce a higher signal strength at the receiver than if only the direct signal were being received. However, the difference in path lengths will most often be other than half-wavelengths or full wavelengths. In those instances, and because the carrier frequency of the FM signal is constantly changing, there will be constant changes in the instantaneous phase and frequency differences in direct and reflected path signals. The result is a signal at the receiver containing both phase modulated and amplitude modulated components that were not in the original transmitted signal.

Several methods can be used to reduce multipath fading of FM signals to an acceptable level, or to eliminate it. One method is to install a highly directive receiving antenna with a good front-to-back ratio, a sharp frontal lobe, and a minimum of spurious side and back lobes, then use an antenna rotator to aim the antenna at an interference-free direct signal.

One way to minimize or eliminate reflected signals arriving at the receiving antenna from 20° to 30° at either side of the direct signal is to *vertically stack* two identical highly directive high-gain antennas $\lambda/2$ apart. This method sharpens the forward lobe of the antenna and increases the direct signal pickup by 2–3 dB.

Yet another method for multipath fading minimization is *stagger stacking*. In stagger stacking, the antennas are spaced about $0.7\ \lambda$ apart vertically, and the lower antenna is spaced $\lambda/4$ behind the upper antenna. A coaxial cable connected between the upper antenna and a coupler is $\lambda/4$ longer than the coaxial cable connected to the lower antenna. Stagger stacking provides a 10–20 dB increase in the front-to-back ratio plus a substantial increase in forward gain.

7-4   EXTENDING THE COMMUNICATIONS RANGE

Of the many possible ways for extending the range of communications in the VHF, UHF, and higher-frequency bands, only two major techniques have proven reliable enough to warrant general use: tropospheric scatter communications and satellite communications.

Tropospheric Scatter

Tropospheric scatter, commonly called *troposcatter*, is a special case of sky wave propagation used for frequencies higher than those in standard sky wave propagation techniques. These higher frequencies are directed to the troposphere instead of to the ionosphere.

The troposphere is the upper region of the earth's atmosphere, ranging from about 6 to 10 miles above the surface of the earth. Propagation by tropospheric scatter uses the properties of the troposphere as a reflector of UHF signals in much the same manner as the ionosphere is used to reflect lower-frequency signals. However, the troposphere only partially reflects the signal, *scattering* the energy, some extremely small portion of which is scattered in the forward direction or in the direction of the receiver. The reason for the scattering is not entirely understood, but it appears to be similar to the mechanism by which the beams of automobile headlights are scattered by fog, snow, or heavy rain. In any case, the phenomenon does exist and is a very reliable technique for extending UHF signal communications paths beyond the horizon.

The scattering process, illustrated in Figure 7-13, shows directional transmitting and receiving antennas aimed so that their beams intercept in the troposphere. Note that much of the energy is scattered in undesired directions and is lost. The scattering process also creates fading problems, caused by multipath transmissions within the scattering path and by atmospheric changes.

Diversity reception is almost always used to overcome the fading in troposcatter communications. There are two methods: *space diversity* and *frequency diversity*. In space diversity, two or more receiving antennas and an equal number of receivers are used. The antennas are separated by several wavelengths. The receivers are connected to a common output stage and are configured to ensure that the automatic gain control (AGC) of the receiver with the strongest signal cuts off all other receivers. In this fashion, only the strongest signal is passed to the common output stage.

In frequency diversity, a common antenna is used to simultaneously pass two or more frequencies to receivers designed with two or more IF stages. Again, the strongest signal received is passed to the common output stage.

Where conditions are critical, a space diversity system using two or more transmitting antennas and transmitters, arranged in the same manner as at the reception end of the link, is often used. Called *quadruple diversity*, this method ensures that adequate signal strengths are received.

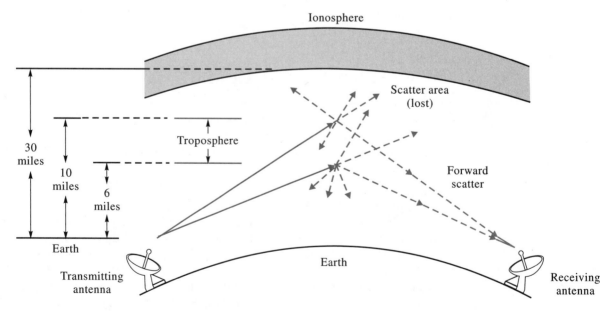

FIGURE 7-13 Tropospheric scatter propagation

The best and most widely used frequencies for the tropospheric scatter technique are 900 MHz, 2 GHz, and 5 GHz. Typically, the distance covered is around 400 miles. When compared to the practical maximum space wave distance of about 60 miles, propagation by troposcatter appears to be ideal. However, because of the minute amount of energy scattered in the direction of the receiver, high-power transmitters, high-gain receivers, and elaborate antenna arrays are required. Thus, the cost of a troposcatter system is quite high. But even with the high costs, these systems are used when reliable long distance communications links are needed, such as across deserts, in mountain regions, to offshore drilling platforms, and between distant islands. Also, this technique is used extensively by the armed services.

Satellite Communications

Satellite communications systems are composed of a communications satellite in stationary orbit approximately 22,000 miles above the earth's surface, an earthbound transmitting antenna, and an earthbound receiving antenna. The stationary satellite is basically a microwave repeater station that receives an *uplink* from the transmitter, amplifies it, and retransmits it on a different *downlink* frequency to the receiver on earth. The satellite is powered by a bank of batteries charged by panels of solar cells.

Because the satellite is in most cases stationary and in such a high position, it provides enormous increases in the communications range for any station within its coverage area. Multiplexing a number of different signals is permitted by using wide bandwidths. Existing satellites are capable of handling 12,500 telephone circuits, 1200 duplex voice channels, 27 television broadcast channels, or some combination of these. A typical satellite communications system is shown in Figure 7-14.

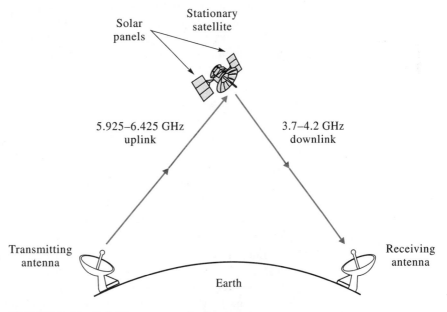

FIGURE 7-14 Satellite communications system

The apparently stationary orbit of the satellite is a result of *geosynchronous* action; that is, the satellite, traveling in the same direction as the earth's rotation, will complete one revolution about the earth's axis in the same time frame as one revolution of the earth. Hence, the satellite appears stationary to a person observing it from earth. For this same reason, a satellite is often referred to as *geostationary*.

Satellite communications systems are examined in more detail in Chapter 14.

SUMMARY

- Three broad classifications for wave propagation are ground, space, and sky waves. Ground wave propagation is used primarily in the LF and MF bands. Energy travels along the surface of the earth. Space wave propagation is used primarily in the VHF, UHF, and higher-frequency bands. Energy travels directly from the transmitting antenna to the receiving antenna. In general, space waves are limited to line-of-sight distances. The radio horizon for space waves is about one third greater than the optical horizon. Sky wave propagation is used primarily in the MF and HF bands. Energy travels upward into space, where it is refracted back to earth by ionospheric layers. Sky waves are used for almost all HF communications and for long distance MF communications at night.

- The ionosphere is composed of ionized layers of the atmosphere between 30 and 250 miles above the surface of the earth. Ionospheric layers are the D layer, 30–60 miles above the earth; the E layer, 60–90 miles above the earth; and the F layer, 90–250 miles above the earth. In the daytime, the F layer splits into F_1 and F_2, each of which varies in depth from summer to winter. Solar radiation is the major influence on the ionosphere.

- Four ionospheric variations are diurnal, seasonal, geographical, and cyclical. The 11-year sunspot cycle has the greatest influence on the ionosphere. The critical frequency for sky wave propagation is about doubled during the highest sunspot activity, greatly increasing communications range and reliability. However, because of hourly changes in the ionosphere, distance and reliability are affected differently in the day than in the night.

- Critical frequency is the highest frequency that returns to earth when radiated upward. Critical angle is the angle above which any given frequency will not be refracted enough to return to earth. Skip distance is the distance from a transmitting antenna that a sky wave returns to earth. Skip zone is a no-signal zone between the end of a usable ground wave signal and the reappearance of a reflected sky wave. Multiple-hop transmission, used to extend the distance of sky wave transmissions, occurs when a sky wave signal is reflected from earth back into the ionosphere, and then is refracted back to earth again. Fading is an annoying variation of a received radio signal caused by multipath reception.

- Maximum usable frequency is the highest frequency for transmission between two points by sky wave. Lowest usable frequency is the frequency below which the RF signal will be totally absorbed by the ionosphere. Optimum usable frequency is the frequency far enough above the LUF and far enough below the MUF to provide reliable sky wave communications.

- Tropospheric scatter (troposcatter) and satellite communications are used to extend the communications range. Frequencies most widely used for troposcatter are 900 MHz, 2 GHz, and 5 GHz, and a typical communications distance is 400 miles.

- Satellite communications uses a stationary satellite with typical uplink frequencies of around 6 GHz and downlink frequencies of around 4 GHz. Satellite communications systems have great capacity and range.

QUESTIONS AND PROBLEMS

1. Briefly describe the three classifications of radio wave propagation and list the primary frequency bands for each.

2. How can ground wave coverage be increased?

3. How can the range of space wave transmissions be increased?

4. Define radio horizon and explain how it occurs.

5. Find the radio horizon distance of (a) a transmitting antenna 65 m high and (b) a receiving antenna 45 m high.

6. Find the radio horizon distance of (a) a transmitting antenna 115 m high and (b) a receiving antenna 70 m high.

7. Find the maximum communications distance for space wave signals using the antennas from (a) Problem 5 and (b) Problem 6.

8. Find the maximum communications distance for space wave signals using the antennas from (a) Problems 5a and 6b and (b) Problems 5b and 6a.

9. Find the required receiving antenna height for a receiver located (a) 75 km from a transmitting antenna 75 m high and (b) 90 km from a transmitting antenna 125 m high.

10. Find the required transmitting antenna height for a transmitter located (a) 45 km from a receiving antenna 50 m high and (b) 55 km from a receiving antenna 35 m high.

11. For long distance communications, which of the wave propagation classifications is used most?

12. Describe each of the layers of the ionosphere.

13. Describe the aurora borealis and explain its effect on communications.

14. Describe (a) sudden ionospheric disturbances, (b) ionospheric storms, and (c) sunspot activity.

15. Define (a) critical frequency, (b) critical angle, (c) skip distance, and (d) skip zone.

16. Explain the concept of multiple-hop transmission.

17. How can skip distance be maximized?

18. What is the maximum distance for a single-hop sky wave transmission under the best conditions? Explain your answer.

19. Define and explain (a) maximum usable frequency, (b) lowest usable frequency, and (c) optimum usable frequency.

20. Describe the effects of multipath reception.

21. Describe the four categories of variations that affect the ionosphere.

22. Describe the effect of sunspot activity on the critical frequency.

23. Describe the effects of day and night hours on (a) ground waves, (b) the VLF and LF bands, (c) the MF band, (d) the HF band, (e) the VHF band, and (f) the UHF, SHF, and EHF bands.

24. How can the range of communications in the VHF, UHF, and higher frequencies be extended?

25. Explain the concept of tropospheric scatter.

26. Briefly describe the concept of a satellite communications system.

CHAPTER 8

Antennas

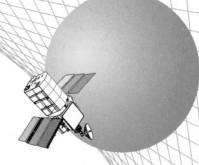

INTRODUCTION

An *antenna* consists of a wire or other conductor, or a collection of wires or conductors, that converts electrical energy into electromagnetic waves for transmission, and electromagnetic waves into electrical energy for reception. Except for their different functions, transmitting and receiving antennas have identical behavior characteristics.

There are three main elements in an antenna system: transmission lines (discussed in Chapter 6), radio wave propagation (discussed in Chapter 7), and the antenna itself. In this chapter, we will discuss the fundamentals of antennas, transmitting antennas, antenna arrays, and receiving antennas.

8-1   ANTENNA FUNDAMENTALS

In a transmitting system, an electrical radio frequency (RF) signal is generated in a master oscillator (MO), amplified by RF amplifier stages, and modulated by a signal from a modulator. The modulated signal is amplified and fed to an antenna, where it is radiated into space as an electromagnetic wave. These electromagnetic waves then propagate (travel) through the atmosphere. In a receiving system, the electromagnetic waves are picked up by an antenna, which converts the waves into alternating currents for use by the receiver, where the modulating signal is extracted, amplified, and applied to an output transducer. Because of high losses in wave travel between the transmitter and the receiver, and to provide adequate signal strength at the receiver, either the transmitted power must be very high or the transmitting and receiving antennas must have high efficiency.

Reciprocity

The *Reciprocity Theorem* describes the behavior of currents and voltages in certain multibranch circuits. This theorem is a direct result of Kirchhoff's and Ohm's Laws.

In a passive, linear, bilateral circuit with two or more separate current-carrying branches, such as the illustration in Figure 8-1, a power supply delivering E_a volts is placed across branch A of this circuit, resulting in a current I_a in branch A and a current I_b in branch B. If the power supply has zero internal resistance, the voltage across branch B is found from the current I_b and resistance R_b according to Ohm's Law: ($E_b = I_bR_b$).

Assume that the power supply with voltage E_a is disconnected from branch A of the circuit shown. A different supply with voltage E_b and zero internal resistance is connected across branch B. The same current (I_b) continues to flow through resistance R_b. According to the reciprocity theorem, under these circumstances, the current through resistance R_a will be the same as before, or I_a. Therefore, the voltage across branch A will be $E_a = I_aR_a$. This results in precisely the same voltage E_a as the original supply.

The reciprocity theorem is important in antenna theory because it can be used to determine currents and voltages in complicated networks.

Except for their different functions, transmitting and receiving antennas behave identically. Thus, the same antenna may be used for transmitting or for receiving. The capability of interchanging an antenna's functional use is called *reciprocity*. Because of reciprocity, our discussion of antennas is from the viewpoint of transmitting antennas, since the same principles apply to receiving antennas.

Antenna Construction

Antennas are constructed from conductors in the form of wires or rods arranged to provide maximum efficiency for the production or collection of electromagnetic waves. Maximum efficiency of operation also requires that the transmitting and

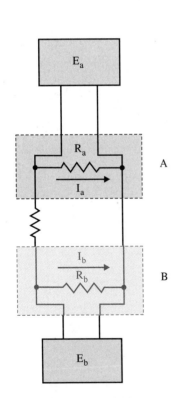

FIGURE 8-1 Equivalent circuit illustrating the reciprocity theorem

receiving antennas have the same polarization; that is, they must both be vertically positioned or horizontally positioned. A vertical antenna transmits or receives best a vertically polarized signal, and a horizontal antenna transmits or receives best a horizontally polarized signal.

Polarization

Polarization of an electromagnetic field is the orientation of the electric flux lines. Polarization is of primary interest in electromagnetic effects. The polarization of an electromagnetic field is considered to be the orientation of the electric flux lines.

An electromagnetic field can be polarized horizontally, vertically, or slantwise. The polarization is generally parallel with the active element of an antenna; thus, a vertical antenna radiates and receives fields with vertical polarization, and a horizontal antenna radiates and receives fields with horizontal polarization.

Electromagnetic fields may have continually changing polarization. This kind of polarization can be produced in a variety of ways. (See Elliptical and Circular Polarization.)

Polarization effects occur at all wavelengths from the very low frequencies to the gamma-ray spectrum. These effects are quite noticeable in the visible-light range. Light with horizontal polarization will reflect well from a horizontal surface (such as a pool of water), while vertically polarized light reflects poorly off the same surface. Simple experiments conducted with a polarized lens illustrate the effects of visible-light polarization.

Information can be impressed on an electromagnetic wave by causing rapid changes in the polarization of the wave. This method of modulation is known as *polarization modulation.* Polarization modulation can be achieved, theoretically, at any wavelength from the very low frequencies through the gamma-ray spectrum. It is most often used at microwave and visible-light wavelengths.

Antenna Polarization

Antennas may be physically erected either vertically or horizontally in relation to the earth. The orientation of the electric field with respect to the radiating antenna—called the *radiation pattern*—is determined by the physical position of the antenna. The radiation pattern for the antenna is an important characteristic. The intended use of the antenna determines the physical position.

Vertical Polarization   *Vertical polarization* occurs when the electric lines of flux of an electromagnetic wave are vertical or perpendicular to the surface of the earth. In communications vertical polarization has certain advantages and disadvantages, depending on the application and the wavelength.

At LF and VLF, vertical polarization is ideal because surface-wave propagation, the major mode of propagation at these wavelengths, requires a vertically polarized field. Surface-wave propagation is effective in the standard AM broadcast band as well; most AM broadcast antennas are vertical.

Vertical polarization is often used at high frequencies because vertical antennas can be erected in a very small space. At VHF and UHF, vertical polarization is used for mobile communications and in repeater communications.

The main *disadvantage* of vertical polarization is that most man-made noise tends to be vertically polarized. Therefore, a vertical antenna picks up more of this interference than a horizontal antenna. At VHF and UHF, vertical polarization results in more flutter in mobile communications, as compared with horizontal polarization.

An antenna erected so that one end points up into the sky and the other end points directly down toward the center of the earth is known as a *vertical antenna* and is said to be *vertically polarized.* Vertical polarization will cause the strongest radiation intensity to be sent out in the horizontal plane, not up and down in the vertical plane.

Standard AM broadcast stations and many other communications users employ vertical antennas. The primary reason is to provide good reception in all horizontal directions—that is, at all points of the compass.

If we measure the strength of the radio waves sent out from a vertical antenna at a certain distance from that antenna, we will find that these waves are of equal strength at that same distance in all horizontal directions from the antenna. The amplitude of this wave is evaluated in terms of the voltage it will induce across a wire. This voltage is known as *field strength,* a measure of the intensity (strength) of an electric, magnetic, or electromagnetic field. The strength of an electric field is measured in volts per meter; a magnetic field in gauss; and an electromagnetic field in watts per square meter, as registered by a field-strength device.

A simple *field-strength meter* is shown in Figure 8-2. One of the simplest types of field-strength meters, this device consists of a microammeter, a semiconductor diode, and a short antenna that serves as the field strength pickup. Although uncalibrated, such a device is useful in estimating the level of RF energy at a particular location.

More sophisticated devices, such as the meter shown in Figure 8-3, use a tuned circuit and an amplifier to measure weak fields. In those devices, the tuned circuit avoids confusion resulting from signals on frequencies other than those desired.

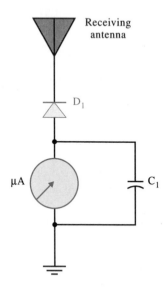

FIGURE 8-2 A simple diode field-strength meter used for rectifying RF energy and measuring field strength

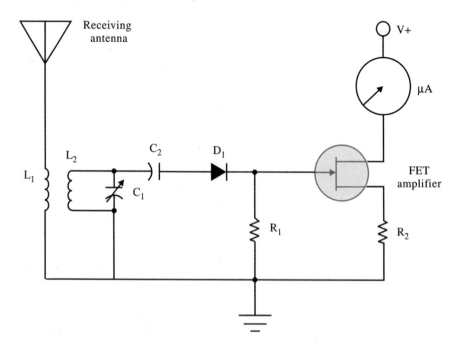

FIGURE 8-3 Amplified field-strength meter with tuned circuit

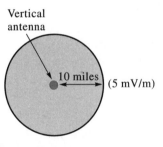

FIGURE 8-4 Field strength radiation pattern for a vertical antenna

The electric field strength from a transmitting antenna, as measured in volts, millivolts, or microvolts per meter, is proportional to the current in the antenna and to the effective length of the antenna. For example, if the field strength of a certain radio wave is measured 10 miles to the north of the radiating antenna and is found to be 5 mV/m, the field strength of the same wave will also be 5 mV/m at distances 10 miles to the south, east, and west. Figure 8-4 illustrates this concept. The circle represents the horizontal radiation pattern of a vertical antenna. As the pattern shows, a vertical antenna radiates equally well in all horizontal (ground plane) directions. Because the vertical antenna radiates to, or receives from, all horizontal directions equally well, it is known as a *nondirectional* or *omnidirectional* antenna.

The farther the electromagnetic wave travels, the weaker it becomes. The field strength at any location depends on two things: (1) the distance from the antenna and (2) the radiated power. The strength varies inversely with distance; that is, if the distance doubles, the field strength is reduced by one half, and if the distance is reduced by one half, the field strength doubles. Likewise, if the radiated power is increased, the field strength at any location will also increase. However, because field strength is evaluated by the voltage induced across a wire, any power increase must be converted to a corresponding voltage increase. From Ohm's law, where power is proportional to voltage squared ($P = V^2/R$), it follows that the increase in field strength is proportional to the square root of the power change. Therefore, it is necessary to quadruple the radiated power to double the field strength at any location.

Horizontal Polarization   When the electric lines of flux of an electromagnetic wave are oriented horizontally, the field is said to be *horizontally polarized*. In communications horizontal polarization has certain advantages and disadvantages at various wavelengths.

At LF and VLF, horizontal polarization is not often used because the surface wave, an important factor in propagation at these frequencies, is more effectively transferred when the electric field is oriented vertically. As previously mentioned, most standard AM broadcast stations, operating in the MF range, employ vertical rather than horizontal polarization.

In the HF part of the electromagnetic spectrum, horizontal polarization becomes practical. Since the polarization is always parallel to the orientation of the radiating antenna element, horizontal wire antennas are simple to install for signals above about 3 MHz. The surface wave is less important at HF than at VLF, since the sky wave is the primary mode of propagation above 3 MHz. This becomes increasingly true as the wavelength gets shorter. Horizontal polarization is just as effective as vertical polarization in the sky-wave mode.

In the VHF and UHF either vertical or horizontal polarization may be used. Horizontal polarization generally provides better noise immunity and less fading than vertical polarization in this part of the spectrum.

An antenna erected in such a manner that it is parallel to the surface of the earth is known as a *horizontal antenna* and is said to be *horizontally polarized*. Maximum horizontal radiation from a horizontal antenna occurs in the directions broadside to the antenna. Minimum radiation occurs from the ends of the horizontal antenna. Figure 8-5 illustrates this concept. Because the horizontal antenna does not radiate to, or receive from, all horizontal directions equally well, it is known as a *directional* antenna. If

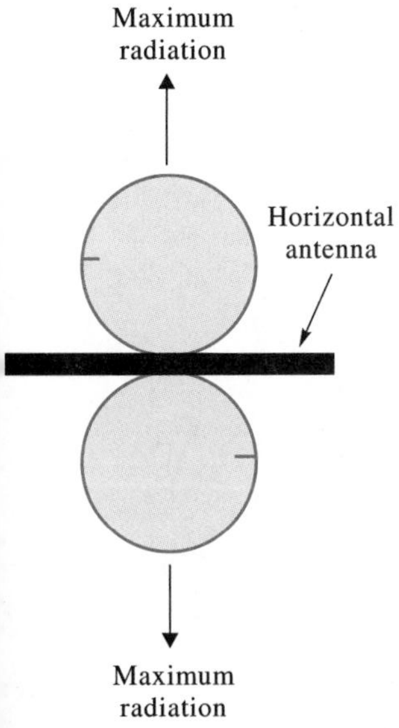

FIGURE 8-5 Radiation pattern for a horizontally polarized antenna

it radiates in one direction only, it is *unidirectional;* if it radiates equally well in two opposite directions, it is *bidirectional.* Television and FM broadcast stations use horizontal antennas because, among other reasons, ignition interference, which causes much trouble at the VHF and UHF ranges, is usually polarized more strongly in the vertical direction.

Elliptical Polarization   The polarization of an electromagnetic wave is the orientation of the electric lines of flux in the wave. While this orientation usually remains constant, sometimes it is deliberately made to rotate as the wave propagates through space. If the orientation of the electric lines of flux changes as the signal is propagated from the transmitting antenna, the signal is said to have *elliptical polarization.*

An elliptically polarized electromagnetic field may rotate either clockwise or counterclockwise as it moves through space, as shown in Figure 8-6. The intensity of the signal might not remain constant as the wave rotates. If the intensity does remain constant as the wave rotates, the polarization is said to be circular. (See Circular Polarization.)

Elliptical polarization is useful because it allows the reception of signals having unpredictable or changing polarization with a minimum of fading and signal loss. Ideally, the transmitting and receiving antennas would both have elliptical polarization, although signals with linear polarization can be received with an elliptically polarized antenna. If, however, the transmitted signal has opposite elliptical polarization from the receiving antenna, there will be considerable signal loss.

Elliptical polarization is generally used at ground stations for satellite communication. In receiving, an elliptically polarized antenna reduces the fading caused by changing satellite orientation. In transmitting, elliptical polarization increases the probability that the satellite will receive a good signal for retransmission.

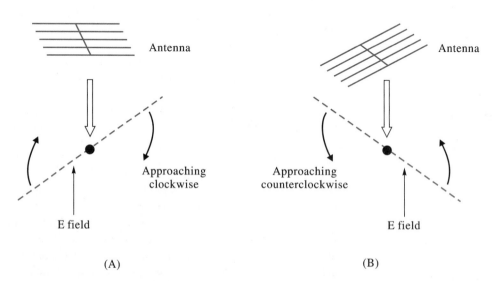

FIGURE 8-6   Representation of elliptical polarization programmed as (A) clockwise and (B) counterclockwise

Circular Polarization As previously stated, polarization can be rotating, either clockwise or counterclockwise. Uniformly rotating polarization is called *circular polarization*. The orientation of the electrical field lines of flux completes one rotation for every cycle of the wave.

Antennas for circular polarization are not turned to produce the rotating electromagnetic field; instead, the rotation is easily accomplished by electrical means. Figure 8-7 shows a typical antenna for generating waves with circular polarization. The antennas are fed 90° out-of-phase by making feed-line stub A a quarter wavelength longer than stub B. The signals from the two antennas thus add vectorially to create a rotating field. The direction of the rotation can be reversed by adding one-half wavelength to either stub A or stub B (but not both).

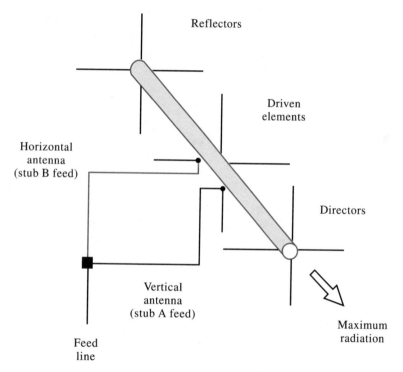

FIGURE 8-7 Antenna configured for circular polarization

Circular polarization is compatible with a 3-decibel power loss, with either horizontal or vertical polarization, or with slanted linear polarization. When communicating with another station also using circular polarization, the rotation directions must be in agreement. If a circularly polarized signal arrives with opposite rotation direction from that of the receiving antenna, the attenuation is about 30 decibels compared with matched rotational sense.

In uniform circular polarization, the vertical and horizontal signal components may have equal magnitude, but this is not always the case. A more general form of rotating polarization, in which the components may have different magnitude, is called elliptical polarization (see Elliptical Polarization).

Antenna Excitation

A transmitting antenna is said to be *excited* when the electrical output of a transmitter is fed into it. Antenna excitation generates antenna current, which in turn causes radio waves to be radiated into the atmosphere. The stronger the antenna current, the stronger the radiated radio waves. If all other factors remain constant, the antenna current will be strongest when the antenna is resonant to the frequency of the antenna current.

In Chapter 2, we learned that the resonant frequency of a circuit is determined by the value of its capacitance and inductance. We assumed that inductance is found only in a coil and that capacitance is found only in a capacitor. However, many circuits contain inductance and capacitance but no coils or capacitors. A coil provides a concentrated inductance, and a capacitor provides a concentrated capacitance. These concentrated values are called *lumped* values. A single straight wire also contains inductance and capacitance, although in nonconcentrated form and with smaller values. The reactive values in a straight wire are called *distributed* values. Distributed inductance and distributed capacitance in a straight wire will produce resonance at a frequency determined by the substitution of these values into the resonant frequency equation. Thus,

$$f_r = \frac{1}{2\pi\sqrt{LC}}$$

(8-1)

where L and C are distributed values.

Maximum Power Transfer

Before discussing power transfer, an understanding of the *decibel* (dB) is necessary. Since the decibel is a means of expressing a ratio between two currents, power levels, or voltages, a reference level is always necessary for the decibel to have meaning.

dBa   The abbreviation *dBa* stands for *adjusted decibels.* Adjusted decibels are used to express *relative levels of noise.* A reference noise level is chosen and assigned the value *0 dBa.* All noise levels are then compared to this value. Noise levels *lower* than the reference level have *negative* values, such as *–3 dBa.* Noise levels *greater* than the reference level have *positive* values, such as *+6 dBa.*

dBd   The abbreviation *dBd* refers to the *power gain* of an antenna, in decibels, with respect to a *half-wave dipole* antenna. The dBd specification is the most common way of expressing antenna power gain.

The reference direction of the antenna under test is considered to be the direction in which it radiates the most power. The reference direction of the dipole is broadside to the antenna conductor. Power gain in dBd is given by the formula:

$$dBd = 10\,\log_{10}\left(\frac{P_{ERP}}{P_d}\right)$$

(8-2)

where P_{ERP} is the *effective radiated power* (*ERP*) from the antenna in question with a transmitter output of P watts, and P_d is the effective radiated power from the dipole with a transmitter output of P watts.

An alternative method of measuring antenna power gain in *dBd* is possible using the actual field-strength values. If E_a is the field strength in microvolts per meter at a certain distance from the antenna in question, and E_d is the field strength at the same distance from a half-wave dipole getting the same amount of transmitter power, then:

$$\text{dBd} = 20 \ \log_{10}\left(\frac{E_a}{E_d}\right) \tag{8-3}$$

dBi The abbreviation *dBi* refers to the *power gain* of the antenna, in decibels, relative to an *isotropic* antenna (discussed later in this chapter). An isotropic antenna, in theory, radiates equally well in all directions. However, measurement is taken from the direction in which the antenna under test radiates the best. The gain of any antenna in dBi is 2.15 dB greater than its gain in dBd. That is:

$$\text{dBi} = 2.15 + \text{dBd} \tag{8-4}$$

Power gain in dBi is given by the formula:

$$\text{dBi} = 10 \ \log_{10}\left(\frac{P_{ERP}}{P_i}\right) \tag{8-5}$$

where P_{ERP} is the effective radiated power from the antenna in question with a transmitter output of P watts, and P_i is the effective radiated power from the isotropic antenna with a transmitter output of P watts.

An alternative method of measuring antenna gain in dBi is possible using actual field-strength values. If E_a is the field strength in microvolts per meter at a certain distance from the tested antenna, and E_i is the field strength from an isotropic antenna getting the same amount of power from the transmitter, then:

$$\text{dBi} = 20 \ \log_{10}\left(\frac{E_a}{E_i}\right) \tag{8-6}$$

Actually an isotropic antenna is not seen in practice, since it is essentially impossible to construct a true isotropic antenna. For other antennas, gain figures in dBi instead of dBd are sometimes used.

dBm The abbreviation *dBm* refers to the *strength of a signal,* in decibels, compared to *1 milliwatt (mW)*, with a load impedance of 600 ohms. If the signal level is exactly 1 mW, its level is 0 dBm. In general:

$$\text{dBm} = 10 \ \log_{10} P \tag{8-7}$$

where P is the signal level in milliwatts.

With a 600-ohm load, *0 dBm* represents 0.775 V, or 775 mV. With respect to voltage in a 600-ohm system, then:

$$\text{dBm} = 10 \ \log_{10}\left(\frac{E}{775}\right) \tag{8-8}$$

where E is the voltage in millivolts. A level of *0 dBm* also represents a current of 1.29×10^{-3} amperes (A) or 1.29 milliamperes (mA). With respect to current in a 600-ohm system:

$$\text{dBm} = 20 \ \log_{10}\left(\frac{I}{1.29}\right) \tag{8-9}$$

where I is the current in mA.

If maximum power is to be delivered by a transmitter to its antenna, the antenna must be resonant to the frequency of the radio wave energy transferred from the transmitter. Stated another way, the antenna must be in resonance with the transmitter, a condition that results in maximum antenna current.

The antenna current is an alternating current flowing back and forth in the antenna at the radio frequency of the transmitter output. The antenna acts as a series resonant circuit, so the inductive reactance and the capacitive reactance at the resonant frequency cancel each other, leaving the resistance of the antenna as the only opposition to the antenna current. Therefore, when the antenna is in resonance with the output of the transmitter, the antenna current is a maximum—the desired condition that causes maximum radiation of radio waves.

Antenna Standing Waves

If we take current measurements along an excited antenna, we find that different current values exist at different points. Figure 8-8 shows an imaginary antenna one wavelength (1 λ) long into which many RF ammeters have been inserted. Assume that the current values shown in the current distribution graph below the antenna are the readings of those meters when the antenna is excited. Note that the antenna current is zero at three different points. These zero points are called *current nodes.* Also, note that the antenna current is maximum at two different points. These maximum points are called *current antinodes,* or *current loops.*

The ammeter readings shown in Figure 8-8 are not merely instantaneous readings that will quickly change in relative values. The meters will continue to give the same readings so long as the antenna excitation strength remains unchanged, because these meters read *effective* RF values, not *instantaneous* values. Thus, the current nodes and loops remain at fixed, or stationary, positions on the antenna. Therefore, we can say that the wave represented by the nodes and loops is an *antenna standing wave.* Of course, the electrons are not actually standing still. They are moving back and forth in

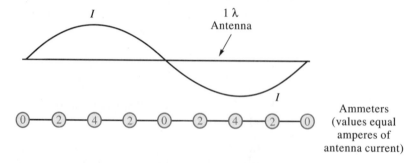

FIGURE 8-8 Current distribution for a full-wave antenna

the antenna at high speed. However, the electron waves are reflected from the ends of the antenna in such a manner that the phase relationships between the reflected wave and the original wave always produce current nodes and loops at the same points.

Because electron flow must cease at the ends of the antenna, current nodes occur at these points. On the other hand, the excess of electrons at the ends of the antenna causes maximum electrical pressure, or voltage, at these points. This means that *voltage loops* occur at the points where current nodes occur; therefore, *voltage nodes* must occur at the same points as current loops. This concept is no different from Ohm's Law, where maximum current always occurs at the same point as minimum voltage, with resistance held constant. Note that these relationships are the same as those discussed for transmission lines in Chapter 6. Figure 8-9 shows the distribution of both current and voltage on an antenna 1 λ long. These current and voltage curves represent the current and voltage standing waves on the antenna and are commonly referred to as the current and voltage *displacement curves*. Because current nodes occur at both ends of an antenna, for best operation the antenna should normally not be electrically less than λ/2 long.

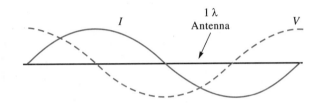

FIGURE 8-9
Current and voltage displacement for a full-wave antenna

Propagation Constant and End Effect   The speed of electrons in a wire is slightly less than the speed of radio waves in space. The speed of radio waves in space is known as the *propagation constant,* and the waves travel at the speed of light (186,000 miles/s or 3×10^8 m/s). Because electric current flows in a conductor at a rate slightly less than the propagation constant, and because of the phenomenon known as *end effect,* an antenna behaves as if it were λ/2 long when its physical length is actually slightly less than that. For example, a half-wave antenna used to radiate a 600 kHz (500 m) signal must be slightly less than 250 m long if it is to have perfect resonance.

Determining Physical Antenna Length   We learned earlier that wavelength (λ) in free space is determined by

$$\lambda = \frac{3 \times 10^8 \text{ m/s}}{f} \tag{8-10}$$

where

λ = wavelength in meters

f = frequency in hertz

However, with an antenna, the wave is developed on a wire, and end effect must be considered. Because of the capacitance between the ends of the antenna, end effect makes an antenna appear to be about 5% longer than its actual physical length. Taking end effect into account, we use the following to calculate physical length, in feet, of a half-wave dipole supported by insulators at each end:

$$\lambda/2 = \frac{4.92 \times 10^8 \times 0.95}{f} \qquad\qquad\text{(8-11a)}$$

where

$\lambda/2$ = half-wavelength in feet

f = frequency in hertz

Simplified, we get a close approximation:

$$\lambda/2 = \frac{468}{f} \qquad\qquad\text{(8-11b)}$$

where

$\lambda/2$ = half-wavelength in feet

f = frequency in megahertz

For a quarter-wave antenna, the approximation equation is

$$\lambda/4 = \frac{234}{f} \qquad\qquad\text{(8-11c)}$$

where

$\lambda/4$ = quarter-wavelength in feet

f = frequency in megahertz

8-2 TYPES OF TRANSMITTING ANTENNAS

Although there are many different types of transmitting antennas available, the two principal categories are the half-wave and quarter-wave. Each has its own special characteristics.

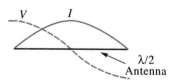

FIGURE 8-10 Current and voltage displacement curves for a half-wave antenna

The Half-Wave Antenna

The values of distributed inductance and distributed capacitance of a straight wire are such that the antenna will be resonant to a frequency whose wavelength is approximately twice the physical length of the antenna. Stated another way, a resonant antenna is approximately $\lambda/2$ long—one half the length of the wave to which it is resonant and one half the length of the wave that it radiates most efficiently. For this reason, the antenna used to transmit radio signals is the *half-wave antenna*. The current and voltage displacements on such an antenna are shown in Figure 8-10.

The Dipole Antenna   A half-wave antenna is variously referred to as a *dipole,* a *doublet,* or a *Hertz* antenna. The half-wave dipole is the most commonly used type of antenna. As the name implies, it is $\lambda/2$ long at its operating frequency. A typical dipole installation is shown in Figure 8-11. The antenna is center-fed (discussed in Section 8-3) with a 75 Ω balanced transmission line, giving a good match to the antenna's 72 Ω input impedance.

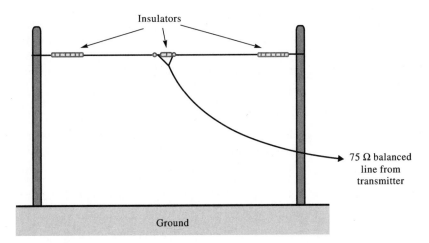

FIGURE 8-11 Typical center-fed dipole installation

A dipole can also be fed with a coaxial transmission line, with the center conductor connected to one side of the dipole and the shield connected to the other. However, some inefficiency will result because the dipole is balanced and the coaxial line is unbalanced. This inefficiency stems from the current and voltage distributions on the antenna being upset, which allows RF current to flow on the coaxial shield and may cause undesirable radiation from the transmission line.

To correct the imbalance between a coaxial line and a dipole antenna, a *balun* must be used. A balun is a *bal*anced to *un*balanced RF transformer. A typical balun is shown in Figure 8-12. This particular balun acts as a center insulator for the dipole. As shown in Figure 8-12(A), the wires connect to the dipole, and the coaxial cable is connected to the RF connector. Baluns are constructed for specific impedance ratios, just like conventional transformers. When 50 Ω or 75 Ω coaxial cable is used to feed a dipole, an impedance ratio of 1:1 is used, as in Figure 8-12(B).

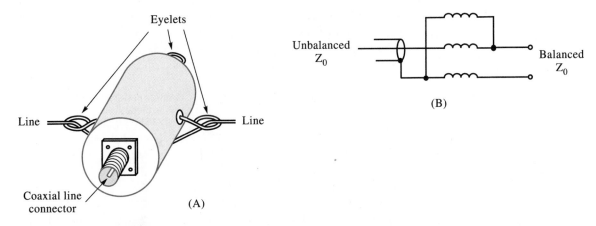

FIGURE 8-12 Typical balun for coaxial cable transmission line connection (A) Connections (B) Schematic for a broadbanded balun with a 1:1 impedance transfer ratio

The Vertical Dipole   When vertical polarization is required, the antenna must be vertical. However, because of antenna height considerations, vertical antennas are usually used at relatively high frequencies. For example, a vertical half-wave dipole operating at 2 MHz would be 233.75 feet high, whereas operating at 10 MHz it would be only 46.75 feet high.

The height restriction at low frequencies can be overcome by constructing a quarter-wave vertical antenna above a perfect ground. The antenna will then have the same characteristics as a half-wave vertical dipole, because a perfect ground will produce a *mirror image* of the quarter-wave, a result of the reflected radio waves as shown in Figure 8-13.

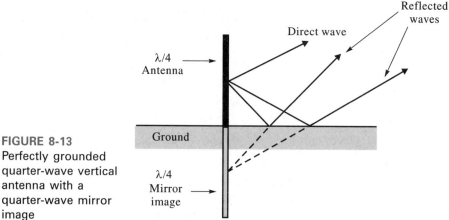

FIGURE 8-13
Perfectly grounded quarter-wave vertical antenna with a quarter-wave mirror image

If the ground is not a perfect conductor, the vertical quarter-wave antenna will lose a considerable amount of power in the resistance of the grounding system. When an antenna is constructed over soil of poor conductivity, such as sandy or rocky soil, an artificial ground system must be used. The artificial system consists of quarter-wave copper wires, called *radials,* extending outward from the base of the antenna in as many directions as possible. Four radials is a practical minimum number, and when only four are used, they should be arranged at 90° angles to one another.

The radiation pattern for a quarter-wave vertical antenna is shown in Figure 8-14. The antenna radiates equally well in all directions in the horizontal plane; that is, it is

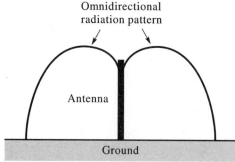

FIGURE 8-14
Omnidirectional radiation pattern for a quarter-wave vertical antenna

omnidirectional. In the vertical plane, the radiation is directed low toward the horizon, thus providing excellent long distance propagation characteristics.

When used with a perfect ground, the input impedance of the quarter-wave antenna is about 36 Ω, which makes an acceptable impedance match for a 50 Ω coaxial transmission line. Also, because one side of the vertical antenna is at ground level, the antenna is unbalanced, so an unbalanced coaxial cable is a good feeder. The center conductor of the coaxial cable is connected to the quarter-wave antenna, which is insulated from ground, and the shield is connected to ground or the radial system.

The Marconi Antenna

The half-wave antenna previously discussed is normally erected in an elevated position and is not connected directly to ground. However, some transmitting antennas are connected to ground at one end. A grounded antenna is known as a *Marconi* antenna.

The actual physical length of a Marconi antenna is normally only λ/4 of the signal to be radiated. However, the earth acts as the *other half* of the antenna, making the overall *effective* length a full λ/2. In other words, the antenna consists of a conductor extending approximately λ/4 above the earth, and the earth acts as an additional λ/4 of the antenna. Electrically, the Marconi behaves as a half-wave antenna.

Figure 8-15 shows the current and voltage displacement curves for a Marconi antenna. Note that both the voltage node and the current loop occur at ground level, which is effectively the center of the antenna. Because antenna current is usually measured at the current loop, current for the Marconi antenna is measured at the base of the antenna—that is, at ground.

The Marconi antenna is used by most standard broadcast stations. A vertical tower is used as the quarter-wave conductor–radiator.

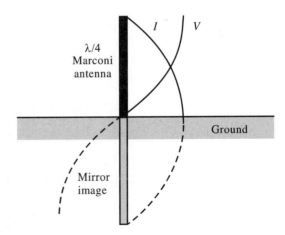

FIGURE 8-15 Current and voltage displacement curves for a quarter-wave Marconi antenna

---

EXAMPLE 8-1

Taking end effect into account, determine the physical length in feet of a half-wave dipole operating at 5 MHz.

SOLUTION

Using Equation 8-11a,

$$\lambda/2 = \frac{4.92 \times 10^8 \times 0.95}{5 \times 10^6}$$

$$= 93.5 \text{ feet}$$

Using Equation 8-11b,

$$\lambda/2 = \frac{468}{5}$$

$$= 93.6 \text{ feet}$$

EXAMPLE 8-2

What is the physical length of a Hertz antenna operating at 88 MHz?

SOLUTION

$$\lambda/2 = \frac{468}{88}$$

$$= 5.3 \text{ feet}$$

Examples 8-1 and 8-2 demonstrate how the physical length of a half-wave dipole decreases as frequency increases.

8-3   SPECIAL CHARACTERISTICS AND CONSIDERATIONS

At first reading, antenna characteristics appear to be almost identical to characteristics of transmission lines. However, many characteristics and considerations are unique to antennas, as this section will illustrate.

Feeding the Antenna

Connecting the RF output of a transmitter to the transmitting antenna is known as *feeding* the antenna. The RF output of a transmitter may be fed directly into the antenna by a coupling transformer or capacitor. However, this method is impractical because the antenna should be in the clear as much as possible, so antennas are usually erected high in the air. A more practical method of feeding the output of the transmitter into the antenna is by RF transmission lines, as discussed in Chapter 6. Although there are many methods of connecting transmission lines to the antenna, the two shown in Figure 8-16 are the most often used for Hertz antennas.

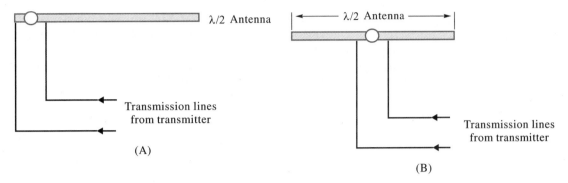

FIGURE 8-16   Hertz antenna feed methods (A) End feed (B) Center feed

Loading the Antenna

Recall that a transmitting antenna operates best when it is resonant to the frequency being transmitted. The process of tuning the antenna to bring it into exact resonance with the transmitter's output frequency is called *loading* the antenna, because the current load carried by the antenna is greatest when it is tuned to exact resonance.

An antenna of fixed length is resonant to a fixed frequency. If the same antenna is to be used to radiate at different frequencies, the antenna's natural frequency must be changed to conform to the new frequency to be radiated. One method of changing the resonant frequency of the antenna is to change the antenna's physical length, but this is usually impractical. Another method is to insert a lumped inductance or a lumped capacitance in series with the antenna. That is, a discrete coil or a discrete capacitor placed in series with the antenna will change the antenna's total inductance or capacitance and thereby change its resonant frequency.

Although most multifrequency transmitters and antennas include loading provisions, not all do. Hence, it is important to know how to change frequencies.

Tuning to a Higher Frequency   Assume that the physical length of a Marconi antenna is such that the antenna is resonant at 2.5 MHz. Further assume that the antenna must be operated at 3 MHz. It is evident that the antenna's resonant frequency must be increased if the antenna is to operate efficiently. We can determine from Equation 8-1 that resonant frequency is inversely proportional to both inductance and capacitance. Therefore, we deduce that f_r can be increased by reducing L or C or both. Recall from basic electronics that inductors in parallel result in a lower overall inductance, so the only way the inductance of the antenna can be reduced is by adding a coil in parallel with the antenna, a decidedly impractical undertaking. Also recall that capacitors in series result in lower overall capacitance. Therefore, the capacitance in the antenna can be reduced by adding a capacitor in series with the antenna at the current loop, thereby making the antenna resonant to a higher frequency.

Tuning to a Lower Frequency   If the antenna discussed in the previous paragraph were operated at 2 MHz, its resonant frequency would have to be reduced. This means that we must increase L or C or both. From basic electronics, we know that capacitors in parallel or inductors in series increase the capacitance or inductance, respectively.

We can see that placing a capacitor in parallel with the antenna is impractical. We therefore elect to place an inductor in series with the antenna at the current loop to obtain the required lower resonant frequency.

Equations for determining the inductance or capacitance required to retune an antenna are derived from algebraic manipulation of Equation 8-1. To find the total capacitance, we use

$$C_t = \frac{1}{4\pi^2 f_r^2 L} \tag{8-12a}$$

To find the unknown series capacitance, we use

$$C_2 = \frac{C_1 C_t}{C_1 - C_t} \tag{8-12b}$$

To find the total inductance, we use

$$L_t = \frac{1}{4\pi^2 f_r^2 C} \tag{8-13a}$$

To find the unknown series inductance, we use

$$L_2 = L_t - L_1 \tag{8-13b}$$

Because the characteristics of capacitance and inductance are seldom known, these calculations are hypothetical. For that reason, lumped inductances or lumped capacitances inserted in series with the antenna normally are made variable so that the antenna may be manually tuned to exact resonance.

The capacitance or inductance can be determined by using other methods, such as the slotted line, or substitution.

EXAMPLE 8-3

An antenna has an inductance of 2 μH and a capacitance of 15 pF. Determine (a) the resonant frequency to which the antenna is tuned, (b) the new capacitance value needed to tune the antenna to 50 MHz, and (c) the value for the required series capacitance.

SOLUTION

(a) From Equation 8-1,

$$f_r = \frac{1}{2\pi \sqrt{LC}} = \frac{1}{6.28 \sqrt{(2 \times 10^{-6})(15 \times 10^{-12})}}$$

$$= 29 \text{ MHz}$$

(b) From Equation 8-12a,

$$C_t = \frac{1}{4\pi^2 f_r^2 L}$$

$$= \frac{1}{(4)(9.86)(2500 \times 10^{12})(2 \times 10^{-6})}$$

$$= 5.1 \text{ pF}$$

(c) From Equation 8-12b,

$$C_2 = \frac{C_1 C_t}{C_1 - C_t}$$

$$= \frac{(15 \times 10^{-12})(5.1 \times 10^{-12})}{(15 \times 10^{-12}) - (5.1 \times 10^{-12})}$$

$$= \frac{7.65 \times 10^{-23}}{9.9 \times 10^{-12}}$$

$$= 7.73 \text{ pF}$$

EXAMPLE 8-4

Assume that the antenna in Example 8-3 is now to be tuned to resonate at 10 MHz. Using the newly calculated value of C, (5.1 pF), determine (a) the new value for L needed to meet the new frequency requirement and (b) the required series inductance.

SOLUTION

(a) From Equation 8-13a,

$$L_t = \frac{1}{4\pi^2 f_r^2 C}$$

$$= \frac{1}{(4)(9.86)(100 \times 10^{12})(5.1 \times 10^{-12})}$$

$$= 49.7 \,\mu\text{H}$$

(b) From Equation 8-13b,

$$L_2 = L_t - L_1$$

$$= 49.7 \,\mu\text{H} - 2 \,\mu\text{H}$$

$$= 47.7 \,\mu\text{H}$$

Antenna Guy Wires

In some cases, antenna erection requires *guy wires* to hold the antenna in position. Guy wires are essential in erecting most vertical antennas.

Guy wire lengths should be varied by using strain insulators in such a manner that the individual lengths will not be resonant to the transmitted frequency. If a guy wire or any part of it is resonant to the transmitted frequency, it can modify the radiation pattern, absorb and dissipate part of the radiated energy, thus reducing the signal strength available at receivers. Also, resonant guy wires reradiate (reflect) some of the RF energy transmitted from the antenna. This reflection is unwanted because it produces variations in the desired radiation pattern. To summarize, insulators in guy wires reduce the tendency of the guy wires to act as unwanted radiators and reflectors to RF energy, thereby reducing the RF losses that may occur in the antenna guy wires.

Figure 8-17(A) shows a guyed vertical antenna. The circles on the guy wires represent strain insulators. The tower itself is the antenna that radiates the radio wave.

The strain insulators used in the antenna guy wires should be of the *compression egg* type, made of glazed porcelain, as shown in Figure 8-17(B). This type of insulator ensures that the separate sections of the guy wire will not part even if the insulator breaks.

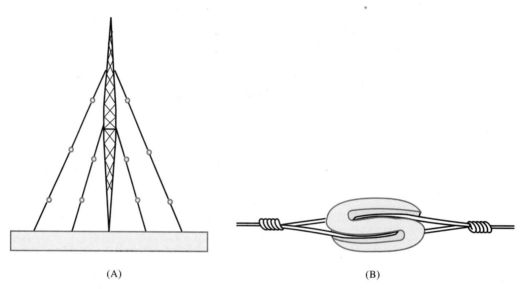

(A)   (B)

FIGURE 8-17 Guying an antenna (A) Vertical antenna with guy wires (B) Compression egg strain insulator

The Dummy Antenna

Sometimes it is necessary to make transmitter adjustments while the antenna is being excited. These adjustments could result in a fluctuation of the carrier frequency and create interference to other signals on the same frequency. Such interference can be avoided if a nonradiating antenna is substituted for the regular antenna. A nonradiating antenna dissipates rather than radiates the power supplied to it; that is, the RF power supplied to a nonradiating antenna is converted into heat rather than into radio waves. A nonradiating antenna designed for such purposes is known as a *dummy, artificial,* or *phantom* antenna.

Interference is not the only consideration for using a dummy load; operating the output of a transmitter into a bad mismatch can possibly destroy the output stage of the equipment. It is especially important to avoid an open circuit. Preliminary tuning of any transmitter should be done with the power reduced to about one half or one quarter of the normal power supply voltage. Power reduction protects the equipment from inadvertent overloads during tuning and decreases the possibility of interference to other stations operating on the same or adjacent channels.

Dummy antennas are helpful in approximating the power output of a transmitter. Power output is determined by using an ammeter in series with the circuit resistance and applying the power equation $P = I^2R$.

A dummy antenna consists of a resonant tank circuit in which a known value of series resistance has been placed. The ohmic resistance of the tank circuit converts the electrical energy from the transmitter to heat energy. A dummy antenna normally is constructed with a capacitive reactance value equal to the reactance of the coupling coil used, and a noninductive resistance equal in value to the impedance of the antenna to be used, as shown in Figure 8-18. In many cases, the *RC* circuit is submerged in an oil bath to ensure adequate heat dissipation and to protect the circuitry from excessive heating.

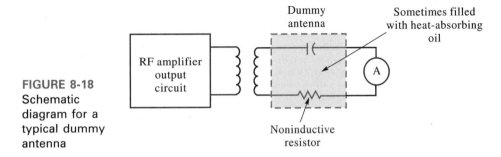

FIGURE 8-18
Schematic diagram for a typical dummy antenna

Radiation Characteristics

An alternating current, known as antenna current, flows back and forth in an excited antenna, producing an alternating electromagnetic field around the antenna. Energy is alternately stored in the magnetic field and returned to the antenna. At low frequencies, all of the energy stored in the magnetic field around a conductor is returned either to that conductor or to some other conductor in the immediate vicinity. However, if a frequency of 10 kHz or higher is used, some of the energy stored in the electromagnetic field around the antenna does not return to the conductor but instead is radiated into space in the form of electromagnetic waves, known as radio waves. As frequency is increased, less energy is returned to the antenna and more energy is radiated into space.

Radiation Resistance

In basic electronics, you learned that pure ohmic resistance is the property of a circuit that converts electrical energy into heat. Also, resistance or impedance is a way of expressing the relationship between voltage and current in a circuit. In a circuit radiating radio waves into space, voltage divided by current will give a value much higher than the pure ohmic resistance of the material that the circuit is made of, even though there is no resistance in the circuit (because it is resonant). There must be some kind of resistance present that we cannot measure with an ohmmeter, because the resistance read on the ohmmeter will be much smaller than the resistance obtained using the equation $R = V/I$. Since the resistance value obtained by Ohm's law is relatively high in a transmitting antenna, it would seem to follow that the high current would cause the antenna to heat to a high temperature. The fact that the antenna does not heat brings up the question of where all the $P = I^2R$ power goes if not to heat the antenna. The answer is that most of the power absorbed by an antenna is radiated into space as radio waves rather than converted into heat in the antenna. Therefore, the special property by which an antenna converts electrical energy into radio waves rather than into heat is known as *radiation resistance* (R_r). The equation for the radiated power P_r is

$$P_r = I^2R_r \tag{8-14}$$

where

$$P_r = \text{power radiated}$$
$$I = \text{antenna current}$$
$$R_r = \text{radiation resistance}$$

Radiation resistance is nothing like pure ohmic resistance, and nothing like ordinary impedance. Radiation resistance is merely a mathematical quantity that expresses the relationship between the antenna current and the power radiated in the form of radio waves.

EXAMPLE 8-5

A 300 Ω antenna is operating with 5 A of current. Determine the radiated power.

SOLUTION

From Equation 8-14,

$$P_r = I^2 R_r = (5^2)(300)$$
$$= 7500 \text{ W}$$

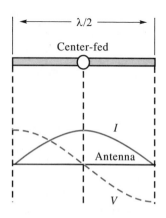

FIGURE 8-19 Current and voltage standing wave patterns for a center-fed half-wave diode

Antenna Radiation

The current and voltage standing wave patterns on a center-fed half-wave dipole are shown in Figure 8-19. Since both ends of the antenna appear as opens, the ends have a current minimum (node) and a voltage maximum (loop). The feed point at the center of the antenna has maximum current and minimum voltage, therefore, the input impedance of the half-wave dipole antenna is low. However, because of the energy lost to radiation, the input impedance is not zero as might be expected from the current and voltage distributions. Because radiated energy is not reflected back to the input, complete cancellation does not occur. For this reason, the input impedance of a half-wave dipole is about 72 Ω.

An excited dipole develops an *electric field* between its ends because of the high voltage potentials at these points. Also, a *magnetic field* develops around the antenna because of the current flow in the antenna. Since the current is maximum at the center of the antenna, so is the magnetic field, as illustrated by the center-fed half-wave dipole in Figure 8-20. The combined electric and magnetic fields radiate into space as an *electromagnetic wave*. This wave travels outward from the antenna and will continue traveling even after the current and voltage are removed from the antenna.

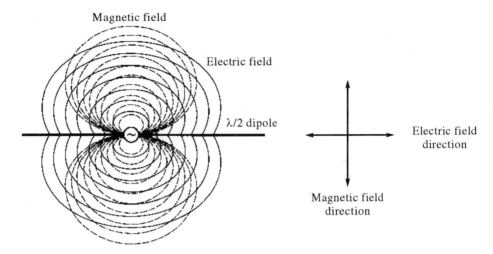

FIGURE 8-20 Electric and magnetic fields developed around a center-fed half-wave dipole

Radiation Patterns The most important property of an antenna is its pattern of radiation. For a transmitting antenna, this pattern is a graphical plot of the field strength radiated by the antenna in different angular directions. This same radiation pattern also indicates the receiving properties of the antenna, because the receiving and transmitting properties of an antenna are reciprocal.

The bidirectional radiation pattern for a horizontal Hertz antenna, as shown in Figure 8-21(A), illustrates how maximum radiation occurs broadside to the antenna and minimum radiation occurs off the ends of the antenna. A polar plot (looking down from above) of the radiation in the horizontal plane is shown in Figure 8-21(B). In three dimensions, this polar plot appears as a doughnut shape.

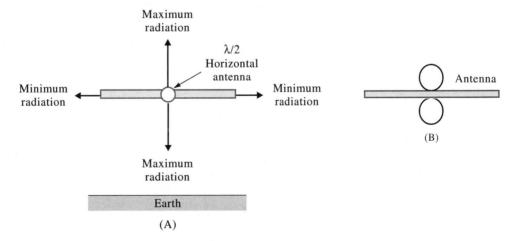

FIGURE 8-21 Bidirectional radiation patterns for a horizontal center-fed Hertz antenna (A) Ground-plane plot (B) Polar plot

The omnidirectional radiation pattern for a vertical Hertz antenna, as shown in Figure 8-22(A), illustrates that maximum radiation occurs in all directions in the horizontal plane and minimum radiation occurs directly above and below the antenna. A polar plot of the radiation in the horizontal plane is shown in Figure 8-22(B). In three dimensions, the polar plot appears as a half-doughnut shape.

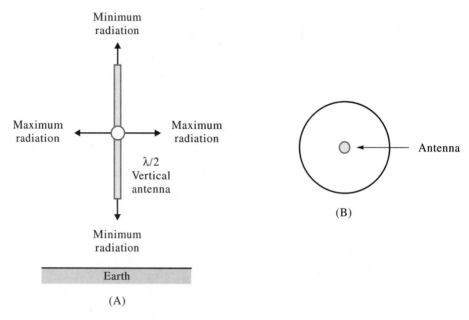

(A)

(B)

FIGURE 8-22 Omnidirectional radiation patterns for a vertical center-fed Hertz antenna (A) Ground-plane plot (B) Polar plot

A Hertz antenna can operate on harmonics of its fundamental frequency. For example, when operating on its second harmonic, the antenna functions as a full-wave dipole. In this case, minimum current occurs at the center of the antenna, indicating that the center of the antenna is a high impedance point. If a low impedance is desired, the antenna feed point must be moved to either of the current maximums, which are located $\lambda/4$ from either end of the antenna. Figure 8-23(A) shows the current distribution for a full-wave dipole. The radiation pattern in Figure 8-23(B) shows four major lobes, a result of the current distribution on the dipole.

(A)

(B)

FIGURE 8-23 Hertz antenna operating on its second harmonic as a full-wave antenna (A) Current distribution (B) Radiation pattern

A Hertz antenna can also operate on its third harmonic, where it functions as a one-and-one-half-wave antenna. The current distribution and radiation pattern for such an antenna are shown in Figures 8-24(A) and (B), respectively. Here, a current maximum occurs at the center of the antenna; hence, the low impedance feed point is at the center of the antenna. In the radiation pattern, four major lobes remain, but two minor lobes are present broadside to the antenna.

The harmonic operation of dipoles can be put to good use by designing an antenna for use on two harmonically related frequencies. For example, an antenna designed for the 7 MHz and 21 MHz amateur radio bands is cut to half-wave at 7 MHz. When operated at 21 MHz, the antenna functions as a one-and-one-half-wave antenna. Because both the half-wave and one-and-one-half-wave antennas have a current maximum at the center, they also have a low impedance center feed point. Therefore, they both offer a good impedance match to a 50 Ω or a 75 Ω transmission line.

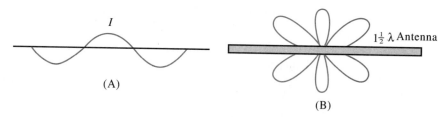

FIGURE 8-24   Hertz antenna operating on its third harmonic as a one-and-one-half wavelength antenna (A) Current distribution (B) Radiation pattern

Effective Radiated Power   *Effective radiated power* (ERP) is an important criterion for determining the effectiveness of a transmitting station. The ERP may be many times the transmitter power. The ERP specification is more often used at VHF than at HF and below.

When a transmitting antenna has power gain, the effect in the favored direction is equivalent to an increase in transmitter power. To obtain a doubling of the radiated power in a certain direction one of two things may be done: (1) the transmitter output can actually be doubled, or (2) the antenna power gain can be increased in that direction by 3 decibels. Effective radiated power takes into account both the antenna and the transmitter output power. It is also affected by losses in the transmission line and impedance devices.

With an antenna that shows zero gain with a dipole (0 dBd), the ERP is the same as the power P reaching the antenna feed point. If the antenna has a power gain of A_p dB, then:

$$A_p = 10 \, \log_{10}\left(\frac{ERP}{P}\right)$$

(8-15)

and the ERP is given by:

$$ERP = P\text{antilog}_{10}\left(\frac{A_p}{10}\right)$$

(8-16)

The greater the forward gain of the antenna, the greater the effective radiated power.

Effective radiated power is usually determined by using a half-wave dipole as the reference antenna. However, *ERP* may be determined with respect to an isotropic radiator.

An *isotropic antenna* is an antenna that radiates electromagnetic energy equally well in all directions. This antenna is a *theoretical construct, and does not actually exist*. However, the isotropic-antenna concept is occasionally used for antenna-gain comparisons.

The power gain of an isotropic antenna is about −2.15 dB with respect to a half-wave dipole in free space. That is, the field strength from a half-wave dipole antenna, in its favored direction, is approximately 2.15 dB greater than the field strength from an isotropic antenna at the same distance and at the same frequency.

The radiation pattern of the isotropic antenna—in three dimensions—appears as a perfect sphere, since the device works equally well in all directions. In any given plane the radiation pattern of the isotropic antenna is a perfect circle, centered at the antenna.

Antenna Power Gain   The *power gain* of an antenna is the ratio of the ERP to the actual RF power applied to the feed point. Power gain may also be expressed in decibels. If the effective radiated power is P_{ERP} watts and the applied power is P watts, then the power gain in decibels (dB) is:

$$\text{Power Gain (dB)} = 10 \, \log_{10}\left(\frac{P_{ERP}}{P}\right) \tag{8-17}$$

Power gain is always measured in the favored direction of an antenna (the azimuth direction in which the antenna performs the best). For power gain to be defined, a reference antenna must be chosen with a gain assumed to be unity, or 0 decibels. This reference antenna is usually a half-wave dipole in free space. Power gain figures taken with respect to a dipole are expressed in dBd. The reference antenna for power-gain measurements may also be an isotropic radiator, in which case the units of power gain are called dBi. For any given antenna the power gains in dBd and dBi are different by approximately 2.15 decibels.

$$\text{Power Gain (dBi)} = 2.15 + \text{Power Gain (dBd)} \tag{8-18}$$

Directional transmitting antennas can have power gains in excess of 20 dBd. At microwave frequencies, large dish antennas can be built with power gains of 30 dBd or more.

With a particular receiving antenna, power gain is the same for reception as for transmission of signals. Therefore, when antennas with directional power gain are used at both ends of a communications circuit, the effective power gain over a pair of dipoles is the sum of the individual antenna power gains in dBd.

Directional Gain   An advantage offered by an antenna system with good directivity is directional gain. *Directional gain* is the ratio of the power required to produce a given field strength at a given location using a reference antenna compared to the power required to produce the same field strength with a directional antenna. Reference antennas are usually half-wave dipoles or quarter-wave vertical antennas.

Because electrically large antenna systems can be constructed in relatively small spaces when operating in the VHF and UHF ranges, very high gains are obtainable.

Directional gain is a result of the radiated power being concentrated within a relatively narrow beam, and the field strength within this beam is much greater than would be obtained from an omnidirectional antenna. Thus, the *effective radiated power* (*ERP*) of the transmitter is increased by the directional gain of the antenna system. If transmitter output power, transmission line loss, and antenna gain are known, effective radiated power can be found:

$$ERP = (P_{out} - P_{L})A_{p}$$ (8-19)

where

P_{out} = transmitter output power

P_{L} = transmission line loss

A_{P} = antenna power gain

An example will demonstrate how the effective radiated power of a transmitter is greatly increased by a directional antenna.

EXAMPLE 8-6

Determine (a) the effective radiated power if transmitter output power is 200 W, transmission line loss is 20 W, and antenna power gain is 5; (b) *ERP* in dB.

SOLUTION

(a) $ERP = (P_{out} - P_{L})A_{P}$

$= (200 \text{ W} - 20 \text{ W})(5)$

$= (180)(5)$

$= 900 \text{ W}$

(b) $dB = 10 \log_{10}\left(\dfrac{P_{ERP}}{P}\right) = 1\left(\dfrac{900 \text{ W}}{200 \text{ W}}\right) = 4.5 \text{ dB}$

8-4   ANTENNA ARRAYS

In some situations, eliminating or reducing interference to or from other stations requires that the direction and bandwidth of a radiated wave be restricted within some specific limits. This situation usually results in more efficient operation, because the radiated power is directed toward a specific receiving station. In this way, very little power is wasted.

When the radiation pattern of an antenna is primarily in one direction, it is unidirectional. Figure 8-25 shows the radiation pattern for a unidirectional antenna. The *beamwidth,* measured from the radiation pattern as an angle between the two points on

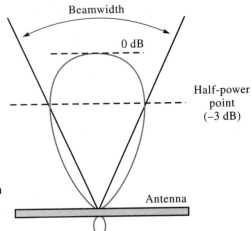

FIGURE 8-25
Unidirectional antenna
radiation pattern
showing directivity
and beamwidth

either side of maximum radiation where the field strength drops to the half-power point (−3 dB), determines the antenna's directivity. In general, the greater the power gain in any given antenna, the narrower the beamwidth.

Two or more antenna elements combined to form an *antenna array* are used to obtain good directivity and directional gain. Depending on the method used to excite the additional elements, antenna array systems are classified as parasitic, driven, or phased.

Parasitic Arrays

A voltage will be developed through induction across an antenna element that is not connected to the transmission line (not driven). Such an element is known as a *parasitic element.* A half-wave dipole with a parasitic element λ/4 removed from the dipole is shown in Figure 8-26. The dipole is connected to the transmission line and is, therefore, the *driven element.* Since both elements are λ/2 long, they are resonant at the operating frequency. Recall that the radiation pattern of the dipole is bidirectional and is maximum broadside to the antenna. The energy traveling toward the parasitic element travels λ/4; therefore, it goes through a 90° phase change before reaching that element. The wave cutting across the parasitic element induces a voltage that is 180° reversed with respect to the wave that induced it. The induced voltage results in current flowing through the element, and the element radiates energy. The radiated field in the direction beyond the parasitic element is opposite to the field from the driven element, resulting in cancellation of the two fields. Therefore, negligible radiation occurs in the *backward* direction. However, the parasitic element also radiates in the direction of the driven element, and, in doing so, it goes through another 90° phase change. The total phase change is 360°; therefore, the energy wave is in phase with the energy from the driven element. For these reasons, radiation toward the driven element is reinforced, resulting in maximum radiation. The direction of maximum radiation is called the *forward* direction, and the parasitic element is called the *reflector.*

For clarification, we will briefly discuss the theory of operation of a parasitic array. The driven element is excited by energy fed to it from a transmission line. If another

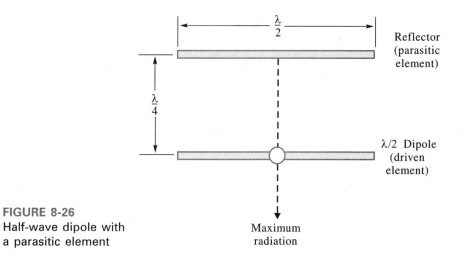

$$\frac{\lambda}{2}$$

Reflector
(parasitic
element)

$$\frac{\lambda}{4}$$

λ/2 Dipole
(driven
element)

Maximum
radiation

FIGURE 8-26
Half-wave dipole with
a parasitic element

antenna element is placed parallel to, and insulated from, the driven element, current will be induced from the driven element into the other element. Thus, the second element is said to be *parasitically excited.*

If the parasitic element is against, but insulated from, the driven element, at any given instant the currents in the two different elements are 180° out of phase (are flowing in opposite directions) because the counter-voltage generated by the current in the driven element is induced by the same field and is therefore in phase with the voltage source for the current in the parasitic element. Now, if the parasitic element is moved λ/4 away from the driven element, the time required for the radiated wave to reach the parasitic element is equal to the time required for the current in the driven element to complete one quarter-cycle. Therefore, the current induced into the parasitic element will be λ/4 (90°) further behind the driven element or, now, a total of 270° out of phase with the driven element current. However, the wave radiated from the parasitic element must travel the distance between the two elements (λ/4 back toward the driven element) before it can affect the current in the driven element. Since the parasitic element current is 270° behind the driven element current, the additional 90° delay gives a full 360° difference between the driven element's original current and the current induced into the driven element from the parasitic element. Since this phase difference is just exactly one full cycle, it has the same effect as no phase difference at all. In this manner, the parasitic element is acting both as a director (increasing the forward radiation) and as a reflector (increasing the backward radiation). Since more energy is being radiated forward and backward, there is less energy radiated to the sides—that is, less energy radiated in the directions to which the ends of the elements point.

To summarize, when a parasitic element is placed λ/4 from the driven element, the radiated wave travels 90° to the parasitic element, is reversed by 180° in the parasitic element, travels 90° back to the driven element, and arrives there in phase with the original wave. In doing so, a gain of approximately 5 dB in both the forward and backward directions (at the expense of the side radiation) is produced.

The radiation pattern obtained with parasitic arrays depends on the magnitude and phase of the current in the parasitic elements. These factors, in turn, depend on the length of the parasitic elements and on the spacing between the driven and parasitic elements. In the foregoing discussion, the parasitic element was used to increase the radiation in both the forward and the backward directions. However, if only forward gain is desired, we can use a *director*—a parasitic element that is resonant at a *higher* frequency than the actual operating frequency. On the other hand, if only backward gain is desired, we can use a *reflector*—a parasitic element that is resonant at a frequency *lower* than the actual operating frequency. For these reasons, *directors must be shorter* than $\lambda/2$, and *reflectors must be longer* than $\lambda/2$. Both elements are tuned for maximum gain when their lengths are approximately 5% different from the operating resonant length. Also, a director gives the greatest gain when it is placed about 0.1 λ in front of the driven element, and a reflector gives the greatest gain when it is placed about 0.2 λ behind the driven element. A three-element (one driven element plus one director and one reflector) antenna can produce a gain of about 8 dB in the forward direction, with a difference of approximately 20 dB between forward and backward radiation. More elements can be added in a progressive arrangement to produce still more forward gain.

The Yagi Antenna A special system by which shortwave antennas can be made highly directional was developed by Hidetsugi Yagi, a Japanese electrical engineer. Any antenna using the principles developed by Yagi is known as a *Yagi antenna.*

The Yagi system uses parasitic elements: directors and reflectors. One or more directors are placed in front of the driven element to direct the wave in a forward direction. A reflector is placed behind the driven element to reflect the wave back to the driven element. The driven element should be exactly resonant—exactly $\lambda/2$ at the operating frequency. The director elements must be shorter than $\lambda/2$, and the reflector must be longer than $\lambda/2$. All elements are mounted parallel to one another.

A three-element Yagi antenna is shown in Figure 8-27(A). The radiation patterns of the Yagi antenna are shown in Figure 8-27(B). Note how the directivity has improved

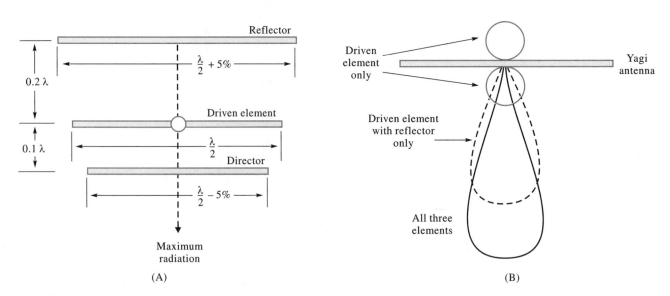

FIGURE 8-27 Parasitic array (A) three-element Yagi antenna (B) Radiation pattern showing improved directivity

by the addition of only one director. Even greater directivity and directional gain would be obtained by adding more directors.

The Folded Dipole   One disadvantage of the Yagi antenna is its low radiation resistance. Adding parasitic elements reduces the radiation resistance of the array, making it somewhat difficult to feed. A center-fed half-wave dipole has a radiation resistance of approximately 72 Ω; a center-fed, closely spaced three-element Yagi antenna has a radiation resistance of approximately 10 Ω. Even though certain impedance-matching systems can be used to feed such an antenna, a simpler solution to the problem is to use a *folded dipole* as a driven element—that is, to use the folded dipole in place of the simple dipole discussed up to this point.

The folded dipole, like the simple dipole, should measure λ/2 from end to end. But in addition to the single conductor used for the simple dipole, the folded dipole has another conductor in parallel with the simple dipole and connected to it at both ends, as shown in Figure 8-28. The radiation resistance of this simplest type of folded dipole is approximately 300 Ω. Higher radiation resistances are obtained by making the folded element larger than the original element or by using an additional folded element.

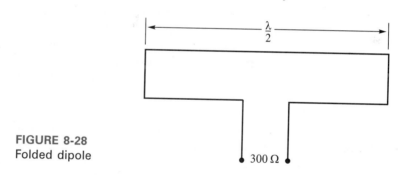

FIGURE 8-28
Folded dipole

Driven Arrays

An antenna system in which all the elements are fed by the transmission line is called a *driven array.* There are no parasitic elements in driven arrays.

Collinear Array   A four-element driven array and its radiation pattern are shown in Figure 8-29. Because all elements are placed in line or end to end, this configuration is called a *collinear array.* The transmission line is connected to each element, and, since the transmission line length to each element is equal, the current in each element is in phase. These collinear arrays are usually vertically oriented to produce omnidirectional coverage. However, the energy is directed down toward the horizon, thus providing greater distance coverage for operation at the VHF and UHF ranges.

Log-Periodic Array   When the element lengths of a driven array are logarithmically related, the antenna system is called a *log-periodic array.* Each element is fed through a special phasing network. This type of antenna provides good directivity and directional gain. Its greatest advantage is its very wide bandwidth, since it can operate over a much wider range of frequencies than can most Yagi arrays. The antenna shown in

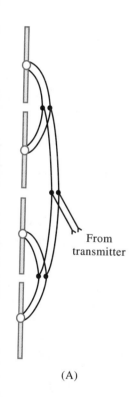

(A)

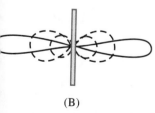

(B)

GURE 8-29   Driven array
) Four-element collinear
ray (B) Radiation pattern

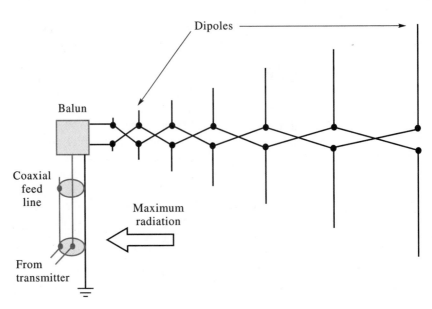

FIGURE 8-30   Log-periodic dipole array

Figure 8-30 offers high directional gain and a VSWR less than 2:1 from 7.5 MHz to 30 MHz. Compare this capability to that of most Yagi arrays, which at best can cover only a 1 or 2 MHz band in this same frequency range.

Phased Arrays

A *phased-array* is an antenna with two or more driven elements. The elements are fed with a certain relative phase, and they are spaced at a certain distance, resulting in a directivity pattern that exhibits gain in some directions and little or no radiation in other directions. Some phased arrays include a reflecting device or parasitic elements for improved directivity and front-to-back ratio.

Phased arrays can be very simple, consisting of only two elements, although more complicated arrangements are sometimes used by radio transmitting stations. For example, several vertical radiators, arranged in a specified pattern and fed with signals of specified phase, produce a designated directional pattern. Stations may use this system to avoid interference with other broadcast stations on the same channel or to reduce radiation to areas where limited reception is likely, such as coastal or rugged mountain areas. Also, by using phased arrays, the radiation pattern can be electronically controlled so that a swept beam can be produced without mechanically moving the antenna.

Phasing   *Phasing* is the technique that gives a phased array a certain directional characteristic. A device that accomplishs this phasing is sometimes called a *phasing harness*.

Depending on the relative phase and the spacing of the elements in a phased antenna, the radiation pattern may have one, two, three, or even four major lobes. Phasing harnesses usually consist of simple delay lines. A transmission line, measuring an electrical quarter-wavelength, produces a delay of 90°. A half-wavelength section of line causes the signal to be shifted in phase by 180°.

Phased Dipoles Two examples of simple pairs of phased dipoles are shown in Figure 8-31. In Figure 8-31(A), the two dipoles are spaced λ/4 apart in free space, and they are fed 90° out of phase. The result is that the signals from the two antennas add in phase in one direction and cancel in the opposite direction, as shown by the arrows. In this particular case, the radiation pattern is unidirectional. However, phased arrays may have directivity patterns with two or more different optimum directions. A bi-directional pattern can be obtained, for example, by spacing the dipoles at 1 λ and feeding them in phase, as shown in Figure 8-31(B).

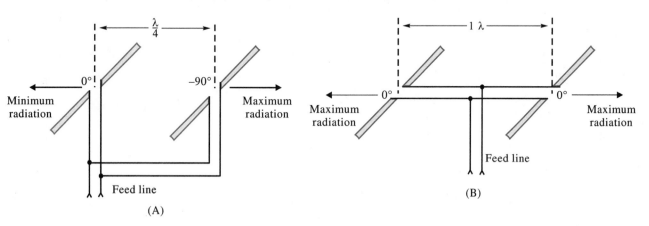

FIGURE 8-31 Phased dipoles (A) Unidirectional (B) Bidirectional

Phased arrays may have fixed directional patterns, or they may have rotatable or steerable patterns. If the wavelength is short enough to allow construction from metal tubing, the pair of phased dipoles in Figure 8-31(A) may be mounted on a rotator for 360° directional adjustability. With phased vertical antennas, the relative signal phase can be varied, and the directional pattern thereby adjusted to a certain extent.

Billboard Antenna A *billboard* antenna (Figure 8-32) is a form of broadside array consisting of a set of dipoles to which a reflecting screen has been added. The reflecting screen is positioned in such a way that the electromagnetic waves at the resonant frequency are reinforced as they return in the favored direction. The more dipoles in a billboard antenna, the greater the forward gain. A billboard antenna has an excellent front-to-back ratio because of the reflecting screen. Billboard antennas are used at UHF and above. Below about 150 MHZ, the physical size of a billboard antenna becomes prohibitively large.

End-Fire Antenna An *end-fire* antenna is a bidirectional or unidirectional antenna in which the greatest amount of radiation takes place off the ends. Although a parasitic array is sometimes considered an end-fire antenna, the term is used primarily for phased array systems. A phased array end-fire antenna consists of two or more parallel driven elements, with all the elements in a single plane.

Figure 8-33 shows a typical end-fire array consisting of two parallel half-wave dipole antennas driven 180° out of phase and spaced λ/2 apart in free space. This

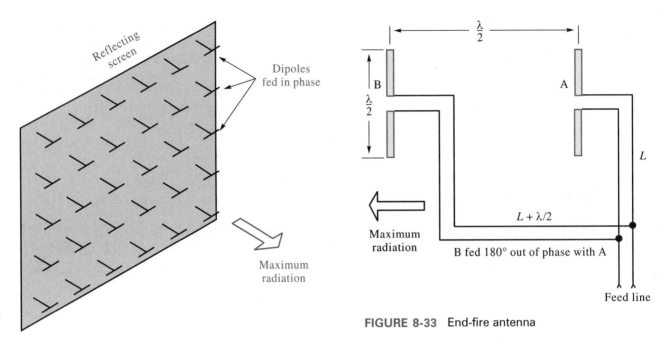

FIGURE 8-33 End-fire antenna

FIGURE 8-32 Billboard antenna

arrangement results in a bidirectional radiation pattern. In the phasing system, the electrical lengths of the two branches of transmission line differ by 180°.

Compared to a single half-wave dipole antenna, end-fire antennas show some power gain in their favored directions. This power gain depends on the number of elements, the phasing, and the spacing: The larger the number of elements, with optimum phasing and spacing, the greater the power gain of the end-fire antenna.

8-5   OTHER ANTENNAS

The antennas discussed thus far are among the major types with which you may come into contact as a technician. However, you should be aware that there are many other types of antennas. Many are simply variations of those types already discussed, but some others have special applications. In this section, we will briefly discuss several of these specialized antennas.

Longwire Antenna

A wire antenna measuring 1 λ or more and fed at a current loop at one end is called a *longwire* antenna. Longwire antennas are sometimes used for receiving and transmitting at frequencies in the MF and HF bands.

Longwire antennas have certain advantages. They offer some power gain and low-angle radiation. The gain depends on the length of the wire: The longer the wire, the greater the power gain. Longwire antennas are inexpensive and easy to install, provided there is sufficient space. For proper operation, the longwire antenna must be as straight as possible.

There are two main disadvantages to the longwire antenna.

1. It cannot conveniently be rotated to change the direction in which maximum gain occurs.
2. A great amount of space is required, especially in the HF spectrum and the longer MF wavelengths. (For example, a 10 λ longwire antenna measures about 1340 feet at 7 MHz. This is more than $\frac{1}{4}$ mile.)

Biconical Antenna

A *biconical* antenna is a balanced broadband antenna that consists of two metal cones arranged so that they meet at or near the vertices, as shown in Figure 8-34. The antenna is fed at the point where the vertices meet. The exact feed-point impedance depends on the flare angle of the cones and the separation between the vertices.

A biconical antenna displays resonant properties at frequencies above that at which the height (h) of the cones is λ/4 in free space. The highest operating frequency is several times the lowest operating frequency. When oriented vertically, as shown, the antenna emits and receives vertically polarized electromagnetic waves.

Often used at VHF and above, the biconical antenna becomes prohibitively large in size at lower frequencies. However, one of the cones can be replaced by a disk or ground plane for the purpose of reducing the physical dimensions of the antenna while retaining the broadband characteristics. If the top cone is replaced by a disk of a certain radius, the antenna becomes a *discone* antenna. If the lower cone is replaced by a ground plane, the antenna becomes a *conical monopole* antenna. Both the discone and the conical monopole are practical at frequencies as low as about 2 MHz.

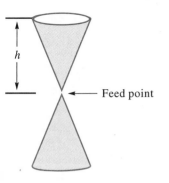

FIGURE 8-34
Biconical antenna

Bowtie Antenna

The *bowtie* antenna is a broadband antenna often used at VHF and UHF. It consists of two triangular pieces of stiff wire or two triangular flat metal plates, as shown in Figure 8-35. The feed point is at the gap between the apexes of the triangles. If a reflecting screen is used, unidirectional operation can be obtained.

The bowtie antenna is a two-dimensional form of the biconical antenna, and it obtains its broadband characteristics according to the same principles as the biconical antenna. The feed-point impedance depends on the apex angles of the two triangles. The polarization is along a line running through the feed point and the centers of the triangle bases.

Bowtie antennas are occasionally used for transmitting and receiving frequencies as low as 20 MHz, but below that frequency the size of the antenna becomes prohibitive.

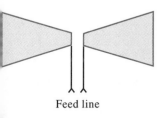

FIGURE 8-35
Bowtie antenna

Dish Antenna

A *dish* antenna is a high-gain antenna used for reception and transmission of UHF and microwave signals. It consists of a driven element or other form of radiating device and a large spherical or parabolic reflector, as shown in Figure 8-36. The driven element is placed at the focal point of the reflector.

Signals arriving from a great distance and in parallel waveforms are reflected off the dish and brought together at the focal point. Energy radiated by the driven element is reflected by the dish and sent out as parallel waves.

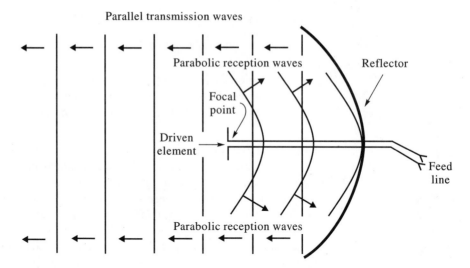

Parallel transmission waves

Parabolic reception waves

Reflector

Focal point

Driven element

Feed line

Parabolic reception waves

FIGURE 8-36 Dish antenna

For proper operation, a dish antenna must be several wavelengths in diameter; otherwise, the waves tend to be diffracted around the edges of the dish reflector. For that reason, the dish is impractical for frequencies below the UHF range. The reflecting element may be made of sheet metal or from a screen or wire mesh, in which case the spacing between the screen or mesh conductors must be a very small fraction of a wavelength in free space.

Dish antennas, which include the *paraboloid* and *spherical* antennas, typically show very high gain. The larger a dish with respect to a wavelength, the greater the gain of the antenna. It is essential that a dish antenna be correctly shaped and that the driven element be located at the focal point. Dish antennas are used in microwave relay systems, radar systems, satellite communications systems, and some television receiving systems.

EXAMPLE 8-7

Assume a 6-foot dish antenna with a mouth diameter of 2 meters, operating on a frequency of 4 GHz. Determine (a) the wavelength, (b) the beam width of the signal, and (c) the gain of the antenna. ($\phi = 70 \lambda/D$, where D = mouth diameter).

SOLUTION

(a) $\lambda = \dfrac{3 \times 10^8}{4 \times 10^9} = 0.075$ meters

(b) $\phi = \dfrac{70 \lambda}{2} = 2.6°$

(c) $A_p = \text{dish size} \left(\dfrac{D}{\lambda}\right)^2 = 4{,}267$, or 36 dB

Whip Antenna

A *whip* antenna is a short-radiator, usually loaded at the base and measuring one quarter physical wavelength or less. Is is often used in mobile communications, especially at high frequencies.

Figure 8-37 shows the construction of a typical whip antenna. A tapered rod of stainless steel forms the radiator. A tapped coil at the base facilitates adjustment of the resonant frequency. A spring allows for wind resistance. A magnetic mounting can be used for small whips. For antennas longer than about 3 feet, a ball mounting is generally used.

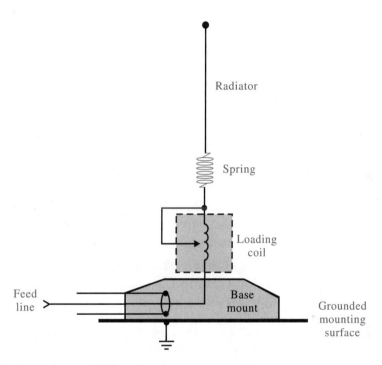

FIGURE 8-37   Whip antenna

The efficiency of a whip antenna depends on the size of the vehicle on which it is mounted: The larger the vehicle, the better the grounding system and the better the antenna performance. In general, efficiency improves as the frequency is increased and worsens as the frequency is decreased.

8-6   RECEIVING ANTENNAS

Any transmitting antenna will function well for receiving purposes within the same range of frequencies. However, receiving antennas need not be as sophisticated or as large as transmitting antennas to function well, especially at VLF, LF, MF, and HF. Because of the nature of electromagnetic waves, tiny antennas can function well at these frequencies, even though they are virtually useless for transmission. At VHF, UHF, and microwave frequencies, large dish antennas may be necessary.

Random Wire Antenna

The simplest type of receiving antenna is the *random wire,* connected to the receiver input terminals either directly or through a tuning network. The antenna may consist of a simple end-fed wire, or a horizontal conductor can be fed by a lead-in wire, as shown in Figure 8-38. Such an antenna is usually made as long as possible. This type of antenna will work well for receiving all radio frequencies into the VHF or even UHF range, but it can provide only marginal performance for transmitting.

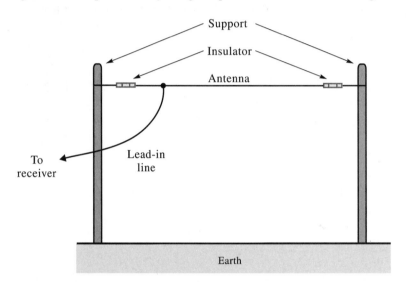

FIGURE 8-38 Random-wire antenna

Tuned Antenna

Tuned receiver antennas provide a high degree of selectivity and some noise suppression. The most common tuned receiver antennas are the *ferrite rod* and the *loop.* These antennas can be used at frequencies from about 30 MHz down to about 10 kHz with excellent results. Despite their physically tiny size, the ferrite rod and the loop perform very well for receiving at the VLF, LF, and MF ranges, with marginal performance up to about 30 MHz. Above that frequency, the ferrite rod develops a high loss, and open loop antennas are not much smaller than a half-wave dipole or a full-wave loop.

At VHF and above, the same antenna is generally used for transmitting and for receiving because physical size is not a major problem. When the same type of antenna is used for transmitting and receiving, the directional characteristics work to advantage in both modes.

Ferrite Rod Antenna   A *ferrite rod* antenna, sometimes called a *loop-stick,* is often used for receiving applications at LF, MF, and HF, up to about 20 MHz. This type of antenna consists of a coil wound on a solenoid ferrite core. A series or parallel capacitor in conjunction with the coil forms a tuned circuit. The operating frequency is determined by the resonant frequency of the inductance-capacitance combination.

These antennas display directional characteristics similar to those for the dipole antenna; that is, the sensitivity is maximum broadside to the coil, and a sharp null occurs off the ends. The null has little effect on sky wave signals, which tend to arrive from various directions. However, the null can be used to minimize interference from local signals and from man-made sources of noise.

The ferrite rod antenna is physically very small. Its small size makes it easy to orient the rod in any direction. Some ferrite rod antennas are affixed to mountings that allow both vertical and horizontal adjustment for best reception.

Loop Antenna Any antenna consisting of one or more turns of wire forming a dc short circuit is called a *loop* antenna. Loop antennas can be categorized as either small or large. Small loops, suitable for receiving signals up to about 30 MHz, have a circumference of less than 0.1 λ at the highest operating frequency. Such antennas exhibit a sharp null along the loop axis. The loop may be electrostatically shielded to improve the directional characteristics.

A popular type of receiving antenna is the *vertical loop* antenna. This antenna consists of one or more turns of wire wound in form of a circle, a square, or a rectangle, as shown in Figure 8-39. The vertical loop antenna will pick up bidirectional signals, or, if electrostatically shielded, it will provide unidirectional reception. Because of its excellent directivity capability, the vertical loop antenna is often used in direction-finding equipment.

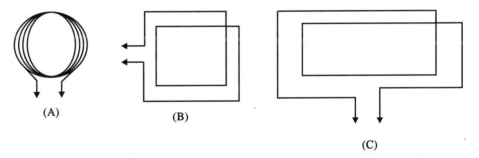

FIGURE 8-39 Typical vertical loop antennas (A) Circle loop (B) Square loop (C) Rectangle loop

Low-Frequency Receiving Antenna Characteristics The frequency, directional, physical, and polarization discrimination characteristics of receiving antennas apply equally to the HF, VHF, and UHF ranges. However, these characteristics are not critical for antennas in the MF and LF ranges. Therefore, the length, location, and polarization of standard AM broadcast receiving antennas are normally of little importance.

8-7 IMPROVING SIGNAL-TO-NOISE RATIO IN RECEIVING ANTENNAS

The basic requirement for a good receiving antenna is that it have a high signal-to-noise (*S/N*) ratio—that is, that it provides a strong signal, with as little noise as possible, to the receiver. A high *S/N* ratio can be attained in four ways: by frequency discrimination, by directional discrimination, by physical discrimination, and by polarization discrimination.

Frequency Discrimination

A high *S/N* ratio is best achieved when the antenna is resonant at the received frequency, because this gives a strong signal response and at the same time attenuates interference and noise at other frequencies. Using a resonant antenna in this manner is referred to as improving the *S/N* ratio of the received signal by *frequency discrimination.*

Directional Discrimination

Another means of improving the *S/N* ratio of the received signal is to use a directional antenna. This antenna provides *directional discrimination,* where the gain for signals coming from the selected direction is increased and interference and noise coming from other than the selected direction are attenuated. The Yagi antenna is an example of using both frequency discrimination and directional discrimination at the same time.

Physical Discrimination

An antenna should be physically isolated from objects that may interfere with reception of desired signals. This setup is known as *physical discrimination.* It is obtained by placing the antenna so that the path of the signal is not interrupted by any object. In some cases, this requirement may make it necessary to place the antenna far from the receiver. If so, a good low-resistance lead-in (transmission) line must be used or the advantage of the better physical location may be overcome by transmission line losses. To prevent serious attenuation of the signal in the line, both the pure resistance and the current in the line must be low. The transmission line current for a given power can be kept low by using a line having a high characteristic impedance.

Figure 8-40 shows an example of physical discrimination.

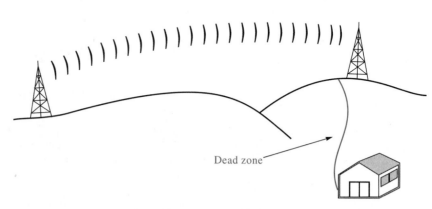

Dead zone

FIGURE 8-40 Example of physical discrimination

Polarization Discrimination

In general, the best reception is obtained when the polarization (physical orientation) of a receiving antenna is the same as that of the transmitting antenna from which the signal is coming. This arrangement is called *polarization discrimination.* For example, if a signal is being transmitted from a horizontal antenna, the best reception will be obtained by using a horizontal receiving antenna. Conversely, if the signal is transmitted from a vertical receiving antenna, the best reception will be obtained by using a vertical receiving antenna.

SUMMARY

- An antenna converts electrical energy into electromagnetic waves for transmission and electromagnetic waves into electrical energy for reception.

- The capability of interchanging an antenna's functional use between transmitting and receiving is called reciprocity.

- Antennas are constructed with conductors in the form of wires or rods. Maximum efficiency occurs when transmitting and receiving antennas for a signal have the same polarization. Polarization is the physical position of an antenna and determines the radiation pattern of the antenna. The intended use of an antenna determines the physical position. A vertically polarized antenna is a nondirectional or omnidirectional antenna and is used primarily for AM signals. A horizontally polarized antenna is a directional antenna and is used primarily for television and FM signals.

- Field strength is the measured signal strength at a specified distance from an antenna. It depends on the distance from the antenna and on the antenna radiated power. It varies inversely with distance, and it is evaluated by the voltage induced across a wire.

- Antenna excitation generates antenna current, which causes radio waves to be radiated into space. Antenna current is strongest, and maximum power is transferred from a transmitter to an antenna, when the antenna is resonant to the frequency of the generated signal.

- Standing waves on antennas have current and voltage loops and nodes. Antenna standing wave current and voltage curves are commonly called displacement curves.

- End effect is the result of capacitance between the ends of an antenna. Because of end effect, the wave developed on an antenna appears to be about 5% longer than its actual physical length.

- The two principal categories of transmitting antennas are the half-wave and the quarter-wave. The half-wave antenna, variously called a dipole, a doublet, or a Hertz antenna, can be vertically or horizontally polarized. Because of a mirror image, a vertical quarter-wave antenna, when constructed with a perfect ground, has the same characteristics as a half-wave dipole. The mirror image causes the ground to act as a quarter-wave to match the quarter-wave above the ground. A grounded quarter-wave antenna is more commonly known as a Marconi antenna.

- Connecting the RF output of a transmitter to an antenna is called feeding the antenna. The process of bringing an antenna into exact resonance with a transmitter's output is known as loading the antenna. An antenna can be tuned to a higher frequency by adding a capacitor in series with the antenna, and to a lower frequency by adding an inductor in series with the antenna.

- Guy wires hold an antenna in position. The wire lengths must be such that they do not reflect any of the RF energy radiated from the antenna. Strain insulators are used to vary the wire lengths.

- A dummy (artificial, phantom) antenna is a nonradiating antenna used for preliminary transmitter tuning and for approximating the power output of the transmitter.

- Radiation resistance—a special property of an antenna that converts electrical energy into radio waves rather than into heat—is a mathematical quantity that expresses the relationship between the antenna current and the power radiated in the form of radio waves.

- An electric field is developed between the ends of an antenna, and a magnetic field is developed around an antenna. The combined electric and magnetic fields radiate into space as electromagnetic waves. The most important property of an antenna is its pattern of radiation.

- A Hertz antenna can operate on harmonics of its fundamental frequency, resulting in full-wave or one-and-one-half-wave antennas. The harmonic operating quality of the Hertz antenna is valuable when two harmonically related frequencies are used on a single antenna.

- The directivity of an antenna is determined by the beamwidth, measured from the radiation pattern as an angle between the two points on either side of maximum radiation where the field strength drops to the half-power point.

- An antenna array is a combination of two or more antenna elements. Arrays offer good directivity and directional gain. Array systems are classified as parasitic, driven, or phased arrays.

- A parasitic element is an element that is not connected to the energy source (not driven). It receives its energy through induction. In a parasitic array, there is one driven element and one or more parasitic elements, called reflectors or directors. Yagi antennas are the most commonly used parasitic arrays, used primarily for television reception.

- A driven array is an antenna system in which all elements are driven—that is, are connected to the energy source. A collinear array is usually vertically oriented to produce omnidirectional coverage. In this array, all elements are placed end to end with each element connected to transmission lines of equal length. In a log-periodic array, the element lengths are logarithmically related, with each element being fed through a special phasing network. This array provides good directivity and directional gain and has a wide bandwidth.

- Phased arrays have driven elements that are fed in a specified phase relationship. These arrays may or may not have parasitic elements.

- Receiving antennas commonly used in radio receivers are the ferrite rod and loop antenna. The vertical loop antenna is bidirectional, but it can be used with a shield to give unidirectional reception. It is often used in direction-finding equipment. Because standard AM broadcast receiving antennas operate in the LF and MF bands, their length, location, and polarization are not important. The *S/N* ratio in receiving antennas can be improved through frequency, directional, physical, and polarization discrimination.

QUESTIONS AND PROBLEMS

1. List the main elements of an antenna system.

2. Define reciprocity.

3. What determines the polarization of an antenna?

4. Briefly describe a vertically polarized antenna and its radiation pattern.

5. Why do standard broadcast stations use vertical antennas?

6. Briefly describe field strength and how it is measured.

7. Briefly describe a horizontally polarized antenna and its radiation pattern.

8. Which communications systems use horizontal antennas? Why?

9. Briefly describe antenna excitation and its result.

10. Explain what condition or conditions must exist to obtain (a) maximum power transfer to an antenna, (b) maximum antenna current, and (c) maximum radiation of radio waves.

11. Describe antenna standing waves.

12. State the relationship between antenna resonance and antenna wavelength.

13. Calculate the wavelength in meters of a 1450 kHz signal propagated into free space.

14. Calculate the physical length in feet of a half-wave dipole operating at (a) 1550 kHz and (b) 100.7 MHz.

15. Describe end effect and how it affects the physical length of an antenna.

16. Taking end effect into account, determine the physical length in feet of a half-wave dipole operating at (a) 10 MHz, (b) 90 MHz, and (c) 108 MHz.

17. Describe how a quarter-wave antenna behaves electrically as a half-wave antenna.

18. Describe a practical method for feeding an antenna.

19. Explain why using a coupling transformer or a coupling capacitor is not a practical method for feeding an antenna.

20. What is the effect of feeding a dipole with a coaxial transmission line? How can this effect be overcome?

21. Explain practical methods for tuning a fixed-length Marconi antenna to (a) a higher frequency and (b) a lower frequency.

22. An antenna has an inductance of 5 μH and a capacitance of 12 pF. Find (a) the resonant frequency to which the antenna is tuned, (b) the new capacitance value needed to tune the antenna to double that frequency, and (c) the value for the required series capacitance.

23. For the original antenna in Problem 22, find (a) the new inductance value needed to tune the antenna to one-half of the original frequency and (b) the value for the required series inductance.

24. Explain how a fixed-length antenna can be used to radiate at different frequencies.

25. Why are guy wires used?

26. Explain why guy wires use strain insulators.

27. What type of strain insulators should be used in guy wires? Why?

28. What is the effect of making transmitter adjustments while an antenna is being excited? How can this effect be avoided?

29. Describe antenna radiation characteristics.

30. Define radiation resistance.

31. A 75 Ω antenna is operating with 3 A of current. Find its radiated power.

32. An antenna operating with 4.1 A of current is radiating 5000 W of power. Find its radiation resistance.

33. A 125 Ω antenna is radiating 10 kW of power. Find the antenna current.

34. A transmitter has an output power of 50 W, a line loss of 5 W, and an antenna gain of 8. Find the effective radiated power.

35. A transmitter has P_{out} = 250 W, P_L = 40 W, and *ERP* = 1000 W. Find A_P.

36. A transmitter has *ERP* = 750 W, P_L = 30 W, and A_P = 3. Find P_{out}.

37. A transmitter has P_{out} = 150 W, A_P = 6, and *ERP* = 900 W. Find P_L.

38. Define (a) parasitic element, (b) director, and (c) reflector.

39. Explain how a Yagi antenna can be made (a) to provide forward gain only and (b) to provide backward gain only.

40. What is a disadvantage of the Yagi antenna? How can this disadvantage be overcome?

41. Describe the folded dipole. What is its normal radiation resistance?

42. Explain the methods used to improve the *S/N* ratio in receiving antennas. What antenna uses two methods at the same time?

43. What are two popular types of receiving antennas?

44. Describe the vertical loop antenna.

45. In what type of equipment is the vertical loop antenna often used? Why?

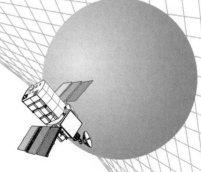

CHAPTER 9

Telephone Systems

OBJECTIVES

When you complete this chapter, you should be able to:

- List the functions of a telephone set.
- Explain how telephone connections are made in the local loop.
- Distinguish between pulse dialing and tone dialing.
- Describe the public switched telephone network.
- Identify and describe the telephone exchange designations for North America.
- Define local network, exchange area network, and long-haul network.
- Explain the types of transmission used for telephone networks.
- Describe and explain the bandwidth, the energy level, and the noise for telephone voice channels.
- List and describe the four methods used for signaling transmission.
- Define and explain the two basic types of multiplexing.

INTRODUCTION

The public switched telephone network in the United States is one of the most sophisticated communications systems in the world. This wireline system provides the ability to interconnect any two out of more than 100 million telephones, usually within a few seconds of the request for connection. Controlled by the world's largest network of interconnected and cooperating computers, the telephones in this network are usable by unskilled operators.

Prior to World War II, about the only type of pulse modulated (digital) communication was telegraphy. Since that time, however, pulse modulated systems have increased at a tremendous pace. Today, signals in many communications systems, and particularly in telephone systems, are in pulse form. The reasons for this increased use of digital techniques for communication are twofold:

1. Large amounts of the information to be transmitted are already in pulse form, so it logically follows that transmitting the information in that form is the simplest technique.

291

2. Large-scale integration (LSI) has produced complex gating circuits in the form of integrated circuits (ICs), which reduce costs and provide ease of operation.

In this chapter, we will examine the telephone network and the technology that has made it possible, and we will consider transmission methods and multiplexing.

9-1 TELEPHONE SYSTEM FUNDAMENTALS

The telephone arrived as a practical instrument on March 10, 1876, as an outgrowth of experiments on a device to send multiple telegraph signals over a single wire. While conducting experiments, Alexander Graham Bell spilled acid on his trousers. He reacted by shouting, "Mr. Watson, come here, I want you." Thomas A. Watson heard these words clearly, carried by electricity, on a receiving set in another room. As a result of this "accidental" discovery, rapid improvement was made on that simple instrument being tested on that day in 1876. The concept has grown into an industry that provides over 100 million telephone sets, profits of several billion dollars a year, and employment for more than a million people.

The Telephone Set

Telephone sets used to originate and receive telephone calls are simple in appearance and operation, yet they perform a number of important functions. The telephone set:

1. Automatically requests the use of the telephone system when a caller lifts the handset.
2. Indicates that the system is ready for use by receiving a dial tone.
3. Sends the number of the telephone to be called to the system when the caller initiates the number by rotating a dial or by pressing number keys.
4. Indicates to the caller the status of a call in progress by receiving tones that indicate ringing, busy, out of operation, or the like.
5. Indicates an incoming call to the called telephone by some type of audible tone.
6. Changes speech of a calling party to electrical signals for transmission to a distant party through the system, then changes the electrical signals received to speech for the called party.
7. Automatically adjusts for changes in the power supplied to it.
8. Signals the system that a call is finished when a caller hangs up the handset.

Rotary Dialer Figure 9-1 shows the block diagram of a telephone set. In the conventional set, a *rotary dialer* is used. The rotary dialer interrupts the telephone circuit a specific number of times for the dialing of a number. This arrangement is called *pulse* dialing. For example, if the dialed digit n is between 1 and 9 inclusive, the rotary dial interrupts the system n times. If $n = 0$, the dialer interrupts the circuit 10 times.

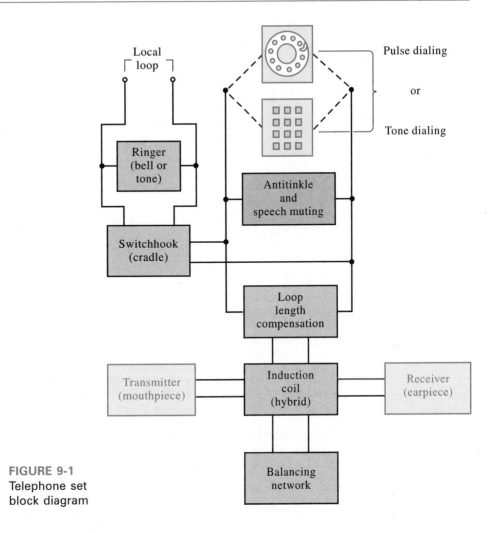

FIGURE 9-1
Telephone set
block diagram

The rotary dialer is a spring-loaded, circular wheel with 10 finger holes for the 10 digits. To dial a given digit, the finger is placed in the appropriate hole and the dialer is turned clockwise until the finger reaches the stop. The dialer then turns counterclockwise by itself, interrupting the circuit at a rate of about 5 Hz.

Touchtone®   In recent years, an increasing number of telephone sets have been using *pushbutton* dialing systems instead of the rotary dialer. The buttons actuate tone pairs that cause automatic dialing of the numbers. This system is called *Touchtone®* dialing (the term is a registered trademark of American Telephone and Telegraph Company, AT&T).

The original Touchtone® dial, still found on most telephone sets, has 12 buttons which correspond to digits 0–9, the star symbol (*), and the pound symbol (#). Some dialers have four additional keys, designated A, B, C, and D, as shown in Figure 9-2. The tone pair frequencies for the 16 designators are listed in Table 9-1.

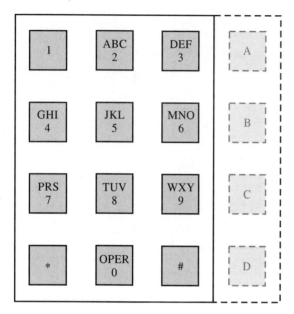

FIGURE 9-2 Extended pushbutton (DTMF) keypad

TABLE 9-1 Tone Pair Frequencies for Pushbutton Telephones

Button Designator	High Group Frequency (Hz)	Low Group Frequency (Hz)
1	1209	697
2	1336	697
3	1477	697
4	1209	770
5	1336	770
6	1477	770
7	1209	852
8	1336	852
9	1477	852
*	1209	941
0	1336	941
#	1477	941
A	1633	697
B	1633	770
C	1633	852
D	1633	941

Although the original pushbutton keypads were called Touchtone pads, other manufacturers have produced similar arrangements known by different names. All arrangements use the common combinations of frequencies listed in Table 9-1 to access the telephone systems. These systems offer what is called dual-tone, multiple-frequency (DTMF) access. Proper operation requires that both tones be present. The DTMF system can be used to remotely control distant objects or electronic equipment.

Tone dialing operation has several advantages over the older rotary dial. Tone dialing is faster and can be done automatically with extreme speed. The tones can be transmitted into a telephone set from an external source, which is not possible with the rotary dial. Many radio repeaters, for example, are interconnected with the telephone lines, so that mobile radio operators equipped with tone keypads in their transceivers can access the lines.

Cellular telephones have become more common over time. These portable systems operate on the following frequency ranges:

824 to 849 MHz (Transmit)

869 to 894 MHz (Receive)

Cellular systems are discussed in detail in Chapter 15.

The Local Loop (Central Office Exchange)

For a telephone to be of any use, it must be connected to another telephone. In the very early days of telephony, telephones were simply wired together with no switching.

This arrangement became impractical as the number of telephones increased, and the *local loop*, or *central office exchange* (CO), was established to handle the switching and other functions.

Today, each subscriber telephone is connected to a central office that contains switching and signaling equipment and batteries that supply dc to operate the telephone. Figure 9-3 shows the simplified circuits for a telephone set and a central office exchange. Each telephone is connected to the central office through a local loop of two wires called a wire pair. One of the wires is designated T (for *tip*); the other, R (for *ring*). Switches in the central office respond to the dial pulses or tones from the telephone to connect the calling telephone to the called telephone. When the connection is established, the two telephones communicate over transformer-coupled loops using the current supplied by the central office batteries.

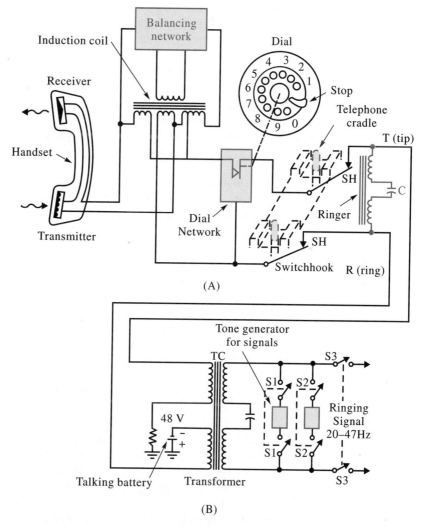

FIGURE 9-3   Simplified circuits (A) Telephone set (B) Central office exchange

Initiating a Call When the handset of the telephone is resting in its cradle, called the *on-hook* condition, the weight of the handset holds the switchhook buttons down and the switches are open. The circuit between the telephone handset and the central office is open; however, as shown in Figure 9-3, the ringer circuit is always connected to the central office. The capacitor *C* blocks the flow of dc from the battery but passes the ac ringing signal. Because of the high impedance presented to speech signals, the ringer circuit has no effect on those signals.

When the handset is removed from its cradle, called the *off-hook* condition, the spring-loaded buttons come up and the switchhook closes. This completes the circuit to the exchange, and current flows in the circuit. The off-hook signal tells the exchange that someone wants to make a call. The exchange returns a dial tone to let the caller know it is ready to accept a telephone number.

Sending a Number Two methods are used by telephone sets to send the telephone numbers: *dial pulses* and *audio tones*. Older conventional telephone sets, using dial pulses, have a rotary dial that opens and closes the local loop circuit at a timed rate. The number of dial pulses resulting from one operation of the dial is determined by how far the dial is rotated before being released.

Most newer telephone sets have a pushbutton keypad and use tone dialing to send the number. Pressing one of the keys causes an electronic circuit in the keyboard to generate two output tones that represent the number (see Table 9-1). Although only central offices equipped with the necessary equipment can process the tones, very few central offices today are not so equipped.

Making the Connection The central office has various switches and relays that automatically connect the calling and called telephones. If the called telephone handset is off-hook when the connection is attempted, a busy tone generated by the central office is returned to the calling telephone set. Otherwise, a ringing signal is sent to the called telephone to alert the called party that a call is waiting. At the same time, a ringback tone is returned to the calling party to indicate that the called telephone is ringing. The signaling arrangement in common use today is the *polarized ringer*, or bell, although some systems use a melodic tone.

When the called party removes the handset in response to a ring, the loop to that telephone is completed by the closed switchhook, and loop current flows through the called telephone. The central office then removes the ringing signal and the ringback tone from the circuit.

Sidetone The transmitter is the part of the handset into which a person talks. It converts acoustical energy (speech) into variations in an electric current (electrical energy) by varying (modulating) the loop current in accordance with the speech of the talker. The receiver is the part of the handset that converts the electric current variations into sound that a person can hear. The signal produced by the transmitter is carried by the loop current variations to the receiver of the called party. Also, a small amount of the transmitter signal is fed back into the talker's receiver. This feedback is called the *sidetone*.

Sidetone enables talkers to hear their own voices from the receivers to determine how loudly they must speak. The sidetone must be at the proper level. Too much sidetone will cause the person to speak too softly for good reception by the called party. Too little sidetone will cause the person to speak so loudly that it may sound like a shout at the receiving end.

The call is ended when either party hangs up the handset. The on-hook signal tells the central office to release the line connections. In some central offices, the connection is released when either party goes on-hook. In others, the connection is released only when the calling party goes on-hook.

9-2   THE PUBLIC SWITCHED TELEPHONE NETWORK

Thus far, the discussion of connecting telephones together has been limited to local loops and a central office exchange. Most central office exchanges can handle up to 10,000 telephones. However, when more than 10,000 telephones or telephones in different cities, states, or countries must be connected, a more complex network of telephone exchanges must be employed.

Exchange Designations

To identify and describe its function, each telephone exchange in North America has two designations: *office class* and *name* (as shown in the network hierarchy in Figure 9-4(A)). Subscriber telephones are normally, but not always, connected to Class 5 End Offices. Toll (long distance) switching is performed by Class 4, 3, 2, and 1 offices. The Intermediate Point, or Class 4X office, is a relatively new class. It applies to all

Key	Class	Name	Abbreviation
1	1	Regional Center*	RC
2	2	Sectional Center*	SC
3	3	Primary Center*	PC
4	4C	Toll Center	TC
4P	4P	Toll Point	TP
4X	4X	Intermediate Point	IP
5	5	End Office	EO
5R		End Office with Remote Switching Unit	
R		Remote Switching Unit	RSU

* May be a "point" rather than a "center." The abbreviation is then RP, SP, or PP.

(A)

(B)

FIGURE 9-4   Telephone network hierarchy (A) Exchange designations (B) North American network

digital exchanges to which remote unattended exchanges, called *Remote Switching Units* (RSUs), can be attached. Class 4X offices may interconnect subscriber telephones as well as other Class 5 and Class 4 exchanges.

All of the 10 Regional Centers (Class 1 offices) in the United States and two in Canada are connected directly to each other by large-capacity trunk groups. A *trunk group* is the logical group of telephone lines entering a telephone office from a specific location. Trunks are classified by the direction (in, out, two-way) or by usage (transit). Each trunk is assigned a unique number. An in trunk terminates at the office to which it is incoming; an out trunk originates at the office from which it is outgoing; and a two-way trunk is used on per call basis in either direction.

Interconnection

The North American network is organized as shown in Figure 9-4(B). Each exchange is optimized for a particular function. A call requiring service that cannot be performed by a lower class exchange is usually *handed off* (forwarded) to the next-higher exchange for further processing.

The Regional Center forms the foundation of the network. The branch levels are the Class 2, 3, 4, 4X, and 5 offices. Most offices are connected to more than one other office, and the interconnections among the various offices are not as simple as shown in Figure 9-4(B). The interconnections depend on the patterns of the traffic arriving and leaving each office.

The network makes connections by attempting to find the shortest path from the Class 5 office serving the caller to the Class 5 office serving the called party. The high-usage interoffice trunk groups that provide direct connection between offices of equal or lower level are used first. If they are busy, trunk groups at the next higher level (called *final groups*) are used. Digital logic circuits in the common control of each exchange make decisions based on rules stored in memory that specify which trunk groups are to be tried and in what order. These rules prevent more than nine connections in tandem and also prevent endless loop connection.

System Structure

Voice signals and signals used to set up telephone connections are carried by transmission systems over paths called *facilities*. These systems are divided into three broad network categories: local, exchange area, and long-haul.

Local Network   The *local network* shown in Figure 9-5 is the means by which telephones in residences and businesses are connected to central offices. The local facilities are almost exclusively wire pairs that fan out from a point called the wire center and extend throughout a serving area. Serving areas vary in size, from an average of 12 square miles in urban locations to 130 square miles in rural areas. More than one central office is often required for an urban serving area whereas one central office is usually sufficient for a rural serving area. An average wire center in an urban area will serve 41,000 subscriber lines and 5000 trunks. Urban exchanges generally have a higher call-carrying capacity than rural exchanges.

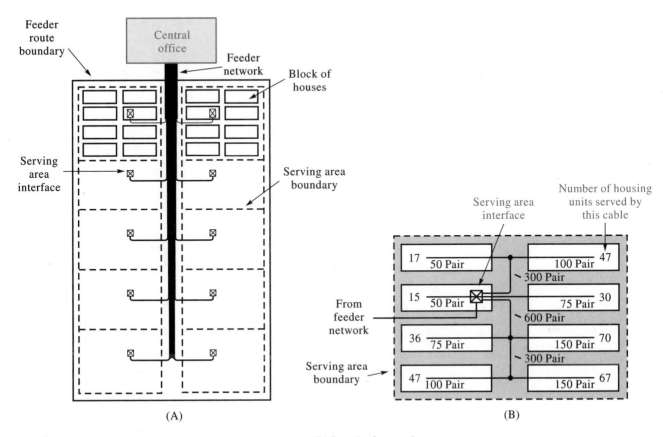

FIGURE 9-5 Local network (A) Local distribution area (B) Detail of a serving area

Exchange Area Network The *exchange area network*, illustrated by the simplified schematic in Figure 9-6, is intermediate between the local network and the long-haul network. Exchanges are interconnected with exchange area transmission systems. These systems may consist of open wires on poles, wire pairs in cables, microwave radio links, or fiber optic cables. The exchange area network normally interconnects local exchanges and tandem exchanges. *Tandem* exchanges make connections between central offices when an interoffice trunk is not available. A tandem exchange is to central offices as a central office is to subscriber telephone sets.

Long-Haul Network In the *long-haul network*, illustrated by the simplified schematic in Figure 9-7, local exchanges are interconnected with toll exchanges. These facilities normally have high capacity per circuit and consist mostly of cable and microwave radio links. In some paths (called *routes*) requiring a great many channels, such as the backbone routes between Boston and Washington, D.C., very high capacity fiber optic cables are being installed. This new Northeast corridor system consists of several fiber optic links, each carrying about 4000 voice channels simultaneously. Such high-capacity links are expected to save operating telephone companies many millions of dollars in construction and operating costs.

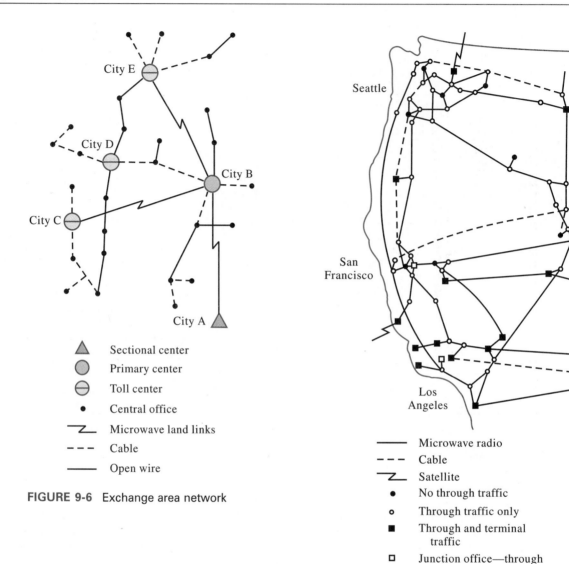

FIGURE 9-6 Exchange area network

FIGURE 9-7 Long-haul network

9-3 TYPES OF TRANSMISSIONS

Spoken messages or voice signals are not the only signals transmitted down a telephone line. The previous discussion about making a connection between the calling telephone and called telephone considered some of these other signals, including dial tone, dial pulses, or key tones used for sending a number, busy tone, and ringback tone. These signals are used to control the switching connections or to indicate the status of the call. Such signals are called *control* signals, or *supervisory* signals. They may be *analog* (tone) signals or *digital* (ON–OFF) signals. Therefore, if you were to examine the signals on many local loops, you would find analog voice signals, analog tone signaling, and digital ON–OFF signaling; that is, you would find a mixture of analog and digital signals.

Analog Voice Transmissions

Recall that signals with continuously and smoothly varying amplitude or frequency are called analog signals. Speech (or voice) signals are of this type; they vary in amplitude and frequency. The typical distribution of energy in voice signals, illustrated in Figure 9-8, shows that the frequencies that contribute to speech extend from below 100 Hz to above 6 kHz. However, it has been found that most of the energy necessary for intelligible speech is found between 200 Hz and 4 kHz.

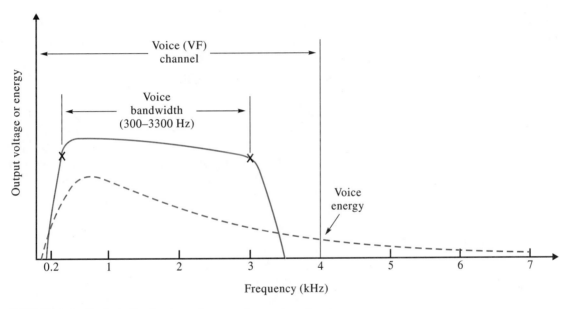

FIGURE 9-8   Typical distribution of energy in a voice signal

Voice Channel Bandwidth   To eliminate unwanted signals (noise) that could disturb conversations or cause errors in control signals, the circuits that carry the telephone signals are designed to pass only certain frequencies, called the *pass band*. Frequencies from zero to 4 kHz represent the pass band of a telephone system voice channel, called the *voice frequency* (VF) *channel*, or the *message channel*. We know that bandwidth is the difference between the upper limit and the lower limit of the pass band; thus, the bandwidth of the VF channel is 4 kHz. However, not all of the VF channel is used for the transmission of speech. The voice pass band is restricted to 300–3300 Hz, as shown in Figure 9-8. Therefore, any signal carried on the telephone circuit that falls within the range 300–3300 Hz is called an *in-band* signal; any signal that does not fall within this range but is within the VF channel is called an *out-of-band* signal. All speech signals are in-band signals. Some signaling transmissions are in-band, and some are out-of-band, as illustrated in Figure 9-9.

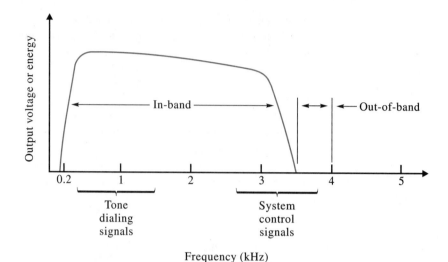

FIGURE 9-9   In-band and out-of-band signaling

Voice Channel Energy Level   The loudness or amplitude of a signal on telephone circuits is usually called the *level* of the signal. The level of a signal is expressed in terms of the power the signal delivers to a load. For example, a pair of telephone wires running together in a cable form a transmission line with an impedance of 600 Ω. The power delivered to a balanced-pair telephone transmission line is

$$P_{load} = \frac{e_s^2}{Z}$$

(9-1)

where

$$P_{load} = \text{power in watts}$$
$$e_s = \text{signal level in volts}$$
$$Z = \text{impedance in ohms}$$

In telephone circuits and audio circuits, the signal reference level is 1 mW of power to the load. So, if $P_{load} = 1$ mW and $Z = 600$ Ω, then

$$1 \text{ mW} = \frac{e_s^2}{600 \, \Omega}$$

or

$$e_s^2 = (600)(1 \times 10^{-3}) = 0.6$$
$$e_s = 0.775 \text{ V}$$

Therefore, a signal level of 0.775 V applied across 600 Ω produces 1 mW of power.

Analog signals that are transmitted at a constant frequency also can have their levels expressed in dB:

$$dB = 10 \log \frac{P_1}{P_2}$$

(9-2)

TABLE 9-2	Power Ratios in Decibels
dB	**P_1/P_2**
40	10,000
30	1,000
20	100
10	10
3	2.0
0	1.0
−3	0.5
−10	0.1
−20	0.01
−30	0.001
−40	0.0001

Table 9-2 lists a number of these power ratios.

Under the condition that 1 mW be used as the reference power P_2, the dB power ratio is measured in dBm—that is, dB referenced to 1 mW. Therefore, from Table 9-2, if $P_2 = 1$ mW, then a signal at 0 dBm will be delivering a power P_1 of 1 mW to a load, because the ratio P_1/P_2 must be 1. Stated another way, Table 9-2 shows that when a signal produces a power P_1 into a load of, for example 600 Ω with a 20 dBm level, it is delivering 100 mW of power (P_1) compared to the reference power of 1 mW (P_2).

In telephone systems, the 0 dBm level is usually set at the output of the switch at the sending end of a transmission line. This point becomes a system reference point called the *zero transmission level point* (0 TLP). Once the 0 TLP is chosen and the 0 dBm level applied at the point, all other power gains and losses in the transmission path between that point and the next switch output can be measured directly with respect to the 0 TLP. If the signal magnitude is measured, then the unit dBm0 is used. If only the relative gain or loss is indicated, the unit dB is used.

Voice Channel Noise   Transmission systems must often operate in the presence of various unwanted signals (noise) that distort the information being sent. Lightning, thermal noise, induced signals from nearby power lines, battery noise, corroded connections, and maintenance activities all contribute to degradation of the signal. Analog speech quality is determined primarily by the absolute noise level on the channel when the channel is *idle*—that is, when no speech signal is present. Speech tends to mask noise, but noise on an idle channel is quite objectionable to a listener. Stringent standards (−69 dBm0 up to 180 miles and −50 dBm0 up to 3000 miles, with −16 dBm0 as speech level) have been set in the U.S. network for this idle channel noise.

Another type of noise that originates in the voice transmission itself is echo. The primary echo is the reflection of the transmitted signal back to the receiver of the person talking. The amount of delay in the echo depends on the distance between the transmitter and the point of reflection. The effect of the delay on the talker varies from barely noticeable to very irritating to downright confusing. Echo also affects the listener, although to a lesser degree. Echos are caused by mismatches in the transmission line impedances, which usually occur at the interface between a two-wire circuit and a four-wire transmission system. The effect of echo is reduced by inserting loss in the lines.

Multiplexing   A local loop can carry only one voice channel conversation at a time. Because this setup is not economical for toll transmission, a method was devised so that a transmission path could carry many conversations simultaneously. This method is called *multiplexing*. For analog signals, several telephone conversations are sent together over one transmission channel but are separated by their frequencies.

The basic principles of analog multiplexing are illustrated in Figure 9-10. In Figure 9.10(A), a voice signal having frequencies within the voice channel bandwidth from 0 to 4 kHz is amplitude modulating the 8140 kHz carrier frequency. Thus, the information in the voice signal is being carried by the changing amplitude of the carrier, and the voice frequencies have been translated to different frequencies.

If different voice signals (different telephone conversations) are placed on different carriers, many conversations may be multiplexed on one transmission path and transmitted to the receiving point. At the receiving point, the different conversations can be

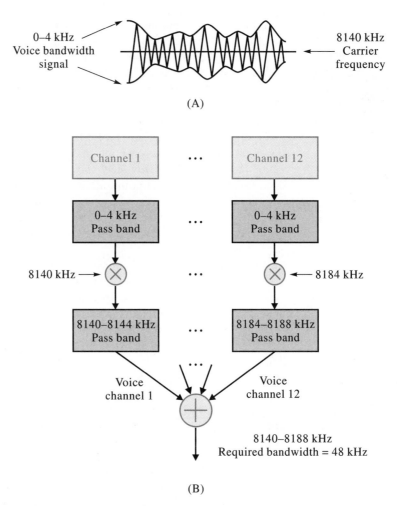

FIGURE 9-10   Analog multiplexing (A) Carrier modulated by a voice signal (B) Multiplexed frequency signals

identified and separated by their unique frequencies, and the original conversation can be recovered from the carrier (demodulated) and sent to the called telephone.

Figure 9-10(B) shows the multiplexing for 12 voice channels. Each channel has a 4 kHz bandwidth, so the 12 channels require a bandwidth of 48 kHz (4 kHz × 12). The lower frequency in this figure is 8140 kHz; therefore, the output multiplexed signal extends from 8140 kHz to 8188 kHz (8140 kHz + 48 kHz = 8188 kHz). In general, as the number of voice channels to be transmitted over a transmission path increases, the required bandwidth of the transmission path must increase.

Signaling Transmission

As stated previously, *signaling* refers to specific signals on the transmission line that are used to control the connection from the calling telephone to the called telephone or to indicate the status of a call as it is being interconnected. In this section, we discuss dc signaling, tone signaling, digital control signals, and common channel interoffice signaling.

dc Signaling   In *dc signaling*, the signaling is based on the presence or absence of circuit current or voltage, or on the presence of a given voltage polarity. The state of the signal indicates on-hook, off-hook, dial pulses, or status of the interconnection. These signals are ON–OFF digital signals.

On local loops, on-hook is indicated by an open circuit and no current flow. Off-hook is signaled by a closed circuit and continuous current flow. Dial pulses consist of current flow interrupted at a specific rate. A potential problem with dc signaling is that dial pulses spaced too far apart may be mistaken by the exchange for an on-hook signal. However, because of careful design, this problem rarely occurs.

A type of dc signaling called *reverse battery signaling* is used between central offices to indicate the status of the switched connection. When the near-end exchange requests service, an idle trunk is seized. A polarity of a given voltage exists on the trunk, which indicates to the near end that the called telephone is on-hook and ringing. The far-end exchange acknowledges and indicates to the near end, by reversing the voltage polarity, that the called party has answered.

Another type of dc signaling is *E & M signaling*, used for the same purpose on long interoffice and short-haul toll trunks. This type of signaling requires two extra wires in the originating and terminating trunk circuits, one for the E-lead (inbound) and the other for the M-lead (outbound). Because separate wires are used for each, the on-hook and off-hook states can be signaled from both ends of the circuit, as shown in Table 9-3. This setup allows on-hook and off-hook signaling to be sent in both directions at the same time without interfering with each other. Sometimes two wires are used for each signal to avoid noise caused by a common ground.

TABLE 9-3   E & M Signaling

State	E-lead (Inbound)	M-lead (Outbound)
On-hook	Open	Ground
Off-hook	Ground	Battery voltage

Tone Signaling   Various *tones* are used for both control and status indication. The tones may be a single frequency or combinations of frequencies. These analog signals are either continuous tones or tone bursts (tones turned ON and OFF at various rates). The call progress tones listed in Table 9-4 are sent by the exchange to the calling telephone to inform the caller about the status of the call. For example, the dial tone is a continuous tone made by combining the frequencies 350 Hz and 440 Hz. The busy signal which tells the caller that the called telephone is busy (off-hook) is a combination frequency tone that appears in bursts of 0.5 s ON time separated by a 0.5 s OFF time. The receiver off-hook warning signal is a combination frequency tone of four frequencies that is ON for 0.1 s and OFF for 0.1 s. This signal is very loud in order to get the

TABLE 9-4 **Call Progress Tones**

Tone	Frequency (Hz)	ON Time (s)	OFF Time (s)
Dial	350 + 440	Continuous	
Busy	480 + 620	0.5	0.5
Ringback (normal)	440 + 480	2	4
Ringback (PBX)	440 + 480	1	3
Congestion (toll)	480 + 620	0.2	0.3
Reorder (local)	480 + 620	0.3	0.2
Receiver off-hook	1400 + 2060 + 2450 + 2600	0.1	0.1
No such number	200 − 400	Continuous, frequency modulated at 1 Hz rate	

attention of someone to "hang up" the handset that has been left off the cradle. All of these tones, as well as the DTMF addressing tones discussed previously, are in-band signaling.

Tone signaling between exchanges may be in-band or out-of-band. The most commonly used single frequency tones are 2600 Hz for in-band and 3700 Hz for out-of-band. E & M signals are converted to a single frequency tone for transmission on carrier systems because the dc signals cannot be transmitted. The tone indicates on-hook when present and off-hook when not present. Multifrequency supervisory signaling uses six frequencies for transmitting telephone number information over toll facilities: 700, 900, 1100, 1300, 1500, and 1700 Hz. The frequencies are used in pairs to represent the digits 0–9, and some control function much like DTMF is used at the telephone set.

Digital Control Signals   Instead of simply interrupting a dc voltage, as in the case of dc signaling, or interrupting continuous tone bursts, control signals also can be *digital codes*. Rather than ON–OFF signals that occur at random times, they are combinations of signals with two levels, 0 and 1, and with a definite time relationship to each other, as illustrated in Figure 9-11. In the telephone system, the binary digit (bit) 1 and 0 levels may be represented by voltage or current levels. Note that the bits occur in a particular time sequence. For example, a binary code of eight bits is shown with bits d_0–d_7 always occurring in the same time slot, t_1–t_8, when transmitted in sequence. For a particular system design, once the time relationship of bits is set, it does not change.

Control information can be contained in the binary code in several ways.

1. All 8 bits may be used as a group to represent a number from 0 to 255. The binary code for the number 136 is shown across the top of Figure 9-11. On the left side, the code is presented in 1s and 0s, and on the right side, the code is presented as voltage levels or pulses.

2. The 8-bit group may represent a letter of the alphabet in a data communication code. The letter *D* in ASCII is shown in the middle of Figure 9-11.

Example	Binary digits (bits)								Time relationship							
	d_7	d_6	d_5	d_4	d_3	d_2	d_1	d_0	t_8	t_7	t_6	t_5	t_4	t_3	t_2	t_1
Binary code for 136	1	0	0	0	1	0	0	0								
ASCII code for letter D	0	1	0	0	0	1	0	0								
Binary code for commands	1	0	1	1	0	1	0	1								

Command X   Command Y   Command Z

Blank

FIGURE 9-11 Serial digital control signals

3. Individual bits or subgroups of the 8-bit code may be used to command different functions. Examples of subgroup codes for the functions X, Y, and Z are shown at the bottom of Figure 9-11.

Common Channel Interoffice Signaling All signaling methods discussed so far send the control and addressing signals over the same circuit as the voice signals. Another method separates the control signals from the voice signals. Here, the control signals are sent over a separate circuit where they are detected and do the control and switching of lines independently from the voice signals. Called *common channel interoffice signaling* (CCIS), this method is illustrated in Figure 9-12. The basic control is by digital computer, and CCIS is a separate network for exchanging control signals among the computers. As the name suggests, CCIS is used on the interconnecting trunks that carry signals between central offices.

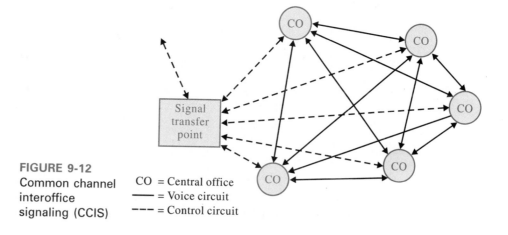

FIGURE 9-12
Common channel
interoffice
signaling (CCIS)

CO = Central office
—— = Voice circuit
--- = Control circuit

9-4 TYPES OF MULTIPLEXING

As stated previously, multiplexing is the simultaneous transmission of several separate information channels over the same communications circuit without interference. For voice communication, this means two or more voice channels on a single carrier. For telephone systems, it means many channels into a single pair of wires or into a single coaxial transmission line. Multiplexing can be accomplished by time division or by frequency division.

Time Division Multiplex

Time division multiplex (TDM) is a means of transmitting two or more information channels over the same communications circuit using a *time-sharing* technique. It is well suited to binary signals consisting of pulses representing a binary 1 or 0. These pulses can be of very short duration and still convey the desired information; therefore, many of them can be squeezed into the time available on a digital carrier channel. The original signal may be an analog wave which is converted to binary form for transmission, such as in speech signals in a telephone network, or it may already be in digital form, such as in a business machine or a computer.

Time division multiplex is a synchronized system that usually involves PCM. A simplified diagram of a TDM system carrying three information channels simultaneously is shown in Figure 9-13. The analog signals are sampled and converted to pulses by PAM, then the samples are coded by PCM. Following this, the samples are transmitted in series over the same communications channel, one at a time. At the receiver, the demodulation process is synchronized so that each sample of each channel is routed to its proper channel. This process is called multiplexing because the same transmission system is used for more than one information channel, and it is called TDM because the available time is shared by information channels.

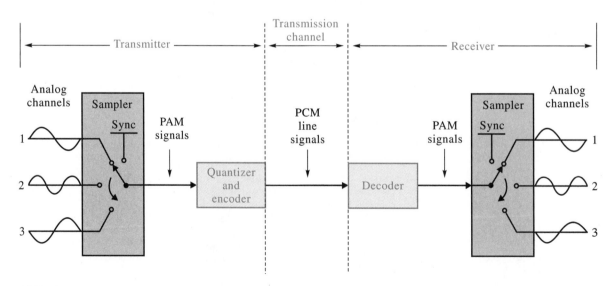

FIGURE 9-13 Simplified diagram of TDM

The modulation and signal preparation portion of the system is called the *multiplexer* (MUX), and the demodulation portion is called the *demultiplexer* (DEMUX). In the MUX, illustrated in Figure 9-14(A), a timed commutator (electromechanical switch) sequentially connects a synchronizing pulse, followed by each information channel, to the output. The combination of this set of pulses is called a *frame*, illustrated in Figure 9-14(B). The synchronizing pulse is used to keep the transmitter and receiver in synchronization—that is, to phase lock the receiver timer with the

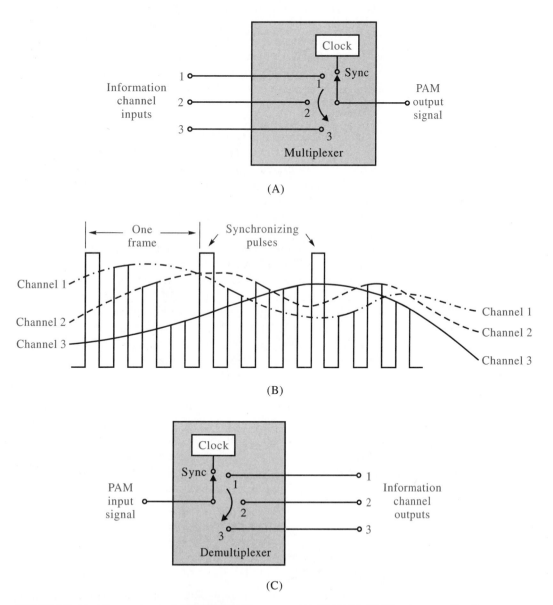

(A)

(B)

(C)

FIGURE 9-14 Three channel MUX-DEMUX system (A) MUX (B) MUX pulse amplitude modulation output waveform (C) DEMUX

transmitter timer. In the DEMUX, shown in Figure 9-14(C), a decommutator routes the synchronizing pulse to the receiver timer, and the information sample pulses to their correct channels for recovery.

An advantage of TDM is that any type of pulse modulation can be used. Many telephone companies use this advantage in their PCM/TDM systems.

Frequency Division Multiplex

Like TDM, *frequency division multiplex* (FDM) is used to transmit several information channels over the same communications channel simultaneously. However, unlike TDM, FDM does not use pulse modulation. In FDM, the frequency spectrum represented by the available bandwidth of a channel is divided into smaller bandwidth portions, with each of several signal sources assigned to each portion. Simply stated, the difference between the two systems is this: In FDM, each channel continuously occupies a small fraction of the transmitted frequency spectrum; in TDM, each channel occupies the entire frequency spectrum for only a fraction of the time.

FDM is used in commercial broadcast radio and television, where many of the stations each use a unique assigned carrier within the overall band to avoid interference. Each of the modulated signals is translated to a much higher and slightly different carrier frequency so it can be transmitted through the common channel (air, vacuum, cable) and discriminated by the receiver.

FDM is also used in telemetry, where many signals must be transmitted from a remote system under test (such as rockets or orbiting spacecraft) to a base station. Sensors are attached to key points of the system being tested. The signals from the sensors are received by the base station and amplified as needed to produce a 0- to 1-V signal used for narrowband FM. The FM signal has an approximate bandwidth of 4 kHz.

Figure 9-15 demonstrates how an FDM system works. In the transmitter (Figure 9-15(A)), the frequencies of each channel are effectively changed through balanced modulators and filters. The filter outputs are then fed to a MUX, where they are placed side by side in a wide-band channel for transmission as a group. At the receiver (Figure 9-15(B)), a DEMUX changes the channels back to their original frequencies by filtering. The filtered signals are then balance modulated and fed through LP filters for recovery.

Frequency division multiplex is still used in some simple data communications systems (Chapter 10) and is extensively used in the long-haul portion of the public telephone network. Although this system is more efficient in terms of bandwidth than digital systems, the problem is that noise is amplified along with the desired signal. That reason, and the fact that the cost of digital electronics has decreased greatly in recent years, has led to the widespread replacement of FDM in telephone systems with TDM.

SUMMARY

- Digital techniques for communication have increased because large amounts of information are already in pulse form and large scale integration has produced complex gating circuits in the form of ICs.

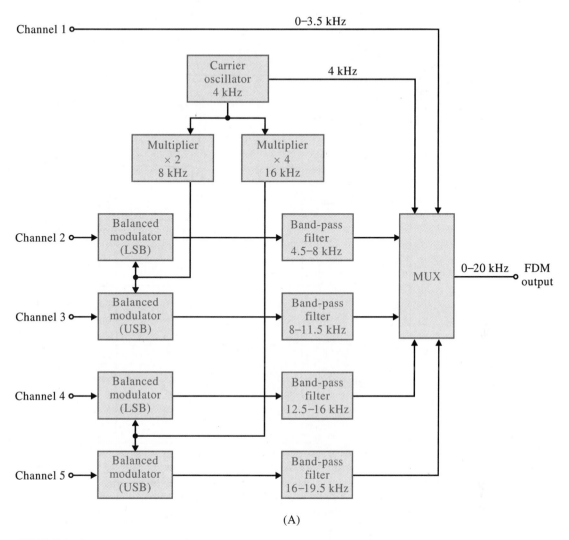

FIGURE 9-15 Five channel FDM system (A) FDM transmitter (B) FDM receiver

- The telephone set is a sophisticated piece of equipment that automatically performs many important functions. Telephone sets use rotary dialers, spring-loaded, circular wheels that produce pulses when dialed, and pushbutton dialers with 12 or 16 buttons that activate tone pairs when depressed. Pushbutton dialing systems offer dual-tone, multiple-frequency (DTMF) access.

- Local loops (central office exchanges) handle telephone system switching and other functions. A telephone call is initiated by going off-hook (lifting the handset from its cradle) and is terminated by going on-hook (replacing the handset in its cradle). Sidetone, a small amount of transmitter signal fed back into a receiver, is needed so that talkers can hear their own voices from the receiver to determine how loudly they must speak.

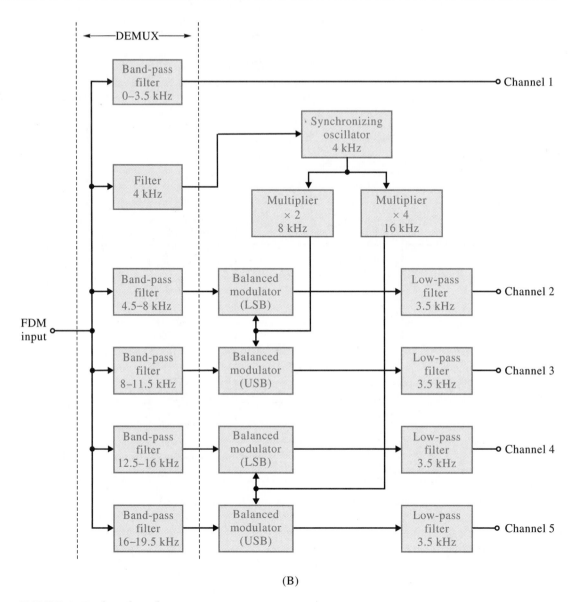

(B)

FIGURE 9-15 (*continued*)

- Public switched telephone network exchanges consist of five classes of offices, listed in Figure 9-4. Facilities are divided into three broad network categories: local, exchange area, and long-haul.

- Telephone system voice channels have a pass band from zero to 4 kHz. Within this bandwidth, the voice pass band is 300–3300 Hz, called in-band signals. Signals in the designated 4 kHz bandwidth but not in the 300–3300 Hz band are called out-of-band signals. Supervisory (control) signals can be analog or digital.

- The loudness or amplitude (level) of a signal on telephone circuits is expressed in terms of the signal power delivered to a load. The signal reference level is 1 mW.

Analog signals transmitted at a constant frequency are also expressed in dB. When the 1 mW reference power is used, the dB power ratio is measured in dBm.

- Analog speech quality is determined by the absolute noise level on the voice channel when it is idle. Noise may be caused by lightning, thermal noise, induced signals from nearby power lines, battery noise, corroded connections, maintenance activities, and echo.

- Multiplexing allows several analog telephone conversations to be sent over one transmission channel, separated by frequency. As the number of conversations transmitted over a path increases, the bandwidth of the transmission path must increase.

- In dc signaling, the signaling is based on the presence or absence of circuit current or voltage, or on the presence of a given voltage polarity. Types of dc signaling are reverse battery and E & M. Tone signals are single frequencies or combinations of frequencies and either continuous tones or tone bursts. Tone signaling between exchanges can be in-band or out-of-band. Digital control signals are combinations of signals that have two levels, 0 and 1, and a definite time relationship to each other.

- Common channel interoffice signaling (CCIS) separates control signals from voice signals and is used on the interconnecting trunks that carry signals between central offices.

- Multiplexing can be achieved by time division multiplex (TDM) or by frequency division multiplex (FDM). In the TDM system, each channel occupies the entire frequency spectrum for only a fraction of the time. In the FDM system, each channel continuously occupies a small fraction of the transmitted frequency spectrum. The TDM system uses pulse modulation for transmission; the FDM system does not.

QUESTIONS AND PROBLEMS

1. Describe the public switched telephone network in the United States.

2. List the important functions of a telephone set.

3. Describe and explain the operation of (a) the rotary dialer and (b) the pushbutton dialer.

4. Define DTMF access, explain how it operates, and state its purpose.

5. List the advantages of pushbutton dialing as compared to rotary dialing.

6. Define local loop and explain how it is used.

7. Define (a) on-hook and (b) off-hook.

8. Define sidetone and explain why it is needed.

9. List the office class and name for each of the telephone exchanges in North America.

10. Define (a) trunk group, (b) final group, and (c) facilities.

11. Give the use of each trunk classification.

12. Explain the operation of (a) a local network, (b) an exchange area network, and (c) a long-haul network.

13. Define supervisory signals and explain their use.

14. What is the pass band of a telephone system voice channel? What is the voice pass band?

15. Give the limits of in-band signals and out-of-band signals.

16. Define level of a signal.

17. What is the standard reference power level in telephone circuits?

18. A signal load of 1.5 V is applied across a load impedance of 300 Ω. Find the load power.

19. Find the signal level for a standard telephone reference power level and a load impedance of 300 Ω.

20. Find the dBm level for a standard telephone reference power level and an output power of 80 W.

21. Find the output power for a standard telephone reference power level and a 15 dBm signal level.

22. List six common causes of voice channel noise.

23. Define (a) idle channel and (b) multiplexing.

24. Define echo and explain its causes and effects.

25. Explain the two methods of multiplexing.

26. Define signaling and list four signaling methods.

CHAPTER **10**

Digital Data Communications

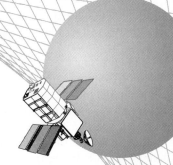

OBJECTIVES

When you complete this chapter, you should be able to:

- Define digital data communications and list the characteristics modern systems must have.
- Define data, teleprocessing, telecommunication, data terminal equipment, and data circuit-terminating equipment.
- Discuss the requirements and objectives of digital data communication systems and discuss how the objectives are met.
- Describe data transmission systems and list six information requirements.
- Describe the two basic types of data transmission systems.
- Draw a block diagram of a basic data communication system and describe the function of each component.
- Define terminal and list and describe the five broad categories of terminals.
- Define front-end processor and list eight functions of the device.
- List and discuss five data communication codes.
- List and describe the three basic modes of operation for data transmission.
- Describe the DTE-DCE interface operation and the three basic types of transmission used.
- Describe how data communication systems are interfaced to telephone networks.
- List the advantages that DTMF dialing has over rotary dialing.
- Define modem and describe modem operation.
- List and describe the steps that occur when two modems are operating in a handshaking mode.
- Describe the telephone channel restrictions on modems.
- Describe the two multiplexing methods used in data communication carrier systems.

INTRODUCTION

Digital data communications is the process of communicating information in binary form between two points. It is sometimes called computer communications because most of the information interchanged today is between computers or between computers and their support (peripheral) equipment. Examples of support equipment are teletype (TTY) units, cathode ray terminal (CRT) units, and printers. The data may be as simple as the binary symbols 1 or 0 or as complex as the characters represented by the keys on a typewriter keyboard. The communication channel itself may be a cable, a switched public telephone line, a radio transmission, or even an optical transmission.

It is important to understand data communications because of its significance in today's world. It is commonly used in business and is being used more and more in homes as well. The telephone channel is the medium by which most data communication systems transfer information from one point to another. In this chapter, we will discuss data communication fundamentals, types of systems, equipment, codes, modes of operation, interfacing with the public telephone system, and system multiplexing.

10-1   DIGITAL DATA COMMUNICATION FUNDAMENTALS

Digital data communications evolved from the union of communications technology and computer technology. The integration of these two technologies makes it possible to transmit data to computers from remote locations. The integration has also brought together professionals from both disciplines, all of whom view data communications from the standpoint of their own training and experience. These differing viewpoints have led to inconsistencies in terminology. Checking a number of references, such as the Consultative Committee for International Telegraph and Telephone (CCITT), the American National Standards Institute (ANSI), and the International Communications Association (ICA), shows that there are no standard definitions for many of the terms used in data communications. Accordingly, the definitions in the following discussion are intended only to clarify the meanings of the terms as used in this text. The terminology you encounter as a technician may be quite different.

Digital data communications can be defined as the movement of encoded data from one point to another by means of electrical transmission systems, including wireline, radio, and optics. A modern data communication system must have two characteristics:

1. The data must be translated into a special code for transmission.
2. The translated code must be transmitted by some electronic means.

Basically, data communications is distinguished from telegraphy by the fact that some form of processing is involved in data communications, either before or after transmission.

The term *data* refers to any representation—for example, letters, numbers, or facts—to which meaning can be ascribed. The term *raw data* is often used to describe unprocessed data, as distinguished from *information,* which usually refers to processed or meaningful data. However, an electrical transmission system cannot differentiate between the terms *data* and *information.* Also, since the purpose of transmitting data is to supply it to the receiving party to whom it will be meaningful, the two terms can be considered synonymous, and they will therefore be used interchangeably in this text.

Teleprocessing

Teleprocessing is a form of information handling wherein data processing equipment is used in conjunction with telecommunications facilities. *Telecommunication* is the process of transmitting information over a distance by electrical or electromagnetic systems. This information may be in voice, data, image, or message form. Teleprocessing includes the transfer of data from one location to another and the processing of those data. Data communications is an integral part of teleprocessing.

Data Communication Systems

A basic requirement of a data communication system is that it be capable of transmitting rectangular pulses at rates of 100–500,000 pulses per second (pps). The most commonly used speeds are 600–50,000 pps. The system may be required to transmit in one direction only (*simplex*), alternately in either direction (*half-duplex*), or simultaneously in both directions (*full-duplex*), with the lowest error rate possible. In the data processing cycle, data must be collected and moved to the processing unit before they can be processed; and before processed data can be used, they must be delivered to the user. The objective of data transmission systems is to provide faster information flow by reducing the time spent collecting and distributing data. Data communication networks facilitate more efficient use of central computers by providing *message switching* capabilities—the routing of messages among three or more locations using either *circuit switching* or *store-and-forward* techniques. If a telecommunications line is available, message switching is accomplished by instantaneous circuit switching. If all lines are busy, store-and-forward procedures are used; that is, messages are accepted and stored in the computer memory until a telecommunications line becomes available, and then forwarded to the next location.

10-2   DATA TRANSMISSION SYSTEMS _____

Communication between computers and their remote terminals usually occurs much faster than the operator can enter the data on a keyboard. Individual characters, consisting of a number of bits, can be transmitted *parallel* or *serial.* For parallel transmission over very short distances, say between a computer terminal and a printer, individual wires, one per bit, are often used. For parallel transmission over long distances, frequencies are assigned to each bit. At the receiver, the individual frequencies are detected, and their presence or absence indicates the binary 1 or 0 condition of the particular bit. Serial transmission requires five, seven, or eight serial bits to transmit one character. Speeding up the rate at which these bits are transmitted makes it possible to achieve the same transmission speed as in the parallel, separate-frequency method.

Data transmission systems are designed to serve a variety of applications; they therefore differ in the way they function. The system used is determined by the information requirements of the user. These requirements include, but are not limited to, the following factors:

1. The quantity of data to be transmitted.
2. Whether immediate action is required.
3. *Response time* (the interval between data input and the system's response to the input).
4. *Delivery time* (the time from the start of transmission at the transmitting terminal to the completion of reception at the receiving terminal, where data are flowing in only one direction).
5. The kinds of input data and the accumulation process employed.
6. The geographical location of the system's users.

There are two basic types of data transmission systems: offline and online.

Offline System

An *offline system* transmits data that do not require an immediate response or that do not require a response at all. Offline means that the data are not transmitted directly to the computer but are stored on magnetic tape, disc, or cards for processing later.

Most offline systems employ *batch-processing* techniques, by which the input records are collected in their original, physical form (time cards, invoices, report cards and the like), accumulated over a period of time, and transcribed onto an input medium that can be read by the computer. The records are then transported to the computer room in groups, or batches, and read into the main computer storage. They are processed in batches, and the output is transmitted to a designated storage area or terminal, or printed as a batch.

In remote batch processing, called *remote job entry* (RJE), data are collected at remote locations and transmitted periodically via remote input/output (I/O) terminals over telecommunications lines to a centralized computer system. The data can be transmitted offline to an auxiliary storage unit and held for input to the computer at a scheduled time. Once the data are processed, the output is transmitted back to the user, who is located at the same remote terminal.

Offline batch processing is comparatively inexpensive. It is used when up-to-the-minute information is not required. The predictable nature of batch-processing requirements permits the best scheduling of computer time; thus, this system is efficient and economical for use in recurring, routine applications.

Examples of offline systems are stock market tickers and national news service agencies. In these systems, the data, in the form of stock market quotes or news, are transmitted from a central location to the subscriber.

Online System

Although offline transmission systems were common in the early days of data communications, the present trend is toward *online systems*. In online systems, devices or subsystems are connected directly to the computer, and data flow directly between the terminal and the computer.

The transmission of online communications can be batched or *real time*. In real-time computing, processing is performed as the operator keys in the data, thus, output is received quickly enough to affect decision making. Although the fast response time of online real-time systems is a great advantage, these systems are expensive to implement and operate. For that reason, some applications use batch-processing techniques even though input devices are online.

In general, real-time means that the data transmission requires an immediate response. Examples of real-time systems are airline reservation and overnight accommodation reservation computers, where an inquiry from a reservation desk anywhere in the world is handled and answered rapidly by the central reservations computers.

10-3 DATA COMMUNICATION SYSTEM COMPONENTS

Data communication systems are networks of components and devices organized to transmit data from one location to another, usually from one computer or computer terminal to another. The data are transmitted in coded form over electrical transmission facilities. This section outlines the basic hardware, controls, and procedures of data communications and describes how they function together as an integrated data communication system.

Basic System Description

The three essential components common to all data communication systems are shown in Figure 10-1. In simple communications terms, we have the source (the transmitter, or originator, of the information), the medium (the channel, or the path through which information flows), and the sink (the receiver of the information). The source and sink in two-way communication systems such as this may switch roles; that is, the same piece of equipment may transmit or receive data.

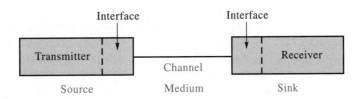

FIGURE 10-1 Typical data communications channel

Remote terminals are the *source* in the system. These are usually devices with typewriterlike keyboards used for entering data. Remote terminals might be in a different department, in a different building, or even in a different geographic location from the mainframe or central processing unit (CPU). The *medium* in the system is the communication link—the facility that links remote terminals to the CPU. The medium could be wire, radio, coaxial cable, microwave, satellite, or light beams. Leased or public switched telephone lines are the most frequently used communication medium. The *sink* in the system is the computer system that receives and processes the data.

Data Terminal Equipment

Data terminal equipment (DTE) is a computer or business machine that provides data in the form of digital signals at its output. It may comprise terminals such as cathode ray tube or teletype terminals, but it may also consist of personal computers, printers, front-end processors for large mainframe computers, or any other device that can transmit or receive data. In general, entries to these devices are made by people. These entries are converted to digital (binary, a series of 1s and 0s) in the devices, then to analog for transmission between points. The receiving-end device returns the analog signals back to digital signals that are then converted to a form understandable to humans. Data terminal equipment can be the source, sink, or both in the system. The information sent between points in the system may be used directly by the DTE, or the DTE may process and display the information in such a manner that it is useful to human operators.

Originally, the word *terminal* meant the point at which data could enter and leave the communications network. In practice, however, the word has become synonymous with *terminal equipment* and refers to any device capable of either input or output to the communication channel.

Terminals provide *interfaces* (connecting points) with computer systems so that people can insert or extract data; they also provide a convenient way for people to exchange data directly. Terminals receive input in coded form and convert it to electrical signal pulses for transmission to the computer. Similarly, terminals at the receiving end transform the electrical impulses into characters that can be read by humans. Thus, terminals serve as *translation* devices for communication codes. Data can be entered into terminals by human operators or by machines that collect data automatically from recording instruments.

Prior to 1968, the major terminal device was a teletypewriter (TTY), and the principal input medium was punched paper tape. These devices may still be found in some applications. Each keystroke on a transmitting teletypewriter (source) produces a sequence of electrical pulses determined by the coding representation for the keyed character. The electrical signals are sent over the communication channel (medium) to a receiving teletypewriter (sink), where they are reconverted to their original form.

Today, there are many different kinds of data communication terminals offered by many manufacturers. Almost all of these devices are combinations that operate as more than one type. We will discuss five broad categories of terminals: keyboard terminals, video display terminals, transaction terminals, intelligent terminals, and specialized terminals.

Keyboard Terminals   The most widely used terminal is the *keyboard* terminal. Resembling a typewriter, it usually has a standard alphanumeric keyboard with special function keys to provide transmission control capability, such as the "line feed" or "bell signal." Keyboard terminals are used on low-speed public or private telephone lines. They operate in two directions: sending and receiving. In the sending mode, they are controlled by an operator; in the receiving mode, they are controlled either by a CPU or by another operator at a distant machine.

There are two types of keyboard terminals: (1) those that are online continuously and (2) those that are accessed on a dial-up basis. Both types of terminals may be *polled* (called up in sequence) by a CPU to request the terminal to transmit a message.

Video Display Terminals   *Video display* terminals consist of a keyboard and a cathode ray tube (CRT), a visual display device resembling a television screen. The keyboard is the input medium; the operator can enter both the data and the control commands that direct the operation of the computer. The CRT provides *soft copy*—a visual display with no permanent record. Display terminals with printers attached enable the operator to print *hard copy*—a permanent record.

Most video display terminals use a standard typewriter keyboard plus control and special function keys, such as "insert," "delete," and "repeat." When the operator types in a character, the character appears on the screen. These terminals have a *cursor* (derived from the Latin *cursus,* meaning "place"), which is usually a blinking symbol that indicates current location on the CRT screen. The operator can move the cursor horizontally and vertically to any desired position. An important advantage of CRT display terminals is their capability for text editing. Errors detected on the screen can be corrected, usually by backspacing and striking over the error. Entire words, lines, or paragraphs can be deleted or repositioned by using special function keys. Changes in input copy are possible because some amount of the copy is held in a *buffer,* or temporary memory, until the user presses a special function key to transfer it to the main memory. Video display terminals with alphanumeric keyboards have become well known through their use in word processing machines.

A special type of video display terminal is the *graphics* terminal, which can display not only letters of the alphabet and numbers but also graphic images (for example, charts, maps, and drawings). These terminals use *matrix* technology, in which many closely spaced dots are connected to draw lines and plot data graphically. Graphic display terminals can accept input from a keyboard, an input tablet, or a light pen (an electronic drawing instrument with a photoelectric cell at its end that allows the user to "draw" designs directly on the display screen). Some graphic display terminals can display charts and drawings in different colors; displayed material can also be reproduced as hard copy with a special printer or plotter.

Video displays are high-speed devices since data output is not slowed by being typed on paper. These terminals are especially useful when quick access to a distant location is needed. They are used by airline reservation systems to determine flight space availability, by hotels and motels to determine room availability, by brokerage firms to transmit stock market quotations, and by insurance companies to access and update policyholder records.

Transaction Terminals   *Transaction* terminals are designed for a particular industry application such as banking, retail point-of-sale, or supermarket checkout. In banking, transaction terminals are used to update customers' passbooks and bank's records. They are also used online for off-hours banking and for processing customer inquiries.

Retail point-of-sale terminals are used to record the details of a sale in machine readable form. Their functions generally include verifying credit, printing sales slips, maintaining a local record of transactions, and updating inventory control records. All of these functions, except credit verification, can be handled by offline terminals, generally by cash register-type machines equipped with special keys to capture transaction data on paper or magnetic tape. Credit verification requires online access to storage files that may be built into the transaction terminal or may be in a central computer.

Transaction terminals used in supermarket checkout lines have the capability to scan or read bar codes printed on the items being sold. As the products pass the check-out point, the codes are read by a recording device or light pen that simultaneously prepares a cash register tape for the customer, records the sale, and updates the store's inventory. Transaction terminals are easy to operate and may be used by persons with little technical knowledge. They have become an integral part of business operations because they help to increase productivity and to control costs.

Intelligent Terminals   Early types of terminals, such as TTY terminals, merely serve as data input and output devices; they perform no processing, editing, or buffering. Such terminals are still used in certain applications, such as simple transaction recording. Because of their limited capabilities, they are sometimes referred to as *dumb* terminals.

The development of microprocessor technology made it possible to incorporate some processing capability into peripheral (support) devices, which greatly enhances their usefulness. Computer terminals equipped with a microprocessor are known as *intelligent* or *smart* terminals. They vary in degree of intelligence. As additional intelligence is incorporated into these machines, more processing can take place at the terminal, lessening the burden on the mainframe computer. The more sophisticated terminals have become small computers in themselves, and they are frequently able to operate independently from the host computer.

Specialized Terminals   Two of the newer types of terminals are *audio response units* and *pushbutton telephones*. Audio response terminals are unique in that their output (response) is verbal rather than printed or visual. The input device may be a keyboard or a telephone. In audio response units, the computer has a built-in synthesizer, enabling the computer to assemble prerecorded sounds into meaningful words. Although some mechanical voices have low-fidelity, robotlike characteristics, the voice quality of audio response units is equivalent to that of the human voice. This synthesized response should not be confused with a response from an answering machine that transmits a recorded message.

The response from an audio terminal is designed for a specific type of message service. Messages are pieced together from sound fragments (called *phonemes*) to produce a reply to a particular inquiry, such as a telephone number request from an information bureau. In this process, an information operator finds the requested number and points with a light wand to that number displayed on a CRT screen. The operator is then released from the call, and the audio response unit synthesizes the message (in this case, the requested telephone number) and transmits it to the customer waiting on the line. The combination of human operators and audio response units saves human effort and thereby improves productivity.

Rotary dial telephones are not generally employed for data transmission since the rotary dial cannot be used as an input terminal. The dial is used to establish a call using dial pulses in digital form. However, once the connection has been established, the dial pulses are ineffective for signaling.

The pushbutton telephone has been used for some time in voice communications, but its use as an input terminal for data transmission is relatively new. Its widespread availability makes it particularly useful. A pushbutton telephone has a keyboard as part of the instrument. Pressing a key on the telephone set transmits a distinctive signal representing a number that the computer uses for the processing operation.

Many banking institutions offer a service that allows customers to conduct certain banking transactions from their homes or offices using pushbutton telephones. For example, where this service is available, a customer can transfer funds from a savings account to a checking account and vice versa or pay bills to a selected list of merchants and utilities by keying a sequence of numbers representing a code into the telephone set. An interesting feature of these transactions is that while the customer communicates with the computer by pressing the appropriate buttons on the telephone, the computer communicates with the customer by using an audio response unit.

Front-End Processors

As terminals are added to a data processing system, the number and complexity of the operations necessary to handle them grow, and the demands on the CPU increase tremendously. To relieve some of these demands, and to control the flow of data and ensure compatibility within the system, a *front-end processor* (FEP), or *communications control unit* (CPU), is employed. Figure 10-2 shows how this processor "front ends" a mainframe CPU by functioning as an auxiliary computer system that performs network control operations. This functioning releases the central computer system to do data processing. Most FEPs are minicomputers. The functional components of the FEP may be freestanding pieces of equipment or combinations of components integrated into one or more equipment units. Functions performed by FEPs include:

1. *Line access* (connecting communication lines to the main computer).
2. *Line protocol* (monitoring line control procedures).

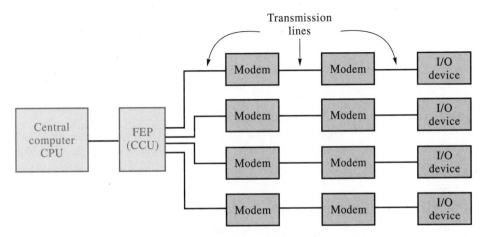

FIGURE 10-2 Data communications system components including the front-end processor

3. *Code translation* (translating the internal code of computer systems into communication codes).

4. *Synchronization* (ensuring that the incoming signals are compatible with the computer requirements).

5. *Polling* (polling terminals to inquire whether they are ready to receive a message or whether they have a message to send).

6. *Error control* (checking the accuracy of the data received using parity techniques).

7. *Path routing* (choosing an alternative path to avoid heavy traffic or an excessive error rate).

8. *Flow control* (controlling the flow of the signal from the processing unit to the destination).

10-4 DATA COMMUNICATION CODES

Humans and machines need different ways to represent information. Humans can quickly and reliably recognize printed characters by their distinctive shapes, whereas machines have difficulty with that. Conversely, machines reliably recognize long strings of two-state signaling elements such as marks and spaces or 1s and 0s, whereas humans cannot easily handle these symbols. Therefore, it is necessary that humans convert (encode) the characters they recognize into symbols that machines recognize.

A *code* is a standard meaning that defines a given set of characters or symbols. *Characters* are the letters, punctuation marks, numerals, and other signs and symbols on an input device's keyboard. Although some characters do not print, they are necessary to control the system. *Symbols* are signaling elements—that is, the representations of characters that are transmitted over transmission lines.

The use of an intelligent device to convert a character or symbol into coded form and vice versa is a characteristic common to all data communication systems. Simplified and standardized binary codes allow information to be encoded and decoded by mechanical or electrical means and make possible automation of data communications. Codes used in data communication systems are built into the equipment. For that reason, a user may never need to deal with codes except when interfacing equipment from different manufacturers. However, to understand data communications systems, we must recognize and understand the basic codes used.

Early Codes
A brief review of two codes (Morse and Baudot) used in early data communication systems will provide insight into how the codes used in today's systems were developed.

Morse Code The Morse telegraph, the oldest form of two-state communications and the first mass communication system based on electric power, was the first electrically based communications system to connect the east and west coasts of the United States, and both sides of the Atlantic Ocean. Figure 10-3 shows a basic telegraph system. When a key at station X is depressed, current flows through the system and a sounder switch at station Y is activated, clicking as it strikes the stop. When the key is released, the electrical circuit is opened and the switch of the sounder is forced to its open position by a spring, striking the other stop with a slightly different-sounding click. If the time between successive clicks is short, it represents a *dot;* if longer, a *dash*. Series of dots and dashes represent characters. The transmitting operator converts the characters in the words of a message into a series of dots and dashes. The receiving operator interprets those dots and dashes as characters. In this manner, the information is transmitted from point X to point Y.

The telegraph channel between the operators is in one of two states. Either current is flowing, or it is not. This illustrates a simple idea that has been repeated many times in the development of modern data communication systems. A two-state system is the simplest, the easiest to build, and the most reliable. The two states can occur in one of several ways: (1) single-current signaling (ON and OFF), (2) double-current signaling (+ and −), (3) sending code by light signals (light and dark), (4) the computer concept (1 and 0), or (5) any other design with only two possible values. A *two-state* or *two-valued* system is referred to as a *binary* system.

The Morse telegraph illustrates the simplicity of a complete data communication system. Much of the terminology that developed around the Morse system is in use today. For example, the terms *mark* and *space* are used worldwide in modern data communications systems. The basic principles of telegraphy have not changed, although the efficiency and implementation of telegraph systems have advanced greatly. For example, although the Morse code is particularly useful for manual, key

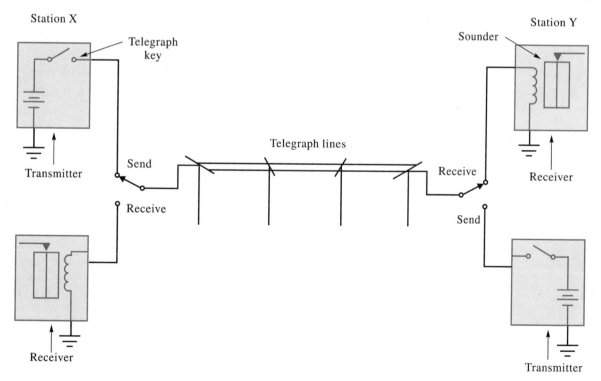

FIGURE 10-3   Basic telegraph system

transmission, it is not suitable for transmission by teletypewriters or other data machines; and, since all modern telegraphy is accomplished by some kind of teletypewriter, the Morse code is no longer used for this purpose. However, a variation of the Morse code was adopted for use by radiotelegraphers and is still used by amateur radio operators the world over.

Baudot Code   The advent of the teletypewriter machine produced the need for codes with a fixed character length and structure. The standard abbreviation TTY is used to represent all teletype operations. In the 1870s, a Frenchman named Emile Baudot developed one of the more successful codes suited to machine encoding and decoding. The Baudot code, shown in Figure 10-4, is a 5-bit code that can generate only 32 combinations, fewer than needed to represent the 26 characters of the alphabet, the 10 decimal digits, the punctuation marks, and the space character. However, by using the letters shift (LTRS) (lowercase) and the figures shift (FIGS) (uppercase), the code set can be made to represent all the characters necessary. The lowercase shift causes all following characters to be interpreted as letters of the alphabet. The uppercase shift causes all following characters to be interpreted as numerals and punctuation marks. In addition to the five bits per character, each character is preceded by a *start* bit, which is a space, and followed by a *stop* bit, which is a mark. The stop bit is approximately 5.5 times longer than the regular data mark. In use for many years, the Baudot 5-bit code is suitable for punched paper tape and standard TTY operation.

Modern Codes

Modern communications required a code that could do much more than early codes such as the Baudot. A modern code:

1. Had to represent all printable characters and still leave room for error checking.
2. Had to permit decoding without reliance on correct reception of previous transmissions.
3. Had to allow decoding by machine.
4. Needed to be expandable.

Although a number of codes were developed in the 1960s, only the three codes discussed in the following paragraphs are common today.

CCITT Standard International Alphabet No. 2   The Consultative Committee for International Telegraph and Telephone (CCITT), based in Geneva, Switzerland, has established international agreements on frequency and code assignments so that all systems using these assignments are compatible. The CCITT Standard International Alphabet No. 2 (shown in the Baudot 5-bit code in Figure 10-4) is still used for TTY transmission.

Since telegraphy signals can be modulated and transmitted in many different forms, a table of standards for data transmission, shown in Table 10-1, was recommended by CCITT in 1961. The standardization described in the recommendation was general and applied to any two-condition transmission, whether over telegraph-type circuits or over telephone-type circuits, making use of electromechanical or electronic devices.

CODE SIGNALS • Denotes positive current							LTRS Shift	FIGS Shift	
Start	1	2	3	4	5	Stop		CCITT Standard International Telegraph Alphabet No. 2 Used for Telex	North American Teletype Commercial Keyboard
	•	•				•	A	—	—
	•			•	•	•	B	?	?
		•	•	•		•	C	:	:
	•			•		•	D	Who are you?	$
	•					•	E	3	3
	•		•	•		•	F	Note 1	!
		•		•	•	•	G	Note 1	&
			•		•	•	H	Note 1	#
		•	•			•	I	8	8
	•	•		•		•	J	Bell	Bell
	•	•	•	•		•	K	(	(
		•			•	•	L	)	)
			•	•	•	•	M	.	.
			•	•		•	N	,	,
				•	•	•	O	9	9
		•	•		•	•	P	0	0
	•	•	•		•	•	Q	1	1
		•		•		•	R	4	4
	•		•			•	S	'	'
					•	•	T	5	5
	•	•	•			•	U	7	7
		•	•	•	•	•	V	=	;
	•	•			•	•	W	2	2
	•		•	•	•	•	X	/	/
	•		•		•	•	Y	6	6
	•				•	•	Z	+	"
						•	Blank		
	•	•	•	•	•	•	Letters shift (LTRS)		
	•	•		•	•	•	Figures shift (FIGS)		
			•			•	Space		
				•		•	Carriage return		
		•				•	Line feed		

Note 1: Allocated internationally; available to each country for internal use.

FIGURE 10-4 Baudot 5-bit code

EBCDIC The Extended Binary Coded Decimal Interchange Code (EBCDIC), shown in Figure 10-5, is a modern code using eight bits to represent 256 characters. It was developed by International Business Machines Corporation (IBM) to provide a standard code for its own products. Although EBCDIC has enough unique characters to allow almost any representation, only IBM and firms that build IBM-compatible equipment adopted it. For that reason, EBCDIC is used almost exclusively for synchronous communication between mainframe computers and peripheral equipment in large IBM-compatible computing systems.

TABLE 10-1 **CCITT Table of Standards**

	Digit 0	Digit 1
	"Start" signal in start–stop code Line available condition in telex switching "Space" element of start–stop code Condition A	"Stop" signal in start–stop code Line idle condition in telex switching "Mark" element of start–stop code Condition Z
Telegraphlike single-current signaling	No current	Positive current
Telegraphlike double-current signaling	Negative current	Positive current
Amplitude modulation	Tone-OFF	Tone-ON
Frequency modulation	High-frequency	Low-frequency
Phase modulation with reference phase	Phase opposite to reference phase	Reference phase
Differential phase modulation	Phase inversion	No phase inversion
Perforation in table	No perforation	Perforation

CCITT Data Transmission Recommendation V.I., 1961 and 1964.

	BIT 87654321		BIT 87654321		BIT 87654321		BIT 87654321
a	10000001	u	00100101	M	00101011	1	10001111
b	01000001	v	10100101	N	10101011	2	01001111
c	11000001	w	01100101	O	01101011	3	11001111
d	00010001	x	11100101	P	11101011	4	00101111
e	10010001	y	00010101	Q	00011011	5	10101111
f	01100001	z	10010101	R	10011011	6	01101111
g	11100001	A	10000011	S	01000111	7	11101111
h	00010001	B	01000011	T	11000111	8	00011111
i	10010001	C	11000011	U	00100111	9	10011111
j	10001001	D	00100011	V	10100111	0	00001111
k	01001001	E	10100011	W	01100111		
l	11001001	F	01100011	X	11100111		
m	00101001	G	11100011	Y	00010111		
n	10101001	H	00010011	Z	10010111		
o	01101001	I	10010011				
p	11101001	J	10001011				
q	00011001	K	01001011				
r	10011001	L	11001011				
s	01000101						
t	11000101						

FIGURE 10-5 EBCDIC 8-bit code

ASCII The American Standard Code for Information Interchange (ASCII), shown in Figure 10-6, is a code defined by the American National Standards Institute (ANSI) in the United States, and by the International Standards Organization (ISO) worldwide. ASCII is a bit code (Figure 10-6(A)) that is in general use today. It can represent 128 characters, not all of which represent printed symbols. Included in the character set are the letters of the English alphabet (both uppercase and lowercase), the numerals

COLUMN	0	1	2	3	4	5	6	7
BITS 4321 765 ⇒	000	001	010	011	100	101	110	111
ROW								
0 0000	NUL	DLE	SP	0	@	P	\	p
1 0001	SOH	DC1	!	1	A	Q	a	q
2 0010	STX	DC2	"	2	B	R	b	r
3 0011	ETX	DC3	#	3	C	S	c	s
4 0100	EOT	DC4	$	4	D	T	d	t
5 0101	ENQ	NAK	%	5	E	U	e	u
6 0110	ACK	SYN	&	6	F	V	f	v
7 0111	BEL	ETB	'	7	G	W	g	w
8 1000	BS	CAN	(	8	H	X	h	x
9 1001	HT	EM	)	9	I	Y	i	y
10 1010	LF	SUB	*	:	J	Z	j	z
11 1011	VT	ESC	+	;	K	[	k	{
12 1100	FF	FS	•	<	L	•	l		
13 1101	CR	GS	•	=	M	]	m	}
14 1110	SO	RS	•	>	N	∩	n	~
15 1111	SI	US	/	?	O	—	o	DEL

Explanation of special control functions in columns 0, 1, 2, and 7:

NUL	Null	DLE	Data Link Escape
SOH	Start of Heading	DC1	Device Control 1
STX	Start of Text	DC2	Device Control 2
ETX	End of Text	DC3	Device Control 3
EOT	End of Transmission	DC4	Device Control 4
ENQ	Enquiry	NAK	Negative Acknowledge
ACK	Acknowledge	SYN	Synchronous Idle
BEL	Bell (audible signal)	ETB	End of Transmission Block
BS	Backspace	CAN	Cancel
HT	Horizontal Tabulation (punched card skip)	EM	End of Medium
LF	Line Feed	SUB	Substitute
VT	Vertical Tabulation	ESC	Escape
FF	Form Feed	FS	File Separator
CR	Carriage Return	GS	Group Separator
SO	Shift Out	RS	Record Separator
SI	Shift In	US	Unit Separator
SP	Space (blank)	DEL	Delete

(A)

3-BIT   4-BIT

7	6	5	4	3	2	1

(B)

FIGURE 10-6   ASCII code (A) ASCII 7-bit code (B) ASCII code word format

0–9, punctuation marks, and special control symbols. ASCII is the standard code set used worldwide in virtually all small computers and their peripherals, and in many large computer systems as well.

The 7-bit ASCII code for each letter, numeral, or control symbol is made up of a 4-bit and a 3-bit group (Figure 10-6(B)). Bit 1 is the least significant bit (LSB), and bit 7 is the most significant bit (MSB). Bits 1–4 are on the right and are derived from the rows of the code. Bits 5–7 are on the left and are derived from the columns of the code. To determine the ASCII code for any character, locate the character in the table. Then using the 3- and 4-bit codes associated with the row and column in which the character is located, determine the 7-bit code.

EXAMPLE 10-1

Write *Your Name* using the 7-bit ASCII code.

SOLUTION

The author's name (Frank Dungan) is used to demonstrate this example:

F	r	a	n	k	SP
1000110	1110010	1100001	1101110	1101011	0100000

D	u	n	g	a	n
1000100	1110101	1101110	1100111	1100001	1101110

The ASCII format is arranged so that uppercase letters can be changed to lowercase letters by changing only one bit—that is, by changing bit 6 from 0 to 1. Another simplifying feature is that the 4-bit groups in rows 0–9 represent the binary coded decimal (BCD) value of the numerals 0–9 in column 3.

Columns 0 and 1 contain the nonprinting control characters which can be used to control the receiving device. These control codes are designed for printing or display devices, although some manufacturers use them for all kinds of special functions. Also, some codes control how a receiving device will interpret subsequent codes in a multiple-character function or command. For example, the two shift characters, designated SI (Shift In) and SO (Shift Out), are used to shift between ASCII and character sets other than those used in English. Other control codes, such as STX (Start of Text) and ETX (End of Text), place limits on text. These codes are used primarily in block or synchronous transmission.

Escape (ESC) sequences are code sequences made up of noncontrol characters that are to be interpreted as control codes. In other words, the ESC character designates that the codes that follow have special meaning. Characters received in an ESC sequence are not interpreted as printing characters but rather as control information to extend the range of the "standard" character set by allowing other definitions. The ESC character makes all character codes available for control of a device. Graphics characters, foreign language character sets, and special applications sets have been developed that are accessible via ESC sequences, permitting a much richer variety of displayed symbols than is possible with any single code.

10-5 MODES OF OPERATION

In this section, we will examine the three basic modes of operation for the transmission of data: simplex, half-duplex, and full-duplex.

Simplex Mode

The *simplex* mode is the simplest and least costly of the three. The earliest use of the simplex mode was for telegraph transmissions. The circuit permits transmission of data in one direction only—that is, from A to B, but never from B to A—as illustrated by the block diagram in Figure 10-7(A). Today, simplex circuits with two-wire telephone circuits are widely used for offline communications.

A somewhat more sophisticated simplex circuit, used for either telephone or telegraph operation, is the *phantom* circuit, in which two separate two-wire telephone lines are connected by means of center-tapped transformers at each end. In this manner, each pair of telephone wires is used for one leg of the simplex circuit. In a phantom circuit, there is no interference between telephone and telegraph operation, but the two telephone circuits must be used to provide one telegraph circuit.

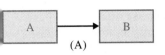

(A)

Half-Duplex Mode

In the *half-duplex* (HDX) mode of operation, illustrated in Figure 10-7(B), transmission of data can occur one way or the other, but not both ways simultaneously, because of the limitations of terminal equipment. This mode is used extensively for realtime communications.

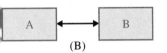

(B)

Full-Duplex Mode

Full-duplex (FDX) operation, shown in Figure 10-7(C), requires two separate communication channels so that simultaneous two-way communication can occur. Generally, this is accomplished by four-wire circuits. However, two-wire circuits can be adapted for FDX operation by using two separate frequencies for transmission and reception. This technique is used in 1200 and 2400 bps FDX two-wire modems, and in most modern telephone systems.

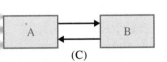

(C)

FIGURE 10-7 Modes of operation for data communication systems (A) Simplex (B) Half-duplex (C) Full-duplex

10-6 DATA CIRCUIT-TERMINATING EQUIPMENT

Data circuit-terminating equipment (DCTE), more commonly known as *data communication equipment* (DCE), provides the functions required to establish, maintain, and terminate a connection, along with the signal conversion required for communication between the DTE and the telephone line or data circuit. In other words, the DCE is the conversion equipment between the DTE and the transmission channel. One type of DCE is the modem, which converts data into tones for transmission over the voice channel.

The DTE-DCE interface consists of the I/O circuitry in the DTE and DCE and the connectors and cables that link them together. The interface conforms to one of the electrical standards, such as the Electronic Industries Association (EIA) RS-232C, RS-422, or RS-423. The RS-232C interface or some variation of the RS-232C specification is by far the most commonly used serial interface in data communications. The

RS-232C standard specifies the rules by which data are moved across the interface between the DTE and DCE; thus, it specifies how data are moved from point to point.

To interface between the DCE and the transmission channel is quite simple. It consists of two wires or four wires. There is no problem with serial sequencing of the electrical signals across this interface.

The electrical characteristics of the transmission channel itself usually conform to specifications published by telephone companies or other carriers. The equipment used for data communications tolerates a wide range of channel characteristics, and the user has very little control over the transmission channel itself. Except for very high-speed data transmission rates, the transmission channel usually does not cause trouble.

How DCE Works

The transmitting computer or DTE sends messages in a sequence of parallel characters to the DCE via the DTE-DCE interface (the RS-232C interface uses from 2 to 24 parallel wires). Each character in the message is serialized for a one-bit-at-a-time transmission. Serial transmission is necessary for a two-wire transmission channel and makes synchronizing the received data easy. Asynchronous, synchronous, and isochronous transmissions are discussed in the next sections.

At the receiving end of the channel, each action of the transmitting end is reversed. The DCE decodes (demodulates) the received signals and extracts the information. The interface produces a serial sequence of pulses that are then assembled into a sequence of parallel characters. These characters are read by the receiving DTE or computer, which acknowledges reception of the message from the sender. Message acknowledgment takes the same path as the data, except that now the receiver becomes the source and the former transmitter becomes the sink, demonstrating the duality of the two-way data communication system.

Asynchronous Transmission

In *asynchronous* transmission, each information character is individually synchronized, usually by the use of *start* and *stop* elements. Asynchronous data usually come from low-speed terminals with data rates below 2 kilobits per second (kbps). In these systems, idle transmission lines are in a *mark* (binary 1) state, as illustrated in Figure 10-8(A). A start bit (transition from mark to *space,* or binary 0) precedes each transmitted character and indicates to the receiving terminal that a character is being transmitted. The receiver detects the start bit first, then the data bits that make up the character. At the end of the transmitted character, one or more stop bits return the line to mark, where it gets ready for the next character. This process is repeated for each character until the entire message is transmitted. The start and stop bits allow the receiving terminal to synchronize its circuitry to the transmitting terminal character by character.

Asynchronous transmission is the most common transmission method in data communications simply because there are more low-speed terminals and small computer applications in which it is used. This is the transmission method used by teleprinters and ASCII terminals. However, there is a large penalty associated with the start-stop bits. For example, if the start-stop bits are used with the 5-bit Baudot code, two out of the seven bits transmitted are for control rather than for information, a 28.6% penalty.

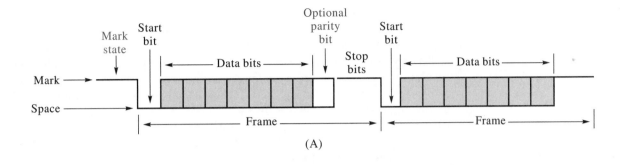

(A)

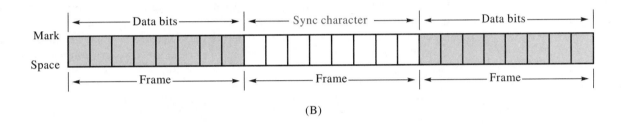

(B)

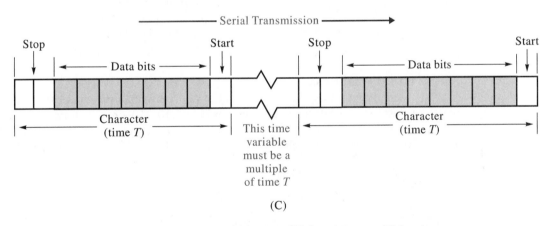

(C)

FIGURE 10-8   Transmission codes (A) Asynchronous (B) Synchronous (C) Isochronous

Even if they are used with the 8-bit EBCDIC code, a 20% penalty exists. For this reason, and others, large systems and computer networks usually use additional transmission methods.

Synchronous Transmission

With *synchronous* transmission, an internal clock synchronizes the communication channel's transmitter and receiver. When one or more synchronization characters are sensed by the receiving terminal, data transmission proceeds character by character without any intervening start or stop bits. A modulated carrier containing *data blocks,* or *frames,* illustrated in Figure 10-8(B), is interpreted with the help of a PLL in the

DCE at the receiving terminal. The PLL produces a receiving clock by locking on transitions in the received data, and then the receiving terminal samples each bit in the data stream using the synchronization provided by the PLL-produced clock.

Bit synchronization depends on transitions between binary 1s or 0s in the serial data input. Therefore, to prevent a long series of 1s or 0s from causing loss of bit synchronization in the PLL, many DCEs include scrambler–descrambler circuitry. At the transmitter, a scrambler changes the data pattern to ensure that there are enough transitions; at the receiver, a descrambler restores the original data pattern.

Isochronous Transmission

Isochronous transmission is a mix between asynchronous and synchronous transmission. Individual characters are framed with a start and stop bit as in asynchronous transmission, but the intervals between characters are time controlled, as illustrated in Figure 10-8(C). So long as it is restricted to multiples of one character time, the time interval may be any length. This technique is used in most modern computer networks.

10-7 INTERFACING TO TELEPHONE NETWORKS

Data communication systems are typically interfaced to public or private telephone networks. All interfacing equipment connected to public utility company lines must be registered with the FCC or use an FCC-approved coupler. Therefore, technicians must know something about how telephone interfacing systems operate.

Rotary Dialing

A simplified rotary dialing telephone system between a *subscriber* (user) and a *central office* (CO) or exchange is shown in Figure 10-9. A 48 V battery at the CO provides the current necessary to operate the handset carbon microphone and various CO relays, which are not shown in this simplified diagram. When the subscriber picks up

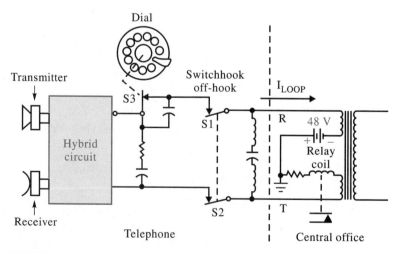

FIGURE 10-9 Simplified rotary dialing telephone

the handset, the hanger switch contacts close, completing a dc circuit from one terminal of the handset microphone, through the hybrid coil and one winding of the CO relay coil, to the negative side of the battery. The other terminal of the handset microphone connects through the hanger contacts and another relay coil to ground and the positive side of the battery. Thus, the microphone is energized, an open operational line in the CO is automatically searched for and selected, and a ±400 Hz dial tone is fed back to the subscriber.

When rotary dialing is used, the subscriber line is sequentially opened and closed, setting up a series of dc pulses that represent numbers. For example, the line is opened and closed three times if a 3 is dialed, five times if a 5 is dialed, and nine times if a 9 is dialed. The line is returned to the closed condition when the dial stops turning. Central office circuits sense the dc pulses representing numbers and, from the first three numbers dialed, connect the initiating subscriber line to the correct CO and then, from the last four numbers dialed, to the called subscriber line.

Dual-Tone, Multiple-Frequency Dialing

Dual-tone, multiple-frequency (DTMF) dialing (more commonly known as *Touchtone®*) was discussed in Chapter 9. For each number pressed on the keypad, a transistor oscillator in the telephone set develops two different tones and feeds these tones across the line. An extended 16-button pad (the additional four buttons are shown as dashed lines in Figure 10-10) is used for some computer and other special applications.

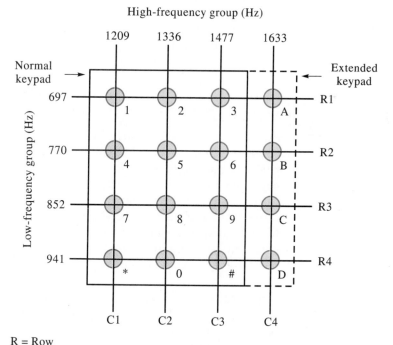

R = Row
C = Column

FIGURE 10-10 Touchtone® pad

Tone Generation Pressing one of the keys causes an electronic circuit to generate two tones: (1) a low-frequency tone for each row and (2) a high-frequency tone for each column. For example, pressing button "8" generates an 852 Hz tone and a 1336 Hz tone.

Tone Detection The tones used have been carefully selected so that the digit receiver or processing circuits in the CO will not confuse them with other tones that may be present. The tones are separated by tuned filters that pass only the frequencies used for DTMF dialing. Timing circuits make sure that a tone is present for a minimum of about 50 ms before it is accepted as a valid DTMF tone. The signals are decoded either to form ON-OFF dc pulses that can open or close relays or to produce binary words (such as 1000 when the "8" key is pressed) that can be made to open or close digital circuits. Decoders may use resonant *LC* circuits and rectifiers to develop the dc pulses.

After a connection is made to an answered telephone, the digit receiver is out of the circuit and the DTMF tones can be transmitted in the same manner as speech. This arrangement permits the use of DTMF tones as data communications for transmitting information to a remote terminal or receiving information from a remote data base.

Although the frequencies and the keypad layout in Figure 10-10 have been standardized internationally, the tolerances on individual frequencies may vary in different countries. The North American standard for frequency tolerances is ±1.5% for the generator and ±2% for the receiver.

DTMF dialing offers the following advantages over rotary dialing: (1) decreased dialing time, (2) solid-state electronic circuits, (3) low-speed data transmission, (4) reduced local CO equipment requirements, and (5) more compatibility with electronically controlled exchanges.

10-8 DATA COMMUNICATION SYSTEM MODULATION AND DEMODULATION

Recall that modulation is the process of changing some property of an electrical wave (the carrier, Figure 10-11(A)) in response to some property of another signal (the modulating signal, Figure 10-11(B)). To transmit data signals over the telephone channel, some property of an ac carrier wave of between 300 Hz and 3300 Hz must be changed in response to a binary (0 to 1) signal from a computer. The properties of the carrier that are available to be changed should be familiar to you; they are the amplitude, the frequency, and the phase. The three methods used for modulation in data communications are illustrated in Figures 10-11(C) through (E). Any one of these methods may be used in modems.

Amplitude Modulation

In amplitude modulation, the amplitude of the carrier is changed in response to the binary modulating signal. As shown in Figure 10-11(C), the amplitude is varied from zero, representing a binary 0, to a maximum value, representing a binary 1. The result is that the 1s and 0s of the data stream are converted to analog waveforms representing tones having frequencies in the 300–3300 Hz range.

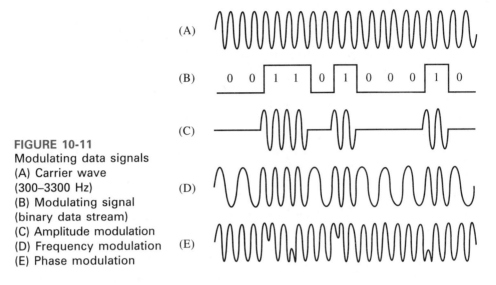

FIGURE 10-11
Modulating data signals
(A) Carrier wave
(300–3300 Hz)
(B) Modulating signal
(binary data stream)
(C) Amplitude modulation
(D) Frequency modulation
(E) Phase modulation

Frequency Modulation

In frequency modulation, illustrated in Figure 10-11(D), the carrier frequency is shifted to a lower frequency to represent a binary 0 and to a higher frequency to represent a binary 1. This technique is sometimes called *frequency shift keying* (FSK). When the carrier frequencies are used in the voice band, the technique is called *audio frequency shift keying* (AFSK). Many modern modems carry digital data signals in the 45–1800 bps range using the AFSK technique.

Phase Modulation

In phase modulation, phase can be visualized as the relative relationship of two waveforms at any given time in their cycle. As illustrated in Figure 10-11(E) the phase is shifted for every occurrence of a 1 bit, but it is not shifted for a 0 bit. Usually, the phase of the signal sent over the transmission medium is not measured absolutely. It is instead measured relative to the phase of the wave during the previous bit interval.

Because the human ear is not very sensitive to the phase of speech or music sounds, the telephone network was not designed to carefully preserve the phase relationships of signals sent through it. For that reason, high-speed modems that use phase modulation to carry data signals usually contain circuits that compensate for disturbances or nonlinearities in the telephone network to help restore the phase linearity of the received signal.

Demodulation

Just as the carrier must be modulated to carry the data, it must be demodulated to recover the original data. In the receiving modem, detectors and filter circuits sensitive to amplitude, frequency, or phase recover the binary 1s and 0s from the modulated carrier. Threshold bias and level-shifting circuits restore the digital signal to common logic levels or to special interface transmission levels specified by the digital signal protocols for interfacing.

10-9   MODEMS AND THEIR OPERATION

As stated previously, data transmission between computers and their remote terminals can be parallel or serial. Both the parallel and the serial methods of transmission require a modulator to encode the original information onto the carrier and a demodulator at the receiver to convert the carrier information into the true digital signal. Each station can transmit and receive; therefore, these devices are incorporated into a single unit called a *modem,* a contraction derived from *mo*dulator and *dem*odulator.

The modem transforms the input into transmittable signals. Two modems are required: (1) to convert data from the source into a form that can be transmitted over telephone lines and (2) to reverse the process at the terminating end. Because modems are capable of both modulation and demodulation, the terminals can alternate as both source and sink.

Consisting of an AF transmitter and receiver, the modem modulates an AF wave to transmit data and demodulates an incoming AF wave. Some modems can also convert parallel data to serial data for transmission, and serial data to parallel data on reception.

Modems are described by their baud rate. *Baud rate* is a measure of the speed of the transmission of a digital code. A baud (a derivative of Baudot) is: (1) a unit of signaling speed in data transmission; (2) a function of bandwidth; (3) a figure indicating the number of times per second that a signal changes; and (4) in an equal-length code, a rate of one signal element per second. In other words, the baud rate is simply the number of code elements or signaling elements transmitted per second. Most often stated simply in baud, it is often represented by bits per second (bps).

However, a *bit per second* is not exactly the same as a baud. Bits per second are: (1) a function of transmission (quality of equipment, lines, modems) and of noise; and (2) the measure of the true bit data transfer rate. Further, the *levels* at which a modem operates can be measured in *dibits* or *tribits*. For example, a 2400-baud modem operating on dibits works at 4800 bits per second (two bits in one baud), and a 3400-baud modem operating on tribits works at 9600 bits per second (three bits in one baud). Bits per second are a function of the means by which the communication is done.

Early low-speed modems operated at speeds of 300, 1200, 2400, 4800, and 9600 bits per second (bps). Modern modems run at much higher rates, such as 14.4 (V.31), 28.8 (V.34), and 33.6 (V.34) Kbps. With high-speed external system connections such as ISDN (*Integrated Services Digital Network*) telephone lines, throughput speeds up to 128 Kbps can be obtained. One example of the newer modems is the ZOOM® V.34I Plus fax/modem. Its technical specifications are:

> *Data Standards:*
> 33.6, 31.2 Kbps—V.34
> 28.8, 26.4, 24.0, 21.6, 19.2, 16.8, 14.4 Kbps—V.34/V.FC
> 14.4, 12.0, 72.0 Kbps—V.32bis
> 9.6, 4.8 Kbps—V.32
> 2.4 Kbps—V.22 bis
> 1.2 Kbps—Bell 212A, V.22, V.22A/B, V.23
> 300 bps—Bell 103, V.21

Fax Standards:
14.4, 12.0 Kbps—V.33, V.17
9.6, 7.2 Kbps—V.29, V.17
4.8, 2.4 Kbps—V.27 ter
300 bps—V.21 channel 2

Terminal-to-Modem (DTE) Data Rates:
115.2, 57.6, 38.4, 19.2, 14.4, 9.6, 7.2, 4.8, 2.4,
1.2 Kbps, 300 bps

The operation of different types of modems is discussed in the following sections. The original pulsed dc (*baseband*) form of information transmitted over the switched telephone network is filtered and amplified in the CO before being transmitted farther. The bandwidth (pass band) of the telephone channel beyond the CO after filtering and amplification is about 300–3300 Hz. Since dc is zero hertz and therefore outside the bandwidth of the channel, transmission of data in their original form as a series of pulses is generally limited to the distance between the subscriber telephone and the CO. Therefore, digital information cannot be transmitted directly over the switched telephone network. The use of modems solves this problem.

The telephone network is designed and optimized for transmission of analog signals in the voice band, so the function of the modem is to convert the pulsed dc or baseband form to analog waveforms resembling tones with frequencies in the 300–3300 Hz range. Thus, a transmitting modem changes the digital signals produced by computers to an analog signal with a frequency band that can be transmitted over the telephone network, and a receiving modem converts the analog signal back to its original form so that the receiving computer can use the data.

Types of Modulation in Modems

There are four basic types of modulation used in modems:

1. Amplitude modulation (AM), or amplitude shift keying (ASK).
2. Frequency modulation (FM), or frequency shift keying (FSK).
3. Phase modulation (PM), or phase shift keying (PSK).
4. Multilevel modulation, also known as quadrature amplitude modulation (QAM).

To a certain extent, the modulation technique used depends on modem speed and the amount of error that can be tolerated. ASK and FSK are used primarily in low- and medium-speed modems (300, 1200, and 2400 bps). Most 4800 bps modems used PSK, and most 9600 bps modems use QAM.

Amplitude modulation operates with two constant amplitude levels: a lower level to represent a binary 0 (or space) and a higher level to represent a binary 1 (or mark). Shifting between these two levels produces ASK. However, ASK suffers from amplitude noise, which can cause the keying of false data.

Phase modulation operates with two phases, on the principle of a 180° phase shift between space and mark, thus producing PSK. But detecting the received data signal is sometimes difficult because no reference phase is transmitted. One method used to overcome this problem is to transmit a 90° synchronizing pulse along with the data. This synchronizing pulse shifts +90° for a mark and −90° for a space, thus producing the 180° required phase shift. This modified approach is called *differential phase shift keying* (DPSK).

Multilevel modulation uses more than two modulation levels. In effect, this method samples two bits at a time and produces four different amplitudes, thus, the designation quadrature amplitude modulation. QAM is especially useful for modems operating at 9600 bps.

Frequency Shift Keying

Frequency shift keying (FSK) is the modulation technique most often used in modems. It uses a specific frequency to represent the ON condition (binary 1) and a different frequency to represent the OFF condition (binary 0). The simplified block diagram in Figure 10-12 shows a typical modem operation. The difference between the two frequencies is the *frequency shift*. In the transmit modem, the generated pulses *key* (shift) the transmitter between the two FSK frequencies, resulting in an FM-like signal that is transmitted over a telephone line. The receiver then uses an FM detector or a PLL to distinguish between the two FSK frequencies and convert them back into the original data pulses.

The audio frequencies used with modems are within the normal telephone line range—that is, 300–3300 Hz. Within this range, many frequency combinations are possible. One standard combination is 1270 Hz for the binary 1 representation and 1070 Hz for the binary 0 representation, providing an FSK signal with a 200 Hz shift.

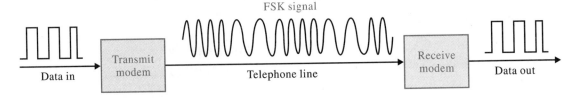

FIGURE 10-12 FSK modem system

Protocols

As the options and intelligence in data communication terminals increased, technology mandated new rules and procedures for efficient operation of the system. These communication controls, called *protocols,* were formulated by equipment suppliers. Protocols may govern lines, types of service, modes of operation, circuit compatibility, or entire networks. This arrangement is analogous to the use of highway traffic rules to control the efficient flow of traffic over city streets and highway networks.

Early protocols were referred to as *handshaking.* However, current usage of this term generally means that a connection has been established and that the communication line is ready for the message. In present-day usage, protocol includes both handshaking and *line discipline,* a term that denotes the sequence of operations involving the actual transmitting and receiving of data. Many suppliers use *line discipline* and *protocol* synonymously. *Line polling* is similar to a two-way conversation between a computer and a terminal wherein each confirms to the other the status of a message.

Error Detection and Correction

A basic problem in using voice facilities for data transmission is the presence of noise and distortion. The resulting errors in transmission require some error control mechanism. Although the subject of error detection and correction has nothing directly to do with the subject of modems, whenever digital data are transmitted over telephone channels using modems, errors are sure to occur. Therefore, the computer systems and terminals that produce and consume the data must take measures to detect and, if possible, correct any data errors.

Recent modem developments combined with the use of microprocessors have allowed the data to be buffered, the error control to be applied, and the data to be checked by the receiver. If the receiver detects an error, it automatically requests retransmission of the original data.

There are a number of methods for detecting errors. The most commonly used techniques add *redundancy*—that is, information in addition to the minimum required to send the original data. The redundant information is related to the original input in some systematic way so that it can be regenerated when it is received. If the regenerated error control matches what was sent along with the data, the transmission is assumed to be error-free.

A classical method of error detection is *parity checking,* which involves the use of a single bit, known as a *parity bit.* A description of parity checking will help you understand how error detection can be built into a coding structure.

Parity describes a condition wherein the total number of 1 bits in each character is always *even* or always *odd,* depending on the parity system used. When a parity system is used, the transmitting equipment automatically adds one information-carrying bit, called the parity bit, or *check bit,* to the characters being transmitted. This addition enables the computer to run its own check on every character it processes.

The redundancy to detect errors in long blocks of data is provided by a class of codes called *cyclic redundancy check* (CRC) codes. Generating a CRC for a message involves dividing the message by a polynomial, which produces a quotient and a remainder. The remainder, which usually is two characters (16 bits) in length, is added to the message and transmitted. The added information is sometimes referred to as a *block check character* (BCC). The receiver performs the same operation on the received message and compares the calculated remainder to the received remainder. If the remainders are equal, the probability is high that the message was received correctly.

Modem Interfaces

A modem has a simple interface to the telephone network consisting of two wires called the *tip and ring.* So long as the modem adheres to the voltage, current, power, and frequency rules of the telephone company, the telephone channel acts like a pipe, used to move analog tones from one place to another.

A more complex interface occurs between the modem and the DTE. The wiring is governed by standards set by EIA. This interface also requires that certain procedures (protocols) be observed in establishing communications between the two ends. First, the DTE and the modem at the transmitting end must establish communication with each other. The DTE indicates to the modem that it wishes to transmit, and this modem polls the modem at the other end of the circuit to see if it is ready to receive. Modems do not store data, so the receiving modem must contact its DTE to see if it is ready to receive. This communication between the equipment is commonly referred to as handshaking. After the transmitting modem knows that the receiving modem and DTE are listening on the line, it notifies the transmitting DTE, which begins passing data to the transmitting modem for modulation and transmission. On the receiving end, the receiving modem demodulates the incoming signal and passes the received data to the receiving DTE.

In HDX transmission, when the transmitter is finished and wants a reply from the other end, the channel must be "turned around." To accomplish this, much of the handshaking must be done again to establish transmission in the opposite direction, and this turnaround handshaking must occur each time the direction is changed. In FDX transmission, the transmission uses two different carrier frequencies; thus, the handshaking is necessary only for the initial setup.

Telephone Channel Restrictions on Modems

Most practical transmission channels have limited bandwidths. There are various causes for this limitation: (1) the physical properties of the channel itself, (2) the electrical signal, (3) noise added to the information signal, or (4) deliberate bandwidth limitations to prevent interference from other sources. From a purely economic point of view, data communication systems should maximize the amount of data that can be sent on a single channel.

Analog Signals   Transmission rates for data communications using analog signals have steadily increased over the years. For a long time, personal computers used low-cost modems with low transmission rates of 300 bps. A 300 bps FDX signal uses two bands of frequencies, each band occupying 300 Hz, for a total of 600 Hz out of the available 3000 Hz bandwidth. This represents an inefficient use of the transmission channel. More modern low-cost modems operate at 1200 bps, four times faster than the earlier 300 bps modems. An FDX 1200 bps modem uses 2400 Hz of the available 3000 Hz bandwidth; therefore, four times as much information can be sent in the same channel in a given time period. It is possible in today's systems, through sophisticated signal processing techniques, to carry up to 56,600 bps using analog signals over a single voice channel.

Digital Signals   The most noticeable disadvantage directly associated with the transmission of digital signals is the necessity for a bandwidth greater than that required for equivalent analog signals. For example, using the standard T1 (paired cable) time division multiplex (TDM) format, the transmission of 24 analog voice channels requires about 96 kHz (24 channels × 4 kHz per channel). Transmitting 24 voice channels in digital form using the same TDM format requires about 772 kHz, eight times as much bandwidth (772 kHz/96 kHz).

The advantages of transmitting signals in digital form more than offset the requirement for greater bandwidth:

1. The telephone network and switching equipment for digital systems can use the same kinds of IC logic as used in digital computers.

2. When both the transmission technique and the switching system are digital, common circuit functions allow transmission and switching integration so that many of the traditionally required interface circuits are eliminated.

3. Digital signals are easier to multiplex, and the filters necessary to separate channels are simplified.

4. Digital signals are represented by pulses of well-defined and uniform shape, so they are easy to reconstruct even if badly distorted by noise.

5. Digital signals are highly resistant to crosstalk.

6. Signals can be mixed; that is, digital transmission channels can handle digital signals from sources other than speech.

In addition to the bandwidth limits, another restriction of the telephone channel that affects modem design is inherent in all analog transmission facilities. The transmission is best at frequencies near the center of the pass band and poorer for frequencies toward the upper and lower limits of the pass band. High-speed modems use almost all of the voice band for one channel; therefore, most high-speed modems in North America utilize a carrier frequency of 1700–1800 Hz because these frequencies are very near the middle of the voice band. Low-speed modems, because of their narrower bandwidth requirements, can use more than one carrier frequency within the voice band and still operate in the "good" portion of the band.

Still another restriction of the telephone channel is that certain frequencies cannot be used. The telephone network uses the transmission channel for passing information and control signals between the switching offices. This process, called *in-band* or *in-channel interoffice signaling,* utilizes tone at frequencies within the voice band. A modem cannot use these same frequencies because the network might interpret them as control tones, with disastrous results to the call.

The basic reason that a digital signal output from a computer cannot be connected directly to a telephone line is the amount of bandwidth available to carry signals over a single telephone channel, which in turn is related to the cost of the channel. The least expensive channels currently available are voice-band channels that have been designed to carry voice signals produced by telephone sets and whose bandwidth extends from about 300 Hz to about 3300 Hz. This means that the channel cannot pass very low frequency signals such as dc, nor very high frequency signals (above 3300 Hz). The signals used in computers are usually unipolar, and the transition from the binary 0 level to the binary 1 level is very fast—that is, at a high frequency. As a result, the signals contain significant frequencies below 300 Hz, even a dc component, and frequencies well above the 3300 Hz limit in a voice frequency telephone channel. Because the telephone channel will not carry these frequencies, the digital waveform must be transformed to a signal compatible with the channel and its bandwidth; that is, the signal must be modulated.

Asynchronous Modem Operation

The AFSK modulation schemes used in most low-speed asynchronous modems are shown in Figure 10-13. As illustrated in Figure 10-13(A), a center frequency (*fc*) of 1170 Hz is frequency shifted to 1070 Hz for a binary 0 (space) and to 1270 Hz for a binary 1 (mark). Every time the serial bit stream for a character is input to the originating

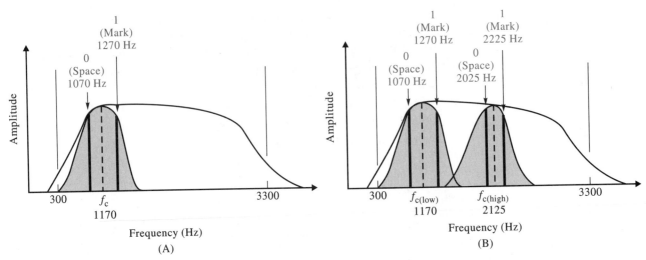

FIGURE 10-13 AFSK modulation schemes (A) Simplex or HDX transmission model (B) FDX transmission model

modem, the modem output is a continuous alternating signal of 1070 Hz or 1270 Hz, depending on whether the input is a binary 0 or a binary 1. This scheme is used for simplex or HDX transmission.

Dividing the available bandwidth into two bands, as shown in Figure 10-13(B), provides FDX transmission. The low band carries data in one direction, and the high band carries data in the opposite direction. The low-band center frequency ($f_{c(low)}$) is 1170 Hz, and the frequency is shifted to 1070 Hz for a space and to 1270 Hz for a mark. The high-band center frequency ($f_{c(high)}$) is 2125 Hz, and the frequency is shifted to 2025 Hz for a space and to 2225 Hz for a mark. These frequency pairs are used in the Bell System 103-series modem which, through general use, has become the U.S. standard.

System Interconnection   Figure 10-14(A) shows the block diagram of the interconnection between the originating and the answering modems in an FDX system. In the U.S. standard system shown, the transmitter at the originating modem and the receiver at the answering modem operate at a center frequency of 1170 Hz. The transmitter at the answering modem and the receiver at the originating modem operate at a center frequency of 2125 Hz. Figure 10-14(B) shows the signal levels versus frequency for the two channels.

U.S. and European Systems Compared   For a similar European standard CCITT system, the transmitter at the originating modem and the receiver at the answering modem operate at a low-band center frequency of 1080 Hz; the transmitter at the answering modem and the receiver at the originating modem operate at a high-band center frequency of 1750 Hz. However, the frequency shifts around these center frequencies representing space and mark are opposite those in the U.S. system. In the CCITT system, a space in the low band is represented by 1180 Hz, and a mark is represented by 980 Hz. In the high band, a space is represented by 1850 Hz, and a mark is represented by 1650 Hz.

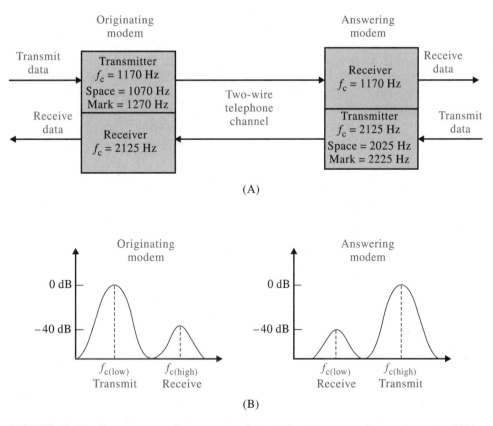

FIGURE 10-14 Frequency assignments and levels for a low-speed asynchronous FDX modem (A) U.S. frequency assignments (B) Transmit and receive levels

Low-Speed Asynchronous Modem A block diagram of a Bell System 103-type low-speed asynchronous modem is shown in Figure 10-15. As illustrated, the two-wire telephone line is terminated in a line-matching transformer. Although the secondary winding of the transformer is connected to both the transmit section output and the receive section input, the received signal has no effect on the transmit section. However, the transmit output could affect the receive section. But, because the transmit and receive frequencies are in different bands, the receive section BP filter prevents the transmit signal from entering the receive section. Also, the receive section BP filter rejects noise and spurious frequencies riding on the signal received from the telephone line. The limiter removes amplitude variations.

The delay detector provides a delayed sample of the signal, compares it to the receive signal, and produces an output proportional to the frequency difference. The slicer circuit (clipper) clips the top and bottom of the detected signal and produces a digital signal at the output with the proper voltage levels for binary 1s and 0s.

For data transmission, the digital serial bit stream is applied to an FSK oscillator, which produces the AF-shifted tones representing the binary 1s and 0s. The transmit section BP filter removes spurious harmonics and passes the signal through the line-matching transformer to the telephone line.

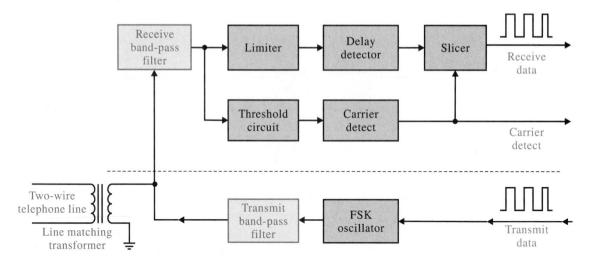

FIGURE 10-15 Low-speed asynchronous modem

Figure 10-15 shows the modem connected directly by wires to the telephone line. Although this is the preferred method of interfacing, some modems are acoustically coupled to the telephone handset. The acoustical coupler has a microphone and speaker that interface with the handset receiver and transmitter, respectively, to couple the transmitted and received signals by sound waves.

In an acoustically coupled system, there are no wire connections between the modem and the telephone line. A telephone call is made to the destination and a connection is established between the sender and receiver in the same way as for a conversation. The handset is then placed in the acoustical coupler for data transmission and reception.

Synchronous Modem Operation

A large amount of design effort has gone into the production of devices that will send the largest number of bits per unit time through standard voice-grade channels. One result of this effort has been *synchronous* signaling.

With the asynchronous modems previously discussed, the time base (clock) for the transmitter and receiver are independent; thus, small differences between the two clocks become increasingly likely to cause errors through sampling of the data at the wrong time. This problem is overcome in *synchronous modems* by deriving the timing information from the received data.

The greater complexity and cost of synchronous modems as compared to asynchronous units are due to the circuitry necessary to derive the timing from the incoming data and to pack more than one bit into one signaling element (*baud*). As shown in the block diagram of Figure 10-16, synchronous modems typically consist of four sections: transmitter, receiver, terminal control, and power supply. We will examine the first three of these sections.

Transmitter Figure 10-17 shows that the transmitter section of a synchronous modem consists of timing (clock), scrambler, modulator, digital-to-analog converter, and

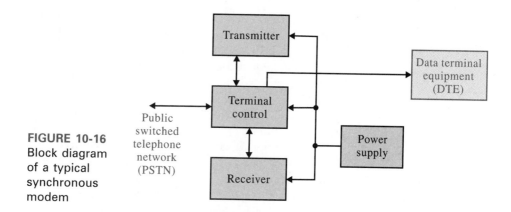

FIGURE 10-16
Block diagram
of a typical
synchronous
modem

equalizer circuits. The *timing* circuit provides the basic clocking information for both the modem and the DTE that is providing the data to be transmitted. Certain data circuit arrangements require that the clocking for the transmitted data be supplied by the DTE (which may be another modem). In these cases, an option is usually provided in the modem to phase-lock the internal clock to an external clock source input through the DTE interface. The internal timing is usually controlled by a crystal oscillator to within about 0.05% of the desired value.

Because the receiver clock is derived from the received data, those data must contain enough changes from 0 to 1 (and vice versa) to ensure that the timing recovery circuit will stay in synchronization. In principle, the data stream provided by the associated terminal or business machine can consist of any arbitrary pattern. If the pattern contains long strings of the same value, the data will not provide the receiver with enough transitions for synchronization. The transmitter must prevent this condition by changing the input bit stream in a controlled way. The *scrambler* performs this function.

Scramblers are usually implemented as feedback shift registers, which may be cascaded (connected in series). They are designed to ensure that each possible value of phase angle is equally likely to occur to provide the receiver demodulator with enough phase shifts to recover the clocking signal. Although scrambling is necessary, it does increase the error rate since an error on one bit is likely to cause an error in subsequent bits. To counteract this problem, some modems encode the input to the scrambler into

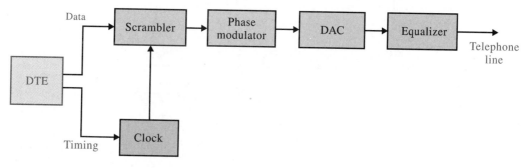

FIGURE 10-17   Block diagram of a modem transmitter

the *Gray* code so that the most likely error in demodulation will cause only a 1-bit error when it is decoded at the receiver. In the Gray code, only one bit changes between any two successive binary numbers.

The *modulator* section of the transmitter converts the bit patterns produced by the scrambling process into an analog signal representing the desired phase and amplitude of the carrier signal. The carrier frequency, baud rate, and number of bits represented by each baud are different for modems of different data rates. The modulator collects the correct number of bits and translates them into a number giving the amplitude of the electrical signal that is correct for the carrier frequency and phase of the carrier at that instant in time. Modulation techniques differ for modems of different speeds and from different manufacturers.

The binary encoded signal from the modulator is fed to a *digital-to-analog converter* (DAC), which produces the actual analog voltage required. This voltage in turn goes through an LP filter to remove frequencies outside the voice band, then through a circuit called an *equalizer* that compensates for transmission impairments on the line.

The transmitter equalizer is set to compensate for the nominal or average characteristics of the transmission medium. It compensates for amplitude distortion in the medium and for the problem called *group delay*. Group delay measures the amount by which a signal of one frequency travels faster in the transmission medium than a signal of a different frequency. Group delay usually is expressed in microseconds at a given frequency.

Receiver The receiver section of a synchronous modem, shown in Figure 10-18, typically consists of the following: adaptive equalizer, clock recovery, demodulator, descrambler, and DTE interface. The receiver equalizer must compensate for the actual errors introduced in the transmission path. This is done by using an *adaptive equalizer*, which measures errors observed in the received signal and adjusts some parameter of the circuit (usually the receiver clock frequency) to track slowly varying changes in the condition of the transmission line.

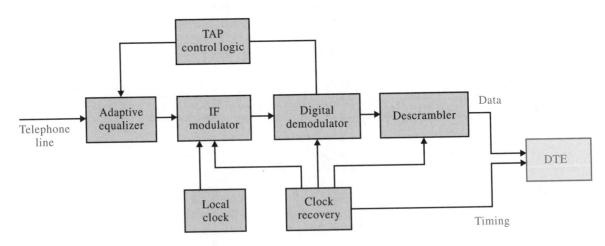

FIGURE 10-18 Block diagram of a modem receiver

Delay distortion has the greatest effect on transmission of an analog signal. Because analog signals of different frequencies travel at different rates through a transmission medium, and since each signaling element contains many frequencies, each signaling element arrives at the receiver over a period of time instead of all at once. Leading frequencies travel faster and arrive earlier than lagging frequencies. The leading and lagging frequencies not only fail to make their proper contribution to the proper signaling element but also cause interference with signaling elements behind and ahead of the proper element. The equalizer must get the parts of each element back together and cancel their effects on other elements.

The adaptive equalizer compensates for delay distortion by temporarily storing the analog signal in a tapped delay line. The signals from each tap, amplified by a different amount as determined by the amount of error detected, are summed to form the corrected signal.

At the receiver, the incoming signal from the line is modulated or frequency-translated using an internal clock. The resulting intermediate frequency is processed to produce a clock signal at the rate at which the data are actually being received. This signal is applied as the reference to a PLL oscillator. The output of this oscillator is a stable signal locked to the incoming line frequency in both phase and frequency.

The operation of the *descrambler* section of the receiver is the inverse of the scrambler operation previously described. If the data have been Gray coded by the transmitter prior to scrambling, they are converted back to straight binary, then applied to the DTE interface circuit.

Terminal Control The control section of synchronous modems must deal with two external interfaces: (1) the telephone line, at one end, and (2) the business machine, at the other. If it is to be used on the public telephone network, the modem must:

1. Sense the ringing signal.
2. Provide line supervision (connecting and disconnecting the modem from the telephone line).
3. Provide a busy indication to the incoming line without tying up telephone CO equipment.

At the other end, it must connect to the associated business machine or DTE. The DTE interface for most modems conforms to one of two standards: EIARS-232C or CCITT Recommendation V.24.

Most synchronous modems support the set of interface circuits shown in Table 10-2. The backward channel circuits (CCITT circuit Nos. 118–122) are required only in modems that provide a reverse or backward channel for control information in the direction opposite normal data flow. The Data Terminal Ready Circuit (CCITT circuit No. 108.2) is required only for modems used on the dial-up connections.

The Clear to Send signal is returned to the DTE in response to assertion of the Request to Send signal but is delayed by the modem by the amount of time necessary for the modem to turn the line around—that is, to send the required data bit stream to the distant end.

Systems that have terminals remote from the terminating modem on the main channel require that the remote be synchronized to the originating modem.

TABLE 10-2 Synchronous Modem DTE Interface Circuits

Circuit Number		Designation
EIA	*CCITT*	
AA	101	Equipment Ground
AB	102	Signal Ground
BA	103	Transmitted Data
BB	104	Received Data
CA	105	Request to Send
CB	106	Clear to Send
CC	107	Modem Ready
CD	108.2	Data Terminal Ready
CE	125	Ring Indicator
CF	109	Received Line Signal Detector (carrier)
CG	110	Signal Quality Detector
CH	111	Data Signal Rate Detector (DTE source)
CI	112	Data Signal Rate Detector (DCE source)
DA	113	Transmitted Signal Element Timing (DTE source)
DB	114	Transmitted Signal Element Timing (DCE source)
DD	115	Received Signal Element Timing (DCE source)
SBA	118	Secondary (backward channel) Transmitted Data
SBB	119	Secondary Received Data
SCA	120	Secondary Request to Send
SCB	121	Secondary Clear to Send
SCF	122	Secondary Received Line Signal Detected (carrier)

Standard 9600 bps modems usually contain internal multiplexers that allow more than one terminal to operate simultaneously over a single channel. In some FDX operations, one or more channels are carried to terminals remote from the first modem terminating the circuit. In such cases, the clock signal for all of the modems in the path must be synchronized to only one of the modems. The external clock input (EIA circuit DA) for the extension modem circuit is fed from the receive clock in the primary modem (EIA circuit DD), causing the clocks of all modems to be slaved to the single originating modem.

Modern Modems

Original modems, such as the Bell System 103, were physically quite large. Using the latest ICs, modern modems are much smaller in physical size. Besides the smaller physical size, modern modems provide many additional performance features. For example, they:

1. Provide switchable data rates up to 115.2 kbps with ISDN telephone system connections. (ISDN is discussed in Chapter 17.)

2. Automatically dial telephone numbers for origination.

3. Automatically answer incoming calls.

4. Return alphabetic characters to the DTE to report on telephone line condition and on the call in progress.

5. Automatically detect the answer tone from a distant modem.

6. Automatically adjust the data rate to match the distant modem's data rate.

Single-Chip Modem The reduced physical size and the significant increase in performance features just discussed are the result of using microprocessor techniques and the reduction in the electronics of the modem to a single IC chip. Figure 10-19 shows a block diagram of Texas Instruments TMS99532, a Bell 103-compatible FSK modem on a single IC chip.

The telephone line is connected through the line matching transformer to the Transmit Analog (TXA) and Receive Analog (RCVA) lines. The digital terminal equipment interface delivers formatted serial bit streams, providing a binary input to the Transmit Digital Data (XMTD) line. The received data, converted to binary format, are output on the Receive Digital Data (RCVD) line. The frequency assignments are as shown in Figure 10-14(A). In the originate modem, digital data are transmitted in the low-band frequencies and received in the high-band frequencies. In the answer modem, the reverse is true. Digital data are received in the low-band frequencies and transmitted in the high-band frequencies.

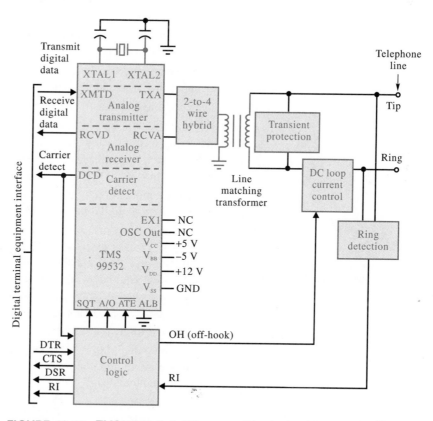

FIGURE 10-19 TMS99532 Bell 103-compatible single-chip modem

The modem chip must be interfaced to the telephone line through a two- to four-wire hybrid circuit and a line matching transformer. Transient protection, on-hook/off-hook control, and ring detection must be provided externally as shown. An external crystal is used to provide accurate control of the frequency of the modem's on-chip oscillator for the timing. Power of +12 V, +5 V, and −5 V and ground must be supplied to the modem.

Handshaking   As previously stated, the steps that occur when two modems are put into operation on the ends of a telephone line are often called handshaking. The discussion that follows refers to the TMS99532 single-chip modem (Figure 10-19), but the steps of the interconnection generally pertain to all asynchronous modems. Similar connections are required for synchronous modems, except that more signals are needed for timing and for transmission rate control.

1. Each modem is connected to a terminal, and power is supplied. The originating modem has the A/O control line set at 1, and the answering modem has it set at 0. No connection is made to EX1 or OSC OUT. For normal operation, the ALB (Analog Loop Back) pin must be grounded.

2. The $\overline{\text{ATE}}$ control line for both modems is set at 1 (inactive) since Bell 103-type modems use the 2225 Hz mark frequency as the answer tone.

3. The SQT input is under logic control. When this input is a 1 (active), the transmitter is off and no signal is output on TXA. Under idle conditions, the input is active and the transmitter is off.

4. At the originating modem, a telephone call is placed to the answering modem number and a ring signal appears at the answering modem.

5. The answering modem detects the ring signal and is enabled by assertion of DTR (Data Terminal Ready) to go off-hook.

6. The answering modem waits 2 seconds (specified by the FCC to provide a billing delay) and, through SQT, turns on its transmitter to send the answer mode carrier (mark frequency 2225 Hz) to the originating modem.

7. The originating modem recognizes the answer mode signal and places a 0 (active) on DCD, the digital carrier detect line.

8. The originating modem turns on its transmitter by placing a 1 on SQT as a result of the DCD carrier detect. It sends an originating mark signal (1270 Hz) down the line. It also starts a 200–350 ms CTS (Clear to Send) time delay to prevent the digital equipment from sending data.

9. The answering modem receives and recognizes the originating modem mark signal and places a 0 on its DCD carrier detect line. Now both modems are sending carriers in both directions.

10. The answering modem, through the DCD carrier detect, puts a 1 (active) on the CTS line to the digital equipment, making the digital equipment clear to send data communications.

11. At the originating modem, the CTS time delay expires, the CTS signal is changed to a 1, and the digital equipment connected to the originating modem is clear to send data communications.

Several techniques are available for disconnect control. A common one negates the DTR signal when sending is complete, which causes the modem to go off-hook, terminating the call.

10-10 DATA COMMUNICATION CARRIER SYSTEMS

In general, data communication carrier systems provide a means of sending signals from more than one source over a single physical transmission channel. The bandwidth available to carry signals in a particular medium can be designated in two ways: (1) by frequency (*frequency division multiplex,* FDM) and (2) by time intervals (*time division multiplex,* TDM).

FDM Carrier System

The principle of FDM is simple: The total available channel bandwidth is divided into smaller bandwidth portions or subchannels, each with its own signal source. Frequency division multiplex is used extensively in the long-haul (long distance calling) part of the public telephone system. It is also used in some simple data communication systems. For example, the standard low-speed modem uses FDM; here, the available voice channel frequency spectrum is divided into two portions, one for transmitting and one for receiving.

Electronic systems that implement FDM are called *analog carrier systems.* In these systems, the carrier is a signal generated by the system that is then modulated by the signal containing the information to be transmitted.

TDM Carrier System

The electronic systems that implement TDM are called *digital carrier systems.* The original signal can be an analog wave (such as the speech signals in the telephone network) that is converted to binary form for transmission, or it may already be in binary form (such as the signal from a business machine or computer).

Time division multiplex uses the technique of dividing the capacity of a transmission channel among several separate signal sources to allocate a very short period of time on the channel in a repeating pattern to each signal. It is well suited to digital (binary) signals consisting of pulses representing a 1 or a 0. These pulses can be of very short duration and still convey the desired information. Therefore, TDM is used in digital carrier systems because many bits can be packed into the very narrow time blocks allocated to each of several signal sources.

SUMMARY

- Data communications is the movement of encoded data from one point to another by means of electrical transmission systems, including wireline, radio, and optics. The most widely used transmission medium is the telephone channel. Some form of processing is involved in data communications, either before or after transmission.

- Data refers to any representation; raw data is sometimes used to describe unprocessed data, as opposed to information, which usually refers to processed data. Telecommunication is the process of transmitting information in the form of voice, data, image, or message over a distance by electrical or electromagnetic systems. Teleprocessing is the use of data processing with telecommunications facilities.

- The objective of a data transmission system is to provide faster information flow by reducing the time spent in collecting and distributing data. Data communication networks make more efficient use of central computers by providing message switching capabilities through circuit switching or through store-and-forward techniques. Data can be transmitted by parallel or serial lines.

- Two basic types of data transmission systems are offline and online. Offline systems transmit data that require no immediate response or no response at all. These systems often employ batch-processing techniques. Examples of offline systems are stock market tickers and national news service agencies. In online systems, devices or subsystems are directly connected to the computer, and data flow directly between the terminal and the computer. Transmissions can be batched or realtime. In realtime, the data transmission requires an immediate response. Examples of realtime systems are airline and overnight accommodation reservation computers.

- Three essential components common to all data communication systems are source, medium, and sink. The source and sink are interchangeable.

- Data terminal equipment (DTE) is a computer or business machine that provides data in the form of digital signals at its output. It may comprise CRT or teletype terminals, personal computers, printers, or front-end processors. It can be source, sink, or both in a system.

- Terminals serve as translation devices for communication codes. They provide interfaces with computer systems, receive input in coded form, and convert the input into electrical signal pulses for transmission to the computer. The most widely used type is the keyboard terminal. Video display terminals consist of a keyboard and a CRT display device. Video displays are high-speed devices used by airline and hotel reservation systems, brokerage firms, and insurance companies. A graphics terminal is a special type of video display terminal. Transaction terminals are used in particular industry applications such as banking, retail point-of-sale, or supermarket checkout. Dumb terminals have limited capabilities and serve as input/output devices only. They perform no processing, editing, or buffering. Intelligent (smart) terminals are equipped with a microprocessor and perform processing, editing, and buffering. Specialized terminals are audio response units and pushbutton telephones. Audio response units respond verbally rather than visually or in print and use voice synthesizers to convert phonemes into intelligible messages. Pressing a key on a pushbutton telephone set transmits a distinctive signal representing a number that the computer uses for the processing operation.

- Front-end processors (FEPs) are used with large mainframe computers to relieve heavy demands on the CPU, to control the flow of data, and to ensure compatibility within the system. Most FEPs are minicomputers.

- A modem is a combination modulator/demodulator used to transmit or receive data. Two modems, one at each end of the transmission medium, are needed for data communications. Modems convert digital data to analog before transmission, and analog to digital after reception.

- Codes used for data communications have standard symbols or characters and are built into the equipment. Codes used are Morse, Baudot, CCITT Standard International Alphabet No. 2, EBCDIC, and ASCII. Data can be transmitted by three modes of operation: simplex, half-duplex, and full-duplex.

- Data circuit-terminating equipment (DCE) provides the functions needed to establish, maintain, and terminate a connection. It provides the signal conversion for communication between the DTE and the data circuit.

- The three data communication modulation methods are amplitude, frequency, and phase, any one of which can be used in modems. Frequency modulation is the most common and uses frequency shift keying (FSK) and audio frequency shift keying (AFSK). All three methods manipulate values between binary 0 and binary 1. Demodulation occurs in the receiving modem, where detectors and filter circuits sensitive to amplitude, frequency, or phase recover the binary 0 and binary 1 from the carrier.

- Protocols are communication controls that establish rules and procedures for efficient operation of the system. In modem usage, handshaking means that a connection has been established and that a communication line is ready for a message. Line discipline denotes the sequence of operations involving the actual transmission and reception of data and is often used synonymously with protocol. Line polling is used between a computer and a terminal to confirm the status of a message. Detection of transmitted errors is made possible by redundancy and parity checking.

- Modems are interfaced to the telephone network through two wires (tip and ring). Interfacing between the modem and DTE, governed by standards set by the Electronic Industries Association (EIA), requires that certain protocols be observed. Telephone channels place several restrictions on modems, including a bandwidth restriction. Microprocessors have reduced the electronics of a modem to a single IC chip.

- Modems may be asynchronous, synchronous, or isochronous, and low-speed or high-speed. In asynchronous transmission (used with low-speed terminals and the most common transmission method), each information character is individually synchronized, usually by the use of start and stop elements. With synchronous transmission, an internal clock synchronizes the communication channel's transmitter and receiver, usually with a PLL clock circuit. In isochronous transmission (a mix between asynchronous and synchronous transmission), the intervals between characters are time controlled.

- Data signals can be multiplexed so that signals from more than one source can be sent over a single physical transmission channel. Multiplexing can be achieved by frequency division multiplex (FDM) or time division multiplex (TDM). FDM uses analog carrier systems; TDM uses digital carrier systems.

QUESTIONS AND PROBLEMS

1. List three examples of computer support equipment.

2. Define data communications and list the characteristics a modern system must have.

3. How does data communications differ from telegraphy?

4. Define (a) data, (b) raw data, (c) information, (d) teleprocessing, and (e) telecommunication.

5. State (a) a basic requirement of a data communication system, (b) the requirements of the data processing cycle, and (c) the objective of a data transmission system.

6. By what means can individual characters of a data communication be transmitted?

7. Describe the techniques used for (a) parallel transmission and (b) serial transmission.

8. List the user information requirements for determining what type of transmission system to use.

9. (a) Describe how offline systems operate and give examples of their use. (b) Describe remote job entry. (c) Describe how online systems operate and give examples of their use.

10. Define realtime computing and give examples of its use.

11. Explain the purpose of each of the three essential elements common to all data communication systems.

12. Define DTE and list five possible DTE devices.

13. Define terminal and state its purpose.

14. List five broad categories of terminals. Which is the most widely used? Describe and explain its operation.

15. Describe and explain the operation and uses of (a) video terminals, (b) graphics terminals, and (c) transaction terminals.

16. Define (a) dumb terminal and (b) smart terminal.

17. Describe two specialized terminals and explain their uses.

18. Define front-end processor and explain its use.

19. List the functions performed by communication control units.

20. Define (a) modem, (b) baud, and (c) baud rate.

21. Define (a) code, (b) character, and (c) symbol.

22. Describe five codes and how each is used.

23. Describe the operation of the three basic methods for data transmission.

24. What is the DCE and what is its function?

25. Explain the interface between the DCE and the transmission channel.

26. Define and explain the operation of (a) asynchronous transmission, (b) synchronous transmission, and (c) isochronous transmission.

27. Explain how data communication systems are interfaced to telephone networks by (a) rotary dialing and (b) DTMF.

28. List the advantages of DTMF over rotary dialing.

29. Explain the three data communication modulation methods.

30. Define and explain (a) FSK and (b) AFSK.

31. Explain how demodulation occurs in data systems.

32. Define (a) protocol, (b) handshaking, (c) line discipline, and (d) line polling.

33. Describe how error detection and correction are achieved.

34. Define redundancy and explain how it works.

35. Define and describe parity checking.

36. Describe how modems are interfaced to a telephone network and to the DTE.

37. Describe the restrictions placed on modems by the telephone channel.

38. Describe the advantages of transmitting signals in digital form.

39. Give the standard operating center frequencies for (a) the U.S. system and (b) the European system.

40. For a Bell System 103 low-speed asynchronous modem receive station, state the function of (a) the BP filter, (b) the limiter, (c) the delay detector, and (d) the slicer.

41. For a Bell System 103 asynchronous modem transmit section, state the function of (a) the FSK oscillator and (b) the BP filter.

42. Describe the operation of an acoustically coupled system.

43. For a synchronous modem transmitter, state the function of (a) the timing circuit, (b) the scrambler, (c) the modulator, and (d) the equalizer.

44. For a synchronous modem receiver, state the function of (a) the adaptive equalizer and (b) the descrambler.

45. Describe the purpose of the terminal control section of a synchronous modem.

46. List the performance features of modern modems.

47. Describe and compare the two data communication carrier systems.

CHAPTER 11

Introduction to Microwave Communications and Components

OBJECTIVES

When you complete this chapter, you should be able to:

- Identify the microwave frequency bands.
- Define microwaves.
- Discuss propagation characteristics and line-of-sight communication.
- Discuss the effects of weather on microwave radiation.
- Describe microwave radiolink systems and how they operate, and list their advantages and disadvantages.
- Define waveguide and describe how it operates.
- Describe how microwaves propagate in waveguide.
- List and describe the three general classes of traveling wave tubes.
- Describe the operation of the magnetron, the carcinotron, and the cavity resonator.
- List the three kinds of klystron tubes and describe the operation of each.
- List the types of microwave electron tubes and describe the operation of each.
- List the types of microwave semiconductor devices and describe the operation of each.
- Describe reflection amplifiers and digital microwave techniques.

INTRODUCTION

The term *microwave* is rather ambiguous, although it enjoys widespread use. Nevertheless, the microwave bands might reasonably be said to extend from ultra-high frequency (UHF) through super-high frequency (SHF) and into extremely-high frequency

359

(EHF). Signals at frequencies from around 1 gigahertz (GHz) to at least 100 GHz encompass these bands. Below 1 GHz, the spectrum has allocations for services such as UHF television, advanced mobile radios, and the like. Well-established technologies based on coaxial lines and lumped reactive and resistive components are available for use in these areas. Above 1 GHz and through millimeter-wavelength bands, design tends to be considerably more difficult, and the product more expensive.

All frequencies above 1 GHz are called microwave frequencies. These frequencies are commonly used for radar and wide-band communications. The radiation in these higher-frequency bands can be directed into very narrow beams of energy. This feature makes these frequency ranges very efficient in utilizing transmitter energy and in minimizing interference between one communication system and another. When the microwave frequency band is from 3 GHz to 30 GHz, the wavelength is from 10 cm to 1 cm; above 30 GHz, it is millimeters in length. These small wavelengths are a significant advantage for transmitting because the antennas can be very small. However, small wavelengths can also be a disadvantage. Transmissions at these frequencies are highly susceptible to weather effects, particularly rain, because each raindrop can become a small antenna, absorbing the energy and causing it to dissipate before reaching its destination.

Important systems that operate at microwave frequencies include the majority of radars, mobile radio systems, satellite communication systems, terrestrial line-of-sight links, and tropospheric scatter links. In this chapter, we will discuss microwave fundamentals, microwave radio, resonators and waveguides, and generating and amplifying devices.

11-1   MICROWAVE FUNDAMENTALS

Microwaves are defined as the portion of the electromagnetic spectrum containing wavelengths between about 1 mm and 30 cm. *Microwave frequencies* range from approximately 1 GHz to 300 GHz.

Microwaves are very short electromagnetic radio waves, but they have a longer wavelength than infrared energy. They travel in essentially straight lines through the atmosphere. Microwaves are not affected by the ionized layers because these layers are so high above the normal line-of-sight transmission of the signals.

Microwave frequencies are useful for short-range, high-reliability radio and television links. In a radio or television broadcasting system, the studio is usually at a location different from the transmitter; a microwave link connects the two. Satellite communication and control are generally accomplished at microwave frequencies. The microwave region contains a vast amount of spectrum space and can therefore hold many wide-band signals.

Microwave radiation can cause heating of certain materials. This heating can be dangerous to humans when the microwave radiation is intense. When microwave equipment is used, care must be exercised to avoid exposure to the rays. The heating of organic tissue by microwaves can be put to constructive use, as demonstrated by the microwave oven, a consumer appliance that will not be considered here.

Propagation Characteristics

As stated previously, microwave energy travels in essentially straight lines through the atmosphere. Although microwaves are unaffected by the ionosphere, temperature inversions can be a problem. The problem arises because, as the air rises, moisture rising with the air causes attenuation of the signal. This is especially bothersome with radar, which also uses microwave signals. The moisture causes false returns to radar receivers.

The primary mode of propagation in the microwave range is line-of-sight. This mode facilitates repeater and satellite communication.

The microwave band is even more vast than the UHF band. In theory, the upper limit of the microwave band is the infrared spectrum. Current technology has produced radio transmitters capable of operation at frequencies of more than 300 GHz.

At some microwave frequencies, atmospheric attenuation becomes a consideration. Rain, fog, and other weather effects cause changes in the propagation path attenuation as the wavelength approaches the diameter of water droplets.

Line-of-Sight Communication

Radio communication by means of the direct wave is sometimes called *line-of-sight* (LOS) communication. The range of LOS communication depends on the height of the transmitting and receiving antennas above the ground, and on the nature of the terrain between the two antennas.

Line-of-sight is the primary mode of propagation at microwave frequencies. Although the range is limited in this mode, propagation is virtually unaffected by external parameters, such as ionospheric or tropospheric disturbances. The LOS communications range is limited to the radio horizon.

Infrared, Visible Light, Ultraviolet, X-Rays, and Gamma Rays

Propagation at infrared and shorter wavelengths occurs in straight lines. The only factor that varies is the atmospheric attenuation. Water in the atmosphere causes severe attenuation between wavelengths of approximately 4500 and 8000 nanometers (nm). Carbon dioxide interferes with the transmission of infrared radiation of wavelengths from about 14,000 nm to 16,000 nm.

Light is transmitted well through the atmosphere at all visible wavelengths. Ultraviolet light at the longer wavelengths can penetrate the air with comparative ease. At shorter wavelengths, attenuation increases.

Rain, snow, fog, and dust interfere with the transmission of infrared radiation, visible light, and ultraviolet light.

X-rays and gamma rays do not propagate well for long distances through the air, primarily because the mass of the air, over propagation paths of great distance, is sufficient to block these types of radiation.

Effects of Weather on Radiation

When transmissions are at frequencies susceptible to weather, water and oxygen in the air absorb the electromagnetic energy, making the air more and more opaque to radiation in the millimeter and shorter wavelengths. Under heavy rain conditions, the loss of energy is particularly severe. The communications system designer must take these losses into account when working in these frequency bands. When microwave frequencies in the SHF band above 10 GHz are used, the distance between transmitter and receiver must be limited to a few miles. Below 10 GHz, the distance can be a few tens of miles as determined by the LOS considerations. If the distance of a microwave system is greater than these minimum distances of a few miles, *repeater* or *relay* stations must be placed along the path as illustrated in Figure 11-1.

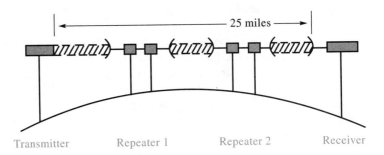

FIGURE 11-1   Typical repeaters required of a 20 GHz system

Microwave Repeater

A *microwave repeater* is a receiver/amplifier/transmitter combination used for relaying signals at microwave frequencies. The signal from the previous repeater is intercepted by a horn or dish antenna, amplified, converted to another frequency, and retransmitted to the next repeater or to the final destination of the information, as the case may be. This concept is illustrated by Figure 11-2.

As the frequencies increase, the distance between repeaters must decrease. The size of the antennas would also decrease proportionately. However, repeaters are not just for the millimeter and centimeter wavelengths. Even the longer-wavelength microwave transmissions may need repeaters if the overall distance between transmitter and receiver exceeds about 25 miles. Repeaters used at these wavelengths are much larger than those used in millimeter-wavelength systems because of the much larger antennas and greater power requirements.

Microwave repeaters are used in long distance, overland communication links. With the aid of such repeaters, microwave links can replace many wire transmission systems.

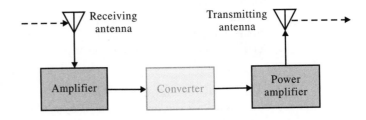

FIGURE 11-2   Block diagram of a microwave repeater

11-2 MICROWAVE RADIO

A *microwave radio system* sends signals through the atmosphere between towers spaced about 20 to 30 miles apart, amplifies the signals, and retransmits them at each repeater station until they reach their destination. Microwave radio operates at the high-frequency end of the radio spectrum. The signals follow a straight-line path, and the relaying antennas must be within sight of each other. One of the principal problems affecting microwave signals is the variation caused by changes in atmospheric conditions. Moisture and temperature conditions can make the radio beam bend, resulting in fading.

The primary advantage of microwave radio is that it can carry thousands of voice channels without physically connected cables between points of communication, thus avoiding the need for continuous right-of-way between points. Further, radio is better able to span water, mountains, or heavily wooded terrain that pose barriers to wire or cable installation. Most long distance links today are coaxial cable or microwave radio.

The tall towers with large horns or dish antennas that can be seen in the country are repeater stations for LOS microwave *relay* systems (sometimes called *radiolink* systems). These systems have become popular for carrying large quantities of voice and data traffic, for several reasons:

1. They require no right-of-way acquisition between towers.
2. They require the purchase or lease of only a small area of ground for installation of each tower.
3. Because of their very high operating frequencies, they can carry large quantities of information per radio system.
4. Because the wavelength of the transmitted signal is short, an antenna of reasonable size can focus the transmitted signal into a beam. This capability provides a much greater signal strength at the receiver without increasing transmitter power.

Radiolink systems are subject to transmission impairments that limit the distance between repeater points and cause other problems. The microwave signals are:

1. Attenuated by solid objects (including the earth), and the higher frequencies are attenuated by rain, snow, fog, and dust.
2. Reflected from flat conductive surfaces (water, metal structures, and the like).
3. Diffracted (split) around solid objects.
4. Refracted (bent) by the atmosphere, so the beam may travel beyond the LOS distance and be picked up by an antenna that is not supposed to receive it.

In spite of these possible problems, radiolink systems are highly successful and carry a substantial part of all telephone, data, and television traffic in the United States. The microwave range of radio frequencies is allocated for various purposes by international treaty. Some of the frequency assignments for the United States are shown in Table 11-1.

Most common carrier radiolink systems carry analog signals, principally FM. There are a few systems, however, that carry digital signals. Two examples in extensive service

TABLE 11-1   Microwave Radiolink Frequency Assignments

Service	Frequency (GHz)
Military	1.710–1.850
Operational Fixed	1.850–1.990
Studio Transmitter Link	1.990–2.110
Common Carrier	2.110–2.130
Operational Fixed	2.130–2.150
Common Carrier	2.160–2.180
Operational Fixed	2.180–2.200
Operational Fixed (TV)	2.500–2.690
Common Carrier and Satellite Downlink	3.700–4.200
Military	4.400–4.990
Military	5.250–5.350
Common Carrier and Satellite Uplink	5.925–6.425
Operational Fixed	6.575–6.875*
Studio Transmitter Link	6.875–7.125
Common Carrier and Satellite Downlink	7.250–7.750
Common Carrier and Satellite Uplink	7.900–8.400*
Common Carrier	10.7–11.7
Operational Fixed	12.2–12.7
CATV Studio Link	12.7–12.95
Studio Transmitter Link	12.95–13.2
Military	14.4–15.25
Common Carrier	17.7–19.3

*Frequencies assigned by international treaty.

in the United States are (1) the AT&T 3A-RDS radio system, which operates in the 11 GHz band; and (2) the AT&T DR-18 radio system, which operates in the 18 GHz band. The 3A-RDS system carries DS-3 digital signals at 44.736 megabits per second (Mbps), and the DR-18 system carries DS-4 digital signals at 274.176 Mbps. The standard hierarchy of digital carrier systems is shown in Table 11-2.

Terrestrial radiolink systems are point-to-point; that is, the signal is transmitted in a beam from a source microwave antenna across the earth's surface to the antenna at which it is aimed. The width of the beam varies between 1° and 5°, depending on the transmission frequency and antenna size. As a result, the transmission is highly directional, which is desirable if the information is intended for only one destination, such

TABLE 11-2   Hierarchy of Digital Carrier Systems

Digital Signal	No. of Voice Channels	Bit Rate (Mbps)
DS-1	24	1.544
DS-2	96	6.312
DS-3	672	44.736
DS-4	4032	274.176

as for a telephone conversation. For many applications, however, the information has multiple destinations, as for television broadcasts, which makes the *satellite* radiolink system more practical and desirable.

11-3 MICROWAVE RESONATORS

At microwave frequencies, transmission lines become waveguides, and resonant portions of transmission lines become cavities with special designs and properties. However, because of the relationship of wavelength to physical dimensions in microwaves, the simplified equations previously used to define electromagnetic fields and waves generally do not apply here.

Cavity Resonators

A *cavity resonator* is a metal enclosure, usually shaped like a cylinder or rectangular prism. Cavity resonators operate as tuned circuits and are practical at frequencies above about 200 MHz.

At its simplest, a cavity resonator is a piece of waveguide closed off at both ends with metallic plates. Figure 11-3 illustrates a cavity resonator consisting of a section of rectangular waveguide.

A cavity has an infinite number of resonant frequencies. When the length of the cavity is an integral multiple of one half wavelength ($\lambda/2$), electromagnetic waves will be reinforced in the enclosure. Thus, a cavity resonator has a fundamental frequency and, theoretically, an infinite number of harmonic frequencies. Near the resonant frequency, or near any harmonic frequency, a cavity behaves like a parallel-tuned inductive–capacitive circuit. When the cavity is slightly too long, it exhibits inductive reactance; when it is slightly too short, it exhibits capacitive reactance. At resonance, a cavity has very high impedance and, theoretically, zero reactance. Cavity resonators can be fixed or tunable.

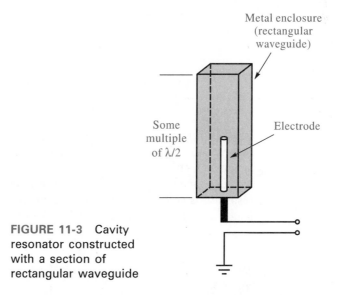

FIGURE 11-3 Cavity resonator constructed with a section of rectangular waveguide

The resonant frequency of a cavity is affected by the dielectric constant of the air inside. Temperature and humidity variations therefore have some effect. If precise tuning is required, the temperature and humidity must be kept constant. Otherwise, a resonant cavity may drift off resonance because of changes in the operating environment.

A length of coaxial cable, short circuited at both ends, is sometimes used as a cavity resonator. Such resonators can be used at lower frequencies than can rigid metal cavities since the cable does not have to be straight. The velocity factor of the cable must be taken into account when designing cavities of this kind.

Tunable Cavity Resonator   In a *tunable cavity resonator,* the physical length is adjustable. One method of adjusting the length is shown in Figure 11-4. The wavelength at resonance is directly proportional to the physical length of the cavity. Therefore, to increase the frequency, the cavity is shortened; to lower the frequency, the cavity is made longer. Tunable cavity resonators can be used (1) in power amplifiers at VHF, UHF, and microwave frequencies; (2) in some types of frequency meters; and (3) to provide selectivity at the front end of a receiver.

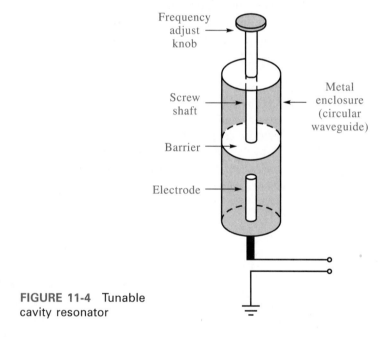

FIGURE 11-4 Tunable cavity resonator

Cavity Frequency Meter   Cavity resonators are sometimes used to measure frequency. A *cavity frequency meter* makes use of a cavity resonator to measure wavelengths at VHF and above. A tunable cavity resonator and an RF voltmeter or ammeter are connected to a pickup wire or loosely coupled to the circuit under test. The cavity resonator is adjusted until a peak occurs in the meter reading. The free-space wavelength is thus easily measured and converted to the frequency according to the following standard equation:

$$f = \frac{300}{\lambda}$$

(11-1)

where

f = frequency in megahertz

λ = wavelength in meters

A cavity resonator is generally tuned to $\lambda/2$ to obtain resonance in the cavity frequency meter, although any multiple of $\lambda/2$ can be used.

11-4 WAVEGUIDES

The two-wire transmission lines and coaxial cables considered in Chapter 6 are efficient devices for carrying RF energy in the frequency spectrum from 30 Hz to about 3000 MHz (3 GHz). At frequencies above 3 GHz, because of losses both in the conductors and in the solid dielectric needed to support the conductors, transmission of electromagnetic waves along transmission lines and coaxial cables becomes difficult. If the frequency of transmission is high enough, the electric and magnetic components of a signal can travel through free space, requiring no solid conductor. However, to avoid interference and loss due to signal spreading, and to be able to route the signal as desired, it is useful to confine these waves to another bounded medium called a *waveguide*. A waveguide is a feed line used at microwave frequencies.

Induced currents in the walls of the waveguide produce power losses, and to minimize those losses the waveguide wall resistance is made as low as possible. There is no need for a center conductor because, at the high frequencies involved, skin effect takes over. The currents tend to concentrate near the inner surface of the waveguide walls because of this skin effect, so the walls are highly polished and sometimes specially plated to reduce resistance. The normal characteristic impedance for a waveguide is 50 Ω.

Waveguides are commonly used from 2 GHz up to 110 GHz to connect microwave transmitters and receivers to their antennas. They are pressurized with dry air or nitrogen to drive moisture from inside because moisture attenuates the microwaves. A waveguide consists of a hollow metal tube, usually having a rectangular or circular cross section, as shown in Figure 11-5. The electromagnetic field travels down the tube, provided that the wavelength is short enough. The metal tube confines the radio waves and channels them to a point where they are released into the air to continue their travel over microwave transmission facilities. Waveguides provide excellent shielding and low loss; thus, they can transmit greater amounts of power with less energy loss than coaxial cables.

Rectangular waveguides have been used for some time to connect microwave transmitting equipment to the microwave towers. Their use is generally limited to distances of less than 1000 feet. The newer and more efficient waveguide is the circular type, which consists of a precision-made pipe about 2 inches in diameter. This type of waveguide can transmit much higher frequencies than rectangular waveguides.

The waveguide is attractive for microwave use because of its wide bandwidth and low-loss transmission characteristics. Deterrents to its use are its critical engineering requirements and very high cost.

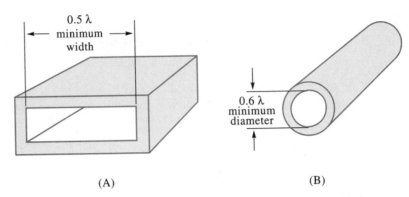

FIGURE 11-5 Waveguide cross section (A) Rectangular (B) Circular

Waveguides have great potential for use in long distance communications. However, the development of optical fibers has resulted in a similar, but somewhat superior, medium at appreciably lower costs.

To efficiently propagate an electromagnetic field, a rectangular waveguide must have sides measuring at least 0.5 λ, and preferably more than 0.7 λ. A circular waveguide should be at least 0.6 λ in diameter, and preferably 0.7 λ or more.

Consider a waveguide operating at 3 GHz. The wavelength is 3.9 inches (10 cm). Thus, a rectangular waveguide must be at least 2.0 inches (5.1 cm) wide, and preferably 2.7 inches (7.0 cm). A circular waveguide should be at least 2.3 inches (6.0 cm) in diameter, but preferably 2.7 inches (7.0 cm). The frequency at which the length of the side or the diameter is λ/2 is known as the cutoff frequency. The waveguide exhibits a high-pass (HP) response.

Waveguide Construction

The most common form of waveguide is rectangular in cross section (Figure 11-6(A)), although round waveguides (Figure 11-6(B)) are also used. The width of the guide must be slightly greater than λ/2 of the signal to be transmitted. Most rectangular waveguides have a height of about one-half the width. A common band of operation, with lower and upper frequencies of 8.2 GHz and 12.4 GHz, respectively, is known as

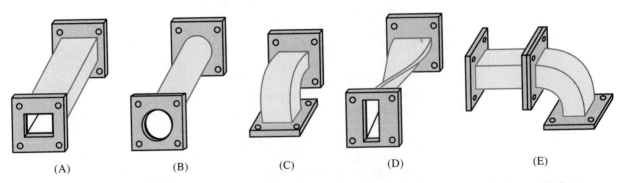

FIGURE 11-6 Waveguides used for transmission of microwave signals (A) Rectangular (B) Round (C) Bent (D) Twisted (E) Straight section connected to a 90° elbow

the *3-cm,* or *X-band.* An X-band waveguide is 2.29 cm wide and 1.02 cm high. Other waveguide bands and sizes are shown in Table 11-3.

Waveguides are made in sections of various length. They may be straight (Figure 11-6(A)), bent to some desired direction (Figure 11-6(C)), twisted to some desired angle (Figure 11-6(D)), or even made flexible. At each end of a waveguide section is a precisely machined flat metal flange: sections can be coupled by bolting the flanges together (Figure 11-6(E)).

Waveguides may be constructed of brass, copper, or aluminum. Because currents on the walls of the waveguide oscillate only on the inner skin, low-loss waveguide sections are silver plated on the inside to reduce skin effect. Energy is transferred within the waveguide by electromagnetic fields; currents and voltages merely aid in establishing those fields. Modes of operation of the waveguide involve two field components: the electric field and the magnetic field.

TABLE 11-3 Waveguide Frequency Bands and Sizes

Band	Frequency Range (GHz)	Waveguide Size (in.)	Waveguide Size (cm)
975	0.75–1.12	9.75 × 4.880	24.80 × 12.40
L	1.12–1.70	6.50 × 3.250	16.50 × 8.26
S	2.60–3.95	2.84 × 1.340	7.21 × 3.40
G	3.95–5.85	1.87 × 0.870	4.75 × 2.21
C	4.90–7.05	1.59 × 0.795	4.04 × 2.02
J	5.85–8.20	1.37 × 0.620	3.48 × 1.57
H	7.05–10.00	1.12 × 0.497	2.84 × 1.26
X	8.20–12.40	0.90 × 0.400	2.29 × 1.02
M	10.00–15.00	0.75 × 0.375	1.91 × 0.95
P	12.40–18.00	0.62 × 0.310	1.57 × 0.79
N	15.00–22.00	0.51 × 0.255	1.30 × 0.65
K	18.00–26.50	0.42 × 0.170	1.07 × 0.43
R	26.50–40.00	0.28 × 0.140	0.71 × 0.36

Propagation Modes   The analysis of propagation in waveguides is based mostly on the electric and magnetic fields in them. For most practical applications, the *transverse electric* field (TE) and the *transverse magnetic* field (TM) are the most important.

A system of letters and subscript numbers is used to describe waveguide nodes. Figure 11-7(A) shows the TE mode of operation of a rectangular waveguide. The letters TE indicate that the electric field (shown as parallel E lines) is in a transverse plane that contains the X and Y axes. The E lines are parallel to the Y axis and are perpendicular to the longitudinal Z axis. Similarly, the letters TM (Figure 11-7(B)) indicate that the magnetic field (composed of closed loops) lies in transverse planes that contain the X and Y axes and are wholly transverse to the guide axis.

Subscripts are used with the TE and TM modes of operation. In rectangular waveguides, the first subscript number states the number of half-wave variations of the

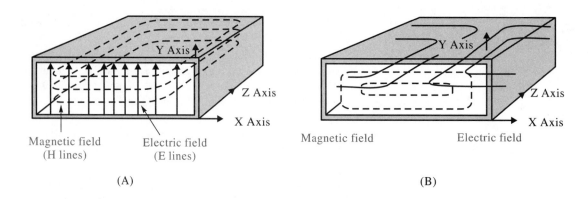

FIGURE 11-7 Modes of waveguide operation (A) Transverse electric (TE) (B) Transverse magnetic (TM)

wide dimension of the guide, and the second number states the number of half-wave variations of the same field in the narrow dimension. The TE_{10} mode, for example, means that the electric field has a half-wave variation along the X axis and none along the Y axis. The TM_{11} mode means that the magnetic field has a one half-wave variation in both the wide and narrow dimensions. The mode with the lowest cutoff frequency for a given waveguide size is called the dominant mode for that waveguide. The dominant mode, TE_{10}, is most commonly used for rectangular waveguides. This mode is easily excited, plane-polarized, and easily matched to a radiator. Moreover, its cutoff frequency depends only on the waveguide width dimension, making system design easier. For circular waveguides, the lowest propagation mode for both TE and TM is the 01 mode.

Directional Waveguide Couplers *Directional waveguide couplers* are produced by welding together two pieces of waveguide and opening one or more holes between them, as shown in Figure 11-8. The larger the holes (or the more there are), the greater the power transfer to the secondary waveguide section. If two holes are $\lambda/4$ apart, the propagation is such that most of the energy induced in the secondary or coupled section is in the forward direction. A dummy load in the backward-direction end absorbs any reflected power, making the coupler a true forward coupler. Couplers are rated in decibels, which specify the fraction of the input signal appearing at the secondary output port.

Waveguide Coupling Methods In addition to the directional coupler, there are two methods for coupling energy into or out of a waveguide, as shown in Figures 11-9(A) and (B). In Figure 11-9(A), energy from a coaxial cable terminates in a single-turn loop connected to the wall of the waveguide. Current in this loop sets up magnetic fields that induce voltages in the waveguide space and currents in the walls, allowing energy to radiate down the waveguide. In Figure 11-9(B), the coaxial cable terminates in a quarter-wave vertical antenna projecting into the waveguide space. Energy radiated from this probe is transmitted down the waveguide.

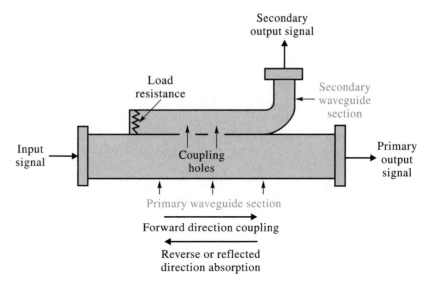

FIGURE 11-8 Directional waveguide coupler

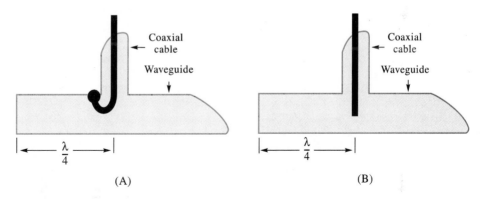

FIGURE 11-9 Coupling coaxial cable to waveguide (A) Loop (inductive) coupling (B) Antenna probe (electrostatic) coupling

With either method, if the coupling devices are some odd quarter-wavelength from the sealed end of the waveguide, the sealed end acts as a parasitic reflector, reinforcing the energy transfer. In waveguides containing coupling devices, the near end will often be adjustable or tunable to ensure maximum reflection from the sealed end.

A coupling horn is used to couple from a waveguide to space. A *horn* is a section of waveguide, flared out in both cross-sectional dimensions, used to change the normal 50 Ω waveguide impedance to the higher impedance of space. Such horns are used with parabolic reflecting antennas.

Other Microwave Transmission Methods

Waveguide is not the only method for transferring microwaves. Within certain limits, coaxial cable, microstrip, stripline, and optical fibers can be used.

Coaxial Cable *Coaxial cable* can be used for microwave transmission, although there are several limitations. In lengthy cable runs, solid dielectric coaxial may have rather high losses at the higher frequencies. In addition, coaxial coupling devices present reflections of energy back up the cable. These reflections produce standing waves on the line and prevent full-power transfer from source to load, as well as high and low voltage points along the cable. On the other hand, if it is only a few inches long, solid dielectric coaxial cable can be used for frequencies up to the X-band.

Microstrip A microwave transmission line called *microstrip* is shown in Figure 11-10(A). Microstrip consists of an insulator, or dielectric material, laid on a flat metal base. A thin metal strip is laid on the insulator. The impedance of this type of transmission line is determined by strip width and the constant and thickness of the dielectric material. Microstrip is used as a printed-circuit transmission line.

Stripline *Stripline* is a transmission line similar to microstrip. The construction technique is the same for both. However, stripline consists of a thin metal strip sandwiched between two insulators laid on two metal bases, as shown in Figure 11-10(B).

The stripline and microstrip transmission techniques are somewhat impractical for transferring signals over long distances. However, the approach is ideally suited to the fabrication of high-Q filters and tuned circuits.

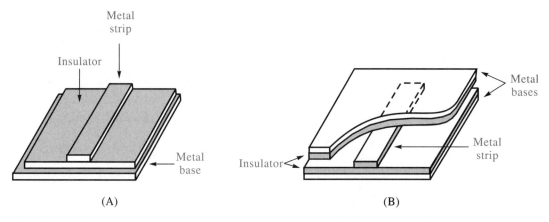

FIGURE 11-10 Other methods for microwave transmission (A) Microstrip (B) Stripline

Optical Transmission Optical transmission of signals is a modern system in which light beams are transferred from one place to another by an *optical fiber* or *fibers*. Light beams can be modulated at frequencies into the hundreds of megahertz, and by many different signals at the same time.

An optical fiber is a thin, transparent strand of material, usually glass or plastic. It contains the light energy inside its walls because of high internal reflection. The light rays are reflected from the inner walls as they propagate lengthwise along the fiber, as shown in the simplified illustration of Figure 11-11.

A bundle of optical fibers may be used to transfer many different light beams at the same time. A single light beam can be simultaneously modulated by hundreds, or even

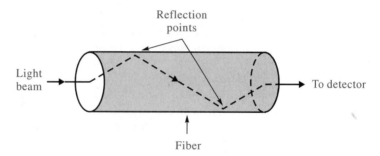

FIGURE 11-11 Optical fiber

thousands, of independent signals. Fiber cables of current design can carry millions of different conversations at one time.

There are several advantages of optical fibers:

1. They are very efficient.
2. They allow the transmission of data at extremely high speeds.
3. They are compact and lightweight.
4. They are essentially immune to electromagnetic interference.

Fiber optics and optical communications are discussed in detail in Chapter 16.

Principles of Guided Waves

The electric field parallel to and near a perfect conductor must be zero. Therefore, assuming that the walls of a waveguide or resonator are perfect conductors, the electric lines of force are at right angles to these walls. The magnetic field at a perfect conductor is either zero or infinite, and surface currents will flow. The depth of these surface or skin currents becomes important because it determines the losses in the walls. The skin depth of the current decreases at higher frequencies.

Reflection of Waves

Most of the principles that apply to optical reflection apply to microwave signals as well. For example, the angle of reflection from a surface is the same as the angle of incidence; that is, the angle of the wave leaving the surface is the same as the angle of the wave striking the surface. The polarity of a reflected electric field gets reversed. Reflective losses depend on the characteristics of the reflecting material and on the depth of penetration. In many instances, the reflection principles described for parasitic antennas in Chapter 8 must be added to the optical reflection qualities of microwaves to account for some of the special effects obtained.

Wave Propagation in a Waveguide   An electromagnetic wave can travel down a waveguide in various ways. If all of the electric lines of flux are perpendicular to the axis of the waveguide, the propagation mode is called *transverse electric* (TE). If all of the magnetic lines of flux are perpendicular to the axis, the mode is called *transverse magnetic* (TM). The electromagnetic field can be coupled into the waveguide via the electric or the magnetic components.

In a waveguide, electromagnetic fields tend to circulate in eddies. Depending on the frequency of the applied energy, there will be one or more eddies. In general, as the frequency increases, the number of eddies in the cross section increases.

Waveguide Operation   When a waveguide is used, it is important that the impedance of the antenna be purely resistive and that the resistance be matched to the characteristic impedance of the waveguide. Otherwise, there will be standing waves on the line, and there will be an increased loss as compared with the perfectly matched condition. The waveguide behaves exactly like other types of transmission lines in this respect.

The characteristic impedance of a waveguide varies with frequency. In this sense, waveguide differs from coaxial cables or parallel-wire lines. In most cases, matching transformers are needed to achieve a low standing wave ratio on a waveguide, because few antennas present impedances that are favorable to a waveguide if direct coupling is employed. A quarter-wave section of waveguide, coaxial cable, or parallel wires or rods can be used for matching.

Because of their HP characteristics, waveguides can be used as filters and attenuators. They are very effective for this purpose because of their high efficiency.

The interior walls of waveguide are highly polished, and it is important that the interior be kept clean and free of condensation. Even a small obstruction can seriously degrade performance.

11-5 MICROWAVE ELECTRON TUBES

In spite of the rapid proliferation of solid-state devices in recent years, there is still a need for vacuum tube devices in applications involving high power levels, such as the output stages of transmitters. At microwave frequencies, the changeover point is at mean power levels of 1–10 W or more, and at pulsed power levels of 10–100 W or more; solid-state devices cannot be manufactured with active regions capable of absorbing the energy lost due to inefficient operation at these levels. However, conventional tube devices (which operate by varying the number of electrons passing through the device by varying the voltages applied to control grids) begin to suffer from the effects of interelectrode capacitance, lead inductance, and electron transit time between electrodes being appreciable fractions of an RF cycle at around 100 MHz. By careful design, the frequency range of these valves can be extended to around 3 GHz for a triode, but the power output and bandwidth performance degrade rapidly with increasing frequency. An important selection of microwave tubes is described in this section.

Planar Triode

The construction and equivalent circuit of a *planar triode* concentric line amplifier are shown in Figure 11-12. The circuit may be changed to an oscillator simply by inserting a feedback probe between the two resonators. For efficient operation at high frequencies, the cathode current density must be high. Planar construction offers the following advantages:

1. Provides ideal terminations for coaxial line resonators.

2. Provides reduced radiation and lead inductance losses.

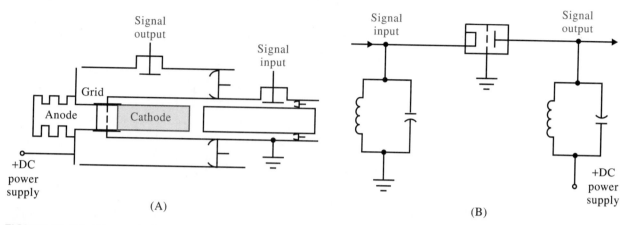

FIGURE 11-12 Planar triode microwave tube (A) Construction (B) Equivalent circuit

3. Permits the use of small electrode areas, which reduces interelectrode capacitances.

4. Enables closer electrode spacing, reducing the loss associated with transit time effects.

Gyrotron

The *gyrotron* depends on electrons rotating at a cyclotron frequency while under the influence of a strong magnetic field. This device was developed in the late 1970s to produce very high microwave power at frequencies up to a few hundred gigahertz. Several hundred kilowatts of *continuous wave* (CW) power have been developed at 60 GHz; and 1000 MW (1 GW), 70 ns pulses have been produced at 10 GHz.

Reflex Triode

Under pulsed conditions, the *reflex triode* is competitive with the gyrotron, at least in the X-band (8–12 GHz). This device causes an oscillating dipole of electrons to be set up, and output energy is extracted from these dipoles. A reflex triode has produced 3 GW pulses at 10 GHz, and the U.S. Army, for example, has these devices under intensive development for battlefield applications.

Pencil Triode

Figure 11-13 shows a *pencil triode,* or *lighthouse tube,* placed in a tunable cavity that is actually a triaxial structure with two simultaneously tunable cavities. The inner cavity is the grid–plate cavity, and the outer cavity is the cathode–grid cavity. The ends of each of the conductors are in direct contact with the electrodes of the tube, which are brought out from the tube in a coaxial manner. The movable short circuits that provide the tuning to each of the cavities must not make ohmic contact with the walls, because this would short out the cathode to the grid, and the grid to the plate. These short circuits have to be of the capacitive or resonant choke type.

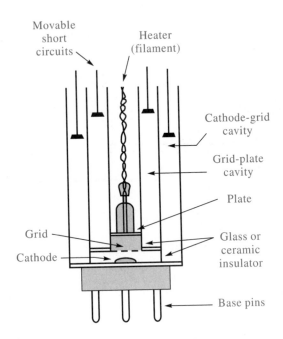

FIGURE 11-13
Lighthouse tube
(pencil triode)

As frequency requirements kept increasing, lighthouse tube oscillators became inadequate at higher frequencies because the electron transit time from the cathode to the plate was too long. However, this kind of oscillator is still being made for use in the lower end of the microwave spectrum.

Parametric Amplifier

Parametric amplifiers, or simply *paramps,* are used primarily at microwave frequencies above about 30 GHz, but low-cost versions may be designed for operation at lower microwave frequencies, where they can be cooled to provide good low-noise performance. The simplified schematic in Figure 11-14 shows the main elements of a paramp.

Nearly all paramp circuits are coupled using ferrite circulators and have power transferred to the signal by enabling this signal to extract power from a varactor diode. The extra power required is delivered to the varactor diode by an oscillator called the *pump.* The varactor diode is maintained under reverse bias.

In the simplest terms, energy that has been stored in the depletion capacitance by means of the pump source is transferred to the signal by means of the circulator, thus providing a net signal gain. Extra circuitry such as tuning and matching is required for proper operation, and the design is critically dependent on the correct choice of varactor diode, the pump frequency (which is different from the signal frequency), and the power.

Note that the only active device in the signal path is the varactor diode. Therefore, the noise performance is dependent only on thermal noise from the varactor and the circulator and any noise from the pump source. The pump source can be designed for the lowest practicable noise, and its output will be well filtered. This explains the superior noise performance of paramps even when they are uncooled; cryogenic cooling yields a further dramatic noise reduction.

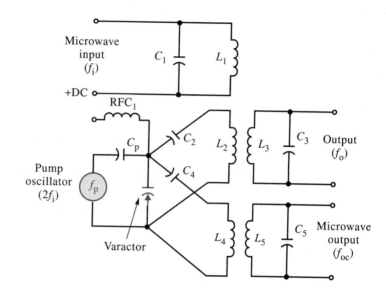

FIGURE 11-14
Simplified
parametric
amplifier

With suitable tuning, the paramp arrangement can be used to convert the signal to one centered on a lower or a higher frequency. These circuits are called *down-converters* and *up-converters,* respectively.

Electron Beam Tubes

An *electron beam tube* is a vacuum tube that uses an electron beam generator rather than a typical cathode. The beam power tube, klystron tube, magnetron tube, cathode ray tube, and photomultiplier tube are examples of electron beam tubes.

The electron beam tube, in contrast to the conventional vacuum tube, operates using a focused and directed beam of electrons. This arrangement can create amplification or oscillation in the same way as with conventional tubes, but according to parameters not available with other kinds of tubes. Electric and magnetic fields are used in electron beam tubes to control the paths of the beams. In an ordinary vacuum tube, the intensity of the electron beam is regulated, but the directional properties are not.

Klystron Tubes

The *klystron* is a form of electron tube used for generation and amplification of microwave electromagnetic energy. It is a linear-beam tube that incorporates an electron gun, one or more cavities, and apparatus for modulating the beam produced by the electron gun. There are several different types of klystron tubes, the most common of which are the two-cavity klystron, the multicavity klystron, and the reflex klystron.

Two-Cavity Klystron A simplified illustration of a *two-cavity* klystron is shown in Figure 11-15. This device is used as an amplifier or oscillator at moderate power levels. In the first cavity, the electron beam is velocity modulated, causing variations in the density of the electron beam in the second cavity. The output is taken from the second cavity. The signal fluctuations are greater in the second cavity than at the input point. This differential results in amplification.

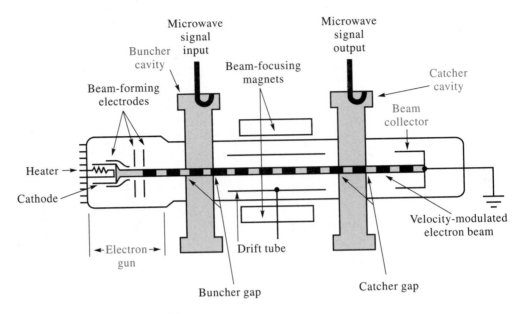

FIGURE 11-15 Two-cavity klystron

Multicavity Klystron The *multicavity* klystron, shown in Figure 11-16, contains one or more cavities between the input and output cavities. The intermediate cavity or cavities produce enhanced modulation of the electron beam, and therefore the amplification factor is greater. Multicavity klystrons are used at high power levels, which in some cases may be as high as 1 MW.

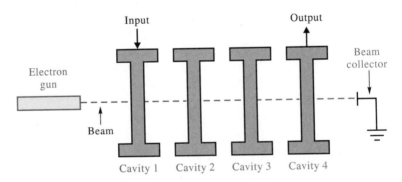

FIGURE 11-16 Simplified multicavity klystron

Reflex Klystron In the *reflex* klystron, shown in Figure 11-17, there is only one cavity and no input termination. Noise on the electron beam acts as the initial input signal. The electrons become velocity modulated and drift into a retarding field which causes the electron beam to reverse direction. This phase reversal allows large amounts of energy to be drawn from the electrons.

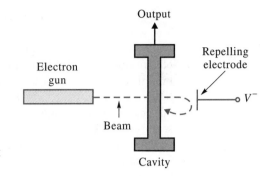

FIGURE 11-17
Simplified reflex
klystron

The reflex klystron is used as an oscillator at low and medium power levels. It can also be used as a frequency-modulating device. This device can produce only a few watts of RF output power in the microwave frequency range. Therefore, it has been largely replaced by semiconductor devices for microwave oscillator applications.

11-6 TRAVELING WAVE TUBES

One of the most difficult problems in electronics is amplifying RF voltages at frequencies above 1 GHz. Special tubes such as the pencil triode will amplify signals up to about 3 GHz. Above this frequency, there has been, until recently, no means of amplifying the RF output other than to build bigger magnetrons and klystrons as oscillators and couple their output directly to the antenna. On the receiving end, it has been necessary to heterodyne feeble signals with a local oscillator and depend on increased IF amplification for results. The limiting factors of noise, interference, loading, and the like, could be reduced greatly if high-powered wide-band amplifiers could be developed. *Traveling wave tubes* (TWTs) seem to be the answer to this problem. These tubes are being used increasingly, as high-powered wide-band amplifiers of microwave signals.

Classes of Traveling Wave Tubes

There are three general classes of TWTs: (1) the linear-beam forward wave tube (traveling wave amplifier), (2) the crossed-field backward wave tube, and (3) the crossed-field forward wave tube. To help distinguish the three general classes, the following brief definitions are provided:

1. *Forward wave*—a wave whose group velocity is in the same direction as the electron stream motion in a TWT.

2. *Traveling wave amplifier*—an amplifier that uses one or more TWTs to provide useful amplification of signals at frequencies of the order of several gigahertz.

3. *Backward wave tube*—a TWT in which the electrons travel in a direction opposite to that in which the wave is propagated.

Construction of Traveling Wave Tubes

The schematic cross section and external views of a TWT are given in Figure 11-18. At the left is an electron gun, similar to the type used in CRTs, that produces a beam of electrons about 1/8 inch in diameter. The anode accelerates the electrons. The long, thin glass tube houses a wire helix having 50 turns per inch. The external connection to the helix is at the far right. The spiral wound on the glass of the tube forms the input and output couplings. Also at the far right is a collecting electrode. The whole assembly may be 12 to 15 inches long, with the helix about 9 inches long. The diameter of the helix is about 11/64 inch, and the glass tube supporting it is slightly less than 3/8 inch in diameter.

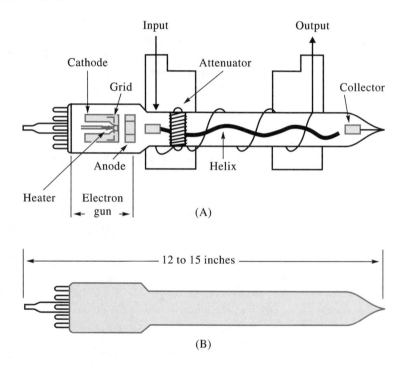

FIGURE 11-18 Traveling wave tube (A) Schematic cross section (B) External view

Power Supplies Figure 11-19 shows a schematic diagram of the connection of the TWT elements to associated power supplies. The power supplies include a 7.5 V_{ac} heater supply, a 0–450 V_{dc} anode supply, and a variable high-voltage (300 V–33 kV) supply for the helix.

The electron gun produces a pencil beam of electrons having uniform thickness. To enable the electron beam to impart energy to the RF traveling wave, the speed of the traveling wave and the speed of the electrons must be about the same. Actually, the speed of the electron beam is slightly greater. The function of the helix is to slow down the traveling wave to slightly less than the speed of the electrons.

A traveling wave in space or in a waveguide normally travels at about the speed of light (186,000 miles per second, mi/s). The helix contains shunt capacitance between

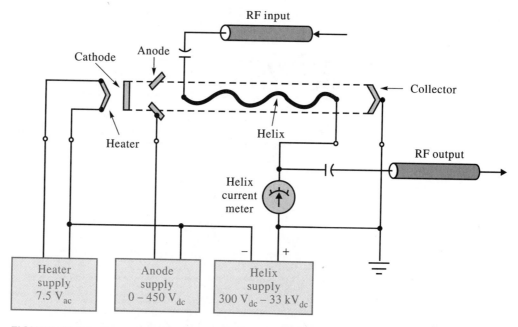

FIGURE 11-19 Power supplies for a traveling wave tube

turns and series inductance within turns, so it corresponds to a delay line having lumped inductances and capacitances. The helix may slow down the speed of the traveling wave by a factor of as much as 30. For example, if the electron-accelerating potential is 1500 V, the electron velocity will be about 15,500 mi/s, or 1/12 the speed of light, and the traveling wave will have a velocity of slightly less than 15,500 mi/s. As the slowed wave travels the length of the tube, the electron beam moves through the center of the tube in the same direction as the traveling wave.

Function of the Magnetic Field   The wave has both electric and magnetic field components. The magnetic component is not useful but rather tends to scatter the electrons in the pencil beam. To counteract this effect, the tube is contained in a focusing solenoid that establishes a magnetic field around the tube. This action is accomplished by a long, thin solenoid or with a structure of permanent magnets spaced along the length of the tube. Traveling wave tubes with permanent magnets are called *permanent magnet* TWTs.

Function of the Electric Field   The electric field component of the traveling wave is useful in interacting with the electron beam, as illustrated in Figure 11-20(A). The arrows indicate the direction in which the field will accelerate negative charges. Some electrons will be accelerated; some will be slowed down under the influence of the traveling wave. The beam of electrons will be made alternately more or less dense.

Electrons are speeded up as they approach a more positive area and slowed down as they approach a more negative area. Arrows pointing to the left represent a positive field and accelerate electrons; arrows pointing to the right represent a negative field

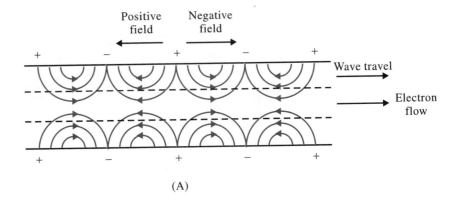

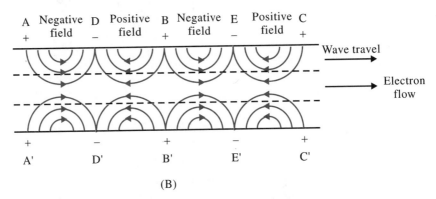

FIGURE 11-20 Electric field components (A) Pattern of the traveling wave at any instant (B) Bunching action of the electron beam

and slow down electrons. As the electrons move to the right slightly faster than the wave, bunching occurs where the field is changing from positive to negative, as illustrated in Figure 11-20(B).

Function of the Collector   When an electron is slowed down, it gives up some of its kinetic energy to the traveling wave that slowed it down. Thus, the wave is increased in strength. The greater the bunching action, the greater will be the amplification. The traveling wave moves slightly slower than the electron beam, so energy is given up from the electron beam to the traveling wave during the time the wave travels down the tube.

By changing the collector voltage (Figure 11-19), the average speed of the electrons in the beam can be changed and the bunches of electrons formed can be positioned so that they are always in a retarding field (arrows pointing to the right in Figure 11-20(B)). This action transfers energy from the electron beam to the wave and makes the wave stronger as it speeds down the tube. The bunches at A, B, and C induce positive charges at A′, B′, and C′, which in turn causes increased negative charges at D and E. Thus, as the bunching increases down the tube, more energy is transformed to the traveling wave. To maintain this action, the electron bunches appear at all times in a retarding field; that is, the electron speed is slightly greater than the speed of the field. By Lenz's law, the induced field opposes the action producing it and thus slows down the electron bunches as they move down the tube to the collector.

Function of the Attenuator Traveling wave tubes can have a gain as great as 70 dB when employed as broadband RF *low-noise amplifiers* (LNAs). If a traveling wave is reflected at the end of the tube, oscillations may be set up by the addition of energy to the input from the output (positive feedback). This action occurs in backward wave oscillators, called *carcinotrons,* to be discussed later. Traveling wave tubes designed only for amplification prevent the reflected wave from reaching the input by means of an attenuator placed (Figure 11-18(A)) around the helix. The attenuator may be a split graphite cylinder or may be provided by spraying a resistive film on the helix and tube envelope at the proper location. In a high-power tube, the attenuation is concentrated near the center of the tube, whereas in a low-power tube, it is usually not more than one-third the distance from the cathode. The necessity for an attenuator capable of handling large average powers is one of the major restrictions on the output power of the TWT.

Function of the Grid The grid (Figure 11-18(A)) of the TWT can be used to turn the electron beam ON or OFF, or to modulate the beam, thus controlling its density and its ability to transfer energy to the traveling wave. Therefore, the grid may be used to amplitude modulate the output. The grid or gate operates like the grid of a CRT for modulation of beam intensity. Thus, the TWT can be used as a modulator at microwave frequencies.

Coupling the TWT to a Waveguide Various methods are employed to couple the TWT to the waveguide. At higher frequencies, the waveguide can be coupled directly to the tube; at lower frequencies, a coaxial cable is used, and the center conductor is connected directly to the helix.

Handling of the Permanent Magnet TWT *Permanent magnet* TWTs can be damaged easily if certain precautions are not observed. These TWTs are generally shipped in a small box packed inside a large primary shipping container. Most damage to these devices can be avoided by following these simple procedures:

1. Do not remove the small box containing the TWT from its primary shipping container until the device is to be installed.

2. Use only nonmagnetic tools to open the container. A large bronze screwdriver is recommended.

3. Do not place the TWT on or near other magnetic material, such as metal benches.

4. Use nonmagnetic screwdrivers to install the TWT.

5. During storage, make sure that both the small box and the primary shipping container are used. The small box alone will not protect the TWT.

6. Observe the printed instructions contained in the "Magnetic Material Proximity Restrictions" located inside the shipping container.

Magnetron The *magnetron* is a form of TWT used as an oscillator at ultra-high and microwave frequencies. Most magnetrons contain a central and a surrounding plate, as shown in Figure 11-21. The plate is usually divided into two or more sections by radial barriers called cavities. The RF output is taken from a waveguide opening in the anode.

The cathode is connected to the negative terminal of a high-voltage source, and the anode is connected to the positive terminal. This arrangement causes electrons to flow outward from the cathode to the anode. A magnetic field is applied in a longitudinal direction causing the electrons to travel outward in spiral, rather than straight, paths. The electrons tend to travel in bunches because of the interaction between the electric and magnetic fields. This bunching results in oscillation, with the frequency being somewhat stabilized by the cavities.

Magnetrons can produce continuous power outputs of more than 1 kW at a frequency of 1 GHz. The output drops as the frequency increases. At 10 GHz, a magnetron can produce about 10–20 W of continuous RF output. For pulse modulation, the peak power figures are much higher.

In a magnetron tube, oscillation sometimes occurs at unwanted frequencies. This effect can be prevented by connecting metal strips between the resonator segments.

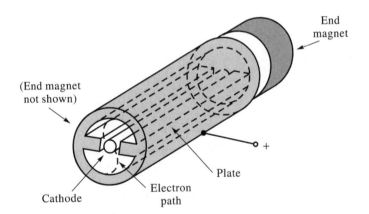

FIGURE 11-21
Magnetron

End-Plate Magnetron The *end-plate magnetron* is a special kind of magnetron tube. In the ordinary magnetron, electrons are accelerated into spiral paths by means of a magnetic field, as shown in Figure 11-22(A). The greater the intensity of the magnetic field, the more energy is imparted to the electrons, up to a certain practical maximum.

With the addition of an electric field between two electrodes at the ends of the cylindrical chamber, as shown in Figure 11-22(B), the electrons are accelerated longitudinally as well as in an outward spiral. The radial velocity component remains the same and is the result of the magnetic field. The longitudinal component, imparted by the electric field between the two end electrodes, gives the electrons extra speed. The electron speed gained by the added end electrodes produces an increased output amplitude.

Carcinotron The *carcinotron* is a *backward wave oscillator* (BWO) used to generate microwave energy by the effects of electric and magnetic fields on an electron beam.

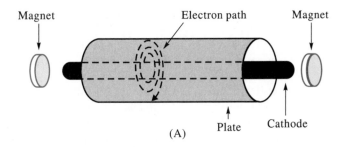

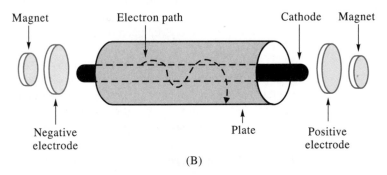

FIGURE 11-22 End-plate magnetron (A) Simple magnetron configuration (B) End plates added

Electrons traveling through the tube interact with the electromagnetic fields induced internally, producing oscillation. The carcinotron may be a linear tube or a circular tube. It operates in a manner similar to the magnetron.

Figure 11-23 is an illustration of a carcinotron. The electric and magnetic fields between the sole and the slow-wave structure cause oscillation in the electron beam as it travels down the tube.

The carcinotron can produce several hundred watts of power up to frequencies of more than 10 GHz. It will produce several milliwatts at frequencies greater than

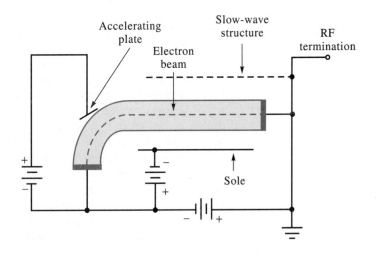

FIGURE 11-23
Carcinotron

300 GHz. This device tends to produce a large amount of noise along with the desired signal.

The carcinotron is a development that grew out of the traveling wave concept, but, in contrast to the TWT, it is inherently an oscillator.

O-Type Backward Wave Oscillator The *O-type* BWO is a development of the TWT with built-in feedback, where an electron beam delivers energy to a wave traveling backward along the slow-wave structure. The amplified backward wave velocity modulates the beam and also provides the output energy of the oscillator from a termination at the electron gun end of the tube. The forward wave is absorbed at the collector end by a nonreflecting termination and plays no further part in the action. These tubes have been largely replaced by solid-state devices.

M-Type Backward Wave Oscillator The *M-type* BWO is similar to the O-type except that the slow-wave structure and the electron trajectories are circular. It can be operated so that it produces a very high noise output over a wide frequency range and thus is used for microwave jamming. It is also frequency-agile, can be noise modulated, and typically has a 40% electronic tuning range.

Crossed-Field Amplifiers

The electron stream formation mechanism in a *crossed-field amplifier* (CFA), more generally known as a *crossed-field tube* (CFT), is quite different from the conventional transparent grid method used for klystrons and TWTs in that it relies on the motion produced in an electron under the influence of perpendicular electric and magnetic fields. Rather than an accelerating drift in the direction of the electric field, circular motion perpendicular to the magnetic field is combined with a constant drift perpendicular to both fields to produce an enhanced circular motion. The CFT construction, then, is simply an anode and a cathode with a magnetic field applied parallel to them, producing an average electron drift parallel to the plates. As in TWTs, this electron stream is allowed to interact with a traveling RF wave, in this case by making the anode a planar version of a slow-wave structure.

The RF field, when traveling at the same velocity as the electron drift velocity, upsets the balance of the forces on the electrons, and spikes of the electrons travel to the anode. As the spikes pass from cathode to anode, they interact with the RF field in such a way as to transfer potential energy due to the static electric field into RF power, causing amplification of the wave. Since the potential energy available depends only on the electric field, and drift velocity only on the ratio of electric and magnetic fields, RF power may be extracted from the tube without interfering with the velocity matching of beam and wave, thus providing improved efficiency. The slow-wave structure design is complicated, however, by the need to dissipate heat generated by the electrons striking the structure.

The planar CFT is not the most commonly used form; it lacks the advantages of compactness that characterize the cylindrical version. In the circular version, the design of amplifiers must ensure that electron spikes decay away in the zone between output and input to avoid feedback effects. Oscillators may easily be produced in this configuration by deliberately violating this design restriction; in the magnetron, for example, the slow-wave structure is replaced by a set of resonant sections. In this device, the rotating "wheel spoke" bunches of electrons are similar to the electron bunches of the amplifier configurations, but the resonant sections have a standing wave electromagnetic field distribution rather than a traveling wave.

Maser

The *maser* is a special form of amplifier for microwave energy. The term itself is an acronym for *m*icrowave *a*mplification by *s*timulated *e*mission of *r*adiation. The maser output is the result of quantum resonances in various substances.

When an electron moves from a high-energy orbit to an orbit having less energy, a *photon* is emitted. For a particular electron transition, or quantum jump, the emitted photon always has the same amount of energy and, therefore, the same frequency. By stimulating a substance to produce many quantum jumps from a given high-energy level to a given low-energy level, an extremely stable signal is produced. This action is the principle of the maser oscillator. Ammonia and hydrogen gas (and sometimes rubidium gas) are used in maser oscillators as frequency standards. The output frequency accuracy in the gas maser is within a few billionths of one percent.

A solid material, such as ruby, can be used to obtain maser resonances. This kind of maser, called a *solid-state* maser, can be used as an amplifier or oscillator. When an external signal is applied at one of the quantum resonance frequencies, amplification is produced. The solid-state maser must be cooled to very low temperatures (cryogenic cooling) for proper operation; the optimum temperature is near absolute zero. While satisfactory maser operation can sometimes be had at temperatures as high as that of dry ice, the most common method of cooling is the use of liquid helium or liquid nitrogen. These liquid gases bring the temperature down to just a few degrees above absolute zero.

The *traveling wave* maser is a common device for use at ultra-high and microwave frequencies. The *cavity* maser is another fairly simple device. Both the traveling wave and cavity masers use the resonant properties of materials cut to precise dimensions. Both are solid-state devices, and in both cases the temperature must be reduced to a low level for proper operation. The gain of a properly operating traveling wave or cavity maser amplifier may be as great as 30–40 dB.

Maser amplifiers are invaluable in radio astronomy, communications, and radar. Some masers operate in the infrared, visible light, and even ultraviolet portions of the electromagnetic spectrum. A visible light maser is called an *optical maser* or, more commonly, a *laser* (*l*ight *a*mplification by *s*timulated *e*mission of *r*adiation).

11-7 MICROWAVE SEMICONDUCTORS

The inherent limitations to the application of semiconductors at higher frequencies are the transit time, the capacity, and the geometry of the device. Diodes and transistors are available that operate in the lower edges of the microwave band, generally up to about 1 GHz. At these higher frequencies, special designs for the terminations of diodes and transistors are required to minimize the inductance and capacitance due to the pin connecting arrangements. Two different approaches are widely used for diodes:

1. The diode leads are flat conducting strips that emerge from the diode in opposite directions.

2. A coaxial arrangement is used in which one end of the diode connects to the center, and the other end connects to the outer conductor. These diodes are usually mounted in specially designed coaxial diode mounts and are used for mixing, detecting, and the special functions discussed in the following sections.

Transistor mounts are usually in the form of flat strips emerging at different points in the transistor package. For microwave applications, the terminal strips are often incorporated into a stripline configuration.

Most microwave transistors are state-of-the-art devices that do not comply with any established standards. For that reason, detailed manufacturers' data and application notes are essential when dealing with these devices.

Microwave Diodes

A number of specialized microwave diodes are currently available. In this section, we will examine several of these special devices, including the varactor, the step recovery, the tunnel, the Gunn, the IMPATT, the TRAPATT, and the BARITT.

Varactor Diode All diodes exhibit a certain amount of capacitance when they are reverse biased. This capacitance limits the frequency at which a diode can be used as a rectifier or detector. However, the reverse-bias capacitance properties of a semiconductor diode can be used to advantage in certain applications. A *varactor* diode, also referred to as a *variable-capacitance* diode, or *varicap,* is deliberately designed to provide an electronically variable capacitance. With a specially doped P-N junction, this diode uses the capacitive variation with reverse bias to provide frequency multiplication. Available in silicon as well as gallium arsenide, varactors can produce as much as 15 W at 1 GHz or 1 W at 5 GHz, with appropriately higher power outputs at lower frequencies.

Step Recovery Diode A *step recovery* diode, also called a *charge storage* or *snap* diode, is used chiefly as a harmonic generator. When a signal is passed through a step recovery diode, dozens of harmonics occur at the output. When connected in the output circuit of a VHF transmitter, this device can be used in conjunction with a resonator to produce signals at frequencies up to 10 GHz at very low power levels, usually below 100 mW.

Tunnel Diode The *tunnel* diode can be used as an amplifier or an oscillator at microwave frequencies. Tunnel diodes depend on a voltage-controlled negative resistance effect for amplification at microwave frequencies. These devices are capable of low signal level amplification but cannot be used for large power handling.

Because of technical problems in mass production, the tunnel diode has found only limited application in the microwave field. In recent years, Gunn diodes have largely replaced tunnel diodes as amplifiers and oscillators at microwave frequencies.

Gunn Diode A *Gunn* diode operates as an oscillator in the microwave frequency range. It has replaced both the klystron tube and the tunnel diode in many applications.

Consisting of a single piece of semiconductor material, such as N-type gallium arsenide (GaAs), the Gunn diode is mounted in a resonant enclosure, as shown in Figure 11-24(A). A dc voltage is applied to the device (Figure 11-24(B)). At a certain

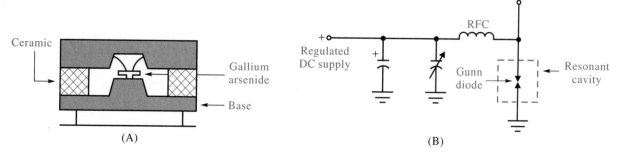

FIGURE 11-24   Gunn diode (A) Cutaway of inside construction (B) Simple oscillator schematic

voltage, the semiconductor develops a negative resistance region and spontaneously generates microwave signals.

Gunn diodes are not very efficient. Only a small fraction of the consumed power results in RF power. These diodes tend to be highly sensitive to changes in temperature and bias voltage. The frequency may vary considerably even with a small change in the ambient temperature, and therefore the temperature must be carefully regulated. The change in frequency with voltage can be useful for frequency modulation, but voltage regulation of some sort is essential. Most Gunn diodes require about 12 V for proper operation. Oscillation can occur at frequencies in excess of 20 GHz, with an output power of 100 mW or greater.

IMPATT Diode   *IMPATT* (*imp*act *a*valanche *t*ransit *t*ime) diodes are common in certain oscillator applications and in the moderate-power stages of reflection amplifiers. They produce substantially greater power than Gunn diodes, but this increased power is largely offset by the considerably greater noise generated. For this reason, IMPATTs tend to complement, rather than compete with, Gunn diodes in many CW, and in a few pulsed, applications.

TRAPATT Diode   *TRAPATT* (*trap*ped *a*valanche *t*ransit *t*ime) diodes operate primarily in pulsed form, and they can deliver considerable microwave power. In fact, TRAPATTs are the leaders in this respect for many microwave systems applications, and hundreds of watts are obtainable for a submicrosecond pulse, at a frequency of a few gigahertz.

BARITT Diodes   *BARITT* (*bar*rier *i*njection *t*ransit *t*ime) diodes are relatively new on the scene and have a rather uncertain future. Although BARITTs are the "quietest" solid-state microwave oscillator currently available, their output power is limited to a few milliwatts. In spite of this low power output, and because of their low noise, BARITT diodes are in strong competition with Gunn diodes. The main disadvantage of the BARITT diode (compared with the Gunn diode) is the high supply voltage required.

YIG Filters

Yttrium-iron-garnet (YIG) filters consist of highly polished spheres of *single-crystal YIG, a ferrite material*. When placed in an RF structure under the influence of a dc magnetic field, this material exhibits a high Q resonance of a frequency proportional to the dc magnetic field. To understand the phenomenon of ferrimagnetic resonance, consider parts A through E of Figure 11-25. In the ferrite with no dc magnetic field applied (part A), there is a high density of randomly oriented magnetic dipoles, each consisting of a minute current loop formed by a spinning electron. Viewed macroscopically, there is no net effect because of the random orientations. When a dc magnetic field, H_o, of sufficient magnitude is applied (part B), the dipoles align parallel to the applied field, producing a strong net magnetization (part C), M_o, in the direction of H_o. If an RF magnetic field is applied at right angles to H_o, the net magnetization

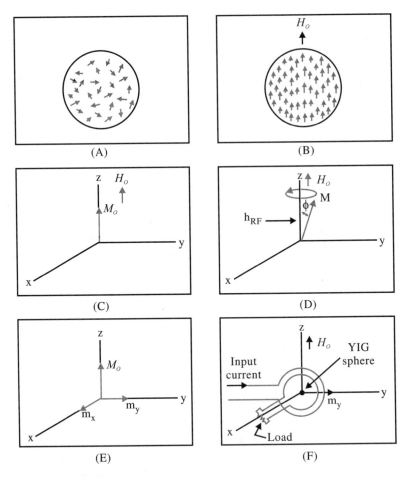

FIGURE 11-25 Illustration of ferrimagnetic resonance in a YIG filter (A) Randomly oriented magnetic dipoles in the unmagnetized ferrite (B) Magnetic dipoles aligned under the influence of a magnetic field (C) Equivalent representation of B showing the combined effect of the aligned dipoles (D) Precession of the net magnetization vector due to the RF magnetic excitation (E) Equivalent representation of precessing magnetization vector (F) Tuned bandpass filter consisting of YIG sphere at center of two mutually orthogonal loops

vector will precess (part D), at the frequency of the RF field, about an axis coincident with H_o. The precessing magnetization vector (part E) is represented as the sum of M_o and two circularly polarized RF magnetization components m_x and m_y. The angle of precession Ø, and therefore the magnitudes of m_x and m_y, will be small except at the natural precession frequency. This frequency, known as *ferrimagnetic resonant frequency,* is a linear function of the dc field H_o.

Figure 11-25(F) shows the basic elements of a YIG band-pass filter. The filter consists of a YIG sphere at the center of two loops whose axes are perpendicular to each other and to the dc field H_o. One loop carries the RF input current, and the other loop is connected to the load. When H_o is zero, there is large input-to-output isolation, since the two loops are perpendicular. With H_o applied, there is a net magnetization vector in the direction of H_o. The magnetic field h_z produced by the RF driving current in the input loop causes the net magnetization vector to precess about the Z-axis. The resulting RF magnetization component, m_y, induces a voltage into the output loop. At frequencies away from the ferrimagnetic resonant frequency, m_y and the voltage it induces are small so input-to-output isolation is high. When the input current is at the ferrimagnetic resonant frequency, Ø and m_y are maximum. There is a large transfer of power from input to output, and insertion loss is low. Thus the filter center frequency is the ferrimagnetic resonant frequency and can be tuned by varying H_o. Commonly, the YIG sphere and RF structure are located between the poles of an electromagnet, and tuning is accomplished by furnishing a control current to the magnet coils.

A typical YIG filter to increase the offband isolation can be made by using two YIG spheres, as shown in Figure 11-26. The heart of the YIG filter is an electronically tunable filter. Two spheres of YIG crystal material in a magnetic field are placed in the path of the RF signal. RF can only pass through the YIG filter when the spheres are at resonance. The frequency of resonance is a linear function of the magnetic field strength produced by the YIG tuning coil. The coil's field is proportional to the tuning current; thus the resonant frequency of the YIG filter is directly proportional to the tuning current supplied to the YIG filter. Insertion loss of the YIG filter at resonance is less than 5 dB. Off resonance, the YIG filter acts like a short circuit reflecting most of the RF energy back to the source.

If a filter like the one just described is connected to the feedback circuit of an oscillator as a tunable band-pass filter, a voltage (or current)-tuned signal source can be

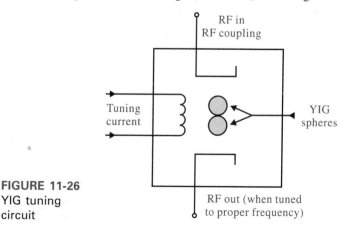

FIGURE 11-26
YIG tuning
circuit

achieved. Tunnel diodes or transistor oscillators can be connected with these filters to provide single-frequency, harmonic-free microwave signal sources.

Other Microwave Diodes

Another group of microwave diodes comprises the so-called *passive* diodes. These devices are referred to as passive only because they are not used in power generation or amplification. They play an active role in mixers and detectors and are used for power control. The passive devices to be discussed here are the point contact, PIN, and Schottky barrier diodes.

Point Contact Diodes   A *point contact* diode is formed by placing a fine gold-plated tungsten wire, called a cat's whisker, in contact with a piece of semiconductor material, such as germanium or silicon. The fine wire is pressed down slightly on the semiconductor material for spring contact. As shown in Figure 11-27, the semiconductor material forms the anode of the junction; the cat's whisker forms the cathode. The semiconductor and the cat's whisker are surrounded by wax to exclude moisture and are placed inside a metal–ceramic housing.

Point contact diodes can be fitted into coaxial or waveguide mounts and are available at frequencies above 100 GHz, although they generate considerable noise at these frequencies. Because the point contact junction cannot carry much power, the devices are limited to low-power applications. They are used as microwave mixers or detectors, although there are some characteristic differences between the two applications.

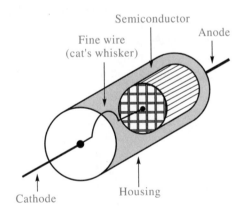

FIGURE 11-27
Point contact diode
construction

PIN Diodes   The *PIN* (*P-i*ntrinsic-*N*) diode, a special diode, with a layer of intrinsic semiconductor material between the P-type and N-type layers, is used for microwave power switching, limiting, and modulation. This device acts more or less like an ordinary diode up to about 100 MHz. Above this frequency, however, it ceases to be a rectifier because of the carrier storage in, and the transit time across, the intrinsic region. At microwave frequencies, the diode acts as a variable resistance.

When negative bias is applied to a PIN diode, the microwave resistance changes to 1–10 Ω from a typical value of 5–10 kΩ when positive bias is applied. Thus, if the diode is mounted across a 50 Ω coaxial line, it will not significantly load the line when it is negatively biased, so power flow will be unaffected. When the diode is forward biased, however, its resistance becomes very low, so most of the power is reflected and little is transmitted; in other words, the diode is acting as a switch. In a similar fashion, the PIN diode may be used as a pulse modulator. Several PIN diodes may be used in series or in parallel in a waveguide or coaxial line to increase power handled or to reduce the transmitted power in the OFF condition.

Figure 11-28 shows a practical coaxial PIN diode switch. As shown, the switch consists of a number of diodes in parallel. This arrangement allows peak powers in excess of 100 kW to be switched. Advantages of the PIN diode switch are (1) long life, (2) reliability, (3) small size, and (4) the ability to remove the initial spike of power coinciding with the beginning of the power pulse. In low-power switching and pulse modulation, PIN diodes are capable of switching times under 10 ns.

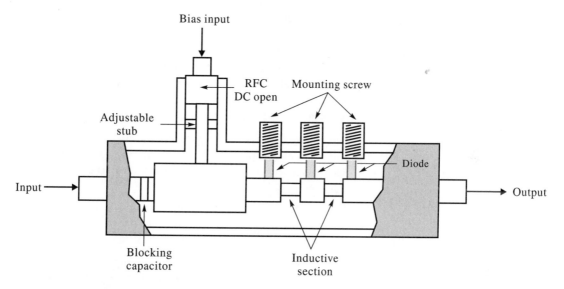

FIGURE 11-28 Coaxial PIN diode switch

Schottky Barrier Diode A *Schottky barrier* diode, also known as a *hot electron* diode, or more simply as a *Schottky* diode, is an extension of the point contact diode. Here, the metal–semiconductor contact is a surface (the Schottky barrier) rather than a contact point. Because of the larger contact area between the metal and the semiconductor, the forward resistance is lower, and so is noise.

Schottky barrier diodes are characterized by extremely rapid switching capability. Used as mixers, harmonic generators, and detectors in microwave applications, they are available for frequencies up to at least 100 GHz.

Microwave Transistors

The following paragraphs present the fundamental device principles and limitations of microwave transistors. As at lower frequencies, there are two general classes of devices, in this case bipolar junction transistors (BJTs) and field-effect transistors (FETs). Likewise, the basic principles of operation are similar to those for lower-frequency devices. The major difference in microwave transistors is their ability to operate at the much higher frequencies of the microwave range.

Bipolar Junction Transistors Almost all *bipolar junction* microwave transistors today are silicon NPN type, principally because of the inherent stability of silicon monoxide (SiO), which is formed as an insulating and protective layer. BJT devices are primarily used as amplifiers and oscillators below 12 GHz. Below 2 GHz, a silicon BJT has higher usable gain, higher transconductance, superior dc threshold uniformity, and consistently lower noise than GaAs FETs, discussed in the next section. BJT devices are used in many microwave systems operating in the C-band and below (less than 8 GHz). They are considered to be very reliable at these frequencies.

Field-Effect Transistors The *field-effect* transistor device differs from the BJT device in both structure and material. An FET is a *unipolar* device, using either electrons or holes as the majority charge carriers. FETs have largely overtaken BJTs, especially for frequencies above about 8 GHz. In particular, FETs are the only transistors that can provide useful gain with low noise at frequencies through the J-band (up to about 20 GHz). Current research is aimed at increasing this frequency to 40 GHz.

The typical material used in the microwave FET is gallium arsenide, and the common name for such devices is GaAs FET. Gallium arsenide allows substantial increase in device performance for two reasons:

1. The charge carriers reach twice the limiting velocity with one third the bias current. This property allows current gain to be reached at a frequency at least twice the frequency characteristic of silicon, and it reduces parasitic resistances.

2. The gallium arsenide substrate has a resistivity of greater than 10^7 Ω per square, compared to silicon, which has typically 30 Ω per square. This property results in lower bonding pad parasitic capacitance.

The GaAs FET is well known as a low-noise amplifying device. It is ideal for use in high-gain LNA circuits at microwave frequencies. It can be used in receiver preamplifiers, converters, and power amplifiers up to several watts.

Special FET devices called MESFETs (*metal semiconductor field effect transistors*) are used in the vast majority of current applications. These devices employ a metal semiconductor contact for the gate. Some GaAs MESFETs achieve a maximum frequency of oscillation of about 50 GHz.

MESFETs are available in single-gate or dual-gate configurations. Dual-gate types are useful in modulators, mixers, and AGC circuits. Power MESFETs producing 2–3 W at 12 GHz are also available.

11-8 REFLECTION AMPLIFIERS AND DIGITAL MICROWAVE TECHNIQUES

In microwave transistor amplifiers, the device that is driven by the RF signal input directly converts available dc power to RF signal output power. Although transistors are available that provide useful gain at frequencies of at least 20 GHz, microwave amplification is still required for communications and radar at frequencies well beyond 20 GHz, even up to hundreds of GHz. Furthermore, the maximum output power that can be obtained from microwave transistors is presently modest.

With a view to filling these and other gaps in system requirements, an alternative amplifying arrangement is often used. In this alternative configuration, a *circulator* is normally employed. Transistors are not necessary to achieve amplification in these configurations. The increase in RF power takes place via *reflection* (negative resistance) from an active two-terminal device connected to the intermediate port of the circulator—that is, between source and load. To emphasize the contrasts between transistor amplifiers and these reflection amplifiers, the general arrangements are shown in Figure 11-29.

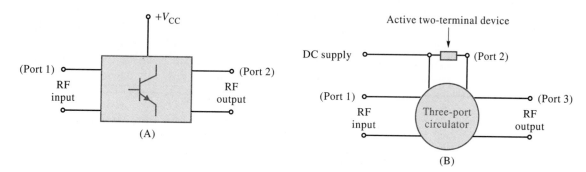

FIGURE 11-29   General arrangement of (A) Transistor amplifier (B) Reflection amplifier

Reflection Amplifier Operation

A schematic diagram illustrating *reflection amplifier* operation is shown in Figure 11-30. The circulator passes RF power, with only a fraction of a decibel loss, from port to port in the direction of the arrow. About 20 dB of attenuation, achieved by means of a magnetized ferrite disc, is offered in the opposite direction. Thus, RF input power enters port 1, travels to port 2 with little loss of power, is reflected from the active two-terminal device at an increased level, and passes to the output on port 3, again with very little loss. In this way, a net power gain is achieved.

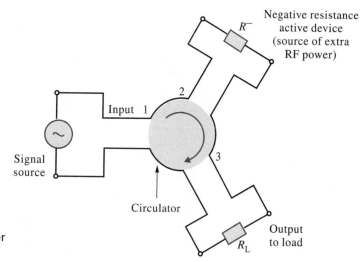

FIGURE 11-30
Schematic diagram
of reflection amplifier
operation

Two-terminal devices discussed earlier (tunnel diodes, Gunn diodes, IMPATT diodes, and BARITT diodes) may be used to provide the negative resistance required for the reflection amplifier. Amplifiers based on these devices are somewhat noisier than amplifiers designed around microwave transistors. However, they can provide considerable gain and power, and they can be designed to operate at much higher microwave frequencies than can transistor amplifiers.

Digital Modulation Techniques

The use of *phase shift keying* (PSK) in *time division multiple access* (TDMA) communication networks is a significant consideration for such microwave communication systems as high-capacity terrestrial microwave links and satellite systems. Table 11-4 shows modulation efficiencies for several levels of PSK, ranging from 0.5 for simple 2-level PSK to 4.0 for 16-level PSK. The complexity of the technology required increases considerably with the increased order of PSK introduced.

TABLE 11-4 Modulation Efficiencies for PSK Methods

Method	*Modulation Efficiency*
2-level PSK	0.5
4-level PSK (QPSK)	Approximately 1 to 1.5
QPSK staggered by one halfbit in time (SQPSK)	Approximately 1.8
8-level PSK	Approximately 2.0
16-level PSK	4.0

As the table shows, 4-level PSK (called *quantary* PSK, QPSK) offers improvement by a factor of 2 to 3 over the simpler 2-level PSK. Also, it can be seen that, to achieve this level of improvement again, 16-level complexity is required. Some actual Intelsat TDMA testing has been carried out with an 8-level PSK system, and 16-level systems have been examined in the laboratory. However, these levels represent such considerable jumps in complexity that most existing systems are oriented toward QPSK.

A series arrangement providing QPSK modulation of a microwave carrier is shown in Figure 11-31. The carrier source provides a pure sinusoidal signal that is relatively stable and noise-free. This microwave signal is passed through the first and second circulators and, in the absence of the circuitry connected to the intermediate ports, would simply pass on to be amplified by the final stage. In this situation, the output to the antenna from the reflection amplifier would be simply the unmodulated carrier, at increased power.

The combination of the first and second circulators produces QPSK of the carrier. In the first circuit, containing the quarter-wave line, this combination transforms the short circuit to an open circuit at the diode. The length of the line from the diode to the circulator is an integral number of $\lambda/2$. Thus, when a 0 is applied to the diode, the diode does not conduct, and the signal passing through the circulator suffers no change of phase. However, when a 1 is applied and the diode does conduct, the signal experiences an effective short circuit in the plane of the diode, and a 180° phase shift is now evident. Thus, this first circuit produces PSK of the 0°, 180° signal.

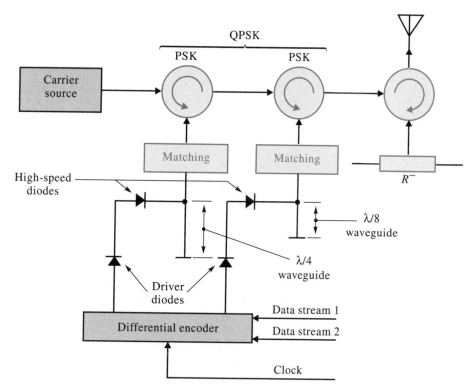

FIGURE 11-31 Series-type QPSK modulator

The second circuit, under similar principles, also produces PSK of the 0°, 180° signal. This second circuit provides the essential intermediate 90° phase shift, by means of the λ/8 line. The overall QPSK angles are thus the conventional 0°, 90°, 180°, and 270°.

PIN diodes act as effective and reliable high speed diodes at the C-band (4–8 GHz), where final-stage IMPATT diode reflection amplifiers yield 10–30 W of CW energy. In the 19–21 and 28–29 GHz bands, GaAs Schottky-barrier diodes must be used to cope with the high modulation rate.

SUMMARY

- Microwaves are the portion of the electromagnetic spectrum where the wavelength falls between about 1 mm and 30 cm. Microwave frequency bands extend from about 1 GHz to 300 GHz and are useful for short-range, high-reliability radio and television links. Systems operating at microwave frequencies are radar, mobile radio, satellite communications, terrestrial line-of-sight, and troposcatter links.

- Microwaves travel in essentially straight lines through the atmosphere, are not affected by the ionized layers, and are affected very little by temperature inversions and scattering. However, weather effects limit the distance between transmitter and receiver to a few miles. Repeater stations placed along the propagation path extend the distance over which communications can occur. A microwave repeater is a receiver/amplifier/transmitter combination used for relaying microwave signals in long distance, overland communication links. As frequencies increase, distance between repeaters must decrease.

- The primary mode of propagation in the microwave range is line-of-sight (LOS). The range of LOS communication depends on the height of the communicating antennas and the nature of the terrain between them. The LOS range is limited to the radio horizon. LOS microwave repeaters (radiolink systems) carry large quantities of voice and data traffic and are subject to transmission impairments, thus limiting the distance between relay points. Most radiolink systems carry analog signals, principally FM, although a few carry digital signals. Terrestrial radiolink systems offer highly directional, point-to-point transmission of voice and date signals. Satellite radiolink systems offer multiple-path transmissions for such signals as television broadcasts.

- Frequencies in the microwave spectrum are normally transmitted through waveguides, the walls of which are highly polished and sometimes specially plated to reduce resistance and, consequently, power losses. Waveguides may be constructed of brass, copper, or aluminum and are made in sections of various lengths. These sections may be straight, bent, twisted, or flexible. The most common form of waveguide is rectangular in cross section, although round waveguides may be used. Within certain limits, microwaves can be transferred by coaxial cable, microstrip, stripline, and optical fibers. Transmission of light-modulated signals through optical fibers is the modern method.

- Because waveguides provide excellent shielding and low loss, they can transmit greater amounts of power with less energy loss than coaxial cables. To be an effective transmission line, a rectangular waveguide must have sides measuring at least 0.5 λ, and a circular waveguide must be at least 0.6 λ in diameter. Microwaves can be propagated in waveguide in the transverse electric (TE) or the transverse magnetic (TM) mode. Interior walls of waveguide must be kept clean and free from condensation.

- Electron tubes are needed in microwave systems where high power levels are involved. Types of microwave electron tubes include planar triodes, gyrotrons, reflex triodes, pencil triodes, parametric amplifiers, and electron beam tubes. Electron beam tubes include the electron beam power tube, the klystron, and the magnetron. Klystron tubes include the two-cavity, the multicavity, and the reflex.

- A cavity resonator is a cylindrical or rectangular metal enclosure that operates as a tuned circuit. The resonant frequency of the cavity is affected by temperature and humidity. Cavity resonators can be fixed or tunable. A cavity frequency meter, consisting of a tunable cavity resonator, an RF voltmeter or ammeter, and a pickup wire, is used to measure wavelengths at VHF and above.

- Traveling wave tubes (TWTs) are used as high-powered wide-band amplifiers of microwave signals. The three general classes of TWTs are the linear-beam forward wave, crossed-field backward wave, and crossed-field forward wave. Permanent magnet TWTs require special handling to prevent damage. The magnetron is a form of TWT. Special kinds of magnetrons are the endplate magnetron, carcinotron, and

O-type and M-type backward wave oscillators (BWOs). The maser is a special form of microwave amplifier. Types of masers are traveling wave and cavity.

- Special construction methods permit semiconductor devices to be used in many microwave applications, particularly at the lower end of the microwave frequency range. Special microwave active diodes include varactor, step recovery, tunnel, Gunn, IMPATT, TRAPATT, and BARITT. Passive diodes include point contact, PIN, and Schottky barrier. Both bipolar and field-effect transistors capable of operating at high frequencies are used in microwave applications. GaAs FETs and MESFETs are used in many special applications.

- A YIG filter consists of highly polished spheres of a ferrite material. A YIG bandpass filter consists of a single YIG sphere at the center of two loops. A YIG filter used for increasing offband isolation consists of two spheres of YIG crystal material placed in a magnetic field in an electronically tunable filter. The resonant frequency of this filter is directly proportional to the tuning current supplied. Off resonance, the filter acts like a short circuit, reflecting most of the RF energy back to the source. Connecting this filter to the feedback circuit of an oscillator as a tunable band-pass filter can result in a voltage or current tuned signal source. Single-frequency, harmonic-free microwave signal sources can be achieved by connecting tunnel diodes or transistor oscillators with these filters.

- An alternative to the transistor amplifying arrangement is the reflection amplifier that uses a circulator and negative resistance from an active diode to produce amplification. Digital modulation techniques for microwave communication use phase shift keying (PSK) in time division multiple access (TDMA) with reflection amplifiers.

QUESTIONS AND PROBLEMS

1. Which range of frequencies do microwaves cover?

2. Describe microwaves and their uses.

3. What precautions must be observed when working with microwaves? Why?

4. State the primary mode of microwave propagation.

5. What factors determine the range of line-of-sight communication?

6. Define microwave repeater and explain its purpose and operation.

7. List the transmission impairments that limit the distance between relay points.

8. Explain terrestrial radiolink systems.

9. What are satellite radiolink systems used for?

10. What is a cavity resonator?

11. How does a cavity behave (a) near resonance, (b) if it is slightly too long, and (c) if it is slightly too short?

12. Describe the operation of (a) a tunable resonator and (b) a cavity frequency meter.

13. A cavity frequency meter is used to measure a wavelength of 7.3 cm. What is the frequency?

14. An applied frequency is 22 GHz. What wavelength is measured with a cavity frequency meter?

15. Define waveguide and explain its structure.

16. For efficient propagation of an electromagnetic field, what measurements must a waveguide have if it is (a) rectangular and (b) circular?

17. Explain the principles of guided waves and describe how microwaves are reflected.

18. Describe the waveguide propagation modes for (a) TE and (b) TM.

19. How can a planar triode concentric line amplifier be changed to an oscillator?

20. Describe the operation of a paramp.

21. List five examples of electron beam tubes.

22. Describe the klystron tube and list the most widely used types.

23. Explain how amplification is obtained in a two-cavity klystron.

24. Describe and state the application of (a) the multi-cavity klystron and (b) the reflex klystron.

25. List the three general classes of TWTs.

26. State the function of the TWT (a) magnetic field, (b) electric field, (c) collector, (d) grid, and (e) attenuator.

27. How can the TWT be coupled to a waveguide?

28. What precautions must be observed when handling a permanent magnet TWT?

29. Define, describe the construction of, and state the application for (a) the magnetron and (b) the carcinotron.

30. How can unwanted oscillations be prevented in a magnetron?

31. Describe how an end-plate magnetron differs from an ordinary magnetron.

32. Describe and compare O-type and M-type BWOs.

33. Describe how CFTs differ from klystrons and TWTs.

34. Define maser and explain its operation.

35. List (a) seven active and (b) three passive microwave diodes.

36. What are BJTs used for? At what frequencies?

37. GaAs FETs are ideal devices for what purpose?

38. Describe how a reflection amplifier operates.

39. Where is PSK used in microwave communications?

40. What level of PSK is used in most existing systems?

CHAPTER 12

Television Systems— Transmission

OBJECTIVES

When you complete this chapter, you should be able to:

- State the concepts of a basic TV system.
- List the elements involved in each section of a TV broadcast transmitting station.
- Draw an illustration of the TV transmission spectrum and identify its various components.
- Draw a composite video wave and identify all synchronizing pulses and video information shown.
- Define and discuss the following elements: pickup device, frame, field, interlaced scanning, deflection and focusing, vestigial sideband, synchronizing pulse, blanking, equalizing pulse, composite signal, diplexer, compatible system, luminance, chrominance, and color burst.

INTRODUCTION

Of all the basic communications systems discussed in this book, television (TV) is perhaps the most widely used by individuals, and it is by far the most popular form of entertainment. In today's society, television has become a way of life. Exciting sports events, musical spectaculars, great motion pictures, important news events—all these and more exist right in our homes because of the magic of television. It is no wonder, then, that television has attracted the greatest audience in history.

Simply stated, the TV system consists of aural (audio) and visual (video) pickup devices (transducers), a transmitter, and a receiver. In this chapter, we will consider a basic TV system, the elements involved in a typical TV broadcast transmitting station, color television, and other TV applications.

12-1   BASIC TELEVISION SYSTEM

A TV system may consist of a simple camera connected to a TV receiver or monitor, or it can be an expensive complex of highly sophisticated electronic equipment geared to the needs of a broadcast network or a space program. Whatever the application, the system should be tailored to both the present and future needs of the user, but it should not be extravagant. In this section, we will examine a basic TV system and the elements involved, all of which are fundamental to all TV systems. We will delay detailed discussion of these elements until we examine the TV broadcast transmitting station.

The Television Camera

Although TV cameras assume many different physical and electrical configurations, they can be divided into two basic groups: (1) self-contained cameras, and (2) two-unit systems that employ separate camera heads driven by remotely located camera control units. Television cameras can also be classified according to resolution capabilities, bandwidth, scanning ratio, environmental abilities, and the like.

Self-Contained Camera   The *self-contained* camera illustrated in Figure 12-1 contains all the elements necessary to view a scene and generate a complete TV signal. All we need is to supply power (usually ac) to the unit and provide a means for routing the video signal to a monitor (TV screen). All power supplies, scanning circuits, and video amplifiers are contained within the unit. Self-contained cameras may also incorporate a synchronizing (sync) generator or have the capability of being driven by a remotely located master sync generator.

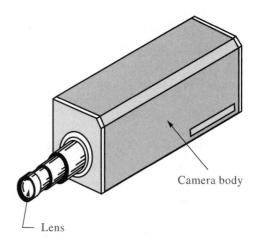

Camera body

FIGURE 12-1
Self-contained
camera

Lens

Two-Unit Camera   Figure 12-2 illustrates the *two-unit* concept of a TV camera that is popular with industrial and scientific users of closed circuit television (CCTV) systems. The camera head and its associated camera control unit (CCU) may contain essentially the same electrical components as the self-contained camera, but the bulk of the circuitry is usually within the CCU, which is connected to the camera head by a multi-conductor cable. This configuration provides some real advantages:

1. The camera head may be placed in locations where environmental problems such as heat, dust, shock, and vibration are severe without endangering the major portion of the circuitry.

2. The remote camera head is generally smaller than a self-contained unit and thus can be used in "tight-fit" areas.

The remote camera head usually contains only the photosensitive pickup device, a video preamplifier, and some filtering circuits. Some cameras may also employ small motors to provide a means of remotely focusing the lens. The exact configuration depends on the application and the manufacturer of the equipment.

When a remote CCU is used, most of the electrical operating and setup controls are contained within it. For this reason, the CCU is usually mounted near a viewing monitor so that the results of any adjustments may be easily viewed on the monitor screen. Most remote camera heads require little or no adjustment since they may be located in relatively inaccessible areas. Sometimes the circuitry of more than one CCU is housed in the same chassis.

The cable that connects the camera and CCU carries the control and other voltages necessary for the camera, as well as video, and other signals. Cable length may vary from a few feet to several thousand feet, although lengths in excess of 2000 feet are rare.

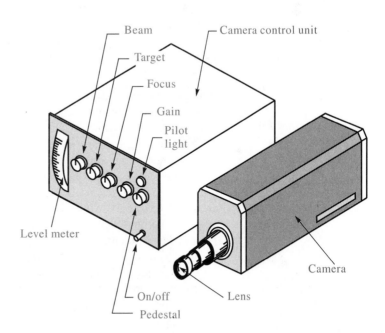

FIGURE 12-2
Two-unit camera

Camera Pickup Devices

The television camera pickup device may be thought of as the eye within the body of a TV system. The tube must possess characteristics that are similar to the human eye—that is, it must have a sensitivity to visible light, it must have a wide dynamic range with respect to light intensity, and it must have an ability to resolve detail when viewing a multielement scene.

Modern camera devices not only fulfill these stated requirements, but in many cases, far exceed the capabilities of the human eye. For example, some devices in use today are so sensitive that they present a useful output under conditions that the human eye interprets as complete darkness. This represents a considerable advancement over early camera pickup tubes that required as much as 1000 foot-candles of illumination to produce a usable picture.

Early Pickup Devices   The earliest pickup devices were tubes. The first TV camera tube was called the *iconoscope*, derived from "icon" (meaning image) and "scope" (meaning to observe). The iconoscope was relatively large and insensitive to light and, because of secondary emission, had a tendency to generate false signals.

The *image orthicon* (IO) replaced the iconoscope. Compared to the iconoscope the IO was somewhat smaller and more sensitive to light. The IO was designed with a slow-speed scanning beam to avoid the creation of secondary emission, the cause of false signals in the iconoscope.

The *vidicon* came into general use in the early 1950s and gained immediate popularity because of its small size (originally 1-inch diameter, later reduced to 1/2 inch), ease of operation, and its comparatively low cost. The vidicon enjoyed great success in the CCTV field and advances in vidicon technology made it attractive for use in many broadcast applications. It incorporated the basic operating fundamentals of the more refined camera tubes such as Plumbicon, Vistacon, Saticon, Newvicon, and others. These tubes were all smaller, lighter, less expensive, simpler in design than the IO, and more light sensitive.

Modern Pickup Devices   Advanced technology has largely eliminated the pickup tubes discussed in the previous paragraphs. Today, the more common pickup device is the *charge coupled device* (CCD). In its basic form, a CCD is a dynamic shift register capable of analog signal processing and storing digital data.

CCD analog signal processing applications include use as video and audio delay lines and as communications and secure communications filters.

Linear imaging devices (LIDs) are formed by pairing linear arrays of photodetectors with CCDs that are used as transport registers. Free electron packets are generated on individual photodetectors in direct proportion to the incident radiation on the chip. If the incident radiation pattern is a focused light image from optics viewing a scene, the free electron charge packets created in the detector array will faithfully reproduce the scene projected on its surface.

After an appropriate exposure time, the charge packets in each photodetector of the array can be simultaneously transferred by charge coupling to a parallel CCD analog transport shift register. The transfer is carried out by a single, long transfer-gate electrode between the line of photodetectors and the transport register. Each charge packet corresponds to a *picture element* (pixel). The CCD analog transport register is then rapidly clocked to deliver the picture information, in serial format, to the device output circuit. LIDs sense and deliver information a line at a time and are electronically scanned in one dimension. Thus, LIDs are also called *line-scan devices*.

Area imaging devices (AIDs) are two-dimensional X-Y arrays of photodetecting elements capable of sensing an area image. They have both vertical (Y axis) and

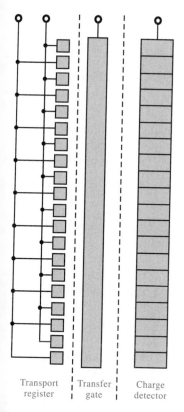

Transport   Transfer   Charge
register   gate   detector

FIGURE 12-3 Simplified schematic of a linear CCD image sensor

horizontal (X axis) transfer gates and transport registers. An entire field of video information can be delivered after each exposure period as a series of lines of video signal.

Charge Coupled Array Camera Charge coupled device cameras contain arrays of evenly spaced, solid-state photosensitive elements, as shown in the simplified diagram in Figure 12-3. Light striking these elements creates a charge that is scanned out sequentially cell by cell. The camera output is a sequence of voltage levels representing individual cell brightness.

A subsystem block diagram in Figure 12-4 shows the major units of a CCD camera. One application for CCD cameras is in precision, noncontact industrial optical inspection of objects and optical data acquisition. The application includes measurement of position, size, and shape of objects. The cameras can also be used for detection and categorization of defects in objects, or for the sorting of objects by size, shape, and color.

A variety of CCD cameras are available. Some are designed for specific applications while others can provide a combination of applications. For example, one B/W CCD camera unit, with a 1/3-inch pickup tube, features 380 lines of resolution for high picture quality. It has a built-in isolation transformer making it ideal for multicamera systems applications. It has an Electronic Sensitivity Control (ESC) for automatic control of shutter speeds (from 1/60 to 1/50,000 second depending on illumination). Minimum illumination for this device is 0.3 lux at a lens setting of f1.2.

Another device is a 1/2-inch format color camera using a single 410,000 pixel CCD. Designed for low light observation, this unit can be used in light levels as low as 0.05 lux, which is 60 times the sensitivity of standard single CCD color cameras. This outstanding sensitivity is produced by controlling the CCD exposure time using field memory for extended integration of the CCD and standard high video gain. Automatic sensitivity switching, auto iris, and real-time auto-white balance maintain color balance and video level. These features allow unattended operation even in locations with

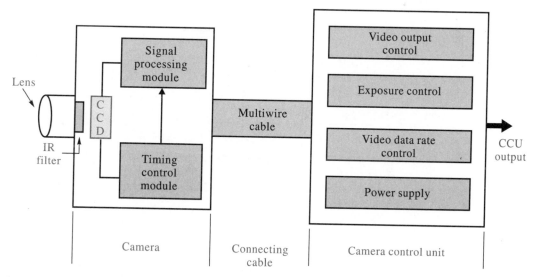

FIGURE 12-4 Functional block diagram of a CCD camera subsystem

extreme changes in color temperature and/or illumination levels. Another CCD camera, designed for security applications, looks like a smoke detector. Its circular housing hides the camera element from view. This high-resolution color image device has backlighting compensation for use under varying light conditions.

Most television stations today use some form of CCD TV camera. These CCD cameras provide standard TV output signals for display of images on monitors or for digital analysis image processing equipment.

The CCD camera typically contains signal processing and timing control modules. The camera is connected by cable to a control unit consisting of a video output control, exposure control, video data rate control, and power supply. The output can be interlaced for compatibility with conventional video equipment.

Deflection and Focusing

Although most modern camera systems use pickup chips instead of tubes, some of the older pickup tube cameras are still being used. With that in mind, this section on deflection and focusing provides the background needed to work on those older systems.

There are two methods by which the beam in a camera tube or picture tube can be deflected and focused: electrostatic and electromagnetic. These methods depend on the properties displayed by the charged electrons moving within electrostatic or electromagnetic fields.

Electrostatic Deflection *Electrostatic deflection* utilizes characteristics exhibited by electrons when placed between two conductive plates that have electrical charges of opposite polarity impressed across them. The electron beam can be deflected up or down by two *vertical deflection plates* in the neck of the tube, as shown in Figure 12-5. How much the beam deflects is determined by the magnitude of the voltage on the deflection plates. For example, in Figure 12-5 the beam will be deflected upward a distance proportional to the voltage applied. Conversely, if the polarity of the voltages

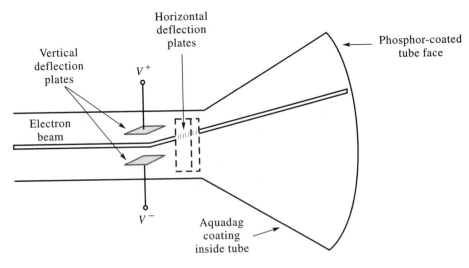

FIGURE 12-5 Electrostatic deflection

were reversed and the magnitude of the voltages remained the same, the beam would be deflected downward a like distance.

A tube also has a pair of *horizontal deflection plates* (shown by dashed lines in Figure 12-5), at right angles to the vertical deflection plates. This pair of plates is closer to the screen, so they will deflect the beam less than will the vertical pair; therefore, the horizontal deflection plates have a lower sensitivity. Voltages applied to the horizontal plates deflect the electron beam horizontally.

Because the deflection plates are within the tube envelope, some vidicon tubes use electrostatic deflection to good advantage. However, electrostatic deflection has a disadvantage in that the electron beam tends to become defocused as the angle of deflection increases, and correction of this tendency is complex and costly. For that reason, electrostatic deflection is used less often in TV systems than electromagnetic deflection.

Electromagnetic Deflection   Practically all TV cameras and display devices use magnetic energy as the principal means of electron beam deflection. A direct similarity exists between a current-carrying conductor located in a magnetic field and an electron beam projected through a magnetic field. In both instances, the electron current is at right angles to the direction of its own magnetic field, and we can assume that it is at right angles to the externally produced field. An analogy can be drawn between the current-carrying conductor in the magnetic field of an electric motor and the electron beam in the magnetic field of a camera tube or picture tube.

Figure 12-6 illustrates how the principle of the electric motor applies to the deflection of the electron beam. In the electric motor, an external magnetic field is produced by current passing through coils wound on a core of magnetic material. In the TV camera pickup tube or the TV receiver picture tube, the external magnetic field is produced by suitable currents passing through deflection coils, only one of which is shown here.

Examining the directions of the lines of force in the two fields, we see that in each case the lines of force aid each other above the current and oppose each other below the current. Where they aid each other, the result is a strengthened field; where they oppose each other, the field is weakened. In both cases, the stronger field is above the

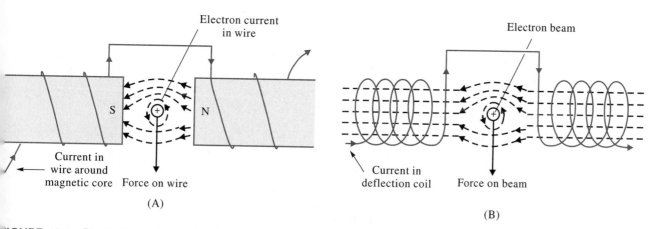

IGURE 12-6   Similarities between (A) An electric motor and (B) An electron beam

current, and the weaker field is below. When a current-carrying conductor or an electron beam is immersed in a nonuniform magnetic field, the field exerts a force that moves both the conductor and the beam from the area of the stronger field toward the area of the weaker field. As shown, both the conductor and the beam are forced, or deflected, in a downward direction. Note that in both cases the deflection is at right angles to the magnetic field through which the conductor and beam move, and not toward either magnetic pole. In the case of the electron beam, we refer to this force as *electromagnetic deflection*.

Two horizontal and two vertical coils form a *deflection yoke*. Figure 12-7 shows the placement of the coils around the neck of the tube. Note that the horizontal coil is placed vertical to the neck, and the vertical coil horizontal to the neck. This placement is dictated by the fact that the lines of force of the magnetic fields are at right angles to the direction of the winding turns.

By using both horizontal and vertical deflection coils, we can move the beam to any position on the tube face by applying the proper polarity and current strength to the two sets of coils. Deflection of the electron beam is the result of one force acting on the beam in a horizontal direction and another force acting on the beam in the vertical direction at the same instant. When current flows through the horizontal coils, a vertical magnetic field is produced across the neck of the tube. When current flows through the vertical coils, a horizontal magnetic field is produced across the neck of the tube. In both cases, the amount of deflection depends on the strength of the magnetic field, and the direction of deflection depends on the polarity of the field.

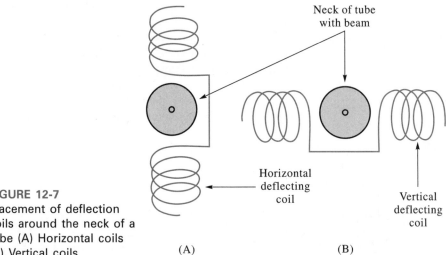

FIGURE 12-7
Placement of deflection coils around the neck of a tube (A) Horizontal coils (B) Vertical coils

Neck of tube with beam

Horizontal deflecting coil

Vertical deflecting coil

(A)   (B)

Beam Focusing   The electron beam in a camera tube or picture tube can be focused electrostatically or electromagnetically. With electromagnetic focusing, as shown in Figure 12-8, a wire carrying a steady current is coiled around the neck of the tube. The lines of force of the magnetic field produced are parallel to the axis of the neck of the tube. Electrons in the beam converge when they pass through the field of parallel magnetic lines. The strength of the field is determined by the current flowing through

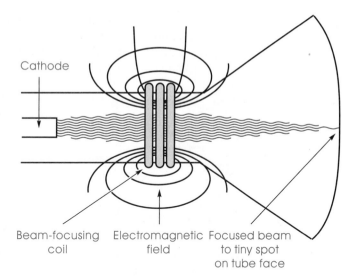

FIGURE 12-8
Position of an electromagnetic focusing coil around the neck of a tube

Cathode

Beam-focusing coil

Electromagnetic field

Focused beam to tiny spot on tube face

the coil, and this field strength determines where the electrons in the beam will come to a point of focus.

Synchronizing the System

To generate a meaningful presentation on the *raster* (lighted screen) of a monitor, some means is needed to synchronize the scanning systems of both the camera and the monitor. A specialized generator achieves this task.

Synchronizing Generator   The scanning system in a monitor or receiver must reproduce exactly the sequence of events that takes place in the camera scanning system. To accomplish this, it is necessary that the camera and monitor/receiver be synchronized in a precise manner. Horizontal and vertical scan and retrace in both camera and monitor/receiver must occur at identical intervals with respect to the video information to obtain a useful presentation. The timing for this action is controlled by a complex piece of equipment called a *synchronizing pulse (sync) generator*.

Another important function of a sync generator is to maintain a rigid phase relationship between the horizontal and vertical scanning systems to ensure the stable 2:1 interlace required for broadcast television. To achieve this, countdown circuitry is generally incorporated to provide a vertical drive signal that is an exact submultiple of the horizontal frequency.

The sync generator should not be considered a part of the deflection system. It serves as a source of accurate timing signals that are used to trigger or drive the deflection circuits in the camera and the monitor/receiver, and it provides horizontal and vertical blanking signals to be inserted onto the video waveform. The sync pulses and the sync waveform are discussed in Section 12-3.

Many cameras have sync generators built into them. Such units may vary considerably in configuration, and some CCTV cameras may generate only a rudimentary signal for use at the monitor. However, when a stable picture that equals or exceeds broadcast quality is desired, it becomes necessary to use a somewhat more sophisticated device, either within the camera or as an accessory unit. Such a system develops signals necessary to drive the deflection systems in the camera and, at the same time, inserts a waveform onto the video output signal that will cause the monitor scanning system to lock to the camera.

Multicamera Synchronization   When several cameras are to be used, it is desirable to have them all synchronized by a single sync generator. This arrangement offers the following advantages.

1. It is often more economical because a separate sync generator is not required for each camera.
2. All cameras are scanning "in time" with one another instead of operating with individually generated sync signals.

Broadcasting applications provide a good example of synchronous multicamera systems. In these systems, when the scene is shifted from one camera to another, the sync waveforms are in phase, so the monitor or home receiver is not interrupted in its scanning process. But if each camera had its own sync system and the scene was switched from one camera to another, the monitor or receiver would have to readjust its scanning procedure for each camera, and the picture might "roll" momentarily.

One way to drive multiple cameras from a single sync generator is the *loop-through* method illustrated in Figure 12-9. (Another way to drive multiple cameras is through distribution amplifiers, to be discussed later.) The outputs from the sync generator go directly to camera 1, then are connected to all other cameras by looping through to each succeeding camera. In each camera, the sync waveforms are sampled by circuits that have such high impedance that they do not affect the waveform. Each camera then uses this sample waveform to develop drive signals for the deflection oscillators. When the sync signal is looped out of the last camera, it is terminated by a resistor

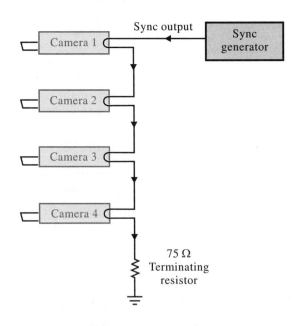

FIGURE 12-9
Loop-through
synchronization
system

connected to ground. The terminating resistor is generally a 75 Ω resistor because the output of most TV camera equipment is designed to work into a 75 Ω load. In all cases, the termination must occur at the extreme end of the line or the waveforms may become severely distorted.

Monitors

Common terminology associates the term *monitor* with a closed circuit or studio TV display device that contains no provisions for receiving broadcast signals; that is, the device contains no RF modulator. Monitors rely on a direct input of unmodulated video signals. Although a home receiver may also be called a monitor, monitors are usually associated with CCTV, and receivers with broadcast television.

The home receiver is a familiar part of a TV system. Although the home receiver is seldom thought of as a component of the system, in a very real sense it is the most important part of the largest, most complicated, and most sophisticated TV system in the world. Television receivers in the United States are all designed to operate at a fixed scan ratio and generally have the same approximate bandwidth and resolution capabilities. On the other hand, TV monitors often have different scan rates and widely varying bandwidth and resolution capabilities. Because closed circuit monitors are wired directly to the camera systems, the bandwidth need not be limited to broadcast regulations. Many TV monitors have bandwidths of 30 MHz or greater.

Several monitors may be used to display the scene being viewed by a single camera. Figure 12-10 shows how the video signal can be looped through monitors, just as in sync distribution, with the last monitor in the series terminated in a 75 Ω resistor. Most monitors are equipped with a termination switch that provides a 75 Ω termination of the input video line in one position and leaves the input line unterminated in the other position.

Distribution Amplifiers

If monitors are to be operated at considerable distances from one another, such as in different rooms or different buildings, it becomes impractical to connect them by looping. In these cases, *distribution amplifiers* (DAs) are used to route the video properly.

A DA converts a single signal line to multiple distribution paths without excessively loading or distorting the original signal. A four-output DA being used to drive four terminated monitors is shown in Figure 12-11. The input to the DA is terminated in 75 Ω to provide proper loading for the camera output. Looping inputs can be used with DAs to allow the addition of more amplifiers to provide as many feed lines as desired. As in all cases, the last unit must be terminated.

All outputs from a DA must be properly terminated to provide correct operation. All outputs are essentially identical to the input, so any of them can be used to drive a series string of monitors, as shown in Figure 12-11, where monitors 4–7 are driven from one output.

Distribution amplifiers are classified into two general categories: (1) *video distribution amplifiers* (VDAs), used to route the camera output signals; and (2) *pulse distribution amplifiers* (PDAs), used to route signals from sync generators and similar equipment that produces pulse waveforms having fast rise and fall times. In general,

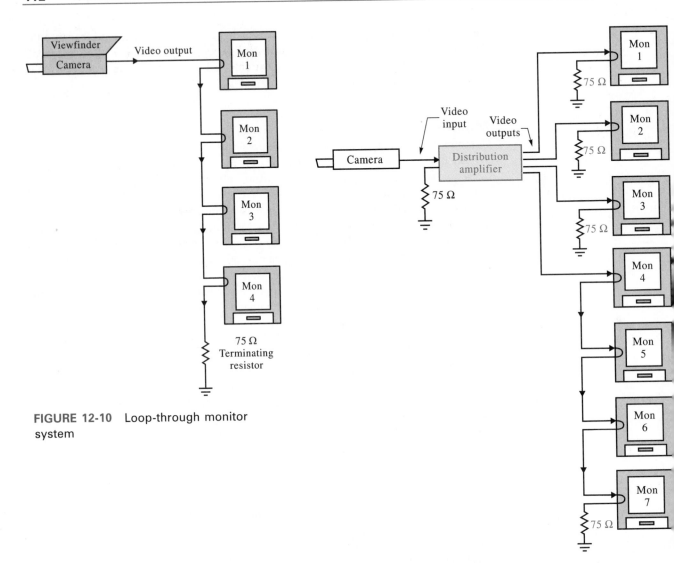

FIGURE 12-10 Loop-through monitor system

FIGURE 12-11 Four-output distribution amplifier

DAs are used only when there are many different using destinations or when the using destinations are too far apart to allow series looping connections.

Video Switchers

It is often desirable to provide a means for viewing the output of several cameras on one monitor. This is accomplished with a *video switcher*. A simple video switching circuit is shown in Figure 12-12(A), where the input to the monitor is switched from one camera to the other with a simple toggle switch. Note that proper terminating resistors for the cameras are used whether or not the camera is being viewed. Simple switchers such as this have the advantages of being very small and inexpensive, requiring no power, and being extremely reliable and simple to maintain.

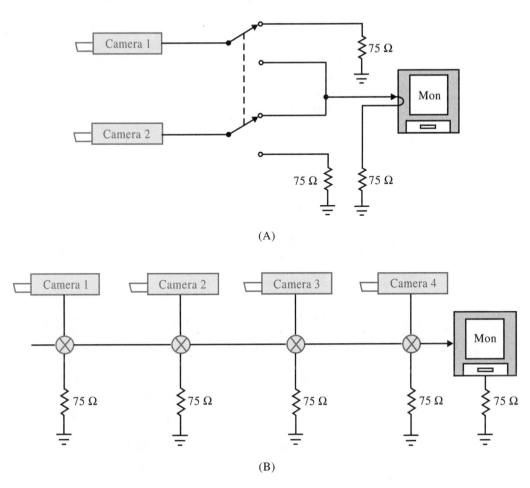

FIGURE 12-12 Video switching systems (A) Simple toggle switch system (B) Industrial-type pushbutton crosspoint system

From an operator viewpoint, pushbutton-operated video switchers are more desirable than toggle-operated video switchers. A simple industrial type switching system is shown in Figure 12-12(B). The circled crosspoints indicate pushbutton switches that, when depressed, connect the associated camera to the monitor. Switches not depressed connect terminating resistors to the remaining cameras. During operation, all switches are interlocked so that only one camera can be connected to the monitor at any time. The switcher is constructed in such a manner that depressing one switch releases all others. Therefore, the outputs of various cameras can be switched onto the monitor by simply depressing the proper switches in a desired sequence.

Remote Controls

Many types of remote controls are associated with TV equipment. For example, the elements of camera lenses can be motor driven to provide focus or zoom control from a remote location. Or it may be desirable to remotely control various camera functions even if the camera is inherently self-contained. Video operators (operators of CCUs) may wish to adjust camera parameters such as video gain, camera sensitivity, blanking level, video polarity, and the like. This adjustment capability is especially important in an educational or broadcast situation where the amplitude of the video signal must be maintained within very close tolerances.

In many instances, especially in closed circuit applications, it is desirable to be able to remotely pan (horizontal rotation) and tilt (vertical action) the camera to view different sections of a scene. Pan-tilt units typically provide up to a 360° rotational capability and allow ±90° tilting action.

Other Equipment

Much additional equipment may be used as part of, or in conjunction with, the basic TV system. Such items include lighting systems, microphones, intercom systems, audio amplifying and recording systems, distribution systems, video tape recorders, microwave relay systems, and RF modulators for displaying information on standard VHF/UHF TV receivers. Most of this equipment will be found in a TV broadcast transmitting station, discussed in the next section.

12-2   TELEVISION BROADCAST TRANSMITTING STATION

Broadcasting is the most familiar use of television. The millions of TV receivers in use around the world attest to its popularity. Broadcast television has been credited with changing the social patterns of a large segment of the human race. Certainly, television influences many people's social behavior, buying habits, political views, and the like.

Transmitting Station Fundamentals

The TV camera and associated equipment used in TV broadcast stations are usually sophisticated and expensive. The broadcaster must be concerned not only about the technical capabilities of these systems but, since the systems are used as an art form, also about presenting a picture whose subtle qualities make it pleasing to the viewer. For this reason, TV stations require extensive lighting facilities and equipment with special effects capabilities to add polish and glamour to the televised image.

Television broadcasters in the United States must use equipment that meets standards set by the FCC. The video signals generated and broadcast must not deviate from specified tolerances, and they must be monitored continuously for conformance. Such requirements, plus the broadcaster's inherent commercial interests, result in the use of equipment that is much more elaborate and expensive than that used for CCTV applications.

The TV cameras used in broadcast studios are relatively large compared to most closed circuit cameras. The greater physical size of studio cameras stems from the specialized circuitry not generally required in closed circuit cameras, and the need for high physical stability.

The advent of color television created an additional dimension of realism. Because of the great commercial success of color television, *monochrome* (black-and-white, or B/W) television for commercial purposes has become rare. The changeover from B/W to color resulted in increased complexity and cost for TV stations.

The wide variety of equipment used to support the TV camera in a studio may take on many configurations. Additional items—such as video tape recorders (VTRs), telecine cameras for display of films and slides, lighting fixtures, audio systems, microwave transmitters and receivers, test equipment, monitors, and switching systems—crowd broadcast studios and control rooms. Items such as the video switcher, which is used to route the various video signals to different locations, are generally quite complex and versatile and have the capacity to provide special effects and professional production techniques. The following section will discuss the equipment normally found in the various sections of a typical TV broadcast station.

Although broadcast television is generally considered high in quality, it is limited because of restricted bandwidth. The bandwidth of the channel allocations in the United States is determined by the FCC, and all broadcasters must conform to FCC requirements. On the other hand, CCTV is not broadcast and therefore may use much wider bandwidths to produce better resolution.

Transmitting Station Equipment

A TV transmitting system involves many of the radio circuits and systems discussed in earlier chapters. Every TV broadcast consists of two separate signals: an aural signal and a visual signal. Each of these signals has its own pickup device, signal processor, and transmitter. The simplified block diagram in Figure 12-13 illustrates a typical TV broadcast transmitting station, which includes a studio, a control room, a video tape and telecine room, a terminal equipment room, and a transmitter room. In Section 12-4, we will discuss additional requirements for color TV systems.

Studio   Television studios vary in size according to application. Some are small, some large. Many have ceiling heights of 30 or 40 feet, with a railed catwalk installed about 20 feet above the studio floor to provide an area for lighting technicians to operate. A maze of pipes from which hang an assortment of lights crisscross the ceiling. Scattered around the room are various sets (scene mock-ups); studio cameras mounted on wheeled tripods (dollies); and microphones, some of which may be boom mounted. Because the microphones may pick up unwanted sounds, many are directional; that is, they pick up sound only from the direction in which they are aimed. To prevent unwanted sounds, a floor director uses hand signals, printed signs, and intercom systems to direct the action. The studio itself is soundproofed to prevent exterior sounds from interfering with studio activities.

The pickup devices in the camera are the transducers for the visual portion of the TV signal. The microphones are the transducers for the aural portion of the TV signal.

Control Room   The control room can rightfully be referred to as the command post of the TV broadcasting station. Operators in the control room monitor and control the functioning of studio equipment. Located in the control room are (1) the CCUs, one for each studio camera; (2) the video switcher for scene selection; (3) the audio mixer console and switcher for sound intensity control and microphone selection; (4) the video monitors, usually one monitor for each camera plus a preview monitor and a line monitor; and (5) the audio monitor speaker. Some control room systems include audio recording equipment.

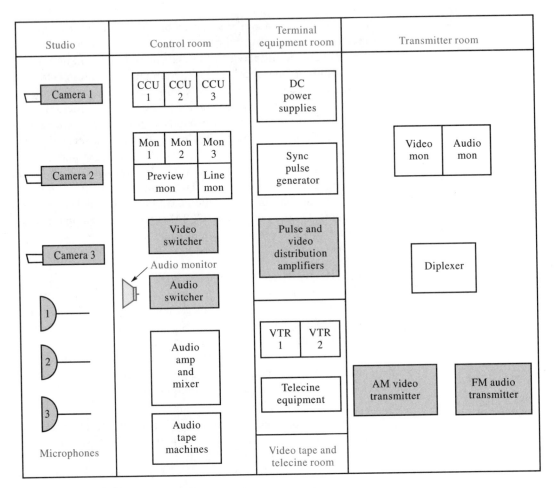

FIGURE 12-13 Typical TV broadcast transmitting station

Video Tape Room Few "live" programs (other than newscasts) are produced today. Instead, almost all TV programs are recorded on video tape before transmission. To maintain top quality, broadcasters employ *video tape recorders* (VTRs) which use high-quality video tape. The VTRs record both video and audio on magnetic tape. The video tape room will normally contain several VTRs so that not only the original program can be recorded, but editing of the program material and commercial inserts also can be made. *Telecine* equipment (film/slide projectors and an optical multiplexer) is often installed in the video tape room, although it may be installed in a separate room.

Terminal Equipment Room *Terminal equipment* includes dc power supplies, pulse-generating equipment, and distribution equipment. In many installations, this equipment is mounted in racks 19 inches wide and 7 feet high. The dc power for all studio, control room, video tape room, and terminal room equipment originates here. The pulses necessary for proper system operation and the amplifiers necessary for proper signal distribution are located here. In some stations, the terminal equipment may be located in the video tape room or in the telecine room.

Transmitter Room The transmitter room contains an AM visual transmitter, an FM aural transmitter, video and audio monitors, and a diplexer. The audio signal is broadcast by an FM transmitter that is essentially the same as the transmitter used for FM radio broadcasting. The major difference is that 100% modulation in the TV audio transmitter is represented by a carrier deviation of ±25 kHz instead of the ±75 kHz deviation required for FM radio broadcast transmitters. The amplifiers still require the 30–15,000 Hz audio capability. Sound signals picked up by the microphone are amplified, monitored, and fed to an FM modulator in the aural transmitter. The FM transmitter signal is monitored and fed to a single high-gain, horizontally polarized antenna through a diplexer, which prevents any FM audio signal from being coupled into the AM visual transmitter.

The video signal is the output of the photosensitive camera pickup device that converts light into an electrical signal, from a video tape machine or from telecine equipment. Since light levels in a visual scene vary over a wide range and at a rapid rate, frequencies range from a few hertz to about 4.5 MHz. The video signals are monitored, synchronizing pulses are added to them, and then the signals are fed to the visual transmitter, where they are amplitude modulated. A *vestigial sideband* filter attenuates most of the lower sideband (leaving only a *vestige* of it) in order to reduce the bandwidth. The signal is then fed to the antenna through the diplexer, which prevents any AM video signal from being coupled into the FM aural transmitter.

Diplexer When more than one receiver or transmitter is connected to a single antenna, the system is called a multiplex or diplex circuit. The *diplexer* allows two transmitters or receivers to be operated with the same antenna at the same time.

The diplexer must pass the signals from the video transmitter and the audio transmitter to the same antenna but prevent either signal from coupling to the other transmitter. A balanced-bridge circuit can accomplish this task. As shown in Figure 12-14, the video signal is coupled through transformer T_1 directly to the center feed points of the antenna

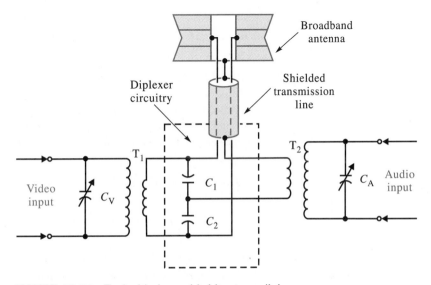

FIGURE 12-14 Typical balanced-bridge type diplexer

through a shielded two-wire transmission line. The shield acts as a center tap of the two wires. The two equal capacitors across the line (C_1 and C_2) also provide a center tap. The audio signal is coupled between the two center taps. Because of the balance of the system, feed-through from transmitter to transmitter is prevented, but the antenna is excited by both transmitters, permitting radiation of both signals.

Transmitting Antenna   Because TV transmissions are in the VHF or UHF ranges, the best site for the transmitting antenna is on a hilltop near the area being served. The transmissions are usually horizontally polarized to reduce *ghosts* (reflected signals). The FCC-allocated channel width for the TV transmitted signal is 6 MHz, thus requiring a broadband antenna. Any beam-type antenna may be used, such as the *turnstile* or the *slotted-cylinder*. Transmission lines for VHF signals can be balanced, shielded two-wire types, or unbalanced coaxial. For UHF signals, waveguide can be used.

12-3   TELEVISION BROADCAST STANDARDS _____

Standardization of all TV broadcast signals is necessary so that any TV receiver anywhere in the country can receive the signals without undue distortion and interference. For that reason, all broadcasters in the United States must use equipment that meets the standards set by the FCC. In the following sections, we will examine the FCC-established TV broadcast standards for the United States.

Synchronizing Standards
Broadcast television requires a sophisticated method of synchronization for several reasons:

1. To provide a stable 2:1 interlace.
2. To provide a precise method of driving the receiver circuits.
3. To generate accurate blanking information to ensure that retrace lines are not visible.

The timing of the TV transmitted lines is controlled by the sync generator. Figure 12-15 illustrates the sync signal for monochrome transmissions as it appears during the vertical retrace portion of the scanning interval. The waveform shows the vertical blanking interval of the composite video waveform and displays the various pulses present. The designation, frequency, and approximate duration for the various pulses are as follows:

1. *Equalizing*—31,500 Hz; 2.7 μs each, with the leading edges separated by 31.75 μs.
2. *Horizontal sync*—15,750 Hz; 5.4 μs, with the leading edges separated by 63.5 μs.
3. *Horizontal blanking*—15,750 Hz; 10 μs.
4. *Vertical sync*—60 Hz; total of 190 μs, but slotted by 31,500 Hz *serration* pulses of 4.4 μs, with the leading edges separated by 27.3 μs.

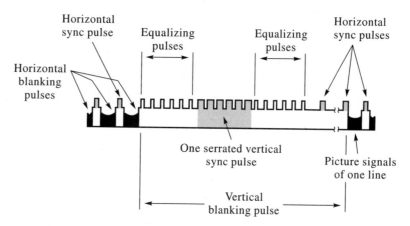

FIGURE 12-15 Synchronizing waveform for monochrome television

5. *Vertical blanking*—60 Hz; 830–1330 μs, depending on how many lines are used in the complete picture.

6. *Driving*—horizontal and vertical; similar to the transmitted sync pulses but slightly out of time to properly synchronize the cameras and equipment in the studio.

As shown, the horizontal sync pulses ride on top of the horizontal blanking pulses. The modulated visual scene is inserted between the 10 ms horizontal blanking pulses, with one full horizontal line between the trailing edges of two successive horizontal sync pulses.

The wide vertical blanking interval occurs at the end of each field. The equalizing pulses and serrations are used to keep the receiver's horizontal oscillator in synchronization during the long vertical blanking interval. An LP filter separates the vertical sync pulse and uses it to synchronize the receiver's vertical oscillator.

A blanking pulse occurs just before the horizontal sync pulse is produced between lines. This pulse cuts off the electron beam in the camera tube of the transmitter and in the picture tube of the receivers. During the blanking pulse interval, horizontal oscillators return the electron beam to the starting position for the beginning of the next line. In receivers, this sweep circuit is produced by a local sawtooth oscillator. The equalizing, vertical, and horizontal sync pulses are not used in the picture pickup of the camera. These sync pulses, transmitted as part of the complete TV signal, are used only to make necessary corrections to the receiver's local sawtooth oscillator frequency to ensure the proper starting time of each line at the end of the blanking pulse.

A block diagram of a basic sync generator is shown in Figure 12-16. The device consists of a master oscillator that generates a signal frequency of 63 kHz, which is fed into a 2:1 digital divider to generate narrow 31.5 kHz pulses. These pulses are then shaped to produce the equalizing pulses used before and after the vertical sync pulse and to serrate the vertical sync pulse into its six sections. The 31.5 kHz equalizing pulses are fed (1) into the sync developing circuits; (2) into divider and gating circuits consisting of consecutive 7:1, 5:1, 5:1, and 3:1 dividers (a total of 525:1), which produces a 60 Hz drive signal and an output to the sync developing circuits; and

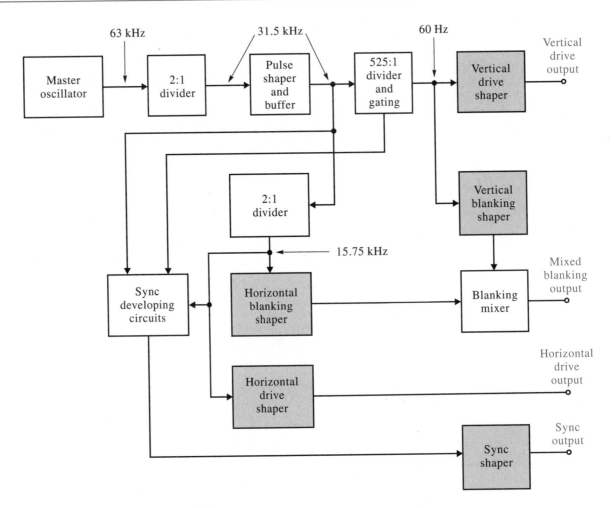

FIGURE 12-16 Basic synchronizing generator

(3) into another 2:1 divider, which produces the 15.75 kHz horizontal sync pulses. The 60 Hz frequency is shaped to provide the vertical drive output. The 15.75 kHz pulses are shaped to provide the horizontal drive output. The vertical and horizontal signals feed through blanking shapers and are mixed to provide a mixed blanking output. The 31.5 kHz equalizing pulses, the 15.75 kHz horizontal pulses, and the 60 Hz vertical pulses are mixed in the sync developing circuits, then shaped to produce the sync output.

Each piece of equipment at the transmitter has its own local sawtooth sweep oscillator. The drive pulses of the sync generator are used to keep these oscillators properly synchronized.

Most commercial sync generators have a crystal oscillator and dividers to produce an accurate 60 Hz ac for operation where local power is not available or when the equipment is to be operated in the field. In the latter case, pulses for the sync generator at the main station transmitter are developed by the remote sync generator. A special input circuit of the sync generator at the main station allows that sync generator to lock in on an external TV signal, such as that provided by a network program, by a remote sync generator, or by field equipment. Digital memory devices called *frame synchronizers* store two complete frames in memory, allowing a shift from one program source to another without loss of vertical sync when the programs are operating from different sync sources.

Two additional timing signals are needed for broadcast TV systems originating compatible color programs: (1) a 3.58 MHz subcarrier frequency and (2) a color burst keying pulse (about eight cycles of the 3.58 MHz subcarrier signal) occurring on the back porch of the horizontal blanking pulse. These additional sync pulses may be derived from circuits used as auxiliaries to standard monochrome sync generators, but modern sync generators incorporate directly the extra circuits needed for color transmission. These signals are discussed in Section 12-4.

Scanning Systems

The deflection process positions the electron beam on the surface to be scanned. The surface may be the target of the IO, the mosaic of the vidicon, or the screen of the TV receiver picture tube. In any case, the beam describes a particular pattern of motion (the *scanning pattern*) across the surface. In line with standards of TV broadcasting employed throughout the world, the pattern used is *interlaced scanning*. This term denotes the way in which the televised image is reconstructed on the screen of the TV receiver, a process that reverses the image's division into parts in the TV camera and its subsequent transmission to the receiver as an electrical signal.

The Scanning Process   We can understand the scanning process used in television by relating it to how we read a printed page. We read the printed page one line at a time, beginning at the left-hand edge and moving toward the right. Having reached the end of the first line, the eye sweeps back across the page and down to the beginning of the next line at the left side, and the horizontal scan is repeated. Upon reaching the bottom of the page, the eye moves rapidly to the left edge of the top line of the next page. This familiar operation is sometimes referred to as scanning, and it closely parallels the method used in television.

What we have described is *progressive* scanning—reading one line after another. However, there are two intervals when no actual reading is done: (1) while the eye is moving from the end of a line to the beginning of the next line and (2) while the eye is moving from the bottom of the page to the top of the next page. These no-reading intervals also occur in electronic scanning.

All TV systems use *rectilinear* scanning, in which two separate scanning procedures occur simultaneously, one moving the beam horizontally and the other moving the beam vertically. Both scans are linear; that is, the movement of the beam, both horizontally and vertically, occurs at a constant rate of speed.

A simplified illustration of the beam path over the photosensitive surface of a camera pickup tube is shown in Figure 12-17. In a TV system, the cathode ray picture tube (CRT) in the monitor or receiver reproduces exactly the beam excursions shown. Note that the beam travels in horizontal and vertical directions simultaneously.

The dashed lines represent *beam retrace*, which occurs very rapidly with respect to normal scan and is usually *blanked* by disabling the beam in the CRT. Consequently,

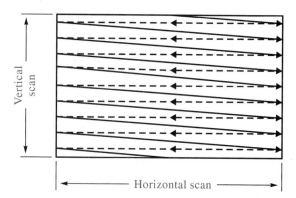

FIGURE 12-17
Simplified scanning
presentation

the CRT develops no output signal during the retrace interval, and visual evidence of the event is eliminated.

Because the horizontal movement of the beam occurs at a much more rapid rate than the vertical movement, we can state that the horizontal occurs at a higher *scanning frequency* than does the vertical. The relationship between the magnitudes of these two frequencies determines the number of horizontal lines that will exist in each scanned pattern, or raster. The greater the number of horizontal lines, the less individually visible they become and the greater the detail in the scene being viewed.

Flicker   A relatively low vertical scan is desirable because it allows the beam to scan a large number of horizontal lines without requiring an unnecessarily high horizontal scanning frequency. However, if the vertical scan rate is too low, the visual presentation will have an objectionable *flicker*, making the scene uncomfortable to watch.

Flicker varies considerably with viewing tube illumination levels, viewing angle, and factors other than frequency. Sensitivity to the effect of flicker also varies from individual to individual. The vertical scanning frequency must be maintained sufficiently above the rate discernible to the average human viewer. For these reasons, a vertical scanning frequency of 60 Hz was chosen.

The 60 Hz vertical scanning frequency standard was selected for several practical reasons. It is the same as the 60 Hz power line voltage supplied by commercial power companies and used by most line-operated electronic equipment in the United States. This relationship reduces the possible effects of equipment power supply ripple and 60 Hz magnetic fields in the final presentation on the monitor or receiver. It also makes the line voltage a convenient source for synchronizing the vertical deflection frequency in simple scanning systems.

Interface and the Standard Raster   A horizontal scanning frequency (*line rate*) of 15,750 Hz is used for broadcast television. There are $262\frac{1}{2}$ horizontal scan lines produced for each 60 Hz vertical deflection period. The raster produced by a single vertical scan is known as a *field*, and the vertical deflection frequency is referred to as the *field rate*.

Since the $262\frac{1}{2}$ lines per field does not constitute a whole number, we can assume that superimposing one field on a succeeding field will not result in an exact mating of horizontal lines. Actually, the number of lines dictates that each field begin $\frac{1}{2}$ line offset with respect to the preceding field. This offset causes an interweaving, or *inter-lacing*, of horizontal lines on each alternate field. As a consequence, the horizontal lines of each of two succeeding fields fall directly between each other, and the total number of horizontal lines is effectively doubled to 525, the standard for broadcast television in the United States.

Two interlaced fields make up one *frame* with a *frame rate* of 60/2, or 30 Hz. The frame constitutes a complete TV picture element, but its low repetition rate does not generate flicker because of the two vertical scans required to complete it and because of a characteristic of the human eye called *persistence of vision.*

The concepts of the preceding discussion describe the 2:1 interlaced standard scan ratio and are summarized mathematically as follows:

line rate = 15,750 (U.S. standard)

field rate = 60 Hz (U.S. standard)

$$\text{frame rate} = \frac{\text{field rate}}{2} = \frac{60}{2} = 30 \qquad (12\text{-}1)$$

$$\text{lines per field} = \frac{\text{line rate}}{\text{field rate}} = \frac{15,750}{60} = 262.5 \qquad (12\text{-}2)$$

$$\text{lines per frame} = \frac{\text{line rate}}{\text{frame rate}} = \frac{15,750}{30} = 525 \qquad (12\text{-}3)$$

In practice, it is not possible to utilize all of the 525 lines of scan generated during each frame. Because of vertical retrace action and its accompanying blanking, or beam cutoff, about 7–8% of such lines are rendered useless. The number of lines lost in blanking can be approximated by multiplying the blanking time (0.07) by the number of lines (525) in the system:

lines lost = (blanking time) (lines in system)

$$= (0.07) (525) = 37 \qquad (12\text{-}4)$$

If we use the 8% figure, we obtain

lines lost = (0.08) (525) = 42

The number of *active* (useful) scanning lines is determined by subtracting the lines lost in blanking from the number of lines in the system:

active lines = lines in system − lines lost

$$= 525 - 37 = 488 \qquad (12\text{-}5)$$

or

active lines = 525 − 42 = 483

These examples show that about 37 to 42 horizontal lines are lost per frame, or 18 to 21 horizontal lines per field, leaving between 484 and 488 useful scanning lines. These are the limits of the tolerance set forth by standardized specifications in the United States.

The details of a raster produced by the 525-line scanning pattern are shown in Figure 12-18. Field 1 begins in the upper left-hand corner and scans across the face of the tube. It is then deflected back across the tube and starts another horizontal line. A vertical deflection signal causes each scanned line to be slightly lower on the CRT and, because of the shape and the amplitude of the vertical deflecting voltage, the horizontal excursions of the beam traverse only the odd-numbered lines in field 1. After $262\frac{1}{2}$ lines are scanned, the vertical deflection signal drives the electron beam back to the top of the picture tube for the start of field 2. Field 2 repeats this process, except that its horizontal lines fall on the even-numbered lines, exactly between the odd-numbered lines of field 1. Even though each frame is displayed as two separate pictures, the human eye allows the viewer to see them as one complete image.

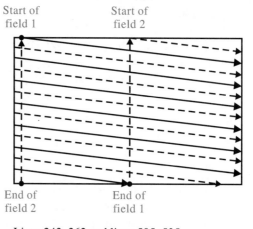

Start of field 1   Start of field 2

End of field 2   End of field 1

FIGURE 12-18
Raster details

Lines 242–262 and lines 505–525 are not part of raster; they are "lost" while beam is shut off during vertical retrace.

Aspect Ratio   The relative width versus the relative height of the generated raster is called the *aspect ratio*. The aspect ratio is 4:3; that is, it is four units wide and three units high, as illustrated in Figure 12-19. The additional width was chosen because most movement occurs in the horizontal plane, producing a slight panoramic effect and resulting in a picture more pleasing to the eye.

Lines of Resolution

The ability of a system to resolve detailed picture elements is termed *resolution*. In the TV medium, the term has a different meaning than in other mediums. In standard photography, for example, four black lines separated by three white lines of equal thickness are defined as four lines of resolution. In television, however, both the blank lines and the white lines are counted, resulting in seven lines of resolution. This difference in definition often causes confusion when resolution capabilities of film and television are being compared. All resolution measurements in television are made with respect to picture height. That is, *lines of resolution* are defined as the number of black and

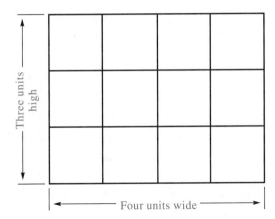

FIGURE 12-19
Standard 4:3
aspect ratio

white lines of equal width that can be contained in the vertical dimension of the TV picture. This definition also holds when we speak of horizontal resolution, even though the horizontal dimension is greater because of the standard 4:3 aspect ratio. Therefore, when we say that a system has a 450-line horizontal resolution capability, we mean that it can resolve resolution elements of such width that 450 of them, placed side by side, would exactly fill the vertical dimension of the TV picture. Vertical and horizontal resolution are calculated differently, as we will see.

Vertical Resolution *Vertical resolution* is the ability of a system to resolve *horizontal* lines in a scene. The vertical resolution capability depends primarily on the number of scan lines used per frame. This number is generally referred to as the *scan ratio* of a TV system. The standard system in the United States has a scan ratio of 525:1, but CCTV uses ratios of 1200:1 or higher. Since the horizontal lines are crowded closer together to achieve the higher ratios of CCTV, vertical resolution will increase accordingly.

There are several limitations to consider in determining the vertical resolution of a particular scan ratio. For computations of a specific ratio, a modifying factor known as the *Kell factor* is used. This is an approximation of utilization and has a value of about 0.7. To approximate the vertical resolution obtainable from a 525-line system, simply multiply the Kell factor (0.7) by the active scanning lines (488) that remain after the vertical blanking action. For standard broadcast television, this results in

$$\text{vertical resolution} = (\text{Kell factor}) \ (\text{active lines})$$

$$= (0.7) \ (488) = 342 \tag{12-6}$$

Most U.S. television equipment manufacturers specify vertical resolution of the standard scan ratio of 350 lines. Since the Kell factor is only an approximation, this 350 lines of vertical resolution is a realistic figure.

EXAMPLE 12-1

A CCTV system employs a 1200:1 scan ratio. Determine (a) the lines lost in blanking (b) the active lines, and (c) the lines of vertical resolution.

SOLUTION

(a) By Equation 12-4,
   lines lost = (0.07) (1200) = 84

(b) By Equation 12-5,
   active lines = 1200 – 84 = 1116

(c) By Equation 12-6,
   lines of vertical resolution = (0.7) (1116) = 781

Horizontal Resolution   *Horizontal resolution* is the ability of a system to resolve *vertical* lines of resolution. Under the assumption that the beam spot in the pickup tube is extremely small, horizontal resolution is determined largely by the bandwidth of the TV system.

Figure 12-20(A) illustrates a series of vertical alternate black and white lines of resolution focused on the target of a pickup tube. The beam horizontally scanning the target produces a series of output pulses, the frequency of which is directly proportional to the speed of the trace and the number of lines the trace crosses. If the video amplifiers in the camera system cannot pass the highest frequency thus generated, the system is said to be *bandwidth limited*, and the horizontal resolution capability will depend on the upper bandwidth limit.

Although calculating horizontal resolution is slightly more involved than determining vertical resolution, simple mathematics still apply. Let us assume a need for a 700-line horizontal resolution capability in a standard 525-line system. A picture element or line of resolution as used in television may be either black or white, so a series of four black lines with four white lines of equal width will represent eight lines of resolution. However, the electrical signal generated by a pickup tube viewing such lines represents an adjoining black and white pair as a single cycle, as shown in Figure 12-20(B), thus producing 350 Hz for a 700-line horizontal resolution capability.

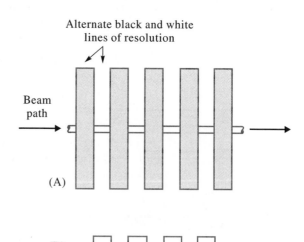

FIGURE 12-20
Horizontal scan of vertical lines of resolution (A) Alternate black and white vertical lines of resolution (B) Output waveform

Remembering that the horizontal frequency of the system is 15,750 Hz, we can determine that one horizontal line is generated in $\frac{1}{15,750}$, or 63.5 μs. However, horizontal blanking time occupies about 17% of this period, so we have about 53 μs of useful time left in the single horizontal line. Also, since the 700 lines of resolution are specified in relation to picture height, a modifying factor of 0.75 must be introduced to compensate for the 4:3 aspect ratio, leaving about 40 μs of actual usable time in the single horizontal line.

Using the information from the preceding paragraph, we can calculate the bandwidth required to present the desired 700 lines of horizontal resolution:

700 lines = 350 cycles of video signal

$$350 \text{ Hz in } 40 \text{ μs} = 350 \times \frac{1}{40 \times 10^{-6}}$$

$$= 8.75 \text{ MHz} \qquad\qquad (12\text{-}7)$$

We can therefore conclude that a bandwidth of 8.75 MHz is necessary to present a 700-line horizontal resolution in the standard 525-line scanning format.

On the other hand, if bandwidth is known, resolution can be calculated. Using the video 4.5 MHz bandwidth of broadcast television in the United States, we have

$$(4.5 \text{ MHz}) \left(\tfrac{1}{40} \text{ μs}\right) = 180 \text{ Hz}$$

$$(180 \text{ Hz}) (2) = 360 \text{ lines of resolution}$$

Now, if we had a 10 MHz bandwidth, we would obtain

$$(10 \text{ MHz}) \left(\tfrac{1}{40} \text{ μs}\right) = 400 \text{ Hz}$$

$$(400 \text{ Hz}) (2) = 800 \text{ lines of resolution}$$

Examination of these calculations shows that the broadcast standard 525-line scanning format cannot produce the 700-line horizontal resolution desired.

Scan Ratio Versus Bandwidth

In TV systems employing scan ratios greater than the standard 525:1, the beam must move faster horizontally. For example, in a 1200:1 scan rate system, the beam must move horizontally more than twice as fast as for the 525:1 ratio. The vertical frequency remains constant at 60 Hz in all cases.

With the beam moving faster horizontally, the frequencies generated by it when it crosses vertical lines of resolution are increased accordingly. Therefore, to achieve a 700-line horizontal resolution capability with a scan ratio greater than 525:1, the bandwidth must be increased.

EXAMPLE 12-2

Determine the bandwidth needed to provide 700 lines of resolution for a 1200:1 CCTV system.

SOLUTION

Determine the horizontal rate.

$$\left(\frac{\text{standard lines}}{\text{line rate (Hz)}}\right)\left(\frac{\text{CCTV lines}}{\text{new line rate (Hz)}}\right)$$

$$= \left(\frac{525}{15,750}\right)\left(\frac{1200}{X}\right) = \frac{18.9 \times 10^6}{525X} = 36 \text{ kHz} \qquad (12\text{-}8)$$

Determine the horizontal interval.

$$\frac{1}{36} \text{ kHz} = 27.8 \text{ } \mu s \qquad (12\text{-}9)$$

Determine the active time less blanking.

$$27.8 \text{ } \mu s - (0.17 \times 27.8 \text{ } \mu s) = 23 \text{ } \mu s \qquad (12\text{-}10)$$

Include the 0.75 modifying factor for aspect ratio.

$$(23 \text{ } \mu s)(0.75) = 17 \text{ } \mu s \qquad (12\text{-}11)$$

The frequency rate for 700 lines is 350 Hz, so

$$(350 \text{ Hz})(\tfrac{1}{17} \text{ } \mu s) = 20.6 \text{ MHz}$$

The preceding examples demonstrate the relationships between resolution capability, scan ratio, and bandwidth. Although it may appear to be advantageous to use extremely wide bandwidths regardless of the scan ratio, excessive bandwidths will amplify noise. For that reason, plus the costs involved, scan ratios and bandwidths should be carefully chosen to fulfill a particular requirement.

12-4   COLOR TELEVISION

Color perception is important to the human eye. Color exerts a great influence on our daily lives. It attracts our attention, increases our perception, and even affects our emotions.

The sensation of color is directly related to the frequency of the light being observed. As the electromagnetic waves that constitute light change in frequency, there is a corresponding change in the color that the eye sees. One frequency will be perceived as blue, another as green, another as red, and so on.

Sunlight contains all the frequencies in the visible spectrum and is considered to be white light. White light is made up of a jumble of frequencies, seemingly without order, yet its various components form the colors seen in everyday life. In the following sections, we will examine the fundamentals of color and how color is developed in the TV system.

Color Fundamentals

White light may be separated into its various components by use of a simple prism, as shown in Figure 12-21(A). The refractive action of the prism delays the higher

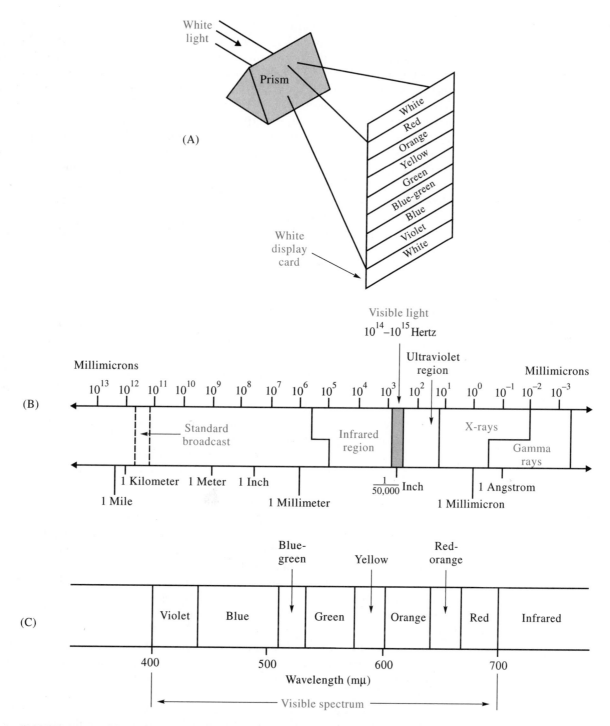

FIGURE 12-21 Visible light spectrum (A) Color separation through refraction action of a prism (B) Relative position of visible light in the electromagnetic spectrum (C) Visible spectrum expressed in millimicrons

frequencies more than the lower frequencies. Thus, the light rays exit the prism at differing angles according to each frequency present. The result is a "stacking" of the frequencies in a sequential order according to their wavelength. When projected onto a flat white surface, the various colors contained in the original white light become visible. As shown, the color red is the lowest in frequency since it is refracted the least. The remaining colors are arranged in order of ascending frequency. Although it may appear otherwise in the illustration, there are no sharply defined borders between the colors. One color blends into another, and it is a matter of individual interpretation where one predominant color ends and another begins.

Figure 12-21(B) shows the relative position of visible light in the total electromagnetic spectrum. Visible light is defined as the portion of the spectrum that stimulates the eye and causes the sensation of seeing light. The color of the light that one sees depends on the wavelengths that reach the eye.

The basic unit of measurement for the wavelengths of light is the *millimicron* (mμ). Figure 12-21(C) illustrates the visible spectrum and defines the wavelengths of the individual color bands in terms of millimicrons.

Color Signal Transmission

Cameras used for color TV broadcasting are much more sophisticated than those used for monochrome systems. Older color cameras use the prism concept just discussed, with a lens system focusing the scene to be transmitted onto and through two special *dichroic glass semimirrors*. In effect, this system separates the color content of the scene into three separate signals. Semimirrors are made to reflect or pass the three primary colors used for television: red, green, and blue. Each semimirror reflects only the color for which it is made and allows the other colors to pass through. The three primary colors are reflected into or passed to three separate pickup tubes, as shown in Figure 12-22. The three pickup tubes, working in unison, scan their respective color scenes and develop modulated-line information to be fed through several stages of processing circuitry.

Modern three-chip CCD color video cameras are lightweight and compact. Easy adaptability allows these cameras to be combined with other equipment, such as VCRs for portable applications using battery packs, or studio viewfinders for studio operations. In the studio, connected to an ac power source and a camera adapter, the camera can be controlled by a camera control unit.

The CCD image sensors produce very few after-images (smears) and offers high resistance to a burned-in image, two of the problems experienced with earlier pickup tubes. Also, the CCDs have no scanning (geometric) distortion, and it is not necessary to adjust the color registration.

In most of these cameras, a built-in electronic shutter enables the production of clear images even when the objects being shot are moving at high speeds. Even under low light levels, when the gain of the video amplifier is at its highest setting, high-quality output is assured.

These cameras have a built-in sync generator, improved color resolution with an R/G mixing detail circuit, and an I/Q encoder through which the standard signal is sent.

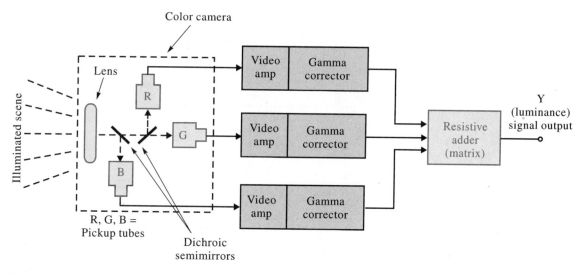

FIGURE 12-22 Block diagram of a tubed color camera and the circuits used to produce luminance signals

At the transmitter, the outputs of the red, green, and blue (R, G, B) pickup devices are amplified by video amplifiers. Because the brightness output of the camera does not correspond to the brightness recognition of the human eye, the amplified signals are fed through *gamma correcters*. (Gamma is a measurement of contrast.) The three gamma-corrected signals are fed to a transmitted *matrix* (resistive adder), where they are combined into a composite signal, called the *luminance* or Y signal. The luminance signal contains a ratio of 59% green, 30% red, and 11% blue (Y = 0.59G + 0.30R + 0.11B). This combination of colors results in a signal that discriminates against noise and produces a good rendition of whites, grays, and blacks when viewed on a monochrome receiver. The composite signal contains video frequencies up to the standard 4.5 MHz, which will produce a B/W picture on any monochrome TV receiver. Such capabilities for color transmissions have been required since 1953 and define a *compatible* system; that is, the color transmission can be received on either a color or a monochrome TV receiver.

Transmitter Color Section   The color section of a TV transmitter is shown in block diagram form in Figure 12-23. The luminance matrix properly proportions the Y signal and produces two outputs, one to the adder, the other to a phase inverter. The +Y signal fed through the phase inverter becomes a –Y signal. The blue signal is added to the –Y signal in the B – Y matrix, and the red signal is added to the –Y signal in the R – Y matrix. The outputs of these two matrices are combined in the IQ matrix, producing an I (in-phase) signal and a Q (quadrature, or 90° out-of-phase) signal. These I and Q output signals are fed through LP filters to balanced modulators, where they modulate the color subcarrier, which is then balanced out (suppressed). The BP filters remove any spurious products of the modulation before the signals are fed into the final adder. The I and Q signals constitute the *chrominance* (color) signal, and each has

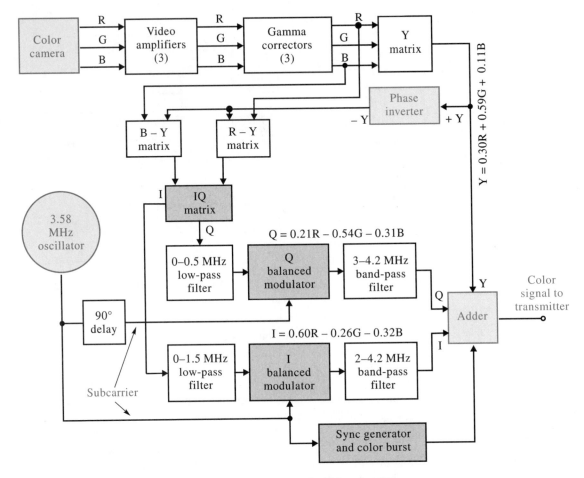

FIGURE 12-23 Block diagram of the color system of a TV transmitter

red, green, and blue voltages with the polarities and relative amplitudes indicated in the illustration.

The +Y signal controls the brightness of the color picture and the visible signal on monochrome receivers. The I and Q signal modulation carries the color information from the transmitter to color receivers, where it adds proper proportions of red, green, and blue to the Y signals. This combination is then fed to the three-color TV screen. The *saturation* (purity) of the colors is determined by the amplitude of the I and Q signals. The *hue* (actual color) produced on the TV screen is determined by the phase developed by the difference in amplitude between the I and Q signals. How these actions in the receiver occur is discussed in Chapter 13.

Color Subcarrier   The chrominance signal is the portion of the composite color signal that represents the color information contained in the televised scene. Its job is to convey the color signals developed by the three individual color pickup tubes in the color camera. If the three separate camera signals were simply added directly to the luminance signal, the result would be an amplitude modulated signal that would be interpreted as distorted luminance information. Therefore, a subcarrier frequency must be provided that can be altered both in phase and in amplitude to convey color information.

The subcarrier frequency chosen had to be such that it would not produce noticeable interference with the luminance signal when displayed on a monochrome TV receiver. The 6 MHz FCC-allocated bandwidth had to carry the luminance signal, requiring 4.5 MHz of sidebands, plus the chrominance information. The sidebands produced by video modulation of the B/W signal are clustered around harmonics of the 15,750 Hz line frequency, resulting in unused spaces between clusters, as illustrated in Figure 12-24(A). The allocated bandwidth must contain both the luminance signal and the chrominance signal, so the color subcarrier frequency had to be positioned

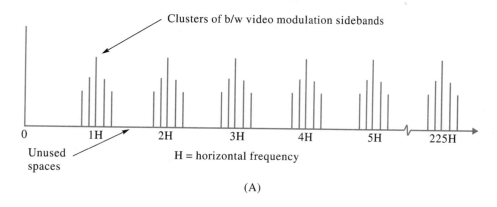

(A)

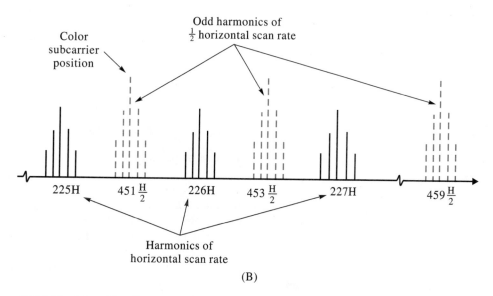

(B)

FIGURE 12-24 Line frequency harmonics and frequency interleaving (A) Video harmonics occur at multiples of the line frequency (B) Color subcarrier generates harmonics that are odd multiples of one half the line frequency

somewhere below the 4.5 MHz limit on the bandwidth curve. It was found that a 3.58 MHz subcarrier (actually 3,579,545 Hz) above the video carrier would produce its video sidebands in the unused spaces between the luminance sidebands. Thus, when this subcarrier frequency is amplitude modulated, the sidebands produced fall between the sideband frequencies produced by the luminance modulation. The result is *interleaving*, illustrated in Figure 12-24(B). In this way, both the luminance and the chrominance modulation can be transmitted within the same 4.5 MHz bandwidth. Figure 12-25 shows the relative positions of the signals contained in the composite color signal.

The color subcarrier frequency was selected to minimize an objectionable beat frequency with the 4.5 MHz sound carrier, but in order to do that, it was necessary to change the line and field frequencies slightly. Recall that the monochrome TV transmitter uses a line frequency of 15,750 Hz and a field frequency of 60 Hz. A color TV transmitter uses a 15,734.26 Hz line frequency and a 59.94 Hz field frequency. However, this slight difference in frequencies does not cause loss of synchronism in TV receivers.

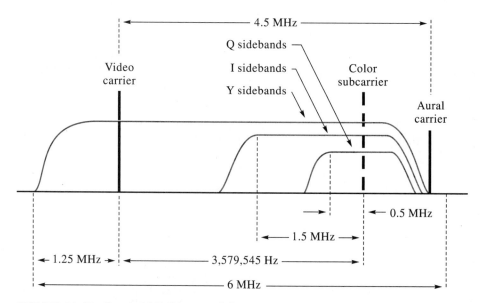

FIGURE 12-25 Bandwidth diagram of the composite color TV signal

Color Burst Remember that the color subcarrier is suppressed in the balanced modulators at the transmitter. Only the sidebands that contain the color information are utilized. Subcarrier suppression reduces the energy content of the chrominance information, which in turn reduces the interference of the chrominance signal on monochrome TV receivers. Because there is no subcarrier to use as a reference for color demodulation, a subcarrier frequency must be generated in the receiver to use for demodulation purposes. To ensure that this signal has exactly the same frequency and phase as the suppressed subcarrier, a short, 8-cycle *burst* of the subcarrier frequency is added to the back porch of the horizontal blanking signal, as shown in Figure 12-26.

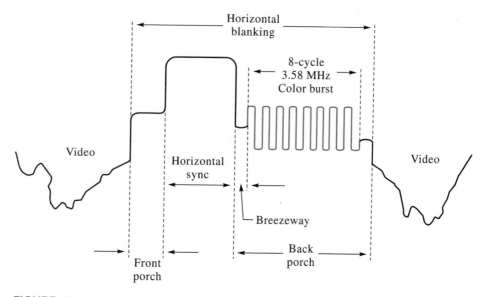

FIGURE 12-26 Color burst rides on the back porch of the horizontal blanking signal

The 3.58 MHz color bursts are used in the color receiver to tightly control the phase and frequency of a crystal oscillator. Generally, the incoming burst signal is compared to the oscillator output, and a dc difference signal is generated when an out-of-phase condition exists. This voltage is then applied to the oscillator as bias and serves to pull it back into the proper phase relationship.

12-5 DIGITAL TELEVISION SYSTEM

Advanced television systems have appeared in recent years, with the concept of *High Definition Television* (HDTV) leading the way. Standardization of HDTV has been an ongoing process for several years. Many proposals were advanced, the most prominant being for an aspect ratio of 16:9 and a scan rate of 1125 lines. Acceptance of a standard for digital TV was finally approved by the Federal Communications Commission (FCC) in January 1997. Approval of these standards paves the way for the delivery of crisp digital signals, CD-quality sound, and much sharper pictures to home viewers by 1998. In effect, these standards result in the format called *high definition*. A digital TV screen can provide about four times the detail of standard screens.

Before broadcasters can provide digital TV, however, the FCC must take two additional steps. (1) It must make slices of the public airwaves available for the service, and (2) it must issue new digital TV licenses. Both steps are expected to be finalized in 1998.

Some stations immediately implemented plans to allow digital broadcasting. These broadcasters will broadcast both the current analog signals and the new digital ones. This dual capability will allow for a gradual transfer to digital TV without rendering the 220 million TV sets nationwide useless.

Broadcasters will simultaneously transmit programs using two TV channels, one for programs in the existing analog format and the other in the new digital format. Digital systems are being created so they can be downgraded to broadcast over analog sets.

The FCC is assigning channels that stations can use to carry digital programs while they continue broadcasting on their existing channels. Most of these channels will be in the lesser-used UHF range (channels 14–69). Stations will be allowed to operate both analog and digital stations for about 10 to 15 years. The FCC will then require that the analog station be shut down.

12-6 OTHER TELEVISION APPLICATIONS

Television has opened many new avenues in the fields of entertainment and news dissemination by its use in public broadcasting. Less well known are applications in education, science, and industry, where the TV camera has contributed greatly to our knowledge of our environment and ourselves.

Because of its ability to relay information instantaneously, the TV camera can best be described as an extension of the human eye. Such *realtime* (see-it-as-it-happens) image transmission has become more important as the complexity and pace of our society have increased. Science and industry require the capability to view events occurring in extremely hazardous locations. The use of television in areas of atomic radiation, in underwater environments, and in the vacuum of space represents only a small sample of the many possible uses. The following sections will examine several applications other than broadcasting.

Education

Probably the greatest challenges and achievements for television are in the field of education. It is here that television's unique ability to relate, inform, and teach finds its ideal application. For example, in a classroom situation, an *educational television* (ETV) or *instructional television* (ITV) system can combine the instructional powers of motion pictures and slide transparencies, a microscope or telescope, and an instructor located many miles away, giving the entire class a simultaneous, unobstructed view of the subject. Or consider the effect on a class of medical students able to watch the movements of a surgeon's hands performing delicate surgery. It is even possible, by the use of flexible fiber optic devices, to enter various internal organs of the body and view their functions directly. And classic operations or unique procedures can be recorded on video tape to be viewed again and again.

Many of the techniques developed by broadcast television are now being applied to ETV in educational institutions. The introduction of the integrated circuit into TV system manufacturing has greatly decreased the size and cost of high-quality ETV systems. The use of these systems is considered by many to be a turning point in educational procedures.

Another important area of education heavily dependent on television is teacher training. Much of a student teacher's time must be spent observing pupils in actual classroom situations. Remotely controlled cameras can be unobtrusively mounted in a classroom and the scene viewed by a group of student teachers at a distant location.

This use of television enriches classroom observations by providing identical viewing situations for all without the disturbing effects of their presence in the classroom.

Medicine

Television is used in many areas of the medical field other than in the education of medical students. It is an effective means of monitoring critically ill patients from a central location within a hospital. Television cameras also view reactions of patients under radiation treatment, making unnecessary the presence of a human observer.

Because most huge medical complexes exist in the larger population centers of the country, patients from outlying communities must usually commute for specialized diagnosis and treatment. High-resolution TV systems can serve as a link between these two areas, allowing a specialist to view x-rays and other pertinent data, and even converse with the patient, without the inconvenience and expense involved in a trip by either. The time saved may be the difference between life and death.

Color television can be an able assistant in many areas of medicine. For example, blood vessel patterns are easily discerned, and various types of eruptions and inflammations can be observed.

Specialized cameras using tubes highly sensitive to the ultraviolet region of the spectrum have been used in cancer research. Also, tubes sensitive to infrared radiation have been used in detection of localized infections.

Surveillance

Surveillance is the most widely used single application of CCTV. Businesses of all kinds use the closed circuit camera to keep an eye on their stock, entrances, warehouses, and parking lots. Banks and financial institutions use the CCTV camera for security purposes. Law enforcement agencies use CCTV for crowd and building surveillance and for monitoring jail and prison corridors. Manufacturers use it to monitor assembly line status.

A major problem with the surveillance application is providing sufficient light for proper camera operation. Lighting conditions are usually adequate during daylight hours, especially for cameras mounted out-of-doors, but after dark, additional lighting must be provided. Most surveillance cameras use a standard vidicon tube as the light-sensitive element. These tubes require a minimum light level of about 5 foot-candles. Low-light-level cameras are available with pickup tubes that can provide acceptable images from light levels lower than 1 foot-candle, but these devices are quite specialized and expensive.

Cameras that are operated unattended usually are equipped with automatic sensitivity control circuitry. This circuitry provides automatic compensation for changes in light level on the scene by adjusting voltages at the camera pickup tube. By providing compensation effective over light variations of 10,000:1, these devices produce a useful image with light levels ranging from bright sunlight for daytime operation to that provided by street lights or floodlights for nighttime operation.

There are other, more specialized, surveillance applications where a CCTV camera is extremely important. For instance, TV cameras in explosion- and fireproof housings are used to monitor rocket fuel loading; they are used at many key points during a space launch to monitor proper operation of the hold-down and release mechanisms;

and they are used in the "white room" to monitor astronauts entering the space capsule, and to monitor capsule door lockdown.

Home Security

A home security system, the WatchCam Security System from Sony, comprises a tiny CCTV camera, a miniature microphone, a 4-inch TV monitor with a tiny speaker, and 67 feet of connecting cable. The camera can fit into the peephole in a door, and the monitor can be placed anywhere in the house within the range of the connecting cable. With this system, a person can see and hear who is calling without opening the door. The system can also monitor any desired spot in the house, such as a baby's crib or the bedroom of a bedridden person.

Data Transmission

The use of CCTV for data transmission has become increasingly popular as the need for instant transfer of information has expanded. Banks use CCTV to approve checks from a remote location. Industrial plants use it at unattended gates to check individuals and identification cards before the gates are opened by remote control. Instant pictorial information of documents, letters, and photographs can be routed to many different locations simultaneously.

The primary consideration in choosing a CCTV camera for data transmission is the camera's resolution—that is, its ability to resolve or reproduce detail. Camera resolution is dictated by the type of information to be televised. A standard typewritten page is often used as a criterion. For the page to be truly legible at the TV monitor, the camera must have high scan ratios and bandwidths. Many high-resolution CCTV cameras are presently available.

Radiation

With the advent of nuclear reactors and radioactive materials, the CCTV camera has become an even more important tool. Televised images of the interior of reactors during the refueling process make such cameras an extremely important part of the total operation. In fact, the ability to quickly restore a reactor to proper operation often depends on the reliability of the camera's operation in areas of high radiation.

Because radiation seriously affects the operation of semiconductor devices, these devices are not normally used in the construction of cameras designed for high-radiation applications. Also, gamma radiation causes changes in the chemical composition of other materials normally used in the construction of TV cameras. For example, the Teflon insulating material may harden or even powder.

Because they are less affected by radiation, vacuum tubes and nuvistors are often employed as the active elements in cameras used for high-radiation applications. The camera pickup tube is usually a vidicon, so its operation is not greatly affected by radiation. However, ordinary optical glass turns brown with the continued exposure to radiation, producing decreased sensitivity of the optical system. Therefore, the optical elements used in such areas are made of a special, nonbrowning glass.

Materials used for camera housings must be impervious to the effects of radiation and should not themselves become radioactive carriers after exposure. Special stainless steel alloys that contain no materials with appreciable half-lives and are highly polished to prevent contaminant retention in small pits or scratches in the surface are used for this purpose. Stainless steel also allows the camera to be decontaminated through dipping or washing in a solution of dilute nitric acid. The camera must be hermetically sealed to prevent the admission of radioactive dust or acid components.

Underwater Applications

Underwater applications generally require miniaturized CCTV cameras enclosed in very sophisticated and durable housings. At depths beyond about 1500 feet, cameras must be able to withstand pressures of several tons per square inch without allowing stress or moisture to be transmitted to the camera circuitry or optics. Operations in a salt water environment require cameras constructed of materials that can withstand the corrosive effects of electrolysis. It is equally important to isolate the camera from its housing; otherwise, the effects of electrolysis will be accelerated. Stainless steel is one of the best materials to use, although where costs might be a factor, hard anodized aluminum may be used in applications at depths up to 1500 feet.

Because of the absence of natural light at great depths and because of the dense medium and the many impurities that may be present, effective video reproduction of the underwater scene requires that high-intensity lighting be used with the camera. Light sources must be able to withstand the same pressures as do the cameras.

Objects under water appear to be about 30% closer than they actually are. To correct this distance distortion, special optically corrected faceplates are used on the camera housing.

Airborne Applications

Motion picture photography has long been used to observe various aircraft functions and armament tests during flight, but in many applications the pilot has no way of determining what is being recorded on film. Mounting a TV camera beside the film camera overcomes this problem. By watching a small TV monitor in the cockpit, the pilot keeps the aircraft oriented to maintain the target in the center of the screen, ensuring a good photographic run. Such systems are widely employed by the military in areas where the use of high-speed, single-seated aircraft is dictated.

Wideband transmission systems are widely used in *drone* (unmanned) aircraft, where the high resolution of the systems permits instrumentation monitoring, reconnaissance, and other airborne applications. Manned aircraft also use the systems for on-board monitoring of critical events.

There are several problems associated with the use of television in an airborne environment:

1. Vibration can cause camera tube microphonics (noise), broken or dislodged electrical components, and intermittent connections.

2. Arcing and corona problems may occur between altitudes of 80,000 and 120,000 feet, depending on the humidity, the voltages used, and so on.

3. The extreme cold encountered at great altitudes has an undesirable effect on the camera circuitry and the photosensitive surface of the pickup tube.

Providing a heater jacket for externally mounted cameras or placing the camera inside the aircraft helps overcome this last problem.

Space

The space program has provided one of the most exciting and noteworthy uses of the TV camera. The TV camera has been on board manned spacecraft from the first suborbital flight, providing valuable information for scientists on earth. It has relayed literally thousands of excellent photographs of the surface of the moon, including our first look at the moon's dark side and our first step onto the moon's surface. Cameras mounted in weather satellites are continuously photographing the cloud cover that moves across the earth, and radio transmitters relay the pictures to ground stations where they provide valuable insight into the earth's weather patterns.

Most of the pictures obtained from spacecraft are a product of *slow-scan* TV systems, where the camera's pickup tube stores scene information until scanned by an electron beam. A shutter assembly, not unlike that used on the familiar family camera, allows a brief exposure of the scene. A slow-moving beam of electrons (about 8.5 s for one complete, noninterlaced frame) then converts the scene into relatively low-frequency electrical signals. The slow-scan process has two advantages:

1. Bandwidth is reduced to 3 kHz.
2. The low-frequency signals are easier to separate from the noise that interferes with radio transmissions from space.

The role of the TV camera in space continues to grow in importance. Television cameras mounted aboard unmanned spacecraft scout future destinations for manned vehicles and relay back pictures of what lies ahead. The space shuttle uses the camera extensively for monitoring cargo bay activities. Voyager craft continue to relay excellent pictures from deep space, long after engineers had predicted system failure.

The superb images received from the summer 1997 Pathfinder landing on Mars are an excellent example of what can be obtained from a slow-scan digital transmission system. As can be imagined, the received signals were quite weak and distorted. Extensive error detection and correction schemes were utilized to produce the almost photo-quality images.

Improvements continue to be made in camera pickup tubes, ICs, and solar power supplies, which will result in cameras with greatly increased capabilities. The potential of television in space applications has so far only been sampled.

Special Applications

Some special application TV cameras can react to light lying outside the visible spectrum. For example, some types of pickup tubes can "see" the radiant infrared frequencies emitted by warm objects. Use of a camera employing such devices allows military personnel to visually detect an enemy vehicle at night by the heat of its engine. Another use is the ability to observe directly on a TV monitor a normally invisible laser beam.

Although not widely used, camera tubes sensitive to ultraviolet light are available. These devices are used to a limited extent to observe biological specimens and experiments being illuminated by high levels of ultraviolet light.

Another type of TV camera has the ability to see in areas where available light is so low that the unaided human eye sees only darkness. Such specialized cameras employ image intensifiers to amplify the light image before viewing it with the photosensitive element of the camera tube. This concept has proven invaluable for military applications.

SUMMARY

- A TV system has audio and video pickup devices, a transmitter, and a receiver. The system may consist of a simple camera and a monitor or of highly complex and sophisticated electronic equipment.

- Television cameras can be self-contained or two-unit systems. The self-contained camera contains all the elements needed to view a scene and generate a complete TV signal. Two-unit systems have a camera head and a separate camera control unit (CCU). Camera pickup tubes must be sensitive to visible light, have a wide dynamic range with respect to light intensity, and be able to resolve detail when viewing a multielement scene. The first TV camera tube was the iconoscope, followed by the image orthicon (IO) and the vidicon. Modern TV cameras use charge-coupled devices instead of tubes as pickup devices. The CCD imaging devices permit easy adaptability for either portable or studio use, and allow shooting scenes even in low light conditions.

- The electron beam in a camera tube or picture tube can be deflected and focused electrostatically or electromagnetically. Electrostatic deflection uses characteristics exhibited by electrons when placed between two oppositely charged conductive plates. Vertical and horizontal deflection plates are used to position the beam. Electromagnetic deflection uses an external current-carrying conductor to set up a magnetic field of the proper polarity to deflect the beam. Two horizontal and two vertical coils form a deflection yoke, placed around the neck of the tube. Beam focusing is most often achieved electromagnetically.

- Synchronizing signals are needed to ensure that the received picture is identical to the original scene; that is, the receiver is "in sync" with the camera. A sync generator develops the necessary pulses and inserts a waveform onto the camera video output signal that causes the monitoring system to lock to the camera. In single-sync multicamera systems, loop-through can be used. In any loop-through operation, the last device in the system must be terminated in 75 Ω.

- Monitors are TV receivers without RF input and audio sections; they rely on a direct input of unmodulated video signals. Loop-through can be used to multimonitor systems.

- Distribution amplifiers (DAs) are used to drive signals for distant operating points. Video distribution amplifiers (VDAs) route camera signals; pulse distribution amplifiers (PDAs) route signals from sync generators and similar equipment that produces pulse waveforms with fast rise and fall times.

- Video switchers are used for switching the output of a number of cameras, one at a time, to a monitor for viewing or for sending out on the line to a transmitter. Crosspoint switching is the most efficient form of video switching.

- Remote controls are used for camera lens focusing; for adjustments for video gain, camera sensitivity, blanking level, and video polarity; and for camera head pan and tilt.

- Television broadcasters must use equipment that meets standards set by the FCC. Generated and broadcasted video signals must be monitored at all times, and they must not deviate from specified tolerances. Television broadcast stations have a studio, a control room, a video tape and telecine equipment room, a terminal equipment room, and a transmitter room. Aural and visual signals are handled separately until they are combined in a diplexer and fed to the transmitting antenna. The studio has cameras, microphones, sets for scenes, and lighting equipment. The control room contains CCUs, video monitors, video switchers and special effects controls, audio mixing and switching equipment, and audio recording and monitoring equipment. The video tape and telecine room contains video tape recorders for recording studio action, for editing programs, and for playback on the air at a later time. The terminal equipment room contains dc power sources for all station equipment, sync generators, and DAs. The transmitter room contains an AM video transmitter, an FM audio transmitter, and a diplexer for combining the video and audio signals for transmission as a composite signal. Synchronization signals are transmitted as part of the composite signal to ensure that all TV receivers are able to receive the transmitted TV signal in the exact timing sequence as the original scene, without vertical or horizontal impairment. Television transmitting antennas must be broadbanded enough to handle the FCC-allocated 6 MHz channel bandwidth. Transmissions are usually horizontally polarized to reduce ghosts.

- Television systems use rectilinear scanning, where the scanning beam moves horizontally and vertically simultaneously. Interlaced scanning involves scanning every other line, retracing, then filling in the untraced lines. The horizontal scanning frequency (line rate) is 15,750 Hz. The vertical scan rate (field rate) is 60 Hz. Two interlaced fields make up one frame with a frame rate of 30 Hz. There are 525 lines per frame, although some lines are lost during vertical retrace. The number of active lines is between 483 and 488. The TV aspect ratio is 4:3.

- Resolution is the ability of a system to resolve detailed picture elements. In television, resolution measurements are made with respect to picture height. Lines of resolution, both vertical and horizontal, are the number of black and white lines of equal width that can be contained in the vertical dimension of the TV picture. Vertical resolution is further defined as the ability of a system to resolve horizontal lines in a scene. Horizontal resolution is further defined as the ability of a system to resolve vertical lines of resolution. Horizontal resolution is determined largely by the bandwidth of the TV system.

- The sensation of color is directly related to the frequency of the light being observed. In color systems, three colors (red, green, and blue) are combined to form the luminance (Y) signal, which equals $0.30R + 0.59G + 0.11B$. Color transmissions are compatible so that the transmitted signal can be received on either a color or a monochrome TV receiver. The color (chrominance) signal consists of the I and Q signals, with color combinations of $I = 0.60R - 0.26G - 0.32B$ and $Q = 0.21R - 0.54G - 0.31B$. Saturation is the purity of colors, determined by the amplitude of the I and Q signals. Hue is the actual color produced on the TV screen, determined by the phase developed by the difference in amplitude between the I and Q signals.

- The 3,579,545 Hz (3.58 MHz) color subcarrier was selected to minimize an objectionable beat frequency with the 4.5 MHz sound carrier. Color systems use a 15,734.26 Hz line frequency and a 59.94 Hz field frequency. The color subcarrier is suppressed at the transmitter, so an 8-cycle burst of the 3.58 MHz subcarrier is added to the back porch of the horizontal blanking signal to use in the receiver for demodulation purposes.

- Television has many uses other than for broadcasting, including education, medicine, surveillance, home security, data transmission, radiation, underwater, airborne, space, and special applications.

QUESTIONS AND PROBLEMS

1. What are the components of a TV system?

2. Describe (a) the self-contained camera system and (b) the two-unit camera system.

3. What characteristics must a camera pickup tube possess?

4. Describe (a) CCD, (b) LID, and (c) AID.

5. How is beam deflection usually achieved in the vidicon?

6. Describe (a) electrostatic deflection and (b) electromagnetic deflection.

7. What is the purpose of the deflection yoke?

8. Define raster.

9. What is the purpose of a sync generator?

10. List the advantages of synchronizing all cameras in a multicamera system to a single sync generator.

11. Explain how multiple cameras can be driven with a single sync generator.

12. What is the ohmic value of the standard TV terminating load?

13. Describe the difference between a TV monitor and a TV receiver.

14. How can several monitors be used to display the scene being viewed by a single camera?

15. What is the purpose of a distribution amplifier?

16. Describe the two types of DAs.

17. Explain the operation and purpose of a video crosspoint switcher.

18. Describe how remote controls are used in a TV system.

19. Which equipment items are required for broadcast television but not for CCTV?

20. List the sections of a broadcast TV station.

21. List the equipment for the TV (a) studio, (b) control room, (c) video tape room, (d) terminal equipment room, and (e) transmitter room.

22. Define (a) telecine equipment and (b) vestigial sideband and explain the use of each.

23. What polarization is used for the TV transmitting antenna? Why?

24. List the parts of the standard sync signal for monochrome TV transmission.

25. Describe the sync generator divider system that produces the frequencies needed for the different sync pulses.

26. What is the purpose of frame synchronizers?

27. Briefly explain the interlaced scanning process.

28. Define beam retrace.

29. Define (a) line rate, (b) field rate, and (c) frame rate and list the frequencies for each.

30. What is the U.S. standard number of scan lines (a) per field and (b) per frame?

31. Define (a) aspect ratio, (b) lines of resolution, (c) vertical resolution, (d) horizontal resolution, and (e) scan ratio.

32. What is the Kell factor and what is its purpose?

33. A CCTV system has an 800:1 scan ratio. Find (a) lines lost in blanking, (b) active lines, and (c) lines of vertical resolution.

34. Explain the operation of a color TV camera.

35. Define (a) luminance signal, (b) compatible TV system, and (c) chrominance signal.

36. Briefly explain how the color section of a TV transmitter works.

37. What are the color ratios for (a) the Y signal, (b) the I signal, and (c) the Q signal?

38. Define (a) saturation and (b) hue.

39. What is the exact color subcarrier frequency?

40. Define color burst and explain its purpose.

41. In a color TV transmitter, what are the exact frequencies for (a) the line frequency and (b) the field frequency?

42. Define realtime transmission.

43. List 10 applications for television other than broadcasting.

CHAPTER 13

Television Systems—Reception

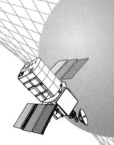

OBJECTIVES

When you complete this chapter, you should be able to:

- State the requirements for good TV signal reception.
- Distinguish between monochrome and color receivers.
- Draw a block diagram of a monochrome TV receiver and discuss each block.
- Draw a block diagram of a color TV receiver and discuss each block.
- Describe in general terms the operation of a color TV receiver.
- Describe the safety precautions to follow when handling TV receivers.

INTRODUCTION

Modern TV reception systems are the result of many years of research. They represent an extension and combination of known radio and electronics principles. The developments that led to these ingenious systems included the simple "cat's whisker" crystal radio; the superheterodyne AM and FM radio receivers; the monochrome TV receiver; and the present culmination, the color TV receiver. The monochrome TV receiver is essentially a superheterodyne radio receiver with circuits added to provide a picture. The color TV receiver is a monochrome receiver to which color circuits have been added.

As explained in the preceding chapter, the signal sent out from a TV station is quite complex. Fundamentally, it consists of two modulated carriers. There is a frequency modulated audio carrier, and there is an amplitude modulated video carrier. The modulation information for the monochrome picture carrier has two basic parts: (1) video information, which contains the monochrome or luminance information, and (2) synchronizing information, which makes it possible for the receiver to assemble the transmitted picture signal in proper order. In color systems, a third element—chrominance (chroma) information, the color signal—is included in the video information. In either case, the RF picture carrier frequency is always 4.5 MHz below the sound carrier frequency. In color systems, the RF picture carrier frequency is modulated by the sidebands of the 3.58 MHz color subcarrier.

Television receivers combine many components to give picture and sound. In this chapter, we will examine these components and the circuitry associated with them.

13-1   THE TELEVISION RECEIVER

The superheterodyne principle (Chapter 3) is also used in the TV receiver. However, the TV receiver is much more complex than the radio receiver. It must:

1. Separate and channel the audio, video, and sync signals into their respective areas within the receiver.
2. Contain circuitry capable of handling video, sync signals, and audio signals.
3. Generate high voltages for the picture tube.

Figure 13-1 shows a simplified block diagram of a TV receiver. The following sections will expand and examine each of the various blocks.

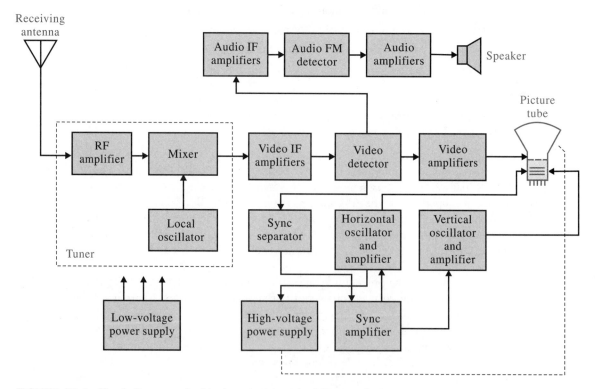

FIGURE 13-1   Block diagram of a black and white television receiver

The Front End

The *front end* or *RF tuner* receives and amplifies the RF signals broadcast by TV stations and provides a means by which individual stations can be selected for viewing. The tuner must:

1. Provide the correct impedance match between the antenna and the tuner.
2. Tune all VHF and UHF and cable television channels.
3. Have low inherent noise to prevent *snow* (white dots in the picture that are the video version of noise).
4. Provide RF amplification.
5. Isolate the local oscillator signal from the receiving antenna to prevent feedback and possible radiation interference.
6. Convert the RF frequencies to IF frequencies.

TV receivers include tuning provisions for 67 TV channels (12 VHF and 55 UHF) as shown in Table 13-1. However, selection of VHF and UHF signals is accomplished by separate tuners, so we will examine each of these tuners individually.

TABLE 13-1   Television Channel Frequencies

VHF Channel	Frequency Limits (MHz)	Channel	Frequency Limits (MHz)	Channel	Frequency Limits (MHz)
2	54–60	23	524–530	47	668–674
3	60–66	24	530–536	48	674–680
4	66–72	25	536–542	49	680–686
5	76–82	26	542–548	50	686–692
6	82–88	27	548–554	51	692–698
7	174–180	28	554–560	52	698–704
8	180–186	29	560–566	53	704–710
9	186–192	30	566–572	54	710–716
10	192–198	31	572–578	55	716–722
11	198–204	32	578–584	56	722–728
12	206–210	33	584–590	57	728–734
13	210–216	34	590–596	58	734–740
UHF Channel	**Frequency Limits (MHz)**	35	596–602	59	740–746
		36	602–608	60	746–752
14	470–476	37	608–614	61	752–758
15	476–482	38	614–620	62	758–764
16	482–488	39	620–626	63	764–770
17	488–494	40	626–632	64	770–776
18	494–500	41	632–638	65	776–782
19	500–506	42	638–644	66	782–788
20	506–512	43	644–650	67	788–794
21	512–518	44	650–656	68	794–800
22	518–524	45	656–662	69	800–806
		46	662–668		

Two popular mechanical arrangements for tuners were used prior to modern digital systems. The *turret* or *drum* tuner (Figure 13-2) had a drum like structure with a separate tuning stick or strip for each channel. The strip of the desired channel was brought into contact with stationary terminals on the tuner by rotating the drum. The second and more popular type used a *wafer switch* arrangement (Figure 13-3) which had one wafer for each tuned circuit in the tuner. The tuners both had 13 positions: 12 VHF channels and one position for UHF.

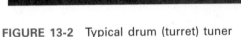

FIGURE 13-2 Typical drum (turret) tuner

FIGURE 13-3 Typical wafer switch tuner

Modern TV receivers use all-electronic tuners (Figure 13-4). The tuning element is a *varactor,* and each tuner tank circuit contains a varactor. Tuning is accomplished by changing the voltage across the varactor with a potentiometer. Pushbutton arrangements (Figure 13-5) on all-electronic switching circuits are used to select a preset potentiometer. In this arrangement, the selection mechanism is separate from the electrical circuit, permitting a wide choice of styling and remote control systems.

Modern receivers use transistor tuners to obtain high gain and low noise in the RF stage. With transistors, an additional concern is cross-modulation. Early transistor tuners used common-emitter RF stages that were highly susceptible to cross-modulation. These tuners were sometimes provided with attenuator pads that could be switched into the antenna circuit to reduce the signal strength and thus eliminate cross-modulation. If such pads were not built in, they could be added externally in areas of strong signals.

RF amplifiers using common-base configurations are less subject to cross-modulation and have only slightly less gain than common-emitter RF stages. Modern circuits use metal oxide field-effect transistors (MOSFETs) as RF amplifiers. Early types of MOSFETs were easily damaged by static charges on the antenna. Modern FET devices using an insulated gate (IGFET) provide immunity to such damage.

The VHF Tuner   The tuner in a TV receiver performs channel selection, amplification, and frequency reduction of the incoming station signal. The block diagram of Figure 13-6 shows the three stages normally found in a tuner: (1) the *RF amplifier,* which is tuned to the frequency of the incoming signal and amplifies it; (2) the *local oscillator* (LO), which generates a continuous wave (CW) signal whose frequency is unique for each channel; and (3) the *mixer,* which receives the amplified RF channel signal and beats this signal with that from the LO.

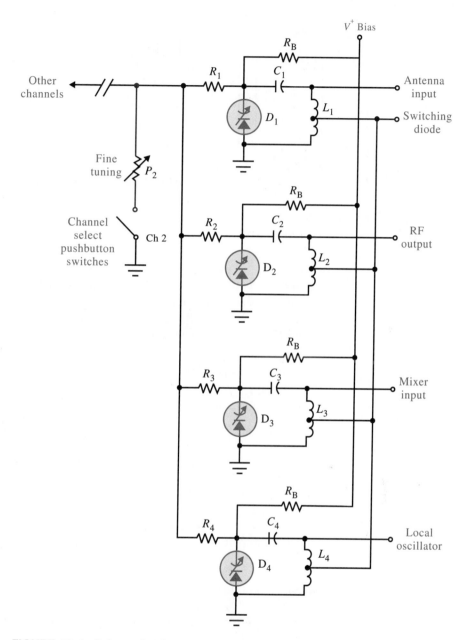

FIGURE 13-4 Schematic of a varactor tuner

The new signal frequencies created by this heterodyning process include the sum and difference frequencies and harmonic combinations of the oscillator signal and the incoming channel signal. The output circuit of the mixer is tuned to the difference, or intermediate frequency (IF), and passes this signal to the IF amplifiers. This is the same process used in superheterodyne radio receivers. VHF tuners are designed to select

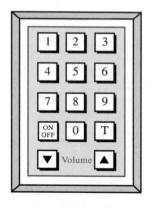

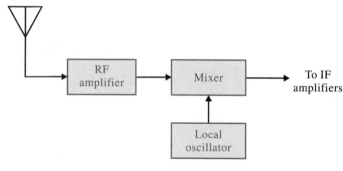

FIGURE 13-6 Block diagram of a tuner

FIGURE 13-5
Pushbutton arrangement

each of the 12 VHF channels (channels 2–13). (In many front ends, a channel 1 position changes the VHF tuner into an IF amplifier for UHF reception from the UHF tuner.)

The frequency characteristics of the tuner circuits can best be understood by considering the tuned circuits for a specific channel. The signal frequencies broadcast by channel 2 (54–60 MHz) are constituted as follows:

> low side = 54 MHz
> picture carrier = 55.25 MHz
> sound carrier = 59.75 MHz
> high side = 60 MHz

When the tuner is set for channel 2, the RF circuits (input to RF amplifier, output of RF amplifier, and input to mixer) must be tuned to accept this 6 MHz band of frequencies. Ideally, the response of the RF circuits should be "flat" over this frequency range. In practice, the RF tuned circuits are more broadly tuned than the required 6 MHz, and their response is not always flat on every channel.

The LO is generally tuned above the frequency of the incoming signal. The relationship among the frequencies of the LO (f_{lo}), the RF picture carrier (f_p), and the IF picture carrier (IF_p) is, for any specific channel,

$$f_{lo} = f_p + IF_p \tag{13-1}$$

In this relationship, f_{lo} and f_p are unique frequencies for each channel, while IF_p is constant for all channels. Applying Equation 13-1 to a receiver in which the IF picture carrier is at 45.75 MHz, the LO frequency for channel 2 must be

$$f_{lo} = 55.25 \text{ MHz} + 45.75 \text{ MHz} = 101 \text{ MHz}$$

Physically, the tuner can be wired as a subassembly on a chassis separate from, but mounted on, the main TV chassis. The IF signal from the tuner is brought out to the first picture IF amplifier on the main chassis. An RF automatic gain control line and the B^+ voltage are connected from the main chassis to the tuner chassis. (B^+ is used instead of V_{CC} to represent active device source voltages in TV receivers.) All these connections are usually brought out at the top of the tuner through feed-through capacitors.

The oscillator frequency is controlled by the position of the channel selector and is chosen so that the mixer output, which is a product of the selected RF signal and the oscillator frequency, constitutes an IF of approximately 45 MHz.

In the color receiver, the requirements of the RF tuner are more stringent than in monochrome receivers. Each channel position must present a flat response over the entire band so that the chrominance information is not attenuated. In addition, stability of the oscillator is an important factor in the reception of color signals.

The UHF Tuner The UHF portion of the 67 available TV channels extends from channel 14 (470–476 MHz) to channel 69 (800–806 MHz). Selection of these channels is made by a separate tuner. The UHF front end typically consists of two *preselector tuned stages,* a *crystal mixer,* and a *transistor oscillator.* Electromechanical UHF tuners do not use RF amplifiers. In varactor UHF tuners, however, an RF amplifier is located between the two preselector circuits. Varactor UHF tuners use the same number of tuned circuits as electromechanical UHF tuners. There is no gain in electromechanical, and very little gain in varactor, UHF tuners.

The UHF tuner is used together with the VHF tuner, which is converted to an additional IF amplifier, to equalize the gain between UHF and VHF. Typically, for UHF operation a transistor VHF tuner such as that shown in the schematic in Figure 13-7 is set on position 1. The antenna is grounded when channel 1 is tuned in. The input to the

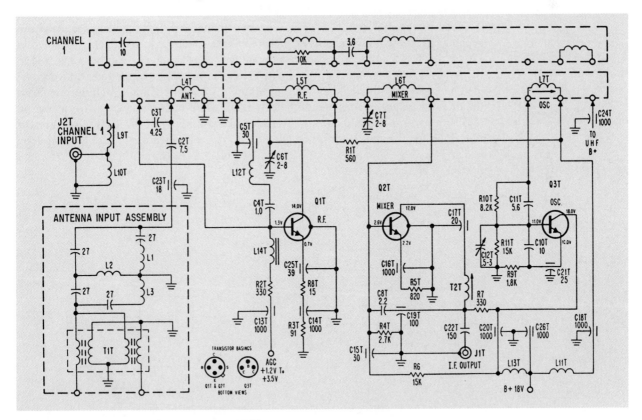

FIGURE 13-7 Schematic of a transistor VHF tuner

RF amplifier (Q_{1T}) then comes from the UHF input cable jack (J_{2T}). The two coils that replace L_{5T} and L_{6T} are tuned to the IF frequency, and Q_{1T} and Q_{2T} now act as a two-stage IF amplifier. The oscillator (Q_{3T}) is disabled because the base has no bias connection. Finally, the UHF tuner is supplied with +18 V through C_{24T}. This arrangement prevents any interference from the UHF tuner when the VHF tuner is on, because there is no dc supply for UHF except when channel 1 is tuned in.

The UHF tuner is in a shielded case to prevent oscillator radiation from interfering with other receivers. It is mounted in the same general area as the VHF tuner. In some receivers, the two tuners are mechanically linked to permit tuning from the same set of shafts. This arrangement is called an *integral tuner,* and the same shaft is used for UHF and VHF fine tuning.

Since the UHF band covers so many channels in a continuous tuning arrangement, the tuning shaft must provide a large gear ratio (approximately 20:1), or some provision for two-speed tuning must be included. In either case, the *dial* must follow the *shaft* of the tuner rather than the knob operated by the user. Two-speed drives depend on a cam, clutch, and gear system.

In 1976, the FCC established the *comparable tuning rule,* which requires that tuning be as simple for UHF as for VHF. In the simplest implementation of this rule, a click detent is provided for each UHF channel, and at least every other channel number appears on the dial. However, in receivers using channel windows, all channel numbers must appear.

Good reception depends on proper antenna connection. Most modern receivers use separate UHF and VHF antennas connected to their respective terminals, as shown in Figure 13-8. If a single all-channel antenna is used, the antenna can be connected to the receiver by a commercially available crossover network. A crossover network can also be easily constructed, as shown in Figure 13-9. The UHF and VHF terminals on the receiver are then connected to their respective terminals on the crossover network.

The most important factors to consider in the circuits of the UHF tuner are the ultra-high frequencies that must be handled. At these frequencies, a lead 1/4-inch long

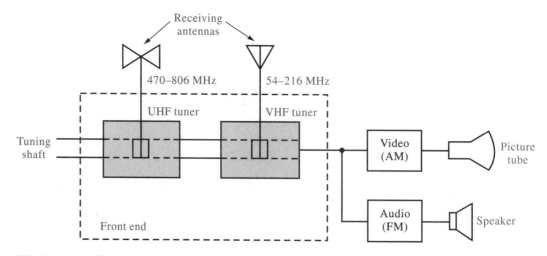

FIGURE 13-8 Block diagram of antenna connects

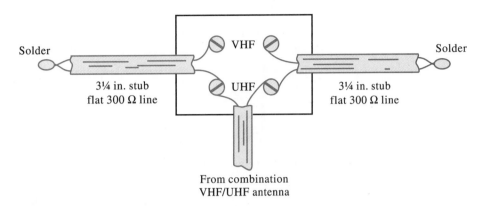

FIGURE 13-9 Antenna crossover network

offers a significant amount of inductance. Changing the position of a component with respect to ground by a similar distance will change the effective capacitive reactance. It is therefore critical that, in any work done inside the UHF tuner, the physical layout of components not be changed. The shape and position of each lead, component, and ground shield or point are integral parts of the electrical circuit.

The IF Section

The *intermediate frequency* (IF) amplifiers amplify the output of the mixer and provide a 6 MHz pass band, from 41 MHz to 47 MHz. In this band of frequencies, the sound carrier falls on 41.25 MHz, and the video carrier falls on 45.75 MHz—an exact 4.5 MHz separation. Although these requirements apply equally for monochrome and color receivers, the color receiver is somewhat more demanding because the IF section must maintain a flat frequency response to the high-frequency end of the video for proper processing of the chrominance information.

Modern TV receivers are called *intercarrier* receivers because of the "double duty" performed by the IF section—that is, amplifying both the video and sound carriers and their respective sidebands. In early *conventional* TV (sometimes called *split-carrier*) receivers, separate IF amplifiers were used for the sound and video signals. However, even slight variations of the LO frequency in those early receivers could detune the sound signal and produce distortion. The intercarrier receiver eliminates that problem by heterodyne action between the video and audio carriers, which are transmitted with the video carrier 4.5 MHz below the sound carrier. This relationship is reversed in the receiver. Figure 13-10 illustrates these relationships.

In the transmitted signal for channel 2 (54–60 MHz), shown in Figure 13-10(A), we see that the sound carrier is on 59.75 MHz and the video carrier is on 55.25 MHz. With the receiver set to receive channel 2, the LO frequency is 101 MHz, which provides an IF (Figure 13-10(B)) of 41 MHz at the high-frequency channel limit and an IF of 47 MHz for the low-frequency channel limit. The reversal occurs in the mixer, and the resultant 4.5 MHz separation is fed to the IF amplifier section.

Video IF Tuning

The IF sections in modern TV receivers have two to five amplifier stages, with AGC applied to all of them. To produce the desired 6 MHz bandwidth, the stages are *stagger*

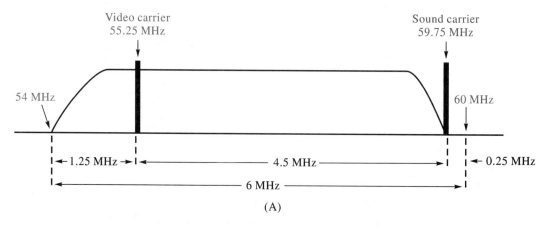

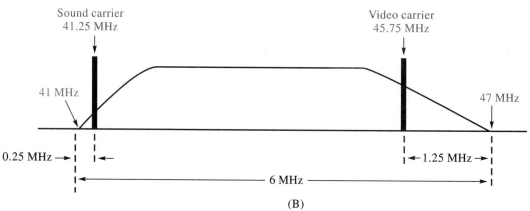

FIGURE 13-10 Transmitted carriers and sidebands of channel 2 TV signal (A) Transmitted signal (B) Inverted relationship in the receiver

tuned. In the stagger-tuned amplifier string, each stage is tuned to a higher or lower frequency than the stage immediately preceding or following it. Stagger tuning:

1. Broadens the BP response.
2. Provides steep skirts.
3. Reduces the possibility of interstage oscillation.

As illustrated by three-stage tuning in Figure 13-11, the use of a lower-Q tuned circuit in the middle stage provides a flatter overall response. Systems with four or five stages provide even better BP response.

The simplified block diagram of a typical IF section given in Figure 13-12 shows three IF amplifiers and the detectors that follow. Note that AGC is applied to the IF circuitry as well as to the RF amplifier in the tuner. The AGC is applied to the base of the second IF amplifier to control its relative gain. The dc voltage that results at the emitter of the second IF amplifier is routed to the base of the first IF amplifier as a gain control. Thus, both the first and second stages of the IF section are gain-controlled by the application of a variable dc potential to the base of the second IF amplifier.

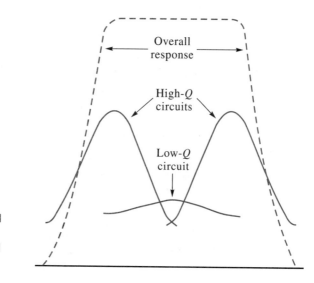

FIGURE 13-11
Three-stage
stagger-tuned
response

In a color receiver, provision must be made for the sound subcarrier to be attenuated at some point before the video detector stage. As shown, the output of the third IF
amplifier is fed into the video detector and the 4.5 MHz beat oscillator. The sound subcarrier must not be present in the detected video signal or a beat signal will be produced between the 4.5 MHz sound subcarrier frequency and the 3.58 MHz color
subcarrier. The beat signal of approximately 920 kHz that would be produced would
result in objectionable interference patterns on the picture tube. Therefore, filtering is
performed in the video detector to eliminate any 4.5 MHz signal that might be present.

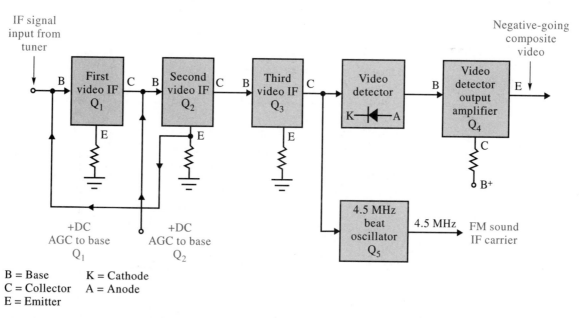

B = Base   K = Cathode
C = Collector A = Anode
E = Emitter

FIGURE 13-12 Typical IF section

A diode acts as the video detector. A 41.25 MHz trap in the detector stage eliminates the sound subcarrier, and the video therefore contains no 4.5 MHz signal to beat with the 3.58 MHz chrominance signal. The IF output to the 4.5 MHz oscillator is taken from the collector of the third IF amplifier before any filtering of the 41.25 MHz signal takes place. To prevent feedback from the output stages becoming part of the input, capacitive networks are used in the collector circuits of the first two IF stages, and a small capacitor is connected to the bottom end of the input transformer primary to the base of the first amplifier to neutralize this stage.

Video Detector

The video detector is an AM detector that demodulates the picture IF carrier. Its operation is the same as for the AM detector in a broadcast band AM radio receiver, except that it must pass the wide band of frequencies associated with the composite video signal (60 Hz to approximately 4.5 MHz). Peaking coils and low-resistance loads are used to extend the frequency response of video detectors. For example, Figure 13-13 shows peaking coil L_2 used in the output of the detector D_1, and the 3900 Ω load resistor R_2.

Signals from the last IF stage are fed to a diode video detector which rectifies them to a varying dc having a waveform that includes the sync and blanking pulses as well as the video information. By mixing the 45.75 MHz video carrier and the 41.25 MHz aural carrier in the diode, a beat frequency of 4.5 MHz is produced. This 4.5 MHz difference frequency is FM modulated by the aural carrier sidebands and AM modulated by the video carrier sidebands. If these modulated signals were allowed to be fed to the following video amplifier, both would be amplified, and the aural signals would produce black sound bars on the picture tube. To prevent or minimize this occurrence, a 4.5 MHz trap can be placed in the coupling stage to the video amplifier. A secondary winding coupled to this trap provides a takeoff point for the 4.5 MHz audio IF stages. This method is used in many monochrome receivers.

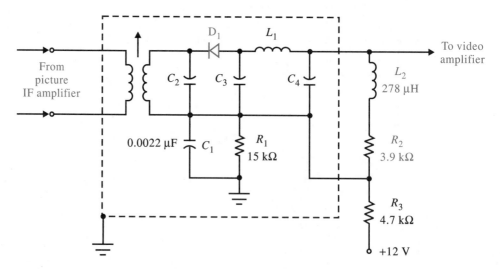

FIGURE 13-13 Video detector

An alternative method of extracting the 4.5 MHz audio IF is by capacitive coupling from the collector of the last IF stage to a separate diode mixer. The output of this mixer is then fed to the audio IF stages. This is the preferred method in color receivers.

Automatic Gain Control

Signal strength varies from station to station and may even vary on a given station due to fading. The *automatic gain control* (AGC) system is designed to eliminate or minimize these variations so that they will not be visible in the picture. In the simple AGC system shown in Figure 13-14(A), the negative dc output of the detector is used to bias the IF and RF amplifiers. As the signal increases, the negative dc detector output increases, which in turn increases the negative bias on the amplifiers. The negative bias reduces the gain and keeps the detector output relatively constant. This arrangement is similar to that used for automatic volume control (AVC) in radios.

The simple AGC system is not satisfactory for TV receivers because the dc component of the video signal depends not only on the RF carrier level but also on the brightness of the scene. Thus, a dark scene produces a more negative dc detector voltage, whereas a bright scene causes a less negative dc voltage. One way to overcome this problem is to use *peak detectors* that respond to the peaks of the video signal. Since the top of the sync pulses in the composite video signal reflects only the RF carrier level, in a noise-free signal a peak detector would develop an AGC voltage directly proportional to the sync pulse amplitude, dependent only on the RF carrier level. This arrangement would make an excellent AGC system for strong signals, but in weak signal areas the noise might be many times larger than the video signal. In the latter condition, the peak detector would give an output dependent on noise rather than on signal.

To perform the peak detector function, the capacitor in the circuit must hold the charge between sync pulses. This long time constant means that the circuit cannot follow rapid changes in signal strength such as that caused by overflying airplanes, and *airplane flutter* results. For the reasons stated, therefore, this simple AGC system is not practical and thus is not used in modern TV receivers.

Delayed AGC   To provide the most noise-free picture possible, low-noise circuits are used in TV receivers. However, even carefully designed circuits will produce some noise in each stage. The effect in a receiver is cumulative, and in a series of high-gain stages the stage noise is emphasized, resulting in a noisy picture. This problem can be overcome by high amplification of the incoming RF signal so that the desired signal will "swamp out" the effects of stage noise as it is processed by the receiver. It is therefore important to have the RF amplifier operate at maximum gain, particularly for low-level RF signals, thus providing a relatively high-level signal for the following stages.

We showed in the simple AGC system how the negative dc voltage was fed back to the RF amplifier as well as to the first two IF amplifier stages. This action resulted in an undesirable reduction in gain of the RF amplifier even for low-level RF signals. This undesirable effect is overcome by the *delayed AGC* system shown in Figure 13-14(B).

In this system, the AGC voltage to the RF amplifier is delayed for low-level RF signals. Diode D_1 is used to achieve the delay. The positive voltage applied to the diode through Resistor R_L causes the diode to conduct in the absence of AGC voltage. Under this condition, the voltage at point B will be about 0.7 V, the drop across the diode junction, thus providing maximum RF amplifier gain for weak signals. However, when

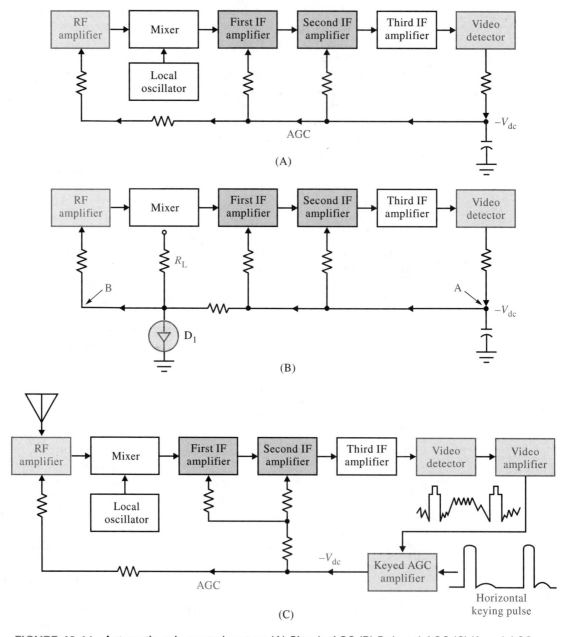

FIGURE 13-14 Automatic gain control system (A) Simple AGC (B) Delayed AGC (C) Keyed AGC

a strong signal is received, the negative AGC bias voltage developed at point A cuts off the delay diode, and the AGC bias is fed to the RF amplifier. The low negative AGC bias for weak signals, although not sufficient to cut off the delay diode, still provides bias to the first two IF amplifier stages.

Keyed AGC   The development of a voltage that will control the gain of the RF amplifier and the IF section is the function of the *keyed AGC* system. In this system, the AGC is developed only during sync pulse time. During this time, the detector can be an average detector, avoiding the long time constant of the peak detector. The noise between sync pulses has no effect on the circuit.

As in the case of the simple AGC system, the controlled stages are the RF amplifier and the first two IF amplifiers, as shown in Figure 13-14(C). Since the input circuits derive their bias from the AGC line, their gain depends on the AGC voltage.

In the keyed AGC system, the control voltage is developed by the AGC amplifier. The base of this amplifier receives a positive-going composite video signal and a dc voltage variation that is dependent on the RF signal level. The collector of the amplifier has no dc voltage applied but receives a narrow positive-going pulse from the horizontal output system at the TV line frequency (15,750 Hz).

Under normal operation, the keying pulse and horizontal sync pulse appear at the amplifier at the same time. The amplifier conducts only when the horizontal keying pulse appears at the collector. The AGC voltage developed is dependent on the sync pulse amplitude in the composite video signal and on the dc bias on the base. Any noise that arrives in the video signal between sync pulses cannot cause the AGC amplifier to operate because there is no keying pulse at its collector at that time. Hence, the noise signal does not affect the AGC output. The circuit is immune to noise between horizontal sync pulses.

Video Section

The output circuit of the video detector contains the composite video signal consisting of video, blanking, and sync signals, as shown in Figure 13-15. The video amplifier raises the level of the composite video signal to an amplitude sufficient to drive the picture tube cathode or grid for proper beam modulation. This signal level may vary

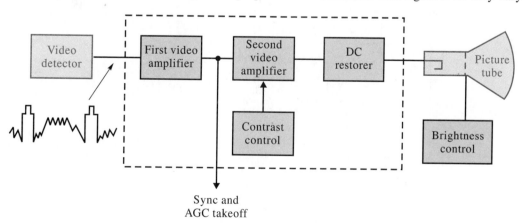

FIGURE 13-15   Video section with composite video signal

from 10 V to 100 V, depending on input signal level, contrast control setting, and type of picture tube. These voltages control the intensity of the beam striking the face of the picture tube. In addition to simple amplification, the video amplifier must have good high-frequency characteristics to faithfully reproduce any fine picture detail elements in the video signal. Since detailed resolution depends on the bandwidth of the amplifying system, the bandwidth must be broad enough to pass the frequencies representing the smaller picture elements desired. This capability is accomplished by special resonant peaking coils in the collector circuit of the video amplifier.

Modern TV receivers are designed with video bandwidth characteristics somewhat less than 5 MHz because limitations of the transmitting bandwidths of TV stations limit video signals at the receiver to about 4.5 MHz. Any additional bandwidth in a receiver would only amplify noise in the area above the 4.5 MHz frequency.

dc Restorer Unless the video amplifier is directly coupled to the picture tube, it is necessary to add a circuit known as a *dc restorer,* which adds a dc bias voltage that remains just at the blanking level but that follows any variation of sync pulse peak amplitude caused by fading. This arrangement prevents variations of signal voltages in light and dark scenes from affecting the bias to which the video signals are added. The result is that light scenes remain light and dark scenes remain dark, with varying shades of gray between the two extremes.

Brightness and Contrast *Brightness* of the displayed picture is controlled by the amount of cathode-grid bias applied to the picture tube. *Contrast* between the light and dark signals on the tube is controlled by the video amplifier signal gain circuit.

Audio Section

The variations in audio circuitry involve the audio takeoff and the type of audio detector used. In color receivers, the preferred audio takeoff is from the final IF stage, as shown in Figure 13-16(A). The audio is extracted here to prevent or minimize the 920 kHz beat between the 4.5 MHz audio carrier and the 3.58 MHz color subcarrier, which would otherwise occur in the output of the video detector. The audio detector

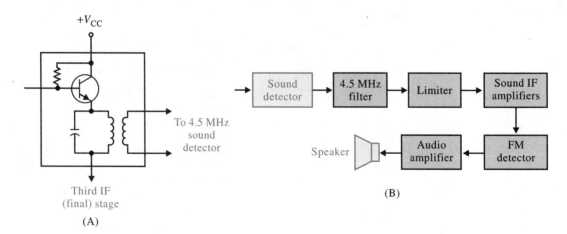

FIGURE 13-16 Audio section for a color TV receiver (A) Takeoff point (B) Block diagram

output is filtered to provide the 4.5 MHz FM intermediate-frequency signal. The remaining stages of the audio section are essentially the same as the equivalent stages in a broadcast FM receiver—that is, limiting, IF amplification, FM demodulation, audio output power amplification, and speaker output—as shown in Figure 13-16(B). The types of FM demodulators found in TV receivers include the ratio detector and Foster-Seeley discriminator in discrete component systems, the dual-control quadrature detector in vacuum tube systems, and PLLs and quadrature detectors in IC systems. In many of today's modern TV receivers, all the audio circuitry including the audio output power amplifier is contained in a single IC.

13-2 SYNC CIRCUITS

The composite video signal in the output of the video detector is fed to the video amplifiers and then to the *kinescope*, or *kine* (CRT or picture tube), where its intensity modulates the beam. However, it is also coupled to the sync circuits, the subject of the following sections.

Sync Separation

As shown in Figure 13-17, the signal is stripped of its video information in a process known as *sync separation.* Filter networks then separate the vertical and horizontal

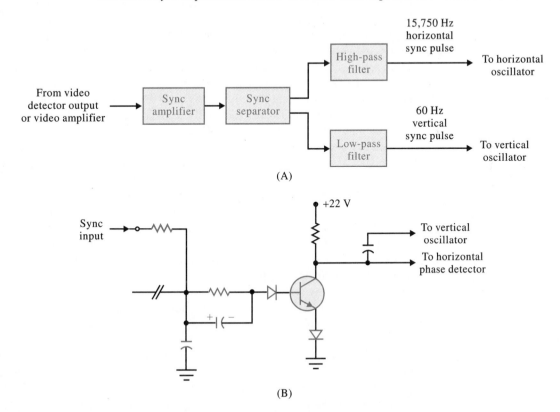

FIGURE 13-17 Sync circuits (A) Block diagram (B) Schematic of a sync separator circuit

sync pulses from each other. This process is called *intersync separation.* The vertical and horizontal sync pulses synchronize the vertical and horizontal sweep oscillators so that the intensity modulated scanning beam in the receiver is in exact synchronism with the camera scanning beam at the transmitter. Normally, there are two types of circuits in the sync separation stages: an amplifier circuit and a limiter circuit.

Sync Amplifier Stage   A *sync amplifier* may not always be used before a sync separator. The setup will depend on the type of transistor used as a separator and on the polarity of the sync pulse at the output of the video detector. If the video detector output is taken from its anode, the signal is negative-going. The negative-going signal may be used as is, if its amplitude is sufficiently high, by feeding it to a PNP-type separator. This arrangement eliminates the need for a sync amplifier before the separator. However, if an NPN-type sync separator is used, it is necessary to reverse the polarity of a negative-going video signal at the output of the detector before the signal can be coupled to the separator. In that case the output of a common-emitter video amplifier may be used as the input to the sync separator. On the other hand, to prevent loading of the video amplifier, a common-emitter sync amplifier may be used to feed the separator.

Sync Limiter Stage   The limiter stage is also known as the *sync clipper,* or the *sync separator.* This stage is biased to cut off in the absence of a sync pulse. When sync pulses of the proper polarity appear at its output, it turns on and goes into saturation. The input to the separator is the composite video signal, which includes video information and vertical and horizontal sync pulses. The separator is self-biased to follow variations in signal amplitude. It has two *RC* circuits in its input, with time constants such that the circuit responds only to the peaks of the input signals, which belong to the vertical and horizontal sync pulses. Thus, the separator does not "see" the video information in its input and responds only to the sync pulses. The output of the separator contains only the vertical and horizontal sync pulses.

Once the sync separator has clipped the sync pulse from the composite signal, the sync pulses are coupled to an *integrator* (LP filter) and a *differentiator* (HP filter). The output of the integrator is the 60 Hz vertical sync pulse. The output of the differentiator is the 15,750 Hz horizontal sync pulse.

The vertical sync pulse is fed to the vertical oscillator. The vertical oscillator generates the required 60 Hz waveform, which is amplified and applied to the vertical windings of the deflection yoke. A free-running relaxation oscillator is used in the vertical circuit. This oscillator may be one of three types: a blocking oscillator, a multivibrator, or a complementary-pair oscillator.

The horizontal sync pulse is not used directly but is fed to the horizontal automatic frequency control (AFC) system. Horizontal AFC systems are part of the horizontal oscillator section and will be discussed later.

Intersync Separation   Separated sync consists of both vertical and horizontal sync pulses. In cases of very weak signals, a large amount of high-frequency thermal noise is also present. Because the vertical sync pulse is a low-frequency (60 Hz) signal, it can be separated from the horizontal sync and from the noise by an LP filter. The horizontal sync pulses are high-frequency (15,750 Hz) signals, and an HP filter can separate them from the vertical pulses. However, this network cannot eliminate the noise from the horizontal sync pulses. To remove both the noise and the vertical signal from the horizontal pulses, AFC systems are used in the horizontal circuits. Low-frequency vertical pulses are removed by some differentiation, and the noise is removed by the *averaging* (fly-wheel) effect of the AFC system.

A three-section integrating network (Figure 13-18) is commonly used in TV receivers. This filter acts on a train of sync pulses. The time constant of the filter is chosen so that it is longer than the 5 μs horizontal sync pulse. As a result, during the horizontal sync interval, C_1 charges up to a small percentage of the amplitude of the applied pulses. However, since the time between the horizontal sync pulses is about 58 μs, much of the charge on C_1 can leak off during this interval. But with three sections in the integrating network, the capacitor charge effect is cumulative, and the leakage effect is negligible during the interval between the horizontal sync pulses.

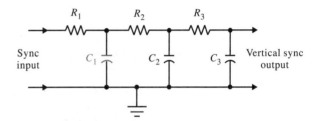

FIGURE 13-18
Three-section
integrator

Noise Inverters   In addition to sync amplifiers and separators, the sync stages usually contain some form of sync noise immunity (*noise inverter*) circuit. The purpose of the noise inverter is to prevent loss of sync due to impulse noise, such as that caused by an automobile ignition system. This type of noise signal voltage is usually much larger in amplitude than the sync pulses in the video signal.

Figure 13-19 shows a typical noise inverter circuit. The noise inverter, Q_1, offsets the effects of noise pulses in the following manner. Because of the long time constant of C_1R_5, diode D_2 acts as a peak detector, which rectifies the normal incoming positive-going composite video signal. The resulting dc voltage developed across C_2 is equal to the peak of the sync pulses. This positive voltage is applied to the cathode of D_1, cutting off that diode under normal conditions. Thus, no normal signal can get through D_1, and the base of Q_1 remains without signal under normal conditions. Also, since the base-emitter bias of Q_1 is zero volts, Q_1 is normally cut off. When a sudden noise spike appears, the voltage across C_1 cannot change quickly, and D_1 will couple the spike through C_2 to the base of Q_1. The noise inverter transistor will conduct heavily and effectively short the collector circuit to ground, shorting out the entire input signal to the sync separator, including video information and sync pulses, during the time of the noise spike. As soon as the noise pulse disappears, however, Q_1 cuts off again, and the next video signal and sync pulses will be unaffected.

The problem caused by noise pulses such as ignition noise is the loss of many sync pulses if the capacitor at the base of the sync separator is allowed to charge up. The function of the noise inverter is to eliminate the input signal at the time of a noise pulse. An occasional sync pulse may be lost due to coincidence with a noise pulse, but

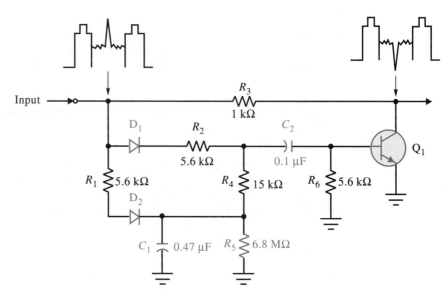

FIGURE 13-19 Noise inverter circuit

this is unimportant because of the flywheel effect which prevents the loss of sync for a prolonged period. The circuit in Figure 13-19 is self-adjusting in that the signal develops the bias that must be exceeded by the noise pulse to cause cancellation. Only noise pulses larger than the signal will be canceled. Noise pulses smaller than or equal to the normal signal will not be canceled because they cannot cause excessive charge on the base capacitor of the separator.

In some older TV receivers, the threshold voltage of the noise inverter is user-adjustable. This adjustment has the effect of canceling some sync pulses on strong stations when advanced too far, and of not providing adequate noise immunity when not adjusted far enough. Therefore, care must be taken in making the adjustment so that good signal performance is not sacrificed.

Vertical Sweep Circuits

The vertical sync output of the integrating network may supply a sync voltage to any type of vertical oscillator found in TV receivers. The type of oscillator and the point of sync injection will determine the polarity of sync pulse required.

The output amplifier in the vertical sweep section of a TV receiver is a power amplifier. Solid-state receivers use a high-current transistor. A power amplifier is required because the deflection yoke winding is a current-driven coil. Deflection is accomplished by the changing magnetic field around the yoke, which results from a changing current through the yoke windings. For linear deflection, a linearly changing magnetic field is required around the yoke. A linearly changing magnetic field is produced by a linearly changing current through the yoke. The condition for linear deflection, therefore, is that a linear current flow through the yoke during the trace period.

The way in which synchronization is achieved is similar regardless of the type of oscillator. As an example of the process, Figure 13-20 shows how a sync pulse is applied to a vertical blocking oscillator. The input pulse is applied to a separate winding

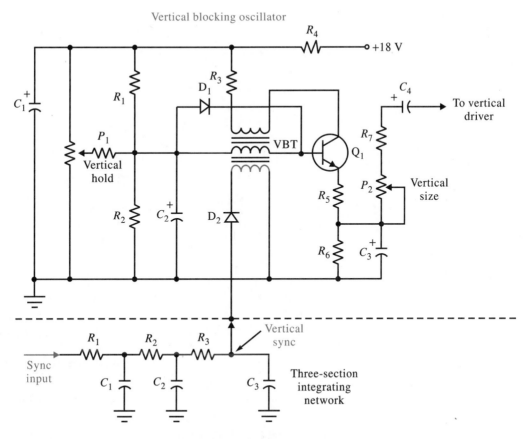

FIGURE 13-20 Vertical blocking oscillator circuit

on the vertical blocking transformer (VBT). Depending on which side of the winding is grounded, either polarity pulse can be used. As shown, a positive pulse is fed to this circuit. The vertical sync pulse is separated from the composite sync signal by a three-section integrating network. The effect of the sync pulse, at sufficient amplitude, is to bring the oscillator into conduction a brief interval before conduction would normally occur, thus providing synchronization.

Vertical Deflection

Figure 13-21 shows a typical vertical drive/vertical output system for vertical deflection. A sawtooth ac is developed in the RC network formed by capacitors C_1–C_2 and resistors R_3–P_2. This sawtooth current is fed to the base of vertical drive transistor Q_1, the output of which drives vertical output transistor Q_2. Adjustment of P_2 controls the sawtooth amplitude, thereby controlling the picture height.

The vertical output signal is fed to the vertical deflection yoke coils. Small resistors are paralleled with each of the coils to decrease inductive effects during the *flyback* (retrace) periods. During flyback periods, a voltage from the vertical yoke is fed to the picture tube cathode or grid to blank out the screen, preventing any retrace line from the bottom to the top of the screen from showing.

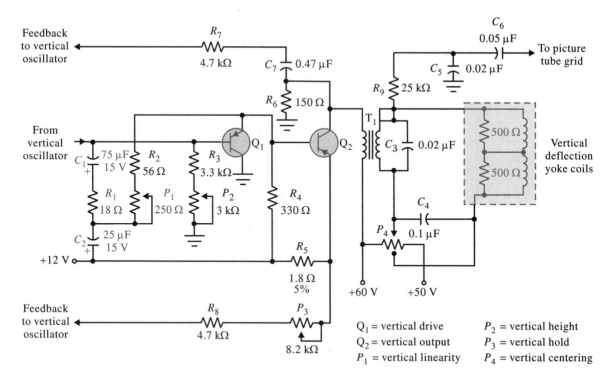

FIGURE 13-21 Typical vertical drive/vertical output circuit for producing a sawtooth ac for vertical deflection

Horizontal Sweep Circuits

The horizontal sweep circuitry is generally considered different from the circuitry used for vertical sweep, primarily because of the higher frequencies involved in the horizontal circuits. Heavy demands are made on the horizontal output transistors because of the high current levels necessary and the large-amplitude retrace voltages generated by the collapsing magnetic field at the end of each horizontal scan. In addition, most TV receivers derive their high voltage potentials by using a portion of the horizontal output signal to drive a high-voltage transformer.

The vertical output stage sees the deflection yoke as primarily a resistive load, but the higher horizontal scanning frequency forces serious consideration of the inductive effects of the horizontal windings of the yoke and other inductive components in the circuit. Output transistors with fast switching speeds are necessary to minimize power dissipation within the semiconductor elements during the actual switching period, and adequate heat sinks must be provided to conduct generated heat away and maintain a stable temperature.

Some early TV receivers triggered the horizontal oscillator directly by the horizontal sync pulses, but the circuitry was susceptible to false triggering by noise spikes passing through the sync separator circuitry. For this reason, the horizontal oscillator in most modern TV receivers is controlled by an AFC system that compares the phase of the horizontal output to that of the incoming sync pulses. If any deviation occurs between the two, a dc voltage is generated and applied to the horizontal oscillator to alter its frequency so that a constant phase relationship is maintained. This type of system is relatively immune to spurious signals, and a more stable mode of operation is achieved.

Horizontal Oscillator

The horizontal oscillator generates the sawtooth sweep current to drive the horizontal output amplifier. It is synchronized to operate at the 15,750 Hz line rate (15,734 Hz for color). Among the most popular oscillator circuits used in the horizontal sweep section are the blocking oscillator, some form of sine wave oscillator (usually a modified Hartley oscillator), and IC systems.

The sawtooth current that accomplishes the horizontal sweep is stepped up and rectified to produce the high voltages required by the picture tube. A small portion of the horizontal flyback voltage is used to key AGC and AFC circuits into operation. In color receivers, some of the output voltage may be shaped and used for convergence correction.

Figure 13-22 is a horizontal sweep oscillator with a blocking oscillator, Q_2, together with its AFC circuit. Frequency control of Q_2 is achieved by a dc correction voltage developed by phase detector CR_1–CR_2. The phase detector compares the frequency and phase of a pulse voltage derived from the horizontal output stage (not shown) and shaped by sawtooth generator Q_3 to those of a sync pulse delivered to the base of phase splitter Q_1. The dc output of the phase detector is used for direct control of the blocking oscillator.

The basic blocking oscillator components are transformer T_1 and capacitor C_8 together with the *hold* control and *range* control. When current flows in the primary (pins 2 and 5), a signal is fed back to the base through winding 4 and 6. This action causes the base of Q_2 to go positive, increasing the base and collector current and discharging C_8. When saturation is reached, the feedback stops and the negative voltage developed on C_8 cuts off the transistor. This result could cause a high pulse on the collector, but a protection circuit, diode CR_3, is in the circuit to clamp this pulse and prevent Q_2 from being damaged. Capacitor C_8 now charges through the hold and range controls from the 30 V source. When the voltage reaches about +0.6 V, the transistor starts to conduct and the cycle is repeated. The output of the oscillator is taken from a separate winding (pins 1 and 3) and is a 17 V_{p-p} positive pulse.

Automatic Frequency Control   The *automatic frequency control* (AFC) circuit in Figure 13-22 consists of the phase detector (CR_1–CR_2) and the circuits that feed it, the phase splitter (Q_1), and the sawtooth generator (Q_3). The base of Q_3 receives a 32 V_{p-p} positive pulse from the horizontal output stage. A sawtooth voltage is developed at the collector of Q_3, which is coupled by C_9 to the anode of CR_1 and the cathode of CR_2. At the same time, a positive horizontal sync pulse from the sync section of the receiver is applied to the base of the phase splitter Q_1. The phase splitter generates two sync pulses—a positive pulse at its emitter and a negative pulse at its collector. The collector pulse is coupled by C_2 to the cathode of CR_1; the emitter pulse is coupled by C_3 to the anode of C_2. If the oscillator is exactly on frequency, the comparison sawtooth voltage is going through zero volts when the sync pulse arrives. In this condition, the collector and emitter pulses from Q_1 turn on CR_1 and CR_2 equally. As a result, the output at test point 1 (TP_1, the junction of R_7 and R_8) will be zero, and the oscillator requires no correction for this condition.

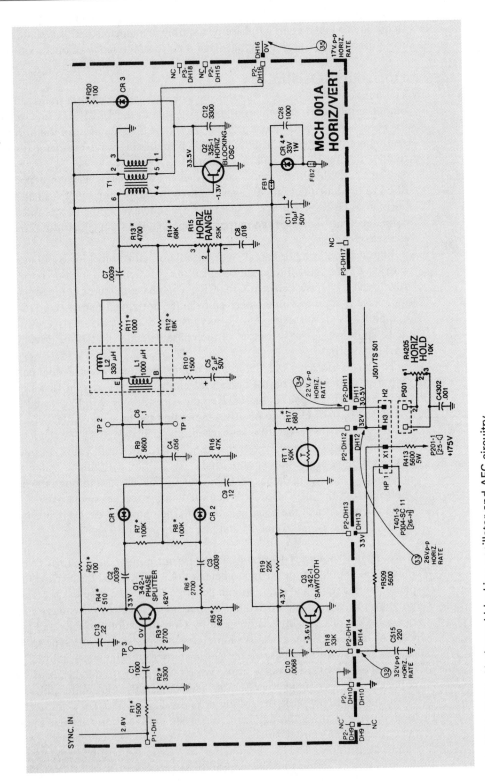

FIGURE 13-22 Horizontal blocking oscillator and AFC circuitry

If the oscillator frequency is low, the sawtooth voltage at R_{16} will be positive when the sync pulse arrives. Diode CR_1 will therefore conduct more heavily than diode CR_2, resulting in a positive correction voltage at TP_1. This voltage causes the base of oscillator Q_2 to go into conduction earlier than it otherwise would, increasing the oscillator frequency as required. If the oscillator frequency is too low, the opposite effect will take place; that is, CR_2 will conduct more heavily than CR_1. The correction voltage at TP_1 will be negative, thus slowing down the blocking oscillator.

The range control permits adjusting the horizontal oscillator to the correct frequency at the center of the hold control. In this manner, the user-operated hold control has a limited range, making operation easy. Coil L_1 and capacitor C_6 form a tank circuit at 15,750 Hz. This circuit is known as a *ringing* or *stabilizing* circuit and combines the stability of an *LC* tuned circuit with the ease of control of a blocking oscillator.

Another example of a horizontal sweep oscillator is shown in Figure 13-23. The multivibrator type of oscillator is followed by the driver stages needed to provide the proper drive signals to the horizontal output stage.

Transistors Q_1 and Q_2 form a multivibrator whose natural frequency of operation is approximately the same as the horizontal frequency of the incoming signal. A manual control of frequency is achieved by applying bias to the base of Q_1 through *horizontal*

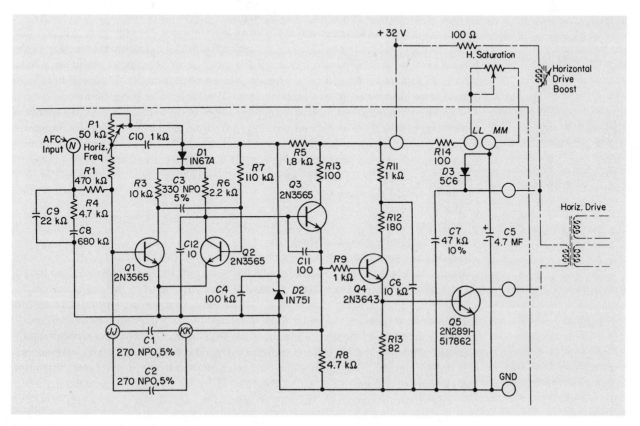

FIGURE 13-23 Horizontal multivibrator oscillator system

frequency control P_1. This voltage is obtained from a source that is regulated by the action of zener diode D_2, thus ensuring stable operation. An input for AFC voltage is also provided through resistor R_1 to the base of Q_1, allowing the oscillator to maintain the stable relationship with the horizontal sync as previously discussed.

Capacitors C_8 and C_9 serve as filtering elements for the AFC voltage. Transistors Q_3, Q_4, and Q_5 are current amplifiers to provide sufficient drive to the horizontal output stage. Transistors Q_3 and Q_4 operate as emitter followers to provide isolation for the oscillator and to allow sufficient current gain to control transistor switch Q_5. Transistor switch Q_5 has a transformer primary as its load that couples the output pulse to the horizontal output transistors.

Horizontal Output and Deflection Circuits   Sweep current is developed in the horizontal output circuits. It is supplied to the horizontal windings of the CRT yoke, which moves the scanning beam from left to right across the CRT screen. The energy in the horizontal output system is also the source of the highest voltage required by the CRT accelerating anode and of the relatively high dc voltages required in other circuits of the receiver.

A simplified schematic of a horizontal output stage is shown in Figure 13-24. The two horizontal output transistors are parallel driven through the action of the transformer secondaries connected to their respective base terminals. The transistors are connected in a series arrangement that allows them to share the voltage transitions that occur during retrace, thereby dividing the power-handling capabilities necessary in the individual transistor elements. Although advances in semiconductor technology have largely eliminated the need for series-connected output stages, their use here is to demonstrate how the high scan rates necessary for large screens are obtained.

The two horizontal output transistors are series-connected switches that open and close simultaneously. The current buildup in the yoke is not instantaneous, however, owing to its inductive nature. The current increases in a relatively slow manner, forming a sawtooth current waveform. The capacitive and inductive components in the circuit are chosen to create a time constant sufficiently long to allow utilization of only the initial part of the sawtooth curve. This setup ensures use of the most linear portion of the waveform and, since the spot position of the CRT electron beam is almost directly proportional to the current in the deflection yoke, results in a sweep pattern with good linearity characteristics.

During the scan, capacitor C_1, the yoke, and the series inductors form a resonant circuit. The first half of the scan is produced by energy stored in the yoke from the retrace action of the previous scan excursion. The current from the yoke charges C_1 through diode D_2. During this period, the current is increasing in the yoke and its series inductors, causing the beam to be deflected through the first half of its scan.

The second half of the scan is produced by discharging C_1 through the yoke with Q_2 and Q_3 completing the circuit as a closed switch. The voltage across the yoke and series inductor elements is maintained, so the current continues to increase. When the switch transistors open at the end of the scan, the yoke inductance, now paralleled by the series inductor, resonates with C_2 at approximately 70 kHz. The energy stored in the inductances causes the circuit to *ring* for half a cycle to produce retrace and reverse the current in the yoke. Diode D_2 is reverse biased during the first half-cycle of the ringing waveform but comes into conduction when the waveform starts to go positive, damping any further ringing.

The reverse current through the yoke causes the beam to retrace rapidly, as a function of the high ringing frequency. When retrace has been accomplished and the damper diode begins to conduct, the series inductance is again in the circuit, and the resonant frequency of the circuit is reduced to its original, lower value to allow normal scan operation.

High-Voltage Generation The high voltages necessary to operate a CRT are generally derived from the horizontal output signal. A *flyback transformer,* found in most TV receivers, transforms the horizontal retrace pulse that appears across its primary winding into the high potentials necessary. A diode in the secondary circuit rectifies the transformer output, and a capacitor-resistor network is usually used as filtering. Since the current drain is quite small, large values of capacitance are necessary.

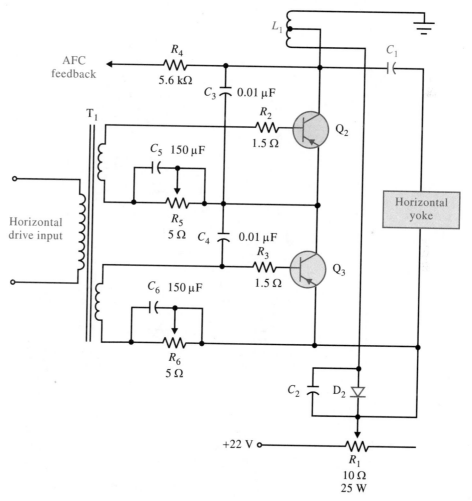

FIGURE 13-24 Simplified schematic of a dual-transistor horizontal output stage

Figure 13-25 shows the circuitry necessary to form the high voltages and illustrates the interaction between the horizontal output and high-voltage sections. The heavy lines indicate the path that the retrace current takes as the yoke discharges during the retrace interval. Note that the primary of transformer T_2 is in series with the yoke.

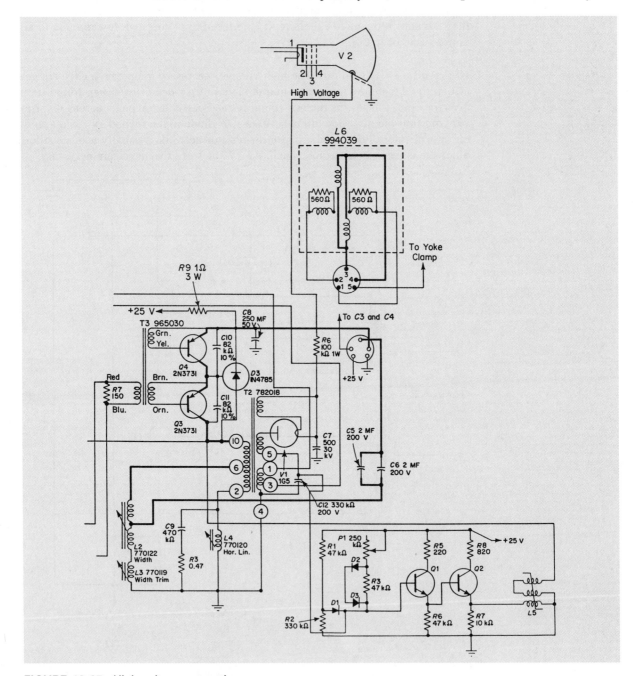

FIGURE 13-25 High-voltage generation

During retrace, energy is coupled into the secondaries, which are wound to effect a voltage step-up. Diode V_1 rectifies the secondary output and applies high voltage to the CRT anode through R_6, with filtering accomplished by capacitor C_7. Other windings of the transformer yield the lower voltages necessary for the accelerating and focusing grids.

High-Voltage Regulation   High-voltage regulation is necessary to eliminate variations in picture quality as brightness or contrast controls are changed or as wide variations in overall brightness levels are encountered in televised scenes. One method of regulation employs a *saturable reactor* (L_5) in parallel with the flyback primary winding. In operation, as the CRT anode current increases (as it would with higher brightness levels), the regulator assembly senses the increased current through the bottom of the high-voltage winding. Transistors Q_3 and Q_4 amplify the current to drive the control winding of the saturable reactor. As the control winding current increases, the saturable winding inductance is reduced, allowing the winding to store more energy during trace time. This stored energy, delivered to the flyback transformer during retrace time, produces additional voltage at the plate of the high-voltage rectifier V_1, thus maintaining the CRT anode voltage at a constant level.

Linearity and Width Control   Figure 13-25 also illustrates a novel means of obtaining control over the horizontal width of the CRT raster. Horizontal *width* control L_2 uses a series and a parallel coil coupled with a movable core. Moving the core changes the inductance in the device, and, since one half of the coil is in series with the yoke, the effective inductance of the entire horizontal circuit is changed. This change in inductance alters the current waveform during the scan and causes a corresponding change in horizontal size. As the width is varied by moving the core, the impedance of one half of the L_2 winding is increased while the impedance of the other half is decreased. This action allows a variation in width while presenting a constant load to the flyback and, therefore, maintaining a constant high voltage.

Horizontal *linearity* is achieved with a resonant circuit in series with the flyback primary (L_4 and C_9). This circuit adds a sawtooth and a parabolic component to the sawtooth current in the yoke. Coil L_4 determines linearity by controlling the amount and shape of the correcting voltage waveform.

13-3   SIGNALS UNIQUE TO COLOR RECEIVERS ___

Color TV receivers in the United States use the modulated National Television Systems Committee (NTSC) color video signal standard to achieve a color presentation. The NTSC system of color transmission stipulates compatibility; that is, the generated composite color signal must be able to produce a B/W presentation in a standard monochrome TV receiver with minimum disturbance from the color information.

The NTSC color TV signal is composed of two elements: (1) the luminance (Y) signal, also called the brightness signal; and (2) the color (I-Q) signal, also called the chrominance signal. In the following sections, we discuss how these two signals are handled in color receivers.

Luminance Signal

The *luminance* signal in color television is identical to the video signal commonly associated with monochrome television, and it is processed in a similar manner. Figure 13-26(A) shows a simplified flow path for the luminance signal, together with the associated noise, sync, and AGC blocks discussed earlier. Note the delay line between the first and second video amplifiers. The delay line delays the luminance signal to compensate for the delay that the chrominance signal experiences in the bandwidth-limited chroma circuits.

The two video amplifier blocks and the delay line are shown schematically in Figure 13-26(B). The first video amplifier, Q_1, a standard common-emitter amplifier, provides drive through the delay line to the second video amplifier, Q_2, an emitter follower whose output contains the contrast control. From this point, the video is routed to final drive amplifiers that provide drive to the picture tube.

The noise takeoff shown in both the block diagram and the schematic feeds a sample of the video signal to the noise separator, where the noise is removed from the video and inverted in the noise inverter. The inverted noise is then added to the emitter of Q_2 via the takeoff line connected to that point, where it tends to cancel the original noise. The signals used to drive the sync separator and the AGC circuits are also taken from the emitter of Q_2.

Chrominance Signal

The *chrominance* (chroma) signal is the portion of the composite color signal that represents the color information contained in the televised scene. This signal modulates a 3.58 MHz color subcarrier which is suppressed at the transmitter. The color burst signal riding on the back porch of the horizontal blanking pulse is used in the color receiver to phase-lock with a locally generated 3.58 MHz signal. As in any modulated signal, the receiver must demodulate the received signal for the signal to be useful.

Chroma Demodulation   Many types of demodulator circuits can be used to remove the chrominance information from the composite color signal. However, because of the nature of the color signal, demodulator circuits must be sensitive to a change in the phase of the input signal. They must provide an output signal that varies in amplitude according to the amount of phase shift. A phase shift is detected by comparing the chrominance information with the locally generated 3.58 MHz signal that is phase-locked to the color burst signal on the composite video waveform.

Recall from Chapter 12 that the I and Q signals generated at the transmitter were developed by using two 3.58 MHz signals 90° out of phase with each other. These two signals were then combined in varying amounts of amplitude, resulting in the chrominance portion of the composite color video waveform. Therefore, demodulation of this signal is accomplished by again using two signals that are 90° out of phase with each other.

A simplified block diagram of a chroma demodulator is shown in Figure 13-27. The output of the chroma amplifier is fed into two demodulator stages, the I detector and the Q detector.

The locally generated, crystal-controlled 3.58 MHz signal is applied directly to the I detector, and through a 90° delay to the Q detector. The detector outputs are a product

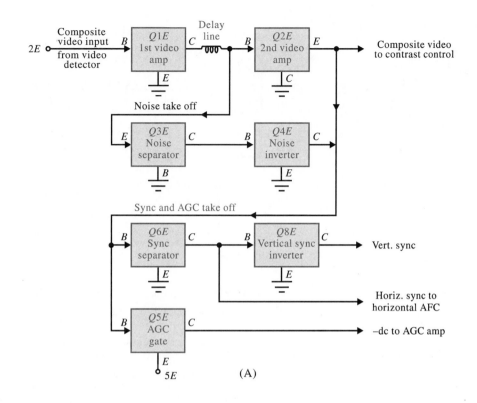

(A)

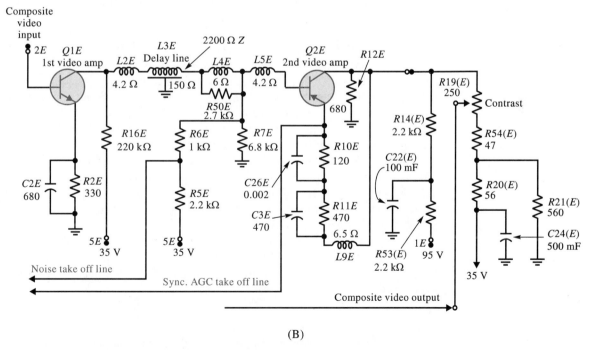

(B)

FIGURE 13-26 Luminance signal path (A) Block diagram (B) Schematic

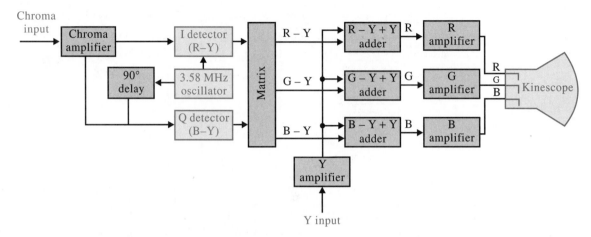

FIGURE 13-27 Simplified chroma demodulator

of the phase difference between their two inputs, but without the 3.58 MHz subcarrier, which is effectively filtered from the resultant video. These outputs are then matrixed together.

The matrix provides three amplitude modulated signals: R − Y, G − Y, and B − Y. Once these three major components of the color signal are reproduced, the luminance (Y) signal is added to each to obtain the red, green, and blue signals that were originally generated by the three respective camera pickup tubes at the transmitting station. These signals are then amplified sufficiently to drive the three color guns in the picture tube.

Color Killer   In the NTSC compatible system, color receivers must be capable of receiving monochrome signals without disturbance from color signals. For that reason, a circuit called a *color killer* is included in color receivers. The function of this circuit is to disable the chrominance circuits when a B/W signal is being received. This disablement ensures that only luminance information will reach the picture tube.

A schematic of a typical *automatic color control* (ACC) amplifier is shown in Figure 13-28. Any 3.58 MHz color burst signal at the input of the circuit will be rectified by diode D_1, filtered, and applied to the base of Q_1. Transistor Q_1 acts as a switch, turning ON when the burst-generated 3.58 MHz signal is present at the input. As a result, the switching action of this stage controls the color processing circuits, turning them ON when a color signal is present at the input and turning them OFF when only a B/W signal is being received.

Many variations of this circuit are used, but they all have one thing in common. They sample the incoming video signal for the presence of a burst. If no burst is present, the chrominance circuits are disabled and a monochrome presentation is obtained.

Color Picture Tubes
Two basic types of color picture tubes (CRTs, or kinescopes) are used in modern color TV receivers: (1) the three-gun, shadow-masked tricolor kinescope and (2) the

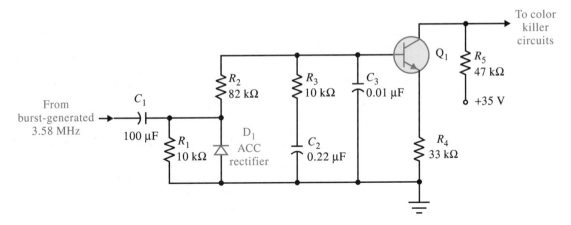

FIGURE 13-28 Color killer circuit

one-gun kinescope. A discussion of the very different characteristics of these two CRTs follows.

Three-Gun CRT In the *three-gun, shadow-masked tricolor* CRT (Figure 13-29(A)), color is generated by using three beams of electrons to bombard tiny phosphor dots that cover the face of the tube. The dots covering the face of the tube are arranged in groups of three (*triads*), consisting of red, green, and blue (Figure 13-29(B)), with each color energized by its respective electron beam. There are hundreds of thousands of these phosphor dot triads on the faceplate of modern color CRTs. The dots are so small that the human eye cannot discern each as a specific element of red, green, or blue. Instead, the eye integrates them into areas of colored light, the color of which is determined by the relative intensities of the three primary colors being emitted by the dots.

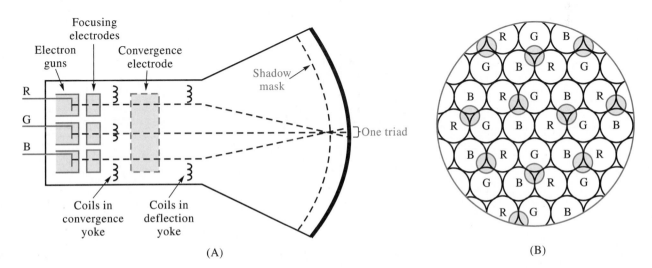

FIGURE 13-29 Three-gun, shadow-masked picture tube (A) Components (B) Color triads

The three high-velocity electron beams are generated by electron gun assemblies positioned in a triangular configuration 120° apart at the rear of the CRT. All of the assemblies are tilted slightly toward the centerline of the tube to cause all three beams to converge near the faceplate. The three *focusing electrodes* have separate high-voltage potentials (about +6 kV) to focus their own beams to a tiny spot on the faceplate.

The *convergence electrode* located at the front of the gun assembly is a magnetic shield that allows each beam to be magnetically shifted in position without appreciably affecting the others. A *convergence yoke* (not the deflection yoke) is positioned on the tube exterior to place the red, green, and blue convergence coils and magnets directly above their respective beams.

Directly behind the phosphor dot screen is the *shadow mask,* a thin metal sheet perforated with tiny holes approximately 0.01 inch in diameter, one hole for each triad on the faceplate. Thus in order for the electron beams to reach their respective phosphor dots, all three beams must converge at this shadow mask, pass through the tiny opening, and deconverge. Proper convergence of the beams and correct placement of the mask ensure that each beam will strike only the particular color of phosphor dot.

One-Gun CRT   The *one-gun* color CRT is quite different from the three-gun CRT. It has only one gun and a simplified focusing system. Instead of a shadow mask with holes, it has a *vertical-stripe metal grille* behind *vertically striped phosphors.* It also has a much simpler convergence system.

Although this CRT has only one gun, it has three separate cathodes, as shown in Figure 13-30. The red, green, and blue signals are applied to the cathodes. Grid G_1 is a single sheet of metal with three openings through which electrons from the three cathodes pass. Grid G_2 is a high-potential accelerating anode, which pulls the electrons out past the three openings in G_1 and accelerates them on to the focusing electrodes G_3

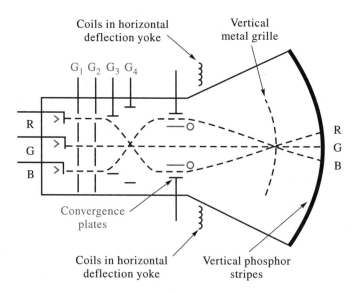

FIGURE 13-30 One-gun picture tube

and G_4. The focusing electrodes converge (focus) the three beams to a point here, but the beams diverge as they move forward. The two inner convergence plates are zero-charged, and the green beam moves straight through, bending only when under the influence of the deflection yokes. The positively charged red and blue beams approach the outer negatively charged convergence plates and are repelled back toward each other. The three beams are rejoined and pass through an opening in the vertical metal grille. The red and blue beams then deconverge and strike phosphor stripes on the faceplate of the tube.

No vertical convergence circuitry is required. Proper horizontal convergence to produce the correct landing of the beams out at the edges of the screen requires a slight parabolic convergence voltage added to the horizontal sweep. Positioning the deflection yoke slightly forward or backward on the neck of the tube helps achieve good horizontal convergence.

The Complete Color Receiver

A block diagram of a typical color receiver is shown in Figure 13-31. It illustrates the interaction among the various circuits discussed in this chapter. As previously stated, many of the blocks have counterparts in monochrome receivers. However, flat BP characteristics, signal delay times, and circuit stability are much more important and stringent in color receivers. Also, the setup, operation, and maintenance of color receivers are much more complex than for monochrome receivers.

13-4   REMOTE CONTROL SYSTEMS

One convenience of modern TV receivers is remote control operation. Different systems are used by different manufacturers. The simplest—a remote ON-OFF switch, extension cord, and power receptacle assembly—can be purchased as an accessory. Its line cord is plugged into a wall socket. The receiver's line cord is plugged into the receptacle of the remote control ON-OFF assembly. In this system, the receiver ON-OFF switch must be ON for remote operation.

This simple system can only turn the receiver ON or OFF. More complex systems permit remote control of ON-OFF switching, volume level, color level, tint adjustment, channel selection, and screen display of channel number and time of day. In some systems, contrast and brightness can also be adjusted. Some simpler systems permit only volume muting rather than full-range volume control.

In modern remote systems, *infrared* (IR) is the controlling link between the remote device and the receiver. Infrared refers to a band of electromagnetic radiation with wavelengths that lie within the range of 770 nanometers (nm) to 1 millimeter (mm). The infrared wavelength is longer than that of visible light; the shortest IR wavelengths occur in the range just outside the visible spectrum; the shortest border on the microwave part of the electromagnetic spectrum.

The remote system has three main components: (1) a remote transmitter; (2) a remote receiver; and (3) devices within the receiver to effect the change ordered by the remote command.

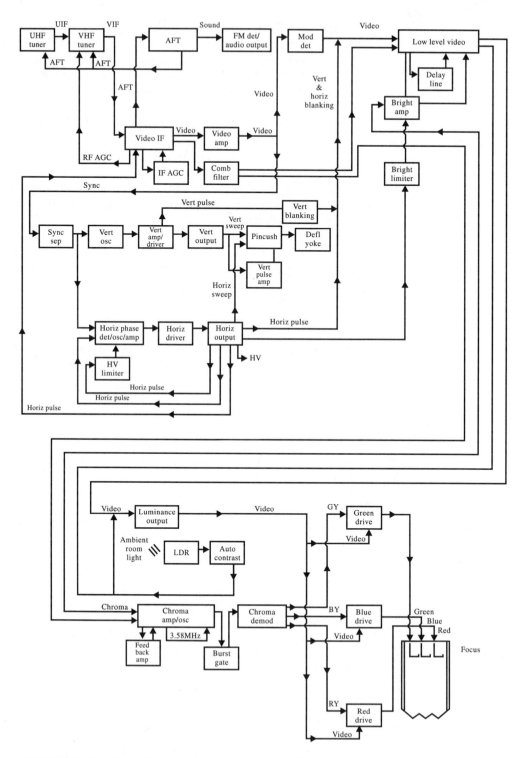

FIGURE 13-31 Block diagram of a complete color TV receiver

Remote Control Transmitter

A remote transmitter must send a command signal to the receiver. For each function, a different signal frequency is transmitted.

Figure 13-32 shows the layout of a typical infrared remote control transmitter. This unit is dedicated to a single TV receiver equipped with an IR remote control sensor. More sophisticated units are the so-called *universal* remote control transmitters. The universal units can be used to control a number of components (each equipped with IR remote sensors), such as audio, video, or satellite components, or any combination thereof.

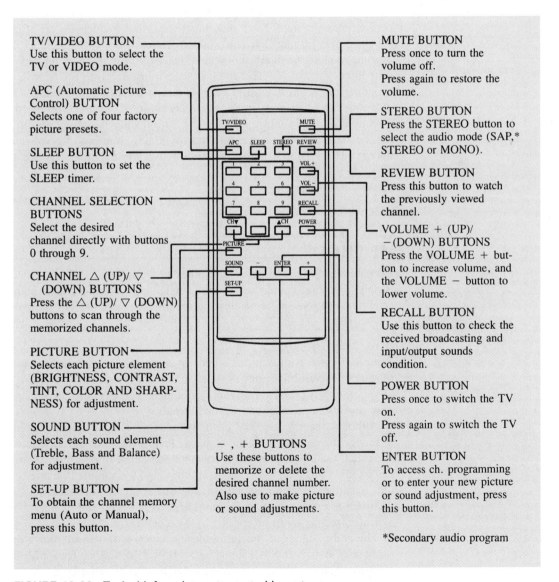

TV/VIDEO BUTTON
Use this button to select the TV or VIDEO mode.

APC (Automatic Picture Control) BUTTON
Selects one of four factory picture presets.

SLEEP BUTTON
Use this button to set the SLEEP timer.

CHANNEL SELECTION BUTTONS
Select the desired channel directly with buttons 0 through 9.

CHANNEL △ (UP)/ ▽ (DOWN) BUTTONS
Press the △ (UP)/ ▽ (DOWN) buttons to scan through the memorized channels.

PICTURE BUTTON
Selects each picture element (BRIGHTNESS, CONTRAST, TINT, COLOR AND SHARPNESS) for adjustment.

SOUND BUTTON
Selects each sound element (Treble, Bass and Balance) for adjustment.

SET-UP BUTTON
To obtain the channel memory menu (Auto or Manual), press this button.

–, + BUTTONS
Use these buttons to memorize or delete the desired channel number. Also use to make picture or sound adjustments.

MUTE BUTTON
Press once to turn the volume off. Press again to restore the volume.

STEREO BUTTON
Press the STEREO button to select the audio mode (SAP,* STEREO or MONO).

REVIEW BUTTON
Press this button to watch the previously viewed channel.

VOLUME + (UP)/ – (DOWN) BUTTONS
Press the VOLUME + button to increase volume, and the VOLUME – button to lower volume.

RECALL BUTTON
Use this button to check the received broadcasting and input/output sounds condition.

POWER BUTTON
Press once to switch the TV on. Press again to switch the TV off.

ENTER BUTTON
To access ch. programming or to enter your new picture or sound adjustment, press this button.

*Secondary audio program

FIGURE 13-32 Typical infrared remote control layout

Remote Control Receiver

The remote control receiver (IR sensor) is installed within, and interconnected with the main TV chassis. Figure 13-33 shows a typical IR remote TV receiver. The two possible inputs to this system are the remote IR signal and the pushbuttons mounted on the front panel of the receiver.

The remote IR signal is received by the IR sensor, which converts the IR signal into an electrical signal. The electrical signal is amplified and fed to internal modules for processing. The pushbuttons on the TV front panel will operate the modules directly.

Many variations of remote receivers are manufactured. The electrical signal that is the input of the IR sensor is amplified in one or more stages of broadband amplification. The band is wide enough to cover the frequencies for all functions of the particular remote system. Next, individual command frequencies are separated to operate the functions. In many systems, this separation is done by a number of tuned circuits. In some modern receivers, a digital IC is used to count the number of pulses in a given time interval, say $\frac{1}{60}$ s. The count is then translated into the function, and the corresponding pin of an IC will change its output to initiate the desired function. In most of these sophisticated systems, noise immunity is achieved by performing the count a number of times and operating the output only if the counts for equal intervals are the same. At least three, and usually more, counting intervals are used to verify the frequency of the command signal. All these operations are performed by a single IC.

13-5 OTHER TELEVISION RECEPTION METHODS

Reception of TV programs is not limited to the individual TV antenna. Other methods include reception via pay television, coaxial cable system, master antenna system, and satellite earth station. The following sections examine each of these methods.

Subscription or Pay TV

Some UHF television stations broadcast *scramble* TV signals that require a subscriber to rent a descrambling unit in order to receive the signal. This system is called *subscription* or *pay* television because only those persons (subscribers) who pay to rent a *descrambler* can use the transmitted signals.

Although there are several methods for scrambling a signal, one simple approach is to reduce the amplitude of the blanking levels, the sync pulses, and the color burst to about half the black level of the transmitted picture information. Normal TV receivers cannot maintain sync under this condition, and a scrambled picture results. The rental unit, installed between the antenna and the receiver, regenerates the required amplitudes for blanking, sync, and color burst. This action causes a normally blanked and synchronized signal to be applied to the receiver.

The audio signal of the scrambled program is transmitted as a double-sideband, suppressed-carrier (DSSC) signal on a 31.5 kHz subcarrier (twice the horizontal sweep frequency), modulating the aural carrier frequency. This DSSC transmission is done to further prevent the use of the transmitted program by nonsubscribers. The sound

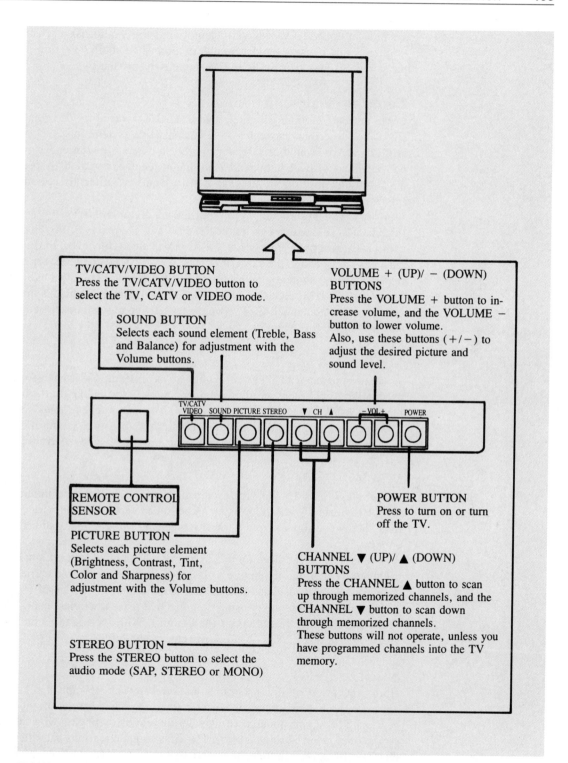

FIGURE 13-33 Typical TV receiver layout with infrared remote control sensor

systems in normal TV receivers cannot hear this audio. Except for the subcarrier frequency that is used, the descrambler employs an audio detection circuit similar to the stereo channel detector of an FM stereo transmission.

Cable or Community Antenna TV

Cable or *community antenna* television (CATV) is used to feed relatively clean TV signals to subscribers through a 75 Ω coaxial cable system. In some systems, the cable is strung above ground on telephone poles; in others, it is run underground. The cable company may provide network TV programs, locally generated TV programs, satellite TV programs, motion pictures, sports programs, weather, time, and bulletin board information with FM music.

Most CATV systems carry the normal 12 commercial VHF television channels (channels 2–13), which can be received by any TV receiver. In systems where more than 12 TV programs are to be relayed or translated (converted) to the VHF television channels on the cable, the subscriber must rent a special converter for the receiver. The rented converter switches the programs being transmitted on other cable frequencies to one of the 12 VHF television channels. This switching also blanks out any other programs from that channel. CATV companies may employ any of the contiguous 6 MHz channels in the following bands: 5.75–47.75 MHz, 120–174 MHz, 216–450 MHz.

Master Antenna TV

A variation of the CATV system is the *master antenna* TV (MATV) system, which is used in large apartment buildings, hotels, motels, and the like. In the MATV system, a master TV antenna picks up all receivable TV signals. These signals are then applied to distribution amplifiers. In multilevel buildings, distribution amplifiers are generally located on each floor. The amplified signals are then fed to receivers in each room.

Wireless Cable TV

The *wireless cable* TV industry, established in 1986, is still in its infancy. This new approach to home TV reception is a competitor to cable TV. The wireless system transmits microwave signals from a tower to beam TV programming directly to a housetop antenna.

Wireless cable can have up to 33 channels. Compared to traditional cable, wireless cable has a better quality picture, is less expensive to set up, and carries a stronger signal. A disadvantage is that—because the system operates on line-of-sight—hills, trees, and buildings impede the transmission. For that reason, wireless cable TV presently is found in relatively flat areas of the country. With increased technology, however, greater use of this medium can be expected in the future.

Satellite TV

Home reception of TV programs via satellite has become a relatively standard approach, particularly in areas where standard systems cannot provide acceptable reception. The key to understanding satellite television begins with the satellite itself (sometimes referred to as a *bird*). The first communications satellite, *Telstar 1,* was launched by American Telephone and Telegraph Company (AT&T) on July 10, 1962. It could carry one TV signal or 12 telephone calls at a time.

Today, satellites can carry up to 24 TV channels. They hover 22,300 miles above the earth in a *geosynchronous orbit,* which means that they orbit at the same speed that the earth turns on its axis so that they always can be found at their specific locations in the sky. This band of satellites is called the *Clarke belt* in honor of science fiction writer Arthur C. Clarke, who first proposed, in 1945, that satellites could be maintained in space in just such a fashion.

Television satellites vary in design, but they all have the same basic parts. They contain control and propulsion systems that can be manipulated from earth should they stray from their orbit, and they are equipped with solar panels that produce the electrical power to operate their systems. Some newer satellites contain 24 *transponders,* or relay systems, which are similar to the channels of your TV receiver. However, some of the satellites launched earlier have only 12 transponders. Each transponder can carry one color TV program or up to 2400 one-way voice circuits.

The TV signal begins at the TV studio, where it is fed to an *uplink* that beams the signal up to the orbiting satellite. The satellite, in turn, beams the signal back down to earth via a *downlink.* The downlink signal is very weak, only about as strong as the signal projected by a CB radio, so its strength is not the same for everybody. Most returning signals are aimed at the Midwest, and the intensity of the signal drops as the signal radiates outward from that point. However, with the notable exceptions of the extreme Northeast and the southernmost tip of Florida, all areas of the nation can expect quality reception, and even viewers in those two areas can get a good picture with the right equipment.

The equipment required at an earth receiving station to pull in the satellite TV signal includes a motor-driven 6–20 foot *dish* (parabolic antenna), made of highly reflective materials to gather and focus the signal to a *feedhorn* assembly at the center of the dish; a 3.7–4.2 GHz broadbanded *low-noise amplifier* (LNA), which boosts the strength of the signal without contributing noise; and a *down-converter,* to lower the gigahertz signal down to a 70 MHz TV intermediate-frequency signal to be fed to a *satellite receiver.*

With a 30 MHz bandwidth FM signal centered on one of twelve 40 MHz wide channels, both the video and the audio signals of the TV program are made to frequency modulate the earth station transmitter. The audio signal frequency modulates a carrier at about 6.5 MHz, the resulting signal of which is added to the 0–4.5 MHz video signal.

The transponder in the satellite then beams back down to earth a 30 MHz bandwidth FM signal containing both the 0–4.5 MHz video information and the 6.5 MHz FM audio carrier. The discriminator output at the receiver is an AM-type 0–4.5 MHz video signal and a 6.2 MHz FM carrier modulated by the audio information. Although the video can be fed directly to a video monitor for display, the FM audio information must be detected with a 6.5 MHz discriminator before being fed to an audio amplifier system.

If a home TV receiver is used to display the TV program, the video is made to amplitude modulate a video oscillator, and the detected audio is made to frequency modulate an audio oscillator, usually for TV channel 2 or 3. The resultant modulated signal is fed to the antenna input of the TV receiver to be modulated and displayed when the receiver is tuned to channel 2 or 3.

Despite the advantages of owning a home satellite TV reception system, the size of the receiving antenna has had a limiting effect on purchasers. To receive the satellite signals and deliver the high-quality audio and video reception that a home system provides, the home-use reflectors had to be at least 12 feet in diameter.

With improved technology, reflectors have been down-sized from a standard 12 foot diameter in 1982 to a 7 foot diameter in 1992. The new 7 foot reflector provides the same or better reception as the earlier, bigger ones. Even smaller reflectors are being developed.

Digital Satellite System (DSS)

Replacing cable with sophisticated *digital satellite systems* (DSS) is becoming commonplace, especially in rural areas. Marketed under several names, such as DirecTV, this service provides a small (18-inch) receiving dish and a set top box with remote control. Such systems can receive up to 175 channels, some of which include commercial-free music channels with digital quality sound. The actual number and types of channels depend upon the subscriber's desire and ability to pay.

Basic installation of these digital systems requires exterior mounting of the dish with an unobstructed view toward satellite locations (depending on what part of the country installation is being made). The dish is connected to the set top translation device by RG-6 or equivalent coaxial cable. The connection requires grounding according to the National Electrical code.

High Definition Television

Advanced digital television systems have appeared in recent years, with the concept of *High Definition Television* (HDTV) leading the way. HDTV was in the standardization process for several years, requiring review of many competing proposals. Acceptance of a standard (an aspect ratio of 16:9 and a scan rate of 1125 lines) for digital TV was approved by the Federal Communications Commission (FCC) in January 1997. This standard will allow the delivery of crisp, digital signals, CD-quality sound and much sharper pictures to home viewers by 1998. In effect, the result is a format called *high definition TV*. A digital TV screen can provide about four times the detail of standard screens.

Before broadcasters can provide digital TV, the FCC must make slices of the public airwaves available for the service and issue new digital TV licenses. Both steps are expected to be finalized in 1998.

Some broadcasters immediately implemented plans to allow digital broadcasting, in which they will broadcast both the current analog signals and the digital ones. This dual capability will allow for a gradual transfer to digital TV without rendering the 220 million TV sets nationwide useless. This simultaneous broadcast will transmit programs using two TV channels, one for programs in the existing analog format and the other in the new digital format. The digital systems are being developed in such a manner that they can be downgraded to be received by analog sets.

Most of the assigned channels the broadcasters will use to carry digital programs while they continue broadcasting on their existing channels will be in the lesser-used UHF range (channels 14–69). Stations will be allowed to operate both analog and digital stations for about 10 to 15 years. Then the FCC will require that the analog station be shut down.

Although Japan is already broadcasting HDTV (they call it Hi-Vision), their system is based on analog transmission from satellite to home. The standard to be adopted in the United States will be digital—a more computerized system that allows for greater precision and can be pushed to uses not yet envisioned.

The HDTV image has approximately twice as much luminance definition horizontally and vertically as do the present 525-line NTSC systems. This means that the total number of picture elements (pixels) in the image is four times as great, and the wider screen adds one quarter more.

The increased vertical definition is achieved by using more than 1000 lines in the scanning patterns. A video bandwidth about five times that used in conventional systems provides the increased luminance detail in the image. Additional bandwidth is used to transmit the color values (chrominance) separately, so the total bandwidth (in a color system) is six to eight times that used in existing color TV systems.

The aspect ratio of the HDTV image of the proposed HDTV systems is wider than conventional images—that is, the HDTV aspect ratio is about 16:9 (width to height) compared with the conventional 4:3 ratio.

What this all means to the technician is a rethinking of how to work with television. But because of the high cost of any new system, it will be several years before large numbers of HDTV sets are in use, and by then technical information will be in ample supply.

13-6   SAFETY PRECAUTIONS

Manufacturers of TV receivers spend much engineering effort, time, and money to make their products as safe for consumers and service technicians as possible. Therefore, it is important that technicians not defeat any of the built-in safety features due to carelessness, ignorance, or use of improper replacement parts. In the sections to follow, we will discuss some of the problems that may arise from improper service techniques.

Hot Chassis

Line-connected (*hot chassis*) receivers must have insulation so that no chassis point is available to the user, because the ac line can be lethal. Therefore, it is important that after servicing, all insulators, such as nylon bushings, knobs, and fishpaper insulating barriers, be replaced as found originally. Make sure that all replacement parts have the proper insulators, such as the nylon shaft on potentiometers. Mounting screws also must be insulated, usually by mounting into nylon bushings.

All of these precautions are for protection of the consumer. To protect themselves, technicians must use an *isolation transformer* whenever a line-connected receiver is being serviced.

B^+ and High Voltage

Coming into contact with the B^+ and HV sources in the receiver can be dangerous. A sudden jerking away from an unexpected jolt of voltage can result in bruises or severe cuts to the hand. Technicians should always stand or sit on an insulated surface and use only one hand when probing into a receiver.

Most modern receivers use an *interlock* with the back cover of the receiver so that any time the back cover is removed, no voltage is applied to the set. The interlock and back cover must be replaced in their original position to ensure that the receiver's high-voltage points are not accessible to the user. *RC* networks connecting the cabinet and the CRT mounting parts to the chassis must be installed properly to prevent the buildup of charge on any floating part of the cabinet, which can cause electrical shock.

CRT

The possibility of *implosion* is always present with a CRT because the inside of the device is a vacuum. If the envelope is damaged, the glass may shatter violently and fly great distances with great force. Older receivers were protected by safety glass or plastic windows placed in front of the CRT. In modern receivers, the faceplate of the CRT is shatterproof and thus provides integral protection. However, to maintain the protection originally built into the receiver, and because a number of different systems are used, it is important that only exact replacement of the CRT be made.

Because of high voltages and the possibility of implosion, extra precautions are necessary when handling CRTs. The technician should always wear safety goggles and gloves. Before removing and installing a picture tube in a cabinet, disconnect the line cord and discharge the CRT high-voltage capacitor by shorting a jumper from the chassis to the anode (*ultor*) button of the CRT. The capacitor formed by the aguadag coating and the outer graphite coating on picture tubes will redevelop a charge as the dielectric stresses in the glass are relieved. Even if the CRT is removed and left stored for some time, you may receive a shock if your finger touches the anode button. It is therefore a good practice to leave the short on for a few minutes and to recheck for charge before handling the CRT at some later time.

Radiation of X-Rays

Although a minimal danger, internal misadjustment of a receiver can cause x-rays to be radiated. Possible sources of this radiation are the CRT, the high-voltage rectifier, and the high-voltage shunt regulator. For that reason, special glass is used in some of these devices to prevent radiation. To maintain safe operation, these tubes must be replaced with types having the exact specific designations.

X-radiation also depends on the level of high voltage in the receiver. Therefore, it is imperative that the high voltage be maintained at the design level. Any voltage increase above the design level can cause radiation, and any drop below the design level will deteriorate picture quality and brightness.

In some receivers, the shielding is designed to limit the radiation of x-rays outside the receiver. Therefore, all shields, especially around the high-voltage and regulator area, must be in place before the receiver can be operated. They also should be kept in place during servicing to protect the technician.

Fire Hazard

The hazard of fire is always present with any electrical appliance. Technicians must be especially careful not to introduce a fire hazard in the repair process. To avoid overheating, correct replacement parts should always be used. Correct replacement is particularly important in circuits where high power is available, such as the power supply and the output circuits of the audio, vertical, and horizontal stages. In the horizontal section, particular care must be taken with the horizontal output transformer. Not only is power consumed in the horizontal output circuit, but high voltage is present, which can cause *arcing,* a possible fire source. Another reason for using exact replacement parts is that special materials are used in many receiver components to ensure that the components are *self-extinguishing*—that is, that they will not continue to burn when the original flame or heat source is removed.

An important consideration in receiver fire safety is lead and component dress. A number of hot components in the receiver must not touch any wires or other possible flammable materials such as plastics, tapes, or paper. The hottest parts are the tube envelopes, output transistors, and high-wattage resistors. Every effort must be made to maintain lead and component dress exactly as arranged originally by the manufacturer. All wire ties and all cabling should be reinstalled in original positions if they are disturbed during servicing.

SUMMARY

- Television receiver circuitry is capable of handling video, audio, and sync signals; generating high voltages for the picture tube; and separating and channeling the video, audio, and sync signals into their respective areas. The RF tuner receives and amplifies RF television signals and selects the desired TV channel to the exclusion of all others. Tuners can receive 12 VHF and 55 UHF television channels. VHF and UHF are received on separate tuners within the same TV set. VHF channels 2–13 range from 54 MHz to 216 MHz. UHF channels 14–69 range from 470 MHz to 806 MHz.

- Television receiver IF amplifiers provide a 6 MHz pass band, from 41 MHz to 47 MHz. The sound carrier is 41.25 MHz, and the picture carrier is 45.75 MHz—an exact 4.5 MHz separation. Modern intercarrier receivers amplify both the video and audio carriers and their respective sidebands. Stagger tuning broadens the BP response, provides steep skirts, and reduces the possibility of interstage coupling by tuning each IF stage to a higher or lower frequency than the stage immediately before or after it.

- In a color receiver, the audio subcarrier must be attenuated at some point before the video detector stage to eliminate a 920 kHz beat frequency between the 4.5 MHz audio subcarrier and the 3.58 MHz color subcarrier. The video detector demodulates the picture IF carrier and operates in much the same manner as the AM detector in a broadcast band AM receiver, except that it must have a 4.5 MHz pass band. Its output circuit contains the composite video signal.

- Peak detectors can be used in AGC systems to overcome the problem of varying brightness levels in the video signal. Delayed AGC provides maximum RF amplifier gain for weak signals. Keyed AGC develops a voltage that controls the gain of the RF amplifier and the IF section. In the keyed AGC system, the AGC signal is developed only during sync pulse time.

- The video amplifier raises the level of the composite video signal to an amplitude sufficient to drive the picture tube cathode or grid for proper beam modulation. The dc restorer adds a dc bias voltage that remains just at the blanking level but that follows any variation of sync pulse peak amplitude caused by fading. The dc restorer is not required in receivers where the video amplifier is directly coupled to the picture tube. Picture brightness is controlled by the amount of cathode–grid bias applied. Picture contrast is controlled by the video amplifier signal gain current.

- Except for the method of audio takeoff and the IF involved, the audio stages in a TV receiver are essentially the same as equivalent stages in a broadcast FM radio receiver.

- Sync separation strips the video information from the composite video signal. Vertical and horizontal sync pulses are separated from each other by filters in intersync separation. Sync amplifiers might not always be used before a sync separator. The sync separator output signals are coupled to an integrator, which produces the vertical pulse, and to a differentiator, which produces the horizontal pulse. The vertical oscillator generates a 60 Hz waveform that is amplified and applied to the vertical windings of the deflection yoke. The horizontal sync pulse is not used directly but is fed to the horizontal AFC system, which is part of the horizontal oscillator section. Sync stages usually contain some form of sync noise immunity or noise inverter circuit that prevents loss of sync due to impulse noise. The noise inverter eliminates the input signal when the noise pulse is larger than the signal.

- The vertical sweep output stage is a power amplifier. Deflection is achieved by the changing magnetic field around the yoke, which results from a changing current through the yoke windings. The horizontal sweep circuitry handles much higher frequencies than does the vertical sweep circuitry. The output stages have high current levels, and large-amplitude retrace voltages are generated by the collapsing field at the end of each horizontal scan. High voltage potentials are derived by utilizing a portion of the horizontal output signal to drive a high-voltage flyback transformer. A small portion of the horizontal flyback voltage is used to key AGC and AFC circuits into operation. The horizontal sweep current, developed in the output circuits, is supplied to the horizontal windings of the CRT yoke, which moves the scanning beam from left to right across the CRT screen.

- The NTSC color signal must be compatible; that is, it must be able to produce a B/W presentation in a standard monochrome TV receiver with minimum disturbance from the color information. The NTSC color signal contains the luminance (Y) and the chrominance (IQ) signals. The luminance signal in color television is identical to the video signal commonly associated with monochrome television, and it is processed in a similar manner. The chrominance signal represents the color information contained in the televised scene. It modulates the 3.58 MHz color subcarrier. An 8-cycle color burst riding on the back porch of the horizontal blanking pulse is used in the receiver to phase lock with a locally generated 3.58 MHz signal.

- Color signal demodulators must be sensitive to a change in the phase of the input signal. They must provide an output signal that varies in amplitude according to the amount of phase shift. Demodulation is achieved by using two signals that are 90° out of phase with each other. A color killer circuit in a color receiver disables the chrominance circuits when a B/W signal is being received. The three-gun, shadow-masked tricolor picture tube contains three electron gun assemblies, a shadow mask, and tiny triads of phosphor dots covering the face of the tube. Color is generated by bombarding the color triads with high-velocity electron beams from the electron guns. The one-gun picture tube uses a vertical-stripe metal grille behind vertically striped phosphors covering the face of the tube. Electrons passing through the grille and bombarding the vertically striped phosphors generate the color.

- Remote control systems are a modern convenience that can be very simple or quite complex. Infrared energy is the controlling link between the remote transmitter and the remote receiver. The remote system consists of a remote transmitter, a remote receiver, and receiver circuitry and devices to effect the change ordered by the remote command. Remote transmitters are infrared. The remote receiver is installed within and interconnected with the main TV chassis. Most receiver systems allow operation either by the remote IR signal or by pushbuttons mounted on the front of the TV set.

- Reception by other than an individual TV antenna includes pay TV, coaxial cable systems, master antenna systems, wireless cable, and satellite earth stations.

- Safety for equipment, the user, and the technician should be the primary consideration when working with a TV system. It is important that technicians not defeat any built-in safety feature due to carelessness, ignorance, or use of improper replacement parts.

QUESTIONS AND PROBLEMS

1. What components are contained in the modulation information for the monochrome picture carrier?

2. What element must be included in the visual information for color TV systems?

3. List tuner functions.

4. How many TV channels can a tuner receive?

5. What are the channels and frequency ranges in (a) the VHF band and (b) the UHF band?

6. Describe the types of mechanical tuners used for TV sets.

7. Explain the operation of an all-electronic tuner.

8. What devices are used as RF amplifiers in modern TV tuners?

9. Explain the functions of the three stages in a VHF tuner.

10. What are the receiver frequencies for the picture and sound carriers for channels (a) 5, (b) 7, and (c) 11?

11. What are the LO frequencies for the three channels in Problem 10?

12. What type of UHF tuner does not use an RF amplifier? What type does?

13. Describe the comparable tuning rule and how it is implemented.

14. Describe the construction of a simple crossover network for use with an all-channel antenna.

15. Describe what precautions must be taken when doing repair work on the UHF tuner, and explain why.

16. What frequencies are used in video IF amplifiers? What is the IF bandwidth?

17. What is the frequency bandwidth separation between the video and audio IF carriers?

18. What are the video and audio IF carrier frequencies?

19. Explain how the video carrier that was below the audio carrier at the transmitter is above the audio carrier in the receiver.

20. Describe stagger tuning, explain its purpose, and state how it works.

21. Explain why the sound subcarrier in a color receiver must be taken off before the video detector stage is reached.

22. Describe the function and operation of the video detector. What is the output signal?

23. Explain the preferred method for extracting the 4.5 MHz audio TV signals in color receivers.

24. Briefly explain each of the three AGC systems.

25. Describe (a) the functions of the video amplifier, (b) how brightness and contrast are controlled, and (c) the TV audio section.

26. Define dc restorer and explain under what conditions it is not needed.

27. What types of audio demodulators are used in TV receivers?

28. Describe (a) the intersync separation process, (b) the sync amplifier operation, and (c) the sync clipper operation.

29. Explain under what conditions the sync amplifier might not be used.

30. Describe (a) how vertical deflection is achieved and (b) the operation and purpose of the noise inverter.

31. Describe (a) how horizontal deflection is achieved and (b) the function of the horizontal oscillator.

32. Describe how AFC is achieved in the circuit in Figure 13-22.

33. Describe how the high voltages necessary to operate a CRT are derived in a TV receiver.

34. Describe the function and operation of the width and linearity controls in Figure 13-25.

35. Define (a) luminance signal and (b) chrominance signal.

36. Describe the chroma demodulation process.

37. Define color killer, explain its function, and describe how it works.

38. Describe (a) the three-gun CRT and (b) the one-gun CRT.

39. How can good horizontal convergence be achieved with the one-gun CRT?

40. List the main components of a remote system.

41. Define infrared and explain its purpose.

42. Describe how an infrared transmitter performs its functions.

43. Describe how the remote receiver is installed in a TV receiver and how inputs are achieved.

44. Explain the operation of the remote receiver activated by an IR signal.

45. List four methods of receiving TV programs other than through the individual antenna.

46. Describe briefly (a) subscription television, (b) CATV, and (c) MATV.

47. List and describe the functions of the equipment required at an earth station.

48. What important factor must a technician observe when repairing a TV receiver?

CHAPTER 14

Satellite
Communications

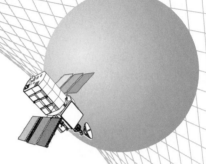

OBJECTIVES

When you complete this chapter, you should be able to:

- Define satellite and satellite communication system.
- Describe a basic satellite communication system.
- Define uplink, downlink, geosynchronous orbit, transponder, footprint, bird, and earth terminal.
- State the advantages and limitations of satellite communications.

INTRODUCTION

From the 1920s to the early 1960s, most long-range communications had been propagated via the HF band. As the number of users continued to increase tremendously each year, the HF band rapidly became overcrowded, and "free" frequencies were at a premium. Although the communications needs of the world had been met in the past, large-scale improvements had to be made to satisfy present and future requirements. Communication via satellite is a natural outgrowth of this continuing demand for greater-capacity, higher-quality communications. Experience with satellite communications has demonstrated that such systems can satisfy many requirements for reliable, secure, and cost-effective communications. In this chapter, we will discuss the basic concepts of satellite communications, the role satellite systems play in today's world, and the advantages and limitations of satellite systems.

 ## 14-1 BASIC CONCEPTS

The earth is literally surrounded by artificial satellites. Many of these satellites carry repeaters and are used for communications. In recent years, satellites have been placed in synchronous orbits, providing continuous communications capability between

493

almost all possible points on the globe. Although we take this situation for granted now, such communications were impossible as recently as the 1960s.

The oldest communications satellite, in terms of the time of launching, is our moon. The moon is a fairly good reflector of radio waves and has been used for long distance communications, mostly by radio amateurs. Unfortunately, it is an inconvenient communications satellite because it is above the horizon only half the time and because it moves across the sky.

In the 1960s, a series of *passive* satellites were launched into orbit around the earth. These devices, called *Echo* satellites, were like large metal balloons that reflected radio waves sent up to them. They were placed in low orbits, since we did not have the technology at that time to put a satellite in synchronous orbit. The area of coverage for each Echo satellite was limited by the low orbit, and access time was brief.

Active communications satellites were developed after the Echo satellites. An active communications satellite is an orbiting repeater with broadband characteristics. The signal from the ground station is intercepted, converted to another frequency, and re-transmitted at a moderate power level. This setup provides much better signal strength at the receiving end of the circuit, as compared with a passive satellite. The first active satellites were placed in low orbits, and thus their users were bothered by the same shortcomings as found with the Echo satellites. Finally, active satellites were placed in *synchronous* orbits, making it possible to use them with fixed antennas, with a moderate level of transmitter power, and at any time of the day or night.

Today, there are so many synchronous satellites orbiting our planet that it is becoming difficult to find room for more. Almost everyone has some need for a synchronous satellite. These satellites are used for TV and radio broadcasting, telephone and data communications, weather forecasting, business communications, and military operations.

Satellite Systems Defined

To define what satellite systems are, we must look at their basic features and historical development. Satellite systems are a relatively new development in the history of communications since their use had to await our entry into the space age in the late 1950s. Satellite systems depend heavily on technology, including rocketry, space mechanics, solid-state electronics, high-frequency electronics and radiation, and modern communication networks. Some of the problems that must be solved to ensure a successful satellite system can be understood by examining the basic features of such a system.

For use in communications, a *satellite* may be defined as a man-made vehicle that orbits the earth. Thus, a *satellite communication system* is a system that uses orbiting vehicles to rely radio transmissions between earth terminals. There are two types of communications satellites: passive and active. A *passive* satellite simply reflects radio signals back to earth. An *active* satellite acts as a *repeater*; that is, it amplifies the signals received and then transmits them back to earth. An active satellite thus provides a much stronger signal at the receiving terminal than that provided by a passive satellite.

A typical operational link involves an active satellite and two earth terminals. One station transmits to the satellite on a frequency called the *uplink* frequency. The satellite amplifies the signal, transfers it to a frequency called the *downlink* frequency, and then transmits it back to earth, where the signal is picked up by the receiving terminal.

When a passive satellite is used, the downlink frequency and the uplink frequency are the same. The basic concept of a satellite communication system is illustrated in Figure 14-1, which shows several types of earth terminals.

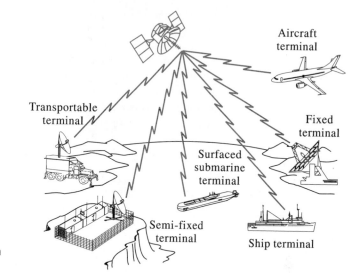

FIGURE 14-1
Satellite
communication
system

Before proceeding in our discussion, let us consider a point worth remembering: Before successful satellite communications can take place between two earth terminals, the communications satellite must be within the view plane of *both* the transmitting and the receiving terminal. This concept is known as a *mutual visibility window* and is illustrated in Figure 14-2. Whether a satellite is within this mutual visibility window of earth terminals depends to a great extent on the orbital pattern of the

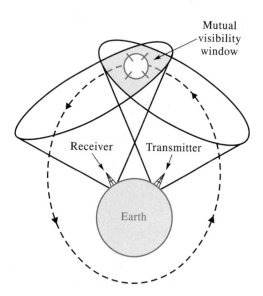

FIGURE 14-2
Mutual visibility
window of two
earth terminals

satellite. Basically, all communications satellites orbit the earth in either an elliptical or a circular orbit.

Elliptical Orbit

The physical shape of an *elliptical* orbit is an ellipse, as shown in Figure 14-3. The actual orbital shapes of individual communications satellites are determined by the initial launch parameters and therefore may vary in degree of elongation.

To better understand elliptical orbits and associated technical terms, we will briefly discuss the basic method of placing a satellite in orbit. As the satellite is launched into space, thrusters stop at a point in space that we will call a temporary apogee. The satellite is then angled into the desired orbital path.

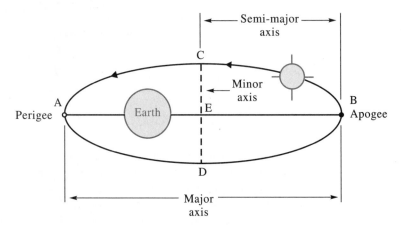

FIGURE 14-3 Elliptical satellite orbit

The foregoing discussion is an oversimplification of the orbital process; however, it should help you to better relate the following technical terms associated with elliptical orbits:

1. *Apogee*—point in the orbit of a satellite at the greatest distance from the center of the earth.

2. *Perigee*—point in the orbit of a satellite nearest to the center of the earth.

3. *Major axis*—distance between perigee and apogee. Half this distance is known as the semi-major axis.

4. *Minor axis*—line perpendicular to the major axis and halfway between perigee and apogee. Half this distance is known as the semi-minor axis.

5. *Angle of inclination*—angle between the earth's equatorial plane and the orbital plane of the satellite, as shown in Figure 14-4. The inclination of the orbit determines the geographical limits of the projection of the path of the satellite over the earth's surface. The greater the angle of inclination, the greater the amount of the earth's surface covered by the satellite, as illustrated by the shaded areas in Figure 14-5.

An elliptical orbit may be inclined any number of degrees up to 90° from the earth's equatorial plane. A satellite orbiting in a plane that coincides with the earth's equatorial plane (very little or no inclination) is considered to be in an *equatorial* orbit. A satellite orbiting in an inclined plane with an angle of 90° is in a *polar* orbit.

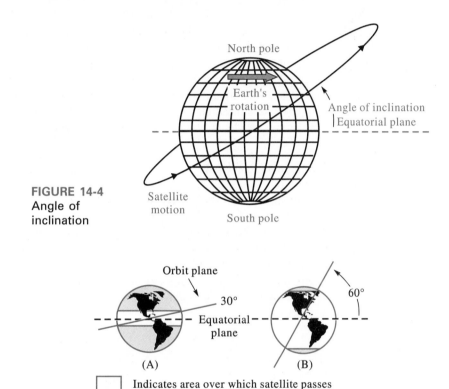

FIGURE 14-4
Angle of
inclination

FIGURE 14-5 Effect of orbit plane inclination
on satellite coverage (A) 30° angle of inclination
(B) 60° angle of inclination

Circular Orbit

In a *circular* orbit, the major and minor axes are equal or approximately equal and the earth is located at the intersection of the axes. The mean height above earth, rather than perigee and apogee, is used to describe a circular orbit. In the following paragraphs, we will be discussing several basic types of circular orbits. Before proceeding, however, we provide a brief overview of the factors that relate to the speed and rotational periods of communications satellites in order to help you understand the different types of circular orbits.

When a satellite is in orbit, the force due to gravity must be exactly balanced by the centrifugal force on the satellite. Centrifugal force is proportional to distance and velocity. Gravitational force is proportional to distance. The resulting balance of gravity and centrifugal force results in a satellite velocity being proportional to its distance

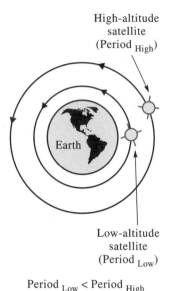

High-altitude
satellite
(Period $_{High}$)

Earth

Low-altitude
satellite
(Period $_{Low}$)

Period $_{Low}$ < Period $_{High}$

FIGURE 14-6 Orbital periods of low- and high-altitude satellites

from the earth—the closer to the earth, the faster the orbital velocity, the farther from the earth, the slower the orbital velocity. We can therefore deduce that a low-altitude satellite must travel at a higher speed to maintain orbit than satellites at higher altitudes. Low-altitude satellites will therefore complete one orbit of the earth in less time than high-altitude satellites since the distance to be covered is considerably less and the velocity is greater, as illustrated in Figure 14-6.

The point of this discussion is that the orbital period will increase with increases in height and decrease with decreases in height. There is a height above the earth, however, where the orbital period of the satellite coincides with the rotational speed of the earth. This height is about 22,300 statute miles above the earth's surface. At this altitude, a communications satellite will appear to be motionless in the sky.

There are three basic types of circular orbit: synchronous, near-synchronous, and medium-altitude. A brief discussion of each type follows.

Synchronous Orbit When a satellite is in a circular orbit approximately 22,300 statute miles above the earth, it is in a *synchronous* (commonly called *geosynchronous*) orbit since its orbital period is the same as the earth's rotational period. Although *inclined* synchronous orbits are possible, the term *synchronous*, as commonly used, refers to a synchronous *equatorial* orbit. It is in this latter type of orbit that the satellite appears to hover motionless in the sky. Figure 14-7(A) shows how one synchronous satellite can cover almost one half of the earth's surface. Three of these satellites can provide coverage over most of the earth's surface (except for the extreme north and south polar regions), as shown in Figure 14-7(B).

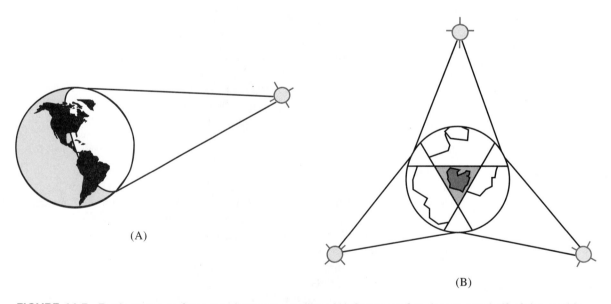

(A)

(B)

FIGURE 14-7 Earth coverage from synchronous satellites (A) Coverage for almost one-half of the earth's surface by one satellite (B) Worldwide coverage from three satellites as viewed from the North Pole

Near-Synchronous Orbit   A satellite in a circular orbit within a few thousand miles of the 22,300 statute miles above the earth is in a *near-synchronous* orbit. If the near-synchronous orbit is *lower* than 22,300 statute miles, the satellite's orbital period is less than the earth's rotational period. Consequently, the satellite will move slowly around the earth from west to east against the sky's background. This type of orbit is called *subsynchronous*. If the orbit is *higher* than 22,300 statute miles, the satellites's orbital period is greater than the earth's rotational period. The satellite will then appear to have a *retrograde* (reverse) motion, from east to west. In actuality, the satellite is still moving in an easterly direction, but the rotational speed of the earth produces the retrograde effect. This effect is similar to the illusion produced when a speeding car overtakes a slower-moving car and the slower car appears to be moving backward. Although inclined and polar near-synchronous orbits are possible, common usage of the term *near-synchronous* implies an equatorial orbit.

Medium-Altitude Orbit   A satellite in a circular orbit at heights from approximately 2400 to 14,300 statute miles above the earth is in a *medium-altitude* orbit. The orbital period of a medium-altitude satellite is considerably less than the rotational period of the earth; therefore, such satellites move rather quickly across the sky from west the east.

Low Earth Orbit Satellite (LEOS) Communication Systems   Recent development of small-satellite technology, highly integrated electronics, and low cost launch options provide new opportunities for commercial use of space. Communication systems based on constellations of small satellites deployed in *Low Earth Orbits* (LEO) present increasing interest. Several companies are producing constellations of small satellites for mobile communications.

Low Earth Orbiting Satellites (LEOS, sometimes designated LEOSAT) are small satellites which can be launched either by small launchers, or by large launchers in a multiple launch configuration. The primary features of the LEOS spacecraft family enable them to support a multisatellite constellation with data message communications, position reporting missions, and voice communications for mobile users. For example, the *Low Earth Orbit Communication* (LEOCOM) system is a store-and-forward communication system. The use of simple design concepts, addressed only to fulfill the essential mission requirements and making use of already existing devices, have lead to relatively low-cost systems of particular interest to government and commercial users in a period of tight finances.

Propagation Effects on LEOS Communication Links   Communication links between LEOS and mobile or portable terrestrial stations can be affected by several fading mechanisms. These mechanisms contribute to degradation of link performance in excess of those usually associated with radio wave communication links. Fading mechanisms include multipath scattering and signal blockage, and their effect on communication links is essential to the design of an effective communications system. The signal behavior due to these mechanisms can have severe degrading effects on communication system performance.

Small-Satellite Power Source   The growth in small-satellite technology has prompted the development of nickel-hydrogen batteries in the 2 to 20 ampere-hour capacity range. The battery designed for LEOS systems utilizes a 2.5-inch diameter, hermetically sealed, stainless steel vessel. The internal structure of this vessel stacks slices of material together in a common pressure vessel configuration that produces a 6 ampere-hour 2.5 volt capacity. Ten of these vessels are sleevemounted to a thermalplate to produce a 6 ampere-hour, 25 volt power source. This package provides the long life and high reliability required for the power source in LEOS systems. The power package also includes strain gauge, heater, voltage, and temperature sensing capabilities.

Confusion About LEOS   There is a possibility of confusion surrounding the designation of LEOS. The IEEE has designated LEOS to represent *Lasers and Electro-Optics Society*. The field of interest of the society is lasers, optical devices, optical fibers, and associated lightwave technology and their applications in systems and subsystems in which quantum electronic devices are key elements. The society is concerned with the research, development, design, manufacture, and application of materials, devices, and systems, and with the various scientific and technological activities which contribute to the useful expansion of the field of quantum electronics and applications.

In some references, the small-satellite systems are designated LEOS, and in some, LEOSAT. Therefore, electronics technicians must be aware of the different LEOS designations, and make sure the reference is appropriate.

An Operational System

The essential system components of a basic operational communication satellite system are (1) an orbiting vehicle with a communication receiver/amplifier/transmitter (*transponder*) installed and (2) two earth terminals equipped to transmit signals to, and receive signals from, the satellite. The design of the overall system determines the complexity of the various components and the manner in which the system operates.

Figure 14-8 illustrates the basic components of a typical satellite system. The satellite is orbiting the earth and receives its energy from the sun through solar cells. It has one or more antennas that receive radiation from earth and send radiation back to earth. One earth station (point A) transmits information to the satellite at a specified carrier frequency, typically in the 6 GHz band. This is the uplink frequency. The satellite receives this transmitted radiation and information and repeats and reinforces the information by transmitting it to the earth on a different carrier frequency, typically 4 GHz. This is the downlink frequency. The satellite is operating in the 6/4 GHz frequency bands.

In this illustration, the satellite antenna has been designed to provide radiation to all parts of the earth visible to the satellite. For a satellite operating at a distance of 22,300 statute miles from the earth, approximately one third (30%) of the earth's surface is exposed to this radiation. With a single satellite in such an orbit, earth station A can send information to any other earth station, including itself, within this 30% area of the earth's surface. If the information is to be sent from station A to station B, nothing prevents station C from receiving it unless the information is coded in such a way that station C cannot understand it without the proper key. In fact, in some applications, station A may want to send information to stations B and C as well as to a large number of other stations over the earth's surface.

Distribution of TV programs from a single central studio to cities and towns on the same or different continents is one example of such a system, and, for this, the satellite becomes a *broadcast* satellite. A typical application is shown in Figure 14-9, where the TV signal (Viewer's Choice) originates in New York and travels 46,600 miles via satellite to California in 0.2 s. At its reception station in California, the signal is bounced around on microwave and satellite dishes until it is transmitted by cable to the private TV sets of paying customers.

Station Conditions

Two situations occur at the receiving station. First, the information sent may be received and understood by all stations but may be of interest to only one or a few. Second, the information may be intended to be received and understood by only one station, in which case the signal must be encoded (*scrambled*) to prevent other earth stations from receiving and understanding the transmission. This latter description pertains to the situation in Figure 14-9.

One other important point about the satellite broadcast signal is that station A (Figure 14-8) can receive its own information and check it for errors. If it is important that information be sent error-free, this capability can be a useful tool for making sure

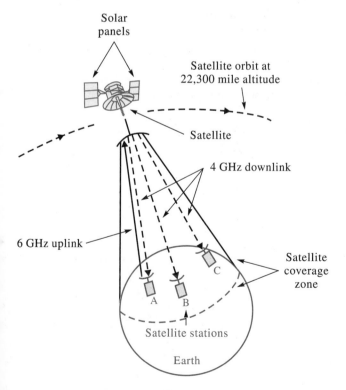

FIGURE 14-8 Basic satellite communications configuration

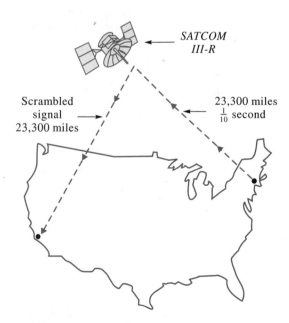

FIGURE 14-9 Viewer's Choice television/satellite system

that the information was correct when it left the satellite. Many techniques can be used at the receiving station for detecting and correcting errors that occur during the transmission of the information.

One other feature illustrated in Figure 14-8 is that the transmitting earth station must provide a very directional beam of radiation that will be received by the satellite being used and not by some nearby satellite, because there are many satellites in orbit. Today's satellites are separated by about 4°; tomorrow's satellites will be separated by only about 2°. Therefore, the earth station transmitter and antenna must be carefully designed to achieve the proper beamwidth. The earth station must also be able to locate and track the position of the satellite so that its beam will not miss the satellite antenna. This situation is simplified if the satellite stays in the same position relative to the earth at all times. Since the earth is spinning on its axis, the only way to achieve a satellite position fixed relative to the earth's surface is to have the satellite orbit match the earth's rotation. An orbit that achieves this match is a geosynchronous orbit. Three geosynchronous satellites spaced at 120° intervals, as shown in Figure 14-7(B), could cover all the world except for some polar regions. Of course, only earth stations in the coverage of two satellites could communicate from one zone to another. Such international stations would be able to relay messages or communications from one zone to another for complete world coverage. Stations within a single zone can communicate only with stations in that zone and are thus considered local stations.

Satellite Channels and Station-to-Station Communications

For stations within the zone of a given satellite (Figure 14-10), any number of independent station pairs can communicate with each other. While the satellite is broadcasting to the entire visible portion of the earth's surface, each station listens only to the signal intended for it. For example, communications can be occurring between station A and B along signal path 1 at the same time communications are occurring

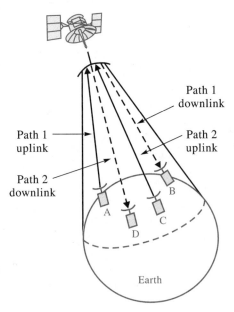

FIGURE 14-10 Multiple communications using the same satellite

between stations C and D along signal path 2. Upon first examination, this arrangement does not seem possible since the satellite antenna is receiving signal frequencies in the 6 GHz range from stations A and C at the same time. In addition, within the same time period, the satellite is broadcasting to the entire visible area below on carrier frequencies in the 4 GHz range. Obviously, some multiplexing must be used to allow these communications to occur simultaneously.

14-2   HOW SATELLITE CAPACITY HAS EVOLVED

To make satellite communications economically competitive, the cost per communication channel must be minimized. Reducing the cost for a given satellite configuration requires that the maximum number of channels possible be placed on a carrier. Therefore, the maximum bandwidth that can be provided by the state-of-the-art technology must be used. The technologies that impact both satellite costs and capacity include launch vehicle technology, solid-state device and system technology, and antenna technology.

Rocket Launch Technology

The amount of satellite weight that can be placed in geosynchronous orbit depends on the rocket used to launch the satellite. The larger the rocket, the more weight that can be placed in orbit and the higher the cost of the launch. The more weight available for the satellite, the more power and channel capacity it can have.

A certain rocket will launch a certain maximum weight into orbit, so the satellites are designed to match these discrete weight options. Early launch vehicles, such as the DELTA series of rockets, launched into orbit a weight of about 1100 pounds; the Atlas/Centaur rocket handled a load of about 2000 pounds; the Titan rocket, about 3300 pounds; and NASA's earth orbiter (shuttle) can carry much greater weights and launch multiple satellites. These heavier satellites have a much higher capacity, with the rate of increase in capacity versus weight much higher than the increased costs due to weight. Some of this capacity has come from advances in electronic technology since the heavier satellites have been launched at later dates. These improvements in technology and the ability to place more weight into orbit have dramatically increased the channel capacity of a satellite. Launch costs are only part of the total system costs, however. The cost of the satellite and the cost of the earth stations must be considered, and cost reduction in these areas depends on advances in electronic and antenna technology.

Satellite Electronic Technology

Rapid advances in electronic technology, such as large scale integration (LSI), have made it possible to construct miniaturized, low-power-drain ICs. These tiny devices have replaced much of the extensive, high-power-drain circuitry of only a few years ago. These advances have reduced system costs, size, and weight and have improved communication capacity and reliability.

Power Sources   One area that depends on modern solid-state devices and that impacts satellite capability is the *solar cell energy source*. The amount of solar energy collected by the satellite depends on the total area of the solar panels that is exposed to the sun and on the generation efficiency of the cell. Significant improvement has been made in solar cell efficiencies; present panels generate about twice the power as early units with the same panel area and weight. This increased power allows an increase in RF energy level radiated from the satellite, which in turn reduces the costs of earth station components.

Early communications satellites were severely limited by a lack of suitable power sources, which in turn severely limited the ability of the satellite transmitter to amplify and transmit signals back to earth. The only source of power available within early weight restrictions was an inefficient panel of solar cells mounted on the surface of the satellite. The major disadvantage of this type of power source was that the satellite was without power when it was in eclipse (located on the opposite side of the earth from the sun). This temporary outage was unacceptable for continuous communications.

Figure 14-11 shows two types of solar cell mountings that have been used for satellites. On the *spin-stabilized* satellite (Figure 14-11(A)), the solar cells are mounted on a cylindrical body and continually rotated so that they face the sun while the antenna is kept pointed at the earth. The spinning cylinder acts as a gyroscope to keep the satellite oriented in space.

The *body-stabilized* satellite (Figure 14-11(B)) is an improvement over the spin-stabilized satellite. On this satellite, a spinning wheel provides the gyroscope; the solar panels and body of the satellite are fixed. The entire solar panel surface is illuminated by the sun at all times, and much larger areas than just the body can be used. Satellite power is increased significantly over that of the spin-stabilized satellite. For example, body-stabilized satellites such as CTS, *Intelsat 5*, and ATS units generate kilowatts of power, whereas early spin-stabilized units generated 40 W (1965) to 400 W (1971).

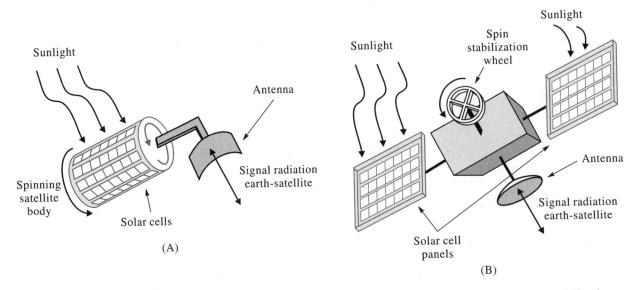

FIGURE 14-11 Types of satellite solar cell power sources (A) Spin-stabilized system (B) Body-stabilized system

Combination of Power Sources A combination of solar cells and storage batteries is a better prime power source for communications satellites than solar cells alone. Although this combination is far from an ideal power source, these systems consist of about 32,000 solar cells, which initially supply about 250 W of power, and a nickel-cadmium battery for back-up power during eclipses. Only about 10% of the sunlight energy that converges on the solar cells is converted to electrical power. The efficiency is further decreased when the solar cells are bombarded by high-energy particles sometimes encountered in space. One approach to overcoming these problems was to arrange the solar cells in the form of wings extending from the satellite, a construction that led to naming satellites *birds*. The position of the satellite with reference to the sun varies throughout the day, and so the wings of solar panels are adjusted to keep them perpendicular to the sun for maximum energy reception.

The use and application of electrical power on board the satellite are controlled in several ways:

1. By a preprogrammed arrangement in which power is supplied automatically.
2. By the use of sensors on board the satellite.
3. By earth control through a telemetry, tracking, and command system. In this way, the weakening or failure of a unit can result in the automatic switching in of a replacement by earth-transmitted instructions.

Electronic Circuitry The electronics used in the receiver, filters, and low-power portions of satellite communications systems have become much more weight efficient in recent years. Figure 14-12 shows a block diagram of a typical satellite. The 6 GHz antenna receives signals from the earth stations, including any command signals. The wideband receiver boosts the signal strength to a level where it can be converted, by mixing with a 2 GHz local oscillator, to the 4 GHz range and drive a traveling wave tube (TWT) preamplifier. This signal is routed through a filter network to the power amplifier TWT that generates the signal for the 4 GHz transmitting antenna for retransmission to earth.

By making the wideband receiver, downconverter, and filters out of microwave ICs and using modern weight efficient TWTs, it is possible to maximize the number of receiver/amplifier/transmitter sections (*transponders*, or *repeaters*) available on a satellite of a given orbit weight. The particular transponder, or *channel*, used is controlled by the signals sent over the control channel. Even with a solar power generator of reasonable size and modern power and weight-efficient electronics, the overall satellite capacity and performance depend on delivering the RF or electromagnetic energy back to earth. Capacity and performance depend on the antennas used in the satellite and at the earth station.

Antenna Technology

One way to get the most out of the available satellite power is to design the antenna so that it delivers a signal to the earth areas that need it and avoids sending to areas having no earth stations. Generally, the larger the antenna, the more precisely focused the radiation beam. The more precisely the radiation pattern is controlled, the closer satellites can be spaced in orbit without interfering with one another. Thus, the total bandwidth available from the synchronous orbit can be increased by using larger, more

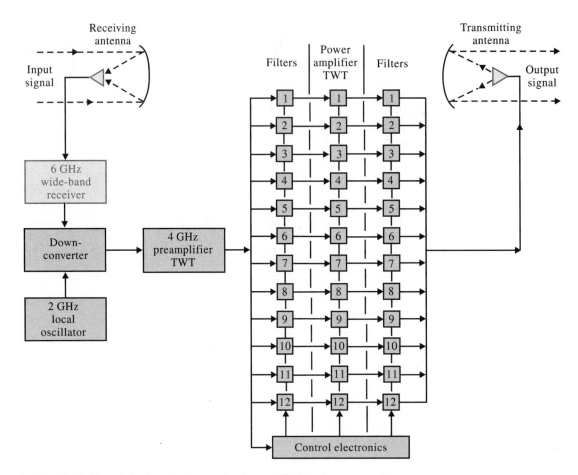

FIGURE 14-12 Typical satellite receive/transmit structure

carefully designed antennas. Also, since the power or energy hitting the earth's surface is increased by a better-focused antenna, smaller earth station antennas are needed. Use of smaller antennas can significantly reduce the costs of earth stations while moderately increasing the cost of the satellite.

Currently, satellite antennas about 30 feet in diameter have been used successfully with the ATS series of satellites. The problem with such large antennas is that special techniques are required to couple energy from the power amplifier to the antenna. At microwave frequencies, energy comes from the power amplifier to the antenna through waveguides. A feedhorn is the end terminator on the waveguide that feeds energy to the antenna. With a large antenna, several feedhorns are required, as illustrated in Figure 14-13(A). Even though the addition of feedhorns increases the weight of the satellite, it has operational advantages. For example, if each feedhorn corresponds to a particular area on the earth's surface, as illustrated by Figure 14-13(B), and if the satellite is transmitting only from area A to area B, horns can be selected so that the power of the satellite can be delivered only to area B, allowing earth stations with small-diameter receiving antennas and inexpensive receivers to receive the information.

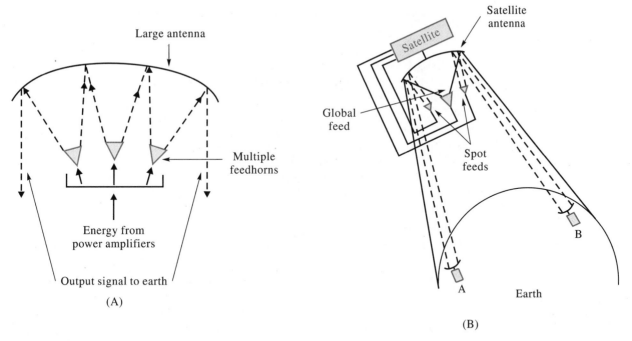

FIGURE 14-13 Multiple-feed satellite antenna (A) Feedhorns for a large satellite antenna (B) Radiation patterns

Another approach is to use separate antennas for global coverage and form localized spot coverage to allow two different transmission paths to use the same frequency channel for separate communications. Such *frequency reuse* will be a common feature in satellites of the future; it is presently in use in cellular radiotelephone systems. Using larger antennas and allowing frequency reuse dramatically increases satellite capacity without significantly increasing costs.

Telemetry

Telemetry is the transmission of quantitative information from one point to another by electromagnetic (radio) means. This process is used extensively by weather balloons, satellites, and other environmental monitoring devices. It is used in space flights, both manned and unmanned, to keep track of all aspects of the equipment and the physical condition of the astronauts.

A telemetry transmitter system consists of a measuring instrument, an encoder that translates the instrument readings into electrical pulses, and a modulated radio transmitter with an antenna. A telemetry receiver system consists of a radio receiver with an antenna, a demodulator, and a recorder. A computer may be used to process the data received.

For satellite systems, telemetry transmits instructions for active satellite operation in duplex; that is, not only are instructions communicated to the satellite, but specific data concerning the functioning of the satellite are transmitted from the satellite to a ground control station. This information monitors all parameters of satellite operation, such as position, fuel supply, condition of the power source system, satellite movement, and temperature.

Satellite Evolution

Table 14-1 uses the Intelsat series of satellites to illustrate the great advances made in the design of typical satellite systems between 1965 and 1980. The table shows that 1980 satellite capabilities include up to 27 transponders, which provide up to 12,500 channels that may be used for voice channels or for computer data at various data rates. Total bandwidth was increased to 2300 MHz. Generally, a given satellite transponder can be allocated as needed to a given communication task. For example, the 36 MHz bandwidth transponder could be used to send a TV channel, 1200 voice channels, or data at a rate of 24 megabits per second (Mbps).

TABLE 14-1   Satellite Evolution of Intelsat Series, 1965–1980

System Parameter	1965–1968 Intelsat 1–3	1971 Intelsat 4	1976 Intelsat 4A	1980 Intelsat 5
Number of voice channels	240–1,200	4,000	6,000	12,500
Total bandwidth (MHz)	50–130	500	800	2,300
Number of transponders	1–2	12	20	27
Transponder bandwidth (MHz)	25	36	36	40, 80, 240
Earth station antenna diameter (m)	25.9	30	30	30, 10
Satellite weight (kg)	38–152	700	790	950

Satellite Positioning

At the present time, satellites in geosynchronous orbit are concentrated most heavily over the United States and the Atlantic Ocean (for trans-Atlantic communication). The orbital path is not yet crowded, with satellites generally spaced about 4° apart. However, as more satellites are placed in orbit, the spacing is expected to be decreased to about 2°. This closer positioning means that each new satellite must have still greater channel capacity and functional capability. To meet this need, development focuses on (1) more efficient networks; (2) improved switching; (3) increased multiplexing; (4) more efficient and smaller antennas, especially at earth stations; and (5) more powerful beams for transmission links.

Earth Stations

At a high enough power level, the earth station antenna can be made smaller, the receiving station electronics can be reduced, and the total cost can be lowered to reasonable levels. The noise characteristics and the focus and directivity of the earth station transmitted beam are affected by the size of the antenna.

Earth Station Antenna   As allowable satellite spacing is narrowed, the size of the earth station antenna becomes more critical in avoiding interference between satellites. In general, the beam width of an antenna is a function of the antenna width to signal wavelength ratio. The higher the ratio, the narrower the beamwidth, and hence the greater the ability of the antenna to differentiate between two closely spaced satellites. For example, a 6 GHz uplink carrier has a wavelength of 5 cm. As a result, the antenna would have to be about 5 m (about 15 feet) in diameter. This size is not excessive for an antenna on top of a large building or on the grounds of a large business plant, but it would be excessive for home or mobile use. If the carrier frequency is changed to 15 GHz, with a wavelength of 2 cm, a 2 m (about 6 1/2-foot) diameter antenna could be used. Increasing the carrier frequency to 30 GHz would allow the diameter to be reduced to 1 m (about 3.3 feet), which is acceptable for home or mobile communications.

Antenna Noise   Noise in an antenna used only for receiving signals varies with the antenna beamwidth, as illustrated in Figure 14-14. If the beamwidth of the receiving antenna includes other satellites so that it detects a wide area of stray electromagnetic radiation, all of this signal appears at the receiver as noise and tends to block out the desired signal from satellite A. Desired signals buried in noise in this manner require a very wide-band FM transmission to effectively detect the signal. Because small antennas with 1–3 m diameters cannot provide as narrow a beam as larger antennas, up to 10 times the normal bandwidth may be required. Thus, a 4.5 MHz TV video signal would take the entire 36 MHz bandwidth of a satellite transponder to reliably transmit

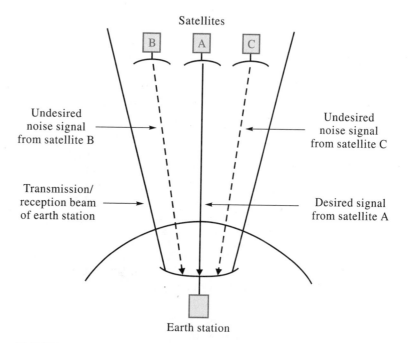

FIGURE 14-14   Effect of earth station antenna beamwidth on signal interference

a high-quality picture. Similarly, a 4 kHz voice channel might require a 40 kHz band-width FM signal to properly transmit it. This would mean that only 100 such voice channels could be sent on a single transponder instead of 1000. The earth station cost is reduced by using a smaller antenna, but the transmission cost is increased significantly because the satellite bandwidth is being used inefficiently. Because of the need to service smaller and smaller earth stations, satellite frequencies continue to increase because system penalties seem to be minimized at the higher frequencies. Table 14-2 shows the frequency bands and frequencies presently used or planned for satellite systems. The 6/4 GHz C band was the band most often used in the past. Increased use of the 14/12 GHz Ku band will continue, especially for inexpensive earth stations, until the band becomes as heavily subscribed as the current C band. Then, the K band use will increase.

TABLE 14-2   Frequency Bands for Satellite Communications

Band	Uplink Frequency (GHz)	Downlink Frequency (GHz)
C	6	4
Ku	14	11/12
K	29/30	19/20

Frequency, GHz	Direction
1.530–1.559	Down
1.6265–1.6605	Up
3.400–4.200	Down
5.850–7.075	Up
7.250–7.750	Down
7.900–8.400	Up
10.70–12.70	Down
12.70–13.25	Up
14.00–14.80	Up
17.30–17.70	Up
17.70–18.10	Both
18.10–20.20	Down
27.00–30.00	Down

Spot and Polarized Beams   A radiation technique that should become more common is the frequency reuse mentioned earlier. Referred to as *spot beaming*, its effect is shown in Figure 14-15. Stations A and B in the west spot can use the same frequencies used by stations C and D in the east spot. The frequencies must simply be multiplexed properly to keep from interfering with each other. Proper switching of transponders will accomplish this.

Another approach allowing satellites to use the same frequencies in parallel without interference is *polarized radiation*, illustrated in Figure 14-16. One pair of satellite

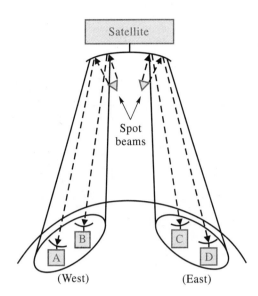

FIGURE 14-15
Reuse of frequencies
through spot beaming

antennas is vertically polarized, and another pair is horizontally polarized. Either vertically or horizontally polarized transmissions are received by the respective antenna and retransmitted in the same polarization. Filter 1 may be 3.7–3.74 GHz for the vertical polarization path, and filter 2 may be 3.72–3.76 GHz without channel interference. This scheme was used in the COMSAT satellite and the Intelsat units from 1976 on.

Multiple Bands   Another way to increase the number of simultaneous communications within a given satellite is to transmit and receive on several of the bands listed in Table 14-2. Communications in the 14/12 GHz bands can occur at the same time as communications in the 6/4 GHz bands, at least doubling the capacity of a given satellite. This approach was used in the *Intelsat 5* unit (Table 14-1). The problem with this technique is that the carrier frequency signals above 10 GHz are heavily attenuated by water droplets in intense rain, thick fog, or heavy cloud cover. One solution is to provide alternate earth stations separated by several miles. It would be unlikely that both stations would be blocked from satellite view at the same time. This approach is called *space diversity*.

Channel Assignments   For a pair of earth stations to communicate via satellite, one of the channels must be assigned to the communication task. Thus, in Figure 14-10, the A–B communication could be assigned to channel 1, and the C–D communication could be assigned to channel 2. In this way, the transmissions do not interfere with each other and can occur simultaneously because different frequencies are being used by the A–B and C–D pairs. Obviously, station A must know that channel 1 is available, and station B must know that station A is transmitting over channel 1. Station B can accept the information coming in on channel 1 and reject any information coming in over the other channels. A similar set of conditions applies to the C–D communication over channel 2.

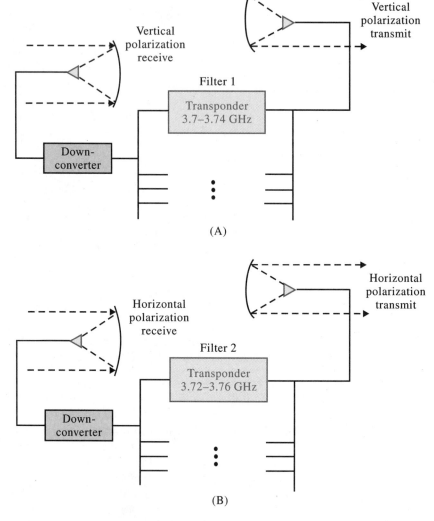

FIGURE 14-16   Receiver structure for polarized radiation (A) Vertical polarization (B) Horizontal polarization

These channel assignments could be made permanently, in which case, if station B detects information on channel 1, the information is being sent to it by station A. If it detected information on channel 2, it would know the information was intended for station D. Of course, this would be a wasteful approach, since the A–B communication might not be continuous for 24 hours a day. For periods during which A was not communicating with B, channel 1 would be idle when it could have been used by other stations.

Multiplexing Satellite Signals

There are two general approaches to sharing a satellite: *frequency division multiplex* (FDM) and *time division multiplex* (TDM). Most satellites to date have used some variation of the FDM approach, although the TDM approach will be used more in the future. A typical FDM approach is shown in Figure 14-17. Each satellite contains a certain number of transponders (repeaters), which are receiver/amplifier/transmitter devices. The receiver of a satellite is a wide-band receiver that covers the entire range of uplink center frequencies shown in Figure 14-17(A). These frequencies are typically from 5.945 GHz to an upper limit that depends on the number of channels the satellite handles. In Figure 14-17(A), which is the FDM structure for the 12-channel Western Union satellite, each channel has a bandwidth of 36 MHz, and channels are spaced 4 MHz apart. Above channel 12 is a 20 MHz command and control or telemetry channel. Thus, the receiver center frequency band extends from 5.945 GHz to 6.425 GHz. The transponders in the satellite convert this frequency range to the downlink channel center frequencies shown in Figure 14-17(B), which cover the 3.72–4.20 GHz band.

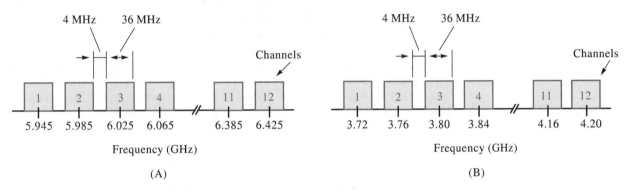

FIGURE 14-17 Frequency division multiplexing in a typical satellite (A) Receive channel (B) Transmit channel

Frequency Division Multiple Access A reasonable approach to channel assignment is to assign available channels to the next station requesting or demanding service and to notify the destination station which channel is being used. Then, once the communication is over, the channel would be released back to the available pool. This approach is called *frequency division multiple access* (FDMA). The command channel can be used to control the allocation of channels. It can also be used to notify the transmitting station which channel is available and to notify the receiving station which channel is being used. This control function can be handled by a central computer or by computers at each earth station. This same strategy would work well if station A (Figure 14-10) were a broadcast station. The transmitter—station A, in this example—would request a channel and state that transmission was to be received by stations B, C, and D. The control system would assign the available channel and notify station B, C, and D to receive information from station A on the assigned channel. With the broadcast and station-to-station capability, the satellite system has all the features required of any information delivery system. Further, it functions without the need for extensive ground repeater/relay systems between the communicating points.

Time Division Multiple Access Yet another way to share the circuitry of a satellite is through TDM. Most of the early satellites had FDM, and many had FDMA. As the use of digital communications increases, TDM is taking over. For TDM (Figure 14-18), all stations use the same single carrier frequency, but their signal occurs at a specific time interval in a specific time slot. As in FDM, the time slot assignment could be permanent, although it is more likely to be assigned by demand. A central control notifies the members of the transmission path which time slot has been assigned, and the receivers look at the information only at those specific times. When the transmission is over, the time slot is released back to the available pool. This approach is called *time division multiple access* (TDMA). TDMA offers some advantages over FDMA in that only a single frequency is used for a station connection, and the connection can be made through digital switches. In the example of Figure 14-18, station A has been assigned the first time slot, over which it sends the message 1001. Station B has been assigned the second time slot, over which it sends the message 0110. Both messages are to be sent to station C. The control system has assigned the other time slots to other communication tasks. The satellite multiplexes the information from A and B into the correct time slots onto the carrier sent to station C. Station C receives the combined TDMA signal, separates the two messages, and deals with them as separate communications.

Individual bits from each of the data signals have been assigned time slots. It is possible to assign a fixed set of bits such as an 8-bit byte or a 128-byte packet to a time frame and time multiplex these frames in a similar manner. In this way, the satellite capacity is increased significantly because several transmissions are sharing the same carrier frequency and the same transponder.

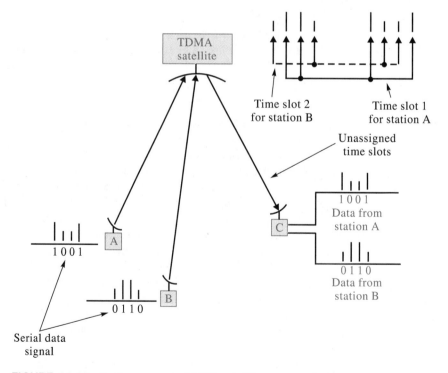

FIGURE 14-18 Basic concept of TDM satellite communications

Whatever the approach, the attempt to provide simultaneous utilization of the same channel of frequencies by several users usually complicates earth station equipment. Radiation polarization calls for the earth station to provide both polarizations in its antenna and power amplifiers. TDMA earth stations must be able to synchronize their digital transmissions with those to or from the satellite, and then demultiplex the signals. Large earth stations must have intricate networks to accumulate information and distribute it over existing telephone or microwave relay networks to the destination.

14-3   GENERAL SYSTEM COMPARISONS

Prior to the satellite system, all transmission systems, with the exception of radio transmission, required a network of wired interconnections between points. Not depending on wired interconnections is one advantage of satellite systems over alternative systems. All but the radio link and the satellite link require repeaters every few miles. Repeaters may not be feasible when the transmission must span oceans or access remote areas. In some of these cases, the radio or satellite link is the only possible solution, and certainly the most economical. The radio link is somewhat unreliable since transmissions over the surface of the earth have noticeable electromagnetic interference from other signals and fading of signals due to ionospheric variations. In addition, the data rate of radio links for worldwide communications is considerably lower than for the satellite system. For high data rates, the bandwidth requires a microwave high-frequency carrier; therefore, microwave waveguide, radio links, or satellite links must be used.

Cryptology (scrambling of information) must be used to make any of the systems secure; otherwise, the transmissions are relatively open to anyone who wants to obtain them. Cable and wire systems rate higher for security than other systems.

Advantages of Satellite Communications

Satellite communications offer distinct advantages over conventional communications for long distance service. Satellite links are unaffected by the propagation abnormalities that interfere with HF radio, are free from the high attenuation of wire or cable facilities, and are capable of spanning long distances without the numerous repeater stations required for line-of-sight or tropospheric scatter links. Satellite communications can furnish the greater reliability and flexibility of service needed to support worldwide communication systems.

Reliability   The propagation of radio waves by satellite is not dependent on reflection or refraction, and waves are reflected only slightly by atmospheric phenomena. Therefore, the reliability of active communication satellite systems is essentially limited only by the reliability of the equipment involved and the skill of operating and maintenance personnel. This improvement in reliability is a remarkable advantage for long distance communications, which in the past have depended mainly on unreliable HF propagation.

Vulnerablity   The destruction of an orbiting vehicle is within the state of the art in rocketry. However, the destruction of a single communications satellite would be difficult and expensive. It would be particularly difficult to destroy an entire multiple-satellite system in which each satellite has a different orbital pattern and period. The earth terminals offer a far more attractive target for destruction, but they can be protected by the same measures taken to protect other vital communications installations. A high degree of invulnerability to jamming is afforded by highly directional antennas at the earth terminals and by a wide-bandwidth system that can accommodate sophisticated anti-jamming techniques.

Flexibility   When satellites were first launched, only very large organizations such as the military and some communications firms could afford them. However, the units that have been placed in space since then have demonstrated that cost-effective communications can be provided. As satellite power, frequency, and channel capability increase, it will be more and more feasible for small businesses and even individuals to utilize satellite systems.

Today's communications satellites offer great flexibility. They can provide multiple-channel TV signals, long distance radio signals, and thousands of voice channels for telephone system use. Satellites allow communication between two stations, simultaneously between many stations pairs, or from one station to many stations, each depending on how the satellite switches and handles the information.

Limitations of Satellite Communications

The limitations of a satellite communication system are determined by the satellite's technical characteristics and its orbital parameters. Active communication satellite systems are limited by satellite transmitter power on the downlink frequencies and, to a lesser extent, by the satellite receiver sensitivity on the uplink frequencies.

Satellite Transmitter Power   The amount of power available in an active satellite is limited by the weight restrictions imposed on the satellite. Early communications satellites were limited to a few hundred pounds because of launch vehicle payload restraints. Additionally, early inefficient power cells were the only feasible power source consistent with these weight limitations. Thus, the RF power output was severely limited, and a relatively weak signal was transmitted on the downlink frequencies. Continued progress in the development of more efficient power sources and the relaxation of weight restrictions have largely overcome these limitations.

Satellite Receiver Sensitivity   Although powerful transmitters and highly directional antennas can be used at an earth station, the wavefront of the radiated signal spreads out as it travels through space. The satellite antenna intercepts only a small amount of the transmitted signal energy, and, because of the antenna's low gain, a relatively weak signal is received at the satellite receiver. Although the strength of the signal received on the uplink frequency is not as critical as that of the signal received on the downlink frequency, careful design of the RF stage of satellite receivers is necessary to achieve satisfactory results. The continued development of stabilized high-gain antennas and improved RF input stages in the receiver will make this problem less critical.

Satellite Availability   The availability of a satellite to act as a relay between two earth terminals depends on the location of the earth terminals and on the orbital pattern of the satellite. All satellites, except those in synchronous circular orbits, will be in the mutual visibility window of any pair of earth stations only part of the time. The length of time that a nonsynchronous satellite in a circular orbit will be in the mutual visibility window depends on the height at which the satellite is orbiting. Elliptical orbits cause the satellite zone of mutual visibility to vary from orbit to orbit, but the times of mutual visibility are predictable.

The Role of Satellite Communication Systems

Within a global communications network, satellite communication systems provide a much needed additional capacity for point-to-point routing of communications traffic. A satellite link is just one of several kinds of long distance communication links that interconnect switching centers at strategic locations around the world. Satellite links are usually in parallel with links that employ the more conventional means of communication—HF radio, tropospheric scatter, and ionospheric scatter, Satellite links provide added capacity between various points in the network, and, since they continue to operate under conditions that often render the more conventional communications means inoperable, they make a significant contribution to reliability.

Pioneer 10: A Communications Success Story

One outstanding example of the technology involved in satellite communications is the saga of Pioneer 10, launched in March 1972. The 570-pound satellite, traveling at 28,900 miles per hour, escaped the solar system in 1983. In March 1992, 5 billion miles from earth and going deeper into outerspace, Pioneer 10 was still providing communication to earth.

The satellite has three antennas: one each of high-gain, medium-gain, and low-gain. Its radio signals take $7\frac{1}{2}$ hours to reach earth. Each signal is produced by just 8 watts of power, the equivalent of a nightlight. By the time this signal reaches the giant antennas of NASA's Deep Space Network, its power measures just 4.2 billionth of a trillionth of a watt.

Intelligent data from this faint radio signal is received once or twice a day in a small control room at NASA's Ames Research Center in Mountain View, California. At least once a day, radio commands are sent to Pioneer 10, taking $7\frac{1}{2}$ hours to reach the satellite.

Originally designed to communicate for just 21 months, Pioneer 10 is expected to continue communicating until at least the year 2000.

14-4   FUTURE SATELLITE COMMUNICATIONS

The capacities of communications satellites have continued to increase thanks to improvements in rocketry, electronics, communications techniques, and antenna design. Individual satellite earth stations using low-cost receivers and small-diameter antennas are possible with today's technology. In the future, satellites will increasingly make use of higher frequencies, multiple frequencies, multiple frequency bands, dual radiation polarization, spot beams, and advanced time and frequency multiplexing techniques to increase their capacity.

As the satellite increased in size and capacity, it has gained the ability to communicate with other satellites directly instead of through the chain of earth stations. The capacity to deliver satellites with the shuttle allows satellite designers to provide, at a reasonable cost, heavier satellites of any desired weight instead of only the discrete weights allowed by rocket launches. This design capability is particularly beneficial to operators of small earth stations, because the operator can become a more active user of satellite communication systems in the future.

Advanced Satellite Technology and Systems Concepts

New technologies and concepts could be major departures or advances beyond today's satellite technology. As we have seen, today's conventional definition of a communications satellite is a three-axis body-stabilized satellite that is positioned in geosynchronous orbit and has a number of antenna beams designed to concentrate RF power and to re-use available frequencies. That definition has expanded to cover LEO satellites in global constellations, LEO store-and-forward satellites, higher powered *direct broadcast services* (DBS) and mobile satellites. Increasingly, communications satellites are moving to higher and higher frequencies despite advances in frequency re-use. These are essentially evolutionary developments driven as much by new applications as new technology.

Non-Conventional Technologies   Questions arise as to whether totally new architectures or experimental concepts will evolve over the next 10 to 20 years which could redefine our understanding of the communications satellite and whether these new ideas are currently being developed and significant progress being made in the United States, Europe, or Japan. We know that possible "break-through" technologies or concepts have been discussed in popular, trade, and scientific journals around the world in the last few years. These ideas can be categorized as follows:

1. Use of high tensile strength, very long tethers either to suspend satellites in lower orbits or to "whip" them into higher orbits.

2. Use of aerostats to establish laser satellite connections above the earth's primary atmosphere.

3. Use of robotic drone rockets to stabilize space communications platforms in low-earth geostationary orbits.

4. Use of ground-based power transmitters to help stabilize extremely low-mass space communications platforms in LEOS. These high-capacity, low-mass satellites would be artificially maintained in low-earth geostationary orbit.

5. Creation of massive global constellations of LEOS (configuration of up to 1,000 spacecraft) to create systems that are able to provide wideband fixed and mobile satellite services. (See Section 14-1.)

6. Satellites that utilize millimeter (mm) wave frequencies in the EHF range.

7. Other advanced concepts or architectures that involve radical departures from the past—such as extremely large space platforms or satellite clusters with many spacecraft configured to operate as a single unit in geosynchronous orbit.

All of these concepts—which imply more departures from today's basic concept of what a satellite is, how it operates, and how much it costs to provide services—could well be critical to the longer term viability of satellite communications and the mix of telecommunications services they provide. New terrestrial wireless technologies and fiber optic cable could indeed require a totally new approach for the 21st century.

In Europe limited work is being undertaken in this area of radical new departures. It is significant to note that the work being done in Europe is for the most part in different areas of technology than in the United States.

Space Station Concepts   The concept of a space station for telecommunications, an orbital antenna farm, is at least 20 years old. The idea is to deploy one or more large antenna reflectors that can be used with a multibeam feed system to create a large number, perhaps up to 50 or even 100 times, of high powered *pencil* beams and a significant amount of frequency re-use. Depending upon the frequency and power employed, such a space station might be used for *fixed satellite services* (FSS), television or radio *direct broadcast services* (DBS) or mobile or *radio determination services* (RDS). This technology might also be employed for new applications—such as *high definition television* (HDTV) broadcasting, wristwatch transceivers (the "Dick Tracy" radio wristwatch), and video on-demand. This approach has not received a great deal of study by NASA and United States aerospace interests in recent years despite a great deal of investigation in the 1970s and 1980s. High costs, technical complexity, and other problems (for example, thermal expansion with large structures, lifting large masses into geosynchronous orbit, and transmission delay from orbit) have tended to reduce United States interest in such projects.

The *European Space Association* (ESA), however, has decided to explore the possibility of large scale antenna reflectors that might be deployed at low cost. If this critical objective were achieved then an entire space station concept for telecommunications might become feasible, particularly for DBS or mobile applications which cannot be easily duplicated by terrestrial fiber systems. In particular, the feasibility of large scale *inflatable antenna structures* with aperture sizes of 12 to 18 meters has been investigated and development undertaken with ESA funding. The European firm *Contraves* carried out the actual work on this inflatable antenna project. *Alenia Spazio* has developed and qualified a large solid reflector (4 meter diameter) with foldable tips for space use. An alternative design for a large *unfurlable mesh antenna* was not pursued when the concept was withdrawn from consideration. A folding deployable large-scale antenna is also under active study. The most mature development of the inflatable antenna is related to the *ARTEMIS* program which was scheduled for 1995 launch. It should seem that overall efforts in the area of large reflectors are well advanced in Europe and may well be ahead of United States efforts.

Millimeter Wave Satellites   Another area of very long range development effort is *millimeter wave* (mm) applications. The need to move to higher frequencies where broad radio spectra can be used for advanced broadband applications seems quite obvious. Applications such as 3-D HDTV, video-on-demand, and virtual reality imaging will, even with digital compression techniques, require a thousand times the bandwidth of fax, medium speed data, or telephone services. Precipitation attenuation, the lack of spectra assignments for operation above 20 GHz, and the lack of an active commercial demand for satellite communications in the mm wave bands has caused development efforts in this range to lag.

However, ESA has begun development in frequencies above 30/20 GHz to look into the true mm wave band. Several other independent efforts have been undertaken in Germany and Italy, suggesting that serious interest in these higher bands is currently being pursued in specifically funded research programs.

The *German Aerospace Research Establishment* (DLR) is pursuing the *PROMETHEUS* project which is examining short distance vehicle mobile communications. This is part of the *EUREKA* industrial research program for Europe and focuses on terrestrial radio communications at 60 GHz. A satellite component is also being examined and funding may be forthcoming to support this effort. Since precipitation attenuation can be very severe at 60 GHz, extensive data on these effects at different elevation angles associated with LEOS have been collected at look angles ranging from a low of 13 degrees and a high of 43 degrees. These have allowed the development of computerized propagation models over this range of conditions.

In Italy, the *ITALSAT* satellites will allow propagation experiments at 50/40 GHz. The first package was launched on the *ITALSAT F1* satellite in January 1991. A similar package was originally planned to be launched on *ITALSAT F2* in 1994, but was later replaced by the *EMS* payload sponsored by ESA. *Alenia Spazio* is currently manufacturing 50/40 GHz propagation measurement terminals for approximately eight sites in Europe. Further, *Alenia* had developed a 44/20 GHz space antenna using a flat dichroic reflector under military funding. *Alenia*'s work covers space antenna development up to 90 GHz as well as large phase arrays up to 10 meters in diameter.

SUMMARY

- Satellites in geosynchronous orbits provide a continuous communications capability between almost all possible points on the globe. Satellite systems are used for TV and radio broadcasting; telephone, data, and business communications; weather forecasting; and military operations. Satellite systems depend heavily on the technologies of rocketry, space mechanics, solid-state electronics, high-frequency electronics and radiation, and modern communication networks.

- A satellite communication system uses orbiting vehicles to relay radio transmissions between earth terminals. A typical operational link involves an active satellite and two earth terminals. The satellite must be within the view plane (mutual visibility window) of both the sending and the receiving terminal. An uplink is the signal from an earth terminal to a satellite. A downlink is the signal from a satellite to an earth terminal.

- Some technical terms associated with elliptical orbits are apogee, perigee, major axis, minor axis, and angle of inclination. An elliptical orbit may be inclined any number of degrees up to 90° from the earth's equatorial plane. In a circular orbit, the major and minor axes are equal or nearly equal and the earth is located at the intersection of the major and minor axes. The mean height above the earth is used in describing the circular orbit. Three basic types of circular orbit are synchronous, near-synchronous, and medium-altitude. A satellite in a synchronous (geosynchronous) orbit is in a circular orbit about 22,300 statute miles above the earth. The term geosynchronous usually refers to an orbit at the equatorial plane. Three orbiting synchronous satellites, properly spaced, can cover most of the earth's surface. A satellite in a circular orbit within a few thousand miles of the 22,300 mile synchronous orbit is in a nearsynchronous orbit. Below 22,300 miles, the orbit is subsynchronous, and above 22,300 miles, the satellite appears to travel in a reverse (retrograde) direction. A satellite in a circular orbit at altitudes from about 2,400 to 14,300 miles is in a medium-altitude orbit. For stations within the zone of a given satellite, any number of independent station pairs can communicate with each other. Communication is achieved by multiplexing.

- LEOS, or LEOSAT, are small satellites which can be launched either by small launchers, or by large launchers in a multiple launch configuration. The primary features of the LEOS spacecraft family enable them to support a multisatellite constellation with data message communications, position reporting missions, and voice communications for mobile users. The LEOCOM system is a store-and-forward communication system. The use of simple design concepts have lead to relatively low-cost systems. Communications links between LEOS and mobile or portable terrestrial stations can be affected by fading mechanisms which contribute to degradation of link performance in excess of those usually associated with radio wave communication links. Fading mechanisms include multipath scattering and signal blockage, and their efforts on communication links is essential to the design of an effective communications system. The signal behavior due to these mechanisms can have severe degrading effects on communications system performance. There can be confusion surrounding the designation of LEOS. The IEEE has designated LEOS to represent *Lasers and Electro-Optics Society*. In some references, the small-satellite systems are designated LEOS, and in some, LEOSAT. Electronics technicians must be aware of the different LEOS designations.

- Power sources for satellites are solar cells backed up by long-life batteries. Weight-efficient electronic circuitry used in satellite wide-band receivers, down-converters, and filters makes it possible to maximize the number of transponders on the satellite. The larger the antenna used, the more precisely focused the radiation beam. The total bandwidth available from the synchronous orbit can be increased by using more carefully designed earth terminal antennas. Frequency reuse is common in satellite communications.

- Telemetry is the transmission of quantitative information from one point to another by electromagnetic means. Telemetry is used in satellite systems to transmit instructions for active satellite operation in duplex; that is, instructions are transmitted to the satellite, and data concerning the functioning of the satellite are transmitted from the satellite to a ground control station. Telemetry is also used by weather balloons and other environmental monitoring devices, and in manned and unmanned space flights.

- The heaviest concentration of geosynchronous orbiting satellites is over the United States and the Atlantic Ocean. The satellites are presently spaced about 4° apart, but future crowding will require spacing of about 2°.

- Polarized radiation allows satellites to use the same frequencies in parallel without interference. Polarization is achieved by vertically polarizing one pair of antennas and horizontally polarizing another pair. The number of simultaneous communications within a satellite can be increased by transmitting and receiving on several different bands at the same time. Space diversity permits alternate earth stations, separated by several miles, to transmit and receive a clear signal during foul weather. Channels must be assigned for communication between two earth stations via satellite. Channels may be assigned permanently, or they may be assigned at the time of need.

- Multiplexed satellite signals can be achieved by frequency division multiplex (FDM) or time division multiplex (TDM). Frequency division multiple access (FDMA) is used to assign available channels to the next station requesting or demanding service. The destination station is notified which channel is being used, and the channel is released back to the available pool upon completion of the communication. Time division multiple access (TDMA) is used for channel assignments on a specified time slot, either permanently or on demand, and is used in digital communications.

- A satellite link is simply one of several kinds of long distance communication links. Satellite links are usually in parallel with links that employ the more conventional means of communications. They are unaffected by the propagation abnormalities that interfere with HF radio, are free from the high attenuation of wire or cable facilities, and are capable of spanning long distances without the numerous repeater stations required for line-of-sight or troposcatter links. Satellite communications provide great reliability and flexibility. Active satellite systems are limited by transmitter power on the downlink frequencies and by receiver sensitivity on the uplink frequencies.

- Future satellites will use higher frequencies, multiple frequency bands. dual radiation polarization, spot beams, and advanced TDM and FDM techniques to increase their capacities. That also have the ability to communicate with other satellites directly instead of through the chain of earth stations.

QUESTIONS AND PROBLEMS

1. What are satellites used for?

2. Define (a) passive satellite and (b) active satellite.

3. Define (a) communications satellite, (b) uplink, (c) satellite communication system, (d) downlink, and (e) mutual visibility window.

4. Define the technical terms associated with elliptical orbits.

5. Define (a) equatorial orbit and (b) polar orbit.

6. Describe the three types of circular orbit.

7. Define subsynchronous orbit.

8. Describe retrograde as it relates to satellites.

9. List the essential system components of a basic operational satellite communication system.

10. Describe (a) how a basic 6/4 GHz communication satellite system works and (b) the Viewer's Choice satellite TV system.

11. Describe geosynchronous orbit.

12. Describe how worldwide coverage can be achieved by satellites.

13. Explain how multiple pairs of earth stations within the zone of a single satellite can communicate with each other.

14. What factors determine the amount of power generated in solar cell sources by the sun's energy?

15. Describe the solar energy sources on (a) spin-stabilized satellites and (b) body-stabilized satellites.

16. Describe (a) how more efficient, more reliable power sources for communications satellites are obtained and (b) how the use and application of electrical power on board the satellite are controlled.

17. Describe how signals are processed in a typical satellite electronic system.

18. Define (a) transponder and (b) frequency reuse.

19. Describe (a) how the total bandwidth available from a synchronous orbit can be increased and (b) the problems encountered in the use of large antennas.

20. Define telemetry and explain its functions in satellite systems.

21. How is the need to provide greater channel capacity and functional capability aboard new satellites being met?

22. What factors are affected by the size of earth station antennas?

23. Describe how antenna noise affects the bandwidth of received signals.

24. Define spot beaming and explain how it is achieved.

25. Describe (a) polarized radiation and (b) channel assignment.

26. Define space diversity and explain its use.

27. Describe the operation of (a) FDM, (b) FDMA, (c) TDM, and (d) TDMA.

28. Define cryptology and explain its use.

29. List the advantages and limitations of satellite communication systems.

30. Define the role of satellite communications.

31. Describe the future of satellite communications.

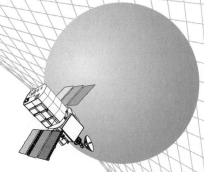

CHAPTER 15

Two-Way Communications

OBJECTIVES

When you complete this chapter, you should be able to:

- List and identify the frequency bands for the various mobile communication systems.
- Describe a basic mobile system.
- Define transceiver, repeater, and VOX, and describe their operations.
- Define walkie-talkie, autopatch, repeater, break-in operation, and duplexer.
- Describe the operation of a cellular telephone system.

INTRODUCTION

Most two-way communications other than direct wireline telephone communication occur between a base station and a mobile unit or between mobile units. The systems may include microwave links to connect control, base, and relay or repeater stations.

Mobile electronic equipment is any equipment operated in a moving vehicle such as a car, truck, train, boat, or airplane. Mobile equipment includes all attendant devices, including antennas, transceivers, microphones, and power supplies.

The rapid development of solid-state technology has made the design and operation of mobile electronic equipment much simpler today than they were just a few years ago. For example, most mobile power supplies provide low-voltage dc power; 13.8 V is by far the most common value. A vacuum tube circuit requires a complex power inverter for operation from a mobile supply, but solid-state equipment can be operated directly from such a supply.

Mobile equipment is more compact than fixed-station equipment. In addition, mobile equipment must be designed to withstand larger changes in temperature and humidity and severe mechanical vibration.

Most two-way mobile communication installations consist of a vertically polarized omnidirectional antenna and a transceiver. The transceiver is a transmitter/receiver combination housed in a single cabinet, an arrangement that saves significant space and expense. Although most mobile systems operate in the VHF and UHF bands, where antenna size is manageable, some mobile operations operate in the low frequency (LF), medium frequency (MF), and high frequency (HF) bands using inductively loaded antennas.

Equipment designed for portable operation may also be used in mobile applications. Generally, however, portable radio transmitters and receivers are less powerful and less versatile than equipment designed specifically for mobile use.

This chapter discusses various services and equipment involved in two-way communications between a base station and mobile units or between mobile units. It also examines the latest two-way system—the cellular telephone system.

15-1 TWO-WAY COMMUNICATION SERVICES

There are many different services offered in two-way radio communications. Each service has its own purpose. Among the basic two-way communication services:

1. *Citizens' band* (CB) *radio* operates at 26.96–27.41 MHz, AM or single-sideband (SSB), using shared assigned channels.
2. *Amateur radio,* the so-called ham operation, covers a broad frequency spectrum; controlled by the FCC.
3. *Aeronautical radio* operates in the HF, MF, and VHF bands, AM or SSB.
4. *General mobile radio service* operates at 460–470 MHz, FM, with a maximum output of 50 W. For use by the general public, with shared assigned frequencies, this is also called a CB service.

The following two-way systems are in the *private land mobile radio services:*

1. *Public safety radio* includes police, fire, highway maintenance, forestry conservation, and local government radio services.
2. *Special emergency radio* includes medical, rescue, disaster relief, school bus, veterinarian, beach patrol, and paging radio services.
3. *Industrial radio* includes power company, petroleum company, forest product, business, manufacturer, motion picture, press relay, and telephone maintenance radio services.

Domestic public landline mobile radio service (DPLMRS) stations are connected to telephone lines (called common carriers) to allow units to make telephone calls while mobile. Users of this service are conventional and cellular radiotelephone systems.

Citizens' Band

The *citizens' radio service,* more commonly called *citizens' band* or *CB,* is a public, noncommercial radio service available to the general population of the United States. There are four classes of citizens' band radio:

1. *Class A* stations are licensed to operate in the 460–470 MHz band and are allowed 60 W maximum input power to the final transmitter stage.

2. *Class B* stations are licensed to operate in the same band as class A stations but are limited to 5 W of input power to the final stage.

3. *Class C* stations are operated in the 26.96–27.23 MHz band, on 27.255 MHz, and in the 72–76 MHz range for radio control purposes.

4. *Class D* stations are allowed to use 40 channels, each 10 kHz wide, in the range 26.965–27.405 MHz for communications between individuals, either for personal or for business purposes. Class D is by far the most popular of the CB radio classes. The 40 channels in the Class D band are listed in Appendix D, Table D-1.

Citizens' band offers a convenient communications system that can be used at home, in a car, in a boat, on an airplane, or even while walking or hiking. In the band from 26.965 MHz to 27.405 MHz, amplitude modulation with double- or single-sideband is permitted for voice, with the frequency 27.255 MHz (channel 23) used mostly for radio control of aircraft, model boats, and garage door openers. In this band, channel 9 is set aside for emergency communications, such as after highway accidents. Channels 1–7 and 16–22 are authorized for communications between stations of a single owner, such as a garage and its tow trucks, local branch offices of a business, or the home and automobile of a private individual.

Voice transmissions are not authorized in the frequency band 72–76 MHz. This band is used, to a limited extent, for radio control systems.

In the UHF region, the band from 462.525 MHz to 467.475 MHz has been authorized for CB voice and control signals, but it is also shared with other services. Most of the recently designed radio-controlled garage door openers operate in this band.

Individual operators of CB equipment need not be licensed, but the transmitting equipment is licensed through the manufacturer's authorization by the FCC. It is the responsibility of the owner of the equipment to file the required license and fee with the FCC in order to obtain a call sign. Citizens' band equipment with less than 100 mW input to the last stage requires no license.

Citizens' band operation requires the transmitter and receiver to be on the same channel, and operators generally use the so-called *10-code*. This code, with some modifications, is also used by police and fire departments. The 10-code is shown in Appendix D, Table D-2.

The transmitter output power for class D stations is legally limited to 4 W for AM and 12 W peak envelope power (PEP) for SSB. The typical maximum communication range is between 10 and 30 miles. Long distance propagation is occasionally observed, since the 27 MHz band is susceptible to the effects of sunspot activity. However, one of the major limitations on CB operation is that, regardless of feasibility, operation beyond a 150 mile radius from the base station is not permitted.

Citizens' band radio enjoyed a great boom in the middle 1970s, when millions of Americans learned how easily they could obtain and operate simple two-way radio equipment. Hundreds of CB clubs exist on the local and national scale. An organization called *Radio Emergency Associated Citizens' Teams* (REACT) monitors channel 9, the officially designated emergency frequency, and provides assistance to motorists in trouble.

Amateur Radio

By definition, the amateur radio operator or *ham* is a duly authorized person interested in radio techniques solely with a personal aim and without financial interest. The FCC issues operator's licenses to U.S. citizens, regardless of age or other considerations, who pass an examination. Part of the examination tests for proficiency in the international Morse code, and part covers electronic theory. There has been a recent change that created one class that has no code test. Several classes of amateur operator licenses are available: they include Novice, Technician, General, Advanced, and Extra. Different licenses have different privileges concerning the use of the frequency bands assigned for amateur operation and concerning the method of operation, such as telegraphy and telephony. Extra Class licensees can operate on any amateur frequency. Advanced Class licensees are restricted from operating on a few frequencies. General Class licensees are still further limited. Technician Class licensees can operate on any Novice frequencies but also have full amateur privileges on all frequencies above 50 MHz. Novice and Technician Class amateurs may operate on only narrow segments on four of the HF bands, using CW (radiotelegraphy) emissions only.

There are over 400,000 amateur radio operators in the United States, and several hundred thousand in the rest of the world. These ham operators converse over distances limited only by the size of the earth and by the power of their stations. They also experiment with new devices and provide emergency back-up communications during times of disaster.

The major organization of ham operators is the *American Radio Relay League* (ARRL). The ARRL operates radio station W1AW, with transmitters on all amateur bands and with special transmissions of interest to ham operators.

Frequency Allocations   Amateur radio operators have many different frequency bands, ranging from 1.8 MHz to the microwave portion of the spectrum. The bands allocated for different types of amateur radio use are shown in Table 15-1, which indicates frequency ranges and meter bands.

The maximum permissible dc power input to the final transmitter stage is 1000 W, with some limitation on this value at specific frequencies and specific locations. For most amateur emissions, the maximum RF power output is limited to 1500 W PEP. One exception is the 200 W PEP limit for all amateurs operating in any Technician/ Novice section of any HF band.

Amateur Stations   All amateur radio stations must be licensed by the FCC, which assigns call letters. Station licenses are given only to licensed operators.

Figure 15-1 shows a basic amateur station consisting of a radiotelegraph transmitter with an LP filter as its output, a key, a receiver with earphones or with a loudspeaker, an antenna, a transmit/receiver (T/R) switch, a standing wave ratio (SWR) meter, and an antenna tuner. Some separate means of checking the transmitter frequency should be available. Many systems use transceivers, with the transmitter and receiver built into a single cabinet.

Some more advanced stations may include equipment to send and receive SSB, AM, FM, radioteleprinter (RTTY), HF slow-scan (3 kHz bandwidth) still-picture TV

TABLE 15-1 **Amateur Radio Bands**

Frequency (MHz)	Band (m)	Frequency (MHz)	Band (m)	Frequency (MHz)	Band (m)
1.800–2.000	160	14.175–14.225		144.1–144.8	
3.500–3.525	80	14.225–14.350		220–225	1.3
3.525–3.700		18.068–18.168	16	420–450	0.7
3.700–3.750		21.000–21.025	15	*(GHz)*	*(cm)*
3.750–3.775		21.025–21.100			
3.775–3.800		21.100–21.200		1.24–1.30	70
3.850–4.000		21.200–21.225		2.30–2.45	or
7.000–7.025	40	21.225–21.300		3.30–3.50	less
7.025–7.100		21.300–21.450		5.65–5.925	
7.100–7.150		24.890–24.990	12	10.0–10.5	
7.150–7.225		28.0–28.1	10	24.0–24.5	
7.225–7.300		28.1–28.2		48.0–50.0	
10.100–10.109	30	28.2–28.5		71.0–76.0	
10.115–10.150		28.5–29.7		165–170	
14.000–14.025	20	50.0–50.1	6	240–250	
14.025–14.150		50.1–50.4		300 and above	
14.150–14.175		144.0–144.1	2		

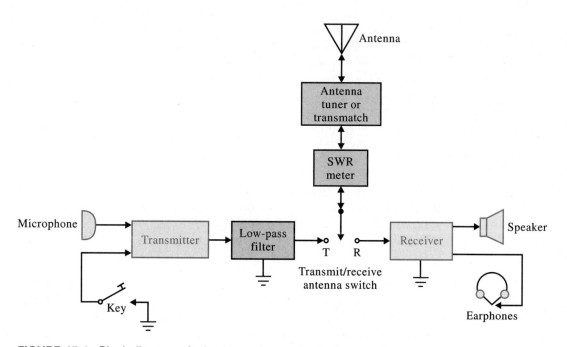

FIGURE 15-1 Block diagram of a basic amateur radio station

transmission (SSTV), VHF or UHF slow- or fast-scan television, and facsimile. Radio-teleprinters may be used on frequencies where frequency shift keying (FSK) or audio frequency shift keying (AFSK) is allowed. Many stations use separate power amplifier stages between a transceiver and the antenna to produce the maximum power output values.

Amateur antennas may be single wires, dipoles, verticals, or fixed or rotatable beams. They may be coupled to transmitters or receivers by transmission lines, possibly using a *balun* or a *transmatch* to match transmitter-to-antenna feed-line impedances. The antennas must conform to local building codes. If they are to be more than 200 feet high, notification must be given to the Federal Aviation Administration (FAA) and the FCC, and lighting may be required. Antenna towers must not be more than 1 foot in height for every 100 feet from a listed airfield having a runway 3200 feet or more in length, 50 feet from other airfields, and 25 feet from listed heliports. Usually, an amateur antenna not exceeding the height of nearby buildings or trees by more than 20 feet has no height requirements.

Codes and Procedures   The most widely used amateur services are CW and tele-phone, with radioteletype operation much less frequent because of the relatively high cost of teletype machines. Amateur TV transmission is generally used only in metro-politan areas at VHF and UHF. The FCC publishes regulations, and the ARRL pub-lishes detailed operating procedures, for each of the amateur services. To keep transmissions as short as possible, it is customary to use certain abbreviations, call signs, and codes. Each station has its own call sign consisting of letters and numerals, and it is customary to use both the calling and the called station call signs when a con-versation is initiated. The code "CQ" is used when the caller wants to speak to anyone who will answer. In telegraphy, the *international Morse code* (Table 15-2) is used.

Q signals are used in CW work as well as in voice communication. Their meanings need to be expressed with brevity and clearness in amateur work. (Q abbreviations take the form of questions only when each is sent followed by a question mark.) Ab-breviations help to cut down unnecessary transmission. However, make it a rule not to abbreviate unnecessarily when working with an operator of unknown experience. With voice operation, the *International Civil Aviation Organization* (ICAO) phonetic alpha-bet is used to spell out proper names. A listing of Q signals, common abbreviations for CW work, and the ICAO phonetic alphabet are in Appendix D, Tables D-3, D-4, and D-5, respectively.

Public Service Operations   In addition to communications between individual ama-teur operators, the ARRL provides a system of networks in which amateur radio op-erators receive and relay personal messages. The *National Traffic System* (NTS) is organized into regional and area networks with specific schedules and station assign-ments to provide this service.

Two services have been organized for emergencies. The *Radio Amateur Civil Emer-gency Services* (RACES) is part of civil defense and is intended to furnish communi-cations to other civil defense activities. The *Amateur Radio Emergency Corps* (AREC) is organized for such civilian emergencies as floods, hurricanes, and other local disas-ters. Both organizations coordinate closely with fire, police, ambulance, and similar services.

TABLE 15-2 **International Morse Code**

Character	Code	Character	Code	Character	Code
A	· —	1	· — — — —	Ü	· · — —
B	— · · ·	2	· · — — —	(OR)	— · — — · —
C	— · — ·	3	· · · — —	"	· — · · — ·
D	— · ·	4	· · · · —	-	· · — — — ·
E	·	5	· · · · ·	=	— · · · —
F	· · — ·	6	— · · · ·	SOS	· · · — — — · · ·
G	— — ·	7	— — · · ·	Attention	— — · —
H	· · · ·	8	— — — · ·	CQ	— · — · — — · —
I	· ·	9	— — — — ·	DE	— · · ·
J	· — — —	0	— — — — —	Go ahead	— · —
K	— · —			Wait	— · · ·
L	· — · ·	.	· · · · · ·	Break	— · · · —
M	— —	;	— · — · — ·	Understand	· · · — ·
N	— ·	,	· — · — · —	Error	· · · · · · · ·
O	— — —	:	— — — · · ·	OK	· · · ·
P	· — — ·	?	· · — —	End message	· — · — ·
Q	— — · —	!	— · — · — —	End of work	· · · — · —
R	· — ·	'	· — — — — ·		
S	· · ·	—	— · · · · —		
T	—	/	— · · — ·		
U	· · —	Ā	· — · —		
V	· · · —	À or Á	· — — · —		
W	· — —	É	· · — · ·		
X	— · · —	CH	— — — —		
Y	— · — —	Ñ	— — · — —		
Z	— — · ·	ö	— — — ·		

Aeronautical Communications

The *Aeronautical Broadcasting Service* (ABS) disseminates information for the purposes of air navigation and communication. These services are provided by an extensive network of transmitting stations, two-way voice and digital communications, radio beacons, radiolocation stations, and radar. Radiotelephone communication for aircraft employs AM and various types of SSB emissions in the HF range in the bands shown in Table 15-3.

VHF Aviation Communication Equipment Although many frequencies in the HF and MF bands have been allocated to aviation, the VHF band is by far the most commonly used in general aviation. Airborne radios are different from automobile and home radios in that airborne radios require two-way conversation. Therefore, aircraft radio systems usually consist of a transmitter, receiver, antenna, microphone, and headset or speaker. In most airborne systems, transceivers are used.

TABLE 15-3 Aeronautical Communications Frequency Allocations

Range	Use	Range	Use
HF (MHz)		*VHF (MHz)*	
2.860–2.990	General purpose	108.0–111.9	Instrument landing systems (ILS)
2.182	General purpose	108.0–117.9	VHF navigation stations (VOR)
3.023	Search and rescue (SAR)	118.0–135.95	Civilian aviation voice communication
3.281	Lighter-than-air	121.5	Simplex, emergency, and distress
3.410–3.460	General purpose	123.1	SAR
4.380–4.990	General purpose	156.8	Distress
5.450–5.680	General purpose	*MF (kHz)*	
8.364	Distress, SAR	410	Direction finding
8.820–8.960	General purpose	457	Working frequency 500 kHz, distress
10.01–11.39	General purpose		
13.27–13.34	General purpose		
17.90–17.97	General purpose		

Two basic methods of communication are used in airborne systems: (1) single-channel simplex, involving the use of the same frequency for transmitting and receiving; and (2) dual-channel or double-channel duplex, involving the use of a different frequency for transmitting than for receiving.

Tuning Most modern VHF radios incorporate crystal-controlled tuning. With this system, the pilot automatically receives the desired frequency once the proper setting on the tuning dial is selected.

Many modern aircraft radios have what is commonly called a $1\frac{1}{2}$ system. The tuning knob on the left side of the radio is normally marked "COM," indicating that both a transmitter and a receiver are provided. This is single-channel simplex communication capability—the "1" part of the $1\frac{1}{2}$ system. The frequency range of the "COM" side of the radio is 118.0–135.95 MHz for a 360-channel radio transceiver and 118.0–126.9 MHz for a 90-channel transceiver.

The right side of the radio has a tuning knob that controls frequencies from 108.0 MHz to 117.9 MHz. This knob tunes only a receiver (the $\frac{1}{2}$ part of the $1\frac{1}{2}$ system) to be used in the reception of VHF omnidirectional radio (VOR) or instrument landing system (ILS) navigation signals or voice reception. The pilot cannot transmit on the right side of the radio.

Special Radio Phraseology English has been adopted as the international aviation communication language. In aviation radio transmissions, the ICAO phonetic alphabet is to be used.

Because of the large number of radio-equipped aircraft, some radio frequencies have become crowded, making it highly desirable to reduce the length of transmissions. For this reason, the special phraseology shown in Table 15-4 was developed.

TABLE 15-4   Special Radio Phrases for Aviation Use

Radio Phrase	Meaning
Acknowledge	Let me know that you have received and understood this message.
Affirmative	Yes.
Correction	An error has been made in this transmission. The correct version is ... .
Go ahead	Proceed with your message.
How do you hear me?	Self-explanatory
I say again	Self-explanatory
Negative	That is not correct.
Out	This conversation is ended and no response is expected. (Used infrequently since it is usually obvious when the conversation has terminated.)
Over	My transmission is ended and I expect a response from you. (Omitted if the message obviously needs a reply.)
Read back	Repeat all of this message back to me.
Roger	I have received all of your last transmission. (Used to acknowledge receipt; should not be used for other purposes.)
Say again	Repeat what you have said.
Speak slower	Self-explanatory
Stand by	I must pause for a few seconds. (If the pause lasts longer than a few seconds, or if the phrase is used to prevent another station from transmitting, the phrase must be followed by the ending OUT.)
That is correct	Self-explanatory
Verify	Check with originator
Words twice	Since communications are difficult, please say every phrase twice. (As a request.)
	Since communications are difficult, every phrase in this message will be spoken twice. (As information.)

15-2   TWO-WAY COMMUNICATION FUNDAMENTALS

A basic two-way communication system consists of a *base station,* or *base,* and one or more *mobile units,* or *mobiles.* The base station with its control or operating point is usually centrally located in a city building. The mobiles communicate with the base by VHF or UHF transmitter/receiver sets, most often configured as transceivers. The base station antenna is erected at a high point near the transmitter, often on top of the building housing the transmitter. The control point and the base transmitter are normally located within 100 feet of each other. The base and the mobiles operate on the same frequency and use common antennas that are switched from transmit to receive by a *transmit/receive* (T/R) relay. The distance between base and mobiles in which reliable communication can occur depends on the location, height, and directivity of the base antenna, and on the power used by the transmitters of the base and the mobiles.

To extend the reliable range from about 25 miles to possibly 100 miles, the control point may remain in place, but the transmitter, receiver, and antenna may be placed in a remote location where the needed height can be gained. These remote locations may be on top of the tallest building in the city or on a nearby hilltop or mountain peak. Transferring control signals to the remote site can be by telephone lines, fixed low-power radio relay systems, or microwave links.

In some industrial systems, in order to allow communications with mobiles by offices located in different parts of a city, it may be necessary to connect one or more dispatch points to the control point. In other systems, such as public safety or special emergency radio, there may be so many mobiles from different departments operating in so many different localities that two or more operating frequencies may be required.

Repeaters are usually located on hilltops or mountain peaks. These repeaters require all units to listen on the same frequency but to transmit on a separate frequency. In the repeater, signals received from mobiles are fed directly to the input of a transmitter and retransmitted to other mobiles. This process provides much better communication between mobiles than is possible between units that are far apart and where signals are blocked by buildings, trees, and so on.

Transceiver

A *transceiver* is a combination of a transmitter and a receiver having a common frequency control and usually enclosed in a single package. Transceivers are used extensively in two-way radio communication at all frequencies and in all modes.

Figure 15-2 shows a simplified block diagram of a transceiver. The principal components are a variable-frequency oscillator (VFO) or channel synthesizer, a transmitter, a receiver, and an antenna-switching device. The simplest transceivers employ direct conversion techniques; this scheme is sometimes used for HF Morse code communication. More sophisticated transceivers use a superheterodyne design.

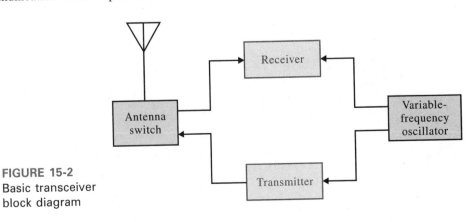

FIGURE 15-2
Basic transceiver block diagram

The main advantage of a transceiver over a separate transmitter and receiver is economy: many of the components can be used in both the transmit and receive modes. Another advantage is that most two-way operation is carried out at a single frequency, and transceivers are more easily tuned than separate units.

The main disadvantage of a transceiver is that communication must sometimes be carried out on two frequencies that differ greatly. Also, duplex operation is not possible with most transceivers. Some transceivers, however, have provisions for separate transmit/receive operation, overcoming these difficulties.

Transceivers are used extensively by amateur and CB radio operators. In both cases, the units are compact and can be used for fixed or mobile operation.

Transmit/Receive Switch

When a transmitter and a receiver are used with a common antenna, some method must be devised to switch the antenna between the two units. It is especially important that the receiver be disconnected while the transmitter is operating. A *transmit/receive* (T/R) switch accomplishes this.

The simplest form of T/R switch is a relay. When the transmitter is keyed, the relay connects the antenna to the transmitter. When the transmitter is unkeyed, the relay connects the antenna to the receiver. This arrangement is used in many transceivers.

Figure 15-3 shows an RF-actuated electronic circuit with a transmit/receive (T/R) switch. In this circuit, the transmitter is always connected to the antenna, even while

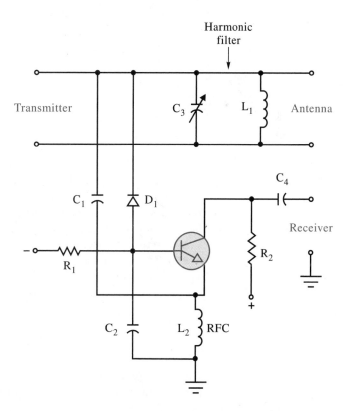

FIGURE 15-3
Transmit/receive
(T/R) switch

receiving. However, when the transmitter is keyed, the receiver is disconnected. This T/R switch allows full break-in operation if CW Morse code emission is used.

Repeater

A *repeater* intercepts and retransmits a signal to provide wide-area communications. Repeaters are generally used at VHF, UHF, and microwave frequencies. They are especially useful for mobile operation. The effective range of a mobile station is greatly enhanced by a repeater. In the 144 MHz amateur band, for example, direct simplex communication between moderate-power mobile stations ranges from 10 to 30 miles; with a repeater, this range may exceed 100 miles. Figure 15-4 illustrates how a repeater located in a high place intercepts and retransmits the signals from a moderate-power or low-power station, extending the range.

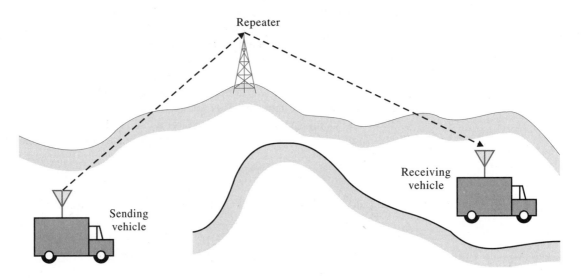

FIGURE 15-4   Typical repeater use

A repeater antenna should be located at the greatest possible height above the ground. A vertical collinear array is generally used to maximize the receiving and transmitting gain. Receiver sensitivity and transmitter power output should be such that repeater coverage is about the same for both reception and transmission.

A repeater consists of an antenna, a receiver, a transmitter, and an isolator (duplexer). The transmitter and receiver are operated at slightly different frequencies. The separation is approximately 0.3–1% of the transmitter frequency. This separation allows the isolator to work at maximum efficiency, preventing undesirable feedback and desensitization.

Duplexer   A *duplexer* is a device in a communications system that allows duplex operation. Duplex operation means that the two operators can interrupt each other at any time, even while one operator is transmitting. Duplex operation is usually carried out using two different frequencies. Notch filters between the transmitter and receiver at each station prevent overloading of the receiver front end. A repeater uses a duplexer to allow the simultaneous retransmission of received signals on a different frequency.

In a radar installation, a duplexer automatically switches the antenna from the receiver to the transmitter whenever the transmitter puts out a pulse. This duplexer action prevents the transmitter from damaging or overloading the receiver. In this application, the duplexer acts as a high-speed, RF-actuated T/R switch. The radar receiver cannot actually operate while the transmitter sends out the pulse.

Simplex Operation In a two-way communication system, both transmitters and receivers are often operated on a single frequency. This operation is known as *simplex operation,* or *simplex.* The two stations communicate directly with each other; no repeater or other intermediary is used.

At VHF and UHF, simplex operation may not provide enough communications range, especially if one or both stations are mobile. To increase the effective range, repeaters are used.

In simplex operation, only one station can transmit at a time, because neither station can receive signals at the same time, and on the same frequency, as it is transmitting. If data must be sent and received simultaneously, two different frequencies are used. This system is called duplex operation.

It is not possible for one station to interrupt the other station in simplex operation. However, interruption of one station by another can be accomplished if CW or single-sideband, suppressed-carrier (SSSC) emissions are used. This arrangement is called break-in operation. It requires a special switch and muting (squelch) system. The receiver is activated during brief pauses in a transmission.

Duplex Operation *Duplex operation,* or simply *duplex*, is a form of radio communication in which the transmitters and receivers of both stations are operated simultaneously and continuously. Each station operator can interrupt the other at any time, even while the other is transmitting. Therefore, a normal conversation is possible, similar to that over telephone lines. There are no "blind" transmissions, and therefore there is no need for exchange signals such as "over."

To achieve full-duplex operation, the transmitting and receiving frequencies must be different. Two stations in duplex operation, such as stations X and Y in the simplified block diagram in Figure 15-5, must have opposite transmitting and receiving frequencies. For example, station X may transmit on 146.01 MHz and receive on 146.61

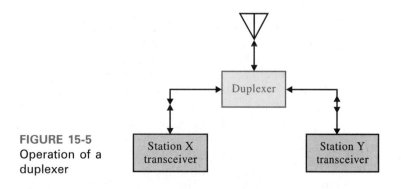

FIGURE 15-5
Operation of a duplexer

MHz; then, station Y must transmit on 146.61 MHz and receive on 146.01 MHz. A duplexer is usually required to prevent the station receiver from being desensitized or overloaded by the transmitter.

In radiotelegraphy, a form of nearly full-duplex operation is sometimes called break-in operation. A fast T/R switch blocks the receiver during a transmitted dot or dash but allows extremely rapid recovery. If the receiving operator desires to interrupt the transmitting operator, a tap on the key will be heard by the sending operator within a few milliseconds. Break-in operation is not true duplex, however, since the sending operator cannot hear the receiving operator interrupt during a transmitted pulse.

Break-in Operation   *Break-in operation* is a form of radio communication in which a transmitting operator can hear signals at all times except during the actual emission of radiation by the transmitter. In break-in operation involving transmission of CW code, the receiver is active between the individual dots and dashes even if the speed is high. In SSB operation, any pause allows reception when break-in is used. True break-in operation is fairly easy to achieve with low-power equipment, but when the transmitter output power is high, sophisticated techniques are needed. Fast break-in operation is impractical with FSK, AM, or FM. However, semi break-in is sometimes used in these modes.

Semi Break-in Operation   *Semi break-in operation* is a switching scheme used in many communications installations. Semi break-in can be used for code, radioteletype, or voice operation. In voice systems, it is called *voice-operated transmission* (VOX).

In semiautomatic code or radioteletype break-in, the transmitter is actuated the moment the operator presses the key. A relay or electronic switch performs the changeover from the receive mode to the transmit mode. When a pause occurs in the transmission, the transmitter stays on for a predetermined time. This delay is adjustable from about 0.1 s to about 5 s. Most code operators do not like to have the receiver click on between the characters or words of a code transmission, and they set the delay according to the speed with which they most often send the code. In radioteletype operation, the delay is not important since the carrier is on all the time.

Semi break-in operation does not allow the code operator to hear between dots and dashes, as full break-in does. However, semi break-in is easier and less expensive to implement and is preferred by some operators.

VOX   Voice-operated transmission (VOX) is a means of actuating a circuit, such as a radio transmitter, using the electrical voice impulses from a microphone or audio amplifier. Such a voice-actuated system may use a relay or an electronic switch. VOX is a form of semi break-in operation.

The schematic in Figure 15-6 shows a VOX circuit. The sensitivity and delay are adjustable. An anti-VOX circuit prevents the received signals from actuating the transmitter if a speaker is used for listening.

VOX circuits are commonly used in communications transceivers, especially in SSB units. VOX eliminates the need for pressing a lever or switch to change from the receive mode to the transmit mode. This feature is especially useful in mobile

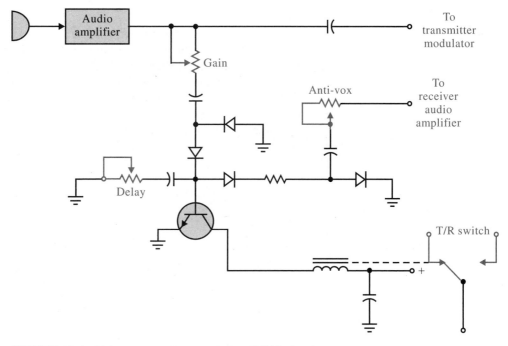

FIGURE 15-6 Voice-actuated transmission (VOX) circuit

operation or when both hands are needed such as for taking notes. The VOX circuit can be disabled, and push-to-talk switching used instead, if the operator desires.

In VOX operation, the transmitter is actuated within a few milliseconds after the operator speaks into the microphone. To prevent unwanted "tripping out" during short pauses, the transmitter remains actuated for a short time after the operator stops speaking.

15-3 MOBILE TELEPHONES

A *mobile telephone* is a radio transceiver designed for access to an autopatch system. Mobile telephones did not originally allow continuous two-way conversation; that is, neither party could interrupt the other. However, some mobile telephones now provide true duplex operation.

Mobile telephones, since they involve radio transmissions, cannot legally be used in some countries without a license from the government. Mobile and portable telephones usually operate in the VHF and UHF radio bands, between about 30 MHz and 3 GHz. This frequency range provides reliable operation within a radius of several miles from the base station or repeater system. Many radio repeaters are interconnected with the telephone lines so that mobile radio operators, equipped with tone (pushbutton) keypads in their transceivers, can gain access to the lines.

A special form of portable telephone consists of a small handheld unit and a base station, with an operating range of several hundred feet. This device can be used by anyone in most countries without a government license.

Autopatch

An *autopatch* is a device for connecting a radio transceiver to the telephone lines by remote control. Autopatch is generally accomplished through repeaters. A simplified autopatch system is illustrated in Figure 15-7.

The repeater is first accessed by transmission of a certain sequence of tone-coded digits. This transmission is done with a telephone-type keypad connected to the microphone input of the transmitter. The autopatch system is then activated by another set of digits. At this point, the operator may dial the desired telephone number. With most amateur radio autopatch repeaters, the repeater is always available for normal communication, but the autopatch must still be activated by a special sequence of digits.

Autopatch telephone conversations are not always the same as ordinary telephone hookups. Interruption of the radio operator is impossible if the receiver is disabled while the operator speaks. A simple autopatch system that works over a limited range and allows duplex operation is the increasingly popular cordless telephone.

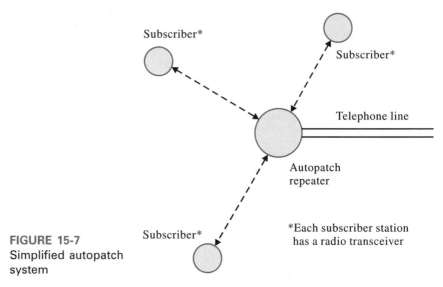

FIGURE 15-7
Simplified autopatch
system

Portable Equipment

Portable communication equipment is any apparatus designed for operation in remote locations. Portable equipment is battery operated and can be set up and dismantled quickly. Some portable equipment can be operated while being carried; an example is the *handy-talkie,* or *walkie-talkie.*

With the improvements in miniaturization and semiconductor technology, portable equipment can have complex and sophisticated features. A few decades ago, mention of a microphone-controlled walkie-talkie would have elicited ridicule. Today, such devices are commonplace.

Portable equipment must be compact and lightweight. Therefore, it always has modest power requirements since a high-power battery is invariably bulky and heavy. In addition to being small and light, portable equipment must be designed to withstand physical abuse such as vibration, temperature and humidity extremes, and prolonged use without complicated servicing.

Portable equipment is sometimes used in mobile applications. For example, a hand-held walkie-talkie unit can be connected to a mobile antenna and used in a car or truck.

Portable Telephone   A *portable telephone* is a radio transceiver designed for access to an autopatch system. Portable telephones resemble handsets with antennas.

There are two types of portable telephones. The short-range variety is called *cordless* or *wireless* telephone. It is designed for use within a few hundred feet of a base unit. The long-range type is similar to a walkie-talkie, and its range is limited only by the access range of the autopatch repeater and transceiver, which is typically 10 to 20 miles. For example, a handheld radio transceiver designed for use on the 2 m (144 MHz) amateur band is equipped with an autopatch encoder, so it can be used as a portable telephone.

Several manufacturers provide newer, lighter-weight portable telephones. The general style and design of portable phones are shown in Figures 15-8 and 15-9. Two specific brands of these telephones are the Bell Atlantic Nynex® *Mobile 232* (manufactured by Nokia) and the Bell Atlantic® *S.E.PC550* (manufactured by Motorola).

FIGURE 15-8
Portable lightweight, handheld telephone

FIGURE 15-9
Portable "flip" telephone

The *232* is one of the smallest micro handheld telephones available. It weighs just 8.7 ounces and uses a battery that provides 110 minutes of talk time and 26 hours of standby time. A travel charger is also included. The following list identifies the most notable features of the *Mobile 232*.

1. A signal strength indicator is a permanent bar graph that indicates the strength of the signal to assist the user in determining whether sufficient signal exists to ensure a quality call. The permanent bar graph also indicates the battery power level to assist in monitoring remaining talk time.

2. The phone can be programmed to identify a telephone number stored in its 100-number memory by associating the number with a name, for easy and accurate retrieval.

3. One-touch dialing allows the storage of the two most frequently dialed numbers in specific locations and allows the user to initiate a call by simply pushing one button. Any digit key on the keypad will answer a call.

4. A call-in-absence indicator feature indicates if a call was missed while out of your vehicle. Used in conjunction with voice mail products, this feature will indicate when messages are in your voice mailbox.

5. A function review acts like an electronic user's guide by displaying which keystrokes to use in order to access the different features of the phone.

6. The phone is capable of having two telephone numbers which permits registration for service in more than one cellular service area to save money on roaming charges. Phone usage can be controlled by imposing restrictions on outgoing calls.

7. Frequency bands for the *232* are 824 to 849 MHz (transmit) and 869 to 894 MHz (receive). Its transmitting power is 0.6 W, +2 dB, −4 dB.

The *550* offers many of the same features as the *232*. It weighs 9.9 oz., and a single battery provides 100 minutes of talk time and 18 hours of standby time. Significant features of the *550* which differ from those of the *232* are:

1. Its one-touch dialing allows the storage of the three most frequently dialed numbers in specific locations.

2. One-touch emergency dialing is offered at memory location "01" which can be programmed for 911, or any emergency number, and accessed when the phone is in the locked mode.

3. The phone can be programmed to lock each time the phone is turned off to help prevent unauthorized use, and phone usage can be controlled by imposing restrictions on outgoing calls.

4. Frequency bands for the *550* are 824.040 to 848.97 MHz (transmit) and 869.040 to 893.97 (receive). Its RF power output is 0.6 W. Input/output impedance is 50 ohms. Channel spacing is 30 kHz, and duplex spacing is 45 MHz. It provides 832/2412 channels.

Both of these phones have been tested to meet or exceed the technical standards established by the *Cellular Telecommunications Industry Association* (CTIA) to ensure quality cellular products.

The Motorola® *Micro T∗A∗C Lite*™ is a *Digital Personal Communicator*™ (DPC) telephone. It weighs just 7.5 ounces with a standard battery, a slim line battery that provides up to 45 minutes of continuous talk time or 8 hours standby time.

Call placement features include memory linking/pause in memory, which enables placement of calls to standard phone numbers that require additional number sequences such as a credit card or pager number. After two rings, the telephone will automatically answer. Super speed dialing provides faster memory dialing with a minimum of keystrokes. Incoming call screening gives the user a visual and audible indication of incoming calls. Its 101 memory locations allow the user to store 99 32-digit phone numbers plus last number redial and scratchpad. It automatically stores a new number into the next available memory location. It automatically attempts to complete a system-busy call for four minutes after the first attempt. A memory overwrite protection warns the user that a memory location is already filled. It offers 832/2412 channel capability (the 2412 capability may not be available in all locations).

Usage control features include an automatic lock that helps control unwanted usage by automatically locking the phone each time it is turned off. Five call timers include Resettable, Individual, Audible, Auto Display, and Cumulative for monitoring phone expenses by tracking airtime usage. Six levels of call restrictions enhance control of phone expenses and security of the numbers in memory.

Cordless Telephone   A cordless telephone uses a radio connection between the receiver and the base, rather than the usual cord. Two small antennas, one at the main unit and the other at the receiver, allow use of the receiver as far away as 600 to 800 feet under ideal conditions.

Cordless telephones are available in various configurations. With some units, calls can be received in a remote place, but outgoing calls must be dialed from the main base. With other units, both incoming and outgoing calls can be controlled entirely from the receiver.

Since cordless telephones are linked by low-power radio, they are subject to interference from other appliances and from stray electromagnetic impulses. However, the technology of the cordless telephone is improving rapidly.

Pager   A *pager* is a shirt-pocket-size, 6 to 10 ounce receive-only device. It is activated by a two-tone signal from the base station that sets off a series of audible beeps to alert the carrier to report to the base station or call the base station by telephone. Frequency bands used by pagers are 30–50, 132–174, 406–420, 450–512, and 929–932 MHz. High-gain antennas are contained within the plastic receiver cases of most pager units.

Some newer pagers have microprocessors to control their capabilities. These devices use subaudible digital codes that may allow several hundred different pagers to be selectively called. These newer pagers may emit audible beeps only, or they may beep and open the AF amplifier to allow the receiver to pass a voice message to the paged person.

Controls vary with different pagers. Depending on the pager type and manufacturer, pagers may have an ON-OFF switch, a volume control, a manual squelch (mute) control, a beep/voice selection switch, or a push-to-hear switch.

A compact tone and voice FM radio pager is shown in Figure 15-10. This device has high receiver sensitivity and a variable volume control to produce loud, clear audio that allows messages to be heard, whether in a large building, an automobile, or a fringe area of operation. The electronic design of the unit reduces the current consumption so that, when the unit is powered by an AA alkaline battery, longer intervals of pager service are gained before battery replacement. An AA rechargeable nickel-cadmium battery, which can be easily recharged in its single-unit charger, can be used.

The pager battery is monitored continuously. When the pager is turned on, it emits a double-interrupted beep to indicate a good battery. If the battery becomes marginal, the pager sounds off with a distinct chirping alert. After this alerting signal, several more hours of reliable operation are available to allow sufficient time for battery replacement or recharge.

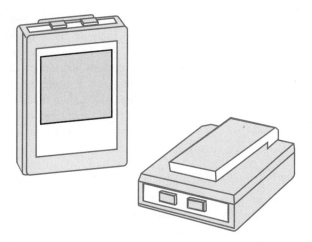

FIGURE 15-10
Motorola BPR 2000
FM radio pager

Motorola® has produced several new pagers. One of those, the *Ultra Express*™, is a numeric pager designed to meet the diverse needs of consumers. It offers a variety of features, such as a protective holster with a hinged belt clip. It provides 16 message slots, shows time of day and date, produces silent vibration, and saves messages when turned off and when changing batteries. This list of features is impressive since it also has an easy-to-read backlit display, front mount display, selectable musical alerts, built-in alarm clock, message lock, selective erase or erase all, programmable on/off, low battery alert, and operates on a single AAA battery.

Another new pager is the *Renegade*™, which provides features such as 10 page message slots, easy-to-use single button operation, musical alert or silent vibration, backlit top mount display, hinged belt clip, low battery alert, reminder alert, and duplicate message detection; and it operates on two AAA batteries.

Conventional Radiotelephone

Motor vehicles outnumber telephones in the United States, and there are more automobiles than houses. Although at least one member of each family drives, less than 10% of all vehicles are equipped with two-way telephone service. In contrast, 97% of all residences and nearly 100% of all businesses have conventional wireline telephone service. Existing conventional mobile telephone service is characterized by (1) high cost, (2) limited range, (3) poor access during busy periods, and (4) long waiting times to obtain service.

The primary reason for the limited availability of mobile telephone service is a lack of radio channel allocations for car telephones. Yesterday's mobile telephone technology cannot offer service to large numbers of subscribers. The use of radio channels for conventional mobile telephone service ties limited spectrum availability to low subscriber numbers. On the bandwidth allocated for one mobile channel, there could be four AM broadcast radio stations; two complete push-to-talk systems for taxicabs; two police, two fire, or two public safety systems; or two private radio systems. This means that with yesterday's technology, there is no efficient way to assign frequency allocations to conventional mobile radiotelephones.

Also, the power requirements for adequate transmission and range mean that when one subscriber is on a radio channel, the use of that channel is prohibited to other users within an area of over 50 miles in diameter. Since conventional systems are usually spaced approximately 70 miles apart, a single conventional mobile telephone conversation consumes 60 kHz of bandwidth over an area of 5000 to 10,000 square miles. This arrangement is obviously an extremely inefficient use of radio channels.

15-4   CELLULAR TELEPHONE SYSTEM

A *cellular telephone system,* sometimes called *cellular radio,* is a special form of mobile telephone developed in recent years. A cellular system consists of a network of repeaters, all connected to one or more central office switching systems. The individual subscribers are provided with radio transceivers operating at VHF and UHF.

The network of repeaters is such that most places are always in range of at least one repeater; ideally, every geographic point in the country would be covered. As a subscriber drives a vehicle, operation is automatically switched from repeater to repeater, as shown in Figure 15-11.

Eventually, most if not all telephone communication may be by cellular radio. Worldwide communications of high quality and low cost, using entirely wireless modes, might be achieved by the end of the twentieth century. Although there are many manufacturers of cellular telephone systems, we will examine just one system—*DYNA T∗A∗C*™, the cellular radiotelephone communication system developed by Motorola. DYNA T∗A∗C is an acronym for *Dynamic Total Area Communications.*

Path of driver

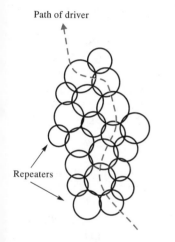

Repeaters

FIGURE 15-11
Cellular telephone system

Solutions Offered by Cellular Systems

Developers of mobile radiotelephone service are in general agreement that an ideal system should have the characteristics of wireline service and also offer efficient spectrum utilization. The solution to keeping the good characteristics of wireline service and overcoming the less desirable characteristics of conventional mobile radiotelephone has been found in cellular radiotelephone.

Good Transmission Quality   *Good transmission quality* has come to mean not only that the voice message is readily understood but also that it is free from annoying static and noise and that it really sounds like the speaker's voice. In mobile telephone, these quality requirements demand (1) a dedicated frequency to minimize co-channel interference, (2) adequate power to ensure a high *S/N* ratio, (3) sufficient bandwidth to transmit voice quality, and (4) frequency modulation to minimize the noise problems that plague most low-power AM radio signals.

Good Service Quality   *Good service quality* has two dimensions: accessibility and usability. *Accessibility* is the ability of a user to obtain an idle transmission channel when he or she desires to place a call. *Usability* refers to ease of operation. Wireline service provides accessibility of giving each subscriber a dedicated pair of copper wires to the local switch point and by providing enough redundancy in the transmission, switching, and control equipment. Most wireline systems are engineered to provide 98% access even during the busiest hour of the day.

Full-Duplex Service *Full-duplex service* is an inherent characteristic of wireline telephone systems, which means that both parties to a conversation can talk at the same time. This capability is one of the great advantages of cellular systems.

Practicality Motorola's cellular radiotelephone system incorporates six 60° directional receiver antennas with 17 dB gain at each cell site. The increased gain of these antennas, along with improved receiver sensitivity, eliminates the need for satellite receivers by balancing the inbound and outbound ranges, thus making portable telephony *practical.*

Frequency Reuse Fundamental to the idea of cellular systems is the concept of short-range *frequency reuse*. Frequency reuse is an answer to limited spectrum availability. It is through this concept that dramatic numbers of subscribers not now being served by conventional radiotelephone systems can be accommodated.

In noncellular radiotelephone systems, the boundary of the system is described by a line on the minimum acceptable signal level. Within the area encompassed by the line, a certain probability exists of receiving a minimum acceptable signal level. A service area is then described as having at least 90% of the locations with greater than the minimum acceptable signal. In cellular systems employing frequency reuse, the coverage, or *cell boundary,* is described as enclosing the area in which 90% of the subscribers will experience at least a 17 dB carrier-to-interference ratio.

Unlimited Area Coverage One of the major differences between conventional mobile telephone service and the cellular approach is distribution of service. Conventional systems use a powerful base station to blanket the service area. Each cell is automatically assigned until all channels are busy. The number of channels is limited, so only a few subscribers may use the system at a given time. The conventional system may serve an area with a radius as large as 50 miles, but the possibility of interference prevents reuse of other channels. Once the conventional system becomes fully loaded, further growth is impossible. In a cellular system, a cell site covers only a small area, but adding cell sites increases area coverage to any size desired.

In summary, an ideal mobile telephone service should meet several objectives that do not pertain to the wireline network:

1. Efficient use of the radio channels because of the increasing demands on the frequency spectrum from competing users.
2. Expandable capacity, the need for which is evidenced by the long waiting lists in many markets.
3. "Roaming" capability (movement from cell to cell), which would, in principle, enable a mobile unit to operate anywhere in the United States.

What Is Cellular?

Cellular systems can meet the demands of both large and small urban areas with the continuing expansion in areas of low population density and along travel routes. This flexibility paves the way to a truly nationwide mobile communications network. Coverage is provided in cellular systems by subdividing the area to be served into a network of coverage areas called *cells*. Each cell contains a low-powered base station (called a *cell site*) and is assigned a set of frequencies. Adjacent cells are assigned different frequency groups to avoid interference. Cells sufficiently far apart may use the same frequency groups simultaneously. The FCC has authorized 666 new channels in the 800 MHz band for cellular communication. Each cell uses about 40 or 50 frequency pairs. Interference between cells does not occur because the frequency pairs are not used jointly in adjacent cells. By limiting the cell site transmitter power levels to 100 W effective radiated power (ERP), channel reuse is possible five cells away. By reusing each radio channel for different conversations many times within a service area, efficient spectrum utilization is accomplished.

Cells ranging from 1 to 12 miles in radius are used in a typical cellular system. At startup of a newly installed system, the use of large cells is economical because it minimizes the cost while providing the desired coverage.

Cell Splitting   One of the big advantages of cellular systems is the ability to expand as subscriber demand grows. New cells may be added to the outside of a system at any time, and existing cells can be subdivided by lowering transmitter power and installing new cell sites. The minimum practical cell radius is approximately 1 mile. Smaller cells allow channels to be reused at smaller distances.

When service demand increases, frequency reuse and a process called *cell splitting* can be used to increase cell capacity. Cell splitting requires the addition of new cell sites between existing cell sites to form a new configuration of smaller cells while using the same number of channels. Since each of the new, smaller cells can serve about the same amount of traffic as the original, larger cell, the capacity of the system is increased. Frequencies are reused many more times with the small-cell configuration than with the large-cell configuration. More simultaneous conversations per given area can be accomplished this way.

Operating Functions   Two main functions must be performed to process a call in a cellular system: *call setup* and *call handoff*. When a mobile unit initiates a call, it sends a request for service and receives a voice channel assignment on one of a set of data channels called *signaling control* channels. In the case of a land-originated call, the mobile is paged, replies to the page, and is assigned a voice channel on the same set of data channels. If the volume of traffic should threaten to exceed the paging capacity of the control channel, the cell site is capable of setting up a separate paging channel and instructing the mobiles to monitor that channel for calls.

15-5   CELLULAR SYSTEM OVERVIEW _____

Cellular systems employ state-of-the-art digital switching base site and subscriber equipment. These modular building blocks allow each system to be customized to the characteristics of an area and the needs of the users. A combination of cells is used throughout the *cellular geographic service area* (CGSA) to provide the composite coverage needed for reliable service.

The capacity provided by an individual cell is directly related to the number of busy-hour subscribers in the area covered by the cell. As the number of users changes, channels are added or removed from specific cells until the available spectrum allocation is fully utilized. The original cell layout of each specific system should take maximum advantage of geographic separation, terrain, and other related factors to facilitate orderly growth with maximum frequency reuse.

After the basic frequency allocation is fully utilized, cells are subdivided or split to increase the total number of channels in the system. By using directional (sector) antennas, cells reusing the same frequencies can be spaced closer together, permitting a significant increase in system capacity. The limits on frequency reuse depend on the terrain and the geographic distribution of subscribers.

Orderly Evolution to Small Cells

The majority of cellular systems are implemented initially with cells that are nominally 11 miles in radius. The standard design technique for cell layout is the definition of a base site, or *cellular grid,* which shows optimum location of cells for most efficient frequency reuse. Sites are located as near as possible on the grid such that they fit the ultimate reuse pattern as it develops.

The channel plan for the system is also prepared with reference to the ultimate layout envisioned for the area. As the system grows from its initial configuration, this layout permits orderly addition of new cells with RF channels which have, in effect, been reserved for that location. This feature permits smooth evolution in system capacity with minimum interruption of existing service. Typically, subscribers will not be aware of any change in system operation during this process.

Nationwide Compatibility

A basic objective of cellular system development is coordinated nationwide service for subscribers with fully automatic operation to the greatest extent possible. To this end, the *Electronic Industries Association* (EIA) has established standards for cellular equipment and systems.

System Interconnection

The cellular system is controlled by an *electronic mobile exchange* (EMX), a fully solid-state pulse code modulation (PCM) switching station, and is capable of interconnecting to a class 5 office, or higher, in the public telephone network. This capability provides full interconnection with subscriber service features comparable to those provided by the most advanced wireline telephone exchanges.

Subscriber Units

Subscriber equipment comprises a complete line of radiotelephone equipment and accessory options, including the portable cellular telephone. These products offer subscribers unrivaled mobility and flexibility, whether their mobility is primarily vehicular based or pedestrian. All portable cellular telephone units have been human engineered to ensure that the features are readily understood by the user.

Remote Diagnostics

Through automatic testing and self-diagnostics, all detectable system failures are alarmed locally and remotely. When an alarm signifying a failed unit is observed by a technician, the system can be accessed and additional tests made. The technician may also determine the replacement boards needed and take them to the site to restore back-up status.

Trunking of RF Channels

Channel use efficiency is maximized through *trunking* of channels within the cells and within the system. A *trunk* is a connection between an EMX system and a central office or another EMX system. Trunks are classified by the direction (in, out, two-way) or by usage (transit). Each trunk is assigned a unique number. All channels are assigned by the system on a demand basis. The system format includes provisions for using channels in adjacent cells in an overflow mode when appropriate. The system is based on trunking of the RF channels; thus, capacity is determined from standard trunking curves for a selected grade of service. Data gathered within the system during operation permit continuous verification of true trunking performance.

Frequency Assignments

The cellular band is divided into two segments to allow two systems to coexist in the same geographic area. Each band occupies one half of the available spectrum. To guarantee nationwide compatibility, signaling channel frequencies are preassigned in each band, and, for convenience, each frequency is assigned a channel number. Table 15-5 shows the assignments by band.

TABLE 15-5   Signaling Channel Frequency Assignments

Signaling Transmitter	Signaling Channel Numbers
Band A base transmitter	313–333
Band B base transmitter	334–354

Channel assignments are made every 30 kHz. Assignments at a particular base site are made every 630 kHz until all channels in a set are assigned. After the full set is used, the next set of channels implemented must be offset by at least 60 kHz.

System Expansion

The methods used to implement frequency reuse determine the system's ability to serve the maximum number of subscribers within a defined area using a fixed amount of spectrum. The two methods of implementing frequency reuse are (1) the use of a 12-cell or a 7-cell site repeat pattern with omni transmit antennas and (2) the use of a 4-cell site repeat pattern with 60° antennas.

Co-channel users are protected from interference by geographic separation and, where used, directional antennas at the base sites. If directional transmit antennas are used in addition to geographic separation, the distance between sites can be reduced. With 60° antennas, each omni cell is divided into six smaller cells in such a manner that a repeat pattern consists of four sites and 24 sector cells. A *sector cell* is defined as an area with a unique frequency set, so that a vehicle leaving that area requires a new channel assignment (handoff). Figure 15-12 reflects these 4-cell and 12-cell patterns.

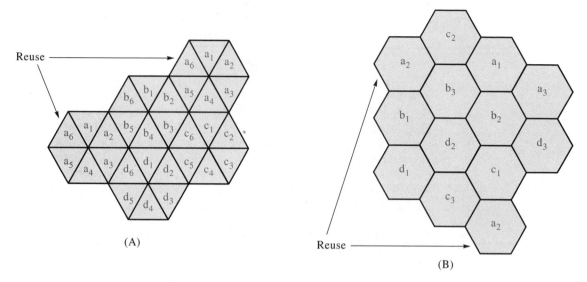

Reuse

(A)

Reuse

(B)

FIGURE 15-12 Cell reuse patterns (A) Sector cells, 4-cell pattern (B) Omni cells, 12-cell pattern

When the system expands sufficiently to require cell subdivision, frequency assignments are allocated using a 4-cell pattern. When the system does not have sector transmit cells, the spectrum is divided into 21 sets. With sector transmit cells, 6 sets per site are required. A disadvantage to implementing frequency reuse with 24 frequency sets (as compared to 12 in omni) is the degradation in trunking efficiency due to the smaller number of channels in any one group.

Sector Sharing

To improve the trunking efficiency, the 4-cell plan employs a technique called sector sharing. *Sector sharing* is a method of assigning channels so that subscribers requesting service can be temporarily assigned a channel from an adjacent sector if all the channels in the desired sector are busy. For example, consider a mobile driving around the cell site such that its path traces an imaginary 60° antenna pattern similar to that shown in Figure 15-13. Along the route, the mean signal strength is constant, because

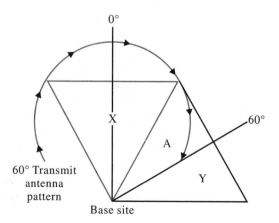

FIGURE 15-13
Sector sharing

as the mobile moves farther from the transmitter, it is also driving toward the center of the antenna beam. This path is the line of constant mean signal strength.

At point A, the user tries to initiate a call. During the signaling time, the signal strength is measured in each sector, and since it is strongest in sector Y, the system tries to assign a channel in sector Y. If all channels in sector Y are busy and the vehicle's mean signal strength is greater than the signal strength for vehicles at the perimeter of sector X, the system will assign the user to a channel that is normally assigned to vehicles in sector X. The area where an assignment to either of two sectors can be made is 59% of the total area. With the signal strength being stronger in sector Y than in sector X, handoff will be initiated when a channel is available in sector Y.

Conversion to Sector Transmit Cells

The omni receive or sector receive cell that is in the omni transmit mode can be converted to sector transmit when frequency reuse considerations require it. The conversion from omni receive/omni transmit to sector receive/sector transmit requires the addition of an omni receive-to-sector receive conversion kit, six directional antennas, and six duplexers. The duplexers provide the isolation between receiver front end and transmitter output to allow the receivers and transmitters to share an antenna.

The sector receive-to-omni transmit conversion requires the addition of six duplexers. The antennas already on-site are used. In addition, the existing firmware in the *base site controller* (BSC) must be replaced whether the cell had previously been omni receive or sector receive, and a new software release is required for the EMX.

Signaling Channel Capacity

As the number of subscribers in a CGSA approaches 75,000, the control channel may become capacity limited in the land-to-mobile direction. To increase the data handling capability of the cell site, a new set of control channels is created, and the data traffic is separated so that only pages are sent on the original control channels, now called *paging* channels, and channel assignments and mobile/portable originated calls are processed on the new control channels, called *access* channels. The call processing is identical except that subscribers start their call-originating and page response sequences on the new access channels. The paging channel numbers are programmed in the number assignment module in the mobile/portable, and the location of the access channel set may be transmitted to the mobile/portable as part of the overhead message transmit so that no modification to the subscriber equipment is required when this change is implemented in the system.

A second set of new control channels may be created, using some channels from the voice channel frequencies if necessary, and this set will be used as a second group of paging-only control channels. The system would then be operating in such a manner that, at any point in the CGSA, one access channel and either of two paging-only channels would be selectable by a mobile/portable unit. New subscribers would then be set up with the second set of paging-only channels, and system signaling capacity would be increased to approximately 160,000 subscribers.

Fortunately, although every cell site requires an access channel, the paging-only control channels need not be implemented at every site. It is necessary that only one paging channel from each group be usable from any point in the CGSA. This condition can usually be achieved with fewer than 10% of the cell sites equipped with paging channel transmitters. Using the same technique, a third set of paging-only control channels could be created and installed (at the same paging sites) to increase system signaling capacity to approximately 240,000 subscribers.

15-6   DYNA T∗A∗C CELLULAR RADIOTELEPHONE SYSTEM

Just one of many cellular systems across the country, the Motorola DYNA T∗A∗C cellular radiotelephone system includes all of the fixed and mobile hardware and software essential for complete system operation. DYNA T∗A∗C offers mobile and portable radiotelephone communications with features and services comparable to those of the public wire-line network. The system complies fully with FCC and EIA guidelines for 800 MHz cellular systems.

How DYNA T∗A∗C Works

In DYNA T∗A∗C, the cell sites are connected by voice and data lines to the EMX, which acts as the interface from the cell sites to the public switched telephone network. Both the cell site base station and the mobile and portable units transmit with reduced power. To complete calls, the system locates the specific mobile or portable telephone subscriber, allocates a free radio channel at the appropriate radio cell site, and connects the subscriber with the wire-line telephone network. The multiple cells form a single-coverage blanket over the area of desired service. Potential interference from portable units operating from tall buildings (or mobile units operating from high elevations) is minimized by automatically adjusting the subscriber's transmitter power to the lowest level necessary for effective communication. This feature ensures that no excess power will be used, thereby minimizing interference with other users and permitting maximum reuse of the frequencies. It also extends the battery life of the portable unit, significantly enhancing the utility to subscribers.

Handoff

When, during the course of a call, the subscriber's vehicle passes one or more radio cell boundaries, the EMX switches the conversation to another frequency channel in the new cell site to maintain strong and clear transmission quality. The vacated channel is then made available for use by another subscriber. This process is called *handoff*. Handoff occurs each time the subscriber passes from one cell to another, roughly as follows:

1. The signal strength of the subscriber unit is monitored at the cell site.
2. As the subscriber approaches the cell boundary, the signal strength of the transmission becomes weaker. When this occurs, the base station sends a message to the EMX.
3. The EMX then orders the station in the cell sites bordering the subscriber's cell area to monitor the signal strength of the subscriber's transmissions.
4. The EMX switches the subscriber's conversation to the cell site base station receiving the strongest transmission from the subscriber. This transmission

is also switched to a different channel, releasing the previous channel for another subscriber. The time required to accomplish this task is only a fraction of a second and is not normally noticeable to the subscriber.

15-7 FUNCTIONAL DESCRIPTION OF EMX

The EMX mobile telephone exchange, illustrated in Figure 15-14, is a distributed microprocessor-managed, completely solid-state system designed to provide maximum system availability under all conditions. Functional redundancy is provided for all key elements, and design modularity accommodates a wide variety of system configurations and architecture. The EMX is accessed by a teleprinter for control of operator functions, and procedures for operation, maintenance, and directory control have been simplified as much as possible.

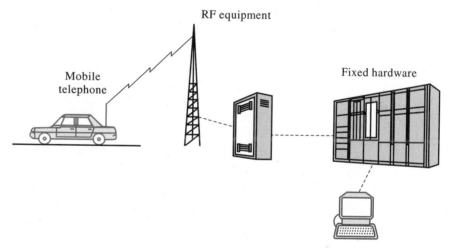

FIGURE 15-14 Electronic mobile exchange (EMX) system

The EMX contains a number of subsystems that work together to produce its total function: (1) dual central processing units, (2) a switch unit, (3) a group multiplexer unit, (4) a tone signaling unit, (5) a maintenance and status unit, (6) numerous voice group units (channel banks which may be customer supplied), and (7) an alarm and status panel. The EMX also utilizes input/output devices such as tape drives, teleprinters, and announcement machines. The tape drives and teleprinters are provided with intelligent interfaces featuring firmware-controlled microprocessors. They are bus switchable between the twin processors of the node with which they are associated.

Figure 15-15 shows an EMX functional block diagram. The primary function of the EMX is to route the input of one port to the output of another. The EMX simultaneously connects any non-busy import port to any non-busy output port with any other signal routing until the maximum port traffic is exceeded. The total port size is calculated as follows. A *voice group unit* (VGU, channel bank) contains 24 voice frequency ports. Each *group multiplexer unit* (GMU) can *voice group interface* (VGI) up to 16 VGUs. This feature provides 24 × 16 = 384 ports per GMU. In its maximum configuration, the EMX can be equipped with four GMUs, providing 64 PCM groups or 1536 voice frequency ports, as shown in Figure 15-16.

Voice or audio signals are applied to the VGU from telephone trunks or lines, base site controllers (BSCs), and announcement machines. These signals are converted to PCM digital data in the VGU through a process of sampling, quantization, and coding. The signals are monitored by various EMX subsystems for (1) serial-to-parallel bit conversion for *time division multiplex* (TDM) transmission, (2) detection of failures, (3) alarming of failures, (4) time switching of PCM voice bytes, and (5) recording of statistical data. Depending on the category of the call (land-to-mobile, mobile-to-land, or mobile-to-mobile) appropriate software routines are called up that control the processing of the signal and the sequence of operations. The voice signal is then converted back to serial bit format, then from PCM form back to audio.

Central Processing Unit

The *central processing unit* (CPU) (Figure 15-15) performs switch control, data acquisition, data base recording, cell coordination (for handoff and frequency reuse), fault management, and general call administration.

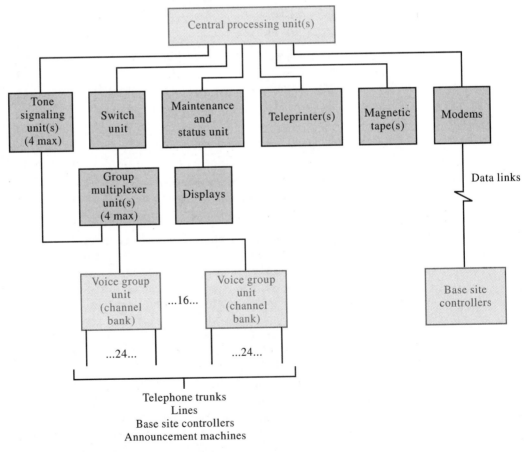

FIGURE 15-15 EMX functional block diagram

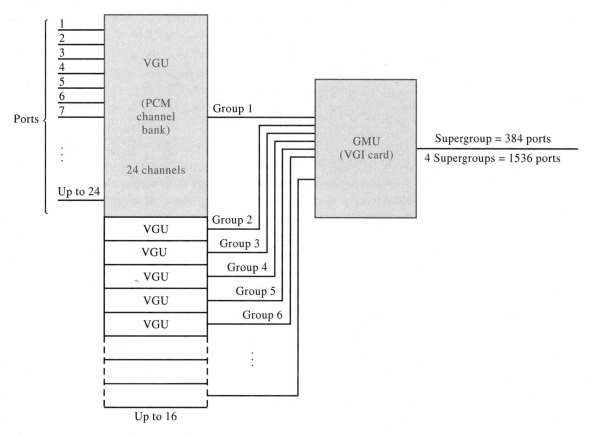

FIGURE 15-16   Supergroup system

Switch Unit

The *switch unit* (SWU) (Figure 15-15) performs the time switching of PCM audio bytes, processes signal bits from the PCM bit stream, and provides an interface with a CPU via a data link. It controls the digitized audio from the VGU and tone signaling units and, except for updating, operates independently of the CPU. Synchronizing and signaling information, which forms part of the PCM bit stream, is extracted in the GMU. Signaling is processed and ultimately sent to the processor. Sync and signal information is also added under processor control to the outgoing PCM bit stream. The SWU "sees" all circuits connecting to its buses as identical, regardless of whether they are telephone trunk, base station channel, or tone signaling circuits.

Group Multiplexer Unit

Each *group multiplexer unit* (GMU) (Figure 15-15) performs the following two functions:

1. Converts, in the group interfaces, the serial bipolar bit streams of up to 16 voice group interfaced VGUs into 8-bit parallel format and time-multiplex transmits this information to the SWU.

2. Receives time-multiplexed 8-bit parallel words from the SWU, converts them to serial format in the group interfaces, sends the resulting PCM bit streams to the VGU, and manages parallel interfaces to the tone signaling unit.

Tone Signaling Unit

The tone sending and receiving circuitry is contained within the *tone signaling unit* (TSU) (Figure 15-15). All voice frequency signaling tones for both transmit and receive directions, and all supervisory tones, are generated within, and/or decoded by, appropriate cards in the TSU. Routing of the tones through the switch is in PCM digitized format, as is the case for all other audio routing. Reception of the digitized signaling through the port of the appropriate card results in the detection, decoding, and subsequent transferral of the decoded information to the CPU. Tone generation, on the other hand, is effected by referring to digital lookup tables or permanent read-only memory (ROM), in which are stored all the data necessary for all signaling and supervisory tones used at the interfaces. This data eliminates the need for frequency or level adjustments at any point other than the external interface card.

Maintenance and Status Unit

The *maintenance and status unit* (MSU) (Figure 15-15) continually monitors itself and the rest of the EMX for failures. Buses are under constant surveillance to prevent device failure from causing total system shutdown. Other portions are periodically tested for failure conditions. If a failure is revealed, the CPU locates the faulty device and isolates it from the remainder of the system.

When the MSU is informed of a failure by the CPU, it strobes out the information to the alarm and status panel. Status for all BSCs is reported to display panels in a similar manner.

The MSU also strobes in the condition of any switch on the alarm and status panel. These switches can be used for such functions as panel lamp tests or audible alarm reset.

Voice Group Units

Voice group units (VGUs) (Figure 15-15) are commercially available PCM channel banks that conform to CCITT PCM Recommendation Q.47. Channel banks are not part of the DYNA T∗A∗C system and may be supplied by the user. They provide analog-to-digital conversion (ADC), digital-to-analog conversion (DAC), and TDM of 24 analog ports that transmit PCM bipolar bit streams to the EMX while receiving bit streams from the EMX—all accomplished on a clock-synchronized basis. Each analog port provides two-way audio (two-wire or four-wire) pulse line signaling. CCITT Recommendation G.711 (North American) types of channel banks are available with up to two line signals in each direction.

The channel banks include independent power and alarm provisions and must be located within 250 cable meters (825 feet) of the EMX unless repeaters are used in the interconnect, in conformance with standard telephone central office PCM practices. For more remote application, any carrier circuit that will support two channel banks back to back may be used to connect a remote channel bank to the EMX.

Alarm and Status Panel

The *alarm and status panel* (ASP) (part of the MSU in Figure 15-15) provides a visual indication of the alarm and/or status situation of the EMX control system and radio channels. Each ASP consists of an *alarm and status electronics board* (ASEB) and up to eight alarm and status modules. The ASP is controlled by the active MSU processor and interfaces the CPU with a display and alarm indicator for each EMX unit.

Input/Output Devices

Input/output (I/O) devices for the EMX consist of teleprinters or magnetic tape drives (Figure 15-15). Teleprinter equipment used in conjunction with the EMX must conform to the following specification or its equivalent: Bell System Technical Reference PUB 41715, "Data Speed 40 Stations for Dataphone Service," December 1973. Magnetic tape read or written by EMX must meet ANSI Standard X3.39 (1973).

Teleprinters   The minimum equipage for any model of the EMX is one *teleprinter* (TTY). The minimum number of TTY ports equipped for a redundant system is six. These ports are split across redundant *serial communication interface peripherals* (SCIPs). Two additional interfaces can be equipped, bringing the total number of TTYs possible to 14. These ports are EIA RS-232C types. Baud rates are switchable between 300 and 1200. Besides these 14 possible RS-232C ports, two more ports are provided via the same SCIPs to drive the MSU.

Magnetic Tape Drives   Up to four *magnetic tape drives* can be provided in a fully redundant EMX. One drive is used to do program loading, and one to collect recent change activity. The other two provide primary and redundant recording of billing and system statistics information.

EMX Software Functional Description

EMX *software* is divided into the following major functional categories:

1. *Call processing software,* which handles all telephone call activity from call origination to call termination.

2. *System administration software,* which updates information in the EMX data base, compiles usage statistics, handles maintenance procedures, and indicates alarm and status conditions.

3. *Common control software,* which controls the execution of call processing and system administration processes, manages hardware and software failures, supports communication of interprocessor messages, and controls the process of loading programs from magnetic tape into EMX memory.

15-8   SUBSCRIBER EQUIPMENT

DYNA T∗A∗C subscriber equipment includes vehicle-mounted units and portable units. Specially designed to operate in 800 MHz cellular systems, these units incorporate the most advanced subscriber-oriented features available. Whether mobile or portable, all subscriber equipment offers the same extensive user performance features. In addition, each product offers unique features, options, and accessories. In the following sections, we will first examine the common features, and then the unique features and options of each product.

Standard Equipment Features

The *mobile* telephone unit is a state-of-the-art, full-duplex radiotelephone designed to operate in the 800 MHz cellular radiotelephone system. It represents significant advances in solid-state electronics, in advanced automated manufacturing technologies, and in logic-based subscriber features. Integrated and externally applied computer-based test procedures provide rapid diagnostics for simplified field service.

Extensive research in 800 MHz technology has culminated in a compact, truly *portable* telephone. The frequency-synthesized transceiver operates automatically over the 666 allocated channels and complies with the EIA Cellular Compatibility Specification. New-found personal mobility and expanded communications control are realizable benefits to the portable subscriber, along with all of the advanced user features cellular radiotelephone has to offer.

Mobile Transceiver An 800 MHz digital programmable divider frequency synthesizer eliminates the need for individual crystal oscillators for each channel. The broadband transmitter power amplifier section is totally microstrip, and it requires no tuning. A power output-leveling circuit ensures constant transmitter power to the mobile antenna.

A newly designed low-loss duplexer provides the superior duplex selectivity of the mobile unit. The duplex selectivity and receiver injection filters are all fixed-tuned, making the unit easy to maintain and adjust.

The digital microprocessor-based logic unit also requires no adjustments. It is encoded by a high-accuracy reference oscillator and encodes and decodes the high-speed signaling information.

The radio package consists of a cast chassis along with rugged top and bottom covers that provide superior RF and environmental protection. Modular construction allows circuit boards to be exchanged easily, and a built-in service mode aids in fast, accurate diagnosis and maintenance.

Portable Transceiver A newly designed frequency synthesizer serves both the transmitter and receiver. This device features divider circuitry to regulate a VCO. A complementary metal-oxide semiconductor (CMOS) is used for the programmable dividers to reduce power consumption. The output of the VCO is stepped and multiplied to generate 666 channels.

Power output of the portable is controlled by a dc feedback network producing six power levels that reduce the output from an EIA standard −2 dBW ERP when the unit is operating close to a base site. A miniature ferrite circulator protects the power amplifier from energy that might be reflected back from the antenna. A high-efficiency antenna ensures maximum ERP.

The portable uses thick- and thin-film hybrid circuit plug-in modules to conserve space yet allow easy maintenance. Custom linear and CMOS integrated circuits are used to reduce parts count and to conserve on battery drain. A CMOS microprocessor IC controls the transceiver call processing, signaling, and logic functions.

An LED display area is used to indicate the various status functions as well as to display the telephone number prior to placing the call. The keypad consists of one piece of silicon rubber, used for tactile response and long life. A custom IC integrates the keypad decoding display and indicator drive, and logic interface, to reduce size.

Full-Duplex Operation   Mobiles and portables operate with simultaneous talk and listen in the same manner as any home or business telephone. In addition, the portable unit is equipped with a VOX mode that conserves the battery by operating the transmitter only when the user is speaking. The voice-operated transmission mode is user-selectable.

Mobile Control Unit   The mobile control unit provides all the controls and indicators necessary for safe, convenient operation by the subscriber. The mechanical design of the package reflects the increasing emphasis on vehicle safety yet offers flexibility in mounting for convenient user interface. The control unit is made of high-impact plastic designed to provide environmental protection from dust, dirt, and splashed water.

All control unit graphics are clearly visible. Pushbuttons are easily distinguished and accessible. All buttons and graphics are illuminated for easy nighttime viewing. The compact size and light weight of the unit allow for versatile and convenient mounting in a large variety of vehicles. The mounting bracket provides a stable support for the control unit while in the vehicle. For added flexibility, the bracket permits users to adjust the position of the control head.

Pushbutton Dialing   A telephone-format pushbutton keypad is located on the *back* of the mobile handset. The keypad is positioned so that the user can hold the handset and select the desired telephone number with one hand.

The pushbutton keypad is located on the *face* of the portable handset. It is positioned so that the user can easily hold the handset with one hand while selecting the desired telephone number with the other hand.

On both units, the pushbuttons and keypad graphics are softly illuminated for convenience of operation at night. Audible feedback tones occur to confirm depression of the pushbuttons, and a series of function buttons is used for various call processing procedures.

Dialed Number Display   On both units, a telephone number digits display is located above the keypad. Whenever a telephone number is selected from the keypad or recalled from a memory location, the digits will appear in the display. The display provides confirmation that the correct telephone number has been entered or recalled before the number is outpulsed. This feature improves the efficiency of call placement and reduces the incidence and frustration of dialing incorrect numbers.

Abbreviated Dialing   Ten frequently called telephone numbers can be stored in either unit for simple, abbreviated recall. These telephone numbers are entered into the memory from the keypad and may be changed or deleted at any time by the user. Storage of these memory locations is protected by a built-in battery.

Preorigination Dialing   All dialing operations are performed prior to attempting to make the call. Telephone numbers entered from the keypad via abbreviated dialing are held in memory and are not transmitted until the appropriate button is depressed. Entering the number before going off-hook permits review of the number prior to outpulsing, thus avoiding errors and conserving air time. For the portable, in the event that the called party cannot be reached, the call attempt is easily terminated by returning the unit to the on-hook mode.

Recall of Last Number Dialed   If the called telephone number is busy or the call cannot be completed for any other reason and the subscriber wishes to try again, the units provide a "last number recall" feature. The last number attempted is held in a temporary storage register and will remain there until it is replaced by another number entered from the keypad. Pressing one button retransmits the desired number.

Electronic Scratchpad Memory   Telephone numbers that the subscriber wishes to note during the course of a conversation may be entered via the keypad into the unit's temporary memory as a form of electronic scratchpad. At the conclusion of the conversation, the new number can be dialed by depressing a single button. A companion feature allows the subscriber to load telephone numbers communicated during a conversation into the more permanent abbreviated dialing memories.

Electronic Lock   The mobile telephone is normally activated in conjunction with the vehicle's ignition. It is possible, however, to electronically lock the mobile or the portable to prevent unauthorized use. The unit is prevented from operating until a unique three-digit security code is entered.

Volume Controls   In the mobile, independent earpiece and speaker volumes are user-adjustable via the control unit. The last level settings are retained until the volume is readjusted.

   In the portable, the earpiece volume and the incoming call alert ring volume are separately controlled from the keypad. Multiple levels of attenuation are selected by a sequential button depression. Memory circuits retain the last level setting until the volume is adjusted.

Function Status Display   On both units, the illuminated display area features a display of status functions. Supplemental audible messages indicate whether a call is being processed properly. On the mobile, in addition to displaying the number to be dialed, the control unit features a display of the following functions: in use, no service, roam, lock, and auxiliary alert.

DTMF End-to-End Signaling   The appropriate dual-tone multiple-frequency (DTMF) tones for digits 0–9 plus * and # are generated over the transmit audiopatch once a call is established to facilitate end-to-end signaling from either unit.

Unique Mobile Features and Options

Vehicular Speaker Phone (VSP) Hands-Free Operation   The mobile control unit includes provisions for VSP hands-free operation using a separate loudspeaker and microphone. When the VSP mode is enabled, the unit will function as a "speaker-phone" allowing the user to carry on a completely hands-free call. Picking up the handset will disable the speaker-phone and activate the normal handset operation.

Electronic Ringer   The subscriber is alerted to incoming calls via an electronically generated ringing signal. The level is adjustable to accommodate the interior noise levels found in the vehicle environment.

Auxiliary Call Alert Function   The auxiliary alert function may be used to trigger (in conjunction with a relay) an external alarm device, such as the vehicle horn or headlights, to notify the user if a call is received while he or she is away from the vehicle.

On-Hook or Off-Hook Call Placement   Because the keypad is located on the handset back, the user can perform the dialing operation with the handset either mounted or handheld, whichever is safer and more convenient. When the call is attempted with the handset mounted, the subscriber uses the accompanying speaker to monitor the progress of the call prior to lifting the handset when the called party answers. If the called party cannot be reached, the call is easily terminated.

Service Mode Provisions   Remote access to key test points along with control contacts to "exercise" the transceiver are provided in a single external connector to facilitate diagnosis and maintenance.

Unique Portable Accessories and Options

The portable radiotelephone is available with a full range of accessories to adapt the basic radio to the subscriber's typical usage patterns. An in-vehicle charger provides convenient vehicular mounting as well as radio unit battery charging. A desktop charger allows spare batteries to be charged separately or accepts the entire portable for easy access while charging.

SUMMARY

- Most two-way communications occurs between a base station and a mobile unit or between mobile units. Systems may include microwave links to connect control, base, and relay stations.

- Mobile electronic equipment is any equipment operated in a moving vehicle and includes antennas, transceivers, microphones, and power supplies. Portable equipment may also be used in mobile applications. Mobile equipment, generally more compact than fixed-station equipment, must be able to withstand large temperature and humidity variations and severe mechanical vibration. Most two-way mobile systems consist of a vertically polarized omnidirectional antenna and a transceiver, and operate in the VHF and UHF bands. Among the two-way radio communication services are CB, amateur, aeronautical, general mobile, private land mobile, and domestic public landline mobile.

- Citizens' band radio is a public, noncommercial radio service available to the general population of the United States. Individual operators need not be licensed, but the equipment is licensed through the manufacturer's authorization by the FCC. Of the four classes of CB stations, class D stations are the most popular, with 40 channels, each 10 kHz wide, in the range 26.965–27.405 MHz. The transmitter output power for class D stations is legally limited to 4 W for AM and to 12 W PEP for SSB. Channel 9 is the officially designated emergency channel, monitored by REACT, an organization that provides assistance to motorists in trouble.

- The amateur radio operator (ham) is defined as a duly authorized person interested in radio techniques solely with a personal aim and without pecuniary interest. The FCC issues operator's licenses to U.S. citizens, regardless of age or other considerations, based on the passing of an examination. Ham license classes are Novice, Technician, General, Advanced, and Extra. Each class has certain privileges and restrictions. All ham stations must be licensed by the FCC, which assigns call letters. Station licenses are given only to licensed operators. The major ham operator organization is the ARRL. The ARRL provides the National Traffic System service, whereby ham operators receive and relay personal messages. Two services are provided for emergencies: RACES for civil defense and AREC for civilian emergencies.

- The ABS provides information for the purposes of air navigation and communication through an extensive network of transmitting stations, two-way voice and digital communications, radio beacons, radiolocation stations, and radar. Radiotelephone communication for aircraft employs AM and various types of SSB emissions in the HF range, plus certain specified frequencies in the MF and VHF bands.

- A basic two-way communication system has a base station and one or more mobiles. The base and the mobiles operate on the same frequency and use common antennas. The distance between base and mobiles in which reliable communication can occur depends on the base antenna location, height, and directivity, and on the power of the base and mobile transmitters. Repeaters located on hilltops or mountain peaks are used to extend the range and reliability of communications between mobiles. These repeaters require all mobiles to listen on the same frequency but to transmit on a separate frequency.

- A transceiver is a combination of a transmitter and a receiver having a common frequency control and usually enclosed in a single package. The principal components of a transceiver are a VCO or channel synthesizer, a transmitter, a receiver, and an antenna-switching device. Transceivers offer both advantages and disadvantages when compared with separate transmitters and receivers.

- A duplexer allows duplex operation, where one of two communicating operators can interrupt the other at any time, even while the other operator is transmitting. Duplex operation normally involves two different frequencies. In duplex operation, the transmitters and receivers of both stations are operated simultaneously and continuously. In simplex operation, both transmitters and receivers are often operated on a single frequency, but only one station can transmit at a time. The two stations communicate directly with each other, with no repeater or other intermediary used. In break-in operation, a transmitting operator can hear signals at all times except during an actual transmitter emission. Break-in provides nearly full-duplex operation. Semi break-in operation is used for code, radioteletype, or voice operation (VOX). VOX actuates a circuit by using the electrical voice impulses from a microphone or audio amplifier. VOX eliminates the need for pressing a lever or switch to change from the receive mode to the transmit mode, thus providing hands-free operation.

- A mobile telephone is a radio receiver designed for access to an autopatch system. Mobile and portable telephones usually operate in the VHF and UHF radio bands, between 30 MHz and 3 GHz. An autopatch is a device for connecting a radio transceiver to the telephone lines by remote control. Autopatch is generally achieved through repeaters.

- Portable communication equipment is any apparatus designed for operation in remote areas. It is battery operated and can be set up and dismantled in minimum time; some can be operated while being carried. A portable telephone resembles a telephone handset with an antenna and can be a cordless or wireless set. Pagers are shirt-pocket-size, 6 to 10 ounce receive-only devices.

- Conventional radiotelephone service is characterized by high cost, limited range, poor access during busy periods, and long waiting periods to obtain service. A cellular telephone system or cellular radio is a special form of mobile telephone consisting of a network of repeaters, all connected to one or more central office switching systems. Individual subscribers use radio transceivers operating at VHF and UHF. The network of repeaters is organized such that most places are always within range of at least one repeater. As a subscriber drives a vehicle, operation is automatically switched from repeater to repeater, with no loss or deterioration of the signal. Cellular radiotelephone systems offer the advantages of good-quality voice transmissions, accessibility and usability, full-duplex service, practicality, frequency reuse, and unlimited area coverage. Area coverage in cellular systems is provided by subdividing the area to be served into a network of coverage areas called cells. Each cell contains a low-powered base station (cell site) and is assigned a set of frequencies. Adjacent cells are assigned different frequency groups to avoid interference. Cells that are sufficiently far apart can use the same frequency groups simultaneously (frequency reuse). The FCC has authorized 666 channels in the 800 MHz band for cellular communication. Each cell uses 40 or 50 frequency pairs. Interference between cells does not occur because the frequency pairs are not used jointly in adjacent cells.

- Cell site transmitter power levels are limited to 100 W PEP, which allows frequency reuse five cells away. A typical cellular system uses cells ranging from 1 to 12 miles in radius. Frequency reuse and cell splitting can be used to increase cell capacity in a given service area. Cell splitting requires the addition of new cell sites between existing cell sites to form a new configuration of smaller cells, using the same number of channels. The cellular system is controlled by an EMX that provides interconnection to the public switched telephone network. Channel utilization efficiency is maximized through trunking of channels within the cells and within the system.

- A sector cell is an area with an unique frequency set, and a vehicle leaving that area requires a new channel assignment, called handoff. Sector sharing is a method of assigning channels so that subscribers requesting service can be temporarily assigned a channel from an adjacent sector if all of the channels in the desired sector are busy. Signaling channel capacity can be increased by creating a new set of control channels and separating the data traffic so that only pages are sent on the original control channels. Channel assignments and mobile or portable originated calls are then processed on the new control channels.

QUESTIONS AND PROBLEMS

1. How do most two-way communications occur?
2. Define mobile electronic equipment and explain what it includes.
3. What distinguishes mobile equipment from fixed-station equipment?
4. On what frequency bands do most mobile systems operate?
5. What distinguishes portable equipment from mobile equipment?
6. List the basic two-way communication services.
7. List the services, and briefly describe the users, of private land mobile radio services.
8. Define DPLMRS, state its purpose, and describe its users.
9. Define CB radio service and briefly describe each class of CB radio.
10. State (a) which class of CB station is the most popular, (b) the frequency range it covers, (c) how many channels it offers, and (d) the channel bandwidth.
11. What CB channel is set aside for emergency communication, and what is its frequency?
12. Describe the licensing arrangement for the CB radio service.
13. What is (a) the legal limit for transmitter output power for class D CB stations and (b) the limit placed on the range of operation from the base station for CB radio?
14. (a) Define REACT and explain its purpose. (b) Define amateur radio operator.
15. Explain the licensing arrangement for the amateur radio service.
16. Briefly describe the privileges and limitations of each class of amateur license.
17. What is the maximum permissible dc power input to the final stage of a ham radio transmitter?
18. What is (a) the maximum RF power output for most ham radio emissions and (b) an exception to this rule?
19. Draw a block diagram of a basic ham radio system.
20. State the rules relating to ham radio antennas.
21. State the purpose of the ICAO phonetic alphabet.
22. Define each of the following and explain its purpose: (a) NTS, (b) RACES, (c) AREC, (d) ABS.
23. Describe two methods of communication used in airborne systems.
24. Describe the $1\frac{1}{2}$ system aircraft radio.
25. What factors determine the distance between base and mobiles in which reliable communications can occur? How may the reliable range be extended?
26. Define transceiver and explain its use.
27. Draw a block diagram of a transceiver.
28. List the advantages and disadvantages of transceivers.
29. (a) Define T/R switch and explain how it works. (b) Define repeater and explain its use. (c) Define duplexer and explain its use.

30. Define the following and describe the working of each: (a) simplex operation, (b) duplex operation, (c) break-in operation, (d) semi break-in operation.

31. Define and explain the operation of (a) VOX and (b) autopatch.

32. Define and briefly describe a mobile telephone.

33. Define portable communication equipment and briefly describe its requirements.

34. Describe the cordless telephone and state its advantages and limitations.

35. Describe pagers and explain their operation.

36. What is the primary reason for the limited availability of conventional radiotelephone service?

37. Define cellular telephone system.

38. Briefly describe each of the features offered by cellular systems.

39. Briefly describe how coverage is provided in cellular systems.

40. Describe the process of cell splitting.

41. Describe the two main functions that must be performed to process a call in a cellular system.

42. (a) Define cellular grid and explain its purpose. (b) Define EMX and explain its function. (c) Define subscriber equipment.

43. What are remote diagnostics?

44. Define trunk and list trunk classifications.

45. What is the purpose of trunking of channels (a) within cells and (b) within the system?

46. Describe how frequency assignments are made in a cellular system.

47. Define (a) sector cell, (b) frequency reuse, and (c) section sharing.

48. Briefly describe how signaling channel capacity is increased when a cell site becomes capacity limited.

49. Describe the handoff process.

50. Briefly describe the EMX mobile telephone exchange.

51. List the subsystems of the EMX.

52. Describe the maintenance and status unit.

CHAPTER 16

Optical Communications

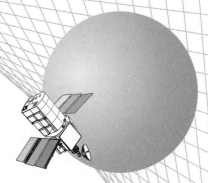

OBJECTIVES

When you complete this chapter, you should be able to:

- Define optical communication, optical fiber, fiberscope, refractive index, modes of propagation, numerical aperture, and acceptance angle.
- Discuss the physics of light and modulated light.
- Describe multimode, single-mode, and graded-index propagation modes.
- List seven advantages of fiber optics over wire.
- List two main sources of bandwidth limitations for fiber optic systems.
- List and briefly describe attenuation losses in optical fibers.
- List three types of optical fibers and describe the composition of each.
- List and describe the light sources used for fiber optic systems.
- List and describe the light detectors used for fiber optic systems.
- Describe wavelength division multiplex.
- List and briefly describe three aspects to be analyzed in the design of a fiber optic system.
- List and describe the types of transmission used in fiber optic systems.

INTRODUCTION

One of the most important developments in communications is the use of optical fibers as replacements for wire transmission lines and microwave waveguide. Optical fiber lines offer several important advantages over wire lines and waveguide. Because light is effectively the same as radio frequency radiation, but at a much higher frequency, theoretically the information-carrying capacity of a fiber is much greater than that of microwave radio systems. Therefore, instead of radiating RF down a microwave waveguide, infrared, visible, or ultraviolet frequency carrier waves are fed into thin, long, clear, round silica or halide glass threads. These thin optical fibers may have less than 2 dB of attenuation per kilometer, and since the fibers are not electrically conductive, they may be used in areas where electrical isolation and interference are severe problems. Furthermore, since the carrier waves are at light and infrared frequencies, each carrier can be modulated by many thousands of voice frequency channels simultaneously (multiplexed), by several TV programs at once, or by very fast computer information. Because of this high information capacity, multiple channel routes may be compressed into much smaller cables, reducing congestion in overcrowded cable ducts.

With present technology, fiber optic communication systems are still considerably more expensive than equivalent wire or radio systems, but this situation is changing. For example, AT&T has a circuit that transmits data over optical fibers 30 times as fast as its previous top-of-the-line model. The service, called Accunet T45, is capable of carrying 45 million bits of data per second (45 Mbps), or the equivalent of 4500 pages of digital information each second. The service is mostly used by large businesses, government agencies, and even telephone companies that lease the lines to provide their own long distance services. For these reasons, fiber optic systems are rapidly becoming competitive with the other systems in price and, because of their many advantages, are increasingly replacing them. In this chapter, we will examine the basic concepts of fiber optics and optical communications.

16-1   BASIC CONCEPTS OF OPTICAL COMMUNICATIONS

Optical communication is the application of light beams for sending and receiving messages. In recent decades, optical communication has become increasingly important. Early sailors used a form of optical communication for sending and receiving signals (flags and flashing lights). By comparison, modern systems are capable of transferring millions of times more information than were those seamen.

Modern optical communication systems use modulated-light sources and receivers sensitive to rapidly varying light intensity. Lasers and light-emitting diodes (LEDs) are the favored methods of generating modulated light, since they have rapid response time and emit light within a narrow range of wavelengths.

Communication can be obtained in the infrared and ultraviolet regions of the spectrum as well as in the visible range, provided that the wavelength is chosen for low atmospheric attenuation. For practical purposes, infrared and ultraviolet communications can be considered as optical communication.

Optical communication can be carried out directly through the atmosphere in much the same way that microwave links are maintained. However, optical propagation is susceptible to the weather. Rain, snow, fog, and dust can obscure the light and prevent the transfer of data. The transmittivity of clear air varies with the wavelength as well. The atmosphere is fairly transparent at visible light wavelengths but considerable attenuation occurs over significant portions of the infrared and ultraviolet regions.

Glass and plastic fibers can transmit light for long distances, in much the same way that wires carry electricity. A single beam of light can carry far more information than can an electric current, however, and optical fibers are used more and more in place of wire systems.

16-2 FUNDAMENTALS OF FIBER OPTIC SYSTEMS

Optical fiber transmission has come of age as a major innovation in telecommunications. Such transmission systems offer the advantages of (1) extremely high bandwidth, (2) freedom from external interference, (3) immunity from interception by external means, and (4) cheap raw materials (silicon is the most abundant material on earth).

Optical fibers guide light waves within the fiber material. They can do this because light rays bend or change direction when they pass from one medium to another. They bend because the propagation of light in each medium is different. This phenomenon is called *refraction*. One common example of refraction occurs when you stand at the edge of a pool of water and look at an object on the bottom of the pool. Unless you are directly over the object, it will appear to be farther away than it actually is. This effect occurs because the speed of the light rays from the object increases as the rays pass from the water to the air. This increase in speed causes the light rays to bend, changing the angle at which you perceive the object.

Optical Fiber

An *optical fiber* is a thin, transparent strand of material, usually glass or plastic or a combination of the two, that is used to carry light beams. A fiber confines the light energy inside its walls because of high internal reflection. The index of refraction of the fiber substance is, for visible light, much higher than the index of refraction for air. Thus, as shown in Figure 16-1, the light rays are reflected from the inner walls as they propagate lengthwise along the fiber.

A bundle of optical fibers may be used to transfer many different light beams at the same time. A single light beam can be modulated simultaneously by hundreds, or even thousands, of independent signals. Fiber cables of current design can carry thousands of different conversations at once.

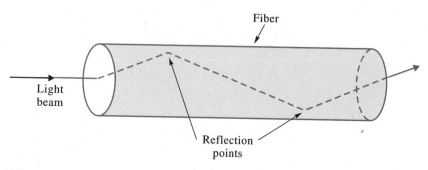

FIGURE 16-1 Fiber for light beam propagation

Fiberscope

A large bundle of optical filaments can carry a picture. The image to be viewed is focused onto one end of the bundle of fibers and is remagnified at the other end. A *fiberscope*, shown in Figure 16-2, is a flexible bundle of optical fibers with a focusing lens at each end that is used for viewing areas not normally observable. This principle makes it possible to look inside the human body. A small light can be placed at the receptor end, and the fiber bundle can be swallowed, allowing medical doctors to examine, for example, the stomach lining.

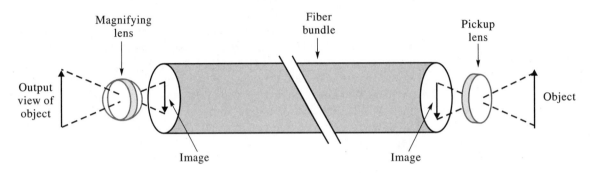

FIGURE 16-2 Fiberscope

The resolution of the fiberscope depends on the size of the lens at the receptor end and also on the number and fineness of the fibers in the bundle. Magnification can be obtained by using a highly convex lens at the receptor end of the bundle. Each fiber carries a small part of the image. Of course, the fibers must be oriented the same way at both ends of the bundle if the image at the display end is to be a faithful reproduction of the image at the receptor end.

Physics of Light

Over the years, it has been demonstrated that light travels at approximately 3×10^8 m/s in free space. It has also been demonstrated that in material denser than free space, the speed of light is reduced. This reduction in the speed of light as it passes from free space into a denser material results in refraction of the light. Simply stated, the light ray is bent at the interface, as shown in Figure 16-3. The degree to which the ray is bent depends on the *index of refraction n* of the denser material. As an equation, we write

$$n = \frac{\text{speed of light in free space}}{\text{speed of light in given material}} \tag{16-1}$$

Values of n for various materials are given in Table 16-1. We are particularly interested in the index of refraction for glass, but the other materials are included for comparison. Although the index of refraction is influenced by light wavelength, and the degree of bending is different for each wavelength, the variation is small enough to be ignored for our purposes.

TABLE 16-1 **Representative Indices of Refraction**

Material	n
Vacuum	1.0
Air	1.0003 (1.0)
Water	1.33
Fused quartz	1.46
Glass	1.5
Diamond	2.0
Silicon	3.4
Gallium-arsenide	3.6

The following three definitions are important to an understanding of light refraction. The *normal* is an imaginary line perpendicular to the interface of the two materials. The *angle of incidence* is the angle of the incident ray to the normal. The *angle of refraction* is the angle of the refracted ray to the normal. These terms are included in Figure 16-3. One other important term is *critical angle*, which is the angle of incidence that will produce a 90° angle of refraction.

Three specific conditions are shown in Figure 16-4. The angle of incidence is labeled A_1, and the angle of refraction is labeled A_2. Material 1 is more dense than material 2, so n_1 is greater than n_2. Figure 16-4(A) demonstrates how a light ray passing from material 1 to material 2 is refracted in material 2 when A_1 is less than the critical angle. Figure 16-4(B) demonstrates the condition that exists when A_1 is at

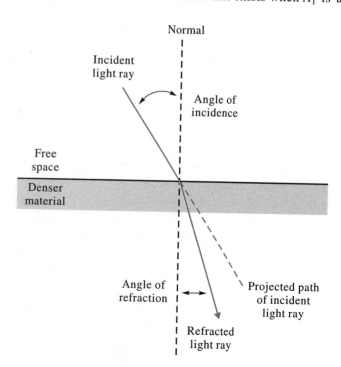

FIGURE 16-3
Light refraction at
an interface

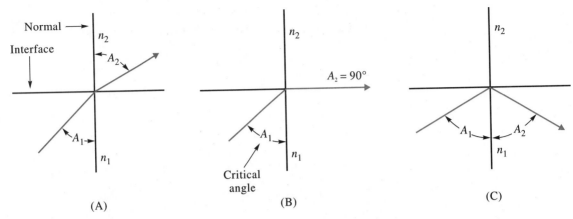

Normal

Interface

FIGURE 16-4 Index of refraction (A) Ray escaptes (B) Ray is absorbed (C) Ray is reflected

the critical angle and angle A_2 is at 90°. The light ray is directed along the boundary between the two materials.

The third condition, Figure 16-4(C), shows that any light ray incident at an angle greater than A_1 of Figure 16-4(B)—that is, greater than the critical angle—will be reflected back into material 1 with A_2 equal to A_1. This ray will be contained in material 1, continuing in a zigzag path, assuming that material 1 is a parallel-sided medium. Figure 16-5 demonstrates the principle of total internal reflection, which forms the basis for light propagation in optical fibers.

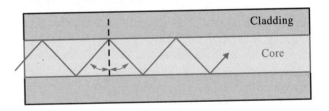

FIGURE 16-5
Total internal
reflection

Light traveling in an optical fiber obeys *Snell's law* of refraction, which gives the relationship between the incident ray and the refracted ray:

$$n_1 \sin A_1 = n_2 \sin A_2 \qquad \text{(16-2)}$$

A light source emits light at many angles relative to the center of the fiber. The angle at which light rays enter an optical fiber determines how the rays will propagate in the fiber.

Numerical Aperture and Acceptance Angle

The *numerical aperture* (NA) of an optical fiber is the measure of the fiber's ability to gather light, much like the maximum f-stop of a camera lens. The optical power accepted by the fiber varies as the square of the numerical aperture, but unlike a camera lens f-stop, the numerical aperture does not depend on any physical dimension, in this case, of the optical fiber.

The *acceptance angle* is the largest angle at which a light ray can enter the fiber and still propagate down the fiber. A large acceptance angle makes the end alignment less critical during splicing and connecting of fibers.

Fiber Composition

An optical fiber is a dielectric (nonconductor of electricity) waveguide made of glass or plastic. As shown in Figure 16-6, it consists of three distinct regions: a *core,* the *cladding,* and the *sheath* (or *jacket*). The index of refraction of the assembly varies across the radius of the cable, with the core being more dense and having a constant or smoothly varying index of refraction, designated n_c, and the cladding region being less dense and having another constant index of refraction, designated n_{cl}.

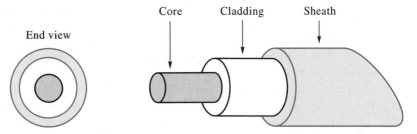

FIGURE 16-6 Optical fiber construction

For a fiber designed to carry light in several modes of propagation (called a multimode fiber), the diameter of the core is several times the wavelength of the light to be carried, and the cladding thickness will be greater than the radius of the core. Some typical values for a multimode fiber might be (1) an operating wavelength of 0.8 µm, (2) a core index of refraction n_c of 1.5, (3) a cladding index of refraction n_{cl} of 1.485 ($n_{cl} = 0.99 \times n_c$), (4) a core diameter of 50 µm, and (5) a cladding thickness of 37.5 µm. The clad fiber would have a diameter of 125 µm.

Lightning and Optical Fiber Cables

Unlike their copper predecessors, optical fiber cables do not use metallic conductors for transmission. However, metallic armoring, twisted copper pairs, and metallic central members are often included in optical fiber cables. Cable designs which incorporate these metallic components are susceptible to damage from lightning strikes both in aerial and buried applications.

Several tests have been proposed for simulating direct lightning strikes on cables. The test that is most widely accepted by the optical fiber cable industry is the sandbox test. The Electronic Industries Association (EIA) has adopted the sandbox test as a *Fiber Optic Test Procedure* (FOTP): FOTP-181, *"Lightning Damage Susceptibility Test for Optic Cables with Metallic Components."* While no test can accurately simulate an actual lightning strike, the sandbox test provides a common basis which can be used to make a relative comparison between different cable designs.

The sandbox test procedure is as follows: A one meter cable sample is placed in a roughly three-by-three foot nonmetallic box filled with pure silica sand. The sand is saturated with water which is allowed to drain through two small holes in the bottom of the box for at least five minutes to ensure that the sand is thoroughly soaked. The cable sample is placed in a channel dug in the sand and covered with a carefully prepared mixture of water and sand. The metallic components in the cable are shorted together and connected to the return terminal of the current pulse generator. The cable sample is subjected to a specified current pulse from an electrode placed one inch away from the sample. A small hole is placed in the jacket to ensure that a proper discharge occurs between the electrode and the sample's metallic components. The cable is rated according to the value of the peak current that it can withstand. It passes the test if all of the fibers are continuous after the test is completed.

Optical cables are rated according to the size of the current pulse they can withstand, with the highest rating being 105 kiloamperes (kA). Studies have shown that 95% of all lightning strikes are less than 100 kA.

Although aerial applications would seem to be very risky in terms of susceptibility to lightning damage, they are actually less risky than buried installations. Aerial cables are often installed on joint-use poles with a large number of metallic conductors such as ground wires, power conductors, and copper telephone cables above them. In these situations, the other cables have a much greater probability of being struck by lightning than the optical fiber cable. Even when the optical fiber cable is installed on a pole by itself, it is generally lashed to a metallic messenger which is much more likely to take the strike.

Many cable manufacturers have also performed tests on cables which simulate aerial lightning strikes. The aerial test setup is identical to the sandbox test except that the sand is removed and the cable is lashed to a metallic messenger. The aerial test is less severe than the traditional sandbox test because the mechanical crushing effect from the vaporization of the water has been removed. Simulated aerial testing has shown that in almost all cases, the lightning strikes the messenger rather than the cable, even when the messenger is on the far side of the cable.

Waterblocked Versus Pressurized Cables

In an outdoor application (aerial, duct, or direct buried), an optical fiber cable must be capable of tolerating a wide variety of environmental conditions which include large temperature variations. It is known that the main cable components will expand and contract in greater proportion with a temperature change than will the fiber itself. Ice crush forces can impact cable performance if water penetrates the cable and is subjected to freezing temperatures. The use of waterblocking materials provides the necessary protection without the need for ongoing maintenance associated with a pressurized cable.

In the loose tube cable design, filling compound is placed in the buffer tubes with the fiber. The fiber is placed with a precise overlength to ensure it remains mechanically decoupled from the cable when the cable expands with temperature change and tensile load. The length is also controlled so that the fiber will not be compressed against the tube wall when the tube contracts with temperature changes. The filling compound provides a mechanical cushion allowing the fibers to "float" within the buffer tube. The filling compound also minimizes the chance of water or moisture penetrating the length of the tube in the event that the tube is breached. A bare fiber exposed to water for a prolonged period of time could also have its strength degraded by a reaction between the water and the glass in the presence of a pre-existing flaw.

Optical fiber cables also typically contain waterblocking materials around the buffer tubes in order to minimize the possibility of moisture penetration and migration within the cable. The primary risks of such penetration and migration are internal mechanical forces such as ice crush. Waterblocked cable cores may be flooded or unflooded. In flooded cores, the interstices between the buffer tubes are filled with a gel-like flooding compound. A *Super Absorbent Polymer* (SAP) may be mixed in the flooding compound. In unflooded cores a SAP, carried by tapes or yarns, swells to block water ingress or migration.

Pressurization is an alternate method of preventing moisture ingress. The only advantage a pressurized cable system has over a waterblocked system is a warning given in the event of sheath penetration; the pressure decreases and an alarm signals that the cable jacket has been breached. However, most disruptions in cable service are catastrophic in nature (for example, a backhoe severing the optical fiber cable). In cases of catastrophic failure, pressurized systems have no advantage over unpressurized systems since the electronics in either system will signal that a disruption in service has occurred. The initial cost and long-term maintenance of pressurized splice closures, pressure blocks, valves, and gauges adds much to the initial expense and ongoing maintenance of the cable system. In addition, the reliability of the system is only as good as the maintenance and attention given to the pressurization equipment. Conversely, waterblocked cable provides passive protection and requires no added cost or maintenance.

Sag and Tension

Appropriate planning for aerial cable installations includes taking into account proper clearances and allowable cable and messenger stress loading. Planning for proper clearances requires knowing the *sag* characteristics of the (planned) installation. Understanding the expected tensile loads placed on an aerial plant is necessary to ensure that legal safety requirements are met and the plant's expected lifetime is not unnecessarily shortened due to exceeding rated tensile loads.

When considering an aerial installation, the following parameters are pertinent: (1) cable size and type, to include diameter, weight, central member type, armored or duct, fiber count, and fibers per buffer tube; (2) span length (distance between poles); (3) allowable sag; (4) messenger type; and (5) geographic location.

Manufacturers have developed a sag and tension algorithm which allows these parameters to be calculated for a variety of messenger/cable combinations and environmental loading conditions. The algorithm can also be used to determine an installation tension based on desired sag or environmental limitations. Additionally, the algorithm can be used to help installers pick the right messenger for the specific application.

The following terms are used with respect to aerial cable installations:

1. *Dedicated Messenger:* The application of installing a steel messenger aerially and lashing one or two cables to it. This application does not include lashing a new cable over an existing messenger and cable.

2. *Overlash:* The practice of lashing a new cable to an existing aerial installation. In this application consideration must be given to the existing installation as well as to the cable being installed.

3. *Span Length:* Horizontal distance between poles.

4. *Tension:* Tensile force applied to the messenger.

5. *Sag:* The vertical distance between the cable and an imaginary horizontal line extending between the points where the cable connects to the poles, assuming zero elevation difference between the poles.

Grounding and Bonding of Optical Fiber Cable in Aerial Applications

Optical fiber cables that contain metallic components are susceptible to an induced voltage when installed in aerial applications in the vicinity of one or more power lines. The grounding and bonding of the metallic components in an optical fiber cable and the supporting metallic messenger is essential to ensure the safety of workers and equipment. The frequency at which the grounding and bonding is performed on the cable plant should be in compliance with documents approved by the *American National Standard Institute* (ANSI).

The *National Electrical Code* (NEC) and the *National Electrical Safety Code* (NESC) are each recognized as ANSI documents and define the grounding and bonding frequency of the optical fiber cable plant. The NEC recommends that noncurrent-carrying metallic members (armor shield, metallic central member, metallic strength member) of optical fiber cables be bonded and grounded at the point of entrance into a building or residence. It should be noted that the NEC recommends that the grounding wire on the outside plant cable at the building entrance should be bonded together with all the building grounding electrodes. This grounding scheme limits the potential differences between the outside cable installation and the building wiring system.

The NESC recommends that the messenger wire employed to support aerial optical fiber cables be grounded at four connections in each installed mile. In addition, the NESC recommends that the messenger wires be grounded at common crossing structures.

Many local and state regulations supersede NEC and NESC recommendations. System designers should contact state and local regulatory agencies to determine if regional regulations exist that differ from the NEC and NESC recommendations. As a minimum, it is recommended that the metallic components of the optical fiber cable be bonded and grounded at each building and cable entry point. The system designers should assess the installation environment to determine the potential of induced voltage from power lines or lightning, or both. The system designer should evaluate the estimated value of the induced voltage to determine if bonding and grounding should be performed at defined frequencies that vary from the NEC, the NESC, or state and local regulatory agencies.

Multimode and Single-Mode Propagation

For optical fibers in which the diameter of the core is many times the wavelength of the light transmitted, the light beam travels along the fiber by bouncing back and forth at the interface between the core and the cladding. Rays entering the fiber at differing angles are refracted varying numbers of times as they move from one end to the other and, consequently, do not arrive at the distant end with the same phase relationship as

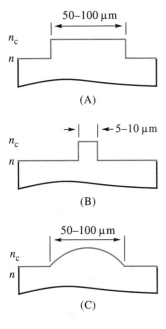

FIGURE 16-7 Refractive index profiles (A) Step index, multimode fiber (B) Step index, single-mode fiber (C) Graded index, multimode fiber

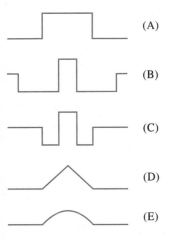

FIGURE 16-8 Different refraction index profiles for optical fibers (A) Step (B) Depressed cladding (C) "W" profile (D) Triangular (E) Graded

when they started. The differing angles of entry are called *modes of propagation*, or simply *modes*, and a fiber carrying several modes is called a *multimode* fiber. Multimode propagation causes the rays leaving the fiber to interfere both constructively and destructively as they leave. This effect is called *modal delay spreading*.

If, on the other hand, the diameter of the core is only a few times the wavelength of the transmitted light, say a factor of 3, only one ray or mode will be propagated, and no destructive interference between rays will occur. Fibers with this characteristic are called *single-mode* fibers. Figures 16-7(A) and (B) show the distribution index of refraction across, and typical diameters of, multimode and single-mode fibers. One of the principal differences between single-mode and multimode fibers is that most of the power in the multimode fiber travels in the core, whereas in single-mode fibers a large fraction of the power is propagated in the cladding near the core. At the point where the light wavelength becomes long enough to cause single-mode propagation, about 20% of the power is carried in the cladding, but if the light wavelength is doubled, over 50% of the power travels in the cladding.

Figure 16-7(C) shows the distribution of another kind of multimode fiber, called a *graded-index* fiber. The index of refraction varies smoothly across the diameter of the core but remains constant in the cladding. This treatment reduces the intermodal dispersion by the fiber, since rays traveling along a graded-index fiber have nearly equal delays. Other refractive index profiles have been devised to solve various problems, such as reduction of chromatic dispersion. Some of these profiles are shown in Figure 16-8. The step and graded profiles are repeated for comparison.

Fiber Optic Systems

The capacity of a transmission system is a direct function of the highest frequency the system can carry; therefore, progress in transmission technology has been measured by the bandwidth of the media available to carry signals. Recent developments in the use of glass fibers to carry binary signals have shown these systems to be extremely well suited to high data rate applications. There are several advantages, in both performance and cost, to be realized by using fiber optics instead of wire.

1. *Greater bandwidth:* The higher the carrier frequency in a communications system, the greater the potential signal bandwidth. Since fiber optics work with carrier frequencies on the order of 10^{13}–10^{14} Hz, as compared to radio frequencies of 10^6–10^8 Hz, signal bandwidths are theoretically 10^6 times greater for fiber optics.

2. *Smaller size and weight:* A single fiber can replace a very large bundle of individual copper wires. For example, a typical telephone cable may contain over 1000 pairs of copper wires and have a cross-sectional diameter of 7–10 cm. A single glass fiber cable capable of handling the same amount of signal might be only 0.5 cm in diameter. The actual fiber may be as small as 50 μm. The additional size is the result of the jacket and strength elements. The weight reduction in this example should be obvious.

3. *Lower attenuation:* Length for length, optical fiber exhibits less attenuation than does twisted wire or coaxial cable. Also, the attenuation of optical fibers, unlike that of wire, is not signal frequency dependent.

4. *Freedom from EMI:* Unlike wire, glass does not pick up or generate electromagnetic interference (EMI). Optical fibers do not require expensive shielding techniques to desensitize them to stray fields.

5. *Ruggedness:* Glass is 20 times stronger than steel, and since it is relatively inert, corrosive environments are of less concern than with wired systems.

6. *Safety:* In many wired systems, the potential hazard of short circuits between wires or from wires to ground requires special precautionary designs. The dielectric nature of optical fibers eliminates this requirement and the concern for hazardous sparks occurring during interconnects.

7. *Lower cost:* Optical fiber costs are continuing to decline while the cost of copper wire is increasing. In many applications today, the total system cost for a fiber optic design is lower than for a comparable wired design. As time passes, more and more systems will be decidedly less expensive with optical fibers.

Bandwidth Limitations

The limitations on bandwidth in fiber optic systems arise from two main sources: modal delay spreading and material dispersion. Modal delay spreading was described earlier and is evident primarily in multimode fibers. *Material dispersion* arises from the variation in the velocity of light through the fiber with the wavelength of the light.

If the light source, such as an LED, emits pulses of light at more than one wavelength, the different wavelengths will travel at different velocities through the fiber. This action causes spreading of the pulses. At a typical LED wavelength of 0.8μm, the delay variation is about 100 picoseconds (ps) per nanometer (nm) per kilometer (km). If the width of the spectrum emitted by the LED is 50 nm, pulses from the source will be spread by 5 ns/km. This will limit the modulation bandwidth product to about 50–100 MHz/km. Fortunately, at certain wavelengths (near 1.3–1.5 μm for some types of fibers) there is a null in the material dispersion curve, giving much better modulation bandwidth performance. Figure 16-9 shows the relationship of loss in doped silicon glass fibers versus light wavelength. Most current development work is aimed at making fibers, light sources, and detectors that work well at the loss nulls at 1.3 and 1.5 μm.

Attenuation

The loss in signal power as the light travels down the fiber is called *attenuation*. Attenuation in the fibers is controlled mainly by four factors: (1) radiation of the propagated light, called scattering; (2) conversion of the light energy to heat, called absorption; (3) connection losses at splices and joints in the fiber; and (4) losses at bends in the fiber.

Scattering Losses *Scattering losses* occur because of microscopic imperfections in the fiber, such as the inclusion of water in the glass. The effect of impurities in the transmission medium is evident when we look up at the sky and see a blue color. In fact, deep space has no color (appears as black), but due to the scattering of sunlight by the dust in the atmosphere, the sky appears as a bright blue.

There is a limit below which scattering cannot be reduced, no matter how perfectly the glass fiber is made, because of irregularities in the molecular structure of glass.

This limit, called the *Rayleigh scattering limit*, is shown graphically in Figure 16-9. For light with a wavelength of 0.8 μm, the scattering limit is about 2.9 dB/km. At a wavelength of 1.3 μm, the value is about 0.3 dB/km, and at 1.55 μm, the limit is about 0.15 dB/km. Commercially available glass fibers exhibit losses of about 3.5 dB/km at 0.8 μm, and 0.7–1.5 dB/km at 1.3 and 1.5 μm. There is less attenuation through 6 m (about 20 feet) of good-quality optical fiber glass than through an ordinary clean windowpane.

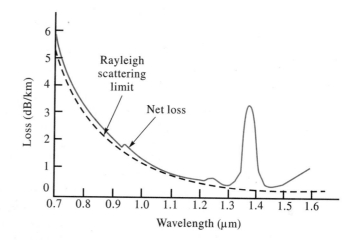

FIGURE 16-9
Net spectral loss curves for a glass core

When light is transmitted down the fiber a percentage of the light is reflected back, or backscattered, to the source. This phenomenon (Rayleigh scattering) is caused by changes in the refractive index on a microscopic level. Because light is reflected back, power is lost from the original input pulse. Rayleigh scattering is one of the major contributors to attenuation; the other is absorption. If the input pulse encounters a refractive index discontinuity, then it is possible to get more light reflected back than occurs just from Rayleigh scattering. This is known as *Fresnel reflection*. Some causes of Fresnel reflections include connectors, couplers, and mechanical splices.

Since the power reflected from the discontinuity is larger than that due to backscatter, a "bump" is seen on the loss trace, as shown in Figure 16-9.

Optical return loss (ORL) is a measure of the power reflected from a discontinuity to power incident upon it. The reflected pulse has the characteristics of the input pulse, accounting for attenuation and dispersion. As an equation:

$$ORL = 10 \, \log_{10} \frac{P_{inc}}{P_{refl}} \qquad \text{(in decibels)} \qquad (16\text{-}3)$$

where

P_{inc} = incident power

P_{refl} = reflected power

Physically ORL represents *how much the reflected power is below the input power.* The larger the magnitude of the ORL the further the reflected pulse is below the input power. This results in less reflected light, which means that an ORL of 50 dB is better than 30 dB.

Reflectance is another expression used to measure refractive index discontinuities. By convention, reflectance is a negative value, usually in dB. Thus −30 dB of reflectance is roughly equivalent to an ORL of 30 dB.

Absorption Losses   *Absorption losses* refer to the conversion of the power in the light beam to heat in some material or imperfection that is partially or completely opaque. This property is useful, as in the jacket of the fiber, to keep the light from escaping the cable but is a problem when it occurs as inclusions or imperfections in the fiber itself. Current fiber optic systems are designed to minimize intrinsic absorption by transmitting at 0.8, 1.3, and 1.5 μm, where there are reductions in the absorption curve for light.

Connection Losses   *Connection losses* are inevitable and represent a large source of loss in commercial fiber optic systems. A relatively large amount of power is lost at every connection point, particularly at repair splices. In addition to the installation connections, repair connections will be required because a typical line will be broken accidentally two or three times per kilometer over a 30-year period. The alignment of optical fibers required for each connection is a considerable mechanical feat, somewhat like threading a needle with your eyes closed. The full effect of the connection is not realized unless the parts are aligned correctly. The ends of the fibers must be parallel to within 1° or less, and the core must be concentric with the cladding to within 0.5 μm. Production techniques have been developed to splice single-mode fibers whose total diameter is less than 10 μm by using a mounting fixture and small electric heater. Mechanical connectors have been developed that allow random mating of fibers with average connection losses of less than 0.3 dB.

Multimode Splice Loss   Splicing is required to create a continuous path for transmission from one fiber to another. Two different methods are used to splice fibers: (1) fusion splicing, and (2) mechanical splicing. Typical attenuation splice loss values are 0.1 dB for fusion and 0.2 dB for mechanical splices. The primary contributors to these splice losses are fiber misalignment, fiber geometry mismatch, numerical aperture mismatch, and mismatch of the backscattering coefficient.

Fiber misalignment can occur with any splice. Likewise, geometry mismatch and mismatch of intrinsic fiber parameters (numerical aperture, backscattering coefficient) must be taken into account. Therefore, fibers can be divided into two groups: (1) Similar fibers—fibers having the same core diameter and similar intrinsic parameters; and (2) Dissimilar fibers—fibers having different core diameters (for example, 50 μm, 62.5 μm, 100 μm) and/or intrinsic parameter mismatch.

(1) *Splices with similar fibers:* When splicing similar fibers, the deviation in the fiber core alignment has the highest influence on attenuation. Even highly sophisticated fusion splicers working with local injection and local detection cannot fully compensate for these misalignments.

Fiber alignment is even more critical for mechanical splices where no optical alignment systems exist. In such splices, transverse as well as longitudinal fiber offset must be considered. Also, the angle and roughness of endfaces influences splice performance. The highest contribution to attenuation is the transverse offset.

(2) *Splices with dissimilar fibers:* To connect two fibers with mismatched parameters, closer consideration must be given to the three main fiber characteristics causing attenuation increase: (1) core diameter; (2) numerical aperture; and (3) scattering coefficient.

Core diameter variations of multimode fibers are not due exclusively to the manufacturing process. Multimode fibers are designed with different core diameters, whereby different core sizes exist. Therefore, two different cases must be considered: (1) splicing a small core fiber to a large core fiber; and (2) splicing a large core fiber to a small core fiber. In the first case, an underfilled (uf) situation exists; all the core power can be transferred from the small core to the large core. In the second case, an overfilled (of) situation exists; only a certain amount of power can be transmitted from the large core to the small core.

The loss due to the overfilled situation can be calculated by:

$$\text{Loss}_{of} = \text{Loss}_d + \text{Loss}_{NA} \tag{16-4a}$$

$$\text{Loss}_d = -10 \ \log \ (d_{re}/d_{tr})^2 \qquad \text{(in dB, and } d_{re} < d_{tr}) \tag{16-4b}$$

$$\text{Loss}_{NA} = -10 \ \log \ (NA_{re}/NA_{tr})^2 \qquad \text{(in dB, and } NA_{re} < NA_{tr}) \tag{16-4c}$$

where

$$of = \text{overfilled situation}$$
$$d_{re} = \text{core of the receiving fiber}$$
$$d_{tr} = \text{core of the transmitting fiber}$$
$$NA_{re} = \text{numerical aperture of the receiving fiber}$$
$$NA_{tr} = \text{numerical aperture of the transmitting fiber}$$

These equations show that the higher the mismatch of numerical aperture and core diameter, the higher the attenuation penalty. *It is important to note that this loss is a truly directional loss, appearing only for the overfilled situation.*

Scattering loss in such a splice is the loss of optical energy due to the molecular structure of the fiber and the embedded impurities. This loss is a relatively small parameter that normally leads to only minor splice attenuation effects.

Bending Losses   *Bending losses* occur because the light rays on the outside of a sharp bend cannot travel fast enough to keep up with the other rays and are lost. Bending an optical fiber is somewhat like playing crack-the-whip with the light rays. As light travels around the bend, the light on the outside of the bend must travel faster to maintain a constant phase across the wave. As the radius of the bend is decreased, a point is reached where part of the wave would have to travel faster than the speed of light, an obvious impossibility. At that point, the light is lost from the waveguide. For commercial single-mode fiber optic cables operating at 1.3 and 1.5 μm, the bending that occurs in fabrication (the fibers are made with the fibers wound spirally around a center stabilizer) and installation does not cause a noticeable increase in attenuation.

Models describing the material strength of glass have been known for many years, and remain virtually unchanged. The specified minimum bend radius for optical fibers is based primarily on the long-term mechanical reliability of the fiber. The *minimum* recommended bend radius is 25 mm (50 mm bend diameter) for standard 125 μm diameter silica fibers. Fibers with other nominal diameters will experience a different stress level for a given bend. The recommended *maximum* bend stress induced on the surface of an optical fiber is available in chart form from equipment manufacturers.

The specified minimum fiber bend radius must be maintained to ensure long term mechanical reliability and optical performance. Placing fiber in a bend induces compression and tension forces on the surface of the glass fiber. Microscopic flaws, present on all glass fibers, can propagate when subjected to tensile forces in the presence of moisture, leading to the growth of the flaw, thus reducing fiber strength. Also, optical performance can be degraded because of bending losses. The minimum recommended bend radius is 25 millimeters for standard 125 μm diameter silica fibers.

Most splicing operations place the fiber in a bend of diameter of 75 to 80 mm. In general, the length of fiber under bend is small compared to lengths proof-tested for strength by the manufacturer. The failure of glass fiber is based on the random distribution of flaws along the length of the fiber, so it is important to look at a strength distribution chart and normalize the probability of failure to the bend situation at hand. Fiber strength distributions are based on the testing of 20 meter gauge lengths to failure.

It had been determined that the bend loss performance of an optical fiber is dependent on the fiber geometry. For example, for single-mode fiber, the smaller the mode-field diameter, the smaller the bending loss will be. This holds true for various fiber design types. However, different fiber types will exhibit different bend loss performance for a given geometry. Current industry standards specify the bend loss performance requirements for single-mode and multimode fiber. Tests are conducted in accordance with EIA-455-62, "*Optical Fiber Macrobend Attenuation,*" usually on mandrel diameters of 75 mm. Most fibers today exhibit little or no increase in attenuation at this diameter, even down to a diameter of 50 mm. However, below a 50 mm diameter (25 mm radius), some fiber types will indicate a definite attenuation increase.

Fiber Versus Cable Bend Radius   Cable manufacturers will specify minimum bend radii for different cable designs which will differ from that for the fiber. Due to the design of the cable, the position of the fiber within the cable, and the characteristics of the materials used in the construction of the cable, other limiting factors may exist. The minimum bend radius of the completed cable should always be observed to preserve both the mechanical and optical performance of the cable structure.

16-3   FIBER OPTIC SUBSYSTEMS AND COMPONENTS

Propagation through optical fibers is in the form of light or, more specifically, electromagnetic radiation in the spectral range of near-infrared or visible light. Since the signal levels to be dealt with are generally electrical in nature (like serial digital logic at standard TTL levels), it is necessary to convert the source signal into light at the transmitter end and from light back to TTL at the receiver end. Several components can accomplish these conversions. In the following sections, we will briefly discuss fiber production but concentrate on light-emitting diodes (LEDs) and injection laser diodes (ILDs) as sources and on photodiodes as detectors.

Fiber Production

Optical fibers are fabricated in several ways, depending on the vendor and purpose of the system. The core and cladding regions of the fiber are doped to alter their refractive indices. This doping is carried out by heating vapors of various substances such as germanium, phosphorus, and fluorine and depositing the particles of resulting oxidized vapor, or "soot," on high-quality, fused-silica glass mandrels, called *preforms*. The preforms are a large-scale version of the core and cladding that is then heated to a taffylike consistency and drawn down into the actual fiber. The core and cladding dimensions have essentially the same relationship in the final fiber as in the preform.

Fiber Types

Fibers may be made of glass, plastic, or a combination of the two. The three main varieties available are (1) plastic core and cladding, or PCP (plastic-clad plastic); (2) glass core with plastic cladding, or PCS (plastic-clad silica); and (3) glass core and cladding, or SCS (silica-clad silica).

All-plastic fibers are extremely rugged and useful for systems where the cable may be subject to rough day-after-day treatment. They are particularly attractive for benchtop interconnects. The disadvantage is their high attenuation.

Plastic-clad silica cables offer the better attenuation characteristics of glass and are less affected by radiation than all-glass fibers. However, the soft clad material should be removed and replaced by a hard clad material for best fiber core-to-connector termination. These cables see considerable use in military-grade applications.

All-glass fibers offer low attenuation performance and good concentricity, even for small-diameter cores. They are generally easy to terminate, compared to PCS. On the other hand, they are usually the least rugged mechanically and more susceptible to increases in attenuation when exposed to radiation. The choice of fiber for a given application will be a function of the specific system's requirements and trade-off options.

As a final statement on fiber properties, it is interesting to compare optical fiber with coaxial cable. Figure 16-10 shows the loss versus frequency characteristics for a low-loss fiber compared with the characteristics of several common coaxial cables. Note that the attenuation of optical fiber is independent of frequency (up to the point where modal dispersion comes into play).

Light Sources

Light sources for fiber optic systems must convert electrical energy from the computer or terminal circuits feeding them to *photons* (particles of electromagnetic radiation in the form of optical energy) in a way that allows the light to be coupled effectively to the fiber. Two such sources currently in production are the *light-emitting diode* and the *laser diode*.

Light-Emitting Diodes   Light is emitted from an LED as a result of the recombining of electrons and holes. Electrically, an LED is simply a P-N junction. Under forward bias, minority carriers are injected across the junction. Once across, they recombine with majority carriers and give up their energy in the process. The energy given up is approximately

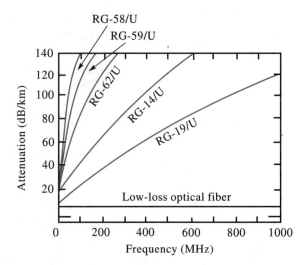

FIGURE 16-10
Comparative attenuation versus frequency for optical fiber and coaxial cable

equal to the energy gap for the material. The same injection/recombination process occurs in any P-N junction, but in certain materials, the nature of the process produces radiation in the visible region; that is, a photon of light is produced. In other materials (silicon and germanium, for example), the process is primarily nonradiative, and no photons are generated.

Light-emitting materials do have a distribution of nonradiative sites, usually crystal lattice defects or impurities. Minimizing these is the challenge to the manufacturer attempting to produce more efficient devices. It is also possible for nonradiative sites to develop over time and thus reduce efficiency. This process gives LEDs finite lifetimes, although 10^5–10^6 hour lifetimes are essentially infinite compared to those of other components of many systems. The principal advantages of LEDs are low cost and high reliability.

A cross section of a *surface* LED is shown in Figure 16-11. The LED emits light over a broad spectrum and, in addition, disperses the emitted light over a large angle. These features cause the LED to couple much less power into a fiber with a given acceptance angle than do other devices. Currently, surface LEDs can couple about 100 μW of power into a fiber with a numerical aperture of 0.2 or more with a coupling efficiency of about 2%.

The *edge-emitting* LED shown in Figure 16-12 emits a more directional pattern than the surface LED. The emitting area is a stripe rather than a confined circular area. The emitted light is taken from the edge of the active stripe and forms an elliptical beam. Although the edge-emitting LED provides a more efficient source for coupling into small fibers than the surface LED, its structure calls for significant differences in packaging. The edge-emitting LED is quite similar to the laser diodes used for fiber optics.

Laser Diodes   The *laser*, also called *optical maser*, is a device that generates coherent electromagnetic radiation in or near the visible part of the spectrum. There are several different methods for obtaining coherent light. Coherent light is characterized by a narrow beam and the alignment of all waveforms in the disturbance.

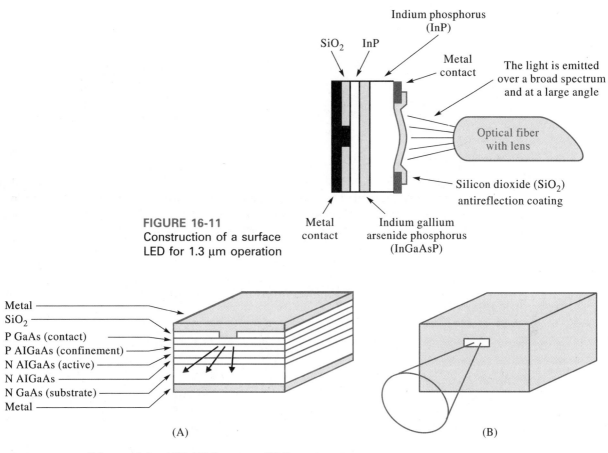

FIGURE 16-11
Construction of a surface
LED for 1.3 μm operation

FIGURE 16-12   Edge-emitting LED (A) Structure (B) Beam pattern

Laser action occurs in many different materials, and a complete description of all possible means of constructing a laser is beyond the scope of this book. The laser process can occur in numerous different ways. Lasers can be described in terms of the following categories: chemical, gas, liquid, metal–vapor, semiconductor, and solid-state.

An early laser, called a *ruby laser*, was constructed from a ruby rod, as illustrated by the simplified diagram in Figure 16-13. A flashlamp surrounds the ruby rod, which has mirrors at each end. The rear mirror is a total reflector, but the front mirror allows about 4–6% of the light to pass through; that is, it reflects only 94–96% of the light. A resonant condition occurs within the ruby rod. If the gain of the system is greater than the loss, oscillation occurs, resulting in a coherent light output at about 700 nm. This output falls into the red part of the visible spectrum.

Low-power lasers can be employed for short-range visible light communications since laser light can be modulated and suffers far less angular divergence than ordinary light.

The *injection laser diode* (ILD) is a simple type of device that is universally available at a low cost. This common form of commercially available laser, composed of

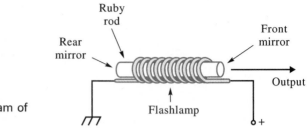

FIGURE 16-13
Simplified diagram of
a ruby laser

helium and neon gas, also produces a red visible beam. The ILD is a form of LED with a relatively large and flat P-N junction, as shown in Figure 16-14. The ILD emits coherent light, provided that the applied current is sufficient. If the current is below a certain level, the ILD behaves much like an ordinary LED, but when the so-called threshold current is reached, the charge carriers recombine in such a manner that laser action occurs.

A cross section of a typical ILD is shown in Figure 16-15. Because of its narrow spectrum of emission and its ability to couple output efficiently into the fiber lightguide, the ILD supplies power levels of 5–7 mW. At present, ILDs are considerably

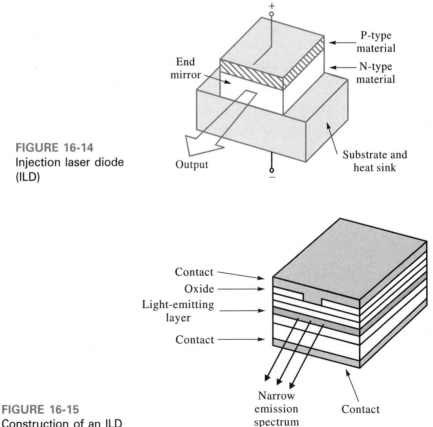

FIGURE 16-14
Injection laser diode
(ILD)

FIGURE 16-15
Construction of an ILD

more expensive than LEDs, and their service life is generally less by a factor of about 10. Other disadvantages of laser diodes are their requirements for automatic level control circuits, controlled laser power output, and protection from power supply transients.

Modulated Light

Electromagnetic energy of any frequency can theoretically be modulated for the purpose of transmitting intelligence. The only constraint is that the frequency of the carrier be at least several times greater than the highest modulating frequency. Modulated light has recently become a significant method of transmitting information. The frequency of visible light radiation is exceedingly high, and therefore modulated light allows the transfer of a great amount of information on a single beam.

A simple modulated-light communication system can be built for a few dollars using components available in hardware and electronics retail stores. The schematic diagram in Figure 16-16 shows such a simple system. This system is capable of operating over a range of several feet. More sophisticated modulated-light systems use laser beams for greater range.

The system shown is an amplitude modulated system, but there are other ways of modulating a light beam. Polarization modulation is one alternative. Lasers make pulse modulation a viable method of transmitting information over light beams. Position modulation is also possible, especially with narrow-beam sources such as lasers. Modulation of the actual frequency of a light beam is difficult to obtain directly, but phase modulation can be achieved with coherent light sources.

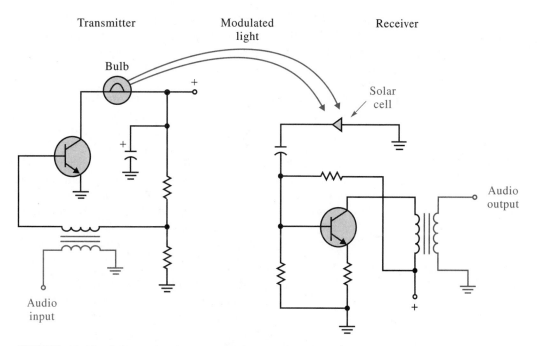

FIGURE 16-16 Schematic of a simple modulated-light system

Light Detectors

Light detectors are used at the receiving end of the optical fiber to detect, amplify, and convert the light signal back to its original electrical form. At the receiving end of the optical communications system, the receiver must have very high sensitivity and low noise. To meet these requirements, there is a choice of three main types of photodiode devices to detect the light beam, amplify it, and convert it back into an electrical signal: (1) the integrated PIN FET assembly, (2) the avalanche photodiode (APD), and (3) the integrated detector preamplifier (IDP).

Detector Fundamentals Just as a P-N junction can be used to generate light, it can be used to detect light. If a P-N junction in a photodiode is reverse biased and under dark conditions, very little current flows through it. However, when light shines on the device, photon energy is absorbed and hole–electron pairs are created. If the carriers are created in or near the depletion region at the junction, they are swept across the junction by the electric field. This movement of charge carriers across the junction causes a current flow in the circuitry external to the diode. The magnitude of this current is proportional to the light power absorbed by the diode and to the wavelength.

Although this type of P-N photodiode could be used as a fiber optic detector, it exhibits three undesirable features:

1. The noise performance is generally not good enough to allow its use in sensitive systems.
2. It is usually not fast enough for high-speed data applications.
3. Because of its depletion width, it is not sensitive enough.

In a typical device, the P anode is heavily doped, and the bulk of the depletion region is on the N cathode side of the junction. As light shines on the device, it penetrates through the P region toward the junction. If all the photon absorption takes place in the depletion region, the generated holes and electrons will be accelerated by the field and will be quickly converted to circuit current. However, hole–electron pair generation occurs from the surface to the back side of the device. Although most of the pair generation occurs within the depletion region, enough does occur outside this region to cause a problem in high-speed applications. Carriers that are generated outside the depletion region are not subject to acceleration by the high electric field. They tend to move through the bulk by diffusion, a much slower type of travel. Eventually, the carriers reach the depletion region and are speeded up. This effect can be eliminated, or at least substantially reduced, by using a PIN structure, shown in Figure 16-17. In addition to response time improvements, the high-resistivity intrinsic (I) region gives the PIN diode lower-noise performance.

There are several critical parameters for a PIN diode in a fiber optic application:

1. *Responsivity*, usually given in amperes per watt at a particular wavelength, is a measure of the diode output current for a given power launched into the diode. The designer must be able to calculate the power level coupled from the system to the diode.
2. *Dark current* is the thermally generated reverse leakage current in the diode. In conjunction with the signal current calculated from the responsivity and

incident power, it gives the designer the ON-OFF ratio to be expected in a system.

3. *Response speed* determines the maximum data rate capability of the diode. In conjunction with the response of other elements of the system, it sets the maximum system data rate. Device capacitance also has an influence on the system data rate.

4. *Spectral response* determines the range, or system length, that can be achieved relative to the wavelength at which responsivity is characterized.

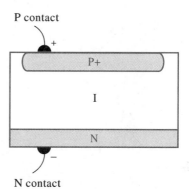

FIGURE 16-17
PIN diode structure

Integrated PIN FET Detector   In the *integrated PIN FET* device, a photodiode with unity gain (the PIN) is coupled with a high-impedance front-end amplifier (the FET). This device combines operation at low voltage with low sensitivity to operating temperature, high reliability, and ease of manufacture.

Avalanche Photodiode   The *avalanche photodiode* (APD), shown in Figure 16-18, produces a gain of 100 or more. However, it also produces noise that may limit receiver sensitivity. The APD devices require high voltage bias, which varies with temperature. Receivers using APDs are so sensitive that they require as few as 200 photons to be detected at the receiver per bit transmitted at data rates of 200–400 Mbps.

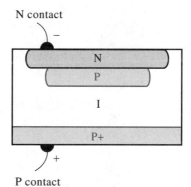

FIGURE 16-18
Avalanche photodiode

Integrated Detector Preamplifier The PIN photodiode is a high output impedance current source. The signal levels are usually on the order of tens of nanoamperes to tens of microamperes. The signal requires amplification to provide data at a usable level (such as TTL). In noisy environments, the noise-insensitive benefits of fiber optics can be lost at the receiver connection between the diode and amplifier. Proper shielding can prevent this. An alternative solution is to integrate the follow-up amplifier into the same package as the photodiode, as shown in the schematic in Figure 16-19. This device is called an *integrated detector preamplifier* (IDP).

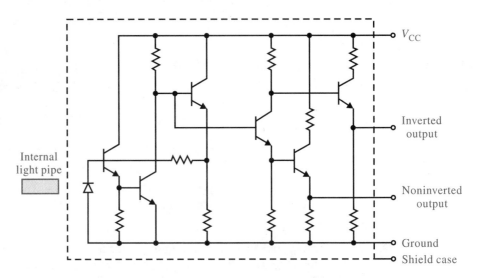

FIGURE 16-19 Schematic of an integrated detector preamplifier (IDP)

Incorporating an intrinsic layer into the monolithic structure is not practical with present technology, so a P-N junction photodiode is used. The first two transistors form a transimpedance amplifier. A third-stage emitter follower provides resistive negative feedback. The amplifier gives a low impedance voltage output which is then fed to a phase splitter. The two outputs are coupled through emitter followers.

The MFOD102F IDP has a responsivity greater than 230 mV/µW at 820 nm. The response rise and fall times are 50 ns maximum, and the input light power can go as high as 30 µW before noticeable pulse distortion occurs. Both outputs offer a typical impedance of 200 Ω.

The IDP can be used directly with a voltage comparator or, for more sophisticated systems, to drive any normal voltage amplifier. Direct drive of a comparator is shown in Figure 16-20.

Wavelength Division Multiplex

A combination of single-mode fiber (low dispersion by the transmission medium), narrow output spectrum (power concentration at a single frequency), and narrow dispersion angle (good power coupling) from ILDs makes possible the extreme bandwidth-distance characteristics cited for systems at the beginning of this chapter. The narrow ILD emission spectrum also makes it possible to send several signals from different

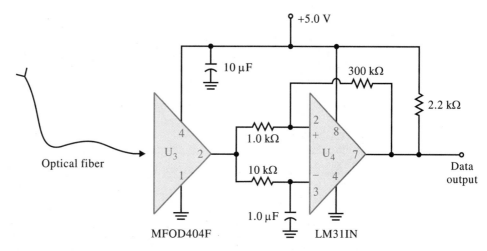

FIGURE 16-20 Simple fiber optic data receiver using an IDP and a voltage comparator

sources down the same fiber by a technique called *wavelength division multiplex* (WDM). As illustrated in Figure 16-21, WDM at optical frequencies is the equivalent of frequency division multiplex (FDM) at lower frequencies. Light at two or more discrete wavelengths is coupled into the fiber, with each wavelength carrying a channel at whatever modulation rate is used by the transmission equipment driving the light source. Thus, the information capacity of each fiber is doubled or tripled.

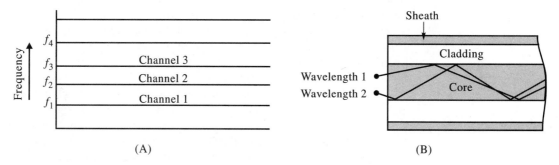

(A)   (B)

FIGURE 16-21 Comparison of multiplexing techniques (A) Frequency division multiplex (FDM) (B) Wavelength division multiplex (WDM)

A Fiber Optic Link

In fiber optics, a *link* is the assembly of hardware that connects a source of a signal with its ultimate destination. The items that form the assembly are shown in Figure 16-22. As the figure indicates, an input signal—for example, a serial bit stream—is used to modulate a light source, typically an LED. A variety of modulation schemes can be used. Although the input signal is assumed to be a digital bit stream, it could just as well be an analog signal, perhaps video.

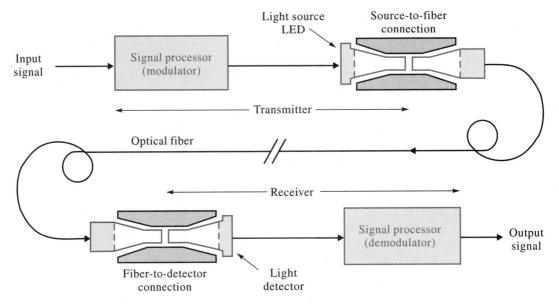

FIGURE 16-22 Fiber optic link

The modulated light must then be coupled into the optical fiber. This is a critical element of the system. Based on the coupling scheme used, the light coupled into the fiber could be two orders of magnitude less than the total power of the source.

Once the light has been coupled into the fiber, it is attenuated as it travels along the fiber. It is also subject to distortion. The degree of distortion limits the maximum data rate that can be transmitted through the fiber.

At the receive end of the fiber, the light is coupled into a detector element (such as a photodiode). The coupling problem at this stage, although still a concern, is considerably less severe than at the source end. The detector signal is then reprocessed or decoded to reconstruct the original input signal.

A link like that described in Figure 16-22 can be fully transparent to the user. That is, everything from the input signal connector to the output signal connector can be prepackaged. Thus, the user need be concerned only with supplying a signal of some standard format and level (such as TTL) and extracting a similar signal. Such a TTL in/TTL out system removes the need for a designer to understand fiber optics. However, by analyzing the problems and concepts internal to the link, the user is better prepared to apply fiber optic technology to the system.

16-4 TRANSMISSION SYSTEMS

Fiber optic transmission systems may be *simplex* (one direction only), *half-duplex* (one direction or the other but not both simultaneously), or *full-duplex* (both directions simultaneously) as shown in Figures 16-23(A) through (C) respectively. They may be *short-link* (usually not more than 1 km in length), *long-link* (greater than 1 km in length), or *networks* (involving more than two terminals).

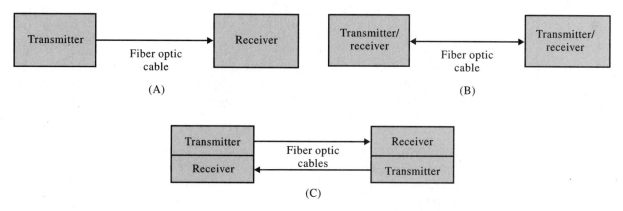

FIGURE 16-23   Methods of transmission (A) Simplex (B) Half-duplex (C) Full-duplex

Three main types of networks are bus, ring, and star systems. Networks may be classified in several ways. For example, a system may be a half-duplex, short-link ring network, or it may be a full-duplex, long-link star network. In the following sections, we will examine three typical network systems.

Bus System

A fiber optic *bus* system, consisting of an optical fiber line, several passive taps, and a number of terminals, each containing a transmitter and receiver, is shown in Figure 16-24(A). In this system, all terminals are connected together so that any one

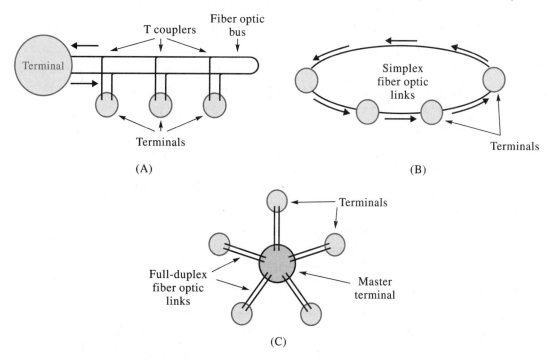

FIGURE 16-24   Transmission networks (A) Bus system (B) Ring system (C) Star system

terminal can communicate with any of the other terminals. Since all terminals can receive the message, a sending terminal is effectively broadcasting its message over the line. How messages for different terminals are identified, or the method of deciding the order in which terminals may transmit, differs from system to system. In some systems, the terminals may listen to the line before transmitting to determine when the line is free. In others, addresses may be added to the message so that the intended terminal will pay attention to that message.

Ring System

A simple *ring* system, consisting of a single optical fiber line connecting several terminals in a ring pattern, is shown in Figure 16-24(B). It may look simple, but, overall, it is more complex than the bus system. Because messages in this system travel in one direction only, they must be passed from terminal to terminal until the addressed terminal is reached. Therefore, the terminals must not only listen to the line at all times, but they must also retransmit every message that comes down the line. This latter requirement places the complexity of the system on the terminals, since the terminals must handle all of the functions of sorting and retransmitting messages. The fiber links themselves are simply a group of simplex links placed between terminals.

Star System

A *star* system, shown in Figure 16-24(C), consists of a central master terminal and several slave terminals. Every slave terminal is directly connected to the master in a point-to-point, full-duplex configuration. In this system, the master can broadcast to any or all of the slaves, but the slaves can communicate only with the master. Messages from one slave to another slave must be routed through the master for retransmission.

16-5   DESIGNING A FIBER OPTIC COMMUNICATION SYSTEM

Now that the basic concepts and advantages of fiber optics and the active components used with them have been discussed, we will consider the design of a specific system. The system discussed will be a simple point-to-point application operating in the simplex mode. It will be analyzed for three aspects: (1) loss budget, (2) rise time budget, and (3) data encoding format.

Loss Budget

If no in-line repeaters are used, every element of the system between the LED and the detector introduces some loss into the system. By identifying and quantifying each loss, the designer can calculate the transmitter power required to ensure a given signal power at the receiver or, conversely, the signal power to be received for a given transmitter power. The process is referred to as calculating the system *loss budget*.

Our sample system will be based on the following:

> 1. *Transmitter*—MFOE107F/MOFE108F, shown in the data sheet in Figure 16-25.

2. *Fiber*—Silica-clad silica fiber with a core diameter of 200 μm; step index mode; 20 dB/km attenuation at 820 nm; NA of 0.35; 3.0 dB bandwidth at 5.0 MHz/km.

3. *Receiver*—MFOD404F, shown in the data sheet in Figure 16-26.

The system will link a transmitter and receiver over a distance of 1000 m (1 km) and will use a single section of fiber (no splices). Some additional interconnect loss information is required:

1. Whenever a signal is passed from an element with an NA greater than the NA of the receiving element, the loss incurred is given by

$$\text{NA loss} = 20 \log \frac{NA_1}{NA_2} \tag{16-5}$$

where NA_1 is the exit NA of the signal source and NA_2 is the acceptance NA of the element receiving the signal.

2. Whenever a signal is passed from an element with a cross-sectional area greater than the cross-sectional area of the receiving element, the loss incurred is given by

$$\text{area loss} = 20 \log \frac{d_1}{d_2} \tag{16-6}$$

where d_1 is the diameter of the signal source (assumes a circular fiber port) and d_2 is the diameter of the element receiving the signal.

3. If there is any space between the sending and receiving elements, a loss is incurred. For example, an LED with an exit NA of 0.5 will result in a gap loss of 1.5 dB if it couples into a fiber over a gap of 0.15 mm.

4. If the source and receiving elements have their axes offset, there is an additional loss. This loss is also dependent on the separation gap. For an LED with an exit NA of 0.5, a gap with its receiving fiber of 0.15 mm, and an axial misalignment of 0.035 mm, there will be a combined loss of 1.8 dB.

5. If the end surfaces of the two elements are not parallel, an additional loss can result. If the nonparallelism is held below 2–3°, this loss is minimal and can generally be ignored.

6. As light passes through any interface, some of it is reflected. This loss, called *Fresnel loss*, is a function of the indices of refraction of the materials involved. For the devices in this example, this loss is typically 0.2 dB per interface.

The system loss budget can now be calculated. Figure 16-27 shows the system configuration. Table 16-2 presents the loss contribution of each element in the link. Note that in Table 16-2 no Fresnel loss was considered for the LED. This loss, although present, is included in specifying the output power in the data sheet.

In this system, the LED is operated at 100 mA. The data sheet in Figure 16-25 shows that at this current, the instantaneous output power is typically 1100 μW. This

SEMICONDUCTORS

PO BOX 20912 • PHOENIX, ARIZONA 85036

MFOE107F
MFOE108F

FIBER OPTICS

HIGH-POWER
AlGaAs LED

AlGaAs FIBER OPTIC EMITTER

...designed for fiber optic applications requiring high-power and fast response time. It is a AlGaAs LED spectrally matched to the minimum attenuation region of most fiber optic cables. Motorola's ferrule package fits directly into a metal fiber optic system cable. Applications include industrial controls, military and other applications.

● Fast Response - 15 ns Typ
● Ferrule Package - Small and Rugged
● Integral Heatsink Improves Reliability
● Prepolished Optical Port Enhances Coupling Efficiency
● Complements All Motorola Fiber Optic Detectors
● Compatible With AMP Connector #227240-3 and Amphenol Connector #905-135-5000
● 200 μm Diameter Fiber Core Optical Port With 0.5 NA

MAXIMUM RATINGS

Rating	Symbol	Value	Unit
Reverse Voltage	V_R	3.0	Volts
Forward Current-Continuous	I_F	200	mA
Total Device Dissipation @ T_A = 25°C Derate above 25°C	P_D	400 4.0	mW mW/°C
Operating Temperature Range	T_A	–30 to+85	°C
Storage Temperature Range	T_{stg}	–30 to+100	°C

THERMAL CHARACTERISTICS

Characteristics	Symbol	Max	Unit
Thermal Resistance, Junction to Ambient	θ_{JA}	250 175*	°C/W

*Installed in compatible metal connector housing.

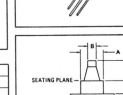

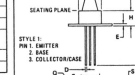

STYLE 1:
PIN 1. EMITTER
2. BASE
3. COLLECTOR/CASE

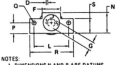

NOTES:
1. DIMENSIONS N AND R ARE DATUMS.
2. ⊤ IS GAUGE PLANE.
3. POSITIONAL TOLERANCE
   LEADS AND MOUNTING HOLES
   | ⊕ | ⌀ 0.36 (0.014) Ⓜ | T | N Ⓜ | R Ⓜ |
4. DIMENSIONING AND TOLERANCING PER ANSI Y14.5, 1973.

DIM	MILLIMETERS		INCHES	
	MIN	MAX	MIN	MAX
A	6.86	7.11	0.270	0.280
B	2.54	2.64	0.100	0.104
C	10.93	11.55	0.430	0.455
D	0.41	0.48	0.016	0.019
E	0.64	0.89	0.025	0.035
F	5.72	5.96	0.225	0.235
G	2.54 BSC		0.100 BSC	
H	3.94	4.44	0.155	0.175
K	12.70	--	0.500	--
L	10.41 BSC		0.410 BSC	
N	5.97	6.22	0.235	0.245
P	3.05 BSC		0.120 BSC	
Q	1.15	1.29	0.045	0.051
R	12.83	13.08	0.505	0.515
S	4.32	4.57	0.170	0.180

CASE 338D-01

© MOTOROLA INC., 1981

FIGURE 1 — SPECTRAL OUTPUT versus WAVELENGTH

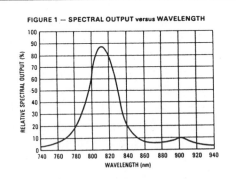

MFOE107F/MFOE108F DATA SHEET

DS2553

FIGURE 16-25 Data sheet for MFOE107F/MFOE108F LED transmitter

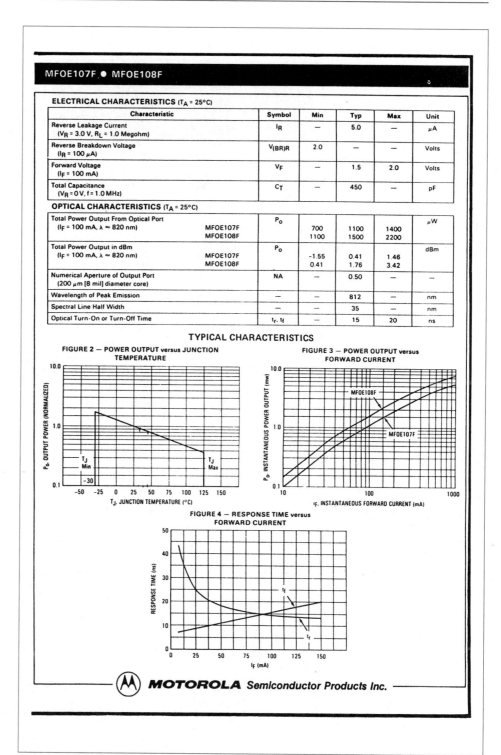

MFOE107F ● MFOE108F

ELECTRICAL CHARACTERISTICS ($T_A = 25°C$)

Characteristic	Symbol	Min	Typ	Max	Unit
Reverse Leakage Current ($V_R = 3.0$ V, $R_L = 1.0$ Megohm)	I_R	—	5.0	—	μA
Reverse Breakdown Voltage ($I_R = 100 \mu$A)	$V_{(BR)R}$	2.0	—	—	Volts
Forward Voltage ($I_F = 100$ mA)	V_F	—	1.5	2.0	Volts
Total Capacitance ($V_R = 0$ V, $f = 1.0$ MHz)	C_T	—	450	—	pF

OPTICAL CHARACTERISTICS ($T_A = 25°C$)

Characteristic		Symbol	Min	Typ	Max	Unit
Total Power Output From Optical Port ($I_F = 100$ mA, $\lambda \approx 820$ nm)	MFOE107F	P_o	700	1100	1400	μW
	MFOE108F		1100	1500	2200	
Total Power Output in dBm ($I_F = 100$ mA, $\lambda \approx 820$ nm)	MFOE107F	P_o	-1.55	0.41	1.46	dBm
	MFOE108F		0.41	1.76	3.42	
Numerical Aperture of Output Port (200 μm [8 mil] diameter core)		NA	—	0.50	—	—
Wavelength of Peak Emission		—	—	812	—	nm
Spectral Line Half Width		—	—	35	—	nm
Optical Turn-On or Turn-Off Time		t_r, t_f	—	15	20	ns

TYPICAL CHARACTERISTICS

FIGURE 2 — POWER OUTPUT versus JUNCTION TEMPERATURE

FIGURE 3 — POWER OUTPUT versus FORWARD CURRENT

FIGURE 4 — RESPONSE TIME versus FORWARD CURRENT

Ⓜ **MOTOROLA** *Semiconductor Products Inc.*

FIGURE 16-25 *(continued)*

MOTOROLA SEMICONDUCTORS

PO BOX 20912 • PHOENIX, ARIZONA 85036

MFOD404F

INTEGRATED DETECTOR/PREAMPLIFIER FOR FIBER OPTIC SYSTEMS

. . . designed as a monolithic integrated circuit containing both detector and preamplifier for use in medium bandwidth, medium distance systems. It joins Motorola family of Straight Shooter devices packaged in the Fiber Optic Ferrule case. The device fits directly into AMP Incorporated fiber optic connectors which also provide excellent RFI immunity. The output of the device is low impedance to provide even less sensitivity to stray interference. The MFOD404F has a 200 μm (8 mil) fiber input with a high numerical aperture.

- Usable for Data Systems up to 10 Megabaud
- Dynamic Range Greater than 100:1
- RFI Shielded in AMP Connector #227240-1
- May be Used with MFOExxx Emitters
- Ferrule Package — Small and Rugged
- Fiber Input Port Greatly Enhances Coupling Efficiency
- Prepolished Optical Port

FIBER OPTICS

INTEGRATED DETECTOR PREAMPLIFIER

MAXIMUM RATINGS (T_A = 25°C unless otherwise noted)

Rating	Symbol	Value	Unit
Supply Voltage	V_{CC}	7.5	Volts
Operating Temperature Range	T_A	–30 to +85	°C
Storage Temperature Range	T_{stg}	–30 to +100	°C

FIGURE 1 — EQUIVALENT SCHEMATIC

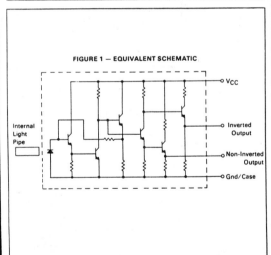

STYLE 1:
PIN 1. $-V_{OUT}$
2. $+V_{OUT}$
3. GND/CASE
4. $+V_{CC}$

NOTES:
1. ⊥ IS SEATING PLANE.
2. POSITIONAL TOLERANCE FOR LEADS:
   ⊕ ⌀ 0.36 (0.014) Ⓜ T
3. DIMENSIONING AND TOLERANCING PER Y14.5, 1973.

DIM	MILLIMETERS		INCHES	
	MIN	MAX	MIN	MAX
A	6.86	7.11	0.270	0.280
B	2.54	2.64	0.100	0.104
C	10.16	10.80	0.400	0.425
D	0.40	0.48	0.016	0.019
E	3.94	4.44	0.155	0.175
G	2.54 BSC		0.100 BSC	
K	12.70	–	0.500	–
M	45° BSC		45° BSC	
N	6.22	6.73	0.245	0.265

CASE 338B-01

Patent applied for.
Straight Shooter is a trademark of Motorola Inc.

MFOD404F DATA SHEET

© MOTOROLA INC., 1980

DS2544

FIGURE 16-26 Data sheet for MFOD404F IDP receiver

ELECTRICAL CHARACTERISTICS (V_{CC} = 5.0 V, T_A = 25°C)

Characteristics	Symbol	Conditions	Min	Typ	Max	Units
Power Supply Current	I_{CC}	Circuit A	3.0	3.5	5.0	mA
Quiescent dc Output Voltage (Non-Inverting Output)	V_q	Circuit A	0.5	0.6	0.7	Volts
Quiescent dc Output Voltage (Inverting Output)	V_q	Circuit A	2.7	3.0	3.3	Volts
Output Impedance	z_o		—	200	—	Ohms
RMS Noise Output	V_{NO}	Circuit A	—	0.4	1.0	mV

OPTICAL CHARACTERISTICS (T_A = 25°C)

Characteristics		Symbol	Conditions	Min	Typ	Max	Units
Responsivity (V_{CC} = 5.0 V, P = 2.0 μW*)	λ = 900 nm	R	Circuit B	20	30	50	mV/μW
	λ = 820 nm			—	35	—	
Pulse Response		t_r, t_f	Circuit B	—	35	50	ns
Numerical Aperture of Input Core (200 μm [8 mil] diameter core)		NA		—	0.70	—	—
Signal-to-Noise Ratio @ P_{in} = 1.0 μW peak*		S/N		—	35	—	dB
Maximum Input Power for Negligible Distortion in Output Pulse*				—	—	30	μW

RECOMMENDED OPERATING CONDITIONS

	Symbol	Conditions	Min	Typ	Max	Units
Supply Voltage	V_{CC}		4.0	5.0	6.0	Volts
Capacitive Load	C_L		—	—	100	pF
Input Wavelength	λ		—	900	—	nm

*Power launched into Optical Input Port. The designer must account for interface coupling losses.

FIGURE 2 — TYPICAL PERFORMANCE OVER OPERATING TEMPERATURE RANGE

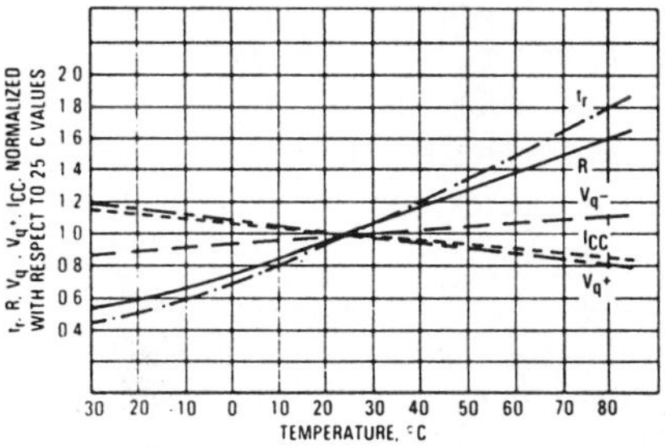

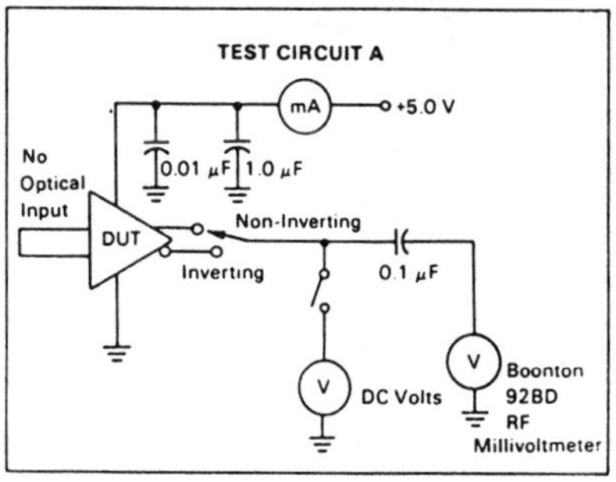

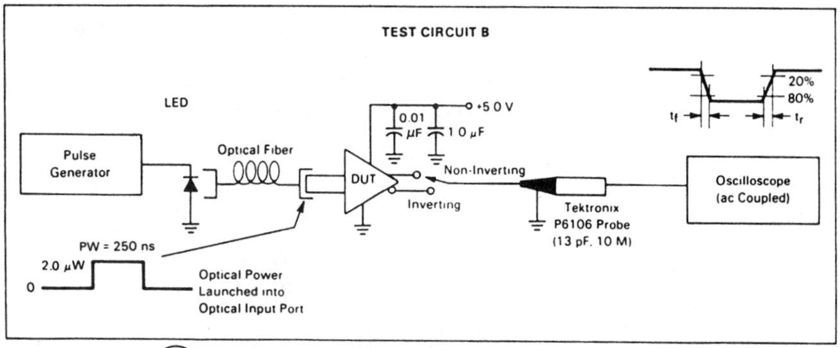

TEST CIRCUIT A

TEST CIRCUIT B

MOTOROLA *Semiconductor Products Inc.*

FIGURE 16-26 *(continued)*

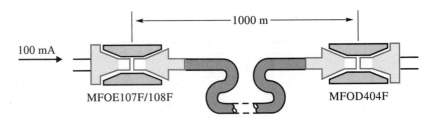

FIGURE 16-27 Simplex fiber optic point-to-point link

TABLE 16-2 Fiber Optic Link Loss Budget

Loss	Loss Contribution (dB)
MFOE107F-to-fiber NA loss	3.10
MFOE107F-to-fiber area loss	0
Transmitter gap and misalignment loss	1.80
Fiber entry Fresnel loss	0.20
Fiber attenuation (1 km)*	20.00
Fiber exit Fresnel loss	0.20
Receiver gap and misalignment loss	1.20
Detector Fresnel loss	0.20
Fiber-to-detector NA loss	0
Fiber-to-detector area loss	0
Total path loss	26.70

Note: Fiber attenuation is directly related to frequency of operation. The loss here would indicate operation at a high frequency of about 100 GHz. Losses this high are acceptable for short cable runs, but for long cable runs, lower frequencies are required. Single-mode fibers can operate at an attenuation loss of less than 1 dB/km at frequencies up to about 50 GHz; graded-index multimode fibers operate at around 1 dB/km up to about 1 GHz; step index multimode fibers operate at around 1.2 dB/km up to about 50 MHz. Above these frequencies, attenuation loss increases dramatically.

measurement assumes that the junction temperature is maintained at 25°C. The output power (P_{out}) from the LED is then converted to a reference level relative to 1.0 mW:

$$P_{out} = 10 \log \frac{1.1 \text{ mW}}{1.0 \text{ mW}} \tag{16-7}$$

$$= 0.41 \text{ dBm}$$

The power received (P_r, in dBm) by the MFOD404F is then calculated:

$$P_r = P_{out} - \text{total path loss} \tag{16-8}$$

$$= 0.41 - 26.70$$

$$= -26.29 \text{ dBm}$$

We then convert this reference level back to absolute power:

$$-26.29 \text{ dBm} = 10 \log \frac{P_r}{1.0 \text{ mW}}$$

$$-2.629(1.0 \text{ mW}) = \log P_r$$

$$P_r = \text{antilog} -2.629$$

$$= 0.0024 \text{ mW} \tag{16-9}$$

Based on the typical responsivity (R) of the MFOD404F from the data sheet in Figure 16-26, the expected output signal will be

$$V_{out} = (35 \text{ mV/}\mu\text{W}) (2.4 \text{ }\mu\text{W}) = 84 \text{ mV} \tag{16-10}$$

As shown in Figure 16-26, the output signal will be typically 200 times greater than the noise level.

In many cases, a typical calculation is insufficient. To perform a worst-case analysis, assume that an *S/N* ratio at the MFOD404F output must be 20 dB. The data sheet shows that the maximum noise output voltage is 1.0 mV. Therefore, the output signal must be 10 mV. With a worst-case responsivity of 20 mV/μW, received power must be

$$P_r = \frac{V_{out}}{R} = \frac{10 \text{ mV}}{20 \text{ mV/}\mu\text{W}} = 0.50 \text{ }\mu\text{W} \tag{16-11}$$

$$P_r = 10 \log \frac{0.00050 \text{ mW}}{1.0 \text{ mW}} = -33 \text{ dBm}$$

It is advisable to allow for LED degradation over time. A good design may include 3.0 dB in the loss budget for long-term degradation. The link loss was already performed as worst-case, so

$$P_{out(\text{LED})} = -33 \text{ dBm} + 3.0 \text{ dB} + 26.29 \text{ dB} \tag{16-12}$$

$$= -3.71 \text{ dBm}$$

Converting to absolute power, we obtain

$$P_{out} = \text{antilog} -0.371 = 0.426 \text{ mW}$$

$$= 426 \text{ }\mu\text{W}$$

Based on the power output versus forward current curve in Figure 16-25, it can be seen that the drive current (instantaneous forward current) necessary for 426 μW of power is about 30 mA. The data sheet also shows a power output versus junction temperature curve which, when used in conjunction with the thermal resistance of the package, enables the designer to allow for higher drive currents as well as variations in ambient temperatures.

At 30 mA drive, the forward voltage will be less than 2.0 V worst-case. Using the 2.0 V will give a conservative analysis for total device power dissipation (P_D):

$$P_D = (30 \text{ mA}) (2.0 \text{ V}) = 60 \text{ mW} \tag{16-13}$$

This value is well within the maximum rating for operation at 25°C ambient. If we assume that the ambient temperature will be 25°C or less, the junction temperature can be conservatively calculated. When an LED is installed in a compatible metal connector,

$$\Delta T_J = (175° \, C/W)(0.06 \, W) \approx 11° \, C \tag{16-14}$$

If we are transmitting digital data, we can assume an average duty cycle of 50%, so ΔT_J will almost always be less than 6°C. Therefore,

$$T_J = T_A + \Delta T_J \tag{16-15}$$

$$= 25° \, C + 6° \, C = 31° \, C$$

The output power derating curve shows a value of 0.9 at 31°C. Thus, the required dc power level must be

$$P_{out(dc)} = \frac{426}{0.9} = 473 \, \mu W \tag{16-16}$$

As Figure 16-25 indicates, increasing the drive current to 40 mA would provide more than 500 μW power output and increase the junction temperature only 1°C. This analysis shows the link to be more than adequate under the worst-case conditions.

Rise Time Budget

The cable for this system was specified to have a bandwidth of 5.0 MHz/km. Since the length of the system is 1.0 km, the system bandwidth, if limited by the cable, is 5.0 MHz. Data links are usually rated in terms of a *rise time budget*. The system rise time (t_{RS}) is found by taking the square root of the sum of the squares of the individual elements. In this system, the only two elements to consider are the LED and the detector. Thus,

$$t_{RS} = \sqrt{(t_{R(LED)})^2 + (t_{R(detector)})^2} \tag{16-17}$$

Using the typical values from Figures 16-25 and 16-26, we have

$$t_{RS} = \sqrt{(15)^2 + (35)^2} = 38 \, ns$$

Total system performance may be impacted by including the rise time of additional circuit elements.

Data Encoding Format

In a typical digital system, the coding format is usually *non-return to zero* (NRZ). In this format, a string of 1s would be encoded as a continuous HIGH level. Only when there is a change of state to a 0 will the signal level drop to zero. In *return to zero* (RTZ) encoding, the first half of a clock cycle would be HIGH for a 1 and LOW for a 0. The second half would be LOW in either case. Figure 16-28 shows NRZ and RTZ waveforms for a binary data stream. Note that between A and B, the RTZ pulse repetition rate is at its highest. The highest bit rate requirement for an RTZ system is a string

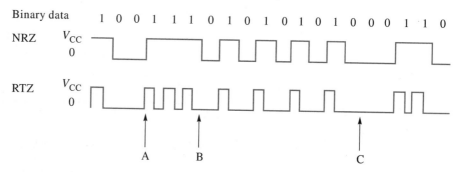

FIGURE 16-28   NRZ and RTZ encoded data

of 1s. The highest bit rate requirement for an NRZ system is alternating 1s and 0s, as shown from B to C. Note that the highest NRZ bit rate is half the highest RTZ bit rate; thus, an RTZ system would require twice the bandwidth of an NRZ system for the same data rate.

However, to minimize drift in a receiver, the receiver will probably be ac coupled, but if NRZ encoding is used and a long string of 1s is transmitted, the ac coupling will result in lost data in the receiver. With RTZ encoding, data are not lost with ac coupling since only a string of 0s results in a constant signal level; but that level is also zero. However, in the case of both NRZ and RTZ, a continuous string of either 1s or 0s for NRZ or 0s for RTZ will prevent the receiver from recovering any clock signal.

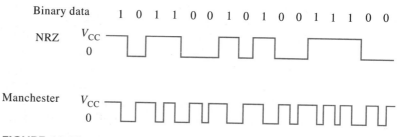

FIGURE 16-29   NRZ and Manchester data encoding

Another format, called *Manchester* encoding, solves this problem. By definition, in Manchester the polarity reverses once each bit period regardless of the data. This encoding action is shown in Figure 16-29. The large number of level transitions enables the receiver to derive a clock signal even if all 1s or all 0s are being received.

In many cases, clock recovery is not required. It might appear that RTZ would be a good encoding scheme for those applications. However, many receivers include automatic gain control (AGC). During a long stream of 0s, the AGC could increase the receiver gain; and when 1s begin to appear, the receiver may saturate. A good encoding scheme for these applications is *pulse bipolar* encoding, shown in Figure 16-30. The transmitter runs at a quiescent level and is turned ON harder for a short time during a data 0 and is turned OFF for a short time during a data 1.

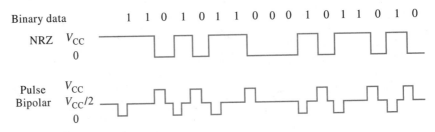

FIGURE 16-30   NRZ and pulse bipolar encoding

16-6   DISPERSION SHIFTED FIBER FOR DIGITAL TELECOMMUNICATIONS

Two primary fiber parameters are critical when selecting a single-mode fiber for digital telecommunications system applications. *Optical fiber attenuation* and *chromatic dispersion* must be considered, since these parameters may limit the system's operating speed or may dictate the maximum distance between amplification or repeater sites. Currently, most telecommunications systems employ conventional single-mode fiber that is optimized for operations in the 1310 nanometer (nm) window. Conventional single-mode fiber offers the user an efficient path for broadband transmission of voice, data, and video information. With the commercialization of *erbium-doped fiber amplifiers* (EDFAs) and the development of more cost effective and improved transmitters, users may begin to explore operation of systems in the 1550 nm window. System operation in the 1550 nm window offers lower attenuation than the 1310 nm window. As a result the number of repeaters in the system may be reduced, effecting an increase in system reliability. The chromatic dispersion of conventional single-mode fiber at 1550 nm, however, may be nearly six times higher than that experienced at 1310 nm. The higher dispersion at 1550 nm may limit the system's operating speed or reduce the spacing between repeaters. Although dispersion compensation techniques are available to increase the operating speed, another option—dispersion-shifted fiber—is available. Dispersion-shifted fiber can be deployed to avoid the requirement for dispersion compensation techniques while capitalizing on the lower attenuation performance of the 1550 nm window.

Chromatic Dispersion

The two fiber parameters that have the greatest effect in limiting digital transmission over optical waveguides are attenuation and pulse spreading. In single-mode fibers, pulse spreading is caused by chromatic dispersion. Attenuation attracted most of the attention in the early years of single-mode fiber because it was generally the limiting factor in determining maximum repeaterless span lengths. As the required information-carrying capacity of optical systems was increased (higher bit rates), the maximum allowable system-pulse spreading decreased. This fact, in conjunction with lower fiber attenuation, eventually resulted in systems that were dispersion-limited instead of attenuation-limited.

Dispersion causes a light pulse to spread in time as it travels through a fiber. This spreading increases the potential for interference between sequential pulses. Pulses launched close together (high bit rates) that spread too much (high dispersion) result in bit errors.

Dispersion for a single-mode fiber is more precisely referred to as chromatic dispersion and consists of material dispersion and waveguide dispersion. Chromatic dispersion is determined by the fiber's material composition, structure and design, and by the light source's operating wavelength and spectral width. Chromatic dispersion is measured in units of picoseconds per nanometer and per kilometer (ps/nm-km): picoseconds (10^{-12} seconds) of light pulse spread per nanometer (10^{-9} meters) of laser spectral width and per kilometer of fiber length (10^{-3} meters).

Material Dispersion

Material dispersion is caused by molecular and light energy interactions. It occurs in every medium except a perfect vacuum. The amount of material dispersion varies with the composition of the glass and is a function of the wavelength of the light source. Material dispersion occurs because the speed of light in a medium is a function of wavelength. No transmitter produces a pure light source of only one wavelength. Instead, sources produce a range of wavelengths around a center wavelength. Light at these wavelengths travels at slightly different speeds, resulting in pulse spreading that increases with distance.

Waveguide Dispersion

Waveguide dispersion is a function of the structure of the refractive index profile of the fiber and varies as the percentage of light in the core and cladding changes. These percentages are a function of the operating wavelength. Material dispersion is changed only slightly by altering the glass composition, but its effect can be changed significantly by using a better light source. Changes in waveguide dispersion can be achieved by altering the internal structure (index profile) of the fiber. The current single-mode fiber is made up of a core with a high index of refraction and a cladding with a lower index. This simple step-index profile yields a zero dispersion wavelength (where the material and waveguide components of dispersion cancel) near 1310 nm. By changing the internal structure of the fiber (variations in refractive index and core geometry) designers can shift the zero dispersion point. Some manufacturers have designed a segmented core fiber which has a larger waveguide dispersion than the standard single-mode fiber. This shifts the zero dispersion wavelength to approximately 1550 nm.

How Chromatic Dispersion Affects an Optical Pulse The dispersion associated with an optical fiber causes the transmitted pulse to spread out as it travels along the length of the fiber. A typical optical source emits light that contains a range of wavelengths. Therefore, an optical pulse will contain a range of spectral components. Since the index of refraction is a function of wavelength, each wavelength of light in a pulse will travel at a slightly different speed. As the pulse spreading increases, there is a potential for overlapping of adjacent bits of information. This effect directly impacts the *bit error rate* (BER) performance of the system.

Shifting the Zero Dispersion Wavelength How is the zero dispersion wavelength of the fiber shifted? As previously discussed, chromatic dispersion is composed of two components. Material dispersion may be altered by changing the glass composition but the effect is typically minimal. However, by changing the refractive index profile of the fiber, the waveguide dispersion can be altered significantly. Conventional single-mode fiber has a step-index profile design that yields a zero dispersion wavelength near 1310 nm. By changing the design to a segmented-core structure, greater wave-guide dispersion is introduced that offsets the material dispersion. This has a net effect of shifting the zero dispersion wavelength to the 1550 nm window.

Types of Dispersion-Shifted Fiber   Two types of dispersion-shifted fiber are available: conventional dispersion-shifted fiber and non-zero dispersion-shifted fiber. The theory behind each dispersion-shifted product is the same—chromatic dispersion is minimized in the 1550 nm window, the same window in which attenuation is minimized. The major difference between the two fibers is the location of the zero dispersion wavelength. The zero dispersion wavelength of conventional dispersion-shifted fiber is nominally centered in the operating window of the fiber. The zero dispersion wavelength of non-zero dispersion-shifted fiber lies outside the operating window of the fiber. Thus, a small quantity of dispersion is assured throughout the operating window of the non-zero dispersion-shifted fiber, which inhibits a non-linear effect known as four-wave mixing. Four-wave mixing is a signal-degrading effect that can occur when two or more optical channels are operated on the same fiber. As a result, conventional dispersion-shifted fiber is optimized for single channel, high speed, long distance operation at 1550 nm. Non-zero dispersion-shifted fiber is optimized for multiple channel (wavelength division multiplexed) high speed, long distance operation. As a result, the long term plans of the system must be considered before selecting a dispersion-shifted fiber.

Tradeoffs   Dispersion-shifted fiber will theoretically exhibit slightly higher attenuation performance due to the increased dopant concentration of the fiber core. This increase in attenuation is on the order of 0.02 dB/km at 1550 nm. Dispersion-shifted fiber also has a mode field diameter that is smaller than that of standard fiber at 1550 nm. The nominal mode field diameter of conventional dispersion-shifted fiber is approximately $8.10 \pm 0.65 \mu m$ at 1550 nm as compared to $10.50 \pm 1.00 \ \mu m$ at 1550 nm for conventional single-mode fiber. The mode field diameter of non-zero dispersion-shifted fiber is $8.40 \pm 0.50 \ \mu m$ at 1550 nm. The smaller mode field diameter may lead to slightly higher splice losses, approximately 0.02 dB, but offers the advantage of improved bend resistance. The real advantage to dispersion-shifted fiber becomes apparent when maximum link distances are considered. Assuming that the attenuation performance for conventional single-mode and dispersion-shifted fibers are relatively the same, as stated here, the difference in dispersion performance becomes the limiting factor.

16-7   OPERATING SYSTEMS

More and more, fiber optic systems are replacing wire systems. A 1978 study by AT&T of a digital lightwave transmission system indicated it would cost less than traditional systems because of differences in terminal multiplexing equipment requirements and cost. In Canada, the Saskatchewan Telecommunications System is installing a 20,000 mile link that will connect most cities in Saskatchewan Province together. In the United States, two large systems have been installed, and undersea fiber optic cable systems are being constructed. In the following sections, we will examine two such systems.

Local and Intercity Systems—The FT3C System

The AT&T FT3C Lightwave System was devised to provide the most economical digital transmission system possible with then-current state-of-the-art fiber optics. It uses wavelength multiplexing techniques to send three 90 Mbps signals over the same fiber, giving over 240,000 digital channels at 64,000 bps in a cable containing 144 optical fibers. The first applications of the system have been the Northeast Corridor Project by AT&T, between Boston and Washington, D.C., and the North/South Lightwave Project on the West Coast by Pacific Telesis, between Sacramento and Los Angeles. A map of the Northeast Corridor system, which contains 78,000 fiber-kilometers of lightwave circuits, is shown in Figure 16-31(A). Figure 16-31(B) shows a map of the North/South Lightwave Project. These two systems were placed in service in 1983. More recently AT&T, MCI, and Sprint have greatly expanded fiber optic systems.

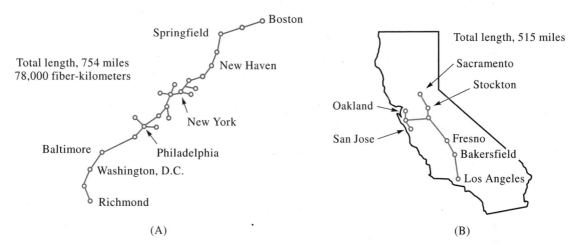

FIGURE 16-31 Local and intercity systems (A) Northeast Corridor Project (B) North/South Lightwave Project

International Systems—The SL Underwater Cable

Undersea optical fiber cables can provide a data communications channel with a carrying capacity equal to that of satellite systems, but with greater security, less interference, less noise, and lower cost. However, undersea cable systems have some understandably difficult environmental requirements. The environment includes pressures of 10,000 pounds per square inch (psi) at depths of 7300 m, salt water, and the possibility of mechanical damage from anchors and earth movement in shallow waters. An important requirement for these systems is that the regenerator spacing be as wide as possible to cut down on the system failure probability and the power requirements, since power must be fed from the ends of the cable.

A schematic of the SL Undersea Lightwave System is shown in Figure 16-32. It comprises (1) a high-voltage power supply, (2) a supervisory terminal, (3) a multiplexer

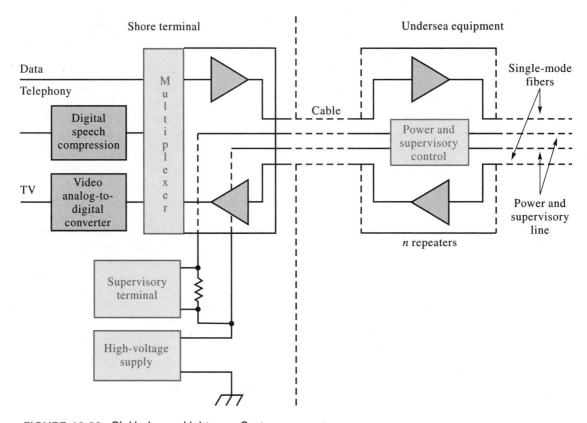

FIGURE 16-32 SL Undersea Lightwave System

with inputs for several types of information, (4) the cable light source, (5) the cable itself, and (6) the repeaters. The cable is composed of a central core and a surrounding support as shown in Figure 16-33. The core has an outside diameter of 2.6 mm and consists of 12 optical fibers wound helically around a central copper-clad steel wire called a *kingwire*, all of which is embedded in an elastomeric substance and covered with a nylon sheath. This assembly, in turn, is covered with several steel strands, a continuously welded copper tube, and, finally, low-density polyethylene for electrical

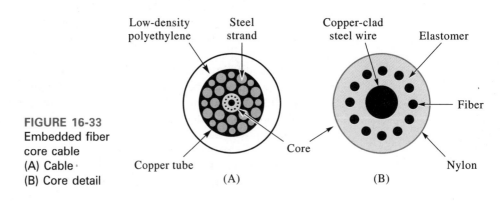

FIGURE 16-33
Embedded fiber
core cable
(A) Cable
(B) Core detail

insulation and abrasion resistance. The outside diameter of the completed cable is 21 mm (about 0.8 inch).

The fibers are single-mode optical lightguides operating at 1.312 μm. The data rate on each fiber is 280 Mbps, and repeaters are spaced every 35 km. The total capacity of the system is over 35,000 two-way voice channels. Inputs from binary data sources are multiplexed directly into the stream. Analog signals are first converted to binary by the adaptive delta modulation technique, then processed by a digital circuit (Digital TASI) that interleaves inputs from a number of input speech channels onto a smaller number of output channels. The light sources for the system are ILDs operating near 1.3 μm with an average output power of 1 mW (0 dBm). Light detectors are idium gallium arsenide (InGaAs) PIN diode receivers followed by silicon bipolar transimpedance amplifiers. Three spare laser diodes, which can be switched in remotely if a failure occurs, are provided for each circuit.

SUMMARY

- Optical communication is the application of light beams for sending and receiving messages. Modern systems use LEDs and lasers as modulated-light sources, optical fibers as lightguides, and photodiodes as detectors. A fiber optic link is the assembly of hardware that connects a source of a signal with its ultimate destination. Optical systems offer extremely high bandwidth, freedom from external interference, immunity from interception by external means, and cheap raw materials (silicon).

- An optical fiber is a nonconductive waveguide made of glass, plastic, or a combination of the two that confines light energy inside its walls. The fiber consists of a core, a cladding, and a sheath. A fiberscope is a flexible bundle of optical fibers with a focusing lens at each end that is used to view areas not normally observable. Fiberscope resolution depends on the size of the lens at the receptor end and on the number and fineness of the fibers in the bundle. A highly convex lens at the receptor end of the bundle provides magnification.

- When light passes from free space into a denser material, the light is bent at the interface. The bend angle depends on the refractive index of the denser material. The refractive index is the ratio of the speed of light in the denser material. Snell's law ($n_1 \sin A_1 = n_2 \sin A_2$) describes the relationship between the incident and reflected light.

- Modes of propagation are the differing angles at which light rays enter a fiber. If the diameter of the fiber core is limited to about three times the wavelength of the transmitted light, only one ray will be propagated, no destructive interference between rays occurs, and we have single-mode fibers. In multimode fibers, the diameter of the core is many times the wavelength of the light transmitted, and the light beam travels along the fiber by bouncing back and forth at the interface between the core and the cladding. Rays entering the fiber at differing angles are refracted a varying number of times as they move along the fiber and arrive at the distant end with different phase relationships than when they started. Multimode propagation causes rays leaving the fiber to interfere both constructively and destructively as they leave the end of the fiber. In graded-index fibers, the index of refraction varies smoothly across the diameter of the core but remains constant in the cladding, thus reducing intermodal dispersion by the fiber.

- Advantages of fiber optics over wire systems are greater bandwidth, smaller size and weight, lower attenuation, freedom from EMI, ruggedness, safety, and lower cost. Bandwidth in fiber optics is limited by modal delay spreading and material dispersion. Attenuation in fibers is controlled by scattering, absorption, connection, and bending losses.

- The numerical aperture is the measure of a fiber's ability to gather light. The optical power accepted by the fiber varies as the square of the numerical aperture but does not depend on any physical dimension of the fiber. The acceptance angle is the largest angle at which a light ray can enter the fiber and still propagate down the fiber. A large acceptance angle makes the end alignment less critical during splicing and connecting of fibers.

- Typical light sources for optical fiber systems are surface LEDs, edge-emitting LEDs, and injection laser diodes (ILDs). The narrow emission spectrum of the ILD makes it possible to send several signals from different sources down the same fiber by wavelength division multiplex.

- Typical detectors are integrated PIN FET assemblies, avalanche photodiodes, and integrated detector preamplifiers. Critical parameters for PIN diodes used in a fiber optic system are responsivity, dark current, response speed, and spectral response.

- Fiber optic transmission systems can be simplex, half-duplex, or full-duplex; and short-link, long-link, or networks. Networks are bus, ring, or star.

- The design of a fiber optic system requires analysis of loss budget, rise time budget, and data encoding format.

- Three typical ICs used for interfacing the various elements of a fiber optic system are a Manchester biphase-mark encoder and LED driver, a Manchester biphase-mark decoder, and a transimpedance amplifier. The encoder/LED driver IC encodes either TTL or ECL data and outputs current at the LED driver outputs. The decoder recovers data and clock signals from the encoder/LED driver. The decoder locks onto incoming data, recovers the clock, and decodes the data. The transimpedance amplifier acts as an interface between a detector diode and a decoder in a fiber optic receiver system.

QUESTIONS AND PROBLEMS

1. Define optical communication and briefly describe the operation of a modern optical communication system.

2. Describe (a) an optical fiber and (b) a fiberscope and explain the use of each.

3. Define (a) refractive index, (b) critical angle, (c) Snell's law, and (d) modes of propagation.

4. Define (a) single-mode fiber, (b) multimode fiber, and (c) graded-index fiber and briefly explain the operation of each.

5. What are the advantages of fiber optics over wire systems?

6. In regard to optical fiber systems, briefly describe (a) the causes of bandwidth limitations and (b) the controlling factors of attenuation.

7. Describe the Rayleigh scattering limit.

8. Define (a) numerical aperture and (b) acceptance angle.

9. Briefly describe optical fiber fabrication.

10. Describe the advantages and disadvantages of the three main varieties of fibers.

11. For fiber optic systems, what is the purpose of (a) a light source and (b) a light detector?

12. For fiber optic systems, briefly describe the operation of the three principal (a) light sources and (b) light detectors.

13. Briefly discuss the critical parameters for a PIN diode in a fiber optic application.

14. Describe how multiplexing is achieved in fiber optic systems.

15. Define fiber optic link and briefly explain its operation.

16. Define fiber optic transmission systems.

17. Describe three main types of networks.

18. Briefly discuss the three major aspects involved in designing a fiber optic system.

19. Briefly describe two operating optical fiber communication systems.

For Problems 20 through 23, refer to Figure 40 in Appendix A.

20. Assume a TTL data input. Explain the interconnection (pin numbers and power supply connections) required for an output at the large LED driver.

21. Repeat Problem 20 for an ECL data input and a small LED driver output.

22. What pins must be connected to produce (a) small driver output of 25 mA; (b) large driver output of 150 mA; and (c) large driver output of 80 mA.

23. Explain how the LED drivers are enabled.

24. Refer to Figure 41(B) in Appendix A. Explain how an input signal is decoded.

25. Explain how the biphase-mark decoder can be used in systems operating over a wide range of data rates without false frequency lock.

26. Explain (a) the purpose for the transimpedance amplifier and (b) how it operates.

CHAPTER 17

Introduction to the Information Highway

OBJECTIVES

When you complete this chapter, you should be able to:

- Understand the benefits and problems involved with the Telecommunications Act of 1996.
- Explain why it is more difficult for consumers to complain about high cable rates.
- Define the possible penalties for violation of the Communications Decency Act.
- List and define the many terms used in the Internet system.
- Define ISDN.
- List the various types of modems and their speeds.

INTRODUCTION

On February 8, 1996, the Telecommunications Act of 1996 was signed into law. As a result of this new law:

1. Local and long-distance telephone, cable, and broadcast companies can compete head-to-head in each other's businesses, which previously had been outlawed by the Communications Act of 1934.
2. Small cable companies may now set their own prices.
3. Media companies can control a bigger share of broadcast and newspaper companies in a specific region.
4. It is illegal to transmit "indecent" material, including speech, over the *Internet* (discussed in Section 17-2) to children under the age of 18.
5. New TV sets, 13 inches or larger, must be equipped with a chip that allows parents to block out violent programs.

6. Non-profit organizations such as libraries, schools, and hospitals can connect to telecommunications networks at a discount rate.

In this chapter, we examine some of the benefits and problems the new law presents, and the Internet, an important part of worldwide communications, is extensively covered.

17-1   THE TELECOMMUNICATIONS ACT OF 1996

The Telecommunications Act of 1996 is a revision of the Communications Act of 1934. These revisions promise a very different future for communications companies as well as electronics technicians and consumers. Those changes likely mean greater opportunities for you, the technician, and for the consumer, as competition increases for communications dollars.

For example, consider a single phone number that follows you wherever you go. You may be able to subscribe to a complete package of communications services from one company, including long distance, local calling, paging, cellular, Internet access, and satellite and/or cable TV, at rates lower than you would pay if you purchased the services individually.

If the Telecommunications Act of 1996 lives up to expectations, you may indeed see such changes soon—or perhaps immediately, as is the case with communications companies offering both local and long-distance service in New York, California, Illinois, and other states. In early 1996, the time and politics were right to bring about a major renovation and modernization of telecommunications policy and law in the United States. After months of negotiations and political discussion, Congress in effect rewrote the 62-year-old Communications Act of 1934, an outdated statute that had perpetuated restrictions on competition in the communications industry. Despite a much-altered business environment in the intervening years, only a few revisions to the law—including the 1984 court-ordered breakup of AT&T which ended a service monopoly and gave consumers a choice of long-distance carriers—had been made. (For example, industry competition as a result of that breakup has forced long-distance per-minute, off-peak rates to drop from about 18 cents per minute to 10 cents per minute in the last 12 years.)

The 280-page Act of 1996 knocked down barriers that for decades had shaped the form and development of telecommunications in this country. Consequently the Act has encouraged greater competition, and the ground rules have changed for virtually every sector of the communications industry, including local and long-distance telephone services, cable television, broadcasting, and equipment manufacturing.

Barriers Are Removed

Clearly, the Act adopts competition as the basic charter for all telecommunication markets. Over the years the technological and operational differences between cable, local telephone, and long-distance companies have become smaller. With the new Act most competitive barriers have been eliminated, giving each more freedom to move into new markets by offering a wide range of communications services. Specifically, the Act's provisions address five major areas:

1. Telephone service
2. Telecommunications equipment manufacturing
3. Cable television
4. Radio and television broadcasting
5. The Internet and online computer services

In each of these areas, cross-market entry barriers have been eliminated, concentration and merger rules have been relaxed, and massive new implementation obligations have been placed on the *Federal Communications Commission* (FCC) and state regulators. Though unleashing competitive market forces was Congress's motivation for many of the Act's provisions, other areas such as television violence and "indecent" online communications were addressed to reflect Congress's current views of appropriate social and moral behavior.

Communications Decency Act   The Communications Decency Act, a part of the Telecommunications Act of 1996, makes forms of free speech that are protected by the Constitution of the United States illegal if transmitted over the Internet. The Act makes it illegal to transmit "indecent" material, including speech, over the Internet to children under the age of 18 years. Violation of the Act can be severe—violators may face up to $100,000 fines and two years in prison. Also, the violator's *Internet Service Provider* (ISP) may be liable. Users of the Internet and their service providers believe that this is a form of censorship and have resisted implementation. A moratorium was declared while this part of the Act is disputed. Results of such action may not be known for years.

Consumer Impact

Proponents hailed the Telecommunications Act as a victory for deregulation and consumers, with promises of more service options and lower prices for communications products and services. Such optimism is due to a provision in the Act which allows cable, long distance, local phone service, Internet, and wireless phone companies to compete in each other's businesses—competition that is expected to reduce prices. Further, the Act's supporters anticipated increased product development and innovation as a result of the removal of restrictive barriers. In truth, however, it may be years before consumers feel the complete impact. Changes which have swept into every corner of the communications industry are sure to affect consumers in profound ways, but today no one is completely sure how. Still, some interesting prospects are already on their way to becoming reality. For example, in the future you should be able to purchase Internet access service from a local phone company, a cable company, or a long-distance carrier. In addition, you'll choose local phone service from a host of communications providers, including Sprint, AT&T, and MCI, as well as from established local phone companies such as the so-called *Baby Bells*. At the same time, the regional Bell companies can begin offering long-distance service in their areas as soon as they meet FCC requirements proving they have fairly opened competition in their own markets. Cable television services or video programming may even be available from your telephone company.

What to Expect

This may seem confusing to you, but you are not alone. It will take time to sort out the options and make sense of the Act's provisions. What seems clearest, for the moment at least, is that some confusion will remain until regulators and legislators on the federal and state levels determine the final impact on consumers. Even now legislators are scrambling to interpret how the new regulation will affect communications entities in their states. In the near term, you are likely to witness an onslaught of restructuring, mergers, and deal-making as companies race headlong into each other's businesses. Most states seem eager to open their local markets to greater competition; some already have done so.

On their part, consumers are hopeful that at least the promise of lower rates will be fulfilled when the dust has settled. As communications companies bundle packages of desirable communications services prices should be pushed down, with low-cost services offsetting higher-cost ones. The U.S. Department of Justice estimates that consumers should save $12 billion a year on communications services as a result of the legislation.

Beyond better pricing, however, consumers should look for better service from providers. Competition for dollars should push providers to reconsider traditional service levels and elect to take a more competitive posture. The poor reputation of some communications service providers will not play well with consumers. Of necessity providers will need, at a minimum, to offer service convenience, operational efficiency, and rewards for loyalty.

In short, competition promoted by the Telecommunications Act may give most companies the incentive they need to step up to the challenge of serving a more demanding consumer in a new area of technological sophistication. What this means to you, the technician, is greater opportunities for a broader job market.

Baby Bells   There is a formidable arsenal of phone services at our beck and call. One feature continuously dials a busy number until the call is completed. Another blocks calls placed from specific telephones. We also have *Call Waiting,* three-way calling, and a device that displays a caller's name as well as number. The extras fatten our phone bill each month—and that is just the point. The Baby Bells and other local phone companies charge stiff monthly fees of up to $8 for each new service, reaping lush profit margins in the process. All told, the Bells and GTE Corporation rake in more than $4 billion a year on these new services, and the take is growing.

For the better part of a century, phone service was simple and unadorned. Even the industry dismissed it as *Plain Old Telephone Service* (POTS). Thanks to new technology, the dull dial-tone network is quietly becoming a web pulsing with a newfound "intelligence." Its copper and fiber tentacles reach into every home and business. It can recognize your voice; it can follow you wherever you go; it knows which callers you like and which you hate. And that, the phone companies assure us, is only a taste of what is to come. It's the beginning of a host of telephone services that can be used to identify the caller in one way or another. Consumers are losing the privacy of their phone numbers. Many of the newer services read like a wish list for the paranoid. *Selective Call Acceptance* lets only calls from certain preset numbers come through, blocking the rest. *Call Block* lets all calls through *except* those from specified phones. *Call Return,* activated by pressing 69, redials the number of the last incoming call. (Sometimes users can even retrieve the caller's number via a recorded message, although this feature is banned in some states as an invasion of privacy.)

Other add-ons offer convenience rather than surveillance. One service lets a single-line home have up to six phone numbers, one for each family member, with a distinctive ring for each unit. Another service uses the BellSouth Corporation service, *ZipConnect*, to let customers in five states call a single toll-free number and automatically reach the outlet nearest them. Yet another service is GTE's *In Contact*, with which one forwards home, work, or cellular calls to wherever one is on a particular day.

The new services all depend on more intelligent software. In the past, offering a new service required the installation of new software at thousands of individual *Switches,* the powerful computers that route calls. Now the Bells install new programming in a handful of master computers, known as *Service Control Points,* which then relay directions to the switches.

Thus, carriers now can introduce some services in one year instead of three. Once the software is up and running, it costs a carrier virtually nothing to add a new customer. And the profit margins of 70 percent or more far exceed the less-than-10 percent profit that regional carriers typically get on basic phone service. Revenues from the new "smart" services could grow more than 50 percent in the next five years.

The most popular (and perhaps most Orwellian-like service) is *Caller ID*. With a special device that can cost up to $200, plus a monthly fee of about $6, a user can see a caller's number or name on a tiny screen. Long-distance callers, shielded in the past, are now included in the service under recent changes in federal rules. Even unpublished numbers show up.

17-2   THE INTERNET

There have been many books written specifically about the Internet. This section presents the basic concepts and answers the question, "What is the Internet?" It is beyond the scope of this text to try to cover every aspect of the Internet. Some of the more commonly used terms and expressions are defined, with some practical suggestions on how to best use and enjoy the Internet system.

What Is the Internet?

The *Internet* is the name for a group of worldwide information resources. These resources are so vast as to be well beyond the comprehension of a single human being. Not only is there no one who understands *all* of the Internet, there is no one who even understands *most* of the Internet.

The first experimental network using Internet-like technology involved four computers and was built in 1969. This was 13 years prior to the development of the first IBM personal computer.

The roots of the Internet lie in a collection of computer networks that were developed in the 1970s. It started with a network called the *Arpanet* that was sponsored by the United States Department of Defense. The original Arpanet has long since been expanded and replaced, and today its descendants form the global backbone of what we call the Internet.

It would be a mistake, though, to think of the Internet as a computer network, or even a group of computer networks connected to one another. For our purposes, consider the computer networks as simply the media that carries the information. The beauty and utility of the Internet lie in the information itself. This is how you should think of the Internet: not as a computer network, but as a huge source of information.

You should also develop an appreciation of the Internet as a people-oriented society. Put simply, the Internet allows millions of people all over the world to communicate and to share. You *communicate* by either sending and receiving *electronic mail (e-mail),* or by establishing a connection to someone else's computer and typing messages back and forth. You *share* by participating in discussion groups and by using the many programs and information sources that are available for free.

What this means is that the Internet resources will become as important to you as your telephone and your post office. In learning how to use the Internet, you are learning about a world in which people from many different countries and cultures cooperate willingly and share their time, their efforts, and their products. The Internet is much more than a computer network or an information service. The Internet is living proof that human beings are able to communicate freely and conveniently.

Computers are important because they do the work of moving data from place to place, and executing the programs that let you access the nation and the world. The information itself is important because it offers utility, recreation, and amusement.

Overall, what is most important is the people. The Internet is the first global forum and the first global library. Anyone can participate, at any time; the Internet never closes. Moreover, no matter who you are, you are always welcome.

One might say that the reason the Internet works so well is that there are no leaders. There is some truth to that reasoning, because nobody actually *runs* the Internet. Nobody is *in charge* and no single organization pays the cost. The Internet had no laws (at least, not until the Decency Act was signed) and no police. There are no real ways to hurt another person. Perhaps, under the circumstances, it is only natural for people to learn how to get along. Nevertheless, this does not stop people from arguing.

What we choose to believe is that, for the first time in history, unlimited numbers of people are able to communicate with ease, and we are finding it is in our nature to be communicative, helpful, curious, and considerate.

That is the Internet.

Using the Internet

Using the Internet means sitting at your computer screen and accessing information. You might be at work, at school, or at home, using virtually any type of computer (including a PC or a Macintosh).

A typical session might begin with you checking your e-mail. You can read your messages, reply to those that require a response and, perhaps, send a message of your own to another city.

You might then read a few articles in some of the worldwide discussion groups, jokes from one of the humor groups, or perhaps recipes for a dinner. Maybe you are following a discussion about philosophy or literature or aviation or—hopefully—electronics.

Understanding the Internet

In order to use the Internet well, you need to understand something of what it is and how it works. In this section, we will start with the idea of computer networks, introduce a few basic ideas and terms, and then move on to the Internet itself.

After explaining the basic services upon which the Internet depends, we will survey the large variety of resources you will be able to use. If you have been wondering what the Internet can do for you, this section will explain it all.

If you hang around the Internet at all, you will from time to time hear people talk about something mysterious called *TCP/IP.* At the end of this section we will discuss TCP/IP and show you what it really is.

The Network   The term *network* refers to two or more computers connected together. There are a number of reasons to connect computers into networks, but the two most important are:

1. To allow human beings to *communicate*

2. To *share* resources.

Once you start using the Internet, you can send messages to anyone else on the Internet. You can even send messages to people who use other networks that are connected to the Internet.

As for sharing, computer managers arrange networks so resources that are expensive or difficult to maintain can be used by anyone on the network. For instance, a manager might attach a costly printer to a network so that everybody who needs it can use the same printer. On the Internet, we share information resources rather than pieces of hardware.

A *local area network* (LAN) is a network in which the computers are connected directly, usually by some type of cable. When we connect LANs together, we call it a *wide area network* (WAN). Most wide area networks are connected via leased telephone lines, although a variety of other technologies, such as satellite links, are used as well. The wide area connections for most of the Internet travel over some telephone system or another. Indeed, the bottleneck in establishing Internet service within developing countries is usually due to the lack of a reliable telephone system.

Here is a typical example of a network. Imagine yourself sitting in a room full of computers in the Computing Facility at a major university. Your computer is connected by a LAN to all the other computers in the room and to the computers within administrators' private offices throughout the building. This arrangement is shown in Figure 17-1.

Computer room

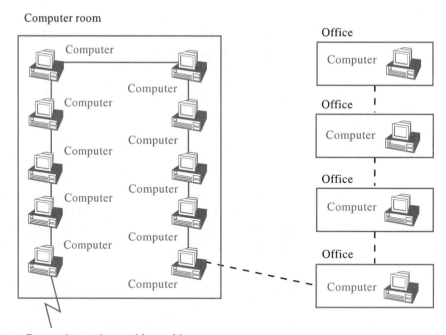

Connection to the outside world

FIGURE 17-1 Local Area Network (LAN)

There are a number of other LANs on the campus. For example, the Psychology department has its own network of computers, as does the Math department, the Computer Science department, and so on. Each of these LANs is connected to a high-speed link, called a *backbone,* to form a campus-wide WAN. This is illustrated in Figure 17-2.

Although we have used a university as an example, many types of organizations use similar arrangements: companies, governments, research facilities, other types of schools, and so on. If an organization is small, it may have only a single LAN. Large organizations may have multiple LANs connected into one or more complex WANs. Such organizations usually have a full-time staff to care for and feed the networks.

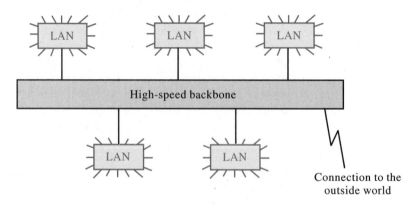

FIGURE 17-2 Wide Area Network (WAN)

The LANs are connected by special-purpose computers called *routers.* The job of a router is to provide a link from one network to another. Routers are used to connect LANs (to form WANs) and to connect WANs (to form even larger WANs). In other words, you can consider the computers within the Internet to be connected into LANs and WANs by a large number of routers.

Hosts, Nodes, and Terminals   There are two meanings for the word *host* that you should know about. First, within the Internet, each separate computer is called a *host.* For example, you might tell someone to find the information needed by connecting to a host in Switzerland. If your computer is connected to the Internet, then it too is a host even though you may not share any resources with the rest of the Internet.

You may also see a computer referred to as a *node.* If you draw a diagram of points and lines to represent the connections within a network, each computer will be a point and each connection will be a line. In the part of mathematics that deals with such diagrams, each such point is called a node. Network specialists have borrowed this term to refer to any computer that is connected to a network. Thus, *node* is a technical synonym for *host.*

The second meaning of the word host has to do with how certain computer systems are set up. In general, there are two ways you might use a computer. You might have it all to yourself, or you might share it. For example, when you use a PC or a Macintosh, you are the only person using the computer. *Single-user* computers, especially the more powerful ones, are often called *workstations.*

Some computers, however, are made to support more than one user at the same time. These *multi-user* systems are often referred to as *host computers.* A large mainframe computer, for example, can act as a host for hundreds of users at the same time. More commonly, a smaller computer, perhaps one that looks no larger than a PC, will act as a host for a small group of users.

When you have your own computer, you interact by using the keyboard, screen, and generally a mouse. These devices are part of the computer. With a multi-user computer, each person has his or her own *terminal* to use. A *terminal* has a keyboard, screen, perhaps a mouse, and not much more. All the terminals are connected to the host, which provides the computing power for everybody. This arrangement, called a *time-sharing system,* is shown in Figure 17-3.

Thus, there are two meanings for the word host. Within the Internet, each computer is called a host. Within a time-sharing system, the main computer that supports each user on a separate terminal is also called a host. If such a computer were connected to the Internet, it would be both a time-sharing host and an Internet host.

Client/Server Systems   One of the principal uses of a network is to allow the sharing of resources. Much of the time, this sharing is implemented by two separate programs, each running on different computers. One program, called the *server,* provides a particular resource. The other program, called the *client,* makes use of that resource.

For instance, say that you are working with a word processing program that is running on your own PC. You tell the program that you want to edit a particular file that is stored on another computer on your network. Your program will pass a message to that computer asking it to send the file. In this case, your word processing program is

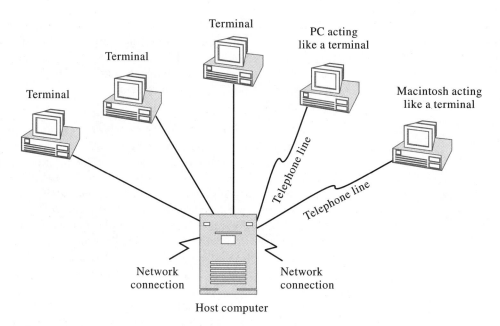

FIGURE 17-3 Time-sharing system

the client while the program that accepts the request and sends the file is the server. More precisely, it is a *file server*.

On local area networks, where the hardware is close by and visible, it is common for people to use the word *server* to refer to the actual computer that runs the server program.

On the Internet, hardware is normally not visible, and the terms *client* and *server* usually refer to the programs that ask for and provide services.

There is another important Internet service. Many Internet sites provide a service called a *Gopher* (discussed in more detail later). Briefly, a Gopher allows you to select items from menus. Each time you select an item, the Gopher performs the required task. For example, if the item describes a particular piece of information (such as *News of the Day*), the Gopher will retrieve this information and display it for you.

When you use a Gopher, two different programs are involved. There is one program that provides your *interface* (your connection between systems). This is the program that interprets your keystrokes, displays the menus, and generally makes sure your requests are carried out. It is called the *Gopher client*.

The other program is the one that supplies whatever the Gopher client has requested on your behalf. This program is called the *Gopher server*.

The beauty of this system is that the client and server programs do not necessarily run on the same computer. Indeed, more often than not, the client and server programs reside on different computers. For example, you may be sitting in front of a PC in California, using a Gopher to read the *News of the Day* at the National Security Agency in Virginia, three thousand miles away. In this case, the Gopher client is a program running on your PC, while the Gopher server is a program running on a supercomputer on the other side of the country.

All of the Internet services make use of this client/server relationship. Learning how to use the Internet actually means learning how to use each of the client programs. In order to use an Internet service, you must understand:

1. How to start the client program for that service
2. How to tell the client program which server to use
3. Which commands you must use with that type of client.

Your job is to start the client and tell it what to do. The client's job is to connect to the appropriate server and to make sure that your commands are carried out correctly.

Each type of Internet client has its own commands and conventions. The commands you use with a Gopher client are different from the commands you use with an *Archie* client (another Internet service, discussed later). Fortunately, there are only a handful of basic services you need to learn.

As you learn about Internet services, you will find that the client programs—such as *Gopher* or *Archie*—have many different commands. However, you will also find that you actually need to learn only a few of the most important commands.

The Internet client programs all come with their own built-in help functions. Thus, if a situation arises in which you require an esoteric command that you have not yet learned, you can use the built-in help to find what you need.

Four Important Services Provided by the Internet

The software that supports the Internet provides a large number of technical services upon which everything else is built. Because most of these services operate behind the scenes, you do not need to understand them.

However, there are four important Internet services we do need to talk about. You don't need to know the details, but you need to know they exist. In this section, we will describe these services. In the following section, we will discuss the many Internet resources available to you that depend on these services.

1. The *mail service:* This service reliably transmits and receives messages. Each message is sent from one computer to another on its way to a final destination. Behind the scenes, the mail service ensures that the message arrives intact at the correct address.

2. The *Telnet:* This service allows you to establish a terminal session with a remote computer. For example, you can use Telnet to connect to a host on the other side of the world. Once the connection is made, you can log in to that computer in the regular manner. (Of course, you will need a valid user account and password.) Telnet also allows two programs to work cooperatively by exchanging data over the Internet. The word *telnet* is often used as a verb.

3. The *File Transfer Protocol* (FTP): FTP allows you to transfer files from one computer to another. Most of the time you will use FTP to copy a file from a remote host to your computer. This process is called *downloading*. However, you can also transfer files from your computer to a remote host. This is called *uploading*. In addition, should you find it necessary, FTP will allow you to copy files from one remote host to another.

4. The *general client/server facility:* We discussed this earlier in the chapter. A client program can connect to another computer and ask for the help of a server program. The Gopher system we mentioned earlier works in just this way. Your Gopher client displays a menu for you. When you make a selection from the menu, the client connects to the appropriate server—no matter where it is on the Internet—and procures the service you requested.

Other Terms

You will encounter many different terms while using the Internet or reading literature about new devices. Some of these follow:

- *AGP* (Accelerated Graphics Port) is a special interface to speed up graphics-intensive applications like 3-D software. The AGP interface will let graphics cards access a PC's main memory directly to accelerate 3-D functions such as Z-buffering. Systems with AGP will be available in 1998.

- *DVD* (Digital Video Disc) is a storage format that can pack about 4.5 gigabytes of data on a disc that looks like one of today's CDs. That's over 7 times the capacity of current CD-ROMs. DVD drives will work like today's CD-ROM drives and be able to read the old CD format. They became available at the end of 1996 for about $350.

- *MMX* (MultiMedia extensors) are a set of new instructions that Intel has added to its Pentium CPUs. These instructions speed up functions like video playback, videoconferencing, and more. High-speed Pentiums, such as the P-200, incorporated MMX first, in late 1996. Other chips received the MMX boost in 1997.

- *SDRAM* (Synchronous Dynamic Random Access Memory) is the next step in PC memory. It's faster than the currently popular *EDO* (Extended Data Out) DRAM, but it's also more expensive. Vendors like Dell and Gateway 2000 have adopted SDRAM in their newest high-end Pentium PCs.

- *USB* (Universal Serial Bus) is now being incorporated into many PCs. This means you will be able to connect multiple peripherals—as many as 126—to your system through one port. Data transfer rates also improve. Intel shipped motherboards with USB at the end of 1996.

- *Spams* (mass mailings of e-mail sales promotions) are a growing problem. Both America Online and CompuServe have taken, or are considering taking, legal action against companies that have been mass mailing to their subscribers. The *Telephone Consumer Protection Act* (TCPA) of 1991 may be their weapon. Among other restrictions, the TCPA prohibits marketers from sending faxes to customers to pitch products.

A Quick Tour of the Internet Resources

Having covered the basic framework of the Internet, let's take a look at all it has to offer. There are many Internet resources, and new ones are added whenever some clever person can figure out a new way to use the network.

In this section, we will examine all the important Internet resources. In later sections, we will discuss each resource in some detail, showing you how it works and the best way to use it. Many of these resources demand an entire chapter to themselves, which is beyond the scope of this text.

The purpose of this section is to give you an idea of exactly what the Internet has to offer you. You will see that each one depends on one or more of these resources. Table 17-1 contains a summary of each of the Internet resources.

TABLE 17-1   Resources and Descriptions

Resource	Description
Electronic Mail	Send and receive messages
Remote Login	Connect to and use a remote host
Finger Service	Show information about a user
Usenet	Vast system of discussion groups
Anonymous FTP	Public access to data archives
Archie Servers	Search Anonymous FTP archives
Talk Facility	Converse with one person
Internet Relay Chat	Converse with a group of people
Gopher	Menu-based information
Veronica, Jughead	Search for Gopher menu items
White Pages Directories	Search for a user's address
Wais Servers	Search indexed databases
World-Wide Web	Access hypertext information
Mailing Lists	Information distributed by mail
Electronic Magazines	Magazines, journals, newsletters
Internet BBSs	Share information and messages

Electronic Mail   As an Internet user, you can send and receive messages from anyone else on the Internet. Moreover, you can do the same for other mail systems—such as CompuServe, America Online, Prodigy, or MCI Mail—that have connections with the Internet.

However, mail does not mean simply personal messages. Anything that can be stored in a text file can be mailed: computer (source) programs, announcements, electronic magazines, and so on.

When you need to send a binary file that cannot be represented as regular text, such as compiled computer programs or graphical pictures, there are facilities for encoding the data into text. At the other end the recipient simply decodes the textual data into its original format. You can mail virtually any type of data to anybody, anywhere in the world. The Internet mail system is the backbone (and original motivation) for the network itself.

Remote Login   As we described in the previous section, you can *telnet* to a remote computer anywhere on the Internet. Once you have established a connection, you can *log in* (as long as you have a valid account on that computer).

The name by which an account is known is called a *userid* (pronounced "user-eye-dee"). The secret code you must enter to prove that it is really you is called your *password.* As long as you have a valid userid and password, you can log in to any computer on the Internet.

As a public service, many Internet systems are set up to allow anybody to log in using a special guest account. For example, in the United States, there is a system that will display weather reports from around the country. Anyone can log in to this system and check out the weather.

Finger Service   Most Internet computers offer a facility that allows you to ask for information about a particular user. This service is known by the descriptive name of *finger.* As you will see, people on the Internet are often known by their userid (user-ID). You can use finger to find out the name of the person behind the userid.

Depending on how the finger service has been set up on the computer you contact, you may be able to find out other information about the person such as a phone number, office address, and so on. In addition, some finger systems will tell you when the person last logged in and if they have unread mail. This can come in handy when you need to check if someone has received an important message.

There is also a way for you to customize part of what people see when they finger your userid. You can specify certain information that you want displayed. A professor might specify his office hours. Somebody giving a party might give directions to her house. You can display this information whenever you want just by fingering that person's userid.

You can also finger a computer rather than a userid. In this case, the computer will respond by showing you a summary of all the userids currently logged in.

Finally, some systems use finger to support a public request for certain specific information. For example, there is a particular userid and computer at the University of Washington at Seattle that you can finger to display information about recent earthquakes.

Usenet   *Usenet* (a contraction of *User's Network*) is one of the main reasons people use the Internet. Usenet itself is not an actual network. It is a system of discussion groups in which individual articles are distributed throughout the world. Usenet has literally thousands of discussion groups, so there is always something for everyone.

At each Internet site the system administrator decides whether or not to carry the Usenet discussion groups; therefore, Usenet is not available everywhere. Moreover, even sites that do provide this resource will not carry every conceivable discussion group.

Anonymous FTP   As mentioned previously, the FTP service allows you to copy files from one computer to another. *Anonymous FTP* is a system by which an organization makes certain files available to the general public. You can access such a computer by using a userid of anonymous. No special password is required.

Anonymous FTP is one of the most important Internet services. Virtually every possible type of data is stored somewhere, on some computer, and it is all available to you for free. For example, many of the programs used on the Internet are created and maintained by individuals who then distribute the programs worldwide via Anonymous FTP. You can also find electronic magazines, archives of Usenet discussion groups, technical documentation, and much, much more. As a full-fledged Internet user, you will come to depend heavily on Anonymous FTP. Anonymous FTP is important for three reasons:

1. Anonymous FTP allows you to download virtually any type of information. Until you are an experienced Internet user, it is difficult to appreciate how important this is. It provides access to the largest library of information ever accumulated. Moreover, it is a library that is always growing, never closes, covers every conceivable topic and, best of all, is free.

2. Anonymous FTP is the principal way in which software is distributed on the Internet. The reason that the Internet can exist at all is that people use programs that provide standardized services using standardized protocols. Many of these programs are distributed via Anonymous FTP and, hence, are available to anyone who wants to set up an Internet host. For example, in order for you to access the *Usenet* discussion groups, your system manager must have installed the Usenet software on your computer. Where does your system manager acquire this software? Anonymous FTP.

3. The final reason Anonymous FTP is so important is that it is used to archive and disseminate the technical information that defines the Internet itself. As was explained earlier, the Internet is based on a large number of protocols and conventions. Each such protocol is explained in a technical publication called a *Request for Comment* (RFC). Think of an RFC as a technical explanation of how something is *supposed* to work. RFCs are also used to offer generally useful information on a specific topic. Each RFC is given a number and is made freely available to anyone who wants to read it. For example, RFC #1325 is a long list of answers to questions that are commonly asked by new Internet users.

Understanding FTP   The term *FTP* comes from *File Transfer Protocol,* the underlying set of specifications that support Internet file transfer. What we mean when we refer to FTP is more than a dry set of specifications. FTP is a service that allows us to copy a file from any Internet host to any other Internet host. As such, FTP provides a large part of the spiritual glue that holds the Internet together.

Like most Internet services, FTP uses a client/server system. You use a client program, named *ftp,* to connect to a server program on the remote computer. Conceptually, the idea is simple. Using your client program, you issue commands that are sent to the server. The server responds by carrying out whatever commands you sent. For example, you might send a command that requests the server to send you a copy of a particular file. The server responds by sending the file. Your client program receives the file on your behalf and stores it in your directory. (We discuss files and directories later in the section.)

When you copy files from a remote computer to your own, we say that you are *downloading* files. When you copy files from your computer to the remote one, you are *uploading* files. Your computer is called the *local host* and the other computer is called a *remote host.* The FTP client program allows you to upload and download files to and from a remote host.

Archie Servers   The name *Archie* was coined from the idea of an *archive server.* In deference to the well-known propensity of computer people to engage in whimsical personification, we usually refer to Archie as if it were human.

There are thousands of Anonymous FTP servers around the world offering a vast number of files. The role of the *Archie servers* is to make the whole system manageable by helping you find what you need. For example, suppose you want a particular file, a program you have heard about. You can use an Archie server to tell you what Anonymous FTP sites store that file. Once you know the names of the sites, it is a simple matter to use FTP and download the file.

If you consider the world of Anonymous FTP as an enormous, worldwide library that is constantly changing, you can think of the Archie servers as the catalog. Indeed, without the Archie servers, most of the Anonymous FTP resources would be unreachable.

Talk Facility   The *Talk facility* establishes a connection between your computer and someone else's. You can then use this connection to type messages back and forth. The great thing about the Internet Talk facility is that it makes it possible to hold a conversation with someone no matter how far away they are. The other person sees what you type as you type it, and you can both type at the same time without your messages getting mixed up.

Internet Relay Chat   *Internet Relay Chat* (IRC) is like a Talk facility for more than one person at the same time. As you might imagine, IRC is used heavily and offers a lot more than simple conversation.

You can take part in public conversations with a large number of people. These conversations are loosely organized around various topics or ideas. Alternatively, you can use IRC to arrange a private conversation with people of your own choosing, much like a telephone conference call.

Gopher   *Gophers* provide a series of menus from which you can access virtually any type of textual information, including that provided by other Internet resources. There are many Gopher systems around the Internet, each one administered locally. Each Gopher contains whatever information the local Gopher people have decided to share.

While some Gophers are stand-alone systems, most Gophers are set up to connect to other Gophers. Assume you are using a Gopher in California. With a simple menu selection, you can connect to another Gopher in Africa or in South America. What makes Gophers so powerful is that, no matter what Gopher you are using and no matter what information you are using, the interface is always the same simple menu system.

Veronica and Jughead   Nobody really knows how many Gophers there are in the world. Suffice it to say there are a great many, all of which have their own series of menu items offering information and services.

Veronica is a tool that keeps track of many Gopher menus from around the world. You can use Veronica to perform a search and look for all the menu items that contain certain keywords (whatever you specify). *Jughead* does the same thing for a specific group of Gopher menus.

The result of a Veronica or Jughead search is a custom menu, containing whatever items were found. Selecting any item from this menu automatically connects you to the appropriate Gopher, wherever it may happen to be. In fact, unless you specifically ask, you will not even know what computer you are using or what country it is in.

Wais Servers   The name *Wais* stands for *Wide Area Information Service* and is pronounced "Wayz". Wais servers provide another method of finding information that is spread around the Internet. Wais is able to access any of a large number of databases. To start, you tell Wais which databases you want to search. Next, you specify a list of one or more keywords to search for. Wais will search every word in every article in all the databases that you specified.

The result of a Wais search is a list of the articles, culled from the various databases, that are likely to be of interest to you. Wais presents them as a menu, with the most relevant items first. From this list, you can ask Wais to display articles you are interested in.

World Wide Web   The *World Wide Web* (*the Web*) is a *hypertext-based* tool that allows you to retrieve and display data based on keyword searches. The Web is probably the most-used Internet access system today. What makes the Web so powerful is the idea of *hypertext: data that contains links to other data.*

For example, suppose as you are reading some information, you notice that certain words and phrases are marked in a special way. You can tell the Web to jump to one of those words. It will follow the *link,* find the relevant information, and display it. In this way, you can jump from place to place, following logical links in the data.

White Pages Directories   Within the often overwhelming world of the Internet, nothing is more important than a person's electronic address. Once you know someone's address, you can send mail, have a Talk conversation, or even find out more about the person by using finger.

When you want to contact someone but you don't know his or her address, you use one of the *White Pages Directories* (special-purpose servers). The very name itself reminds us of the standard telephone book.

However, the electronic counterpart is actually much different, principally because there is no single Internet directory. After all, nobody actually runs the Internet. With no single person or organization in charge, it is not surprising that there is no central source of names and addresses.

There are a number of different White Pages Directories that you can search for a name. For example, if you lose that note on which you carefully wrote someone's name during your last vacation, there is a chance that a White Pages Directory may help you track down your quarry's electronic address if the person is an Internet user.

Electronic Magazines   The Internet is host to a variety of magazines that are published electronically. The articles are stored as text files that are accessible to one and all. Some *electronic magazines* (in computereze, called *'zines*) are scholarly journals of interest mainly to specialists. Other magazines are of general interest.

There are two ways in which electronic magazines are distributed. Some maintain a mailing list. When a new issue comes out, it is sent to you as a mail message. Other magazines are stored in well-known Anonymous FTP sites. You can download copies, including back issues, whenever you want.

Aside from the ease of distribution, electronic magazines have two important advantages over their conventional paper and ink counterparts.

1. Electronic magazines do not contain advertisements showing people who are richer and better looking than you, having more fun than you will ever have.

2. Electronic magazines do not contain irritating subscription cards that fall out when you turn the page.

Mailing Lists   A *mailing list* is an organized system by which a group of people are sent messages pertaining to a particular topic. The messages can be articles, comments, or whatever is appropriate to that topic.

All mailing lists, and there are thousands of them, have someone in charge. You can *subscribe* or *unsubscribe* to a list by sending a message to the appropriate address. Many mailing lists are *moderated,* which means that someone decides which material will be accepted. Other lists will accept and send out messages from anybody.

Internet BBSs   A *Bulletin Board System* (BBS) is a repository for messages and files, often devoted to a particular topic. To use a BBS, you connect to it and select items from a series of menus.

Typically, a BBS will be maintained by a single person or by an organization. There are countless BBS systems in the world, most of them reachable by telephone. The Internet has many BBSs that you can reach by the more refined method of Telnet.

What is TCP/IP?

In order to complete this section and our overview of the Internet, we need to talk about TCP/IP. As you know from our previous discussion, the Internet is built on a collection of networks that cover the world. These networks contain many different types of computers, and somehow, something must hold the whole thing together. That something is *TCP/IP.*

To ensure that different types of computers can work together, programmers write their programs using standard protocols. A *protocol* is a set of rules that describes, in technical terms, how something should be done. For example, there is a protocol that describes exactly what format should be used for a mail message. All Internet mail programs follow this protocol when they prepare a message for delivery.

TCP/IP is the common name for a collection of over 100 protocols that are used to connect computers and networks. We have already mentioned two of the TCP/IP protocols, Telnet and FTP *(file transfer protocol).*

The actual name *TCP/IP* comes from the two most important protocols: *TCP (Transmission Control Protocol)* and *IP (Internet Protocol).* Although you don't need to know the details, it is useful to have an appreciation for what these protocols are and how they hold the Internet together.

Within the Internet, information is not transmitted as a constant stream from host to host. Rather, data is broken into small packages called *packets*. For example, you send a long mail message to a friend on the other side of the country. TCP will divide the message into packets. Each packet is marked with a sequence number and with the address of the recipient. In addition, TCP inserts some error control information.

The packets are then sent over the network, where it is the job of IP to transport them to the remote host. At the other end, TCP receives the packets and checks for errors. If an error has occurred, TCP can ask for that particular packet to be resent. Once all the packets are received correctly, TCP will use the sequence numbers to reconstruct the original message. In other words, the job of IP is to get the raw data (the packets) from one place to another. The job of TCP is to manage the flow and ensure that the data is correct.

Breaking data into packets has several important benefits. First, it allows the Internet to use the same communication lines for many different users at the same time. Since the packets do not have to travel together, a communication line can carry all types of packets as they make their way from place to place. As a comparison, think of a highway in which separate cars all travel on a common road even though they are headed for different places.

As packets travel, they are routed from host to host until they reach their ultimate destination. This means the Internet has a lot of flexibility. If a particular connection is disrupted, the computers that control the flow of data can usually find an alternate route. In fact it is possible that, within a single data transfer, various packets might follow different routes.

This also means that as conditions change the network can use the best connection available at the time. When a particular part of the network becomes overloaded, packets can be routed over other, less busy lines.

Another advantage of using packets is faster data transmission. When something goes wrong only a single packet may need to be retransmitted rather than the entire message. This greatly increases the overall speed of the Internet.

All of this flexibility makes for high reliability. One way or another, TCP/IP makes sure the data gets through. The Internet runs so well that it may take only a few seconds to send a file from one host to another, even though they are thousands of miles apart and all the packets must pass through multiple computers.

Therefore, there are several answers to the question, "What is TCP/IP?" The technical answer is that TCP/IP is a large family of protocols that are used to organize computers and communication devices into a network. The two most important protocols are TCP and IP. IP *(Internet protocol)* transmits the data from place to place, while TCP *(transmission control protocol)* makes sure that it all works correctly.

The best answer, though, is that the Internet depends on thousands of networks and millions of computers, and TCP/IP is the glue that holds it all together.

17-3   CONNECTING TO THE INTERNET _____

It is stimulating to read about the wonderful things you can do once you have access to the Internet. But how do you get that access? In this section, we will show you how to connect to the Internet.

The Internet Connection

To have access to the Internet means you are using a computer that is a part of a network connected to the Internet. In practical terms, this means you can use the Internet resources we described in the previous section. When you are using such a computer, you are on the Internet. In common usage, the Internet is often referred to as *the Net*.

It is important to realize that a great many people who do not have full Internet access are able to communicate with Internet users. For example, many people use some type of non-Internet mail service that can exchange messages with the Internet. It is also common to find people who are not on the Internet who have access to the Usenet discussion groups.

What you should understand is that while mail and Usenet are important, they are not everything. You want to have access to all the Internet resources, including Telnet and FTP. In other words, you want to be able to use a computer that has a full TCP/IP connection with the Internet.

Types of Internet Connections   Before we discuss the various ways in which you might obtain access to the Internet, we need to talk about the two different types of Internet connections.

First, you might use a computer that is directly connected to the Internet. You might be using a PC or Macintosh or workstation that is part of a network connected to the Internet. In such a case, your computer will be a full-fledged Internet host, with its own electronic address.

The other way to connect to the Internet is by using a terminal that is connected to an Internet host. In this case, the terminal itself is not on the Internet. You simply use the terminal to access a computer that is on the Internet.

For example, imagine that you are being taken on a tour of a building occupied by many Internet users. First, you are taken to the Computer Room, in which you see 40 different PCs, connected in a network. You are told that this network is connected to the Internet, so that all the PC users can access the Internet directly. Each PC user in the Computer Room has his or her own Internet host. Each PC has its own Internet address and is a self-sufficient, stand-alone system.

Next, you are taken to another room, the Terminal Room. Here you are shown 40 terminals, all of which are connected to a time-sharing computer in a closet at the back of the room. This computer is also connected to the Internet. In the Terminal Room, life is different. Each person is actually using the same computer, the one in the closet. These users access the Internet by logging in to the time-sharing system that provides Internet access. Thus, they all share one computer that has a single Internet address.

Telephone Connections   In the previous section, we explained that you can use either a computer or a terminal to access the Internet. The example we used described two rooms. One contained 40 PCs connected into a network. The other contained 40 terminals connected to a time-sharing host. What they had in common was that all the devices—computers and terminals alike—were connected directly using some type of cable. This type of arrangement is called a *hardwired connection*.

The main advantage of a hardwired connection is its permanence. All you have to do is turn on your PC or terminal, and your connection is ready to use. The disadvantage, of course, is the lack of flexibility. If you want to move the computer or terminal to another location, you must deal with the cables.

A more flexible system is one in which the computer or terminal uses a *dial-up connection* over a telephone line. In such cases, you can work anywhere you want, as long as you have access to a phone line.

To use a dial-up connection, you need a hardware device to convert computer signals to telephone signals and back again. In technical terms, the computer signal is *digital* and the telephone signal is *analog*. A device that converts from digital to analog is called a *modulator*. A device that converts from analog to digital is called a *demodulator*. When we connect computers over a phone line, we must be able to send data in both directions. For this, we use a *modem*, a contraction of modulator-demodulator. Various types of modems are discussed in the next section.

A modem can be either external (a separate box that attaches to your computer, usually via a "serial" cable), or internal (an adapter board that is installed in the computer). Figure 17-4 compares a hardwired connection to a dial-up connection. Notice that a dial-up connection requires a modem at each end of the telephone line. Once a dial-up connection is established with the remote host, the communications program will emulate a terminal.

There are many different types of terminals, and most programs will give you a choice as to which one you want to emulate. The standard choice is the *VT-100*. This is a terminal that used to be made by the *Digital Equipment Corporation*. It has been a long time since Digital actually made a VT-100, but nevertheless, it has become a standard. Many host computers assume that, by default, all dial-up connections use some variation of a VT-100, such as *VT-102*.

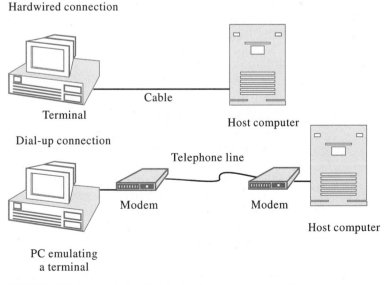

FIGURE 17-4   Hardwired/dial-up systems compared

If you want to use the Internet from your home, you need four things: a computer, a modem, a communications program, and the telephone number of a remote host to which you can connect.

There are two things that will make your work with a remote host a lot easier:

1. Choose a communications program capable of storing the information that is displayed on your screen. As new data is displayed, the previous data will scroll off the top of the screen. It is a great convenience to be able to ask your communications program to recall previous lines of output that otherwise would be gone for good. Sometimes such a facility is called a *scroll buffer.*

2. Use a computer system that will run more than one program at the same time. For example, with a PC, running your communications program under Microsoft Windows or OS/2 allows you to keep an Internet session in its own window while you do other work. Moreover, such systems enable you to cut and paste from one part of the screen to another, which can be a great advantage. Let's say that you are reading a Usenet discussion group in which someone has mentioned the address of a computer that has a great new service. You would like to enter a command to connect to that computer to try out the service. Normally, you would type the command, including the computer's address. However, Internet addresses can be long and must be typed exactly right. Instead, you can *cut* the address from one part of your screen and *paste* it back as part of the command. That way, you do not have to worry about making a spelling mistake. Once you become a veteran Internet user, you will find that such situations arise frequently.

Connecting a Computer over a Telephone Line   Most of the time, using a computer that emulates a terminal is the best way to connect to an Internet host over a telephone line. But remember, your computer is merely acting like a terminal. It is not on the Internet itself. For example, you might work for a small company that cannot afford a network with a dedicated full-time Internet connection. You may still want your own Internet host. In such cases, there is a way to establish a full Internet connection over a telephone line.

To do so, you must first arrange for some other Internet host to act as your connection point. You then install a set of programs on your computer called *Point-to-Point Protocol* (PPP). Once a telephone line connection is made between the two computers via modems, PPP will endow your computer with TCP/IP capabilities. This enables your computer to be a real Internet host with its own official electronic address. You will also hear about an older system called *Serial Line Internet Protocol* (SLIP). Both PPP and SLIP will work, but if you have a choice, use PPP.

When you use these systems, there are two ways to set up the telephone line. First, you might use a regular dial-up line with a standard modem, just as we described in the previous section. Although your computer will be considered an official Internet host, it will not be connected at all times. When such a system is set up, you must arrange for the computer that acts as your connecting point to save mail messages that arrive when you are not connected.

The alternative is to use a *dedicated* phone line. As the name implies, a dedicated phone line is always connected. This, of course, costs a lot more than a regular line and may not be feasible for an individual user. However, for a small company, a dedicated phone line using PPP can be a relatively economical way to establish Internet access. This connection can then provide Internet access for the rest of the computers in the company. However, if there is to be a fair amount of Internet traffic, you are probably better off with a faster connection.

Later in the section, we will explain how certain companies are in the business of providing commercial access to the Internet. Many such companies offer not only a host computer to which you can connect with a PC, but PPP or SLIP service for customers who require a full Internet connection.

If you want to access the Internet from your home, use a communications program and emulate a terminal. Establishing a PPP or SLIP connection sounds nice, but it is expensive and requires a fair bit of technical knowledge. When you use a terminal emulator to connect to a remote host, someone else is taking care of all the details for you. When you use PPP or SLIP, you will have to manage the system yourself.

Using the Internet for Free

If you do not already have Internet access, there are several ways to get it. Let's start with those that are free.

Virtually all colleges and universities and many companies have access to the Internet. To get free access, all you have to do is convince somebody to give you an account on one of the computers. If you belong to such an organization, find out who is in charge of computer accounts. Then find out what criteria one has to meet to get an account. In some colleges and universities any student who asks will be given a computer account, but you have to ask. In other schools you must be taking a particular class or working on a particular project. Sometimes if you work for a professor you can get an account. Similarly, some companies give Internet access to any employee. Other companies give it only to qualified employees, such as researchers.

If your school does not have access to the Internet, but there is a college or university nearby, you may be able to get someone to give you free access to one of the university computers.

Community Computing and the Freenets   An alternative to using a local university is to look for a free community computing service. Such services offer computer-mediated communication facilities to anyone for no charge. All you need is a computer with a modem. Using your modem, you connect to a host computer over a telephone line. The best known community computing services are the *Freenets* (sometimes spelled *Free-Nets*). There are a number of Freenets in the United States and Canada that offer all sorts of information services as a general service to the community. The original system is the Cleveland Freenet. Be aware that not every such system has the word Freenet in its name.

The Freenet system has given rise to an organization called the *National Public Telecomputing Network* (NPTN). The NPTN is a nonprofit organization dedicated to providing free public access to a large variety of information.

Community computing services usually have an electronic mail system that can communicate with the Internet. Most of them also carry the Usenet discussion groups. Moreover, they have a number of unique information offerings, including special programs for schools. Perhaps their best feature is the price: they are free. However, there are a number of significant disadvantages.

First, these systems do not have full Internet access. A Freenet user usually cannot connect to a remote Internet host or use a Gopher.

Second, unlike the Internet, community computing services have Someone-In-Charge, just like the commercial services such as CompuServe, America Online, or Prodigy. This means that there are rules and there is some censorship (although it is usually not heavy-handed).

The Freenets are typically connected to a local college or university computer system. This is how they exchange mail messages with the Internet and how they access Usenet. Although their users cannot connect to remote Internet systems, Internet users can use a Freenet by using Telnet.

Public Internet Access   When all else fails, you can always get Internet access by paying for it. There are many companies or organizations that will provide such access. Typically, you must pay a certain amount per month, along with a fee for connect time, that is, the actual time you are using the service. Here are some suggestions on how to find such a service and how to make a good choice.

First, try a local college or university. Some colleges and universities offer Internet access to the general public. Even if you have to pay, the service will probably cost a lot less than from a commercial provider. If this is not available, you may find that some companies offer services to a large area—for example, the entire United States. Other companies are regional in scope.

Start by looking for a company that offers a phone number that is a local call for you. You can expect to spend many hours on the Internet, so it is best not to have to pay long-distance charges. You will find that some companies have toll-free numbers that you can use from anywhere in the country. However, if you compare closely, you will find there is a surcharge for using such a number. This surcharge may amount to more than paying for the long distance yourself.

After you have found a few likely Internet providers, call them and ask:

1. What are your rates?
2. Do you offer full Internet service?
3. Do you offer Telnet and FTP?

Do not settle for only electronic mail and Usenet. You want the whole set of resources that we described earlier.

You should plan on spending many hours a month on the Net. Therefore, look for a company that offers as many hours of connect time as you need for a flat fee. If you are not yet an experienced Internet user, the Internet will consume a lot of your time, and paying an hourly rate can get expensive. You may find that paying a long distance charge to call a flat-fee service is cheaper than using a local service that charges by the hour. This is especially true if you call at night or on the weekend when the rates are lower.

If you have access to Usenet and to electronic mail, there is a discussion group specifically to help people find Internet access in their area. This is the place to send a query asking if someone can tell you where to find public Internet access in the (your hometown) area. If you do not have access to Usenet, perhaps you can find a friend to ask the question for you.

Finally, there is a large, up-to-date list of companies that provide public dial-up Internet access. This list is posted regularly to several Usenet groups and is also available by Anonymous FTP and by mail. If you are just starting, you might prevail upon an experienced Internet person to help you download this list.

17-4   INTEGRATED SERVICES DIGITAL NETWORK (ISDN)

As new as ISDN (*Integrated Services Digital Network*) appears to be, it's really more than 10 years old, about the same age as cellular phones, though most people have only been able to buy it since about 1995. Before that, when the Internet wasn't such a general fact of life, there was little reason to buy so much bandwidth. Now, in the days of big commercial Web sites and multimegabit downloads, we need all the bandwidth we can get. At this writing, that means ISDN.

However, it can take up to six months between the time you request an ISDN line and the time you actually get the line to work. (This is not atypical; a well-known anchorman at CNN had to wait four months.) It can take nearly two months to confirm whether ISDN is available even in a densely urban neighborhood. Another two months can be lost in setting an installation date and getting the telephone company to show up at the appointed time. And finally, it can take another month or so to fix a provisioning problem that keeps the line from working correctly.

The point of this discussion is not to belabor the many possible problems, but to explain why ISDN is great—and why it might take many more years to catch on.

The why-it's-great part is easy. ISDN is fast, very fast. It's one thing just to know that an ISDN adapter running at 128 kbps is almost five times faster than a 28.8-kbps modem. But actually seeing what a difference that makes is a great experience. Graphics on Web pages leap up to greet you. FTP sessions fly. Simultaneous sessions aren't painful, and multitasking online becomes a reality. And there's none of that annoying modem whine. Although ISDN is great, you may not want it, or if you want it, you may not get it.

Ordering ISDN is not something you want to do if you don't have enough time and patience to follow through. *One of the many ISDN books on the market takes 80 pages to work through just the process of ordering service.* If you want ISDN, you're going to have to learn some *telco-speak* (telephone language). You will have to learn about ordering codes, terminal adapters, and many other such things. Although the industry is trying to simplify matters, it has a very long way to go.

Consider how unpopular ordinary modem communications would have become if you had to go through ISDN-style problems. You would have needed a special modem line, different from the one you used for voice calls. You would not have been able to plug in any handy phone to make sure the line was working. You would have to tell the phone company what brand modem you wanted to use—just to be sure it was compatible—and be ready to learn the modem you spent many dollars on might not work.

Every one of those restrictions is true about ISDN. You cannot plug a regular phone into an ISDN wall jack. Not every ISDN adapter works with every ISDN line. Sales representatives are severely undereducated about the system.

And there's more. For ISDN to work properly, you must be located less than three miles from your telephone company's switch, and that switch has to have the ISDN-capable software installed. In cities that's not much of a problem, but in the suburbs or rural areas the three-mile limit presents a large problem. One of the major applications of ISDN is *telecommuting*—dialing into corporate networks (*intranets*) from home offices. Home offices tend to be in the suburbs. Pacific Bell, for example, has found that some 70 percent of its requests for ISDN service comes from the suburbs.

Typically, the situation is that the most affluent potential customers of ISDN will be beyond the reach of a service they need. If you're a big believer in ISDN, or if you just need its bandwidth, this is a serious barrier.

Conspiracy theorists believe that the reason ISDN remains such a hassle a full decade after it was first rolled out is because telephone companies really aren't interested in ISDN. They would rather be selling access over coaxial cable through their cable TV franchises.

Like most conspiracy theories, there's some truth to it. *Telcos* (telephone companies) have wanted to carry cable TV for years, but were prohibited by law from doing so until the Telecommunications Act of 1996 was signed into law. Cable TV has lots more bandwidth than normal twisted-pair phone wire, so telcos may well have been reluctant to sink so much money into upgrading their phone networks for ISDN when they could do the same thing over cable TV.

Also, another new technology is just around the corner. *Asymmetrical Digital Subscriber Line* (ADSL) will supposedly allow speeds up to 10 times faster than ISDN over normal phone wiring. Between cable TV and ADSL, it might be smart for telcos to not be especially aggressive about rolling out ISDN.

Ironically, ADSL will have the same three-mile limit as ISDN. And it will take time for telephone companies to buy cable companies, let alone build competing ones or install the kind of infrastructure that two-way switched data transmission requires.

Even with the problems it has, ISDN is just about the only mid-bandwidth service that the telcos have to sell right now. It's either ISDN or V.34—128 kbps or 28.8 kbps.

Before ISDN becomes real to enough people, telcos need to make it more friendly. Ordering and using ISDN has to become as simple as it is with an analog line. And while charging a premium price for high throughput is right and appropriate, the current difference between analog and ISDN is much too great.

The net you see through an ISDN adapter is fundamentally different from the one you see through even a 28.8 modem; it is faster, more interactive, and more involving. It's worth the wait; it's worth the aggravation; and it's worth the money. As with all technology, ISDN will be cheaper tomorrow. But the time to make the move is now.

ISDN Terminology   For the typical PC user, one of the biggest stumbling blocks to ISDN technology is the special terms that come with the territory. The following is a list of the most common terms you'll see, along with their meanings:

- *Analog port:* The standard RJ-11 jack that analog devices such as phones, fax machines, and modems plug into.

- *BONDing:* From *Bandwidth on Demand,* this describes a proprietary protocol implemented in the hardware that is used to combine both *B* channels to form a single 128-kbps connection. ISDN modems at both ends of the connection must support BONDing to achieve the combined throughput.

- *BRI:* The *Basic Rate Interface* is the most common ISDN service for individuals and small offices. It comprises two *B* channels, each of which can carry up to 64 kbps of voice or data, and one *D* channel that carries 16 kbps of signaling information.

- *Dynamic Bandwidth Allocation:* A feature that drops one of two *B* channels being used for a data connection and automatically gives that channel to an incoming voice call or an outgoing call without disrupting the data connection. When the call is completed, the second *B* channel is then reassigned to the data connection.

- *ISDN:* The *Integrated Services Digital Network* uses two or more 64-kbps *B* channels to carry data, digitized voice, or video across digital phone lines at high speeds.

- *Multilink PPP: Multilink Point-to-Point Protocol* is a multiconnection protocol that lets you bind the two *B* channels together for a combined throughput of 128 kbps on a single call. Unlike BONDing, Multilink PPP is implemented in software and requires only that your communications software support the protocol.

- *NT 1:* The *Network Terminator* is a device that stands between the ISDN network and your PC. Many ISDN modems include an integrated NT 1, while others require an external unit. If you need to hook additional devices to your ISDN line—such as multiple PCs or a digital phone—then purchasing a separate NT 1 device can make it easier.

- *PPP:* The *Point-to-Point Protocol* governs the way computers connect to networks over a direct link such as a phone line. PPP is the standard for communications with your Internet service provider, as well as for many remote networking clients.

- *PRI:* The *Primary Rate Interface* consists of 23 *B* channels and a single *D* channel. PRI is geared for large companies that may have multiple users calling a network server or phone system at one time.

- *SPID:* The *Service Profile ID* is a unique number assigned by the phone company to an ISDN modem. Depending on the type of service and your provider, a separate SPID may also be assigned to each *B* channel.

- *Switch:* This is the device used by your local ISDN service provider to connect callers to circuits. There are four common switches in use in North America today: *AT&T* 5ESS, *Northern Telecom DMS 100, National ISDN-1,* and *Siemens* WSD.

- *V.120:* A protocol that allows devices that communicate at speeds of less than 64 kbps to adapt their rates to fill an entire 64 kbps *B* channel. Many PC communications products require support for the V.120 protocol in order to connect.

17-5 TYPES OF MODEMS

Modem manufacturers have a tendency to exaggerate the speeds of their products. For example, it is commonly claimed that a 14.400 bps modem, using *data compression* (designated as *bis*), can achieve an overall performance of 57,600 bps, (four times the nominal speed). Similarly, it is said that a 9,600 bps modem with data compression will perform at 38,400 bps. These numbers are highly inflated theoretical maximums that are completely divorced from reality.

When you buy a modem for your own computer, you will encounter a bewildering number of technical terms. Both analog and digital modems are available today. However, all you have to do is remember the following three guidelines:

1. *The speed of a modem is measured in bits per second (bps).* Fast analog modems transmit at speeds up to 33,600 bps (often abbreviated as 33.6K). Slower modems transmit at 9,600 bps (9.6K), 14,400 bps (14.4K), or 28,800 bps (28.8K).

2. *Buy no less than a 28.8K modem.* If you buy anything slower, you will be disappointed. No matter what anyone tells you, do not buy one of the very slow 2.4K modems (the standard just a few years ago).

3. *You will see a number of strange terms that describe various functions:* V.32bis, V.34I, V.42, V.42bis, x2, ISDN, and so on. *Don't worry about them.* All modern 28.8K and 33.6K modems come with the right stuff. Just buy no less than a 28.8K modem and you will be okay.

Analog Modems   The standard analog modem installed in new PCs has steadily increased in speed and capability. Each upgrade provided about a four-fold increase. Just a few short years ago, the 2,400-bps modem was standard. Next came the 9,600-bps; then the 28,800-bps modem. Now, 33,600-bps has become standard. These later upgrades, in general, provide data communications at 33.6K, with fax capabilities of 14.4K.

Newer modems, such as U.S. Robotics *x2*, are appearing on the shelves. These devices claim communications between the customer and the ISP at speeds up to 33.6K, depending on the modem and line conditions, with download speeds of up to 56 kbps (using V.42bis compression) over ordinary telephone lines. However, the actual performance will be somewhat less (about 53 kbps) until the FCC changes some rules for digitally-connected server equipment. Some servers (AOL and CompuServe) have suggested they will provide support for the x2 technology. Also, several of U.S. Robotics existing 28.8K and 33.6K modems can be software upgraded to the new speed.

ISDN Modems   ISDN modems give you fast Internet access (standard speeds of 115 kbps with capability of up to 128 kbps) and quick connections to office networks, but not without headaches.

This is the age of information, and whether on the road or at home, you can get late-breaking news over the Internet or download up-to-the-minute profit reports from your corporate server. But you're probably using nineteenth-century technology to access that information: the telephone. Most homes, hotels, and small businesses use aging analog lines tied to a digital telephone system. Designed for simple voice communications, analog lines start to choke when you push lots of data through them. It's like having a highway one mile from your house with a washboard dirt road leading to it.

ISDN technology is far from ideal, but for now it is the best solution for paving a data onramp between the PC on your desk and the wealth of information online. The basic ISDN service gives you throughput up to 128 kbps; the typical analog modem tops out at 33.6 kbps. ISDN is digital, which makes perfect sense for linking computers: Your PC is digital; the phone network is digital; any computer network you'd connect to is digital. Shouldn't your modem also be digital?

To answer this question, consider this: Although there is a wide range of prices and features, you will discover that many of these devices are quite difficult to install.

The existing 3Com *Impact* ISDN modem is easy to use. But the manufacturer plans a new model that will feature, among other things, an automated wizard to ease installation and use.

The Motorola *BitSurfr Pro* and the U.S. Robotics *Sportster ISDN 128K* are excellent choices for their ease of use and great performance. The *Sportster* features *TurboPPP* that provides compression up to 512 kbps. It also has an onboard analog port that lets you connect a standard phone or fax, allowing you to talk or send/receive a fax simultaneously with data transmission.

A close contender is the U.S. Robotics *Courier I-Modem*, which offers both digital and analog access. ISDN modems from Adak, Adtran, Digi International, IBM, and ISD-Ntek are difficult to install and flounder in remote access operations.

Preparing for the ISDN Connection   Unlike an analog modem, which you can typically just plug in, you will have to do some homework before you purchase an ISDN modem.

- An ISDN modem can connect only with another ISDN modem, so you will have to make sure that the users you plan to contact have them. If you want the faster Net access provided by ISDN, ask your Internet service provider whether it offers ISDN service, and also find out which ISDN modems it currently supports.

- Make sure your communications software supports ISDN. Most terminal emulation programs, such as *HyperTerminal*, and remote access software, including *pcAnywhere* and *LapLink*, support ISDN.

- Call your local phone company and make sure ISDN is available in your area. This is a problem that has dogged the technology for years. If you live outside a major metropolitan area, chances are that your phone company cannot provide ISDN service. If it does offer ISDN, request the *Basic Rate Interface* (BRI) service, which provides two 64-kbps *B* channels for voice and data. ISDN BRI is ideal for the home or small business because you can use one line for data and the other for phone and fax. You can expect to pay between $30 and $200 for installation, and monthly charges can run anywhere from $30 on up; the exact price will vary based on where you live.

- ISDN modems also require a *Network Terminator* (NT 1). This unit mediates between the ISDN line and your PC. In the past, ISDN modems required a separate

NT 1, but most now come with the NT 1 device built into the unit. Of the modems mentioned before, only the IBM *WaveRunner Digital Modem* requires a separate NT 1.

Problems with ISDN Modems   Installing an ISDN modem is not for the squeamish or casual user. Under Windows 95, none of the modems just discussed come with the files needed to just plug and play. In nearly all cases it is necessary to track down device drivers, configuration files, and command strings to get the modems up and running.

Your work is not finished even when the units are installed. You then have to get the modem and ISDN phone service on speaking terms. To do so, you have to configure the modems using information supplied by the phone company. This includes specifying the type of switch used at the phone company's central office and entering the *Service Profile ID* (SPID) assigned to each data channel.

Clearly, ISDN modems have a long way to go before they match the ease of use of analog modems. But, also clearly, you can make up for lost productivity fairly quickly with the performance you will gain. In situations where you can control the hardware and software at both ends of a connection, such as telecommuting, you will benefit from ISDN. In situations where you don't have as much control, such as when you connect to the Internet, improvement will not be as dramatic, but an ISDN modem will still surpass the current transfer rates of an analog modem. In any case, plan to spend plenty of time getting your ISDN line installed and your hardware set up before you can reap the benefits of the new digital frontier.

Cable Modems   Manufacturers of *cable* modems are moving slowly in putting their products on the market. Initial installations of these devices are proceeding so slowly because the makers wish to avoid the mistakes of ISDN or conventional dial-up systems.

While initial costs are relatively low (about $150 for installation and about $35 per month), the cost advantage is mitigated by technical issues. For example, data transmission over cable is subject to interference, and providing technical support for net access is more complicated than for cable TV service.

Concern over such interference has led some cable modem manufacturers to temporarily provide hybrid modems—devices that send the outgoing data over a phone line, while providing a high-speed return channel via cable.

The cable industry realizes it cannot yet handle the technical sophistication of computer users' support questions. For that reason, the cable industry is trying to team up with Internet service providers and other computer specialists to upgrade its service to handle computer users. This is an opportunity for ISPs to band together with the cable companies.

As of the writing of this text, the percentage of cable systems that can support cable modems was less than 10 percent, and the percentage that can support two-way service was smaller than that. Depending upon where you live, it may be a long time before you have such access. It is estimated that only one-half of the nation's 62 million cable households will be able to access the Internet via cable by the year 2004.

Internet Appliances

Computer companies are rushing to bring out low-cost, easy-to-use Internet devices. Products range from bare-bones *Net PCs,* to *Network Computers* (NC), to pocket telephones, which comply with an Internet appliance specification. The NC is intended to be no bigger than a cable TV set-top box, with a simple operating system, a modest amount of RAM, and little or no local storage. Data storage and applications reside on the network. Because an NC can use a TV as a monitor, it's expected to cost less than $500. One unit available today plugs into a TV, uses a remote control or wireless keyboard, and runs a Web browser. Future systems will bring TV channel and Web surfing together in Internet TVs.

Internet Phones   Internet phones concern traditional telecommunications companies because, for the price of a local call on either end, two net accounts, and a little advance planning, two people anywhere in the world can talk as often and for as long as they want. Compare that to international telephone rates that climb upward at the cost of $1 or more per minute, and net phones start to look pretty good, even with all the logistical and operational difficulties.

In fact, the *America's Carriers Telecommunication Association* (ACTA), which is made up of 130 small long-distance telephone companies, has filed a petition before the FCC. ACTA wants the FCC to regulate Internet telephone products in the same way that all long-distance telephone companies are now regulated, and to treat ISPs as local telephone companies. As of the end of 1996, the issue had not been addressed by the FCC, though public comments were being accepted.

You will need to pay attention to your hardware setup if you plan to use an Internet phone product. Telephone software handles streaming files, playing uncompressed audio while processing incoming compressed information. To send out your voice while simultaneously processing the incoming phone call, your sound card needs to support full duplexing. If your card supports only half duplexing, you won't be able to talk while listening to the incoming voice. The experience is similar to using two-way radio—or a bad speakerphone.

There's more to a phone application than merely chat. Some let you send more than just voice; you can transmit e-mail, or even files. Several packages feature multiple lines, which allow you to put someone on hold while chatting with another person. However, you should have some sort of encryption that prevents other people from listening in on your conversations.

This is a field that's changing rapidly. Netscape is building a net telephone component into Navigator 3.0, endorsing technology developed by Voxware Inc. The endorsement comes in the form of *Metavoice,* a patent-pending technology that Voxware claims is the closest to replicating the entire sound band of the human voice while incorporating compression ratios of 53-to-1.

Another issue slowing down the growth of the net telephone arena is that competing products cannot connect to each other. In theory, an open standard proposal that has been endorsed by Intel Corp. and Microsoft would solve the problem. Although more than 100 companies have signed up to support this new specification, no products have even been announced that support it—much less any implementations. Basically, every vendor will position its own technology as the industry standard and will try to force other vendors to adopt it.

Overall you may be pleasantly surprised by the products' sound quality, though diminished expectations may have something to do with your level of acceptance.

AT&T PocketNet Phone   AT&T's *PocketNet* phone is an integrated cellular phone and wireless Internet appliance for accessing Internet information and two-way messaging services and displaying the information on the phone's screen. The phones communicate via AT&T's *Cellular Digital Pocket Data* wireless IP network in markets where the service is available.

ISDN for Notebook Computers   While travelers may not find ISDN in many hotels, notebook computers are enjoying increased usage, in some cases replacing desktop computers. For these notebooks, a new class of ISDN PC cards are available. One of the first, Angia Communications Inc.'s *I-bahn*, provides both an ISDN active terminal adapter and a V.34, 28.8K analog fax/modem. Using the PC card port avoids bottlenecks on older slower serial ports. I-bahn exceeds the 115.2-kbps of most serial ports, allowing full 128-kbps ISDN throughput. The unit includes a 115.2-kbps 16550 UART to facilitate the analog modem's communications.

17-6   INTRANETS

Intranets are mini Internets, internal networks at small and large companies, organizations, colleges, and universities. While bandwidth limitations challenge Internet Web designers, intranets are using sound bytes, real-time video, virtual 3-D worlds, and other audiovisual enticements. Indeed, if you want a glimpse of what the Internet may look like when every household and business has ISDN or cable modems, check out the multimedia-rich intranets at any communications-savvy company. The browser Netscape estimates that 90 percent of all Fortune 1000 companies already have intranets up and running. The reason is simple: Information managers find that pictures and sounds convey messages more effectively, and Web-based intranets offer a simpler way to deliver these messages and updated information on demand. Best of all, since the audience is internal, managers do not have to worry about protective security measures (*fire-walls*) and other complex security schemes.

Although we humans speak in words, we instinctively rely on audiovisual aids to communicate. Even mathematicians and physicists who play with numbers all day long use sensory-oriented terms like "big bang," "superstrings," and "black holes" to describe their discoveries. Multimedia will inevitably become an integral part of intranet and Internet applications because sound and motion are our preferred ways of getting ideas across. The broader bandwidth and more robust nature of intranets offers the perfect opportunity, particularly for today's electronic technicians, to get in on the ground floor thereby benefiting from the current fledgling multimedia technologies while preparing for the worldwide applications of the future.

In this section, the emphasis is on what you can and can't do with multimedia on an intranet, what benefits an organization can gain from specific types of multimedia, and what's involved and what you need to implement the latest technologies.

Why Intranets?

Intranets allow for archiving essential information that can be maintained by those who are responsible for it, with little support from the *information services* (IS) staff. That means if you have some simple multimedia authoring tools and a $10 microphone or $99 eyeball videocam, plus capture software, you can post audio and video updates on your intranet Web page as often as you like. There's no waiting for IS to get around to it, nor do you have to worry about something getting lost in the translation.

Intranet-based multimedia frees you from the limitations of bandwidth, so you can use as much multimedia content as you need. *Webmasters* (designers of Web pages) on the Internet, faced with many users at overtaxed 28.8-kbps or slower servers, try to keep graphics-heavy pages down to 200 KB or so and video or audio files to around 2 MB—all of which may translate into only a few seconds of material. An intranet, in contrast, can count on a minimum of 10-Mbps transfer rates and up to 100 Mbps on newer, faster networks. These fast conduits can handle all the traffic multimedia-rich content can generate.

Multimedia intranets are easy to implement; their internal nature allows for a uniform set of browsing and plug-in tools across an enterprise. If your users need a new browser and a certain set of plug-ins to access embedded audio and video material on your multimedia pages, it's easy to make sure everyone has them. On the same note, providing a *WYSIWYG* (*w*hat you *see* is *w*hat you *get*) Web authoring tool for internal information providers is easy.

Working with multimedia on an intranet also has a less obvious advantage: It provides you with the opportunity to gain experience and get all the bugs out of using multimedia while it's out of the eye of public scrutiny. In this way, you will be better prepared when bandwidth bottlenecks are solved and you're ready to place your multimedia pages onto the Internet.

Why Multimedia?

Multimedia on an intranet has two compelling advantages:

1. Multimedia is a more precise and understandable way to deliver information efficiently. For example, employees quickly can grasp how their company's new product works by actually watching it in action on a video clip. A *Chief Executive Officer* (CEO) can express sweeping policy changes more clearly with a personal message instead of depending on workers to read between the lines of a dry memo. Groups working on collaborative projects can get a better idea of what their counterparts are doing if they can see and hear each other, rather than drawing the information from reports, summaries, and charts.

2. Studies have shown that, because multimedia is a painless way of learning, workers are more likely to watch a video than to read a manual, and probably will remember what they saw a lot longer, too.

What Intranets *Can* Do   The rapid adoption of multimedia over intranets has been largely application-driven. Companies are using video and sound because they work and are cost-effective. Here are some of the ways multimedia is being used:

- *Video conference rooms,* complete with chat windows, video insets, and shared *whiteboards* (white areas, similar to a chalk board, upon which text and/or art work can be shown). Several users can work on the same spreadsheet or word-processing document simultaneously. If it is difficult for everyone to get away from their desks for a meeting, they can stay in their offices and meet in a virtual conference room instead.

- *Employee "business card" databases.* Some fast growing companies may add employees at a 50 to 100 percent clip for their first few years. It makes sense to videotape a brief interview with each new hire, packed with background information, a job description, and how he or she fits in with the organization. Other workers can access the interviews to match names with faces and jobs, and then refer to them in the future as they renew dealings with an individual.

- *Memos and reports.* Even routine memos can become more memorable when they're delivered as voice messages attached to e-mail or embedded in intranet Web pages. Add video when the message warrants it (for example, a CEO's pep talk to the employees) and you've increased the effectiveness even more.

- *Product introductions.* Nobody but the launch teams really knows everything about new products, and in larger companies there may be a tight veil of secrecy that prevents the average worker from knowing anything at all until the public announcement. It is probably better not to keep your own employees in the dark once the product is unveiled. Modest multimedia productions shared over intranets can provide employees with the same information you may be paying thousands of dollars to distribute to the media and potential customers.

- *Training.* You do not need to spend a lot of money to create computer-based training programs, when a well-organized intranet Web page using video clips and audio can do the job for much less. Employees (or students) can download the pages and graphics to their own (or school) computers, refer to them using a browser as needed, and then erase them when finished.

- *Benefits databases.* Make benefits information more understandable with multimedia clips that explain difficult topics, or ease employee concerns about their rights with a comforting talk from a top manager.

- *Job and class schedule listings.* A Web-based job listings page for in-house positions can include video or sound clips that describe openings and quickly bring your workers up-to-date on internal opportunities for advancement. In an educational institution, class offerings and schedules can help students determine their own schedules.

- *Newsletters.* It is not necessary to fill a corporate newsletter, for example, with bowling scores when you can display actual bowling clips. Not only can multimedia enhance a newsletter, it can be opened to submissions from anyone with a video camera or tape recorder.

What Intranets *Can't* Do   While multimedia-rich intranets are great for enhancing information with sound and video, they currently do not have the same collaborative tools as traditional groupware. Intranets do not have the built-in ability to create new shared databases using simple user commands. Nor do intranets have the replication capabilities of groupware; you can't easily download intranet databases to your laptop, review them during a plane trip, and automatically resynchronize your files with the network.

The smart thing to do for an organization heavy into groupware is to retain its existing environment and supplement it with an intranet that can manage and distribute user-maintained text, graphics, and multimedia.

How to Do It   Sound, video, and animation are the core of intranet-based multimedia. With access to a microphone, a camcorder, a video capture board, and authoring software, you are able to mount the multimedia world. But before we go further, let's take a closer look at the lastest technologies involving sound, video, Java, ActiveX, and Shockwave—all of which are defining multimedia today.

Sound   Today, sound is the easiest multimedia component to transfer over an intranet, since file sizes are small and the bandwidth required is only 1 percent of live video. Small sound clips are easy to embed in Web pages, and users need only a Web browser with built-in sound support or a helper application to play back .wav, .azcf, and .aifjr sound files. Windows and Mac operating systems have built-in sound recorder *applets* (small applications programs) that can record brief messages with a microphone or convert prerecorded productions into one of these low-fidelity sound formats.

With the introduction of new plug-ins for net browsers, FM-quality sound or better is also easy. These plug-ins offer a streaming audio (prerecorded material that is displayed on the client machine as it is received over the intranet) delivery method using feeds of digital audio that go into a buffer on the client computer for smooth playback even before the entire file has been received. You could set up a corporate streaming audio system to transmit daily company radio news programs on user demand or direct everyone to listen to an announcement at the same time. Audio clips can substitute for memos and be embedded in HTML pages as tips or advice.

If you want to send voice messages as e-mail attachments, a full-featured TCP/IP mail client can record and transmit voice messages using handy recorder/speaker icons. You will need a microphone and sound card in your PC, of course, but the audio is attached as a MIME-format file, so anyone on your intranet (or the Internet) equipped with the right retrieval software can listen to the voice messages.

Video   Video is perfect for intranets; even the smallest of intranets can boast personal video, corporate or educational television stations, and phone conferencing. Quality video requires just a few seconds to transmit over an intranet, whereas it can take minutes to deliver over external Internet connections. For example, a 2-MB video file that would take only 15 seconds to display in real time on an intranet can take up to 10 minutes to download at 28.8-kbps on the Internet.

The trend today is away from expensive videoconferencing hardware solutions that use hardware compression/decompression boards at each end of a connection. Instead, video is squeezed down at the transmission end and viewed by the user using a browser plug-in. Intranet video can take several forms: prerecorded files, streaming video, or live video feeds. These files are captured, stored on the server, and downloaded to the client workstation for viewing in .avi, QuickTime, or MPEG (Moving Picture Experts Group) format.

Limited to Windows machines, .avi is compressed using software rather than hardware, so it's slow. It provides only 15 frames-per-second images that often exhibit synching problems between the video and audio tracks.

The more advanced technology of QuickTime can handle playback of MPEG movies with compression boards that sell for $1,000 or less. Software for editing and manipulating QuickTime movies also is comparably inexpensive. Although MPEG provides the highest quality video while still maintaining reasonably small file sizes, it must be encoded, a time-consuming process that requires expensive hardware. Such equipment can cost as much as $5,000 or more if you want to squeeze the image down in real time. Fortunately, vendors have brought the product costs down to $1,000 or less, offering encoding at one-third the playback speed with promised upgrades to real time in the future.

Video also can be distributed over intranets in a streaming format. As with streaming audio, this type of video is often prerecorded material that is displayed on the client machine as it is received over the intranet. Streaming can be made available as files that are downloaded and viewed. The very ambitious can even view them as live video.

Live video can be quite expensive unless you are willing to settle for CU-SeeMe-type connections. Developed by Cornell University, *CU-SeeMe* is distributed as freeware, and also is available in a commercial version costing about $99 for up to eight video inputs. CU-SeeMe doesn't limit the number of audio participants, and it features chat windows, whiteboards for group collaboration, and videoconferencing support.

To use CU-SeeMe in business, every network must have special server software installed. For group conferences, *Reflector* software must be installed as well; this software accepts multiple CU-SeeMe connections and reflects the video, audio, and data to up to 100 participants at once.

More ambitious projects, like corporate TV channels, call for higher quality images and sound than what's provided by CU-SeeMe-type connections. That means you'll need *MPEG-1* (full-screen VHS quality) or even *MPEG-2* video (broadcast TV quality—720 samples per line by 480 lines by as many as 30 frames per second). MPEG is an open standard developed for CD-ROMs, video games, and other high-quality video display systems. MPEG files are created by discarding redundant data within a video stream and producing a file size that is typically 200 times smaller than the raw digital video alone (which typically amounts to 30 MB per second, uncompressed).

Video Options   Corporate TV often calls for special Webcasting packages on the server that can send compressed signals to all or just selected clients on the network. Live, full-motion video is possible using two available systems, each of which requires expensive server hardware and software packages, capture cards, and encoder and workstation hardware. Each can work only with its own proprietary video streams. A third system is available that uses MPEG-I without the need for special hardware or software. It also supports .avi, .mov, and other formats.

Key Technologies

It is not necessary to immerse yourself in technical jargon, nor to understand all the fine points of intranets to work with custom multimedia on an intranet, but some familiarity with the leading technologies will help you understand what your options are and how they can work for you. The key providers in the multimedia world are programming and scripting languages like Java, APIs such as Microsoft's ActiveX, and applets, plug-ins, and controls that put both to work.

Java *Java* is a C++-type language developed by Sun Microsystems that turns the Web from an information-delivery system to an interactive environment. If you become a developer of an intranet with multimedia content, you'll be interested in Java chiefly if you require specialized applications, particularly interactive programs and animation that need user feedback. Java applets can run automatically on any platform (including Mac and Windows) that has a browser with an interpreter capable of interpreting Java code. Incoming Java applets can be played by the Java interpreter included in Netscape 2.0 and Microsoft Internet Explorer 3.0

Java takes the workload from the server and puts it on the client workstation, which can be a significant benefit with graphics-intensive multimedia applications. It can easily be used to create animated sequences, scrolling text, slide shows, and other multimedia content.

Java programmers can take advantage of class libraries, like the QuickTime libraries announced by Apple in May 1996, to drop sets of functions into applets without coding them all from scratch. Sun plans for a new set of APIs, code-named *Java Beans*, that allows building Java applets from reusable components and should make creating applications even easier.

Microsoft Microsoft's plans for the Net are anything but small; its Internet Explorer Web browser supports a variety of media formats, scripting languages, and programming environments. The most important of the programming environments is *ActiveX*. It started life as an *OLE* (Object Linking and Embedding) control and has been repackaged as a multimedia-savvy way for programmers to build controls with improved performance. ActiveX controls use animation sequences that are dropped into "container" applications to run. For example, Microsoft has stated that Internet Explorer will eventually just be a tight ActiveX container full of objects that carry out Web browsing and content playback functions. Microsoft is also including Java in future releases of Windows 95 and NT. It also has announced *Visual J++,* a full implementation of Java that incorporates support for ActiveX.

Microsoft's *ActiveMovie* API is an ActiveX control that offers playback of MPEG, QuickTime, .avi, MPEG audio, .wav audio, and the new Active Movie streaming format. ActiveMovie can repackage any of these into content that can be transmitted as a continuously playable stream over intranets.

Microsoft also offers a developer tool kit that includes an editor, a software developer's kit for creating ActiveMovie OLE controls using Visual Basic and Visual C++, and sample files.

One advantage that ActiveMovie has (apart from being made an integral part of the operating system in the future) is that developers can specify the maximum data transmission rate that will be used to deliver content appropriate to the user's method of linking to the site. Intranet hookups could be provided with full-motion video and sound, while the same ActiveMovie tools can be used to create slide show/sound multimedia productions with stills and synchronized voice.

Netscape LiveMedia   Netscape's strategy for real-time audio and video is called *LiveMedia*, which is an additional communications layer between the browser and its plug-ins. LiveMedia makes it easier for third parties to produce audio and video components that work smoothly with Netscape and other LiveMedia-compatible products, including the real-time audio and video applications.

Netscape offers three plug-ins that take advantage of LiveMedia's open architecture. These include *LiveVideo*, a plug-in which plays .avi and QuickTime movies; *LiveAudio*, which plays .midi, .wav, .azif, and .aiff files; and *Live3D*, a plug-in that adds *Moving Worlds* (VRML 2.0) support.

Shockwave   Of all the Netscape plug-ins, the one with the greatest impact on the Web is undoubtedly Macromedia Inc.'s *Shockwave*. Shockwave is a playback plug-in that allows display of "movies" created by Macromedia Director and Authorware. Both products are virtually industry standards for creating large-scale interactive multimedia productions for CD-ROMs. Shockwave is made up of two parts: a compression engine called *Afterburner*, which compresses Director and Authorware movies by as much as 60 percent for easier delivery over the Internet, and a player that includes security, caching, and streaming features optimized for network playback. This results in Inline Director and Authorware presentations that play back within a Web browser window.

Getting Started

The good news about implementing a multimedia-rich intranet is that the start-up costs can be easily justified by any school or business that really needs these applications. Virtually every school, business, and many individuals already have nearly all the hardware they need to create an intranet laced with multimedia add-ons. Software for creating multimedia content and making it available on an intranet is another one-time purchase that can amount to as little as a few thousand dollars.

Hardware   Webmasters and network administrators agree that a modest amount of multimedia intranet traffic on a network doesn't add a lot of overhead or load to an existing network. A lot of extra memory probably will not be required, and a gigabyte of extra hard disk space can handle a large number of HTML pages, graphics, video, and sound clips.

While premium-quality, real-time-speed MPEG encoding can cost $15,000 to $50,000 in hardware and software, or $300-per-minute charges for a service bureau, low-cost video capture hardware can put you into full-motion production for less than $1,000. This type product consists of a capture card with its own processor, intended for a 75-MHz Pentium (or faster) platform with 8 MB of RAM, and at least 500 MB of hard disk space.

On the client side, each workstation will need a browser and the right plug-ins to view the multimedia content you plan to provide. A sound card and speakers are required.

MPEG playback of full-screen, real-time video at 30 frames per second can put some special demands on a computer, but any Pentium 100 (or faster) or Power Macintosh is more than up to the job using only a software playback system. Hardware MPEG playback chips commonly are found on separate MPEG cards, in video cards, or even built into motherboards, but are required mostly for full-screen playback on slower machines.

If you want to do videoconferencing or incorporate video clips from different users, a tiny white eyeball camera will be needed for each input site in your organization. These cameras clip onto a monitor and plug into a parallel or serial port. The cameras cost about $100 for a monochrome model and $200 for a color version. The color camera can capture 24-bit, 640 by 480 dpi (dots per inch) still frames, and 160 by 120 at 24 frames per second.

Software   In addition to the software required to distribute multimedia components over an intranet, you will need at least some software to create your video, sound, and animation.

Animated sequences can be stored in the same formats as video files and are played back over intranets using the same plug-ins. The key difference is that animations must be created from scratch, using sophisticated 3-D modeling and animation software. Video, on the other hand, is captured with a camera, then clipped, edited, merged, processed with special effects, and combined with other elements in a finished production. Simple videos require little more than a camera and someone to edit the images. If you want to create animation, however, it's better to have an art department with some time to spend.

Whether you're working with video or animated sequences, a standard package for taking motion, sound, and graphics elements and creating multimedia programs is required. It should contain modular, customizable tools, a visual interface that's easy to grasp, and drag-and-drop editing for many functions. It should let you collect all the inputs you want—text, objects, film clips, sounds—to use in a window, orchestrate the movements and timing, and then play the whole works back. Inputs can be unlinked from a movie and used in multiple productions, so you can create input libraries with a variety of reusable components.

Internal Resources   Most experts agree it is best to do as much intranet multimedia work internally as possible. If an outside source is required to set up the system, once a multimedia intranet is up and running it should be handled internally. One reason is obvious: The whole point of most intranets is to have information distributed by those responsible for collecting and maintaining it with as little filtering as possible. An outside consultant can't possibly stay on top of everything that is going on in your company. Also, once a multimedia intranet is set up, it won't be that difficult to maintain. Video clips, sound bites, and whole Web pages easily can be created by individuals with a minimum amount of training, and then uploaded to the Web server for distribution.

SUMMARY

- The term *Information Highway* was coined to represent worldwide communications systems utilizing all available means of communication. Part of these systems are telephones of all types, computers tied into the Internet, cable TV, radio, and intranets.

- The Telecommunications Act of 1996 is a revision of the Communications Act of 1934. The new Act opens markets in communication areas not previously available to telephone and cable companies. This allows companies to compete in one another's area of expertise.

- The Communications Decency Act is part of the Telecommunications Act of 1996. In part, it makes it unlawful for anyone to transmit so-called indecent material on the Internet to children under the age of 18. It does, in effect, make forms of speech previously protected by the Constitution illegal if sent over the Internet.

- The Internet is a worldwide system of communication, open to access by any person, anywhere, anytime. To access this system, one must have access to a computer, modem, a method of connection, and access to a server.

- Many types of modems are available. The minimum standard device is the 28.8-kbps analog modem. ISDN modems offer higher speeds, but are expensive and difficult to obtain and maintain. Cable modems are presently available in limited areas and good primarily for receiving downloads from the Internet. New devices that will allow send and receive will not be available for many years.

- Intranets are mini-Internets designed to operate internally within any organization, institution, or corporation, large or small. In most cases an Intranet allows access to the Internet as well.

QUESTIONS AND PROBLEMS

1. List the general terms of the Telecommunications Act of 1996.

2. Explain the possible effects and penalties of violations to the Communications Decency Act.

3. What should you, the consumer, expect as a result of the Telecommunications Act?

4. Define the Internet.

5. Define the terms network, LAN, WAN, router.

6. Define the two meanings of host.

7. Define the terms node and terminal.

8. What do the terms client and server refer to?

9. List and briefly describe the four important services provided by the Internet.

10. List and briefly describe sixteen Internet resources.

11. Define the terms FTP, TCP, and IP.

12. List the ways to connect to the Internet.

13. Explain briefly how you can use the Internet for free.

14. What questions should you ask of an Internet provider before signing on with them?

15. Define ISDN and briefly explain its advantages and disadvantages.

16. What is meant by Internet appliances, and what are some of them?

17. Briefly explain what is required to set up and operate an intranet.

CHAPTER **18**

Troubleshooting and Instrumentation

OBJECTIVES

When you complete this chapter, you should be able to:

- Identify the major categories of faults in electronic communications systems.
- Perform basic manual troubleshooting diagnostics.
- Understand computer-aided troubleshooting methods.
- Make accurate measurements and use those measurements to troubleshoot problems involving antennas, receivers, transmitters, and related components.
- List three hazards of working with lasers and know the appropriate safety precautions specific to each.
- Understand how to perform troubleshooting with the aid of a frequency synthesizer.
- Analyze and troubleshoot a transistor amplifier.
- List four types of undesirable oscillations and explain how to correct and/or prevent each type.
- Give a method of troubleshooting for each of the five major types of power supplies.
- Approach troubleshooting in an analytical manner and be able to use all of the information at hand to devise a solution to correct the problem.

INTRODUCTION

Troubleshooting is the act of locating and repairing, in an electrical circuit, a fault that causes improper operation of the circuit.

Instrumentation is the use of equipment to measure electrical quantities for the purpose of keeping these quantities within prescribed limits to ensure proper system operation.

Several major categories of faults in electronic communications systems are as follows:

1. *Complete failures.* This is usually the easiest kind of problem to correct. A complete failure means that the entire electrical system is inoperative. A complete failure may occur because the power cord is not connected to the wall outlet, or the on-off switch may be faulty, or the electrical power supply within the system is defective.

2. *Poor system performance.* This area can be more difficult to analyze. Poor performance means that the system is not operating within the performance limits of its original design. Thus a communication receiver may not be receiving all of the stations it was designed to receive. A transmitter may not be conveying all of the information it was designed to handle. Poor system performance may be due to a weak component within the system. A knowledge of system operation and proper instrumentation is necessary to locate errors due to poor system performance.

3. *Tampered equipment.* An old saying among communication technicians is, "If it works, don't fix it!" Tampering with equipment for the purpose of experiment, modification, or attempted repairs can leave the system inoperative or performing poorly.

4. *Intermittent fault.* This problem is the most difficult. An intermittent is an inconsistent fault that causes the system to be inoperative or perform poorly. The difficulty of finding the cause of an intermittent is in keeping the problem consistent. Intermittents can be caused by mechanical defects, temperature changes, or erratic electrical behavior of components and connections.

5. *Massive traumas.* These system failures may cause more than one part of the system to fail. A massive trauma is a system failure caused by an outside intrusion into the system that results in an inoperative system, poor system performance, or intermittents. Examples of massive traumas are fire, smoke, dropped equipment, water immersion, lightning damage, and applications of voltages with incorrect polarities or values.

18-1 INSTRUMENTATION

When devices are used to measure electrical quantities, a general procedure is used by the experienced technician. The procedure is as follows:

1. *A decision is made to make the measurement.* This decision may be made after you determine that a system fault exists. This determination usually results from experience with the system to be measured.

2. *A measurement procedure is selected.* The procedure may consist of making sure that all system operating controls have been properly set, or it may be a good visual inspection. The technician may "feel" for the presence or

absence of heat, based on what the technician has experienced as normal. The technician's sense of smell can detect the odor of a burning resistor or transformer.

3. *The measurement is conducted.* Here the actual measurement is made. It is important to record what the measurement indicated. There is another old saying: "The lightest pencil mark is better than the world's best memory." There are usually good reasons for old sayings.

4. *The data are analyzed.* The process of analyzing the data consists of comparing the recorded measurement to some standard. This step is very important. From previous measurements, experienced technicians usually know what the standard should be. If the equipment being measured comes with complete documentation, then electrical values printed on schematics or measurement tables are used as the standard. If none of these are available, then making the same measurement on an identical working system, if available, will yield a standard of reference. Without any of these standards, the technician may have to mathematically analyze the circuit.

5. *Act on the analysis.* If a discrepancy exists between the measured data and the standard, then some action is required, such as troubleshooting to correct the fault, reporting the discrepancy to others, or simply giving the system to someone else for troubleshooting.

18-2 BLOCK DIAGRAMS

Schematics of complex communication circuits that show all the resistors, capacitors, inductors, diodes, and other circuit components are sometimes too detailed. Occasionally, a more simplified representation is desired, especially when trying to determine the signal flow through a communications system. This can be done with *block diagrams.*

The simple AM receiver can be represented by the block diagram in Figure 18-1. The signal flow is from the antenna, through the tuner, through the detector, and to the reproducer. If the detector were not functioning (suppose it were open), then the received signal would still be present in the antenna and tuner, but there would be nothing at the input of the reproducer. Figure 18-2 shows the common types of block diagram representations for different circuits in communication systems. Table 18-1 shows what each block in Figure 18-2 could represent and presents a troubleshooting strategy.

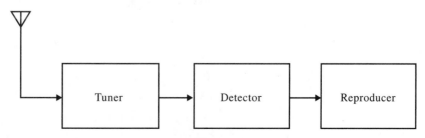

FIGURE 18-1 Block diagram of a simple AM receiver

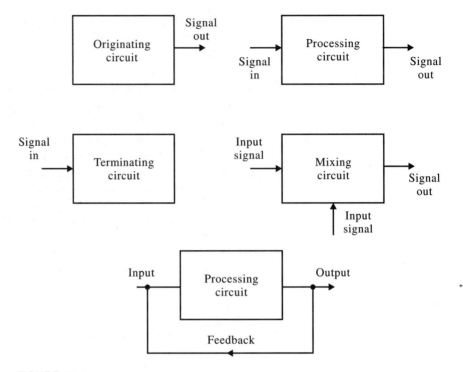

FIGURE 18-2 Typical block diagrams

TABLE 18-1 **Block Diagrams and Troubleshooting Techniques**

Type of Circuit	Function	Troubleshooting Method	Example(s)
Originating	Creates own signal	Check output with scope or meter to see if signal is present	Oscillator
Processing	Changes the signal or impedance matching	Check input and output and compare the two signals for desired results	Amplifier Tuner
Terminating	Performs a final function	If input signal is OK and terminating circuit does not produce desired output, then problem is in terminating circuit	Headset CRT Relay
Mixing	Combines two signals to produce a single output	Check both input signals; then check for desired output	Mixer
Feedback	Output affects the input	Break open feedback loop and test as a processing circuit	Automatic gain control

Each block in a block diagram can be considered as a *stage*. The first step in trouble-shooting a defective system is to use a block diagram and to find the defective stage. Once the defective stage is located, the system schematic can then be referenced to find the exact cause of the problem.

Section 18-3 will introduce you to equipment used for signal injection and signal tracing. Both methods are widely used in troubleshooting communication systems to isolate a defective stage.

18-3   TROUBLESHOOTING BLOCK DIAGRAMS

Troubleshooting with a block diagram is a convenient way of isolating a faulty stage so that a detailed analysis of that stage can be made. In this section, you will analyze the superheterodyne (superhet) receiver from a block-diagram standpoint. You will then see how to troubleshoot it using the two most common methods of isolating a faulty stage: *signal injection* and *signal tracing*.

Superhet Analysis

Consider the block diagram of a superheterodyne receiver in Figure 18-3. Table 18-2 summarizes the purpose of each stage and the input and output signals you should expect.

Signal Tracing

In the troubleshooting examples that follow, assume that the problem with the receiver is no sound from the speaker. When troubleshooting using the signal tracing method, use an instrument like the oscilloscope. No matter which method is used, always check the power supply voltage first, using a voltmeter. The reason is that if the power supply is not functioning properly, then nothing in the receiver will work correctly, since the power supply supplies power to the amplifiers.

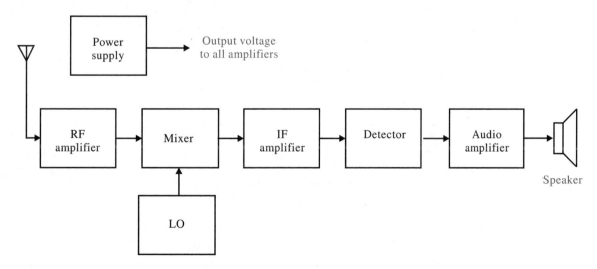

FIGURE 18-3 Block diagram of superheterodyne receiver with power supply

TABLE 18-2 **Superheterodyne Functions**

Stage	Function	Input Signal	Output Signal
Power supply	Provides power to all stages of receiver	AC power 120 VAC, 60 Hz	6 to 12 VDC
RF amplifier	Selects desired frequency and amplifies it	Radio signals	Amplified radio signal
LO	Creates a sine wave to be mixed with the incoming received signal	None	AC that is 455 kHz more than f_R
Mixer	Mixes the LO signal with RF to produce a difference frequency	LO and RF	$f_{LO} + f_R$ $f_{LO} - f_R$ f_{LO} and f_R
IF amplifier	Amplifies the 455-kHz signal from mixer	Same as output of mixer	Amplified 455 kHz AM
Detector	Restores the audio and bypasses the RF	455-kHz AM	Restored audio
Audio amplifier	Amplifies the audio signal from the detector	Audio from detector	Amplified audio
Speaker	Converts the audio waveform into sound	Audio signal	Sound waves

Once you have checked the power supply, go to the "middle" of the receiver and see if a signal is present on the output of the IF amplifier. If there is one, this tells you that all the stages to the left of the IF amplifier output are working. This means that the problem must be in one of the stages to the right of the IF amplifier output. If you do not see a signal here, then the problem must be to the left of the IF amplifier output. See Figure 18-4.

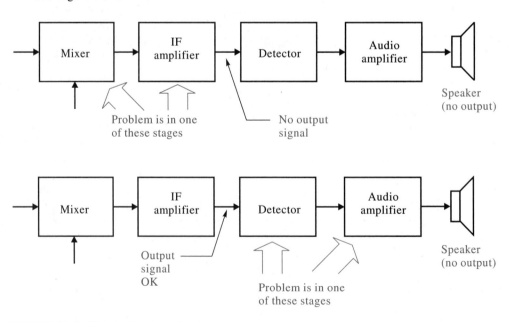

FIGURE 18-4 Using signal tracing in a superhet

Continue this procedure until you find a stage with an input signal but no output signal. That stage is the defective one. To check the LO, simply look for an output signal.

Signal Injection

The signal injection method of troubleshooting involves using an RF generator that can be modulated with an audio tone. The output frequency of the RF generator is variable and can be tuned to represent that of a radio station, LO, or IF amplifier. Most RF generators contain an audio signal output for injecting an audio signal into the audio amplifier.

To troubleshoot a superhet receiver using the signal injection method, inject an audio tone at the input of the audio amplifier. If you hear a tone at the speaker, then the problem must be to the left of the audio amp. If a tone is not heard, then the problem must be from the audio amp to the speaker. As before, always check the power supply first. Figure 18-5 illustrates the signal injection troubleshooting method.

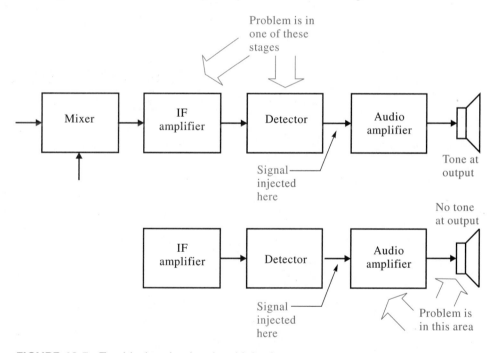

FIGURE 18-5 Troubleshooting by signal injection

18-4   SSB INSTRUMENTATION

The alignment and maintenance of SSB systems is different from other types of communication systems. This section introduces the techniques used in SSB transmission testing.

Carrier Suppression Testing

Recall that for a true SSB transmitter, the carrier should not appear at the output of the transmitter. If the SSB transmitter is not properly maintained, a carrier can appear at the output, with the result that transmitter power is wasted. This condition defeats one of the main advantages of SSB transmission.

A simple test setup to measure the relative degree of carrier suppression is shown in Figure 18-6. The *dummy load* is a device that simulates the transmitting antenna but does not allow any signal to be transmitted. Dummy loads are necessary because you don't want to transmit a radio signal while adjusting the transmitter. That could disrupt other communication signals because your transmitting frequency is changing from the adjustments. A typical dummy load is shown in Figure 18-7.

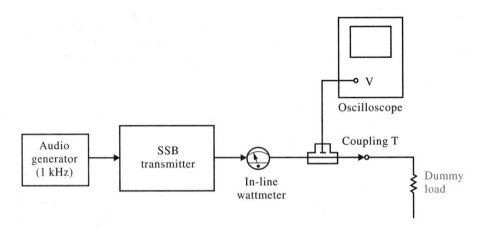

FIGURE 18-6 Basic carrier suppression test

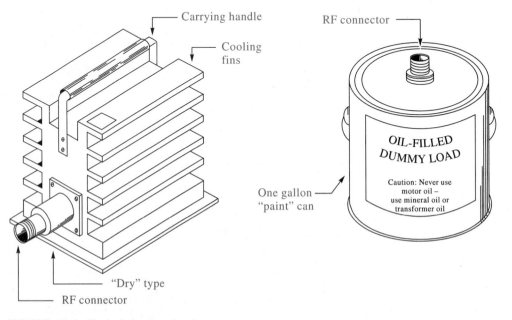

FIGURE 18-7 Typical dummy loads

The *in-line* wattmeter ensures that the output power of the transmitter is adjusted properly. The *coupling T* is an electrical device that extracts a small amount of RF energy for use by the oscilloscope.

The adjustment procedure in the maintenance manual for the particular SSB transmitter should be followed for proper adjustment of carrier suppression. In general, no matter what the adjustment procedure, here is what you look for.

Figure 18-8 shows the scope signal for good and poor carrier suppression. Note that the presence of a carrier begins to produce an output waveform that is similar to the standard AM transmitted wave.

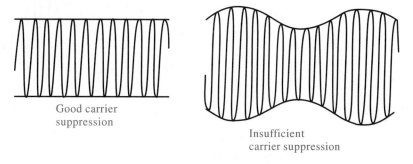

FIGURE 18-8 Carrier suppression scope patterns

Carrier-Suppression Measurement

For some systems, it is impossible to completely suppress the carrier. In these cases, the technician must adjust the transmitter to bring the carrier down to certain specifications. Carrier suppression is usually measured in dB and is defined as

$$S_C \text{ dB} = 20 \ \log \left(\frac{V_{\text{p-p}}}{V_R} \right)$$

where

$S_C\text{dB}$ = Carrier suppression in decibels

$V_{\text{p-p}}$ = Peak-to-peak envelope voltage in volts

V_R = Voltage of envelope ripple in volts

Figure 18-9 illustrates where the measurement is made on the output signal.

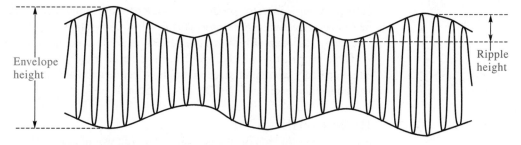

FIGURE 18-9 Measurements for carrier suppression

EXAMPLE 1

For the output signal of the SSB transmitter in Figure 18-10, what is the carrier suppression in dB?

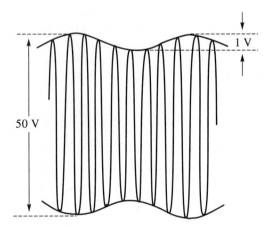

FIGURE 18-10

SOLUTION

Using the formula for carrier suppression:

$$S_C \text{ dB} = 20 \ \log \left(\frac{V_{\text{p-p}}}{V_{\text{R}}} \right)$$

$$S_C \text{ dB} = 20 \ \log \left(\frac{50 \text{ V}}{1 \text{ V}} \right) = 20 \ \log(50) = 20 \times 1.7$$

$$S_C \text{ dB} = 34 \text{ dB}$$

18-5   SWEEP GENERATORS

Sweep generators are used in testing and aligning communications equipment. Basically, a sweep generator produces a range of radio frequencies over and over again—hence the name *sweep generator*. This frequency repetition allows you to quickly test the frequency response of radio frequency RF and intermediate frequency IF amplifiers.

The construction of sweep generators is based upon a special oscillator circuit that can repeatedly change or sweep its output frequency. One such circuit is called a voltage controlled oscillator (VCO); its operation is presented in this section.

Basic Construction

The basic construction of a sweep generator is shown in Figure 18-11. The VCO creates a sine wave whose frequency is directly proportional to the amplitude of the control

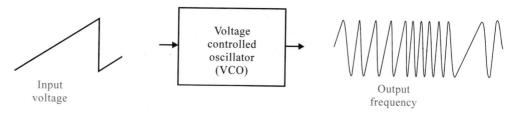

FIGURE 18-11   Basic construction of a sweep generator

voltage. If the control voltage input is a sawtooth waveform, then the output of the VCO will be a sine wave with a sweeping change in frequency that repeats itself according to the frequency of the sawtooth control voltage.

Practical Sweep Generator

It is important that a sweep generator have a constant rate of change of its sweep frequency and a constant output amplitude during the sweep. These qualities are produced by adding the circuits in Figure 18-12.

The *discriminator* circuit does exactly the opposite of a VCO. A discriminator gives an output voltage whose amplitude is proportional to the frequency of the input. The *differential amplifier* has an output voltage proportional to the difference between its two input signals. The purpose of using a discriminator with a differential amplifier is to ensure that the output frequency change of the VCO remains linear.

The automatic level control (ALC) is similar to the automatic gain control (AGC) circuit discussed with AM receivers. The ALC makes the variable gain amplifier maintain a constant output amplitude. The *calibrated attenuator* is used to reduce the output of the sweep generator to an amount specified by the user.

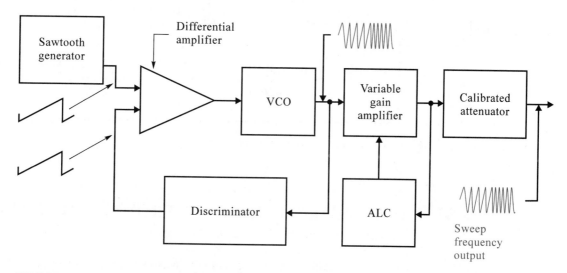

FIGURE 18-12   Practical sweep generator

Sweep Generator Application

A typical application of a sweep generator is shown in Figure 18-13. If the amplifier under test is an IF amplifier, then the display on the oscilloscope will replicate the frequency response curve of the amplifier.

When alignment of RF and IF amplifiers is required, the sweep generator becomes a valuable tool. Be sure to disable any AGC circuits when doing an alignment. If you don't, you will get an inaccurate alignment because the AGC is trying to keep the amplifier gain constant.

The sweep generator is used in the alignment of AM and FM radio receivers as well as TV receivers. It is a valuable tool in any communication repair and service area. Sweep generators come in a variety of frequency ranges, depending on the particular equipment to be aligned. Specialized sweep generators, such as those made for commercial TV alignment, have built-in markers and specific key frequencies to mark picture, sound, and color information.

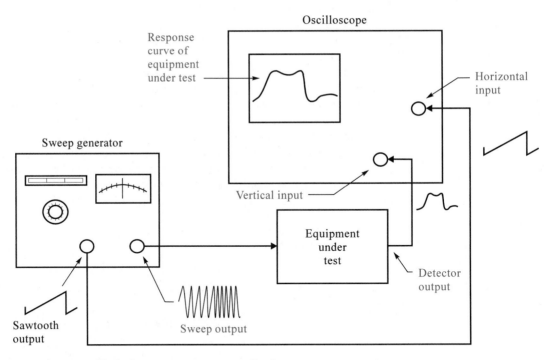

FIGURE 18-13 Typical sweep generator applications

18-6 COMPUTER-AIDED TROUBLESHOOTING

Computer-aided troubleshooting (CAT) consists of using a computer when troubleshooting and repairing electronic equipment. Any personal computer can be used that can accommodate an interface to allow signals to be analyzed. Figure 18-14 shows the basic requirements for CAT.

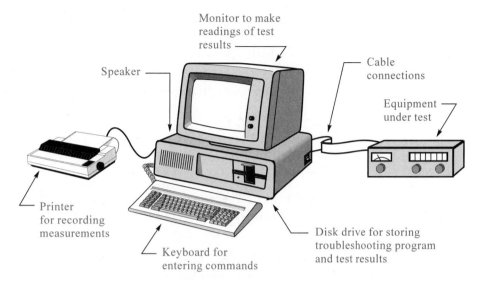

FIGURE 18-14 Basic requirements for CAT

CAT Example

Suppose a technician wants to use CAT to measure the gain of an amplifier. The amplifier under test is shown in Figure 18-15. The system works as follows:

1. The computer interface takes the input and output signals of the amplifier and converts them to digital information.

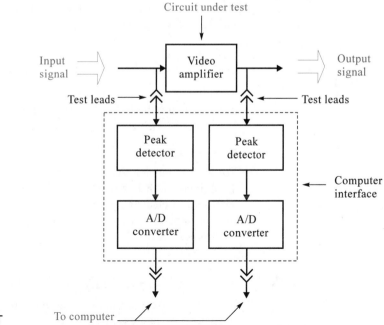

FIGURE 18-15
Example of CAT

2. The *peak detector* outputs the peak reading of its input signal. The magnitude of the peak detector output is then sent to an analog-to-digital converter (A/D converter).

3. The A/D converter inputs digital information into the computer.

4. A program (software) inside the computer records the two readings.

Levels of CAT

You may think of CAT as being divided into different levels. Theoretically, a computer could not only diagnose the problem but also cause a robot arm to replace the defective part.

Level 1 CAT   The technician gives all instructions to the computer, and the computer simply displays the values of the measurements. The technician decides what to do with the data.

Level 2 CAT   The technician gives all instructions to the computer, and the computer displays both measured and calculated values.

Level 3 CAT   The computer recommends troubleshooting procedures to the technician. These procedures are based on previous readings made by the computer. The technician still decides what measurements are to be made.

Level 4 CAT   The computer automatically makes tests and displays results for observation or technical documentation. Voice synthesizers inform the technician of possible difficulties. The technician may override the computer by voice or keyboard commands.

Level 5 CAT   Same as level 4, except the computer directs a robot arm or similar device to change suspected parts.

Level 6 CAT   Same as level 5, but the computer is integrated into parts inventory control. Parts are automatically ordered as needed to maintain proper inventory level.

You can speculate on higher levels of CAT. CAT is used where it is more economical than a technician with hand tools and equipment. Large organizations that frequently service the same type of equipment depend heavily on CAT.

Future of CAT

For the communication service technician, CAT represents exciting new opportunities. Technicians familiar with software and hardware have new employment potentials. The field of electronics has never been stagnant. Its future and the future of all technicians depends on the dynamic changes inherent in the science and applications of electronics.

18-7   PULSE MEASUREMENTS

There is a difference between an *ideal* pulse and a real pulse. The pulses presented in Chapter 5 were assumed to be ideal; that is, they looked like perfect square waves.

In communications, it is often necessary to make measurements with real pulses that will be seen on an oscilloscope. These measurements will tell you how close the measured pulse comes to the ideal.

Real-Pulse Measurements

Figure 18-16 shows a real pulse along with measurements of various properties of that pulse. The definition of each pulse measurement is as follows:

1. *Preshoot.* A change of amplitude of the opposite polarity that precedes the pulse.

2. *Rise Time.* The amount of time it takes for the pulse to go from 10% of its maximum value to 90% of its maximum value.

3. *Nonlinearity.* Any variation from a straight line drawn from the 10% to 90% amplitude points.

4. *Overshoot and Rounding.* Changes that occur in the pulse after the initial transition.

5. *Settling Time.* The time it takes for the pulse to reach its maximum amplitude (not counting overshoot).

6. *Ringing.* Amplitude changes that follow overshooting. They are usually a series of dampened sine waves.

7. *Droop.* A decrease in the pulse amplitude with time.

8. *Pulse Width.* The amount of time between the 50% amplitude points of the pulse.

These pulse measurements are important in data communications devices.

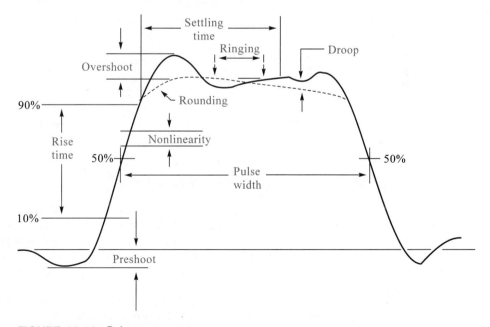

FIGURE 18-16 Pulse measurements

Pulse Spectrum

The frequency spectrum of a pulse changes dramatically as the pulse width or pulse repetition rate changes. See Figure 18-17. Observe that as the width of the pulses decreases, the spectrum changes. In the limit, when the width of the pulse is zero, it is no longer a pulse but an impulse. This condition is impossible, since it implies zero rise time. Note that the spectrum of an impulse consists of harmonics of equal amplitude and infinite bandwidth, which implies than an impulse needs infinite power—another reason why a pulse of zero width is impossible.

From the figure, observe what happens as the duration between pulses increases. In the limit, when the period approaches infinity, the spectral components become indistinguishable from one another and the amplitudes merge into an almost continuous wave.

The actual shape of a pulse and its spectral components are important considerations in pulse communications.

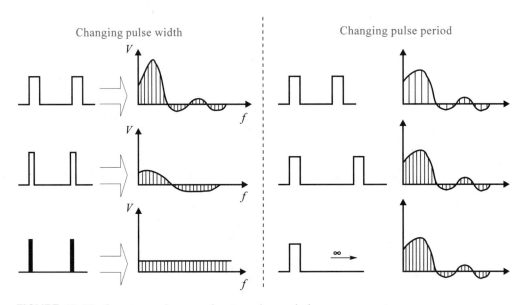

FIGURE 18-17 Spectrum changes due to pulse variations

18-8   DIGITAL METERS

Digital meters are quite popular. A digital multimeter must take an analog quantity and convert it to an equivalent digital readout. This section introduces the essentials of digital meters and then compares them to the conventional analog meters.

Basic Idea

Figure 18-18 illustrates the basic system used in digital meters. The quantity to be measured must first be converted to a form usable by the A/D converter. Since the quantity to be measured may be current, resistance, or some form of ac, the *signal*

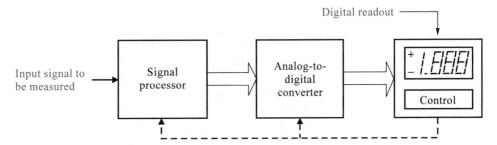

FIGURE 18-18 Basic digital meter system

processor circuit must convert this input to a dc voltage that is within the measurement range of the A/D converter.

Signal Processor

The typical A/D converter used in a digital voltmeter (DVM) is capable of reading a dc voltage in the range of 0 V to 0.1 V. This limited voltage range must actually measure a much wider range of dc voltages. A circuit that will provide such a range is shown in Figure 18-19. An input voltage of 1 V is to be measured by the meter. Note that the total input impedance is greater than 10 MΩ. The input impedance to the amplifier is about 100,000 MΩ; thus the position of the *range* switch will not have any noticeable effect on the DVM input impedance.

This system lets the DVM user select the proper DVM range. Thus, if the input signal to be measured causes the input voltage to the amplifier to be greater than 0.1 V, the digital meter will indicate an *overflow,* which tells the user to change the range switch to a higher setting. This means turning the switch toward R_5.

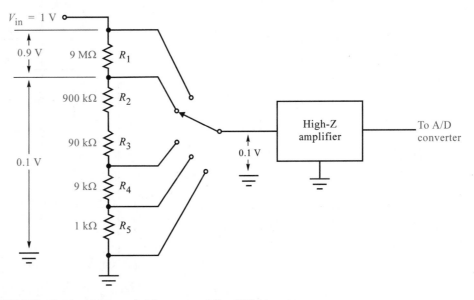

FIGURE 18-19 Voltage divider network for DVM input

Autoranging

Figure 18-20 illustrates the basic concepts of a system for *autoranging*. If the input signal causes the voltage to the A/D converter to exceed 0.1 V, the comparator will cause the binary counter to increment. This in turn will cause the bilateral switch to automatically change the resistance of the voltage divider until the correct input voltage to the A/D converter is produced.

If the input signal is lower than a given *threshold* level, then the second comparator is activated, causing the binary counter to count down until the correct input is achieved.

Autoranging can be used to measure quantities other than just dc voltages. The quantity need only be converted to a dc voltage.

True rms Converter

An ac voltage can be converted to a true rms dc voltage by a *thermocouple*. Recall from your study of basic electricity that a thermocouple produces a dc voltage that is proportional to its temperature. See Figure 18-21. The voltage produced by the thermocouple is proportional to the heat produced by the input ac voltage. Thus, the larger the ac voltage, the greater will be the corresponding dc voltage.

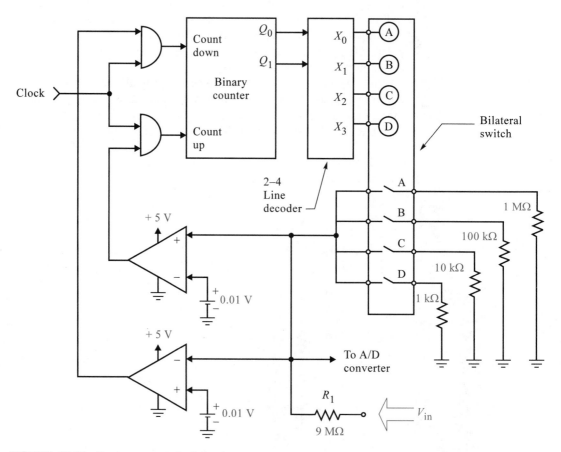

FIGURE 18-20 Basic autoranging circuit

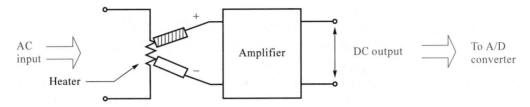

FIGURE 18-21   Thermocouple input to measure rms

Measuring Resistance

For a digital multimeter (DMM) to be a true multimeter, it must be capable of measuring resistance. Since any input quantity to be measured must be converted to a voltage, a system for developing a dc voltage that is proportional to the amount of resistance being measured must be used. Figure 18-22 shows such a basic system. Its basic operation is as follows:

1. The voltage drop across R_X is equal to the product of R_X and the value of the current I_S from the constant-current source (Ohm's law).

2. When the constant-current source is set at 1 mA, the input voltage will be 0 V for 0 Ω and 1 V for 1 kΩ. Any value over 1 kΩ will produce an *over-range* condition.

3. Larger values of resistors are measured by switching the constant-current source to its smaller output current state. This may be done by the operator or an autoranging circuit. This means that the maximum resistance that can be read by the digital ohmmeter is 10 kΩ with the 100-μA source and 1 MΩ with the 1-μA current source.

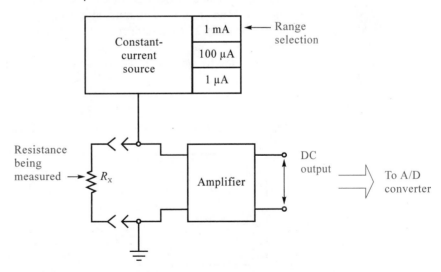

FIGURE 18-22   System for converting resistance reading to a DC voltage

Comparison of Analog and Digital Meters

Table 18-3 gives a comparison of digital and analog meters.

TABLE 18-3   **Comparison of Digital and Analog Meters**

Digital Meters	*Analog Meters*
Display Easy to read. Many displays give the decimal point and the polarity of the reading.	**Display** Must interpret reading on scales that are usually nonlinear. Accurate readings require a great deal of practice and are subject to parallax errors as well as operator reading errors.
Resolution A simple 3-digit display gives a resolution of one part in 1,000, a 4-digit display gives one part in 10,000. These are resolutions of 0.1% and 0.01%, respectively.	**Resolution** At best, an experienced operator can read one part in 100. That is a resolution of 1%.
Reading time An inexpensive DMM makes about 5 readings per second. In many applications, the reading rate exceeds 100 readings per second.	**Reading time** It usually takes about 1 second or more to settle after each reading.
Noise immunity Requires special circuits to reduce input noise interference of the digital reading.	**Noise immunity** Relatively immune to circuit noise. Thus readings are easier in a noisy environment.

18-9   RADIO FREQUENCY GENERATOR IMPEDANCE MATCHING

The *radio frequency* (RF) generator is a necessary instrument for designing, trouble-shooting, and aligning communications equipment. As with any piece of laboratory equipment, the RF generator must be connected properly to the circuit under test. In this section, you will be introduced to the importance and methods of *impedance matching* the RF generator and the circuit under test.

Basic Idea

For reliable testing to occur, the load impedance must equal the generator impedance. This concept is shown in Figure 18-23. The RF generator is usually connected to the

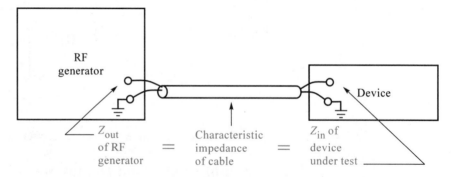

FIGURE 18-23   Impedance-matching requirements of RF generator

circuit under test through a connecting coaxial cable. This coax cable must have a characteristic impedance (Z_0) equal to the output impedance of the generator. Equally important, the cable must be terminated in its characteristic impedance. Any impedance mismatch will cause signal reflections and result in standing waves on the connecting cable.

Resistive Pad

When the RF generator is to be connected to a circuit whose input impedance is different from the impedance of the generator, an impedance-matching device, such as an impedance-matching transformer, should be used. If such a device is not available, then a *resistive pad* can be constructed and used.

The design equations for a resistive pad are

$$R_1 = Z_1 \sqrt{1 - \frac{Z_2}{Z_1}}$$

$$R_2 = \frac{Z_2}{\sqrt{1 - \frac{Z_2}{Z_1}}}$$

where

$\quad R_1$ = Resistor in series with the load

$\quad R_2$ = Resistor in parallel with the source

$\quad Z_1$ = Larger of the load or source impedance

$\quad Z_2$ = Smaller of the load or source impedance

A typical resistive pad is shown in Figure 18-24.

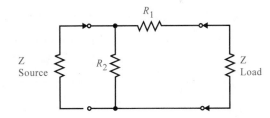

FIGURE 18-24
Typical resistive pad

EXAMPLE 2

Design a resistive pad for connecting an RF generator with a 50 Ω output impedance to a circuit with a 75 Ω input impedance.

SOLUTION

$$R_1 = Z_1 \sqrt{1 - \frac{Z_2}{Z_1}}$$

$$R_2 = \frac{Z_2}{\sqrt{1 - \frac{Z_2}{Z_1}}}$$

Since Z_1 must represent the larger of the two impedances, solving for R_1 gives

$$R_1 = 75 \ \Omega \ \sqrt{1 - \frac{50 \ \Omega}{75 \ \Omega}}$$

$$R_1 = 75 \sqrt{1 - 0.666}$$

$$R_1 = 75\sqrt{0.333} = 75(0.577)$$

$$R_1 = 43.3 \ \Omega$$

Solving for R_2 gives

$$R_2 = \frac{50 \ \Omega}{\sqrt{1 - \frac{50 \ \Omega}{75 \ \Omega}}}$$

$$R_2 = \frac{50}{\sqrt{1 - 0.666}}$$

$$R_2 = \frac{50}{\sqrt{0.333}} = \frac{50}{0.577}$$

$$R_2 = 86.6 \ \Omega$$

The completed and connected pad is shown in Figure 18-25, using practical values of R_1 and R_2.

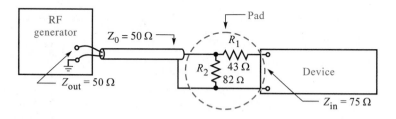

FIGURE 18-25

Pad Analysis

To demonstrate that the source sees 50 Ω looking toward the load and that the load sees 75 Ω looking back toward the source, refer to the steps in Figure 18-26. Note that when the equivalent circuit for the source is solved, the impedance is almost 50 Ω. When the equivalent circuit is solved for the load, the impedance is almost 75 Ω. The

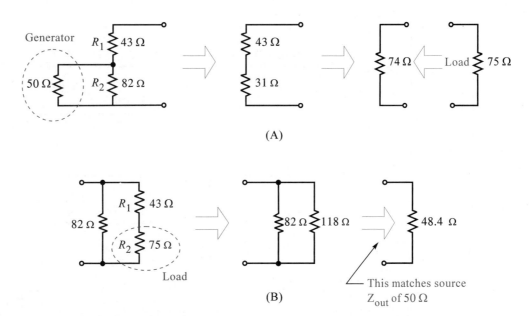

(A)

Load

(B)

This matches source
Z_{out} of 50 Ω

FIGURE 18-26   Analysis of resistive pad (A) Equivalent circuit looking from the load
(B) Equivalent circuit loking from the cable

pad will cause a 4 dB loss, which you must consider when making measurements. Make sure that all RF test equipment is properly terminated in order to obtain accurate readings.

RF Generator Coupling Circuits

Some standard methods of coupling an RF generator to communications equipment are shown in Figure 18-27. For an RF generator with a 50 Ω output, the coax cable is always terminated in a 51 Ω resistor (closest standard value to 50 Ω). For the all-wave receiver, a standard *dummy antenna* recommended by the Institute of Radio Engineers (IRE) is used.

18-10   ANTENNA MEASUREMENTS

There are many ways to measure the resonant frequency of an antenna. To begin these measurements, you should know the approximate resonant frequency of the antenna. You can find it from the physical dimensions of the antenna. The method presented in this section uses the VSWR bridge.

VSWR Bridge

A *VSWR bridge* is a device that separates the forward wave from the reflected wave. See Figure 18-28. With the VSWR bridge, the frequency at which the reflected wave drops to a minimum is considered to be the resonant frequency.

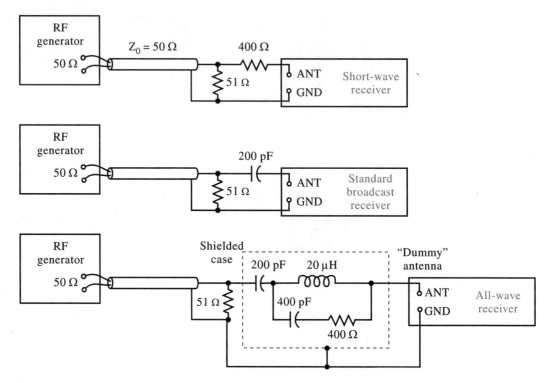

FIGURE 18-27 Coupling an RF generator to receivers

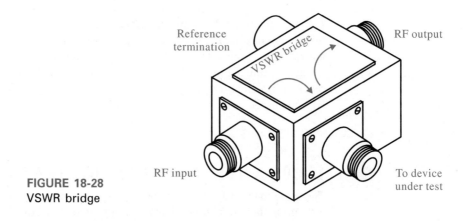

FIGURE 18-28
VSWR bridge

Sweep Generator Method

One method of using the VSWR bridge to determine the resonant frequency of an antenna is illustrated in Figure 18-29. This method produces a scope pattern of the reflected signal amplitude with respect to the applied sweep frequency. The point of minimum response represents the resonant frequency of the antenna. A signal generator can be used to place a marker on the scope pattern to indicate the resonant frequency.

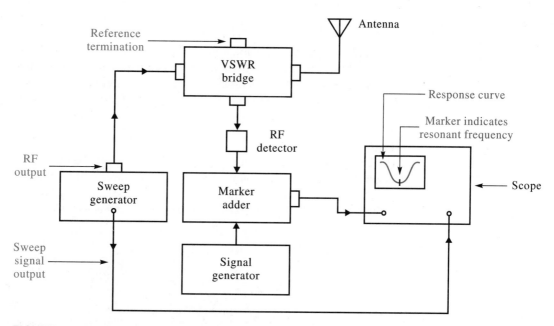

FIGURE 18-29 Sweep generator method of measuring antenna resonant frequency

RF Voltmeter Method

Figure 18-30 illustrates a method of measuring the resonant frequency of an antenna with the VSWR bridge and an RF voltmeter. Before performing this test, make sure that the response of the RF generator and RF voltmeter are reasonably flat over the range of frequencies to be measured. Then vary the output frequency of the RF generator around the expected resonant frequency of the antenna under test. The frequency at which the RF voltmeter shows a minimum reading indicates the resonant frequency of the antenna.

This section presented two methods for measuring the resonant frequency of an antenna. There are other methods, depending upon the type of equipment available and the type of antenna under test.

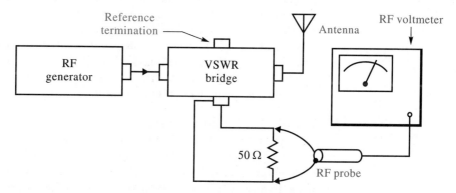

FIGURE 18-30 Using RF voltmeter to measure antenna resonant frequency

18-11 LASER SAFETY

Laser safety is such an important topic that it warrants inclusion in the troubleshooting chapter. Lasers produce an extremely intense light that can cause permanent eye and skin damage if proper precautions are not taken. Laser manufacturers must certify that their laser products are in compliance with federal safety standards. This section presents the potential hazards of working with lasers and precautions to take.

The Human Eye

The energy absorbed by your eye depends on the wavelength of the energy. Wavelengths from 400 nm to 1,400 nm are a potential danger to the retina of the eye because these wavelengths are absorbed by the eye. Figure 18-31 is a schematic representation of the human eye. The eye consists of the *cornea, aqueous, lens, vitreous,* and *retina.*

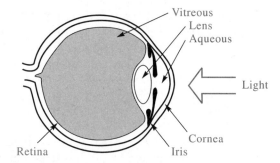

FIGURE 18-31
Schematic
representation of
the human eye

Ultraviolet Damage

Ultraviolet wavelengths of 280 nm or less are absorbed by the cornea of the eye. These are frequencies that you and I cannot see, but can severely burn the cornea of the eye. Damage done to the eye in this manner produces a reddish color of the cornea. The effect is similar to snow blindness or "welder's flash," which occurs from watching arc welding. The eye hurts and feels as if sand were in it. If the damage is not severe, recovery takes a few hours to a couple of days.

Other damage to the cornea can be caused by a photochemical reaction due to ultraviolet radiation. This reaction is a delayed effect and usually occurs several hours after exposure. The cornea takes on a milky hue. This damage can result from eye exposure to a short, intense (3 mJ and above) burst of ultraviolet radiation.

Medical science has strong evidence that cataracts are developed from long-term exposure to ultraviolet radiation. Statistically, people who live near the equator have a higher incidence of cataracts than those living further north. Thus the ultraviolet radiation from a brighter sun at regions near the equator could lead to cataracts.

Visible Radiation Damage

Since the retina absorbs visible radiation (that's what makes the radiation visible), it can be damaged from overexposure. Laser beams whose wavelengths are 400 nm to 1,400 nm can cause this damage. An extra hazard is caused by the eye's amplification of the light by the lens. Thus, a very weak radiation can be magnified thousands of times and cause even more damage.

Skin Damage

The skin is the largest organ of the body. A section of human skin is shown schematically in Figure 18-32. The skin consists of the *epidermis, dermis,* and *subcutaneous tissue.* The epidermis is further divided into the *stratum corneum* and the *basal layer.* The outermost layer of the epidermis consists mostly of dead skin cells that have moved up from deeper parts of the skin. This layer of dead cells protects living cells from water loss, abrasion, dust, air, and radiant energy.

Skin damage from optical exposure depends on many variables. Usually, the lowest detectable level of damage to the skin is called *erythema,* which causes a reddening of the skin. This photochemical change also takes place during a *sunburn.* The problem is that the effect of erythema is usually not noticed until several hours after the exposure. There is some medical evidence that long-term exposure to this kind of radiation can cause skin cancer.

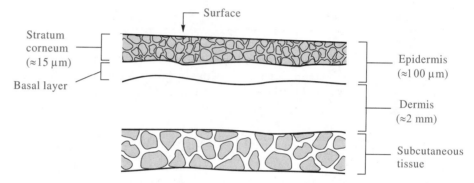

FIGURE 18-32   Schematic representation of human skin

Laser Safety Standards

The two major safety standards for lasers are (1) standard 1040 published by the federal government and (2) standard Z136 published by the American National Standards Institute (ANSI). The main difference between these standards is that the federal standard classifies lasers according to the maximum possible output from the device during *normal* operation. The ANSI standard classifies the laser in terms of the potential hazard according to the maximum exposure a human could receive.

Generally, lasers are classified for safety purposes as shown in Table 18-4.

Typical warning labels for the different classes of lasers are shown in Figure 18-33.

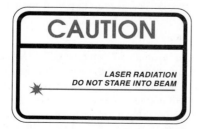 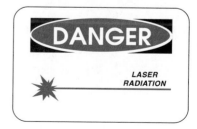

FIGURE 18-33   Warning labels for lasers

TABLE 18-4 **Laser Classification**

Class of Laser	Meaning
I	Not known to have caused human injury
II	Only visible lasers allowed in this class. Can cause eye damage if laser beam strikes the eye directly for 0.25 s or longer. (Note: It takes about 0.25 s for the eye to blink.)
III	Will produce eye injury if laser beam strikes the eye directly. (This is quicker than the eye can blink!)
IV	Exposure to skin is hazardous. Beam is a fire hazard. Even diffused reflections as well as direct eye contact can cause severe eye injuries.

This section should be considered the most important section on lasers. Hopefully, you now have a healthy respect for any laser. Observing the safety precautions that apply to lasers can result in productive applications of the device.

18-12 TELEPHONE MODEM TROUBLESHOOTING

Table 18-5 lists some of the most common problems encountered with telephone modems. As you will see, most of the problems are software-related ones.

Other common problems encountered involve some very simple hardware considerations. For example, telephone modems usually come with two separate telephone

TABLE 18-5 **Common Telephone Modem Problems**

Symptom	Possible Cause(s)
Cannot connect	Usually this means that your baud rates are not the same or the number of data bits are not matched. This is especially likely if you see "garbage" on the screen, especially the { character.
Cannot see input	You are typing in information and it doesn't appear on the monitor screen. However, the person on the other side can see what you are typing. You need to turn your local 1ECHO1 on. In this way, what you type will be echoed back to you and you will now see it on your monitor screen.
You get double characters	For example, if you type HELLO, you get HHEELLLLOO, and, at the same time, what the other computer is getting appears normal. This means that you need to set your local 2ECHO2 off. In this way, you will not be echoing back the extra character. With some systems, *half duplex* refers to local ECHO on, while *full duplex* refers to local ECHO off.

line connectors. As shown in Figure 18-34, the purpose of the PHONE input is to connect a telephone, not the output line from the modem to the telephone wall jack. The PHONE input is simply a convenience. It allows the telephone to be used without having to disconnect the telephone line from the computer to the wall telephone jack. If you mistakenly connect the line from the wall telephone jack to the 3PHONE3 input, you will be able to dial out from your communications software, but then your system will disconnect you. Make sure that the telephone line that goes to the telephone wall jack comes from the LINE output and not the PHONE output jack of your modem.

Another common hardware problem is that there is simply something wrong with your telephone line. This can be quickly verified by simply trying to use your phone to get through to the other party. If you can't do that, then neither can your computer!

A problem that is frequently encountered in an office or school situation involves the phone system used within the building. You may have to issue extra commands on your software in order to call out of the building. In this case, you need to check with your telecommunications manager or the local phone company.

Sometimes your problem is simply a noisy line. This may have to do with your communications provider or with how your telephone line is installed. You may have to switch to a long-distance telephone company that can provide service over more reliable communication lines. Or, you may have to physically trace your telephone line from the wall telephone jack. If the installation is old, your telephone line could be running in the wall right next to ac power lines. If this is the case, then you need to reroute the phone line.

Telephone modems are an increasingly popular attachment to personal computers. It is important that you understand the basic troubleshooting procedures for these important communication devices.

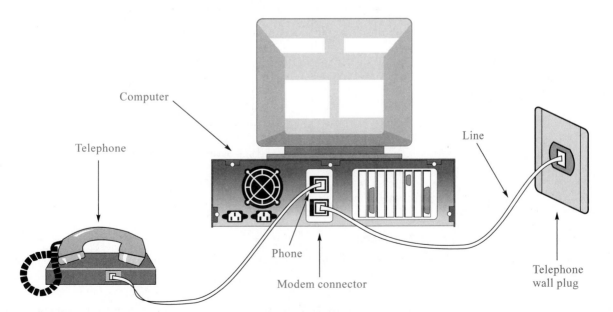

FIGURE 18-34 Telephone connectors for typical telephone modems

18-13   FREQUENCY SYNTHESIZERS

Frequency synthesizers are much used in communication instrumentation. Frequency synthesis provides stable frequencies that are easily selected by the user. Another advantage of frequency synthesizers is that you can easily select output frequencies by digital means. Thus, different frequencies may easily be selected by a computer or other digital control circuits that have been preprogrammed to respond to specified conditions.

Because of the digital selection of frequency synthesizers, music synthesizers are easily constructed using this basic principle. Frequency synthesizers have also found their way into communication receivers and transmitters. As you may suspect, the backbone of the frequency synthesizer is the PLL.

Basic Idea

Figure 18-35 is the block diagram of a basic frequency synthesizer. Note that the output frequency is *higher* than the input frequency. This is important because the accuracy of the reference frequency is passed on the final output frequency. The whole system consists of a PLL with a counter in the feedback loop. In Figure 18-35, the counter is simply a *J-K* flip-flop, which changes states at the leading edge of each input pulse, thus producing an output frequency that is half of the input frequency.

If the switch SW-1 were changed to position ÷16, then a 4-bit binary counter is in the feedback circuit. Thus the output frequency would then be 2^4 or 16 times larger than the input frequency.

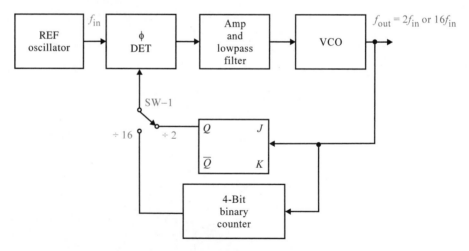

FIGURE 18-35   Basic frequency synthesizer

EXAMPLE 3

Determine the output frequencies of the frequency synthesizer in Figure 18-36.

SOLUTION

Since the feedback circuit consists of a 4-bit binary counter, the output frequency will be a multiple of the binary count selected by the *frequency selector* switch.

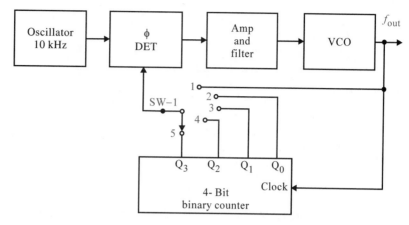

FIGURE 18-36

The resultant output frequencies are listed in Table 18-6.

TABLE 18-6   **Output Frequencies of Frequency Synthesizer**

Selector Switch Position	Binary Counter Output	Output Frequency (kHz)
1	0	10
2	2	20
3	4	40
4	8	80
5	16	160

The preceding circuit is a handy troubleshooting device capable of quickly producing five frequency standards that are just as accurate as the source oscillator. However, just one oscillator is enough.

This is a workable scheme for low frequencies. For frequencies in the 30 MHz to 300 MHz range (VHF and UHF), it isn't practical to use this scheme because these frequencies are out of the range of most ICs.

Synthesizing Communication Frequencies

Figure 18-37 illustrates a method of reducing the incoming frequency to a lower frequency for the PLL. The frequency of the crystal-controlled oscillator is reduced by a divide-by-N counter. Thus a crystal frequency of 40 MHz can be reduced to a workable 4 MHz with a divide-by-10 counter.

Another way to work with higher frequencies in frequency synthesizers is to use the principles of *heterodyning*. Figure 18-38 is a block diagram of a frequency synthesizer using this method. The circuit operation is as follows:

1. The *offset* oscillator is fed into a mixer along with the output of the VCO.

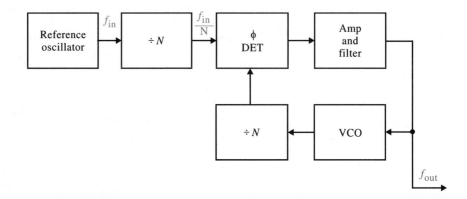

FIGURE 18-37 Producing a lower frequency for the PLL

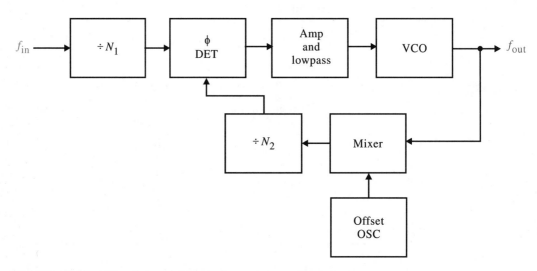

FIGURE 18-38 Using heterodyning in frequency synthesizers

2. The output difference frequency of the mixer is fed into the second divide-by-N_2 counter.

3. The divide-by-N_1 counter output is fed into the phase detector with the lower-frequency output of the second divide-by-N_2 counter.

The advantage of this system is that it can work with a much higher frequency range than other frequency synthesizers.

18-14 TROUBLESHOOTING A TRANSISTOR AMPLIFIER

This section will take you step-by-step through the analysis and troubleshooting of a transistor amplifier.

Transistor Amplifier

A typical transistor amplifier is shown in Figure 18-39. The troubleshooting analysis of this amplifier assumes that any of the transistor resistors can either *increase* or *decrease* in value.

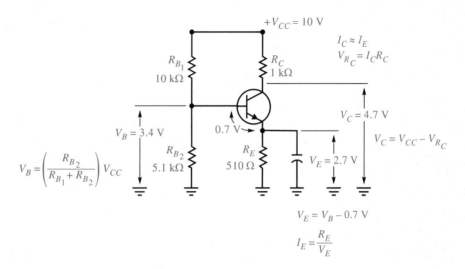

FIGURE 18-39 Typical transistor audio amplifier

Table 18-7 is a troubleshooting summary for the circuit of Figure 18-39. To encourage circuit analyzing, the problems caused by the resistors will be viewed in terms of the circuit voltage readings with respect to ground.

Note: Readings are for a percentage change of less than 100% in resistance values, not a drastic change approaching an open or a short.

Figures 18-40 through 18-43 illustrate the reason for the readings in Table 18-7.

<hr />

TABLE 18-7 **Transistor Amplifier Troubleshooting**

		Effect on		
Component	*Problem*	V_B	V_E	V_C
R_{B_1}	Increase	Decrease	Decrease	Increase
(refer to Fig. 18-40)	Decrease	Increase	Increase	Decrease
R_{B_2}	Increase	Increase	Increase	Decrease
(refer to Fig. 18-41)	Decrease	Decrease	Decrease	Increase
R_E	Increase	None	None	Increase
(refer to Fig. 18-42)	Decrease	None	None	Decrease
R_C	Increase	None	None	Decrease
(refer to Fig. 18-43)	Decrease	None	None	Increase

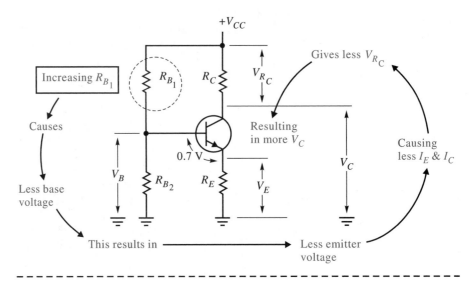

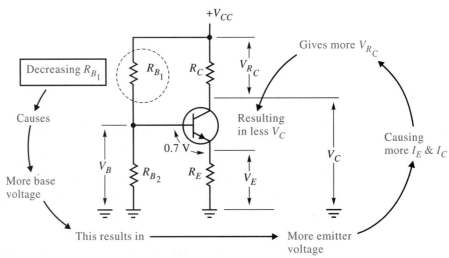

FIGURE 18-40 Effects of changes in R_{B_1}

18-15 OSCILLATORS

Undesirable Oscillations

In electronics, any circuit that meets the criteria for an oscillator will oscillate. What this means is that there are times where circuits, not intended to be oscillators, will act as oscillators. This is especially true of high-gain RF amplifiers whose purpose is to amplify a received signal. Instead of amplifying the received signal, the amplifier oscillates (produces its own signal) and ignores the intended information. This behavior can have disastrous effects on any communication system. Consider the space

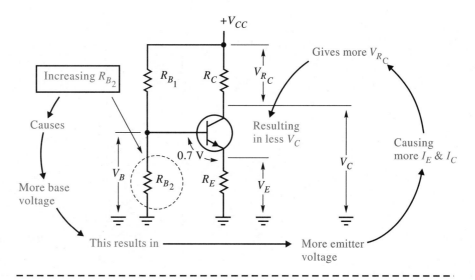

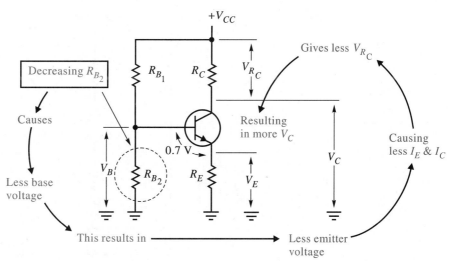

FIGURE 18-41 Effects of changes in R_{B_2}

satellite that can no longer interpret earth signals because of undesirable oscillations in its own circuitry. This section describes the most common causes of undesirable oscillations and what can be done to prevent them.

Types of Undesirable Oscillations
Table 18-8 lists the different types of undesirable oscillations most commonly found in communications equipment.

Correcting Undesirable Oscillations
The oscillations outlined in Table 18-8 can be very difficult to analyze and cure. Some standard methods for preventing these types of oscillations follow.

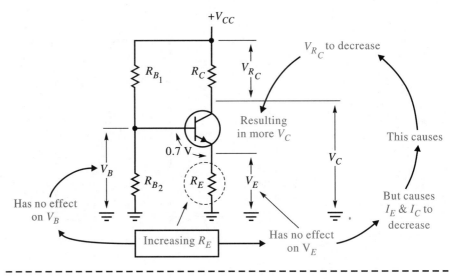

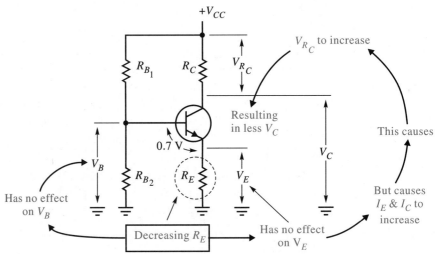

FIGURE 18-42 Effects of changes in R_E

Low-Frequency Oscillations

The cause of low-frequency oscillations is shown in Figure 18-44. The best way to prevent them is to make sure that the power supply for the communication system is well regulated. An output filter capacitor in need of replacement may also cause low-frequency oscillations (motorboating).

Mid-Frequency Oscillations

The cause of undesirable mid-frequency oscillations is shown in Figure 18-45. These types of oscillations are best prevented by using shielding between the circuit output

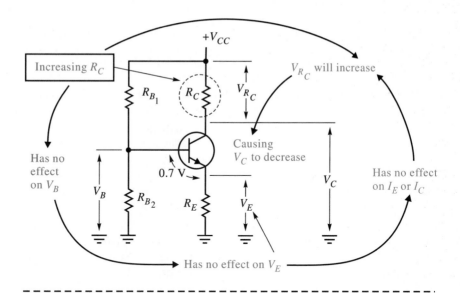

FIGURE 18-43   Effects of changes in R_C

and input to block any feedback path from the circuit output to the circuit input for electromagnetic radiated energy. A typical shielding method is shown in Figure 18-46.

Mid- to High-Frequency Oscillations
Mid-frequency to high-frequency oscillations are usually caused by improper ground connections in the circuit. See Figure 18-47.

The standard acceptable cure for poor ground loops is to connect all ground connections to a single point. This single point is sometimes called a ground plate. Good

TABLE 18-8 **Types of Undesirable Oscillations**

Type	Description of Symptom	Most Probable Cause
Low-frequency (refer to Fig. 18-44)	Low-frequency audio sounds like low-frequency vibrations; sometimes called *motorboating*	Poor power supply regulation
Mid-frequency (refer to Fig. 18-45)	Oscillations at the resonant frequency of the tuned circuits of the IF or RF amplifiers	Feedback from the output to the input of the amplifiers
Mid- to high-frequency (refer to Fig. 18-47)	Oscillations at the resonant frequency of the tuned circuits within the system or some other frequency	Feedback through poorly decoupled ground loops
Parasitic (refer to Fig. 18-49)	Weak oscillations that may not be consistent; usually a high-frequency that may change when test probes are brought near	Stray capacitance or circuit inductance

circuit construction techniques are called for here. A suggested solution is shown in Figure 18-48.

Parasitic Oscillations

One reason for parasitic oscillations could be poor decoupling. See Figure 18-49. There are circuit points that do not represent a low impedance to ground for radio frequencies. For proper circuit operation, these points should be bypassed with capacitors, which creates low-impedance points to ground for radio signals. See Figure 18-50.

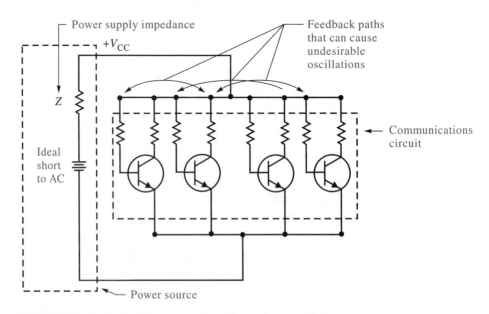

FIGURE 18-44 Probable cause of low-frequency oscillations

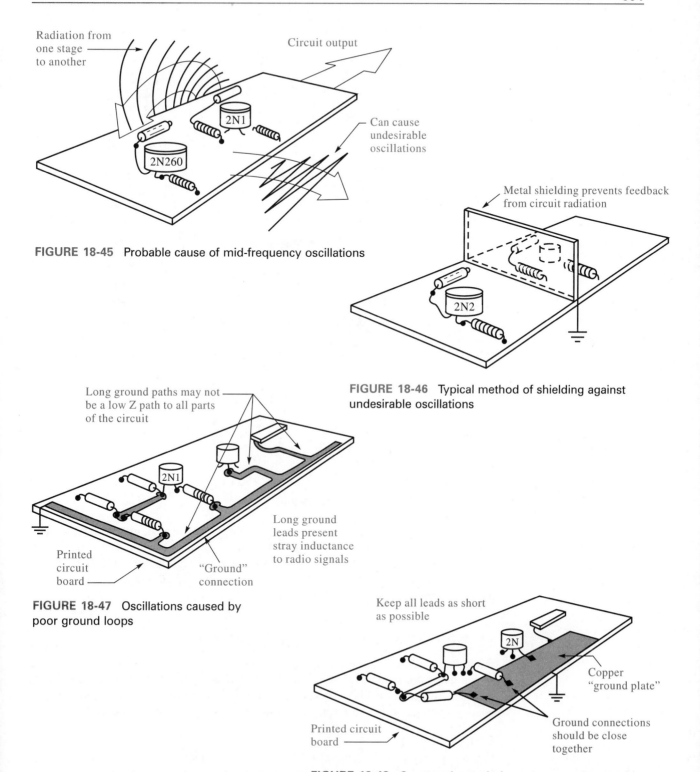

Radiation from one stage to another

Circuit output

2N1

2N260

Can cause undesirable oscillations

FIGURE 18-45 Probable cause of mid-frequency oscillations

Metal shielding prevents feedback from circuit radiation

2N2

FIGURE 18-46 Typical method of shielding against undesirable oscillations

Long ground paths may not be a low Z path to all parts of the circuit

2N1

Long ground leads present stray inductance to radio signals

Printed circuit board

"Ground" connection

FIGURE 18-47 Oscillations caused by poor ground loops

Keep all leads as short as possible

2N

Copper "ground plate"

Printed circuit board

Ground connections should be close together

FIGURE 18-48 Construction techniques to prevent ground loops

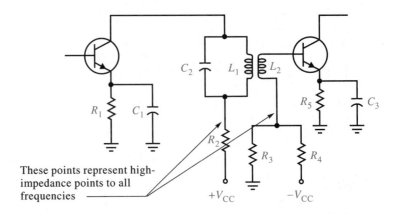

These points represent high-impedance points to all frequencies

FIGURE 18-49 Poor decoupling causing parasitic oscillations

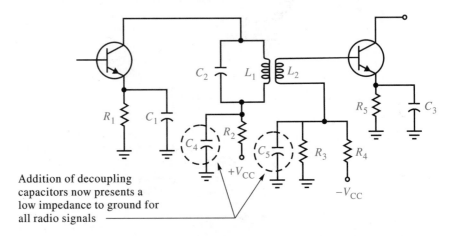

Addition of decoupling capacitors now presents a low impedance to ground for all radio signals

FIGURE 18-50 Proper decoupling

Undesirable oscillations are a very real problem in the design, troubleshooting, and maintenance of communication systems. If the system is designed properly, then, when replacing components, you must use the same lead length and layout of the original component. If you don't, the system may turn into an oscillator and ignore or modify signals it is supposed to process.

18-16 MEASURING FREQUENCY RESPONSE

Often you want to know the frequency response of a communication circuit when a sweep generator is not available. The operation and application of sweep generators were given in section 18-5. This section presents information for measuring the frequency response with a sine-wave generator and a voltmeter or oscilloscope.

General Idea

Figure 18-51 shows the general setup to test the frequency response of an amplifier. What you want to do is measure the amplitude of the amplifier output as you change

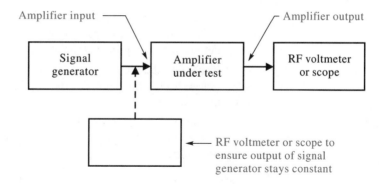

FIGURE 18-51   General setup for measuring frequency response

the frequency of the input. Thus, a voltage reading will be made for a given frequency input. The resultant readings will then be displayed in graphical form as shown in Figure 18-52.

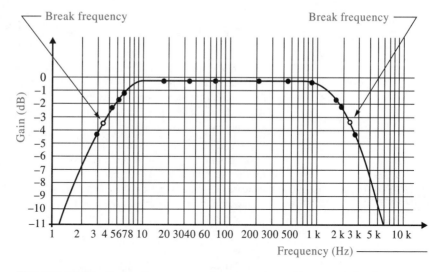

FIGURE 18-52   Recording measurements and resulting graph for frequency-response measurements

At the breakpoints, several frequency readings are made to determine the shape of the curve. At the points where the response is flat, taking readings every octave is satisfactory. Also note from Figure 18-52 that the vertical axis is linear and measured in dB, and the horizontal axis is logarithmic and measured in hertz. This, you should recall, is called a *Bode plot.*

You must observe the following points to obtain accurate measurements:

1. The output of the signal generator must always have exactly the same amplitude as its frequency is changed.

2. The input signal must be made small enough so that the amplifier under test is not saturated. This can easily be checked with a scope connected to the output of the amplifier while a sine wave is inserted at the input.

3. If the circuit contains any AGC, you must break the AGC loop. If you don't, you will read the action of the AGC along with the frequency response of the amplifier.

4. The frequency response of the voltmeter used to measure the output voltage of the amplifier must be flat over the range of frequency measurements to be made.

Performing the Measurements

Table 18-9 lists the steps for making a frequency-response measurement of a typical amplifier.

TABLE 18-9 Steps for Making Frequency Response Measurement

Step	Action	Comments
1	Make sure frequency range of meter or scope is within the frequency range to be measured.	Read owner's manual to see if your voltmeter has the required frequency range. If not, use a scope.
2	Connect the signal generator to equipment under test. Change its frequency from below the amplifier frequency response to above it. Adjust the amplitude as small as possible to prevent saturation of the amplifier.	Make sure that the output amplitude of the generator stays constant. Check this with a scope or voltmeter. Make sure you break the AGC connection.
3	Without adjusting the amount of output signal from the generator, start at the lowest measurable output signal and gradually increase the frequency. Note the change in output amplitude of the amplifier under test.	If the amplitude of the generator does change during these readings, it will be necessary to adjust generator signal amplitude to keep it constant.
4	As you increase the input frequency, the output signal strength of the amplifier should increase.	You are now measuring the low end of the amplifier response.
5	You will arrive at a point where the output signal strength of the amplifier will stay constant as you increase frequency.	You are now measuring the midband of the amplifier response.
6	Keep increasing the frequency until the amplifier output signal starts to decrease.	You are now measuring the high end of the amplifier response.
7	Change the frequency of the generator to find at what frequencies the output voltage is 0.707 of the midband frequencies. Do this for the low end and the high end.	These are the breakpoints or 3 dB points used to measure bandwidth of the amplifier.
8	Transfer your readings to a semilog graph. Be sure to indicate the 3 dB points on the graph.	You will now have a Bode plot of the amplifier.

18-17   POWER SUPPLY TROUBLESHOOTING

Power supply problems can cause many different symptoms in a communication system. Thus always check the power supply first when troubleshooting any electronic system.

This section suggests methods of troubleshooting five major types of power supplies:

1. Unregulated
2. Series regulated
3. Shunt regulated
4. Switching regulator
5. Battery supplied

The first part of this section introduces troubleshooting methods that are common to two or more types of power supplies.

General Procedures

Figure 18-53 shows the block diagram of a typical power supply. There are five major sections that can cause a problem in a power supply system. They are listed in Table 18-10.

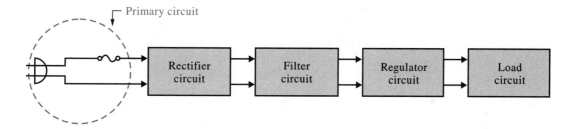

FIGURE 18-53   Block diagram of a typical power supply

TABLE 18-10   **Major Causes of Power Supply Failure**

Circuit	Reason for Failure	Symptom
Primary	Blown fuse or open line cord	No output voltage
Rectifier	Faulty rectifier or open or shorted secondary	Low output or no output voltage
Filter	Open or shorted filter capacitor	Audible hum or no output voltage
Regulator	Open, shorted, or faulty regulator	Poor regulation or no output voltage
Load	Short or open in load	No output for short. Voltage larger than normal output voltage for open.

Figure 18-54 shows a method of isolating a short in a power supply. If the power supply consistently blows a fuse, it usually means a short or overload is in the circuit. The best way to deal with this is to break open the power chain at the indicated points, starting with point A. Then, with the power off, replace the fuse and apply power. Keep repeating this procedure, going next to point B, until the fuse no longer blows. The problem is then in the circuit to the right of the last break you made.

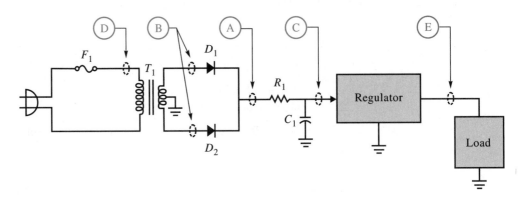

FIGURE 18-54 Method of isolating a short in typical power supply

Series and Shunt Regulated Power Supplies

Be cautious when troubleshooting voltage regulators. Often the problem is not in the regulator but in the load itself. If the load is taking too much current, the voltage regulator will undergo *thermal shutdown* or *current foldback*. In the first case, the output voltage of the regulator will be close to zero. In the second case, the output voltage will be very low. The best way to check for these symptoms is to remove the load from the power source and check the output voltage of the regulator *after it has cooled down.*

IC voltage regulators, because of the built-in protective circuits, are very difficult to destroy. You should suspect these circuits last as a probable cause of power supply problems.

Switching Regulators

A switching regulator can be analyzed with an oscilloscope. One method of troubleshooting is to remove the load from the regulator and replace it with a rheostat in series with a fixed resistor. Be sure that the *power rating of the rheostat is large enough to accommodate the power output of the power supply.* This method is illustrated in Figure 18-55.

Batteries

For the purpose of this section, there are three major types of batteries used as a power source in communication equipment:

1. Dry cell
2. Nickel–cadmium (nicad) battery
3. Lead–acid cell

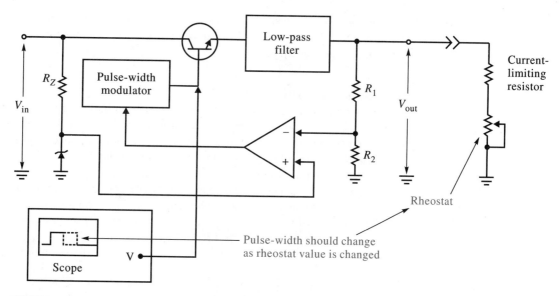

FIGURE 18-55   Method of troubleshooting switching regulator

The problems encountered with each type are summarized as follows:

Dry Cells   Dry cells usually develop a high internal resistance. The symptom here is that the cell will measure an open-circuit voltage that appears to be right, but its output voltage will then decrease the moment it is connected to the load. The rule here is to measure the voltage of these cells when they are connected in the circuit.

Nickel–Cadmium Battery   Nickel–cadmium batteries are rechargeable over a long useful life. These cells produce hydrogen and oxygen gases when being recharged. These two gases are *potentially highly explosive* near an open flame. Charging cells must therefore be well vented and kept away from sparks or open flames.

These cells will usually experience an internal short, indicated by excessive charging current, at which time they should be replaced.

Lead–Acid   Lead–acid batteries are low-voltage high-current types used to start internal combustion engines in vehicles. They are extremely dangerous because of their potentially high current output and corrosive liquid contents.

18-18   FREQUENCY MEASUREMENT

Frequency measurement is necessary in electronic communications. Due to frequency-division multiplexing and the necessary rigorous FCC requirements that transmitters stay on their assigned frequency allocations, frequency measurement becomes very important.

Many methods are used to measure radio frequencies. The method shown here uses the information presented in this section. Frequency counters are commonplace in the communications industry.

Frequency Counter

The block diagram of a frequency counter is shown in Figure 18-56. The more precise the one-shot is, the more accurate the count is. A *one-shot* is a flip-flop that stays true for a given amount of time and then returns to false. Since the one-shot in Figure 18-56 stays true for 1 s, the binary counter will count for the number of cycles of the unknown frequency that occur in 1 s. For the frequency counter shown in the figure, this is only good for frequencies less than 15 Hz. The one-shot could be designed to remain true for 1 ms, and thus the number appearing in the binary counter would be the number of cycles of the unknown waveform that occur in 1 ms. In this case, the highest frequency measured would be 15 kHz.

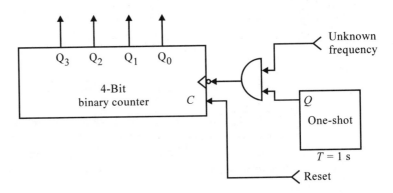

FIGURE 18-56   Block diagram of a frequency counter

Practical frequency counters use a *binary-to-decimal* converter to display the output count in decimal form. This arrangement is shown in Figure 18-57. Observe that the output is now displayed using *seven-segment* readouts. These readouts are commonly found in communication equipment. Many radios use seven-segment readouts to indicate the selected station.

Seven-Segment Readout

Figure 18-58 shows the details of a seven-segment readout. The readout is usually connected in series with a resistor value recommended by the manufacturer. This resistor limits the current in each segment to a safe value. Each segment is nothing more than a *light-emitting diode* (LED). When current flows in the LED, it emits energy in the form of light.

In this section, you saw a practical application of a binary counter. You will find that many other forms of digital electronics are now being used in electronic communications, from satellite communications to digital audio recordings.

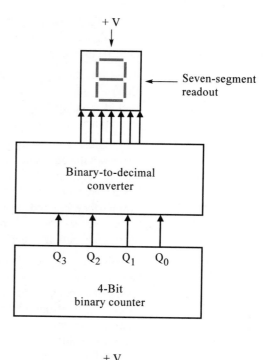

FIGURE 18-57
Using binary-
to-decimal
converters

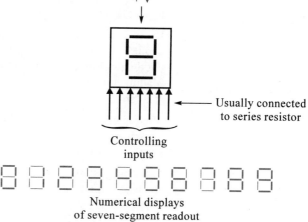

FIGURE 18-58
Typical seven-
segment readout

SUMMARY

- Troubleshooting is the act of locating and repairing a fault that causes improper operation of an electrical circuit.

- Instrumentation is the use of equipment to measure electrical quantities for the purpose of keeping those quantities within prescribed limits to ensure proper system operation.

- Block diagrams are simplified versions of schematics used to trace the signal flow through a communication system. They provide a convenient way of isolating a fault so that an analysis can be made.

- Troubleshooting often involves a trial-and-error process whereby each connection or component is checked in sequence until the fault can be identified and corrected. Two common methods of locating the fault are signal tracing and signal injection.

- Computer-aided troubleshooting (CAT) is dependent upon the use of a computer when troubleshooting and repairing electronic equipment. Any personal computer that can accommodate an interface to allow signal analysis can be used. There are six levels, each requiring more complex instructions than the one before, at which most troubleshooting and repair can be effected. Level one requires considerable technician involvement. As you use higher levels, more decisions will be made by the computer including the use of a robotic device and selection and ordering of parts.

- Troubleshooting and instrumentation are inseparable in many situations. There are many kinds of measurements necessary for evaluation. Many problems require measurements so exact that technicians have to rely on specialized instruments for accuracy. Real-pulse measurements that will be seen on an oscilloscope are important in data communications devices. The shape of a pulse may be critical to successful troubleshooting. Digital meters, signal processors, and radio frequency generators all provide various measurements which are invaluable for successful troubleshooting.

- Laser safety is such an important issue that technicians need to be well informed about the hazards of carelessness and inadequate safety precautions. Lasers produce an extremely intense light that can cause permanent eye and skin damage. Manufacturers must certify that their products are in compliance with federal safety standards.

- Most telephone modem problems are software related. Hardware-related problems are usually quite easy to solve. Table 18-5 presents an easy reference to common problems and their solutions.

- Frequency synthesizers provide stable frequencies that are easily selected by the user. The user can specify output by digital means, thus, different frequencies may be selected by a computer or other digital control circuits that have been pre-programmed to respond to specified conditions. The backbone of the frequency synthesizer is the PLL.

- Undesirable oscillations are a real problem in the design, troubleshooting, and maintenance of communication systems. If the system is designed properly, then, when replacing components, you must use the same lead length and layout of the original component.

- Power supply problems can cause many different symptoms in a communication system. Always check the power supply first when troubleshooting any electronic system.

- Frequency measurement is necessary in electronic communications. Due to frequency-division multiplexing and the necessary rigorous FCC requirements that transmitters stay on their assigned frequency allocations, frequency measurement becomes very important.

- Troubleshooting is becoming a more complex and specialized skill. Computer advances, satellite communications, and networks of all kinds continue to make demands on existing and experimental instrumentation.

 ## QUESTIONS AND PROBLEMS

1. How would you define troubleshooting? Instrumentation?

2. Name the five major categories of troubleshooting.

3. What is usually the easiest category of troubleshooting? What is the most difficult?

4. Define an intermittent. Give an example.

5. Before making a system measurement, of what must you be sure?

6. State a good habit when making measurements.

7. What do you do when analyzing measurement data?

8. What are some actions to be taken when a measurement discrepancy is found?

9. What is a block diagram?

10. How does a schematic differ from a block diagram?

11. When are block diagrams used? When are schematics used?

12. Describe the function of the (A) originating circuit, (B) processing circuit, (C) terminating circuit, (D) mixing circuit, (E) feedback circuit.

13. Give examples of each circuit in question 12.

14. State the troubleshooting procedure for each circuit in question 12.

15. State the function of each section of a superheterodyne receiver.

16. Describe the input and output signals for each stage of a superheterodyne receiver.

17. Describe the signal tracing troubleshooting method. Give an example. What instrument can be used for signal tracing?

18. What is a signal generator?

19. Describe the signal injection method of troubleshooting method. Give an example.

20. Explain the result of having the carrier frequency appear at the output of an SSB transmitter.

21. Describe a dummy load. Why is it needed?

22. Explain a coupling T. Give an example of its use.

23. Describe the oscilloscope pattern of the output wave of an SSB transmitter when no carrier is present.

24. How does the oscilloscope pattern of the output wave of an SSB transmitter appear when the carrier affects the output signal?

25. Describe the basic principle of operation of a sweep generator.

26. Why is the term "sweep generator" used to describe the instrument?

27. Explain the operation of a VCO.

28. What effect does an input control voltage of a sawtooth have on the output frequency of a VCO?

29. Describe the purpose of a discriminator.

30. Explain the basic operation of a differential amplifier.

31. Define ALC. How is it used in a sweep generator?

32. Describe a typical application of a sweep generator. How is an oscilloscope used in this kind of application?

33. List the basic components needed for CAT. What is the purpose of each?

34. Discuss a typical application for CAT. Why is a computer interface circuit needed?

35. List some advantages and disadvantages of CAT.

36. Explain the purpose of an A/D converter.

37. Describe different levels of CAT. What is the highest level you can envision?

38. What is an ideal pulse? Does it exist?

39. Name some factors that characterize a real pulse.

40. How is the rise time of a pulse measured?

41. Explain how pulse width is measured.

42. Describe the changes in the pulse spectrum as the pulse duration is decreased. What happens in the limit (when the pulse becomes an impulse)?

43. Describe the changes in the pulse spectrum as the period between pulses is increased. What happens in the limit (pulse period approaches infinity)?

44. Describe the basic operation of a digital multimeter.

45. What is the purpose of the signal processor circuit used in a DMM?

46. Explain autoranging.

47. Describe the operation of a true rms converter as it applies to a DMM.

48. Explain the process used by a DMM to read resistance.

49. What are some advantages and disadvantages of a DMM compared with a VOM?

50. State the relationship between the RF generator output impedance, characteristic impedance of the connecting cable, and the input impedance of the circuit under test in order for reliable measurements to occur.

51. Explain the use of a resistive pad when an RF generator is used.

52. Discuss what the input impedance of a properly designed resistive pad looks like to the source.

53. Discuss what the output impedance of a properly designed resistive pad looks like to the load.

54. To use the measurements described in this section, state what you should already know about the antenna to be measured.

55. Describe a VSWR bridge.

56. Describe the equipment needed to measure the resonant frequency of an antenna by the sweep generator method.

57. Describe the equipment needed to measure the resonant frequency of an antenna by the RF generator method.

58. What parts of the body are most sensitive to laser damage?

59. Describe the most common type of damage to the human eye by laser radiation. Is this radiation always visible?

60. Describe the most common type of damage to the skin by laser radiation. How long does it usually take this kind of damage to develop?

61. Name the two major standards for lasers.

62. Name the different classes of laser radiation.

63. Describe the meaning of each class of laser radiation.

64. State the most common problems found in using computer telephone modems.

65. Name the two telephone connectors available on most computer telephone modems.

66. Describe the symptom that shows up on the transmitting computer if neither computer has its ECHO ON activated.

67. Describe some advantages of a frequency synthesizer.

68. Name the major sections of a frequency synthesizer.

69. What section of a frequency synthesizer determines the output frequency?

70. Describe a method of synthesizing communications frequencies.

Questions 71–73 refer to the transistor amplifier given in section 18-14.

71. Why are the voltage divider resistors R_{B1} and R_{B2} used?

72. How do the values of R_{B1} and R_{B2} affect the base voltage of V_B?

73. How does the value of R_E affect the base voltage V_B?

74. How does the value of R_E affect the transistor collector voltage V_C?

75. Describe the relationship between the values of R_C and V_C.

76. What are undesirable oscillations?

77. Name three causes of undesirable oscillations.

78. What is motorboating? Describe what usually causes it.

79. Discuss what can be done to prevent undesirable oscillations caused by radiation.

80. Explain the use of a ground plate. How does it serve to prevent undesirable oscillations?

81. Describe what equipment you would use to measure the frequency response of an amplifier if a sweep generator were not available.

82. For the measurement in question 81, what precautions should be taken if any AGC is present in the amplifier?

83. What must you do as the frequency of the signal generator is changed?

84. Describe what results you may get in doing a frequency response while the amplifier under test is saturated. How do you relieve this condition?

85. What is a Bode plot?

86. Name the five major types of power supplies.

87. Describe the five major sections of a power supply. What is the most likely problem and the cause of the problem for each section?

88. How would you isolate a short in a power supply?

89. Discuss the precautions you should exercise when troubleshooting a series or shunt regulated power supply.

90. Describe the major types of batteries in communications equipment.

91. Why is frequency measurement important?

92. Describe the main sections of a frequency counter.

93. Name a main factor that determines the accuracy of a frequency counter.

94. Explain the purpose of a binary-to-decimal converter.

95. Describe the details of a seven-segment readout.

96. What is an LED?

Appendixes

TYPICAL IC APPLICATIONS FOR CHAPTER 2 _____

This section presents a small selection of typical integrated circuits and applications. Many of the applications discussed here and in the following applications can be constructed with single monolithic ICs containing the major circuit components and a minimum of external components. Manufacturers' data books and data sheets provide a wide selection of devices and applications.

Dual-Power Audio Amplifier

The connection diagram in Figure 1(A) shows the LM1877 dual-power audio amplifier, a 14-pin monolithic IC designed to deliver 2 W per channel continuously into 8 Ω loads. The device is designed to operate with few external components and still provide flexibility for use in stereo phonographs, tape recorders, and AM-FM stereo receivers. Each power amplifier is biased from a common internal regulator to provide high power supply rejection and output Q centering. The IC is internally compensated for all gains greater than 10 and contains both internal current limiting and thermal shutdown for device protection. It operates from a 6 to 24 V supply and provides a typical open-loop output gain of 70 dB, with minimum channel separation of 50 dB and minimum power supply ripple rejection ratio (PSRR) of 50 dB.

A stereo phonograph amplifier with bass tone control is shown in Figure 1(B). The frequency response of the bass tone control is essentially flat from 100 Hz to 20 kHz. The circuit voltage gain A_v is 50 (34 dB) when feeding into an 8 Ω load. The circuit in Figure 1(C) is a stereo amplifier with an output voltage gain of 200 (about 46 dB) into an 8 Ω load.

Audio Power Amplifier

The LM380 audio power amplifier, shown in Figure 2(A), is designed for low-cost consumer applications. Gain is internally fixed at 34 dB, the unique configuration of the input stage allows inputs to be ground referenced, and the output is automatically self-centering to one half of the supply voltage. The output is short-circuit protected with internal thermal-limiting protection.

Uses of the LM380 include simple phonograph amplifiers (Figure 2(B)), phase-shift oscillators (Figure 2(C)), intercoms (Figure 2(D)), and bridge amplifiers (Figure 2(E)). Other uses include line drivers, teaching-machine outputs, alarms, ultrasonic drivers, TV sound systems, AM-FM radios, small servo drivers, and power converters.

Battery Operated Low-Voltage Audio Amplifier

A battery operated device is the LM389 low-voltage audio power amplifier with NPN transistor array, shown in Figures 3(A) and (B). The amplifier inputs are ground referenced, and the output is automatically biased to one half of the supply voltage. Figures 3(C) through (F) show four typical applications for the LM389.

The three transistors have high gain and excellent matching characteristics. They are well suited to a wide variety of applications in dc through VHF systems. The transistors are general-purpose devices that can be used in the same manner as other small-signal

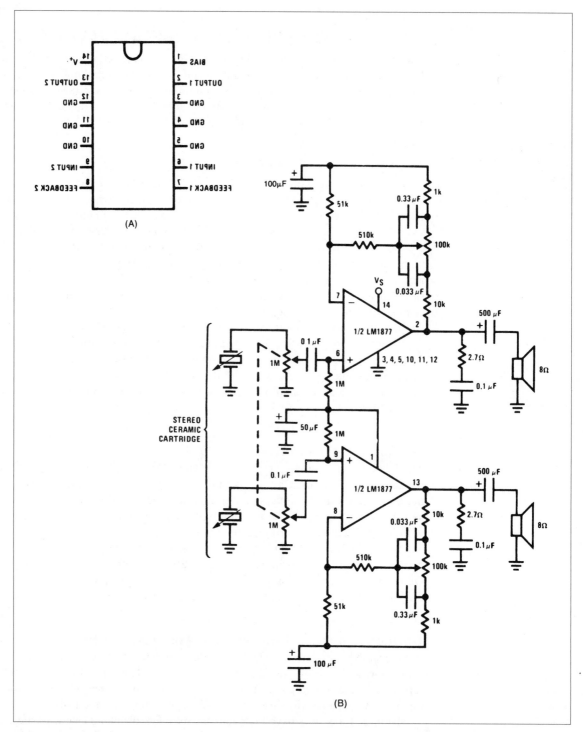

FIGURE 1 LM1877 dual-power audio amplifier (A) Connection diagram (B) Stereo phono amplifier with bass tone control (C) Stereo amplifier with $A_v = 200$

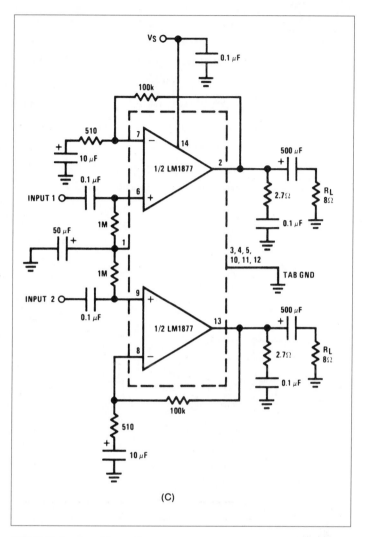

FIGURE 1   *(continued)*

transistors. So long as the currents and voltages are kept within the absolute maximum limitations and the collectors are never allowed to go to ground potential with respect to pin 17, there is no limit to the way they can be used.

The LM389 is made a more versatile amplifier by pins 4 and 12, which provide gain control. With pins 4 and 12 open, the 1.35 kΩ resistor sets the gain at 20. Bypassing the 1.35 kΩ resistor with a 10 µF capacitor between pins 4 and 12 increases the gain to 200. If we put a variable resistor in series with the bypass capacitor, we can set the gain at any level between 20 and 200.

Additional external components can be placed in parallel with the internal feedback resistors to tailor the gain and frequency response for individual applications. For example, we can compensate for poor speaker bass response by placing a series *RC* network from pin 1 to pin 12, thus paralleling the internal 15 kΩ resistor.

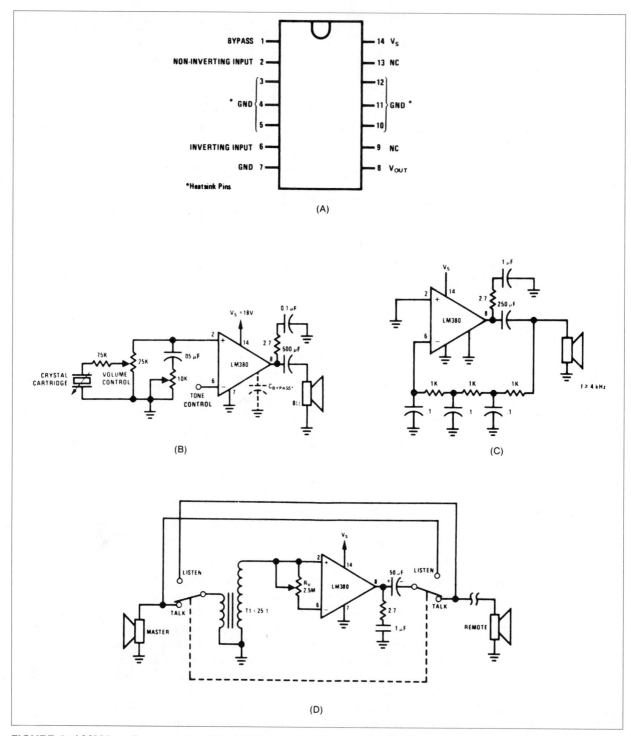

FIGURE 2 LM380 audio power amplifier (A) Connection diagram (B) Phonograph amplifier (C) Phase-shift oscillator (D) Intercom (E) Bridge amplifier

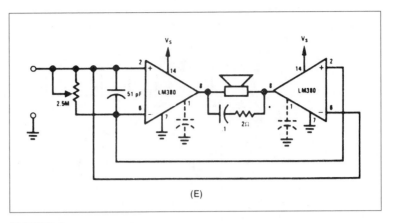

(E)

FIGURE 2 (*continued*)

The schematic in Figure 3(B) shows that both inputs are biased to ground with 50 kΩ resistors. The base current of the input transistors is about 250 nA, so the inputs are at about 12.5 mV when left open. If the dc source resistance driving the LM389 is higher than 250 kΩ, it will contribute very little additional offset (about 2.5 mV at the input and 50 mV at the output). If the dc source resistance is less than 10 kΩ, then shorting the unused input to ground will keep the offset low. For dc resistances between these values, we can eliminate excess offset by putting a resistor of equal value to the dc source resistance from the unused input to ground. All offset problems can be eliminated by capacitively coupling the inputs.

When the LM389 is used with higher gains (bypassing the internal 1.35 kΩ resistor), it is necessary to bypass the unused input, with a 0.1 μF capacitor or a short to ground, depending on the dc source resistance, to prevent gain degradation and possible instability.

IC APPLICATIONS FOR AM FOR CHAPTER 3 ⎯⎯⎯⎯

Integrated circuits have greatly simplified the design and construction of AM radio systems. This section examines a few applications in which these modern ICs are used.

LM1863 AM Radio System

The block diagram of the 18-pin LM1863 IC is shown in Figure 4. This IC is a high-performance AM radio system intended primarily for electronically tuned radios (ETRs). Important to this application is an on-chip stop detector circuit, which allows for a user-adjustable signal level threshold and center frequency stop window. The IC uses a low-phase-noise, level-controlled LO.

A buffered output for the LO allows the IC to drive a phase-locked loop synthesizer directly. The IC uses an RF automatic gain control detector to gain-reduce an external RF stage, thereby preventing overload by strong signals. An improved noise floor and a lower third harmonic distortion (THD) are achieved through gain reduction of the IF stage. Fast AGC setting time and excellent THD performance are achieved by using a two-pole AGC system. Low *tweet* (an audio tone produced by the second and third

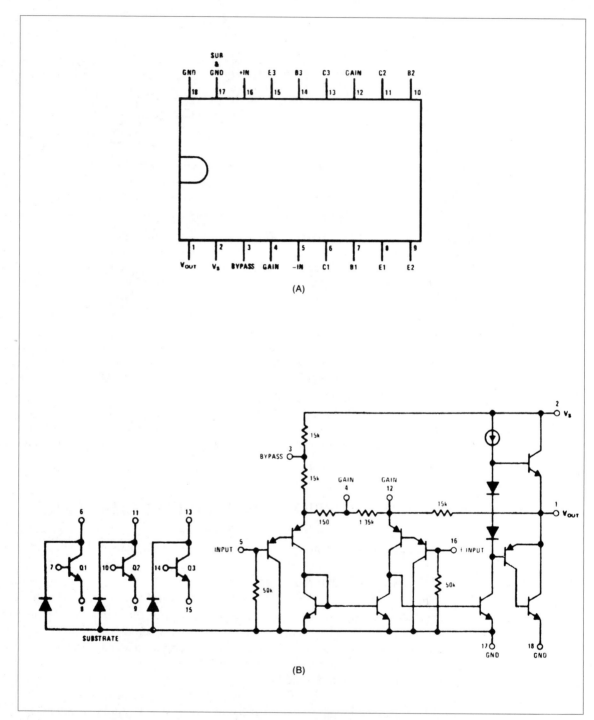

FIGURE 3 LM389 battery-operated, low-voltage audio amplifier (A) Connection diagram (B) Equivalent schematic (C) Ceramic phono amplifier with tone controls (D) FM scanner noise-squelch circuit (E) AM radio (F) Tape recorder

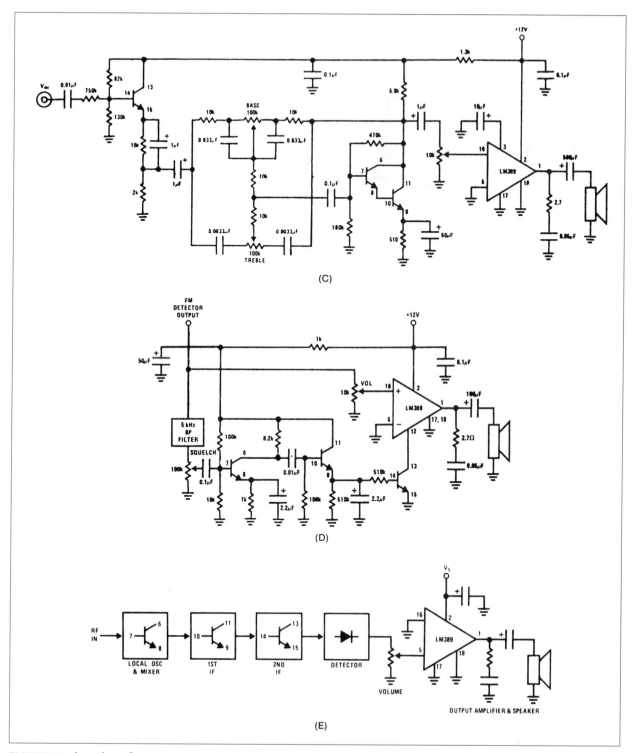

(C)

(D)

(E)

FIGURE 3 (*continued*)

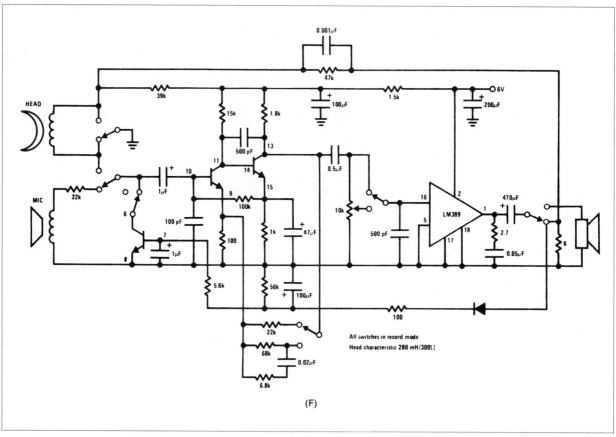

FIGURE 3 (*continued*)

harmonics of the IF beating against the received signal) radiation and sufficient gain are provided to allow the IC to be used in conjunction with a loop-stick antenna.

NE546 AM Radio Receiver Subsystem

The NE546 AM radio receiver subsystem shown in the block diagram in Figure 5(A) is a 14-pin monolithic IC that provides a mixer, RF amplifier, IF amplifier, voltage regulator, LO, and automatic gain control detector in a single chip. The NE546 is used primarily for superheterodyne AM radio receivers, particularly, in automobile radios. A typical application is shown in Figure 5(B).

TCA440 AM Receiver Circuit

The TCA440 AM receiver circuit, shown in the block diagram in Figure 6(A), is a 16-pin monolithic IC developed especially for AM receivers operating at frequencies up to 50 MHz. It includes an RF stage with AGC, a balanced mixer, a separate oscillator, and an IF amplifier with AGC. Because of its low current consumption and internal stabilization, the TCA440 is perfectly suited for battery operated portable, automobile, and home radios.

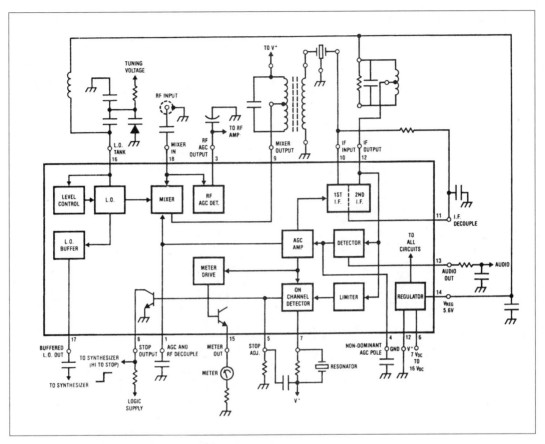

FIGURE 4   Block diagram of the LM1863 AM radio system IC

The TCA440 contains two control loops, independent of each other, that control the RF and IF stages. By AGCing the RF stage, we can obtain excellent signal handling. A voltage of 2.6 V_{p-p} on the IC input can be handled with very low distortion. A push-pull mixer operates multiplicatively, resulting in few harmonic mixing products and whistling points. The oscillator, which is separated from the mixer, is also very adaptable for short waves. From the AGC of the RF amplifier, a voltage is derived that can be connected directly to a tuning meter. The symmetric composition of the circuit provides high stability against oscillation and, at the same time, an AGC range of more than 100 dB. A mixer bridge circuit provides good isolation of the oscillator. A typical AM receiver application for the TCA440, using varicap diodes, is shown in Figure 6(B).

LM3820 AM Radio System

The LM3820, shown in Figure 7, is a three-stage AM radio IC consisting of an LO, mixer, RF amplifier, IF amplifier, AGC detector, and zener regulator. Although originally designed for use in slug-tuned automobile radio applications, it is also suitable for capacitor-tuned portable radios.

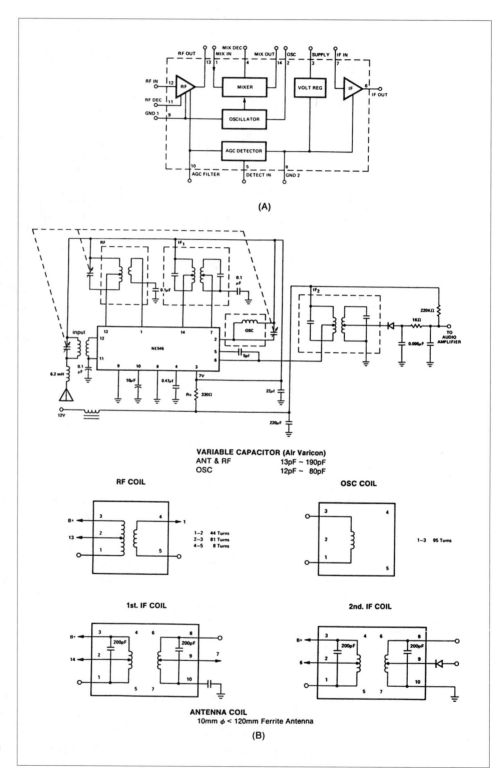

FIGURE 5
NE546 AM radio receiver subsystem
(A) Block diagram
(B) Typical application circuit

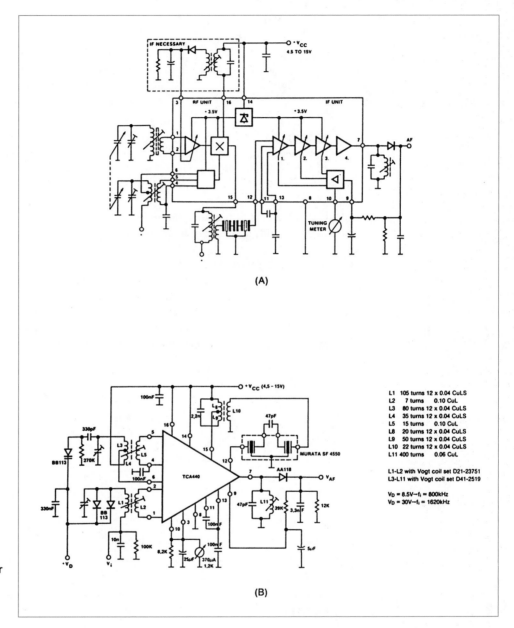

(A)

(B)

FIGURE 6
TCA440 AM receiver circuit (A) Block diagram (B) Typical application circuit

PLL APPLICATIONS FOR CHAPTER 4 _____

The PLL circuit has many applications. It is ideally suited for routine applications such as AM and FM detectors, FSK decoders, signal conditioning, prescalars for frequency counters, tone detectors, and touch-tone decoders. In general, when the circuit is used to control the frequency of an oscillator, such as in radio and television receiver sound systems, it is termed *automatic frequency control* (AFC). In the RF tuner section of

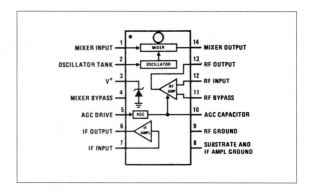

FIGURE 7
Connection diagram
of the LM3820 AM
radio system IC

television receivers, it is called *automatic fine tuning* (AFT), and in the horizontal synchronizing circuit, it is called the *horizontal AFC*.

Frequency Synthesis

An important application is in communications systems where a CCO, such as the MC4024 (Figure 8), is the standard source for a reference. The PLL circuit then makes an oscillator without a crystal have the same frequency stability as the crystal reference oscillator. This procedure is called *frequency synthesis*—that is, putting together (mixing) two frequencies to provide the desired output.

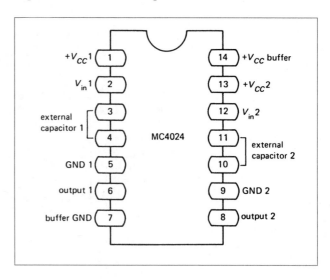

FIGURE 8
MC4024 CCO

Figure 9 shows a basic frequency synthesizer, where a divide-by-N ($\div N$) counter is inserted in the feedback path of a PLL. The output frequency of the synthesizer is

$$f_{out} = Nf_{in} \tag{1}$$

where

N = divider value of $\div N$ counter

f_{in} = input frequency (generally from a CCO)

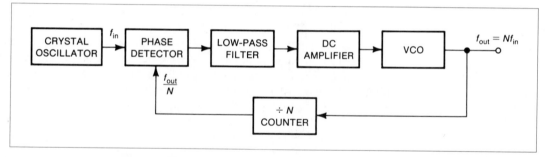

FIGURE 9   Basic frequency synthesizer

The synthesizer phase detector produces a dc control voltage proportional to the phase difference between the input frequency f_{in} and the $\div N$ counter output (f_{out}/N). The counter generates a single output pulse for every N input pulses. The dc control voltage from the phase detector, after filtering and amplification, then controls the output frequency of the VCO. The output signal from the $\div N$ counter is equivalent to the reference input frequency, except for the small phase difference, or

$$f_{in} = \frac{f_{out}}{N} \qquad (2)$$

EXAMPLE 1

A frequency synthesizer has an input frequency of 750 kHz. The $\div N$ counter is set at a count of 50. Calculate the synthesizer frequency output.

SOLUTION

From Equation 1,

$$f_{out} = Nf_{in} = (50)(750 \text{ kHz}) = 37.5 \text{ MHz}$$

EXAMPLE 2

What is the input frequency of the phase detector if the output frequency of a synthesizer is 200 MHz and the $\div N$ counter is set at a count of 400?

SOLUTION

From Equation 2,

$$f_{in} = \frac{f_{out}}{N} = \frac{200 \text{ MHz}}{400} = 500 \text{ MHz}$$

The 560 Series PLL

A group of monolithic analog PLL devices is the 560 series, which uses the double-balanced mixer, the output of which is an average dc voltage proportional to the phase difference between its two inputs. The most popular of the series are the 565 (Figure 10) and the 567 (Figure 11).

The 565 is a general-purpose PLL containing a stable, highly linear VCO for low-distortion FM demodulation and a double-balanced phase detector with good carrier suppression. The VCO frequency is set with an external resistor and capacitor, and a tuning range of 10:1 can be obtained with the same capacitor. The bandwidth, response speed, capture, and pull-in range can be adjusted over a wide range with an

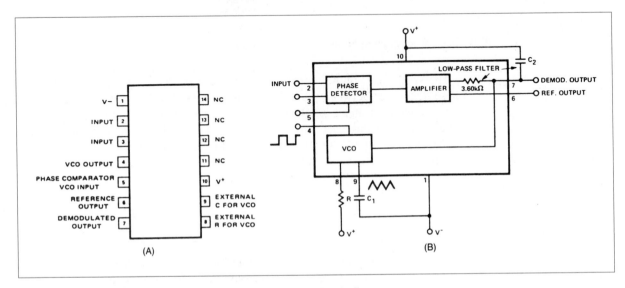

FIGURE 10 SE/NE565 PLL (A) Connection diagram (B) Block diagram

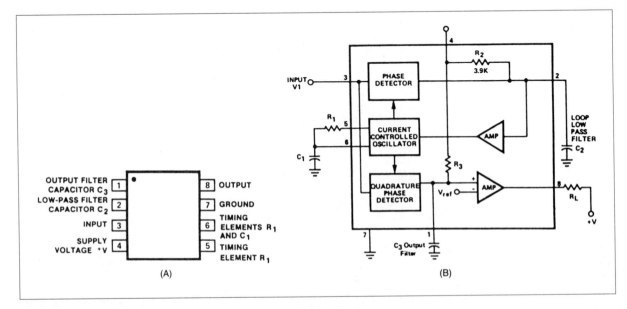

FIGURE 11 SE/NE567 PLL (A) Connection diagram (B) Block diagram

external resistor and capacitor. Inserting a digital frequency divider between the VCO and the phase detector provides frequency synthesis.

Other applications of the 565 are data and tape synchronization, FSK demodulation, FM demodulation, tone decoding, frequency multiplication and division, telemetry receivers, and signal generation.

The 567 is a general-purpose tone decoder designed to provide a saturated transistor switch to ground when an input signal is present within the pass band. The bandwidth, center frequency, and output delay are independently determined by external components. Typical applications are touch-tone decoding, precision oscillators, frequency monitoring and control, wide-band FSK demodulation, ultrasonic controls, carrier current remote controls, communications paging decoders, and 0° to 180° phase shifting.

Four-Block VCO

The XR-2207 in Figure 12 is a monolithic IC voltage-controlled oscillator composed of four functional blocks: a VFO that generates the basic periodic waveforms; four current switches actuated by binary keying inputs; and two buffer amplifiers for simultaneous triangular- and square-wave outputs, available from 0.01 Hz to 1 MHz. The internal switches transfer the oscillator current to any of four external timing resistors to produce four discrete frequencies that are selected according to the binary logic levels at the keying terminals. This device is ideally suited for FM, FSK, and sweep or tone generation, as well as for PLL applications.

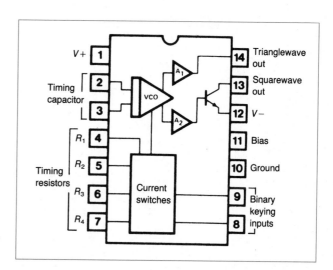

FIGURE 12
Connection and block diagram for the XR-2207 VCO

Frequency-Locked Loops FLL circuits are generally more complicated than phased-locked loops. Frequency-locked loops are used primarily in tuning and control circuits.

Tuning is performed using FLL digital control. The actual tuning frequency is divided by a factor of 64 or 256.

The minimum tuning voltage that can be generated during digital tuning is programmable to prevent the tuner from being driven into an unspecified low tuning voltage region.

LINE IC APPLICATIONS FOR CHAPTER 6 _____

One might think there would be no place for IC applications in transmission line systems, but in a few instances an IC can solve a problem simply and inexpensively. This section examines a few of these ICs and their applications.

Improving Coaxial Cable Performance

The circuit in Figure 13(A) illustrates the connection between the shield of an input coaxial cable and an LM11C operational amplifier (op amp) configured as a voltage follower. This simple circuit reduces cable capacitance, leakage, and spurious voltages from cable flexing. Instability can be avoided with a small (0.01 µF) capacitor on the input.

Figure 13(B) illustrates the coaxial cable connected to a summing amplifier configuration. The summing node is at virtual ground, so it is best to ground the shield of the input cable. The small feedback capacitor (0.01 µF) ensures stability. Figure 13(C) shows the connection diagram for the LM11 op amp.

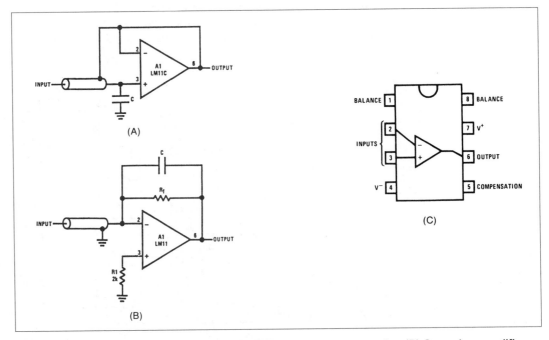

FIGURE 13   LM11 operational amplifier (A) Voltage follower connection (B) Summing amplifier connection (C) Connection diagram

Line Drivers

A balanced line driver (amplifier) using the LM159/359 dual high-speed programmable current mode (Norton) op amp is shown in Figure 14(A). This circuit operates at a 1 MHz bandwidth with a gain of 10 into a 600 Ω load. At full bandwidth, it produces 0.3% distortion, reduced to 0.05% with a bandwidth of 10 kHz. It will drive $C_L = 1500$ pF

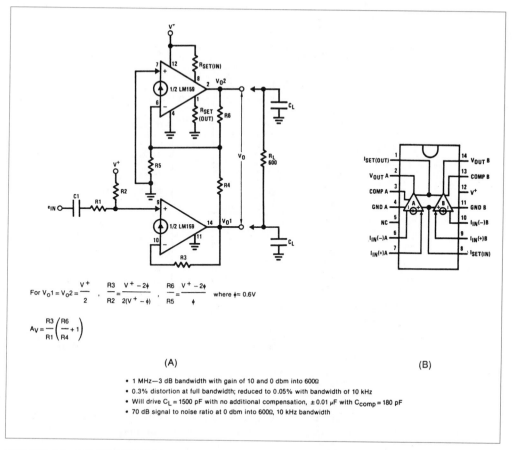

For $V_{o}1 = V_{o}2 = \dfrac{V^{+}}{2}$, $\dfrac{R3}{R2} = \dfrac{V^{+} - 2\phi}{2(V^{+} - \phi)}$, $\dfrac{R6}{R5} = \dfrac{V^{+} - 2\phi}{\phi}$ where $\phi \approx 0.6V$

$$A_V = \dfrac{R3}{R1}\left(\dfrac{R6}{R4} + 1\right)$$

(A)   (B)

- 1 MHz—3 dB bandwidth with gain of 10 and 0 dbm into 600Ω
- 0.3% distortion at full bandwidth; reduced to 0.05% with bandwidth of 10 kHz
- Will drive C_L = 1500 pF with no additional compensation, ±0.01 μF with C_{comp} = 180 pF
- 70 dB signal to noise ratio at 0 dbm into 600Ω, 10 kHz bandwidth

FIGURE 14   LM159/LM359 Norton amplifier (A) Balanced line driver (B) Connection diagram

with no additional compensation. When operating at a bandwidth of 10 kHz into a 600 Ω load, the circuit provides a 70 dB *S/N* ratio. Figure 14(B) shows the connection diagram for the LM159/359 op amp.

The line driver shown in Figure 15(A) can accept an unbalanced high-impedance input and convert it to a balanced output suitable for driving a low-impedance line. It is particularly useful in an environment where magnetically induced hum or noise pickup is a problem.

The outputs of the two LM13080 op amps are of opposite polarity. Therefore, terminating the line with a balanced load (for example, a differential amplifier or a transformer) will cause common-mode interference pickup to be canceled.

This circuit will drive a 20 $V_{p\text{-}p}$ signal into a 50 Ω load for frequencies up to 10 kHz. Above 10 kHz, the output signal is slew rate limited, but the line driver will still supply a 13 $V_{p\text{-}p}$ signal at 20 kHz. The voltage gain of the network is 2, and the low-frequency roll-off is determined by $f_{\text{low}} = 1/2\pi RC$.

If the load is connected directly between the outputs of the amplifiers, the line driver becomes a simple bridge amplifier capable of delivering 2 W into a 16 Ω load.

Figure 15(B) shows the connection diagram for the LM13080 op amp.

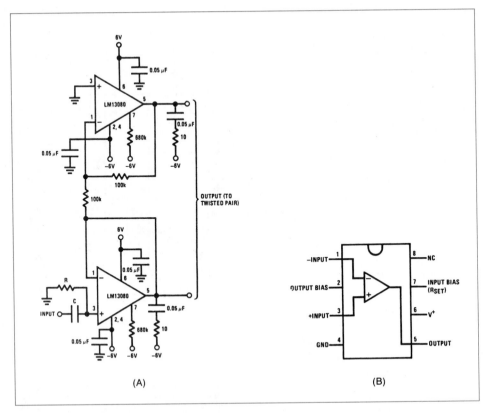

FIGURE 15 LM1380 programmable power op amp (A) Line driver for unbalanced input to balanced ouput (B) Connection diagram

TELEPHONE SYSTEM IC APPLICATIONS FOR CHAPTER 9

Integrated circuits are ideally suited to telephone system applications because of their small size, low cost, flexibility, and reliability. Because there are so many different ICs and applications for telephone systems, we limit this section to a few representative circuits.

Electronic Telephone Circuit

The MC34010/MC34011 electronic telephone circuits (ETCs) provide all the necessary elements of a tone dialing telephone in a single IC. The functional blocks of the ETC (Figure 16(A)) include the DTMF dialer (Figure 16(B)), speech network (Figure 16(C)), tone ringer (Figure 16(D)), and dc line interface circuit (Figure 16(E)). The MC34010 also provides a microprocessor interface (Figure 16(F)) port that facilitates automatic dialing features.

Low-voltage operation is a necessity for telephones in networks where parallel telephone connections are common. An electronic speech network operating in parallel with a conventional telephone may receive line voltages below 2.5 V. DTMF dialers operate at similar low line voltages when signaling through battery powered station carrier equipment. These low voltage requirements have been addressed in these ICs by use of a bipolar I^{2}L (integrated injection logic) technology with appropriate circuit techniques. The resulting speech and dialer circuits maintain specified performance with instantaneous input voltage as low as 1.4 V.

Applications Information Figure 16(G) specifies a typical application circuit for the MC34010 and MC34011. Complete listings of external circuit components, along with nominal component values, are provided in Table 1.

The hook switch and polarity guard bridge configuration in Figure 16(G) is one of several options. If two bridges are used, one for the tone ringer and the other for speech and dialer circuits, then the hook switch can be simplified. Component values should be varied to optimize telephone performance parameters for each application. The relationships between the application circuit components and certain telephone parameters are briefly described in the following paragraphs.

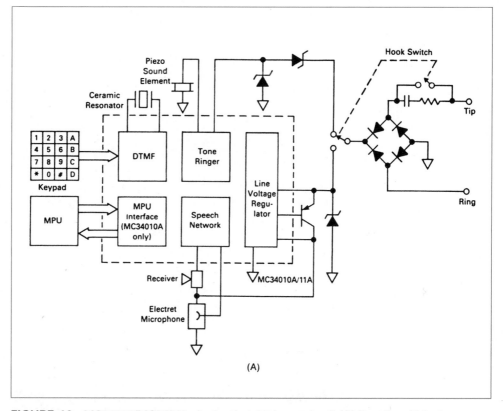

FIGURE 16 MC34010/MC34011 electronic telephone circuit (A) Functional block diagram (B) DTMF dialer block diagram (C) Speech network block diagram (D) Tone ringer block diagram (E) dc line interface block diagram (F) Microprocessor interface block diagram (MC34010 only) (G) Typical application circuit

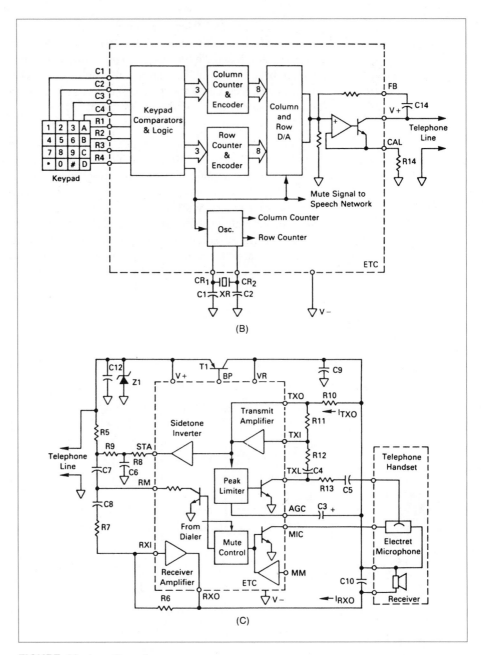

FIGURE 16 (*continued*)

On-Hook Input Impedances—R_1, C_{17}, and Z_3 are significant components for on-hook impedance. C_{17} dominates at low frequencies, R_1 at high frequencies, and Z_3 provides the *nonlinearity* (a measure of the deviation of an analog output level from an ideal transfer curve) required for 2.5 V and 10 V impedance signature tests. C_{17} must generally be ≤ 1.0 μF to satisfy 5.0 Hz impedance specifications.

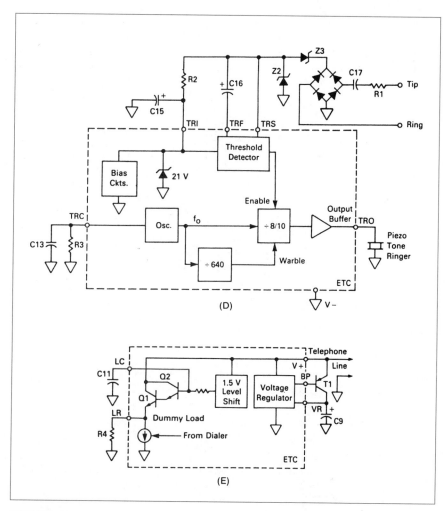

FIGURE 16 (*continued*)

Tone Ringer Output Frequencies—R_3 and C_{13} control the frequency f_o of a *relaxation oscillator* (an oscillator whose operation results from the buildup of a charge and the rapid discharge of a capacitor). Typically, $f_o = (R_3 C_{13} + 8.0\ \mu s)^{-1}$. The output tone frequencies are $f_o/10$ and $f_o/8$. The *warble rate* (a periodic rise and fall in the pitch of the tone) is $f_o/640$. The tone ringer will operate with f_o from 1.0 kHz to 10 kHz. R_3 should be limited to values between 150 kΩ and 300 kΩ.

Tone Ringer Input Threshold—After R_1, C_{17}, and Z_3 are chosen to satisfy on-hook impedance specifications, R_2 is chosen for the desired ring start threshold. Increasing R_2 reduces the ac input voltage required to activate the tone ringer output. R_2 should be limited to values between 0.8 kΩ and 2.0 kΩ.

Off-Hook dc Resistance—R_4 conducts the dc line current in excess of the speech and dialer bias current. Increasing R_4 increases the input resistance of the telephone for line currents above 10 mA. R_4 should be selected between 40 Ω and 120 Ω.

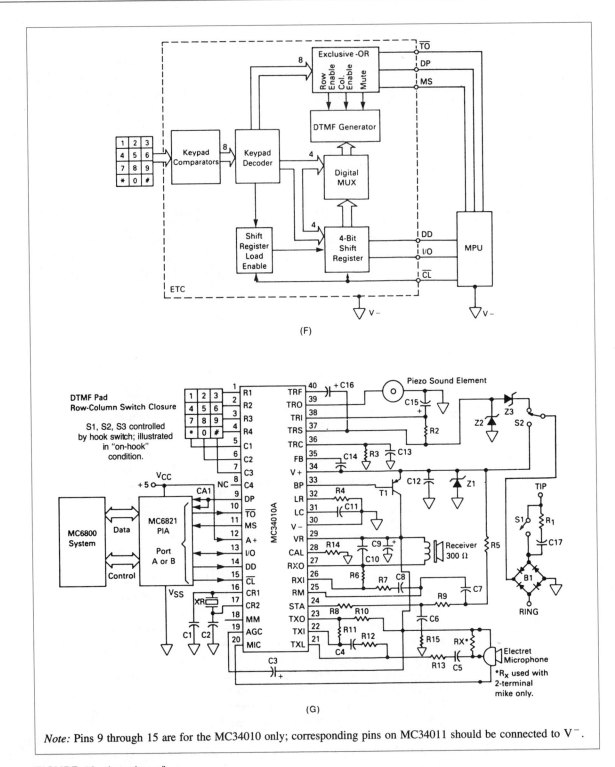

(F)

(G)

Note: Pins 9 through 15 are for the MC34010 only; corresponding pins on MC34011 should be connected to V⁻.

FIGURE 16 *(continued)*

TABLE 1 **External Components for Figure 16(G)**

Capacitors	Nominal Value	Description
C1, C2	100 pF	Ceramic Resonator oscillator capacitors.
C3	1.0 µF, 3.0 V	Transmit limiter low-pass filter capacitor: controls attack and decay time of transmit peak limiter.
C4, C5	0.1 µF	Transmit amplifier input capacitors: prevent dc current flow into TXL pin and attenuate low-frequency noise on microphone lead.
C6	0.05 µF	Sidetone network capacitor: provides phase-shift in sidetone path to match that caused by telephone line reactance.
C7, C8	0.05 µF	Receiver amplifier input capacitors: prevent dc current flow into RM terminal and attenuates low frequency noise on the telephone line.
C9	2.2 µF, 3.0 V	VR regulator capacitor: compensates the VR regulator to prevent oscillation.
C10	0.01 µF	Receiver amplifier output capacitor: frequency compensates the receiver amplifier to prevent oscillation.
C11	0.1 µF	DC load filter capacitor: prevents the dc load circuit from attenuating ac signals on V +.
C12	0.01 µF	Telephone line bypass capacitor: terminates telephone line for high frequency signals and prevents oscillation in the VR regulator.
C13	620 pF	Tone ringer oscillator capacitor: determines clock frequency for tone and warble frequency synthesizers.
C14	0.1 µF	DTMF output feedback capacitor: ac couples feedback around the DTMF output amplifier which reduces output impedance.
C15	4.7 µF, 25 V	Tone ringer input capacitor: filters the rectified tone ringer input signal to smooth the supply potential for oscillator and output buffer.
C16	1.0 µF, 10 V	Tone ringer filter capacitor: integrates the voltage from current sense resistor R2 at the input of the threshold detector.
C17	1.0 µF, 250 Vac Nonpolarized	Tone ringer line capacitor: ac couples the tone ringer to the telephone line: partially controls the on-hook input impedance of telephone.
C18	25 pF, 25 V	Speech equalization coupling capacitor. Prevents dc current flow into SPE terminal. (optional)
C19	5.0 µF, 3.0 V	Sidetone equalization coupling capacitor. Prevents dc current flow into STE terminal. (optional)

Resistors	Nominal Value	Description
R1	6.8 k	Tone ringer input resistor: limits current into the tone ringer from transients on the telephone line and partially controls the on-hook impedance of the telephone.
R2	1.8 k	Tone ringer current sense resistor: produces a voltage at the input of the threshold detector in proportion to the tone ringer input current.

TABLE 1   (*continued*)

Resistors	Nominal Value	Description
R3	200 k	Tone ringer oscillator resistor: determines the clock frequency for tone and warble frequency synthesizers.
R4	82, 1.0 W	DC load resistor: conducts all dc line current in excess of the current required for speech or dialing circuits; controls the off-hook dc resistance of the telephone.
R5, R7	150 k, 56 k	Receiver amplifier input resistors: couple ac input signals from the telephone line to the receiver amplifier; signal in R5 subtracts from that in R9 to reduce sidetone in receiver.
R6	200 k	Receiver amplifier feedback resistor: controls the gain of the receiver amplifier.
R8, R9	1.5 k, 30 k	Sidetone network resistors: drive receiver amplifier input with the inverted output signal from the transmitter; phase of signal in R9 should be opposite that in R5.
R10	270	Transmit amplifier load resistor: converts output voltage of transmit amplifier into a current that drives the telephone line; controls the maximum transmit level.
R11	200 k	Transmit amplifier feedback resistor: controls the gain of the transmit amplifier.
R12, R13	4.7 k, 4.7 k	Transmit amplifier input resistors: couple signal from microphone to transmit amplifier, control the dynamic range of the transmit peak limiter.
R14	36	DTMF calibration resistor: controls the output amplitude of the DTMF dialer.
R15	2.0 k	Sidetone network resistor (optional): reduces phase shift in sidetone network at high frequencies.
R17	600	Speech equalization resistor. Reduces transmit and receive gain when EV terminal switches on. (optional)
R18	5.1 k	Sidetone equalization resistor. Reduces sidetone level when ES terminal switches on. (optional)
R_X	3.0 k	Microphone bias resistor: sources current from VR to power a 2-terminal electret microphone; R_X is not used with 3-terminal microphones.

Semiconductors	Electret Mic	Receiver
B1 = MDA101A, or equivalent, or 4-IN4005 T1 = 2N4126 or equivalent Z1 = 18 V, 1.5 W, IN5931A Z2 = 30 V, 1.5 W, IN5936A Z3 = 4.7 V, 1/2 W, IN750 XR—muRata Erie CSB 500 kHz Resonator, or equivalent Piezo—PBL 5030BC Toko Buzzer or equivalent	2 Terminal, Primo EM-95 (Use R_X) or equivalent 3 Terminal, Primo 07A181P (Remove R_X) or equivalent	Primo Model DH-34 (300 Ω) or equivalent

Off-Hook ac Impedance—The ac input impedance is equal to the receive amplifier load impedance (at RXO) divided by the receive amplifier gain (voltage from V + to RXO). Increasing the impedance of the receiver increases the impedance of the telephone. Increasing the gain of the receiver amplifier decreases the impedance of the telephone.

DTMF Output Amplitude—R_{14} controls the amplitude of the row and column DTMF tones. Decreasing R_{14} increases the level of tones generated at V +. The ratio of the row and column tone amplitudes is internally fixed. R_{14} should be greater than 20 Ω to avoid excessive current in the DTMF output amplifier.

Transmit Output Level—R_{10} controls the maximum signal amplitude produced at V + by the transmit amplifier. Decreasing R_{10} increases the transmit output signal at V +. R_{10} should be greater than 250 Ω to limit current in the transmit amplifier output.

Transmit Gain—The gain from the microphone to the telephone line varies directly with R_{11}. Increasing R_{11} increases the signal applied to R_{10} and the ac current driven through R_{10} to the telephone line. The closed loop-gain from the microphone to the TXO terminal should be greater than 10 to prevent transmit amplifier oscillations.

Note: Adjustments to transmit level and gain are complicated by the addition of receiver sidetone current to the transmit amplifier output current at V +. Normally, the sidetone current from the receiver will increase the transmit signal (if the current in the receiver is in phase with that in R_{10}). Thus, the transmit gain and sidetone levels cannot be adjusted independently.

Receiver Gain—Feedback resistor R_6 adjusts the gain at the receiver amplifier. Increasing R_6 increases the receiver amplifier gain.

Sidetone Level—Sidetone reduction is achieved by the cancellation of receiver amplifier input signals from R_9 and R_5. R_8, R_{15}, and C_6 determine the phase of the sidetone balance signal in R_9. The ac voltage at the junction of R_8 and R_9 should be 180° out of phase with the voltage at V +. R_9 is selected such that the signal current in R_9 is slightly greater than that in R_5. This ensures that the sidetone current in the receiver adds to the transmit amplifier output current.

Hook-Switch Click Suppression—When the telephone is switched to the off-hook condition, C_3 charges from 0 V to 300 mV bias voltage. During this time interval, receiver clicks are suppressed by a low impedance at the RM terminal. If this click suppression mechanism is desired during a rapid succession of hook switch transitions, then C_3 must be quickly discharged when the telephone is on-hook. R_{16} and S_3 provide a rapid discharge path for C_3 to reset the click suppression timer. R_{16} is selected to limit the discharge current in S_3 to prevent damage to switch contacts.

Microprocessor Interface (MC34010 Only)—The six microprocessor interface lines (DP, $\overline{TO}$, MS, DD, I/O, and $\overline{CL}$) can be connected directly to a port, as shown in Figure 16(G). The DP line (depressed pushbutton) is also connected to an interrupt line to signal the microprocessor to begin a read data sequence when storing a number into memory. The MC34010 clock speed requirement is slow enough (typically 20 kHz) so that it is not necessary to divide down the processor's system clock, but rather a port output can be toggled. This facilitates synchronizing the clock and data transfer, eliminating the need for hardware to generate the clock.

The DD pin must be maintained at a Logic "0" when the microprocessor section is not in use, to permit normal operation of the keypad.

When the microprocessor interface section is not in use, the supply voltage at Pin 12 (A +) may be disconnected to conserve power. Normally, the speech circuitry is powered by the voltage supplied at the V + terminal (Pin 34) from the telephone lines. During this time, A + powers only the active pullups on the three microprocessor outputs (DP, MS, and I/O). When the telephone is "on-hook," and V + falls below 0.6 V, power is then supplied to the telephone speech and dialer circuitry from A +. Powering the circuit from the A + pin permits communication with a microprocessor, and/or use of the transmit and receiver amplifiers, while the telephone is "on-hook."

Telephone Tone Ringer

The block diagram for the MC34012-1/-2/-3 telephone tone ringer IC is shown in Figure 17(A). This is a complete telephone bell replacement circuit requiring a minimum of external components, as illustrated by the application circuit in Figure 17(B).

Circuit Description The MC34012 tone ringer derives its power supply by rectifying the ac ringing signal. It uses this power to activate a tone generator and drive a piezo-ceramic transducer. The tone generation circuitry includes a relaxation oscillator and frequency dividers which produce high- and low-frequency tones as well as the tone warble frequency. The relaxation oscillator frequency f_o is set by resistor R_2 and capacitor C_2 connected to pin RC. The oscillator will operate with f_o from 1.0 kHz to 10 kHz with the proper choice of external components. Table 2 lists external components and their uses.

The frequency of the tone ringer output signal at pin RO alternates between $f_o/4$ and $f_o/5$. The warble rate at which the frequency changes is $f_o/320$ for the MC34012-1, $f_o/640$ for the MC34012-2, or $f_o/160$ for the MC34012-3. With $f_o = 4.0$ kHz, the MC34012-1 produces 800 Hz and 1000 Hz tones with a 12.5 warble rate; with $f_o = 8.0$ kHz, the MC34012-2 generates 1600 Hz and 2000 Hz tones with a similar 12.5 warble frequency; with f_o 2.0 kHz, the MC34012-3 will produce 400 Hz and 500 Hz tones with a 12.5 warble rate. The tone ringer output circuit can *source* (occurs when the output terminal is high) or *sink* (occurs when the output terminal is low) 20 mA with an output voltage swing of 20 V_{p-p}. Volume control is readily implemented by adding a variable resistance in series with the piezo transducer.

Input signal detection circuitry activates the tone ringer output when the ac line voltage exceeds a programmed threshold level. Resistor R_3 determines the ringing signal amplitude at which an output signal will be generated at RO. The ac ringing signal is rectified by the internal diode bridge. The rectified input signal produces a current through R_3 which is input at terminal RI. The voltage across R_3 is filtered by C_3 at the input to the threshold circuit. When the voltage on C_3 exceeds 1.7 V, the threshold comparator enables the tone ringer output. Line transients produced by pulse dialing telephones do not charge C_3 sufficiently to activate the tone ringer output.

Capacitors C_1 and C_4, and resistor R_1 determine the 10 V, 24 Hz signature test impedance. C_4 also provides filtering for the output stage power supply to prevent *droop* (the variation or drift caused by the charge leaking out of the holding capacitor) in the square wave output signal. Six diodes in series with the rectifying bridge provide the necessary nonlinearity for the 2.5 V, 24 Hz signature tests.

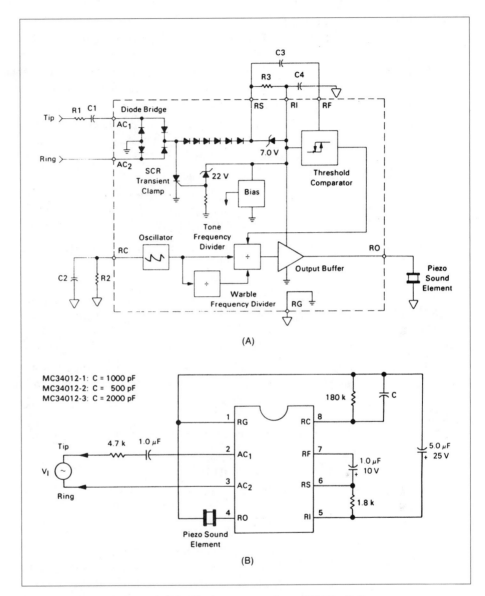

FIGURE 17 MC34012-1/-2/-3 telephone tone ringer (A) Block diagram
(B) Application circuit

An internal shunt voltage regulator between the RI and RG terminals provides dc voltage to power output stage, oscillator, and frequency dividers. The dc voltage at RI is limited to approximately 22 V in regulation. To protect the IC from telephone line transients, an SCR is triggered when the regulator current exceeds 50 mA. The SCR diverts current from the shunt regulator and reduces the power dissipation within the IC.

TABLE 2   **External Components for Figure 17**

R_1	Line input resistor. R_1 controls the tone ringer input impedance. It also influences ringing threshold voltage and limits current from line transients. (Range: 2.0 kΩ to 10 kΩ).
C_1	Line input capacitor. C_1 ac couples the tone ringer to the telephone line and controls ringer input impedance at low frequencies. (Range: 0.4 μF to 2.0 μF).
R_2	Oscillator resistor. (Range: 150 kΩ to 300 kΩ).
C_2	Oscillator capacitor. (Range: 400 pF to 2000 pF).
R_3	Input current sense resistor. R_3 controls the ringing threshold voltage. Increasing R_3 decreases the ring-start voltage. (Range: 0.8 kΩ to 2.0 kΩ).
C_3	Ringing threshold filter capacitor. C_3 filters the ac voltage across R_3 at the input of the ringing threshold comparator. It also provides dialer transient rejection. (Range: 0.5 μF to 5.0 μF).
C_4	Ringer supply capacitor. C_4 filters supply voltage for the tone-generating circuits. It also provides an ac current path for the 10 V_{rms} ringer signature impedance. (Range: 1.0 μF to 10 μF).

Crosspoint Switch

Figure 18 shows the connection diagram for the MC3416 monolithic IC, a $4 \times 4 \times 2$ crosspoint switch. This device, intended for switching analog signals in communications systems, consists of a pair of 4×4 matrices of dielectrically isolated SCRs triggered by a common selection matrix. Dielectric isolation processing provides excellent crosstalk isolation while maintaining minimal insertion loss. The MC3416 features (1) low ON resistance (6 Ω typical), (2) high OFF series resistance (100 MΩ minimum), (3) high breakdown voltage (30 V typical), and (4) selection matrix compatibility with TTL or CMOS logic levels.

Telephone Line Feed and Conversion Circuit

The MC34F19 bipolar thin-film IC, shown in Figure 19(A), is a telephone line feed and two- to four-wire conversion circuit designed to replace the hybrid transformer circuit in central office exchange subscriber carrier equipment. It provides signal separation for two-wire differential to four-wire single-ended conversions and suppression of longitudinal signals at the two-wire input. It provides dc line current for powering the telephone set, operating from supplies up to 56 V. Features of the MC34F19 include (1) external programming of all key parameters, (2) current-sensing outputs that monitor the status of both tip and ring leads, (3) on-hook power below 5 mW,

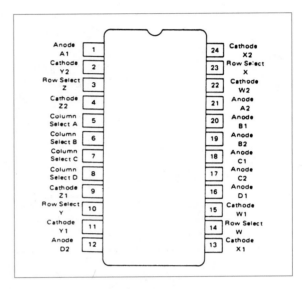

FIGURE 18
MC3416 4 × 4 × 2
crosspoint switch

(4) digital hook status output, (5) power-down input, (6) ground fault protection, and (7) size and weight reduction over conventional approaches. Figures 19(B) through (D) are suggestions for interfacing the MC34F19 with various digital logic levels.

Special Application Devices

Figure 20 shows the block diagram of the MC3417/MC3418, MC3517/MC3518 series of continuously variable slope delta modulator-demodulators (CVSDs), designed for military secure communication and commercial telephone applications. Encoding and decoding functions are contained on the same chip, with a digital input for selection, thus providing a simplified approach to digital speech encoding/decoding. The MC3417/MC3517 devices have a 3-bit algorithm, used for general communications. The MC3418/MC3518 devices have a 4-bit algorithm, used for commercial telephone systems.

DATA COMMUNICATION IC APPLICATIONS FOR CHAPTER 10

A small sampling of the ICs available for data communication systems is included in this section. For those persons interested in further study about data communication IC applications, manufacturers' data books are recommended.

Universal Low-Speed Modem

The MC14412 universal low-speed modem, shown in the block diagram in Figure 21(A), is a 16-pin CMOS integrated circuit containing a complete FSK modulator and demodulator compatible with both CCITT standards and U.S. low-speed data communication networks. It has an on-chip CCO (with external crystal), an on-chip sine wave generator, an echo suppressor disable tone generator, and a postdetection filter. It can operate in the originate or answer mode and offers simplex, HDX, and FDX operation.

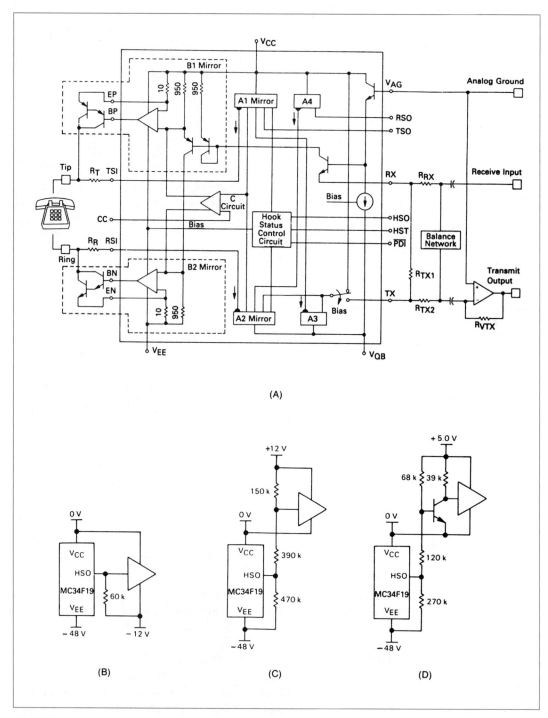

FIGURE 19 MC34F19 telephone line feed and two- to four-wire conversion IC (A) Functional block diagram (B) Interface-to-CMOS using a negative supply (C) Interface-to-CMOS using a positive supply (D) Interface-to-TTL/LS using +5 V for V_{CC}

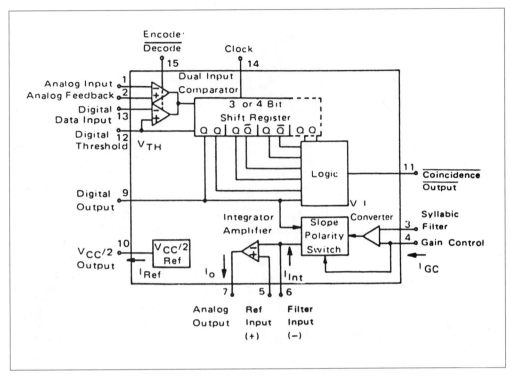

FIGURE 20 Block diagram of MC3417/MC3418, MC3517/MC3518 CVSDs

It provides selectable data rates (0–300, 0–600 bps) and a modem self-test mode. A typical application is illustrated in Figure 21(B).

Modem Band-Pass Filter

The block diagram in Figure 22(A) shows the 18-pin MC145440 300 baud modem BP switched capacitor filter designed to be used with the MC14412 low-speed modem (Figure 21). Features of the MC145440 include (1) low-band and high-band BP filters, (2) a spare op amp, (3) single or split power supply operation, (4) a self-test loopback configuration, (5) an answer or originate mode, and (6) Bell 103 frequency compatibility.

Figure 22(B) shows a typical application using the MC14412 modem and the MC145440 BP filter. These modem/filter combinations fulfill the major requirements of a complete Bell 103 300-baud modem system.

1200 Baud FSK Modem

Figure 23 shows a block diagram of the 22-pin CMOS MC145450 1200 baud FSK modem intended for use in Bell 202 and CCITT V.23 applications. It offers (1) soft turn-off capability, (2) an answer back tone generator (U.S. and CCITT tones) (3) a 0–150 baud reverse channel, (4) a carrier detect input, and (5) a Bell 202-compatible 0–1800 baud main channel. A typical 1200 baud four-wire modem application is shown in Figure 23(B).

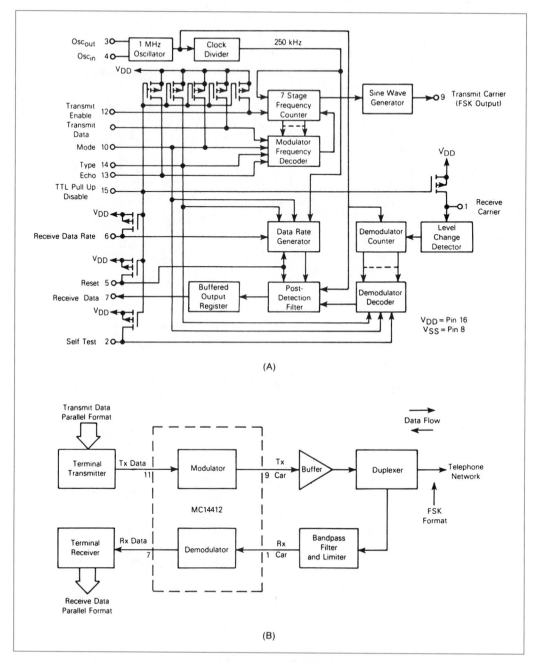

FIGURE 21 MC14412 universal low-speed modem (A) Block diagram (B) Typical application

2400 bps DPSK Modem System

The tremendous growth in data communications has spurred the development of many diverse modems for use on the normal dial-up telephone network and on private leased lines. One of the more prominent modems is the type 201 2400 bps system such as the

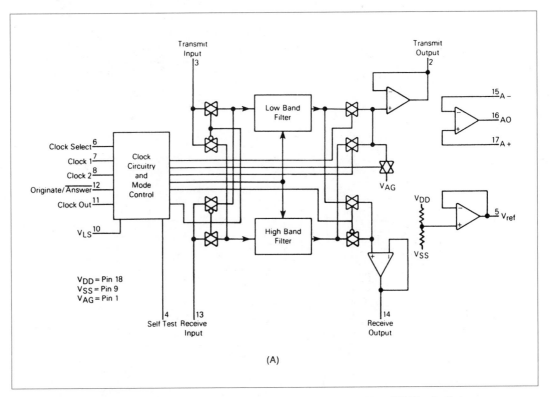

FIGURE 22   MC145440 300 baud modem BP switched capacitor filter (A) Block diagram
(B) Typical application

Bell 201B/C data set. This type of modem uses a technique of modulation called *differential phase shift keying* (DPSK) in which a carrier frequency is phase modulated to represent different information states. Figure 24 shows the MC6172 modulator/MC6173 demodulator chip set, an NMOS LSI subsystem designed to perform the modulate/demodulate and control functions for implementing a DPSK modem. Pin-selectable options permit compliance with either Bell or CCITT requirements and also allow selection of the standard data rate of 2400 bps or a secondary rate of 12 bps.

The system shown in the block diagram is a simplex mode of operation. For HDX operation, an MC6172 modulator and an MC6173 demodulator would be needed at both ends of a two-wire connection. If FDX operation is desired, a four-wire connection would be required.

Extending a Communication Link

Although the maximum recommended length for communication links over uncompensated No. 24 AWG (American Wire Gauge) twisted-pair wires is 4000 feet, a few off-the-shelf ICs, when configured as a line driver, will drive a communication link with a maximum length of over 4 miles. The bandwidth of over 3000 Hz is adequate for remote control, sensing, and even private telephone voice communications.

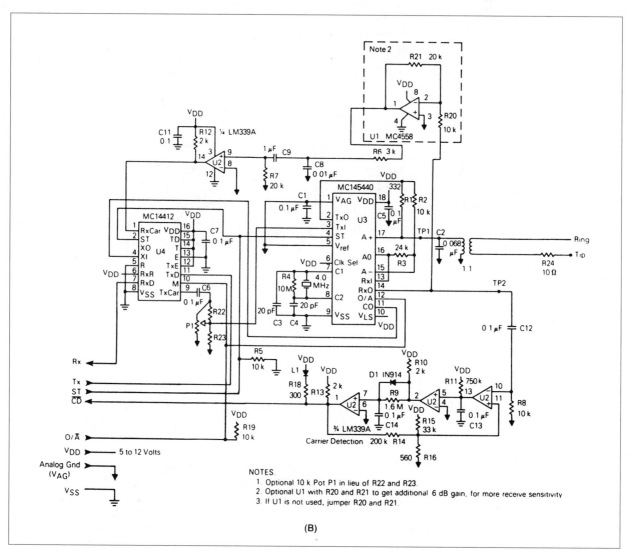

FIGURE 22 (*continued*)

Any subscriber-loop interface circuit (SLIC) can be used to drive the line. In Figure 25(A), an SLIC interfaces with the voltage-to-frequency and frequency-to-voltage converters and a line terminator. Thus, a dc voltage related to a process at a remote site can be measured from a central location and a corrective dc voltage can be sent in the reverse direction to adjust or control the process.

A variety of ICs can be interfaced with the basic driver for signaling or for data or voice transmission. Analog-to-digital and digital-to-analog converters, such as coders/decoders (codecs) and CVSDs (continuously variable slope delta modulators used for voice companding), perform the front-end data conversion for a low-cost computer link. A variety of DTMF encoders and decoders facilitate simple signaling over privately owned twisted pairs.

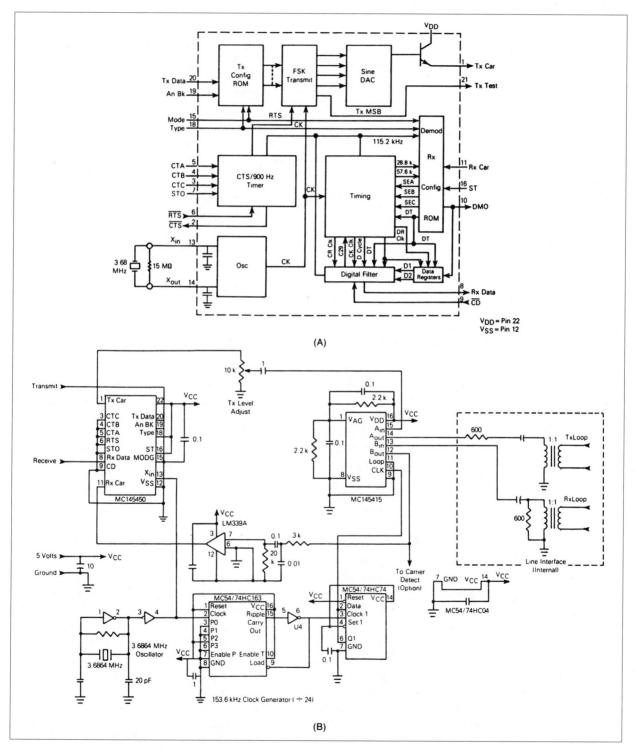

FIGURE 23 MC145450 1200 baud FSK modem (A) Block diagram (B) Typical application

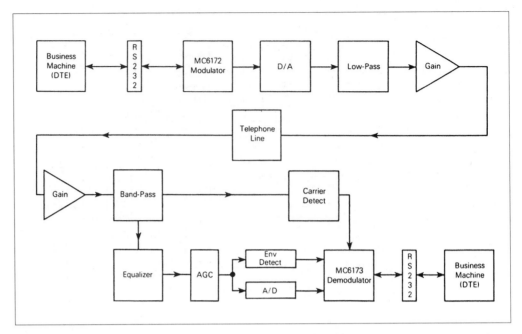

FIGURE 24   MC6172/MC6173 2400 bps modem block diagram

A line driver card, shown in Figure 25(B), exemplifies the simple hardware configuration required. The SLIC uses two Darlington pairs (MJE270 and MJE271) as pass transistors to handle currents up to 120 mA. Changing the value of the 59 kΩ resistor used in the feedback path of the 741 op amp will adjust the transmission output gain. The MDA220 rectifier bridge protects the circuit against lightning damage.

The line terminator card converts a bidirectional two-wire line into two pairs of unidirectional lines, one transmit and one receive, and then amplifies the received and transmitted signals. A simple terminator can be made with the hybrid transformer and two varistors from a telephone handset.

FSK Demodulator/Tone Decoder

The functional block diagram of the XR-2211 FSK demodulator/tone decoder is shown in Figure 26. This is a 14-pin monolithic PLL system designed especially for use in data communications. It operates over a wide supply voltage range (4.5–20 V) and a wide frequency range (0.01 Hz–300 kHz). It can accommodate analog signals between 2 mV and 3 V and can interface with conventional logic families. It offers an adjustable tracking range from ± 1% to ± 80%. The circuit consists of (1) a basic PLL for tracking an input signal frequency within the pass band, (2) a quadrature phase detector that provides carrier detection, and (3) an FSK voltage comparator that provides FSK demodulation. External components are used to independently set carrier frequency, bandwidth, and output delay.

In addition to FSK demodulation and tone decoding, the XR-2211 IC can be used for data synchronization and FM detection.

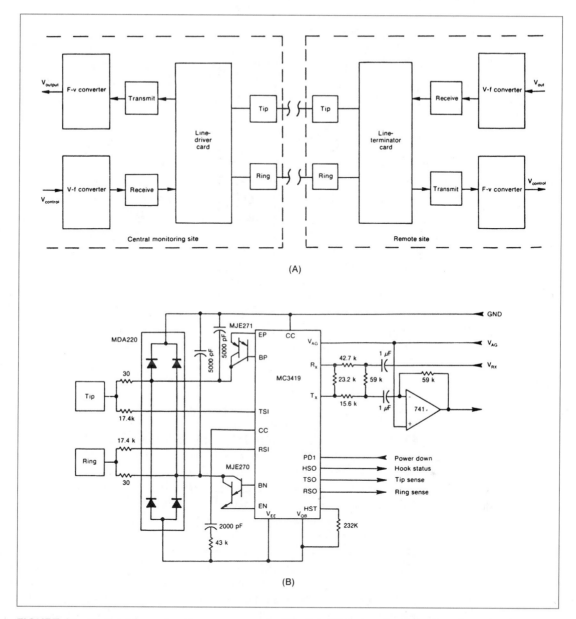

FIGURE 25 Method for extending a communication link (A) Subscriber-loop interface circuit
(B) Line driver card

IC APPLICATIONS FOR TV FOR CHAPTER 13 ____

Many modern TV receivers contain a variety of ICs that simplify set manufacture,
reduce chassis clutter, and reduce heat and power consumption. Maintenance and repair are eased because, in most cases, the repairperson simply replaces an IC. The

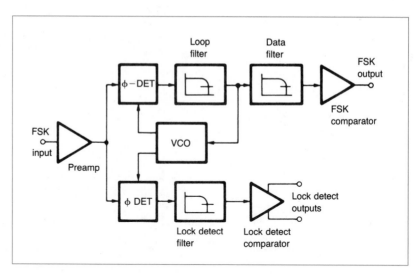

FIGURE 26 Functional block diagram of the XR-2211 FSK demodulator/tone decoder

following discussion examines a few of the more sophisticated devices used in TV receivers and video recording/playback systems.

TV Modulator Circuit

A block diagram of the MC1374 TV modulator circuit is shown in Figure 27(A). This IC includes an FM audio modulator, sound carrier oscillator, RF oscillator, and RF dual-input modulator. It is designed to generate a TV signal from audio and video inputs. The wide dynamic range and low-distortion audio of the device make it particularly well suited for applications in video tape recorders (VTRs), video cassette recorders (VCRs), and video disk players (VDPs). It operates from a single supply (5–12 V) and is designed for channel 3 or 4 operation.

The oscillator components shown in the typical application circuit in Figure 27(B) are selected to have a parallel resonance at the carrier frequency of the desired TV channel. The values of C_2 (56 pF) and L_1 (0.1 μH) were chosen for a channel 4 carrier frequency of 67.25 MHz. For channel 3 operation, the resonant frequency is 61.25 MHz, so the values would be $C_2 = 67.5$ pF and $L_1 = 0.1$ μH. Resistors R_2 and R_3 are chosen to provide an adequate amplitude of switching voltage, while R_1 is used to lower the maximum dc level of switching voltage below V_{CC}, thus preventing saturation within the IC.

The video modulator is a balanced modulator. Sound carrier and video information are applied to pins 1 and 11. The other modulator inputs are internally connected to the RF oscillator. The modulator output appears at pin 9.

In a typical application where the composite video information is dc coupled to pin 11, the bias on pin 1 is set to give the desired modulation characteristics. Minimum carrier occurs when the voltages on pins 1 and 11 are equal—a desirable trait. The minimum permissible voltage on either input is 1.6 V. The maximum voltage should be 1.5 V below the dc voltage on pins 6 and 7. The value for gain-setting resistor R_8

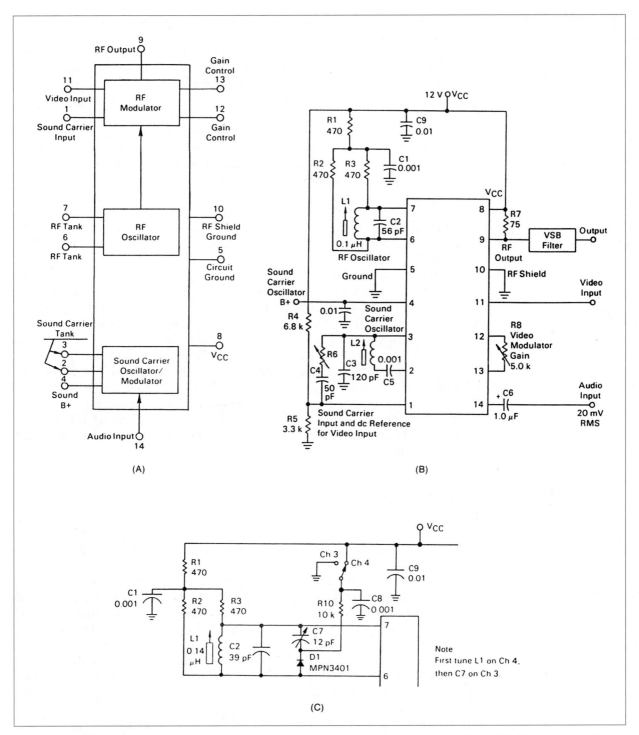

FIGURE 27 MC1374 TV modulator IC (A) Block diagram (B) Channel 4 application circuit (C) Channels 3 and 4 application circuit

between pins 12 and 13, is selected to give the proper modulation depth for the available composite video amplitude.

The modulated RF signal is presented as a current at RF output pin 9. Since this pin represents a current source, any load impedance may be selected for matching purposes and gain selections so long as the voltage on the pin is high enough to prevent output devices from reaching saturation. Lowering the dc voltage on pins 6 and 7 gives increased RF output capability, but at the expense of video input range.

The sound carrier oscillator and audio modulator have an internally set bias. A separate B^+ is supplied to the oscillator through pin 4, so the oscillator may be easily disabled while the RF tank is tuned. The sound carrier frequency is determined by L_2 and C_3. The oscillator feedback is fed to L_2 through dc blocking capacitor C_5, and 4.5 MHz appears at the input to the oscillator, pin 3.

The sound carrier is coupled to the modulator input, pin 1, through variable resistor R_6 and capacitor C_4. The value for R_6 is chosen to give the desired sound carrier amplitude and depends on the Q of L_2 and the values for R_4 and R_5. The RF modulator gain is set by R_8.

Baseband audio is fed to the audio modulator on pin 14, where it directly modulates the sound carrier oscillator for a flat characteristic and very low distortion. The input impedance on pin 14 is nominally 6 kΩ. If the audio available is much greater than necessary for proper deviation, a series resistor may be added to allow a low value for coupling capacitor C_6.

When application calls for tight frequency stability, the sound carrier may be frequency controlled by supplying a dc current to pin 14 from a suitable AFC circuit. The nominal voltage at pin 14 is approximately 3 V. Supplying current to pin 14 increases the frequency, and pulling current out of pin 14 decreases the frequency.

Two-channel operation is possible by switching in a second capacitor to tune the lower channel by means of a PIN diode. Figure 27(C) shows the circuit and components for channel 3 or channel 4 operation.

No-Holds TV Circuit

The LM 1880 no-holds vertical/horizontal circuit shown in Figure 28(A) uses compatible linear/I²L technology to produce a TV vertical and horizontal processing system that eliminates the hold controls. The heart of the system is a precision 32-times horizontal frequency VCO designed to use a low-cost resonator as a tuning element.

The VCO signal is divided in the horizontal section to produce a predriver output that is locked to negative sync by means of an on-chip phase detector. The vertical output ramp is injection locked by vertical sync subject to a sync window derived from the vertical count-down section. A gate pulse centered on the chroma burst is also provided.

A typical application circuit is illustrated in Figure 28(B). Since the LM1880 uses a counter to derive the horizontal frequency, care must be taken to prevent extraneous signals from the horizontal driver and output stages from feeding back to the VCO, where they could cause false counts and consequent severe phase jitter. To accomplish this, keep the VCO feedback capacitor C_L as close as possible to device pins 6 and 7 and limit the lead length on horizontal output pin 8. If a long line is required to the driver base, isolate it with a small (200–300 Ω) series resistor next to pin 8.

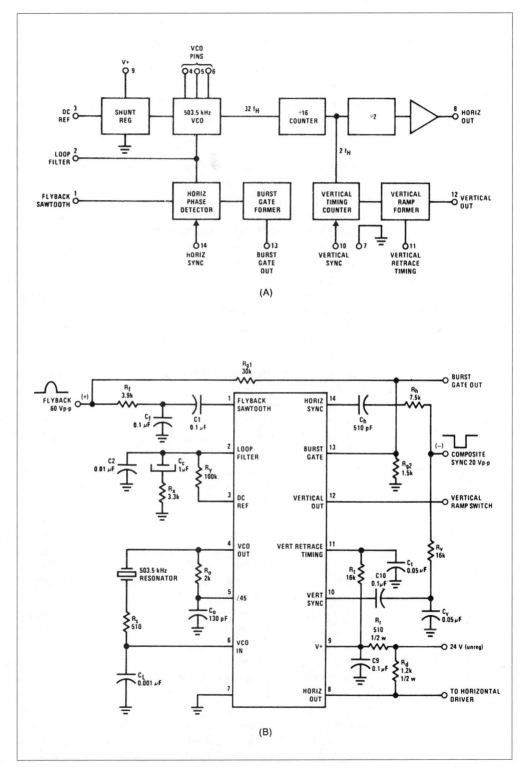

FIGURE 28
LM1880 no-holds
TV circuit
(A) Block diagram
(B) Application
circuit

TV Signal Processing Circuit

Figure 29(A) shows the connection diagram for the TBA950-2 TV signal processing large-scale integration (LSI) IC. This device is designed for pulse separation and line synchronization in TV receivers with transistor output stages. As shown in the block diagram in Figure 29(B), the TBA950-2 IC consists of (1) a sync separator with noise suppression, (2) a frame pulse integrator, (3) a phase comparator, (4) a switching stage for automatic changeover of noise immunity, (5) a line oscillator with frequency range limiter, (6) a phase control circuit, and (7) an output stage. It delivers prepared frame sync pulses for triggering the frame oscillator. The phase comparator may be switched for video recording operation. Because of LSI, few external components are required.

The sync separator separates the sync pulses from the composite video signal. The noise inverter circuit, which needs no external components, in conjunction with integrating and differentiating networks, frees the sync signal from distortion and noise.

The frame sync pulse is obtained by multiple integration and limitation of the sync signal and is available at pin 7. The leading edge of the frame pulse should be used for triggering because of possible pulse duration differences in production of the sync pulses.

The frequency of the line oscillator is determined by a 10 nF polystyrene capacitor at pin 13 that is charged and discharged periodically by two external current sources. The external resistor at pin 14 defines the charging current and, consequently, in conjunction with the oscillator capacitor, the line frequency.

The phase comparator compares the sawtooth voltage of the oscillator with the sync pulses. Simultaneously, an AFC voltage is generated that influences the oscillator frequency. A frequency range limiter restricts the frequency holding range.

The oscillator sawtooth voltage, which is in a fixed ratio to the line sync pulses, is compared with the flyback pulse in the phase control circuit to compensate all drift of delay times in the driver and line output stage. The correct phase position and, hence, the horizontal position of the picture can be adjusted by the 10 kΩ potentiometer connected to pin 11. Within the adjustable range, the output pulse duration (pin 2) is constant. Any larger displacements of the picture (for example, that due to a nonsymmetrical picture tube) should not be corrected by the phase potentiometer, since in all cases the flyback pulse must overlap the sync pulses on both edges.

The switching stage has an auxiliary function. When the two signals supplied by the sync separator and the phase control circuit are synchronized, a saturated transistor is in parallel with the integrated 2 kΩ resistor at pin 9. Thus, the time constant of the filter network at pin 4 increases and, consequently, reduces the pull-in range of the phase comparator circuit for the synchronized state to approximately 50 Hz. This arrangement ensures disturbance-free operation.

For video recording operation, this automatic switchover can be blocked by a positive current fed into pin 8 through a resistor connected to pin 3. It may also be useful to connect a resistor of about 680 Ω or 1 kΩ between pin 9 and ground. The capacitor at pin 4 may be lowered in value to 0.1 μF. These alterations do not significantly influence the normal operation of the IC and thus do not need to be switched back.

At pin 2, the output stage delivers output pulses of duration and polarity suitable for driving the line driver stage. If the supply voltage goes down (for example, when power is turned off), a built-in protection circuit ensures defined line frequency pulses down to $V_3 = 4$ V and shuts OFF when V_3 falls below 4 V, thus preventing pulses of undefined duration and frequency. Conversely, if the supply voltage rises, pulses

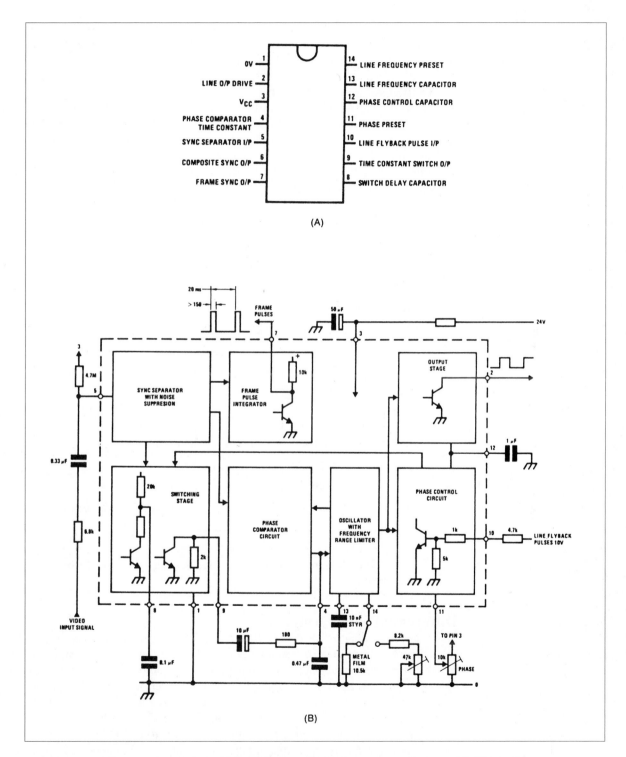

FIGURE 29 TBA950-2 TV signal processing circuit (A) Connection diagram (B) Block diagram

defined in duration and frequency will appear at the output pin as soon as V_3 reaches 4.5 V. In the range between $V_3 = 4.5$ V and full supply voltage, the shape and frequency of the output pulses are almost constant.

IF Amplifier and Detector

The TBA120U/TBA120T shown in the connection diagram in Figure 30 is a monolithic IC designed specifically for audio detection in TV and FM receivers. It incorporates a eight-stage limiting IF amplifier and balanced detector plus a dc-operated volume control. The circuit also provides connection facilities for a VTR. The TBA120T is designed primarily for use with ceramic filters, and the TBA120U is optimized for inductive tuning.

These devices:

1. Require very few external components.
2. Operate over a supply voltage range of 6–18 V.
3. Operate over a frequency range of 0–12 MHz.

Appendix B provides the data sheets with specifications and application circuits for the TBA120U/TBA120T ICs.

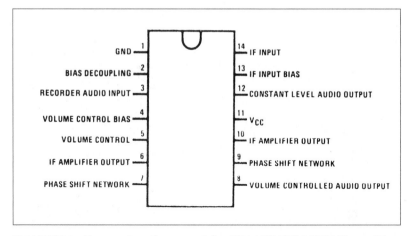

FIGURE 30 Connection diagram of the TBA120U/TBA120T IF amplifier and detector

Differential Video Amplifier

The connection diagram in Figure 31(A) and the equivalent circuit in Figure 31(B) show the μA733 two-stage differential input/differential output video amplifier IC. Internal series-shunt feedback is used to obtain wide bandwidth (120 MHz), low phase distortion, and excellent gain stability. Emitter-follower outputs enable the device to drive capacitive loads, and all stages are current source biased to obtain high power supply and common-mode rejection ratios. The IC offers fixed gains of 10, 100, or 400 without external components and adjustable gains from 10 to 400 by the use of a single external resistor. No external frequency compensation components are required for any gain option.

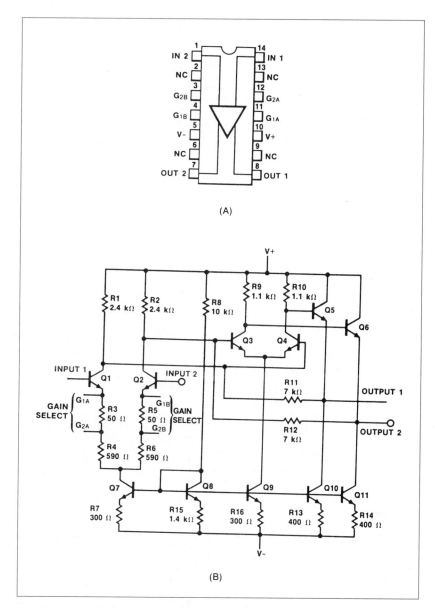

FIGURE 31 μA733 differential video amplifier (A) Connection diagram
(B) Equivalent schematic

The device can be used for general-purpose video and pulse amplifiers.

Chroma Demodulator

The connection diagram in Figure 32(A) and the equivalent circuit in Figure 32(B) show the MC1327 chroma demodulator, a monolithic IC designed for use in solid-state color TV receivers. This is a dual, double-balanced detector with RGB matrix, PAL switch, and chroma driver stages. It offers (1) good chroma sensitivity (0.28 $V_{p\text{-}p}$

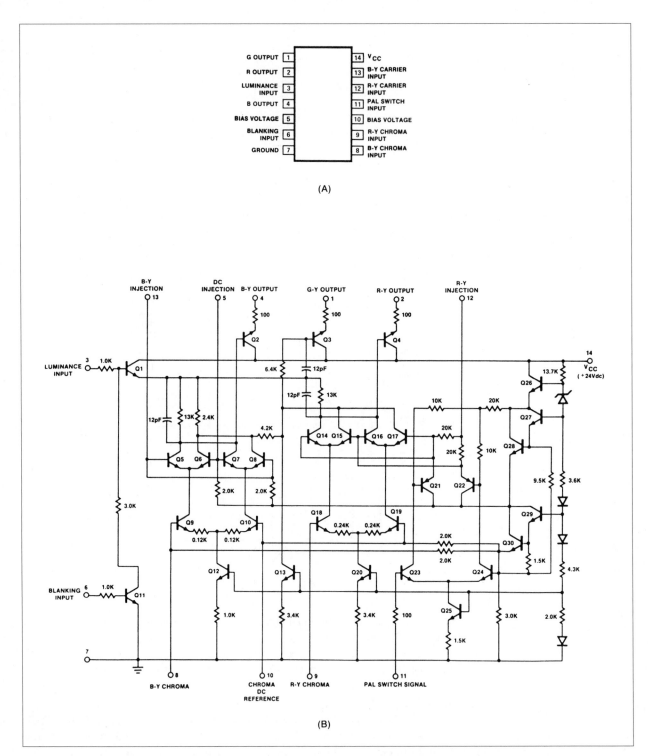

FIGURE 32 MC1327 chroma demodulator (A) Connection diagram (B) Equivalent schematic (C) Application circuit

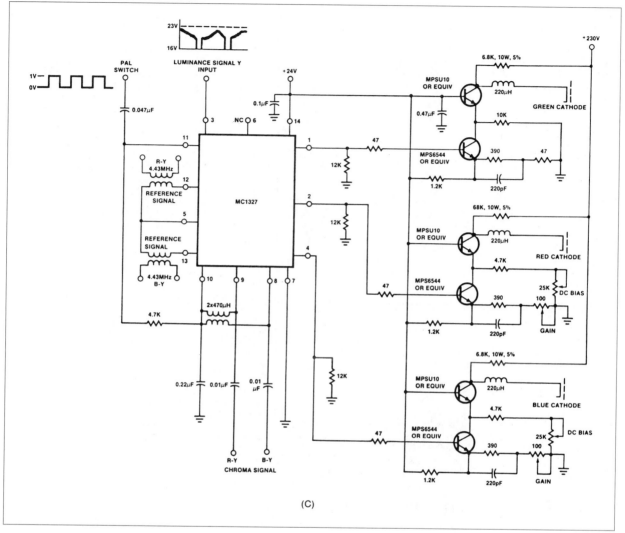

(C)

FIGURE 32   (*continued*)

input typical for 5.0 V_{p-p} output), (2) low differential output dc offset voltage (0.6 V maximum), (3) high blue output voltage swing (10 V_{p-p} typical), (4) a blanking input (pin 6), and (5) a luminance bandwidth greater than 5 MHz. Figure 32(C) shows a typical application circuit for the MC1327 IC.

FLL Tuning and Control Circuit

The SAB3037 shown in the connection diagram in Figure 33 is an FLL tuning and control IC. It provides closed-loop digital tuning of TV receivers, with or without AFC. It also controls up to four analog functions, four general purpose input/output ports, and four high-current outputs for tuner band selection.

Additional applications for SAB3037 are in CATV converters and in satellite receivers.

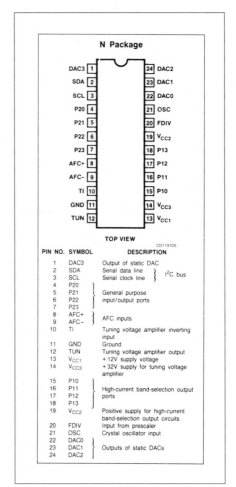

FIGURE 33
SAB3037 FLL Tuning
and control circuit
connection diagram

VHF Mixer/Oscillator Circuit

Figure 34(A) shows the functional block diagram for the TDA5030A VHF mixer/oscillator IC. This IC performs the VHF mixer, VHF oscillator, surface acoustic wave (SAW) filter IF amplifier, and UHF IF amplifier functions in TV tuners.

Features of the TDA5030 IC include a balanced VHF mixer, amplitude-controlled VHF local oscillator, a buffer stage for driving an external prescaler with the local oscillator signal, a voltage stabilizer, and a UHF/VHF switching circuit.

Additional applications for the device are CATV, local area networks (LANs), and demodulation.

Figure 34(B) shows a test circuit for the IC.

Video Amplifier

Figure 35 shows the equivalent circuit for the NE/SA/SE592 video amplifier. This monolithic, two-stage, differential output, wideband video amplifier offers fixed gains of 100 and 400 without external components and adjustable gains from 400 to 0 with one external resistor.

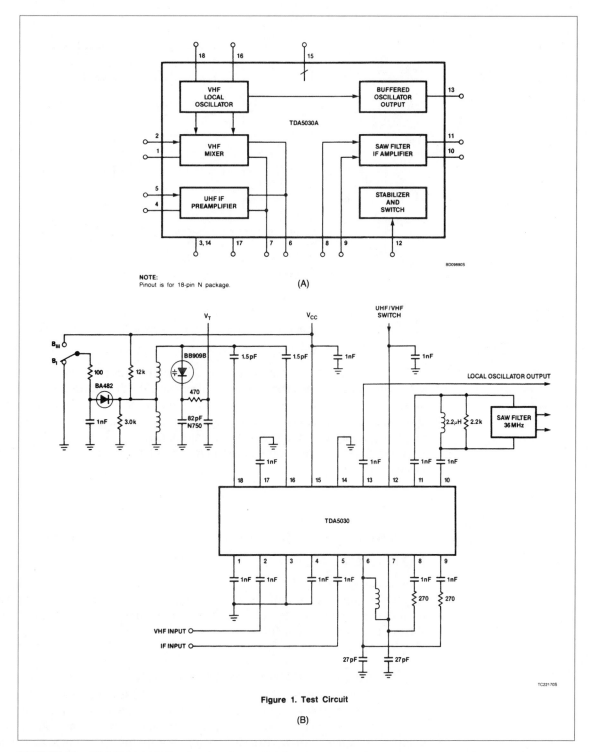

(A)

NOTE:
Pinout is for 18-pin N package.

Figure 1. Test Circuit

(B)

FIGURE 34   TDA5030A VHF mixer/oscillator (A) Functional block diagram (B) Test circuit

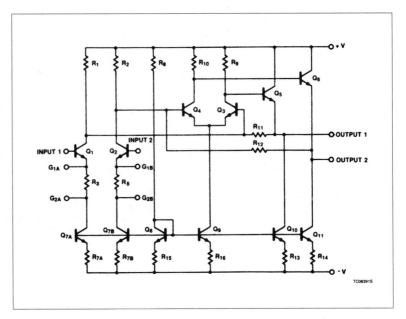

FIGURE 35 Equivalent circuit for the NE/SA/SE592 video amplifier

The input stage has been designed so that with the addition of a few external reactive elements between the gain select terminals, the circuit can function as a high-pass, low-pass, or band-pass filter. This feature makes the circuit ideal for use as a video or pulse amplifier in communications, magnetic memories, display, video recorder systems, and floppy disk head amplifiers.

Appendix C provides application note AN141 for the use of the NE/SA/SE592 video amplifier.

IC APPLICATIONS FOR TWO-WAY SYSTEMS FOR CHAPTER 15

Integrated circuits are ideally suited for use in mobile and portable two-way communication systems. Because of the wide variety of such equipment, it will be necessary to limit the number of devices examined here. Manufacturers' data books and data sheets should be studied if further information is desired.

High-Gain, Low-Power FM IF Circuit

The MC3359 IC illustrated in Figure 36(A) and (B) is a high-gain, low-power FM intermediate-frequency circuit that includes an oscillator, a mixer, a limiting amplifier, automatic frequency control (AFC), a quadrature detector, an op amp, a squelch control, and a mute switch. It is designed primarily for use in voice communication scanning receivers.

The mixer/oscillator combination converts the 10.7 MHz input frequency down to 455 kHz where, after external BP filtering (ceramic filter at pin 3), most of the amplification is done. The audio is recovered using a conventional FM quadrature detector (at pin 8). The absence of an input signal is indicated by the presence of noise after the desired audio frequencies. This "noise band" is monitored by an active filter and a detector. A squelch-trigger circuit (pin 14) indicates the presence of noise (or a tone) by an output (pin 15) that can be used to control scanning. At the same time, an internal switch (pins 16 and 17) is operated. This switch can be used to mute the audio.

The oscillator is an internally biased Colpitts with the collector, base, and emitter connections at pins 4, 1, and 2, respectively. A 10.245 MHz crystal is used in place of the usual coil.

The mixer is doubly balanced to reduce spurious responses. The input impedance at pin 18 is set by a 3.6 kΩ internal biasing resistor and has low capacitance, which allows the circuit to be preceded by a crystal filter. The mixer output at pin 3 has a 1.8 kΩ impedance to match the external ceramic filter.

After BP filtering, the signal goes to the input of a six-stage limiter (pin 5) with an impedance of 1.8 kΩ. The output of the limiter drives a multiplier, both directly and through the quadrature coil, to detect the FM.

The external capacitor at pin 9 can combine with the internal 50 kΩ resistor to form an LP filter for the audio. The audio is delivered through the emitter follower to pin 10, which may require an external resistor to ground to prevent the signal from rectifying with some capacitive loads.

Pin 11 provides AFC. If AFC is not required, pin 11 should be grounded, or it can be tied to pin 9 to double the recovered audio amplitude.

A simple inverting op amp is included, with an output at pin 13 providing dc bias externally to the input at pin 12, which is referred internally to 2.3 V. A filter can be made with external impedance elements to discriminate between frequencies. With an external AM detector, the filtered audio signal can be checked for the presence of either noise above the normal audio band or a tone signal. The result is applied to pin 14.

An external negative bias to pin 14 sets the squelch-trigger circuit such that pin 15 is HIGH, at an impedance level of about 2.5 kΩ, and the audio mute (pin 16) is open-circuit. If pin 14 is raised to 0.7 V by the noise or tone detector, pin 15 will go open-circuit, and pin 16 is internally short-circuited to ground. There is no hysteresis. Audio muting is accomplished by connecting pin 16 to a high-impedance ground-referenced point in the audio path between pin 10 and the audio amplifier.

Gain-Controlled IF Amplifier

The µA757 shown in Figures 37(A) and (B) is a high-performance, gain-controlled IF amplifier intended primarily for use in AM and FM communication receivers. It also has excellent performance when operated in FM receivers as a limiting amplifier.

The IC contains two sections which can be operated independently or in cascade from audio frequencies up to 25 MHz. It offers 70 dB gain and 70 dB automatic gain control (AGC) range at 10.7 MHz, and a 300 mV input signal capability.

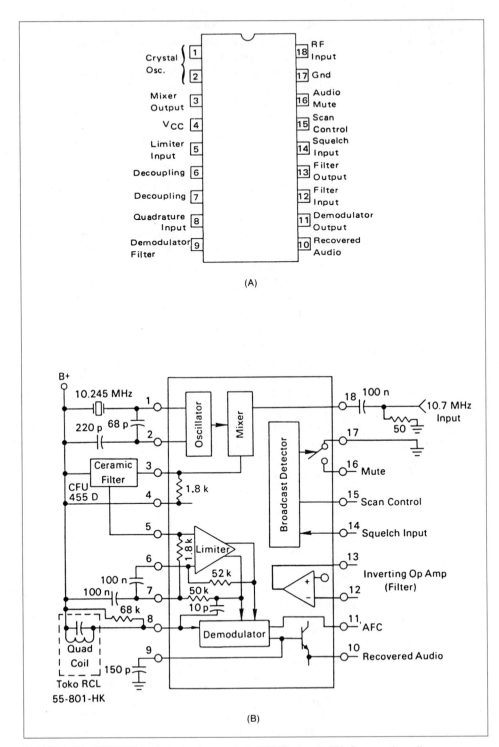

FIGURE 36 MC3359 high-gain, low-power FM-IF circuit (A) Connection diagram (B) Functional block diagram

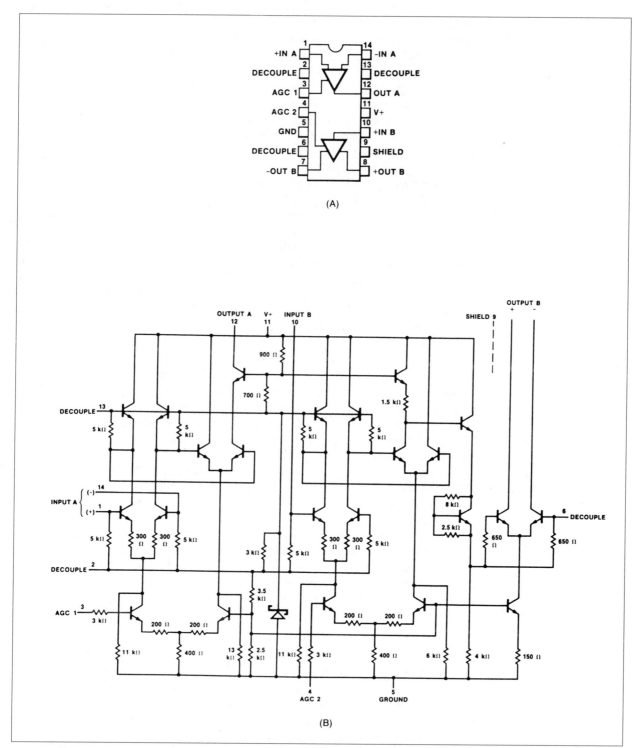

FIGURE 37   μA757 gain-controlled IF amplifier (A) Connection diagram (B) Equivalent schematic

VCO/Modulator

Figure 38(A) and (B) show the MC1376 VCO/modulator IC. This device is ideally suited to cordless telephone applications. It operates over a range of 5–12 V_{dc}, has a useful frequency range of 1.4–14 MHz, has less than 1% distortion, and offers excellent oscillator stability.

Orginally designed for the base station of a cordless telephone system, the MC1376 IC includes a separate, or auxiliary, transistor (pins 2, 3, and 4) suitable for service as an output buffer or amplifier for up to 50 mA. Although the oscillator contains internal phase-shift components that are not accessible, the device still has a wide frequency operating range of 1.4–14 MHz.

The wide frequency operating range makes the MC1376 a good companion to other devices, such as the MC1372 and MC1373, as a 4.5 or 5.5 MHz intercarrier sound modulator for TV signal generation. Also the device can be used as a low-cost FM intermediate-frequency (10.7 MHz) signal source.

Figure 38(C) shows an application for the MC1376—a 1.76 MHz cordless telephone base station transmitter. The oscillator center frequency is approximately the resonance of the inductor (pin 6 to pin 7) and the total capacitance from pin 7 to ground. If the internal capacitance of about 6 pF is included, the circuit strays in the resonant frequency calculations for the higher-frequency applications. For overall oscillator stability, the inductive and capacitive reactances should be kept in the 300 Ω to 1 kΩ range.

Most applications require no dc connection at the audio input (pin 5). However, some performance improvements can be achieved by the addition of biasing circuitry. The unaided device will usually establish its own pin 5 bias at 2.9–3.0 V. This bias is a little high for optimum modulation linearity and results in some modulation distortion. This distortion can be significantly reduced by pulling the pin bias down to 2.6–2.7 V. Temperature and supply voltage factors must also be considered when determining biasing.

Temperature stability can be improved by pulling pin 5 down to 2.6 V through a 27 kΩ resistor. If V_{CC} is well regulated, then a simple 180 kΩ/30 kΩ resistor divider, as shown in Figure 38(D), provides optimum distortion-and-frequency stability versus temperature.

The FM output at pin 7 is usually about 600 mV_{p-p} and has low harmonic content and high (2 kΩ) output impedance. The oscillator behavior is relatively unaffected by loading above 1.0 kΩ. If lower impedance must be driven, the capacitor divider, shown in Figure 38(E), can be used, or the auxiliary transistor can be used as a buffer.

CMOS Frequency Synthesizer

Most modern channelized transceivers use frequency synthesizers to achieve the desired operating frequencies. Figure 39 shows the block diagram of the TDD1742 CMOS low-current frequency synthesizer. The device includes a high-gain phase comparator, using a sample-and-hold technique. It can operate with a low 7 V supply with a maximum input frequency of 8.5 MHz.

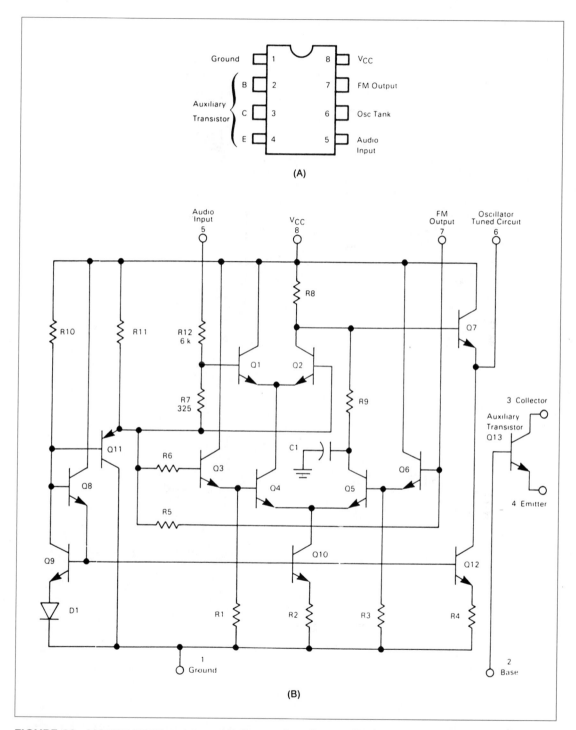

FIGURE 38   MC1376 VCO/modulator (A) Connection diagram (B) Equivalent schematic (C) 1.76 MHz cordless telephone base station transmitter (D) Frequency stability versus supply voltage (E) Capacitor divider output test circuit

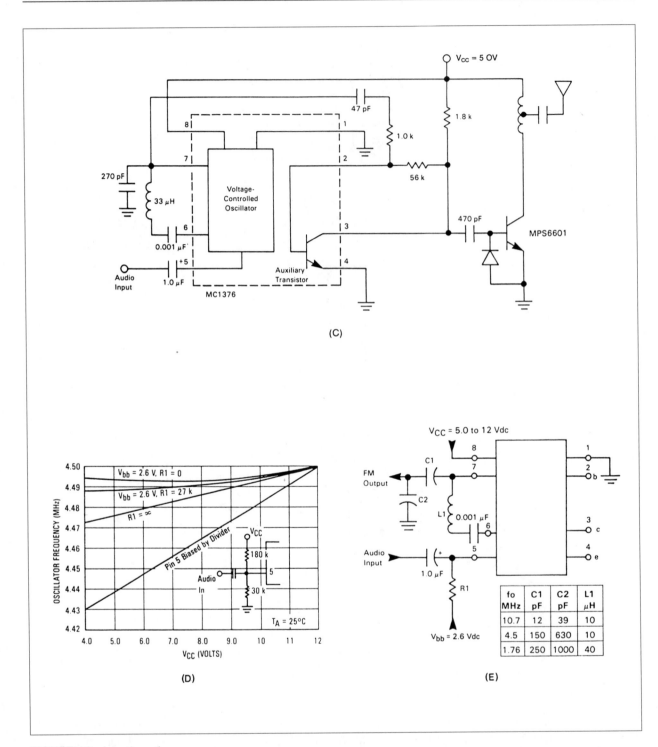

(C)

(D)

(E)

FIGURE 38 (*continued*)

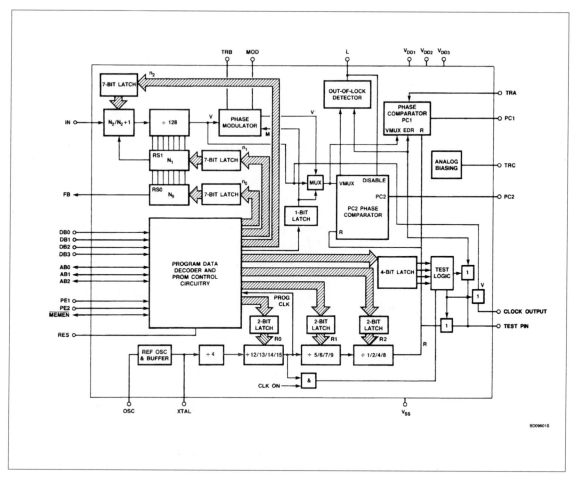

FIGURE 39 TDD1742 CMOS frequency synthesizer

The TDD1742 IC is designed for VHF/UHF portable or mobile transceivers. It offers low phase noise and spurious response, power-on reset circuitry, and is microprocessor controllable.

Applications include cellular radio and digital frequency synthesis.

IC APPLICATIONS FOR FIBER OPTICS FOR CHAPTER 16 _____

A variety of ICs are available for use in interfacing the various elements of a fiber optic system. In this section we look at three such ICs: a Manchester biphase-mark encoder and LED driver, a Manchester biphase-mark decoder, and a transimpedance amplifier.

Manchester Biphase-Mark Encoder and LED Driver

The SP9960 Manchester biphase-mark encoder and LED driver is designed for use in fiber optic links at up to 50 Mbits per second (Mbs). Operating on a single power supply voltage, this IC encodes transistor-transistor logic (TTL) or emitter-coupled logic (ECL) data and outputs as a current at the outputs of either the large or the small LED drivers. The LED driver and the current output are selectable.

Figure 40(A) shows a simplified block diagram for the SP9960. Data arriving at a data input (pin 3 for TTL, pin 6 for ECL) is sampled by the positive edge of the appropriate clock (pins 4 or 6) encoded into a biphase-mark signal, and output as a current at the chosen LED driver (pin 9 for the small driver, pin 10 for the large driver).

When the TTL inputs are used, negative V_{EE} (pin 1) is normally tied to 0 V and the ECL inputs are left open (unconnected). Conversely, when the ECL inputs are used, V_{EE} is set to −5.2 V and the TTL inputs are left open.

The ladder diagram in Figure 40(B) shows the biphase-mark encoding alignment scheme. The input data is sampled by the positive edge of the clock. If the data is HIGH (logic 1), the driver switches to the opposite state (OFF if it was previously ON, or ON if it was previously OFF). If the data is LOW (logic 0), the driver does not switch to its opposite state on the positive clock edge. However, regardless of the state of the sampled input data, the driver always switches to its opposite state on the negative edges of the clock.

This form of encoding ensures a high number of transitions in the signal, thus simplifying the task of clock recovery at a remote detector. And, since the data is encoded in terms of transitions instead of absolute levels, the signal can be given a net inversion without corrupting the information carried.

The LED driver (SD or LD) used is chosen by the V_{EE}: LD pin. This pin should be tied to the negative V_{EE} pin if the large driver is selected, and left unconnected if the small driver is selected.

When ON, the small driver typically outputs 25 mA if the V_{EE}: 10 is connected to V_{EE}, or 15 mA if unconnected.

When ON, the large driver typically outputs 150 mA if both V_{EE}: 35 and V_{EE}: 70 pins are connected to V_{EE}; 115 mA if only V_{EE}: 70 is connected to V_{EE}; and 80 mA if only V_{EE}: 35 is connected to V_{EE}. If neither of these pins is connected, the large driver outputs 45 mA.

When the LED driver is OFF, the current is switched to the dummy load (DL) pin, which is normally connected to the positive supply V_{CC} through a dummy load resistor. This reduces ringing that could otherwise occur on switching relatively large currents.

The LED drivers are enabled if the ODE pin is left unconnected or pulled HIGH; they are disabled by pulling the ODE pin LOW.

Figure 40(C) shows a typical ECL application using the SP9960 IC.

Manchester Biphase-Mark Decoder

Figure 41(A) shows the simplified block diagram of the SP9921, a bipolar monolithic silicon IC designed for clock and data recovery from a Manchester biphase-mark encoded input signal. The device locks onto incoming data, recovers the clock, and

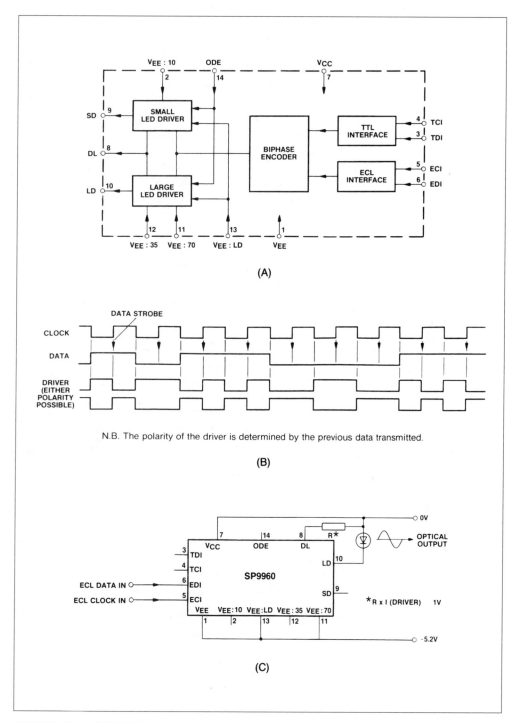

FIGURE 40 SP9960 Manchester biphase-mark encoder and LED driver (A) Functional block diagram (B) Encoding alignment (C) Typical ECL application circuit, with the large driver set at 115 mA (nominal)

decodes the data, making use of a reference clock input at approximately one-fifth of the data rate.

Data is received at differential input pins (12 and 13) of the limiting amplifier. The received signal can be monitored at the amplifier output, pin 7. The signal is also fed into a loop which outputs the recovered clock at pin 6 and the decoded data at pin 5.

Figure 41(B) shows how the input signal is decoded. The Manchester biphase-mark code uses a transition at the center of the bit to indicate a logic 1 and the absence of a transition to indicate a logic 0. There is always a transition at the end of the bit.

The SP9921 can be used in systems operating over a wide range of data rates without false frequency lock. This is achieved using a reference VCO and a recovery VCO.

The reference VCO is phase-locked to the reference clock input (pin 10), thus generating an internal clock at 5 times the frequency of the reference clock input. The output

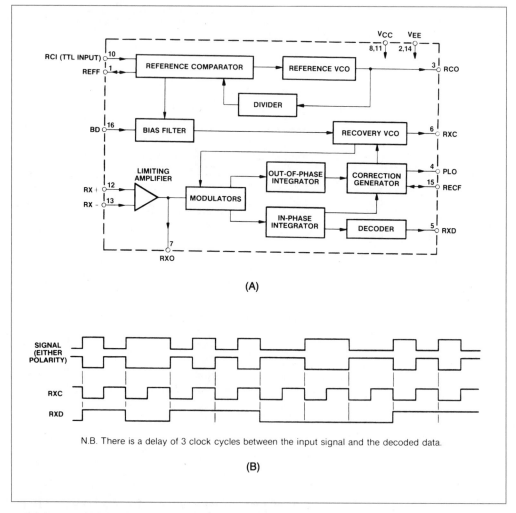

FIGURE 41 SP9921 Manchester biphase-mark decoder (A) Functional block diagram (B) Biphase-mark decoding (C) Typical application circuit

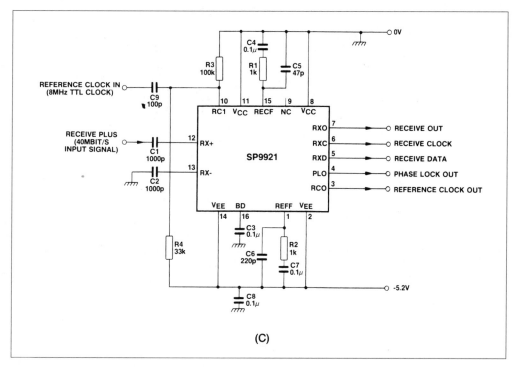

(C)

FIGURE 41 (*continued*)

of the reference VCO is monitored at pin 3. Filtering of the bias control signal to the VCO is performed at the reference filter pin (pin 1).

The bias control signal for the reference VCO is filtered at bias decoupling pin 16 and used to set the free-running frequency of the recovery VCO. The recovery VCO drives receive clock pin 6 and the modulators, which in turn drive the integrators. The integrators analyze the components of the signal which are in-phase and 90° out-of-phase, and thus obtain the recovered data and the correction signal for the loop. The correction signal is filtered at recovery filter pin 15.

The loop also pulls phase-lock output pin 4 LOW for any bits when the output of the in-phase integrator (data) does not exceed the output of the out-of-phase integrator (error) by a set margin. This can occur when there is a loss of data, if there is enough noise on the link (even if no data is corrupted), or if the loop has difficulty locking.

Figure 41(C) shows a typical application circuit for the SP9921 IC.

Transimpedance Amplifier

Figure 42(A) shows the functional block diagram of the SL9901 transimpedance amplifier. This monolithic silicon IC is designed to interface between a detector diode and a decoder in a fiber optic receiver system.

A typical application for the SL9901 IC is shown in Figure 42(B). The photocurrent generated by the PIN diode is converted to a voltage suitable for driving a comparator input stage in a decoder/detector circuit. The device has a 3 dB electrical bandwidth of 50 MHz, enabling NRZ data rates of up to 100 Mbs to be received.

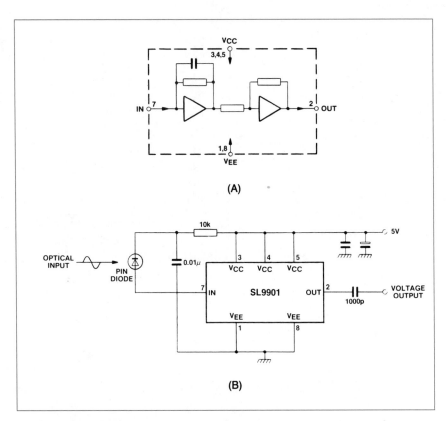

FIGURE 42 SL9901 transimpedance amplifier (A) Functional block diagram (B) Typical application circuit

National Semiconductor

Audio, Radio and TV Circuits

TBA120U, TBA120T IF Amplifier and Detector

General Description

The TBA120U, TBA120T is a monolithic integrated circuit specifically designed for audio detection in TV and FM radio receivers. It incorporates an 8 stage limiting IF amplifier and balanced detector plus a DC operated volume control. The circuit also provides connection facilities for a video tape recorder. The TBA120T is designed primarily for use with ceramic filters while the TBA120U is optimized for inductive tuning.

Features

- Electronic attenuator: replaces conventional AC volume control
- Volume reduction range: 85 dB typ
- Sensitivity: 3 dB limiting voltage 30 μV typ
- Excellent AM rejection 68 dB typ 500 μV
- Wide supply voltage range (6 to 18V)
- Easy video recorder connection
- Very low external component requirement
- Simple detector alignment: one coil

Block and Connection Diagrams

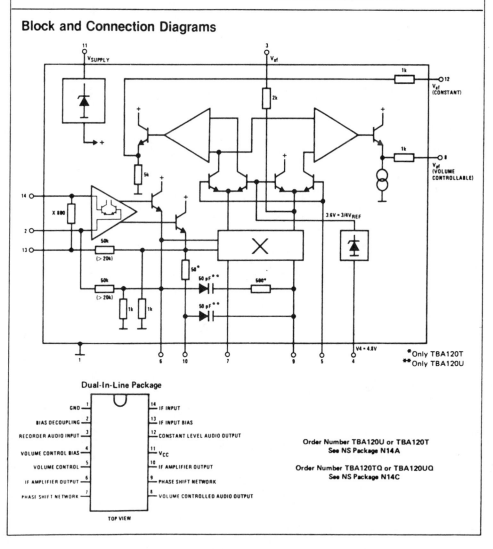

Dual-In-Line Package

GND 1	14 IF INPUT
BIAS DECOUPLING 2	13 IF INPUT BIAS
RECORDER AUDIO INPUT 3	12 CONSTANT LEVEL AUDIO OUTPUT
VOLUME CONTROL BIAS 4	11 Vcc
VOLUME CONTROL 5	10 IF AMPLIFIER OUTPUT
IF AMPLIFIER OUTPUT 6	9 PHASE SHIFT NETWORK
PHASE SHIFT NETWORK 7	8 VOLUME CONTROLLED AUDIO OUTPUT

TOP VIEW

Order Number TBA120U or TBA120T
See NS Package N14A

Order Number TBA120TQ or TBA120UQ
See NS Package N14C

769

Absolute Maximum Ratings

Supply Voltage, V11	18V	Current Pin 4, I_4	5 mA
Operating Temperature Range, T_u	−15°C to +70°C	Operating Frequency Range, f	0 to 12 MHz
Storage Temperature Range, T_s	−40°C to +125°C	Power Dissipation, P_{tot}	400 mW
Voltage Pin 5, V5	6V	Resistor Parallel to Pins 13 and 14	1 kΩ

Electrical Characteristics (V_{CC} = 12V, T_A = 25°C)

PARAMETER		CONDITIONS	MIN	TYP	MAX	UNITS
I_{CC}	Supply Current		9.5	13.5	17.5	mA
G_V	IF Voltage Gain	f = 5.5 MHz		68		dB
V_O	IF Output Voltage (Each Output Limiting)			250		mVp-p
R8	Output Impedance			1.1		kΩ
R12				1.1		kΩ
R3	Input Impedance			2		kΩ
R4	Regulator Impedance			12		Ω
V8	DC Output Level	V_i = 0		4		V
V12		V_i = 0		5.6		V
V4	Regulator Voltage		4.2	4.8	5.3	V
$\dfrac{\text{Vaf max}}{\text{Vaf min}}$	Volume Control		70	85		dB
$\dfrac{\text{Vaf8}}{\text{Vaf3}}$	Video Recorder Output Ratio			7.5		
V_{LIM}	Sensitivity	Vaf − 3 dB, f = 5.5 MHz		30	60	μV
$\dfrac{\text{V8}}{\text{V11}}$	Supply Rejection			35		dB
$\dfrac{\text{V12}}{\text{V11}}$				30		dB
R4−R5	Impedance		1		10	kΩ
$\dfrac{\text{Vaf max}}{\text{Vaf8}}$	Output Ratio	R4--R5 = 5 kΩ R5−R1 = 13 kΩ	20	28	36	dB
TBA120T Only						
Z_i	Input Impedance	f = 5.5 MHz		800/5		Ω/pF
a_{AM}	AM Rejection	f = 5.5 MHz m = 30% Δf = ±50 kHz V_i = 500 μV f_{MOD} = 1 kHz	50	60		dB
Vaf8	A.F. Output Voltage	f = 5.5 MHz f_{MOD} = 1 kHz		900		mV
Vaf12		Δf = ±50 kHz		650		mV
TBA120U Only						
Z_i	Input Impedance	f = 5.5 MHz	15/6	40/4.5		kΩ/pF
a_{AM}	AM Rejection	f = 5.5 MHz V_i = 500 μV f_{MOD} = 1 kHz Δf = ±50 kHz m = 30%	50	60		dB

Electrical Characteristics (Continued) (V_{CC} = 12V, T_A = 25°C)

PARAMETER		CONDITIONS	MIN	TYP	MAX	UNITS
TBA120U Only (Continued)						
Vaf8	A.F. Output Voltage	f = 5.5 MHz f_{MOD} = 1 kHz Δf = ±50 kHz V_i = 10 mV Q_B = 45		1.3		V
Vaf12	A.F. Output Voltage			1.0		V
k	Distortion	f = 5.5 MHz Δf = ±50 kHz f_{MOD} = 1 kHz Q_B = 45 V_i = 10 mV		1		%

Typical Application (5.5 MHz)

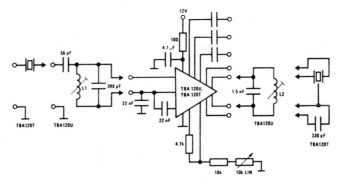

Circuit for Direct Connection to Video Recorders

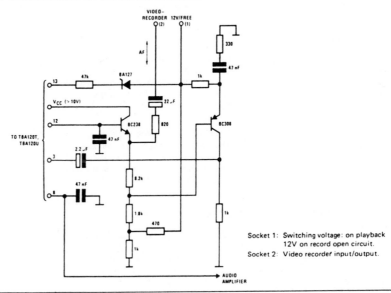

Socket 1: Switching voltage: on playback
12V on record open circuit.
Socket 2: Video recorder input/output.

Using the NE/SA/SE592 video amplifier   AN141

VIDEO AMPLIFIER PRODUCTS

NE/SA/SE592 Video Amplifier

The 592 is a two-stage differential output, wide-band video amplifier with voltage gains as high as 400 and bandwidths up to 120MHz.

Three basic gain options are provided. Fixed gains of 400 and 100 result from shorting together gain select pins G_{1A}-G_{1B} and G_{2A}-G_{2B}, respectively. As shown by Figure 1, the emitter circuits of the differential pair return through independent current sources. This topology allows no gain in the input stage if all gain select pins are left open. Thus, the third gain option of tying an external resistance across the gain select pins allows the user to select any desired gain from 0 to 400V/V. The advantages of this configuration will be covered in greater detail under the filter application section.

Three factors should be pointed out at this time:

1. The gains specified are differential. Single-ended gains are one-half the stated value.

2. The circuit 3dB bandwidths are a function of and are inversely proportional to the gain settings.

3. The differential input impedance is an inverse function of the gain setting.

In applications where the signal source is a transformer or magnetic transducer, the input bias current required by the 592 may be passed directly through the source to ground. Where capacitive coupling is to be used, the base inputs must be returned to ground through a resistor to provide a DC path for the bias current.

Due to offset currents, the selection of the input bias resistors is a compromise. To reduce the loading on the source, the resistors should be large, but to minimize the output DC offset, they should be small—ideally 0Ω. Their maximum value is set by the maximum allowable output offset and may be determined as follows:

1. Define the allowable output offset (assume 1.5V).

2. Subtract the maximum 592 output offset (from the data sheet). This gives the output offset allowed as a function of input offset currents (1.5V-1.0V=0.5V).

3. Divide by the circuit gain (assume 100). This refers the output offset to the input.

4. The maximum input resistor size is:

$$R_{MAX} = \frac{Input\ Offset\ Voltage}{Max\ Input\ Offset\ Current} \quad (1)$$
$$= \frac{0.005V}{5\mu A}$$
$$= 1.00k\Omega$$

Of paramount importance during the design of the NE592 device was bandwidth. In a monolithic device, this precludes the use of PNP transistors and standard level-shifting techniques used in lower frequency devices. Thus, without the aid of level shifting, the output common-mode voltage present on the NE592 is typically 2.9V. Most applications, therefore, require capacitive coupling to the load.

Filters

As mentioned earlier, the emitter circuit of the NE592 includes two current sources.

Since the stage gain is calculated by dividing the collector load impedance by the emitter impedance, the high impedance contributed by the current sources causes the stage gain to be zero with all gain select pins open. As shown by the gain vs. frequency graph of Figure 2, the overall gain at low frequencies is a negative 48dB.

Higher frequencies cause higher gain due to distributed parasitic capacitive reactance. This reactance in the first stage emitter circuit causes increasing stage gain until at 10MHz the gain is 0dB, or unity.

Referring to Figure 3, the impedance seen looking across the emitter structure includes small r_e of each transistor.

Any calculations of impedance networks across the emitters then must include this quantity. The collector current level is approximately 2mA, causing the quantity of 2 r_e to be approximately 32Ω. Overall device gain is thus given by

$$\frac{V_O\ (S)}{V_{IN}\ (S)} = \frac{1.4x10^4}{Z_{IN}(S) + 32} \quad (2)$$

where $Z_{(S)}$ can be resistance or a reactive impedance. Table 2 summarizes the possible configurations to produce low, high, and bandpass filters. The emitter impedance is made to vary as a function of frequency by using capacitors or inductors to alter the frequency response. Included also in Table 2 is the gain calculation to determine the voltage gain as a function of frequency.

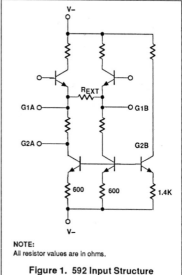

NOTE:
All resistor values are in ohms.

Figure 1. 592 Input Structure

TABLE 1. VIDEO AMPLIFIER COMPARISON FILE

PARAMETER	NE/SA/SE592	733
Bandwidth (MHz)	120	120
Gain	0,100,400	10,100,400
R_{IN} (k)	4-30	4-250
V_{P-P} (Vs)	4.0	4.0

TABLE 2. FILTER NETWORKS

Z NETWORK	FILTER TYPE	$\dfrac{V_o(s)}{V_1(s)}$ TRANSFER FUNCTION
(series R–L network)	LOW PASS	$\dfrac{1.4 \times 10^4}{L} \left[\dfrac{1}{s + R/L} \right]$
(series R–C network)	HIGH PASS	$\dfrac{1.4 \times 10^4}{R} \left[\dfrac{S}{s + 1/RC} \right]$
(series R–L–C network)	BAND PASS	$\dfrac{1.4 \times 10^4}{L} \left[\dfrac{S}{s^2 + R/Ls + 1/LC} \right]$
(R with parallel L–C network)	BAND REJECT	$\dfrac{1.4 \times 10^4}{R} \left[\dfrac{S^2 + 1/LC}{s^2 + 1/LC + s/RC} \right]$

NOTES:
In the networks above, the R value used is assumed to include 2 r_e, or approximately 32Ω.
$S = j\Omega$
$\Omega = 2\pi f$

Differentiation
With the addition of a capacitor across the gain select terminals, the NE592 becomes a differentiator. The primary advantage of using the emitter circuit to accomplish differentiation is the retention of the high common mode noise rejection. Disc file playback systems rely heavily upon this common-mode rejection for proper operation. Figure 4 shows a differential amplifier configuration with transfer function.

Disc File Decoding
In recovering data from disc or drum files, several steps must be taken to precondition the linear data. The NE592 video amplifier, coupled with the 8T20 bidirectional one-shot, provides all the signal conditioning necessary for phase-encoded data.

When data is recorded on a disc, drum or tape system, the readback will be a Gaussian shaped pulse with the peak of the pulse corresponding to the actual recorded transition point. This readback signal is usually 500μV_{P-P} to 3mV_{P-P} for oxide coated disc files and 1 to 20mV_{P-P} for nickel-cobalt disc files. In order to accurately reproduce the data stream originally written on the disc memory, the time of peak point of the Gaussian readback signal must be determined.

The classical approach to peak time determination is to differentiate the input signal. Differentiation results in a voltage proportional to the slope of the input signal. The zero-crossing point of the differentiator, therefore, will occur when the input signal is at a peak. Using a zero-crossing detector and one-shot, therefore, results in pulses occurring at the input peak points.

A circuit which provides the preconditioning described above is shown in Figure 5. Readback data is applied directly to the input of the first NE592. This amplifier functions as a wide-band AC-coupled amplifier with a gain of 100. The NE592 is excellent for this use because of its high phase linearity, high gain and ability to directly couple the unit with the readback head. By direct coupling of readback head to amplifier, no matched terminating resistors are required and the excellent common-mode rejection ratio of the amplifier is preserved. DC components are also rejected because the NE592 has no gain at DC due to the capacitance across the gain select terminals.

The output of the first stage amplifier is routed to a linear phase shift low-pass filter. The filter is a single-stage constant K filter, with a characteristic impedance of 200Ω. Calculations for the filter are as follows:

$$L = \frac{2R}{\omega C}$$

where

$R = characteristic\ impedance\ (\Omega)$
$C = \dfrac{1}{\omega C}$

where

$\omega C = cut-off\ frequency\ (radians/sec)$

The second NE592 is utilized as a low noise differentiator/amplifier stage. The NE592 is excellent in this application because it allows differentiation with excellent common-mode noise rejection.

The output of the differentiator/amplifier is connected to the 8T20 bidirectional monostable unit to provide the proper pulses at the zero-crossing points of the differentiator.

The circuit in Figure 5 was tested with an input signal approximating that of a readback signal. The results are shown in Figure 7.

AUTOMATIC GAIN CONTROL
The NE592 can also be connected in conjunction with a MC1496 balanced modulator to form an excellent automatic gain control system.

The signal is fed to the signal input of the MC1496 and RC-coupled to the NE592. Unbalancing the carrier input of the MC1496 causes the signal to pass through unattenuated. Rectifying and filtering one of the NE592 outputs produces a DC signal which is proportional to the AC signal amplitude. After filtering; this control signal is applied to the MC1496 causing its gain to change.

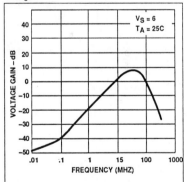

Figure 2. Voltage Gain as a Function of Frequency (All Gain Select Pins Open)

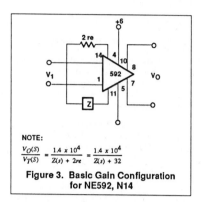

NOTE:

$$\frac{V_O(S)}{V_T(S)} = \frac{1.4 \times 10^4}{Z(s) + 2re} = \frac{1.4 \times 10^4}{Z(s) + 32}$$

**Figure 3. Basic Gain Configuration
for NE592, N14**

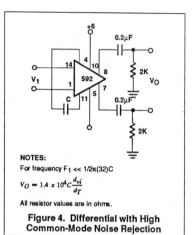

NOTES:

For frequency $F_1 \ll 1/2\pi(32)C$

$$V_O \approx 1.4 \times 10^4 C \frac{d_{vi}}{d_T}$$

All resistor values are in ohms.

**Figure 4. Differential with High
Common-Mode Noise Rejection**

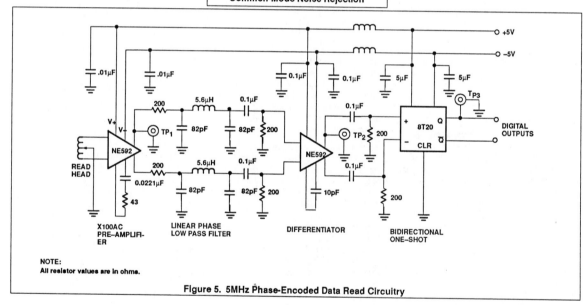

NOTE:
All resistor values are in ohms.

Figure 5. 5MHz Phase-Encoded Data Read Circuitry

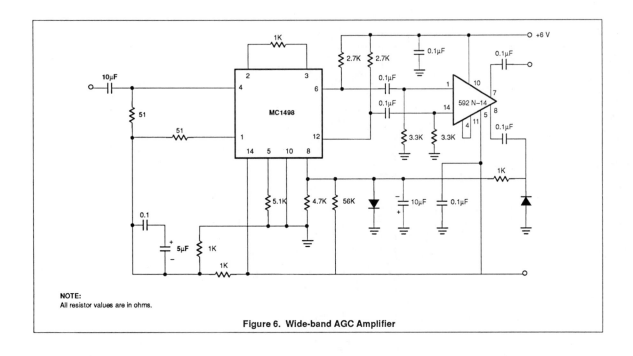

NOTE:
All resistor values are in ohms.

Figure 6. Wide-band AGC Amplifier

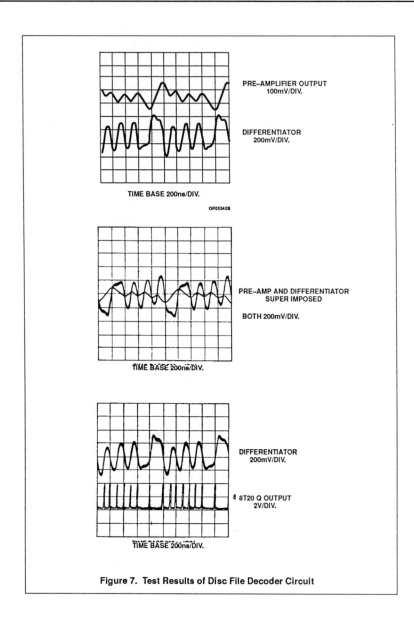

PRE–AMPLIFIER OUTPUT
100mV/DIV.

DIFFERENTIATOR
200mV/DIV.

TIME BASE 200ns/DIV.

OP053408

PRE–AMP AND DIFFERENTIATOR
SUPER IMPOSED

BOTH 200mV/DIV.

TIME BASE 200ns/DIV.

DIFFERENTIATOR
200mV/DIV.

8T20 Q OUTPUT
2V/DIV.

TIME BASE 200ns/DIV.

Figure 7. Test Results of Disc File Decoder Circuit

TABLE D-1 **Class D Citizens' Band Radio Channels**

Channel	Frequency (MHz)	Channel	Frequency (MHz)
1	26.965	21	27.215
2	26.975	22	27.225
3	26.985	23	27.235
4	27.005	24	27.245
5	27.015	25	27.255
6	27.025	26	27.265
7	27.035	27	27.275
8	27.055	28	27.285
9	27.065	29	27.295
10	27.075	30	27.305
11	27.085	31	27.315
12	27.105	32	27.325
13	27.115	33	27.335
14	27.125	34	27.345
15	27.135	35	27.355
16	27.155	36	27.365
17	27.165	37	27.375
18	27.175	38	27.385
19	27.185	39	27.395
20	27.205	40	27.405

TABLE D-2 10-code

Code	Message	Code	Message
10–1	Receiving poorly	10–37	Wrecker needed at _____
10–2	Receiving well	10–38	Ambulance needed at _____
10–3	Stop transmitting	10–39	Your message delivered
10–4	OK, message received	10–41	Switch to channel _____
10–5	Relay message	10–42	Traffic accident at _____
10–6	Busy, stand by	10–43	Traffic tied up at _____
10–7	Out of service	10–44	I have a message for you
10–8	In service	10–45	All units within range please report
10–9	Repeat transmission	10–50	Break channel _____
10–10	Transmission completed, standing by	10–60	What is next message number?
10–11	Speaking too rapidly	10–62	Unable to copy, use phone
10–12	Visitors present	10–63	Net directed to _____
10–13	Advise weather and road conditions	10–64	Net clear
10–15	Pick up _____ at _____	10–65	Awaiting your next message/assignment
10–16	Picked up _____	10–66	Cancellation
10–17	Urgent business	10–67	All units comply
10–18	Anything for us?	10–70	Fire at _____
10–19	Nothing for you, return to station	10–71	Proceed with transmission in sequence
10–20	What is your location? (QTH)	10–73	Speed trap at _____
10–21	Call by landline	10–75	You are causing interference
10–22	Report in person to _____	10–77	Negative contact
10–23	Stand by	10–79	Report progress of fire
10–24	Completed last assignment	10–81	Reserve hotel room for _____
10–25	Can you contact _____	10–82	Reserve room for _____
10–26	Disregard last information	10–84	My telephone number is _____
10–27	Moving to channel _____	10–85	My address is _____
10–28	Identify your station	10–89	Radio repairman needed at _____
10–29	Time is up for contact	10–90	I have TVI
10–30	Does not conform to rules and regulations	10–91	Talk closer to mike
10–32	I will give you a radio check	10–92	Your transmitter is out of adjustment
10–33	**Emergency traffic at this station**	10–93	Frequency check this channel
10–34	**Trouble at this station, help**	10–94	Give a test with voice
10–35	Confidential information	10–95	Transmit dead carrier for 5 seconds
10–36	Correct time (QTR)	10–99	Mission completed, all units secure
		10–100	Nature calls _____ 20/40
		10–200	Police needed at _____

TABLE D-3 Q Signals

Signal	Meaning	Signal	Meaning
QRG	Will you tell me my exact frequency (or that of …)? Your exact frequency (or that of …) is … kc.	QSA	What is the strength of my signals (or those of …)? The strength of your signals (or those of …) is … (1. scarcely perceptible; 2. weak; 3. fairly good; 4. good; 5. very good).
QRH	Does my frequency vary? Your frequency varies.	QSB	Are my signals fading? Your signals are fading.
QRI	How is the tone of my transmission? The tone of your transmission is … (1. good; 2. variable; 3. bad).	QSD	Are my signals mutilated? Your signals are mutilated.
QRK	What is the intelligibility of my signals (or those of …)? The intelligibility of your signals (or those of …) is … (1. bad; 2. poor; 3. fair; 4. good; 5. excellent).	QSG	Shall I send … messages at a time? Send … messages at a time.
QRL	Are you busy? I am busy (or I am busy with …). Please do not interfere.	QSK	Can you hear me between your signals and if so can I break in on your transmission? I can hear you between my signals; break in on my transmission.
QRM	Is my transmission being interfered with? Your transmission is being interfered with … (1. nil; 2. slightly; 3. moderately; 4. severely; 5. extremely).	QSL	Can you acknowledge receipt? I am acknowledging receipt.
QRN	Are you troubled by static? I am troubled by static … (1–5 as under QRM).	QSM	Shall I repeat the last message which I sent you, or some previous message? Repeat the last message which you sent me (or message(s) number(s) …).
QRO	Shall I increase power? Increase power.	QSN	Did you hear me (*or* …) on … kc? I did hear you (*or* …) on … kc.
QRP	Shall I decrease power? Decrease power.		
QRQ	Shall I send faster? Send faster (… w.p.m.).	QSO	Can you communicate with … direct or by relay? I can communicate with … direct (or by relay through …).
QRS	Shall I send more slowly? Send more slowly (… w.p.m.).	QSP	Will you relay to … ? I will relay to … .
QRT	Shall I stop sending? Stop sending.	QSU	Shall I send or reply on this frequency (or on … kc)? Send or reply on this frequency (or on … kc).
QRU	Have you anything for me? I have nothing for you.		
QRV	Are you ready? I am ready.	QSV	Shall I send a series of Vs on this frequency (or … kc)? Send a series of Vs on this frequency (or … kc).
QRW	Shall I inform … that you are calling him on … kc? Please inform … that I am calling on … kc.	QSW	Will you send on this frequency (or on … kc)? I am going to send on this frequency (or on … kc).
QRX	When will you call me again? I will call you again at … hours (on … kc).	QSX	Will you listen to … on … kc? I am listening to … on … kc.
QRY	What is my turn? Your turn is Number … .	QSY	Shall I change to transmission on another frequency? Change to transmission on another frequency (or on … kc).
QRZ	Who is calling me? You are being called by … (on … kc).		

TABLE D-3 (*continued*)

Signal	Meaning	Signal	Meaning
QSZ	Shall I send each word or group more than once? Send each word or group twice (or ... times).	QTH	What is your location? My location is ... .
		QTR	What is the correct time? The time is ... .
QTA	Shall I cancel message number ... ? Cancel message number	**Special abbreviations adopted by ARRL:**	
QTB	Do you agree with my counting of words? I do not agree with your counting of words; I will repeat the first letter or digit of each word or group.	QST	General call preceding a message addressed to all amateurs and ARRL members. This is in effect "CQ ARRL."
QTC	How many messages have you to send? I have ... messages for you (or for ...).	QRRR	Official ARRL "land SOS." A distress call for emergency use only by a station in an emergency situation.

TABLE D-4 **Abbreviations for CW Work**

Abbreviation	Meaning	Abbreviation	Meaning
AA	All after	GA	Go ahead (or resume sending)
AB	All before	GB	Good-bye
ABT	About	GBA	Give better address
ADR	Address	GE	Good evening
AGN	Again	GG	Going
ANT	Antenna	GM	Good morning
BCI	Broadcast interference	GN	Good night
BCL	Broadcast listener	GND	Ground
BK	Break; break me; break in	GUD	Good
BN	All between; been	HI	The telegraphic laugh; high
C	Yes	HR	Here; hear
CFM	Confirm; I confirm	HV	Have
CK	Check	HW	How
CL	I am closing my station; call	LID	A poor operator
CLD; CLG	Called; calling	MA, MILS	Milliamperes
CUD	Could	MSG	Message; prefix to radiogram
CUL	See you later	N	No
CUM	Come	ND	Nothing doing
CW	Continuous wave	NIL	Nothing; I have nothing for you
DLD, DLVD	Delivered	NM	No more
DX	Distance; foreign countries	NR	Number
ES	And, &	NW	Now; I resume transmission
FB	Fine business; excellent	OB	Old boy

TABLE D-4 (*continued*)

Abbreviation	Meaning	Abbreviation	Meaning
OM	Old man	TNX, TKS	Thanks
OP, OPR	Operator	TT	That
OT	Old timer; old top	TU	Thank you
PBL	Preamble	TVI	Television interference
PSE	Please	TXT	Text
PWR	Power	UR; URS	Your; you're; yours
PX	Press	VFO	Variable-frequency oscillator
R	Received as transmitted; are	VY	Very
RCD	Received	WA	Word after
RCVR, RX	Receiver	WB	Word before
REF	Refer to; referring to; reference	WD; WDS	Word; words
RIG	Station equipment	WKD; WKG	Worked; working
RPT	Repeat; I repeat	WL	Well; will
SED	Said	WUD	Would
SIG	Signature; signal	WX	Weather
SINE	Operator's personal initials or nickname	XMTR, TX	Transmitter
SKED	Schedule	XTAL	Crystal
SRI	Sorry	XYL, YF	Wife
SVC	Service; prefix to service message	YL	Young lady
TFC	Traffic	73	Best regards
TMW	Tomorrow	88	Love and kisses

TABLE D-5   **ICAO Phonetic Alphabet**

Letter	Phonetic Tag	Letter	Phonetic Tag
A	Alpha	N	November
B	Baker	O	Oscar
C	Charlie	P	Papa
D	Delta	Q	Quebec
E	Echo	R	Romeo
F	Foxtrot	S	Sierra
G	Golf	T	Tango
H	Hotel	U	Uniform
I	India	V	Victor
J	Juliett	W	Whiskey
K	Kilo	X	X-ray
L	Lima	Y	Yankee
M	Mike	Z	Zulu

Glossary

A-law companding standard for the European telephone network.

absorption loss the conversion of the power in the light beam to heat in some material or imperfection that is partially or completely opaque.

active filter a device constructed with a network of resistors and capacitors around an active device that screens out or passes electric current of certain frequencies.

amplitude modulation (AM) is generated by combining intelligence and carrier frequencies through a nonlinear device called a modulator.

analog pulse modulation includes pulse amplitude modulation and pulse time modulation. Periodic samples of the modulating waveform are taken and only those samples, in the form of a pulse train, are transmitted. If the sampling frequency is high enough, the modulating wave can be recovered at the receiver.

angle modulation variation of the angle of a sine-wave carrier in response to the modulating source.

angle of inclination the angle between the earth's equatorial plane and the orbital plane of the satellite. The inclination of the orbit determines the geographical limits of the projection of the satellite's path over the earth's surface.

antenna array a combination of two or more antenna elements.

antenna a device for transmitting or receiving electromagnetic energy.

apogee the point in the orbit of a satellite at the greatest distance from the center of the earth.

Armstrong system an early system for FM broadcast transmitters requiring a phase shifter and a balanced modulator.

asynchronous transmission sending each information character, individually synchronized by means of start and stop elements. The most common transmission method used with low-speed terminals.

atmospheric noise distortion caused by electrical disturbances in the earth's atmosphere and manifesting itself in static. (Lightning is a visible source of atmospheric noise.)

audio amplifier device designed to accept the range of frequencies extending from approximately 20 Hz to 20,000 Hz. This range includes the human voice and the various frequencies of musical instruments.

audio frequency (AF) a frequency lying within the audible spectrum.

audio frequency shift keying (AFSK) the modulation technique in which the RF carrier is modulated by a tone. Keying is accomplished by changing the tone to another frequency. Used in facsimile and radioteletype communications.

aurora borealis a luminous atmospheric phenomenon (northern or southern lights) ascribed to electricity. It occurs near or radiates from the earth's northern or southern magnetic poles and is visible at different times over varying portions of the earth's surface.

automatic frequency control (AFC) a system that keeps a circuit automatically tuned to the desired frequency.

automatic gain control (AGC) a system that keeps the gain, and therefore the output, of a receiver or amplifier relatively constant, in spite of input-signal amplitude fluctuations.

autopatch a device for connecting a radio transceiver to the telephone lines by remote control.

avalanche photodiode (APD) a diode that produces a gain of 100 or more. However, it also produces noise that may limit receiver sensitivity.

B

backward wave oscillator (BWO) a microwave oscillator tube similar to the traveling wave tube. Electron bunching results from interaction between the electron beam and the RF field, and reflection takes place at the collector. As a result, the wave moves backward from the collector to the cathode. Oscillation is sustained because the wave is in phase with the input.

backward wave tube a traveling wave tube in which electrons travel in a direction opposite to that in which the wave is propagated.

balanced modulator a symmetrical circuit that delivers an output signal containing the frequency sum and frequency difference of two input signals. The key circuit in SSB generation.

balanced ring modulator a circuit using diodes as nonlinear elements and capable of suppressing the carrier by 60 dB.

balun a *bal*anced to *un*balanced RF transformer. Used to correct the imbalance between a coaxial line and a dipole antenna.

band-pass filter passes a band of frequencies between designated values and rejects all others.

band-reject filter prevents a band of frequencies between two designated values from passing and passes those frequencies above and below that band.

band-stop filter See *band-reject filter*.

bandpass response the ability of a receiver to pass a band of frequencies.

bandwidth (BW) defines the band of frequencies occupied by a signal and required for effective transfer of the information carried by the signal. The width depends on the quality (Q) of the tuned circuit.

Barkhausen criterion states that for an oscillator to produce its own input signal continuously, the product of the amplifier gain (A_v) and the fractional feedback factor (B_v) must equal unity ($A_v B_v = 1$).

base bias the least effective way to bias a circuit for operation in the active region of the load line because the Q point is unstable. It should never be used in linear circuits. Also called *fixed bias*.

beat frequency oscillator (BFO) an oscillator used to set up an audible beat with the signal generated by another oscillator.

bending loss the energy loss that occurs because light rays on the outside of a sharp bend in an optical fiber cannot travel fast enough to keep up with the other rays and, consequently, are lost.

Bessel functions advanced calculus functions used for the analysis of an FM signal modulated with a sine wave.

biconical antenna a balanced broadband antenna that consists of two metal cones arranged so that they meet at or near the vertices. The antenna is fed at the point where the vertices meet.

bipolar junction microwave transistors devices of the silicon NPN type and used primarily as amplifiers and oscillators below 12 GHz. Also used in many microwave systems operating in the C-band and below.

body-stabilized satellite an improvement on the spin-stabilized satellite. A spinning wheel provides the gyroscope and the solar panels and body of the satellite are fixed. The entire solar panel is continuously illuminated and power is increased significantly.

break point See *half-power point*.

break-in operation a form of radio communication in which a transmitting operator can hear signals at all times except during the actual emission of radiation by the transmitter.

buffer amplifier circuit used extensively for impedance matching.

bus system a transmission system consisting of an optical fiber line, several passive taps, and a number of terminals (each containing a transmitter and a receiver). All terminals are connected together so that any one terminal can communicate with any of the other terminals.

C

capture range in a PLL, the bandwidth over which capture is possible. It can never be wider than the lock range.

carcinotron See *backward wave oscillator* and *magnetron*.

carrier wave serves as the vehicle for the transmission of information.

cascade two or more amplifier stages connected together to provide a higher overall gain.

cavity resonator a cylindrical or rectangular metal enclosure that operates as a tuned circuit. It can be fixed or tunable. The resonant frequency of the cavity is affected by temperature and humidity.

ceramic filter constructed from lead zirconate titanate, similar in use to a crystal filter but with lower Q values. Used for portable SSB equipment.

ceramic ladder filter ceramic disc having a series and a parallel resonant frequency. Used in applications requiring ruggedness, low cost, and small sized filters.

characteristic impedance also called surge impedance or surge resistance. Special property determined by the distributed inductance (of the conductors) and the distributed capacity (between the conductors) of a line. Expressed in ohms, it is approximately equal to the square root of the ratio of inductance to capacitance, with the ratio taken for any given length of line.

charge coupled device (CCD) a dynamic shift register capable of analog signal processing and storing digital data. Applications include use as video and audio delay lines and as secure communications filters.

chrominance signal the portion of the composite color signals developed by the three individual color pickup tubes in the color camera.

circular polarization condition in which the orientation of the electrical lines of flux completes one rotation for every cycle of the wave. Uniform rotation may be either clockwise or counterclockwise.

citizens' band (CB) radio a public noncommercial radio service available to the general population of the United States. Individuals need not be licensed but the equipment is licensed through the manufacturer's authorization by the FCC.

coaxial lines are either rigid (concentric) or flexible. Rigid lines are limited by high-frequency losses over distance, therefore flexible lines are most commonly used.

codec acronym for *co*der/*dec*oder.

collector-feedback bias although more sensitive than emitter-feedback bias, it is more effective and used in practical linear circuits. The smaller base current offsets the increase in collector current.

collinear array an antenna system that is usually vertically oriented to produce omnidirectional coverage. All elements are placed end to end with each element connected to transmission lines of equal length.

color killer a television circuit that disables the chrominance circuits when a black and white signal is being received. This ensures that only luminance information will reach the picture tube.

common-base amplifier (CB) the base is common to both the input and output signals, with the emitter and collector leads serving as input and output leads, respectively.

common-collector amplifier (CC) the collector is common to both input and output signals, with the base and emitter leads serving as input and output leads, respectively.

common-emitter amplifier (CE) most common configuration for amplifiers having the emitter common to both input and output signals with the base and collector leads serving as input and output leads, respectively.

communications system a collection of individual communications networks consisting of intelligence origination, transmission systems, receiving systems, and intelligence reproduction.

compander acronym for *com*pressor-ex*pander*. The circuit that compresses the dynamic range of an input signal in the coder and then expands it to almost original form in the decoder.

companding reduces the level of quantizing noise during periods of very weak signals. It is essential to quality transmission of digital modulation signals.

comparator a device used to generate dc control voltage. It is used by all PLL systems.

conical monopole antenna See *biconical antenna*. If the lower cone is replaced by a ground plane, the antenna becomes a conical monopole antenna.

connection loss in fiber optic systems, the loss of power at every connection point, particularly at repair splices.

convergence electrode a magnetic shield that allows each beam in a CRT to be magnetically shifted in position without appreciably affecting the other.

cosmic noise an electromagnetic radiation disturbance generated by the stars.

critical angle the angle of radiation, with respect to the tangent at the transmitting point, above which a signal will not be reflected back to earth by the ionosphere.

critical frequency the highest frequency returned to earth when radiated upward in a vertical direction.

crossed-field amplifier (CFA) an electron stream formation mechanism. Generally known as a crossed-field tube (CFT).

crystal filter has very high Q value. Crystal filters produce narrow pass bands and sharply defined skirts to filter out unwanted sidebands.

crystal lattice filter uses two matched pairs of crystals in a bridge circuit.

crystallization the result of bending a line back and forth (often by wind) which causes the metal of the conductors to become brittle. Crystallization increases resistance in the conductors, which in turn increases copper loss.

cutoff frequency the frequency at which the voltage gain of the filter drops to 70.7% of its maximum value. The value of the cutoff frequency determines the bandwidth of the pass band.

D

dark current the thermally generated reverse leakage current in a PIN diode. In conjunction with the signal current, it provides the ON-OFF ratio to be expected in a system.

decade represents a tenfold increase or decrease in frequency.

decibel (dB) a means of measuring relative levels of current, voltage, or power.

deemphasis the process of restoring distorted HF intelligence signals to their proper relative amplitudes in the receiver.

deflection yoke two vertical coils and two horizontal coils placed horizontally and vertically, respectively, which allow the beam to be moved to any position on the tube face by applying the proper polarity and current strength to the coils.

delayed AGC obtained when the generation of the bias is prevented until the signal level exceeds a preset threshold. As the signal strength increases, the delayed AGC provides greater and greater attenuation.

delta modulation (DM) a train of fixed-width pulses is transmitted. Pulse polarity is determined by whether the signal sample is larger or smaller than the previous sample. Also called slope modulation.

demodulation (decoding) the process of stripping the carrier from the modulating signal, thus recovering the original information.

demultiplexer device or circuit that separates the components of a multiplexed signal.

dial-up connection a telephone connection requiring a hardware device only to convert computer signals to telephone signals and back again.

dielectric loss is the I^2R power dissipation due to the heating that occurs in the insulating material between the conductors in a transmission line.

difference frequency the output of the mixer stage (in a superheterodyne receiver between the RF signal and the LO frequency). It is a constant value because of the relation between RF and LO tuning.

digital data communications   the process of communicating information in binary form between two points. Sometimes called computer communications. It can be defined as the movement of encoded data from one point to another by means of electrical transmission systems.

digital pulse modulation (DPM)   transmission of pulse trains in groups representing binary numbers that correspond to the modulating signal voltage amplitudes. Includes pulse code modulation and delta modulation.

digital-to-analog converter (DAC)   a circuit or device that converts a digital quality (unit) into an analog presentation (e.g., position of a rotating shaft).

diode detector   the simplest and most widely used AM demodulator. It provides no amplification.

diplexer   a coupler that allows two transmitters or receivers to be operated with the same antenna at the same time.

directional coupler   See *reflectometer*.

discone antenna   See *biconical antenna*. If the top cone is replaced by a disk of a certain radius, the antenna becomes a discone antenna.

discriminator   also called a demodulator or detector. It converts the phase variations brought into the receiver on the FM or PM wave into audio voltages.

dish antenna   a high-gain antenna used for reception and transmission of UHF and microwave signals. It consists of a driven element or other radiating device and a large spherical or parabolic reflector.

distributed elements (distributed constants) the equivalent resistance, capacitance, and inductance in the wiring of a circuit.

double-sided, suppressed-carrier (DSSC) variation on an SSB. The output of a properly balanced modulator is a DSSC signal.

down-converter   a circuit which heterodynes a signal to one centered on a lower frequency.

drift   an unwanted change in the final output caused as transistor characteristics vary with temperature.

driven array   an antenna system in which all elements are driven.

driven element   an element that is connected to an energy source.

dual-conversion   the use of two IF sections to improve stability. Dual conversion is used at higher frequencies.

dummy antenna   a nonradiating antenna used for preliminary transmitter tuning and for approximating the power output of the transmitter. Also called a phantom antenna.

E

Echo satellite   an early earth-orbiting satellite shaped like a large metal balloon, used to reflect radio waves sent to it. It was in low orbit and access time was brief.

edge-emitting LED   a diode that emits a directional pattern of light. It is more efficient than a surface LED for coupling into small fibers.

effective radiated power (ERP)   important criterion for determining the effectiveness of a transmitting station. The ERP may be many times the transmitting power.

electromagnetic radiation   complex form of energy containing both electric and magnetic fields. An example is radio waves radiated into space.

elliptical polarization   condition in which the orientation of the electric lines of flux changes as the signal is propagated from the transmitting antenna.

emitter-feedback bias   an arrangement whereby increased collector current produces more voltage across the emitter resistor, which reduces the base current and, consequently, the collector current. Theoretically workable but not a practical arrangement.

emitter-follower amplifier   circuit in which the output voltage follows the input voltage. See *common-collector amplifier*.

end effect   the result of capacitance between the ends of an antenna. Because of end effect, the wave developed on an antenna appears to be about 5% longer than its actual physical length.

end-fire antenna a bidirectional or unidirectional antenna in which the greatest amount of radiation takes place off the ends.

end-plate magnetron a specialized form of magnetron tube which produces increased output amplitude.

envelope a depiction of the actual amplitude variations resulting from modulation.

equivalent input noise resistance (R_{eq}) a fictitious resistance provided by manufacturers to simplify shot noise calculations.

error control function which checks the accuracy of data received using parity techniques.

exchange area network the intermediate system between a local telephone network and the long-haul network. It normally connects local exchanges with tandem exchanges.

exciter an amplifier or oscillator which supplies the input signal to the output amplifier.

external noise unwanted disturbance (noise) generated outside a receiver. It may be caused by atmospheric conditions or may be man-made.

F

fading a large variation in signal strength which may result in the signal being lost completely.

ferrite rod antenna sometimes called a loopstick, often used for receiving applications at LF, MF, and HF, up to about 20 MHz. Consists of a coil wound on a solenoid ferrite core. A capacitor in conjunction with the coil forms a tuned circuit.

fiberscope a flexible bundle of optical fibers with a focusing lens at each end that is used for viewing areas not normally observable.

field strength the amplitude of a radio wave evaluated in terms of the voltage it will induce across a wire. It is a measure of the intensity of an electric, magnetic, or electromagnetic field.

fixed bias See *base bias*.

flow control function which controls the flow of the signal from the processing unit to the destination.

flyback in an oscilloscope or TV picture tube, the rapid return (retrace) of the beam to its starting position.

focusing electrode the internal electrode (in a cathode-ray tube) that is used to focus the electron beam.

Foster-Seeley discrimination a phase discriminator requiring a limiter stage preceding the discriminator. Both primary and secondary windings are tuned to the IF center frequency. This greatly simplifies circuit alignment and provides increased linearity.

frequency division multiplex (FDM) signal used to transmit several information channels over the same communications channel simultaneously. Unlike TDM, FDM does not use pulse modulation. In FDM, each channel continuously occupies a small fraction of the transmitted frequency spectrum; in TDM, each channel occupies the entire frequency spectrum for only a fraction of the time. FDM uses analog carrier systems. See *time division multiplex*.

frequency modulation (FM) condition in which carrier amplitude remains constant while carrier frequency varies around the center frequency. The amount by which the instantaneous deviation is made proportional to the instantaneous amplitude of the modulating signal.

frequency multiplier a circuit that produces a whole numerical multiple, or many whole numerical multiples, of a given signal.

frequency shift keying (FSK) the modulation technique used in most models. It uses a specific frequency to represent the ON condition (binary 1) and a different frequency to represent the OFF condition (binary 0).

frequency swing the total frequency deviation above and below the rest frequency.

frequency synthesizer device that generates highly precise frequency signals by means of a single crystal oscillator in conjunction with frequency dividers and multipliers.

frequency-conversion section commonly called the *mixer* stage, where heterodyning of the received RF signal and the LO signal occurs.

front-end processor (FEP) device that functions as an auxiliary computer system to perform network control operations. It releases the central computer system to do data processing.

fundamental frequency the basic frequency, sometimes called the first harmonic.

G

gamma correctors devices which regulate (or correct) the degree of contrast in a TV picture.

geostationary having an orbit requiring an angular velocity the same as the earth's so that the position of the satellite in orbit appears stationary to an observer on the earth.

geosynchronous See *geostationary*.

Gray code a computer code in which the representations of sequential numbers differ in only one bit (for the purpose of minimizing errors).

ground-emitter amplifier circuit in which the emitter is bypassed to ground. See *common-emitter amplifier*.

grounded-base amplifier a common-base amplifier circuit so named because the base is an ac ground.

guard bands 25 kHz frequency bands at the upper and lower ends of a standard FM station bandwidth used to prevent interference from adjacent stations.

Gunn diode operates as an oscillator in the microwave frequency range. It has largely replaced both the klystron tube and the tunnel diode.

gyrotron a device developed to produce very high microwave power at frequencies up to a few hundred gigahertz. Dependent on electrons rotating at a cyclotron frequency while under the influence of a strong magnetic field.

H

half-power points the points defining a band of frequencies lying within 70.7% of the maximum amplitude at resonance. These associated frequencies also provide resonance effects. Also called cutoff frequency points.

half-wave dipole antenna a Hertz antenna whose radiation has been cut to half a wavelength.

hard-wired connection a telephone connection that requires all the devices (terminals, computers, and so on) be directly connected using some type of cable.

harmonic generator a frequency multiplier is often called a harmonic generator because the output of the circuit is a harmonic of the fundamental input frequency.

harmonic a multiple of the fundamental frequency.

Hertz term that means "cycles per second" and is used when referring to the frequency of radio waves.

High Definition Television (HDTV) an advanced television system in which very minute detail can be faithfully reproduced. It will offer CD-quality sound and much sharper pictures by means of digital signals.

high-pass filter rejects lower frequencies and passes only frequencies above a given value.

horizontal deflection plates positioned at right angles to the vertical deflection plates. They are positioned closer to the screen so they will deflect the beam less than the vertical pair. Voltages applied to the horizontal plates deflect the electron beam horizontally.

horizontal resolution the ability of a system to resolve vertical lines of resolution.

horizontally polarized condition of a wave whose electric field is horizontal to the earth.

horn a section of a waveguide, flared out in both cross-sectional dimensions, used to change a normal 50 Ω waveguide impedance to the higher impedance of space. Used with parabolic reflecting antennas.

hot electron diode See *Schottky Barrier diode*.

hue the actual color produced on the television screen, determined by the phase developed by the difference in amplitude between the I and Q signals.

hypertext on the Internet, data that contains links to other data.

I

iconoscope the first camera tube. The name was derived from the terms "icon" (image) and "scope" (to observe). It was large and insensitive to light.

image frequency occurs when input signals on two different frequencies produce an output at the IF. The undesired response frequency is called the image frequency.

image orthicon (IO) replaced the iconoscope. It was somewhat smaller, more sensitive to light and did not create the false signals generated by the iconoscope.

in-band signal any signal carried on the telephone circuit that falls within the range of 300–3300 Hz. All speech signals are in-band signals.

incident wave the original wave moving outward from the source.

independent-sideband (ISB) transmits two independent sidebands, each containing different intelligence, with the carrier suppressed by about 50 dB. Frequency used for HF point-to-point telephony. Also referred to as *twin-sideband, suppressed-carrier* (TSSC).

index of modulation variation in the AM signal compared with the unmodulated carrier. Also called *modulation factor* or *percent modulation*.

injection laser diode (ILD) a simple, low cost device composed of helium and neon gas and producing a red visible beam. It supplies power levels of 5 to 7 mW.

integrated detector preamplifier (IDP) a device which integrates a follow-up amplifier into the same package as a photodiode. This device can be used directly with a voltage comparator or to drive a normal voltage amplifier.

integrated PIN FET a device created from the coupling of a photodiode with unity gain (the PIN) to a high-impedance front-end amplifier (the FET). It combines operation at low voltage with low sensitivity with operating temperature, high reliability, and ease of manufacture.

interlaces scanning the way in which a televised image is reconstructed on the screen of the TV receiver. It is a process that reverses the image's division into parts in the TV camera and its subsequent transmission to the receiver as an electrical signal. Interlaced scanning involves scanning every other line, retracing, then filling in the untraced lines.

intermediate frequency (IF) the radio frequency which lies between the original RF carrier and AF modulating frequencies.

intermediate-frequency (IF) amplifier used in receivers and designed to operate at frequencies above the AF range but below the received RF range.

internal noise random noise created by the passive or active devices inside a receiver. Internal noise consists of thermal noise and shot noise.

international Morse code the international code consisting of dots and dashes or long and short sounds used in radiotelegraphy.

intersync separation the process of separating (by filters) the vertical and horizontal sync pulses from each other.

intranets mini Internets designed to operate internally within any organization, institution, or corporation. Intranets allow for archiving essential information that can be maintained by those who are responsible for it.

ionization the loss or gain of electrons by an atom.

ionosphere the collective layers of atmosphere between 30 and 250 miles above the surface of the earth.

isochronous transmission a mix between asynchronous and synchronous transmission. Individual characters are framed with a start and stop bit, but the intervals between are time controlled. The technique used in most modern computer networks.

J

j operator an imaginary number, mathematically defined as the square root of -1. It is used to represent reactance.

Johnson noise See *thermal agitation*.

K

kinescope (kine) a television picture tube or a videotape made from the screen or taken from the circuit of a TV picture tube.

klystron a form of electron tube used for generation and magnification of microwave electromagnetic energy.

L

laser *l*ight *a*mplification by *s*timulated *e*mission of *r*adiation. See *optical maser*.

leading-edge PDM the trailing edge of the sampling pulses remains fixed and the leading edge varies in accordance with the modulating signal amplitude.

light-emitting diode (LED) a semiconductor device that emits visible light when forward biased.

line-of-sight (LOS) communication radio communication by means of the direct wave. It depends on the height of the transmitting and receiving antennas and on the nature of the terrain between the two antennas. Range is limited to the radio horizon.

linear imaging device (LID) transport register formed by pairing linear arrays of photodetectors with CCDs.

load a device that absorbs or consumes energy fed into it.

local oscillator (LO) an oscillator included in a piece of equipment as opposed to an oscillator located in a distant transmitter.

lock range the frequency range over which the PLL can follow the incoming signal.

log-periodic array an antenna system having element lengths logarithmically related with each element being fed through a special phasing network.

loop-stick antenna See *ferrite rod antenna*.

loops points of maximum strength.

Low Earth Orbiting Satellites (LEOS) small satellites which can be launched by small launchers or in a multiple launch configuration by large launchers.

low-pass filter passes frequencies below a given value and rejects frequencies above that value.

lowest usable frequency (LUF) a frequency below which an RF signal will be totally absorbed by the ionosphere.

luminance signal in television, the monochromatic signal conveying brightness information. Called the Y signal.

M

M-type backward wave oscillator similar to the O-type except that the slow-wave structure and the electron trajectories are circular. See *O-type backward wave oscillator*.

magnetistrictive device constructed with nickel and ferrite. It will lengthen or shorten in the presence of a magnetic field.

magnetron a form of TWT used as an oscillator at ultra-high and microwave frequencies. Most magnetrons contain a central and a surrounding plate.

Manchester encoding a format in which polarity reverses once each bit period regardless of the data.

maser a special form of amplifier for microwave energy. Maser output is the result of quantum resonances in various substances.

maximum usable frequency (MUF) the highest frequency at which a signal radiated into the ionosphere will be refracted back to earth with usable strength.

mechanical filter mechanically resonant device that uses a double-conversion technique for better suppression of unwanted frequencies.

mirror image 1. a response curve that is identical with another. 2. the extra quarter-wave supplied by the earth (in a quarter-wave Marconi antenna).

mixer stage See *frequency-conversion section*.

modulating signal the signal employed to modulate another signal (the carrier).

modulation (coding) the process of impressing information in the form of a low-frequency signal (the modulating signal) onto a high-frequency signal (the carrier).

monimatch an amateur version of a reflected-power meter and the SWR meter.

multiple-hop transmission concept of a sky wave striking a good conducting surface when it returns to earth and being reflected back into the ionosphere. If the signal is strong enough the wave may be reflected (hop) again and again. This principle is used extensively for long-distance communications.

multiplexing the capability of allowing several analog telephone conversations to be sent over one transmission channel, separated by frequency.

muting (squelch) circuit interference-preventing device that automatically shuts off a radio receiver while the transmitter is operating.

mutual visibility window the concept applied to a communications satellite requiring it to be within the view plane of both the transmitting and the receiving terminal.

N

neutralization a method of reducing the possibility of unwanted oscillations using a degenerative (negative) feedback circuit.

nodes points of minimum strength.

noise factor (*nf*) See *noise figure*.

noise figure (*NF*) the ratio R_1/R_2, where R_1 is the signal-to-noise power ratio of an ideal network and R_2 is the signal-to-noise ratio of the network or device being tested. Also called noise factor.

noise inverter a circuit that eliminates the input signal when the noise pulse is larger than the signal. It provides sync noise immunity.

non-return to zero (NRZ) a coding format in which the current flowing in the write-head coils is sustained (it does not return to zero) after the write pulse.

nondirectional (omnidirectional) antenna one which radiates to or receives from all directions equally well.

nonresonant line one of infinite length or one that is terminated with a resistive load equal to the characteristic impedance of the line.

notch filter See *band-reject filter*.

Nyquist rate the minimum sampling frequency of twice the highest modulating signal frequency.

O

O-type backward wave oscillator a development of the TWT with built-in feedback, where an electron beam delivers energy to a wave traveling backward along the slow-wave structure.

octave represents a doubling or halving of a given frequency.

optical fiber a thin, transparent strand of material, usually glass or plastic. The refraction characteristics of the fiber keep the visible light inside. Used for optical communications.

optical maser a visible light maser, commonly called a laser.

optical return loss (ORL) a measure of the power reflected from a discontinuity to power incident upon it.

optimum usable frequency (OUF) a frequency well above LUF but far enough below MUF that it is not appreciably affected by atmospheric changes. An OUF provides relatively reliable sky wave communications.

oscillator 1. a circuit designed specifically to produce electrical oscillation (repetitive motion). All oscillators use the feedback principle. 2. an electronic generator that operates from a dc source, has no moving parts, and can produce ac signal frequencies.

out-of-band signal signal in the designated 4 kHz bandwidth but not in the 300–3300 Hz band.

overmodulation undesirable condition in which the modulation carrier goes to more than twice its unmodulated value and the signal is intermittently turned off, causing interference and distortion and resulting in lost information.

P

parallel resonance occurs at the frequency at which $X_L = X_C$ in a parallel LC circuit. The current is in phase with the source voltage.

parallel-tuned filter a wave trap used to prevent interference from unwanted signals or from harmonics of the desired frequency.

parametric amplifier usually called a paramp. It is used primarily at microwave frequencies above about 30 GHz. Nearly all are coupled using ferrite circulators and have power transferred to the signal by enabling this signal to extract power from a varactor diode.

parasitic element an element that is not connected to the energy source (not driven). It receives its energy through induction.

parasitically excited a state resulting from current induced from a parallel, insulated element into an adjacent element.

pass band the range of frequencies passing through a filter with maximum gain or minimum attenuation.

passive filter a device constructed with resistors, capacitors, and inductors and containing no active device that passes or screens out electric current of certain frequencies.

passive satellite a satellite which reflects a signal from one ground point to another. It does not provide reception, amplification, or retransmission.

peak envelope power (PEP) rating for an SSB transmitter which is the product of rms voltage and rms current. It is equal to the average power of the SSB signal.

peak voltage the highest value reached by a pulse voltage or an ac voltage half cycle.

perigee the point in the orbit of a satellite nearest to the center of the earth.

phantom antenna See *dummy antenna*.

phase modulation (PM) both carrier amplitude and frequency remain constant while carrier pulse is varied.

phase-locked loop (PLL) an electronic feedback loop. Its purpose is to make a variable-frequency oscillator lock in at the frequency and phase angle of a standard frequency used as a reference.

phased array an antenna system having driven elements that are fed in a specific phased relationship.

phasing harness device usually consisting of simple delay lines. Used to give a phased array a certain directional characteristic.

photon a quantum of radiant energy.

pi network a network of filter sections having a configuration resembling the capital Greek letter pi.

piezoelectricity in crystalline substances, the electrical polarity having the properties of generating a voltage across its faces when subjected to mechanical pressure and compressing when a voltage is applied across its faces.

polarization in a radio wave, the direction of the electrical lines.

positive (regenerative) feedback occurs when a small portion of the output signal of an oscillator is fed back to the input of the circuit.

positive feedback signal feedback signal which is in phase with the input signal.

power amplifier (PA) 1. an amplifier that delivers useful amounts of power to a load. 2. the last stage of amplification before an ac signal leaves the AF system.

power gain the ratio of the ERP to the actual RF power applied to the feed point. It is always measured in the favored direction of the antenna.

preemphasis the process of deliberate frequency response distortion of the intelligence signal at the transmitter, where the higher frequencies are boosted to unnaturally high amplitudes as compared to the lower frequencies.

progressive scanning the reading of one line after another in electronic scanning.

propagation velocity the speed at which energy travels from a source.

pulse amplitude modulation (PAM) the simplest form of analog pulse modulation. It is the result of a train of pulses, occurring at a fast repetition rate, that are made to vary in amplitude in proportion to the amplitude of a lower-frequency modulating signal.

pulse code modulation (PCM)   the principal form of digital pulse modulation, it provides greater immunity to noise and interference.

pulse dialing   a form of telephone dialing in which each digit is formed by a series of pulses. Each digit except 0 has the corresponding number of pulses. Zero has 10 pulses.

pulse distribution amplifier (PDA)   device used to route signals from sync generators and similar equipment that produces pulse waveforms with fast rise and fall time.

pulse duration modulation (PDM)   also called pulse width modulation (PWM) or pulse length modulation (PLM), is the result of varying the duration of pulses used for modulation. See *pulse time modulation*.

pulse length modulation (PLM)   See *pulse duration modulation*.

pulse modulation   use of pulses as the modulating signal.

pulse position modulation (PPM)   state in which both the amplitude and width of the pulse remain constant but the position of the pulse varies in relation to the reference pulse, either leading or lagging, in accordance with the polarity of the modulating circuit.

pulse time modulation (PTM)   uses constant-amplitude pulses varying in timing for sampling. The timing may be variable in duration or position.

pulse width modulation (PWM)   See *pulse duration modulation*.

push-pull amplifier   one transistor conducts during one half-cycle, and the other transistor conducts during the other half-cycle. This arrangement results in a class B amplifier circuit with large load and low distortion.

Q

quadrature detector   a device that requires no transformer. Its purpose is to compare the input signal to be modulated with its phase-shifter version.

quadrature   the use of signals whose phases are at right angles to each other.

quality factor (Q)   a numerical value that is the ratio of reactive power (in either the inductor or the capacitor) to resistive power. Q is the ratio of energy stored to energy lost.

quantization noise   distortion resulting from the difference between sampling point amplitudes and quantum levels. So called because the errors are random.

quantum   the nearest standard amplitude.

quiescent (Q) point of operation   the resting, or no-signal, point of the dc load line in a biased circuit.

R

radials   quarter-wave copper wires used to construct an artificial ground system when an antenna is erected on soil of poor conductivity.

radiation loss   the unwanted radiation of energy as radio waves from a transmission line.

radiation resistance (R_r)   the special property by which an antenna converts electrical energy into radio waves rather than into heat.

radio frequency (RF)   an ac frequency which is higher than the highest audio frequency. Termed radio because of the application of such frequencies to radio communications.

radio frequency voltage amplifier   may be used in transmitters to amplify RF signals before modulation, in receivers to amplify receiver RF signals, and as a frequency multiplier.

raster   the lighted screen of a monitor.

reactance modulator   circuit that provides a varying capacitance or inductance in a resonant oscillator circuit to achieve frequency deviation.

receiver   a device or circuit that intercepts a signal, processes it, and converts it into a useful form.

Reciprocity Theorem   a direct result of Kirchoff's and Ohm's laws, based on the theory that the functional use of an antenna can be interchanged between transmitting and receiving.

rectilinear scanning system where the scanning beam moves horizontally and vertically simultaneously. Both scans are linear; that is, the movement of the beam, both horizontally and vertically, occurs at a constant rate of speed.

reflected wave the wave moving back toward the source.

reflectometer (SWR meter) an instrument installed in a transmission line for measuring the standing wave ratio. Some are calibrated in both forward watts and reflected watts. Also called a directional coupler.

reinsertion process of carrier substitution, or the reinserting of a new carrier to replace the old carrier.

resonance state in which the natural response frequency of a circuit coincides with the frequency of an applied signal, or vice versa.

resonant circuit a circuit whose constants are chosen for maximum response at a given frequency.

resonant frequency the frequency (f_r) at which the opposite reactances are equal.

rest frequency (deviation) the amount by which the signal frequency varies above and below the center of the main carrier.

retrograde reverse motion. An orbit from east to west around the earth. Generally, the satellite motion is from west to east.

return to zero (RTZ) a method of coding in which the write current returns to zero after the write pulse.

RF choke (RFC) a low inductance coil used to block RF currents.

ring system a transmission system consisting of a single optical fiber line connecting several terminals in a ring pattern. Messages travel in one direction only; they must be passed from terminal to terminal until the addressed terminal is reached.

ripple amplitude the ratio of maximum to minimum attenuation within the useful pass band of a filter. Also called *peak-to-valley ratio*.

rolloff the per decade or per octave increase or decrease above or below the cutoff frequency in a

filter's response curve. It represents the performance of the filter.

S

saturation the purity of a color or colors.

scan ratio the number of scan lines (in television transmission) used per frame. The standard system in the United States has a scan ratio of 525:1.

Schottky barrier diode a solid-state diode in which a metal and a semiconductor form a pn junction. Because of the larger contact area between the metal and semiconductor, the forward resistance is lower, and so is the noise.

self-induction induction that occurs within a single circuit.

series resonant condition occurs when the current in a series *RLC* circuit is in phase with the voltage across the circuit, the impedance is low at the resonant frequency and all other frequencies are highly attenuated, and inductive resistance is equal to capacitive reactance.

shadow mask in a television, a thin metal sheet perforated with tiny holes approximately 0.01 inch in diameter (one hole for each triad on the faceplate). In order for the electron beams to reach their respective phosphor dots, all three beams must converge at the shadow mask, pass through the tiny opening, and deconverge.

shielded-pair line consists of two separated parallel wires surrounded by solid, low-loss dielectric. The dielectric containing the parallel wires is enclosed in a metal braid, then covered with a flexible coating.

shot effect a condition caused by random variations in the arrival of majority current carriers at the output terminal of an amplifying device.

shot noise the result of shot effect and is present in all amplifying devices. The term is used because excessive random varying noise current, when superimposed on the dc of the output terminal and amplified, sounds like a shower of lead shot dropping onto a metal surface.

sidebands frequencies (signals) occurring above and below the carrier frequency. They are the result of mixing between the carrier and the modulating circuit.

signal-to-noise ratio (*S/N*) the ratio of signal power to noise power at the same point.

signal-to-quantizing-noise (SQR) for a linear system, the ratio of the size of the input signal to .25 the size of the quantization level.

simple AGC a condition in which the AGC bias starts to increase as soon as the received signal level exceeds the background noise level. The receiver immediately becomes less sensitive.

single-sideband (SSB) a form of AM in which only one sideband is transmitted. The remaining sideband and the carrier are suppressed at the transmitter.

single-sideband, controlled-carrier (SSCC) a pilot carrier system in which the controller carrier rises to almost full amplitude during brief inactive periods but is reduced to very low levels during actual modulation, thus keeping the average power output of the transmitter constant.

single-sideband, full-carrier (SSFC) a pilot carrier system that can be used as a compatible AM broadcasting system. The transmitted signal can be received and demodulated by standard AM and SSB receivers.

single-sideband, reduced-carrier (SSRC) a pilot carrier system in which an attenuated carrier is reinserted into the SSB signal, after removal of the unwanted sideband, to facilitate receiver tuning and demodulation.

single-sideband, suppressed-carrier (SSSC) a transmission system commonly identified as SSB. See *single-sideband*.

skin effect type of copper loss in which most of the flow of electrons is diverted to the surface (skin) of the wire, ultimately resulting in increased power losses as frequency increases.

skip distance the distance a signal wave travels over the earth (increase) as the radiation angle decreases.

skip zone the distance between the end of a usable ground wave signal and the reappearance of the reflected sky wave. It is a no-signal zone.

sky wave a radio wave propagated by ionospheric refractions and reflections.

slope modulation See *delta modulation*.

slotted line a section of coaxial transmission line with a lengthwise slot cut into the outer conductor. A probe is inserted into the slot to show that the magnitude of a signal picked up is proportional to the voltage at the conductors at the insertion point.

slow-scan a TV system whereby the camera's pickup tube stores scene information until scanned by an electron beam.

solar flux See *solar radio noise flux*.

solar radio noise flux the amount of radio noise emitted by the sun. Also called *solar flux*.

solid-state maser a solid material (ruby) that can be used as an amplifier or oscillator.

source an energy generator in any form.

spectral response a characteristic of a device that determines the range, or system length, that can be achieved relative to the wavelength at which responsivity is characterized.

spin-stabilized satellite a satellite having its solar cells mounted on a cylindrical body and continually rotated so that they face the sun while the antenna is kept pointed at the earth. The spinning cylinder acts as a gyroscope to keep the satellite oriented in space.

stagger stacking the placement of antennas about 0.7λ apart vertically with the lower antenna spaced $\lambda/4$ behind the upper antenna. A coaxial cable connected between the upper antenna and a coupler is $\lambda/4$ longer than the coaxial cable connected to the lower antenna. This produces an increase in the front-to-back ratio and a substantial increase in forward gain. Used to minimize multipath fading.

standing wave ratio (SWR) ratio of high current points to low current points in a standing wave.

standing wave when the frequency of the energy from the source remains constant, the points at which maximum and minimum currents appear along the line remain at fixed distances back from the open end. They are said to "stand" in a fixed position. Hence the term standing wave.

star system a transmission system consisting of a central master terminal and several slave terminals. The master can broadcast to any or all of the slaves, but the slaves can communicate only with the master.

stop band the range of frequencies attenuated.

surface LED a diode that couples less power into a fiber with a given acceptance angle than do other devices. This occurs because it emits light over a broad spectrum and disperses that light over a large angle.

symmetrical PDM both leading and trailing edges of the sampling pulses are varied in accordance with the amplitude of the sampling points of the modulating signal.

sync separation the process of stripping the video information from the composite video signal.

synchronizing pulse generator complex piece of equipment used to time vertical and horizontal scan and retrace and to maintain a rigid phase relationship between the horizontal and vertical scanning systems in a monitor or receiver.

synchronous detector uses two diodes and has a separate carrier input. The carrier input in this circuit is a signal that is regenerated by earlier stages of the receiver. The receiver-generated carrier must have exactly the same frequency as, and be exactly in pulse with, the original suppressed carrier.

synchronous transmission uses an internal clock to synchronize the channel's transmitter and receiver. When one or more synchronization characters are sensed by the receiving terminal, data transmission proceeds character by character without intervening start to stop bits.

T

tandem exchange telephone exchange which makes connections between central (local) offices when an interoffice trunk is not available.

tank circuit used in series with an antenna and tuned to resonance at the undesired frequency, it presents high impedance to currents at the undesired frequency but allows currents at other frequencies to pass with little or no opposition. See *parallel-tuned filter*.

tapped coil an inductor to which one or more intermediate connections (taps) are made to provide intermediate values of inductance.

telemetry the transmission of quantitative information from one point to another by electromagnetic (radio) means.

thermal agitation noise generated by the random motion of atoms and electrons in any electronic equipment. Also referred to as *white noise* or *Johnson noise*.

three-gun, shadow-masked tricolor CRT a tricolor picture tube containing three electron gun assemblies, a shadow mask and tiny triads of phosphor dots covering the face of the tube. Color is generated by bombarding the color triads with high-velocity electron beams from the electron guns.

time division multiplex (TDM) a means of transmitting two or more information channels over the same communications circuit using a time-sharing technique. See *frequency division multiplex*.

trailing-edge PDM the leading edge of the sampling pulses remains fixed and the trailing edge varies in accordance with the modulating signal amplitude.

transducer (speaker) any device that converts one form of energy into another form.

transfer function of a filter is considered to be its gain or amplitude response.

transmitter a device that produces a signal for communication purposes.

transponder a communication receiver/amplifier/transmitter.

transverse electric (TE) propagation mode in which all of the electric lines of flux are perpendicular to the axis of the waveguide.

transverse magnetic (TM) mode in which all of the magnetic lines of flux are perpendicular to the axis of the waveguide.

transverse wave a wave occurring in a direction or directions perpendicular to the direction of wave propagation. This is the type of wave produced by water.

traveling wave amplifier an amplifier that uses one or more TWTs to provide useful amplification of signals at frequencies of the order of several gigahertz.

traveling wave maser a device for use at ultrahigh and microwave frequencies.

traveling wave tube (TWT) is used as a high-powered wide-band amplifier of microwave signals.

traveling wave a wave of energy moving down a line at the speed of light, setting up an oscillating electron motion in sections of the wires as it moves along.

tropospheric scatter (troposcatter) an atmospheric phenomenon which allows a UHF signal communications path to extend beyond the horizon.

tuned radio frequency (TRF) type of radio receiver used primarily as a fixed-frequency receiver. It offers simplicity and high sensitivity.

tunnel diode device that can act as an amplifier or oscillator of microwave frequencies. It has been largely replaced by the Gunn diode. See *Gunn diode*.

twin-sideband, suppressed-carrier (TSSC) See *independent-sideband*.

twin-T oscillator provides a 180° phase shift from the active device output to its input.

twisted-pair line consists of two rubber-insulated wires twisted together. No spacers are used; the result is line flexibility. Used for short-run, low-frequency use (ac power cables).

two-cavity klystron a device used as an amplifier or oscillator at moderate power levels.

two-wire ribbon line commonly called twin-lead, consists of two parallel wires embedded in a thin ribbon of low-loss dielectric. The ribbon ensures uniform spacing along the entire length of the line.

U

u-law *u* = 255 law companded PCM digital coding standard. Common scheme for analog-to-digital conversions and companding. Standard for North American and Japanese telephone networks.

up-converter a circuit which heterodynes a signal to one centered on a higher frequency.

V

varactor (varicap) voltage-variable capacitance diode in which the junction capacitance is easily varied electronically by changing the reverse bias on the diode.

variable-frequency oscillator (VFO) an oscillator whose frequency is continuously variable.

VCO circuit also termed a voltage-to-frequency converter, it is used for electronic tuning of the oscillator frequency.

velocity of a radio wave that is radiated in space is simply the speed at which the wave travels.

vertical deflection plates in a television tube, two plates used to deflect the electron beam up or down. The amount of deflection depends on the magnitude of the voltage on the plates.

vertical resolution the ability of a system to resolve horizontal lines in a scene. It is dependent on the number of scan lines used per frame.

vertically polarized condition of a wave whose electric field is vertical to the earth.

vestigial sideband filter a filter that attenuates most of the lower sideband in order to reduce the bandwidth.

vestigial sideband a trace (vestige) of the unwanted sideband is transmitted, usually with the carrier and the other sideband left intact. Used for television video transmission.

video distribution amplifier (VDA) device used to route camera signals.

video switcher circuit used for switching the output of a number of cameras, one at a time, to a monitor for viewing or for sending out on the line to a transmitter.

vidicon a camera tube developed in the 1950s which was a vast improvement over earlier TV camera tubes. It was small, easy to operate and of comparatively low cost. It enjoyed great success in the CCTV field.

voice frequency (VF) channel also called the message channel. Limited to carrying frequencies from zero to 4 kHz. This is the telephone system voice channel.

voltage amplifier an audio amplifier specifically designed to amplify low-voltage audio signals.

voltage standing wave ratio (VWSR) ratio of high voltage points to low voltage points in a standing wave.

voltage-to-frequency converter See *VCO circuit*.

W

wave trap a form of band-reject filter (notch filter) sometimes used in antenna systems of receivers or high frequency equipment.

wavelength symbolized by the Greek lambda, λ, is the space occupied by one full cycle of a radio wave at any given instant.

whip antenna a short-radiator, usually loaded at the base and measuring one quarter physical wavelength or less. Often used in mobile communications at high frequencies.

white noise See *thermal agitation*.

Y

yttrium-iron-garnet (YIG) filter consists of highly polished spheres of a ferrite material. It is electromagnetically tuned.

Z

zero-beat condition that occurs when no difference exists between two frequencies being compared (no heterodyne frequency will be produced).

Index

UNIVERSITY OF STRATHCLYDE

15 MAR 2007

UNIVERSITY LIBRARY